SOCIOLOGY

United States Map

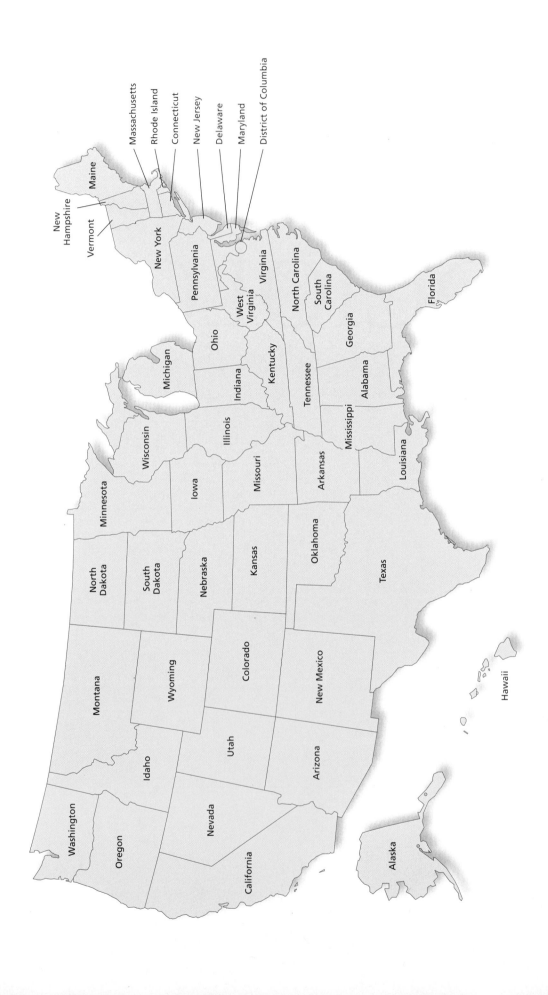

World Map

North America
South America
Europe
Africa
Asia
Australia

Pacific Ocean
Atlantic Ocean
Arctic Ocean
Indian Ocean

Canada
United States
Mexico
Greenland
Iceland
Guatemala
El Salvador
Belize
Honduras
Nicaragua
Costa Rica
Panama
Cuba
Haiti
Jamaica
Dominican Republic
Puerto Rico

Colombia
Venezuela
Guyana
Suriname
French Guiana
Ecuador
Peru
Brazil
Bolivia
Paraguay
Chile
Argentina
Uruguay
Pa

United Kingdom
Ireland
Norway
Sweden
Denmark
Portugal
Spain
France
Italy
Greece
Germany
Poland
Belarus
Ukraine
Finland

Russia
Kazakhstan
Mongolia
China
Japan
North Korea
South Korea
Taiwan
Philippines
Vietnam
Laos
Thailand
Cambodia
Malaysia
Brunei
Indonesia
Papua New Guinea
New Zealand

India
Pakistan
Afghanistan
Uzbekistan
Turkmenistan
Kyrgyzstan
Tajikistan
Nepal
Bhutan
Bangladesh
Myanmar
Iran
Iraq
Turkey
Syria
Lebanon
Israel
Jordan
Saudi Arabia
Yemen
Oman
U.A.E.
Qatar
Bahrain
Kuwait
Georgia
Armenia
Azerbaijan

Morocco
Algeria
Tunisia
Libya
Egypt
Mauritania
Mali
Niger
Chad
Sudan
Eritrea
Ethiopia
Somalia
Senegal
Gambia
Guinea Bissau
Guinea
Sierra Leone
Liberia
Ivory Coast
Burkina Faso
Ghana
Togo
Benin
Nigeria
Cameroon
Central African Rep.
Equatorial Guinea
Gabon
Congo
Democratic Republic of the Congo
Cabinda (Angola)
Angola
Zambia
Malawi
Tanzania
Kenya
Uganda
Rwanda
Burundi
Mozambique
Madagascar
Zimbabwe
Botswana
Namibia
South Africa
Swaziland
Lesotho

1. Netherlands
2. Belgium
3. Luxemburg
4. Switzerland
5. Czech. Rep.
6. Austria
7. Slovakia
8. Hungary
9. Slovenia
10. Croatia
11. Bosnia and Herzegovina
12. Serbia
13. Albania
14. Macedonia
15. Montenegro

SOCIOLOGY

TENTH EDITION

Jon M. Shepard

Virginia Polytechnic Institute
and State University

WADSWORTH
CENGAGE Learning™

Australia • Brazil • Japan • Korea • Mexico • Singapore • Spain • United Kingdom • United States

Sociology, Tenth Edition
Jon M. Shepard

Senior Acquisitions Editor: Chris Caldeira

Assistant Editor: Erin Parkins

Editorial Assistant: Rachael Krapf

Media Editor: Lauren Keyes

Marketing Assistant: Jillian Myers

Marketing Communications Manager: Martha Pfeiffer

Project Manager, Editorial Production: Cheri Palmer

Creative Director: Rob Hugel

Art Director: Caryl Gorska

Print Buyer: Linda Hsu

Permissions Editor, Text: Roberta Broyer

Permissions Editor, Images: Don Schlotman

Production Service: Lachina Publishing Services

Photo Researcher: Pre-PressPMG

Illustrator: Lachina Publishing Services

Cover Designer: Yvo Riezebos

Cover Image: Blend Images/Veer

Compositor: Lachina Publishing Services

For product information and technology assistance, contact us at
Cengage Learning Customer & Sales Support, 1-800-354-9706.

For permission to use material from this text or product,
submit all requests online at **cengage.com/permissions.**
Further permissions questions can be e-mailed to
permissionrequest@cengage.com.

Library of Congress Control Number: 2008937204

ISBN-13: 978-0-495-59901-2

ISBN-10: 0-495-59901-8

Wadsworth
10 Davis Drive
Belmont, CA 94002-3098
USA

Cengage Learning is a leading provider of customized learning solutions with office locations around the globe, including Singapore, the United Kingdom, Australia, Mexico, Brazil, and Japan. Locate your local office at: **www.cengage.com/international.**

Cengage Learning products are represented in Canada by Nelson Education, Ltd.

To learn more about Wadworth, visit **www.cengage.com/wadsworth.**

Purchase any of our products at your local college store or at our preferred online store **www.ichapters.com.**

Printed in the United States of America
1 2 3 4 5 6 7 11 10 09 08

About the Author

While an undergraduate student, Jon Shepard was inspired and nurtured by his sociology professor, Richard Scudder. After graduating from Michigan State University with a Ph.D. in sociology, Shepard taught introductory sociology and the sociology of organizations at the University of Kentucky. For fourteen years, he was Head of the Virginia Tech Department of Management. He is the author of ten books and more than forty professional journal articles. He has received teaching awards at both universities, including the University of Kentucky Great Teacher Award.

To Ava

Contents in Brief

Contents

BOXES

Preface

A Note to Students from the Author

The issues discussed in my freshman introductory sociology class were not the sort of things I usually spent time thinking about: Is divorce more likely or less likely when people have the same social-class background? Are some races inferior to others? What is the social significance of Darwinism?

Suddenly, I began to see human behavior in a different light. I discovered that Richard Wright's classics, *Native Son and Black Boy,* are not merely stories about black youth but rather autobiographical reflections of the black experience in America. Prejudice and discrimination are not just characteristics of individuals; they are part of society as a whole. I learned that football is actually as much a business as a sport. It became apparent that the fraternity I was about to join was not only a brotherhood but also part of the campus social hierarchy.

I began to see social relationships as essential for human survival. And if the world is a stage and all its men and women merely players, these players generally deliver their lines and act out their parts as if they were rehearsed, and with a definite flair for mimicry. Yet, the action that sociologists label "social structure" depends less on the conscious learning of appropriate attitudes, beliefs, and behavior than on unreflective acceptance of our culture and society. In this sense, we are like puppets responding to tugs on the strings that bind us to essential social relationships—relationships in which people, I came to understand, do not usually behave randomly and do not always behave only as individuals. People often think, feel, and behave in rather predictable ways because of what they have been taught and because of the many social pressures to which they have been exposed. At the same time, however, individuals interacting with others create their own understandings of situations. In this sense, human beings are not like puppets, because they have the potential to buck tradition; they are active, thinking creatures even when they are conforming.

Society was demystified for me. I came to value sociology as a tool for understanding the world. In fact, this experience led me to major in sociology and subsequently to obtain my PhD in the field. I have never regretted those choices.

You might not major in sociology. You can, however, enjoy this course and take lifelong benefits from the slant on social life that sociology provides.

A Note to Instructors from the Author

Several specific goals continue to energize me as we introduce the significance and excitement of sociology to students. First, sociology—with its perspectives, concepts, theories, and research findings—offers a window on the social forces that affect us all daily. This perspective is vital for students as they grapple to understand the social factors that promote patterned behavior in themselves and others. Second, the material must be readable and at the same time theoretically and empirically sound. Third, students deserve a textbook design that is dynamic and demonstrates, clearly, the application of a new perspective to their personal lives. Fourth, I want a presentation from which students can not only learn the basics of sociology, but can acquire the ability to pose their own questions about social life. Finally, the catalyst: stimulate students to become more active learners.

Unifying Themes and Features

Sociological Imagination

The study of sociology encourages critical thinking about conventional wisdom through the development of the sociological imagination—the mind-set that enables individuals to see the relationship between events in their personal lives and events in their society. To this end, each chapter opens with a question about some aspect of social life. The answer to each question contradicts a popular or commonsense belief. Sometimes the question will focus on a result that even sociologists may have doubted until a sufficient amount of convincing research had been done. The correct answer to each question is given at the beginning of the chapter and is elaborated on within the chapter itself. Topics covered include the following:

- Suicide (Chapter 1, "The Sociological Perspective")
- Television and violence (Chapter 4, "Socialization over the Life Course")
- Selfishness and human nature (Chapter 5, "Social Structure and Society")
- Date rape (Chapter 7, "Deviance and Social Control")
- Racial inequality (Chapter 9, "Inequalities of Race and Ethnicity")
- Gender income inequality (Chapter 10, "Inequalities of Gender")
- National health care (Chapter 15, "Health Care and Aging")
- Revolutions and social change (Chapter 18, "Social Change and Collective Behavior")

Sociology and You

A new boxed feature called *Sociology and You* now appears in each chapter. It is included to further aid students in developing the sociological imagination. This feature presents interesting, up-to-date examples of the relevance of sociology to the students' personal lives. Here are some examples:

- "Should You Believe Everything You Read?" (Chapter 2, "Social Research")

- "MySpace and Personal Identity" (Chapter 4, "Socialization over the Life Course")
- "Cyber Bullying and Stratification" (Chapter 8, "Social Stratification")
- "Spinning a Web of Hate" (Chapter 9, "Inequalities of Race and Ethnicity")
- "Gender-Based Hierarchy" (Chapter 10, "Inequalities of Gender")
- "Looking for Mr. or Ms. Right" (Chapter 11, "Family")
- "How to Avoid Bigotry in Sport" (Chapter 16, "Sport")
- "Terrorist Attacks and Disaster Myths" (Chapter 18, "Social Change and Collective Behavior")

Of course, opportunities to cultivate a sociological imagination permeate the entire text.

Doing Research: Sociologists as Scientists

A second boxed feature within each chapter, entitled *Doing Research,* presents the theory, methods, conclusions, and implications of significant sociological studies. This feature is intended to demonstrate the ways sociologists use the scientific method in their work.

Several criteria guided the selection of these research studies. Some studies, such as Emile Durkheim's work on the social antecedents of suicide, are sociological landmarks. Others, such as Philip Zimbardo's experiment involving a simulated prison and George Ritzer's analysis of the McDonaldization of higher education, are included because they reinforce a major point in a chapter. Still other studies are included because they illustrate the imaginative use of a major research method. Durkheim's use of existing sources in the study of suicide and Donna Eder's use of a variety of research methods in her study of popularity in middle school are examples.

Given these criteria, it is hardly surprising that many of the studies included are sociological "classics"; they have had a lasting influence on the field and are continuously being cited by other researchers. Like classics in all fields, these pieces of research generally have high interest value. They are innovative in approach and explore important topics in ingenious ways. If read carefully, these detailed accounts of significant sociological studies cannot fail to pique interest in social research and stimulate the sociological imagination. For example, Chapter 16 contains a *Doing Research* entitled "Tough Guys, Wimps, and Weenies," which focuses on sports and masculinity. Here are some other examples:

- "Teenagers in a Cultural Bind" (Chapter 3, "Culture")
- "High School Reunions" (Chapter 4, "Socialization over the Life Course")
- "Adopting Statuses in a Simulated Prison" (Chapter 5, "Culture and Society")
- "Who's Popular, Who's Not" (Chapter 8, "Social Stratification")
- "No Shame in My Game" (Chapter 9, "Inequalities of Race and Ethnicity")
- "Men's Work, Women's Work" (Chapter 10, "Inequalities of Gender")
- "The McDonaldization of Higher Education" (Chapter 12, "Education")
- "Gang Violence" (Chapter 17, "Population and Urbanization")
- "The Withering of the American Dream?" (Chapter 18, "Social Change and Collective Behavior")

Accent on Theory

Each chapter contains a prominent section on the distinctive views of three major theoretical perspectives—functionalism, conflict theory, and symbolic interactionism. Each chapter also presents a table entitled "Focus on Theoretical Perspectives," in which succinct illustrations are offered for the three theoretical perspectives. For example, the following table appears in Chapter 16 ("Sport"):

TABLE 16.2

Social Effects of Sport

This table illustrates how each theoretical perspective might study an issue involving sport. For each assumption, provide a specific example from your own experience or from a team you folllow.

Theoretical Perspective	Concept	Power Elite Model
Functionalism	Social integration	Athletic teams promote togetherness and belonging in a community.
Conflict Theory	Social conflict	Deep social conflict exists within a community and persists despite widespread attachment to athletic teams.
Symbolic Interactionism	Social concept	Participation in a team sport may promote or harm self-esteem depending upon factors such as emphasis on winning and fair play.

Cross-Cultural Perspective

Never has the world been as interconnected as it is now. Gone are the comfort and simplicity of geographically based social isolation. It is now apparent that events in cities, states, and nations have repercussions for each other. Thus, it is imperative that stimulating sociological phenomena be viewed in a larger geographic frame.

To this end, cross-cultural examples and research are presented throughout the text. This cross-cultural emphasis alerts students to our tendency to accept our own culture while rejecting others. By interacting mentally with other cultures within the context of sociological concepts, theories, and research findings, students are better able to apply the sociological perspective to their daily lives. This cross-cultural emphasis is intended to encourage in students a more self-conscious awareness of their own society and a better understanding of other cultures.

In each chapter, a world map feature, entitled *Worldview,* displays a worldwide comparison of a particular social phenomenon. Eight of the maps are new to this edition, six are revised, and four are retained from the Updated Ninth Edition. Here are some examples:

- "Availability of Television" (Chapter 4, "Socialization over the Life Course")
- "Social Classes in World Perspective" (Chapter 8, "Social Stratification")
- "Median Household Income" (Chapter 12, "Education")
- "Olympic Success" (Chapter 16, "Sport")

In addition, a U.S. map is included in each chapter. Each *Snapshot of America* permits a state-by-state comparison of some aspect of American culture. Twelve of the Tenth Edition maps are new to this edition, six are revised, and three are retained from the Updated Ninth Edition. Here are some examples:

- "Rates of Imprisonment" (Chapter 4, "Socialization over the Life Course")
- "Hate Groups in the United States" (Chapter 9, "Inequalities of Race and Ethnicity")
- "Women in Public Office" (Chapter 10, "Inequalities of Gender")
- "Who Are the Biggest Baseball Fans?" (Chapter 16, "Sport")

What is New in the Tenth Edition?

Text Photos

The photo program for this edition is brand new. This keeps the photo presentation as current as text material.

Text Additions

The Tenth Edition contains two new chapters, one a combined chapter covering health and aging and one on sport. Health and aging are so intimately connected

that it makes sense to cover them together. Sport is added for two reasons. It is now a significant institution in the United States, and the subject matter of sport demonstrates sociological principles well for the students. These are the major topics in each new chapter:

"Health Care and Aging" (Chapter 15)

1. Health Care as a Social Institution
2. Theoretical Perspectives and the Health-Care System
3. Health-Care Reform in the United States
4. The Graying of America
5. Health Care and Aging in the United States
6. Ageism and Health Care

"Sport" (Chapter 16)

1. The Nature of Sport
2. Theoretical Perspectives and Sports
3. Sport and Social Mobility
4. Racial and Ethnic Inequality in Sport
5. Gender Inequality in Sport

As noted earlier, a new feature, *Sociology and You,* emphasizes the effects of society on personal life. This revision also includes the latest research, data, and illustrations. The addition of approximately 200 new references reflects the inclusion of new topics as well as the expansion and updating of existing topics. There are 25 new figures and seven new tables. The most current information is added to other figures and tables. The U.S. and world maps included in each chapter are either new or revised with up-to-date data. All of these changes are reflected in the following thumbnail sketches of each chapter.

Chapter 1 introduces students to the sociological perspective, its practical uses, and history. The three major sociological theories (functionalism, conflict theory, and symbolic interactionism) close the chapter. New material is presented on the sociological perspective, symbolic interactionism, and the difference between individual and group behavior. The new *Sociology and You* feature highlights job opportunities in sociology. Health insurance is the subject for a new *Snapshot of America.*

Chapter 2 explores the major research methods used by sociologists. It concludes with a new discussion of ethics in social research. *Sociology and You* covers ways to read the popular press more skeptically. New data refresh *Snapshot of America* and *Worldview.*

Chapter 3 promotes a broader understanding of the pivotal concept of culture. New material is presented on the subject of social sanctions. The new *Sociology and You* focuses on the concept of cultural relativism. Immigration to the United States is illustrated in a new *Snapshot of America.*

Chapter 4 focuses on socialization, the process of learning to participate in society through the acquisition of culture. A new *Sociology and You* explores the place of personal identity in a MySpace world. Both the *Snapshot of America* and the *Worldview* contain new data.

Chapter 5 examines the concept of social structure. It explains and illustrates the major sociological concepts underlying social structure. A new *Sociology and You* looks at ways to reduce role conflict in the modern two-career family. There is a new *Snapshot of America* on unemployment rates and a new *Worldview* on agriculture employment. All figures and tables present new information.

Chapter 6 distinguishes among various types of groups and the basic interactions within them. A new *Snapshot of America* shows membership in unions, and a new *Worldview* displays information on human development. Figures and tables are revised and updated.

Chapter 7 opens with the biological, psychological, and sociological explanations of deviance. New material is presented on the definitions of deviance and the nature of stigma. The last half of the chapter focuses on a discussion of crime in the United States. Then, new material is brought to bear on the death penalty and imprisonment in the United States and the world. Violent crime is highlighted in a new *Snapshot of America,* and the *Worldview* contains new data. In addition to three new figures, other figures and tables on crime are revised and updated.

Chapter 8 takes up social stratification, one of the most important areas of sociology. This edition contains new information on these topics: income inequality; race, ethnicity, and poverty; gender, age, and poverty; societal responses to the problem of poverty; and social mobility. A new *Sociology and You* introduces the idea of cyber bullying and stratification. A new *Worldview* displays information on social classes, and the *Snapshot of America* is updated. There is a new figure on change in average real after-tax income in the United States. Other figures are revised and updated.

Chapter 9 turns to the significance of race and ethnicity and prejudice and discrimination in America. Much new material is presented on these subjects: the biology of race; anti-immigration; inequality of African Americans and Latinos; social mobility of African Americans; and racial and ethnic diversity in the United States. Hate groups is the topic of a new *Snapshot of America*. This chapter contains two new figures as well as updated and revised figures.

Chapter 10 concentrates on gender inequality. There is new information on gender income inequality. Gender-based hierarchy is the subject of a new *Sociology and You.* A new *Snapshot of America* provides a picture of women in public office. Two new figures and one new table are included.

Chapter 11 looks at the family as a social institution. This edition contains new material on the definition of marriage, working mothers, cohabitation, and same-sex domestic partners. "Looking for Mr. and Ms. Right" is the title of a new *Sociology and You.* A new

Snapshot of America contains data on grandparents as caregivers. A new *Worldview* displays the percentage of women aged 15 to 19 who are either newly married or have been married. Two new figures, one new table, and updated figures add currency to this edition.

Chapter 12 turns to the organization of schools, the functions of education, educational inequality, and the transmission of culture in schools. New material is introduced on the relationship between research universities and corporations. A new *Sociology and You* entitled "Are Americans Losing Their Minds?" examines the question of anti-intellectualism and anti-rationalism in modern America. A new *Snapshot of America* and a new *Worldview* contain up-to-date data on household income and literacy rates, respectively.

Chapter 13 covers the political and economic institutions. This edition introduces new material on several topics: contingent employment, offshoring of U.S. jobs, political action committees, the power elite, and government regulation of business. Employee rights is the subject of a new *Sociology and You.* The *Worldview* is updated, and three new figures and two new tables are added.

Chapter 14 delves into sociology's unique perspective on the institution of religion, including new material on religiosity in the United States. A new *Sociology and You* looks at the danger of cults. A new *Snapshot of America* provides nationwide information on public charities. Gender inequality in religion is the subject of a new *Worldview*. Also added are three new figures and one new table.

Chapter 15 is a new combined chapter covering health and health care and the inequalities of age, the contents of which were listed earlier. The *Snapshot of America* is updated, and a new *Worldview* displays data on worldwide population aging. A new figure on rising health-care expenditures is added, and the other figures are updated.

Chapter 16 is an entirely new chapter on sport in the United States. Its major topics are the nature of sport, theoretical perspectives and sport, sport and social mobility, racial and ethnic inequality in sport, and gender inequality in sport. New concepts include stacking, sport subculture, and sport. A new *Doing Research* is titled "Tough Guys, Wimps, and Weenies." A new *Sociology and You* explores ways to avoid bigotry in sports. A new *Snapshot of America* and a new *Worldview* cover baseball and Olympic success, respectively. This chapter includes three new figures and two new tables.

Chapter 17 contains the dual and related topics of population and urbanization. The following concepts are new to this edition: urban area, urban cluster, urbanized area, and combined statistical area. A new *Sociology and You* deals with the uses of demography in business. There is a new *Snapshot of America* on the population under age 18 and a new *Worldview* on urban population. This chapter incorporates seven new figures.

Chapter 18 combines the areas of social change and collective behavior. New material is included on the natural environment as a source of social change and on "smart mobs." A new *Sociology and You* focuses on terrorist attacks and disaster myths. Both the *Snapshot of America* and *Worldview* are revised.

Distinctive Study Aids

Joseph Butler, an eighteenth-century English minister and moral philosopher, wrote:

> [P]eople habituate themselves to let things pass through their minds, as one may speak, rather than to think of them. Thus by use they become satisfied merely with seeing what is said, without going any further. Review and attention, and even forming a judgment, becomes fatigue; and to lay anything before them that requires it, is putting them quite out of their way. (Butler 1983:11; originally published in 1726)

Butler eloquently expresses the elements underlying our approach to learning in this textbook: the SQ3R method and critical thinking. Students continue to express enthusiasm for this approach. It helps them combat passivity and become better active learners, comprehending the material more fully as they become progressively more mentally involved.

SQ3R: A Format for Study

This text is designed with the "SQ3R" study format at its core. Research-tested, this approach helps students identify significant ideas, understand these ideas rapidly, remember important points, and review effectively for exams. As a result, students learn more about sociology more easily while performing better on tests. I recommend that instructors and students go over these steps together.

The letters in SQ3R symbolize five steps in effective reading and learning: survey (S); question (Q); read, recite, and review (3R). The steps in the SQ3R method are built into each chapter.

1. *Survey.* Before reading the chapter, students should read the outline, the introduction, and the summary. This survey, which will give them an overall picture of the chapter content, should take only a few minutes.
2. *Question.* Third-level headings are phrased as questions to help students select and concentrate on the important points. For example, instead of seeing subtopic headings such as "Working Women," they will find such questions as "Have men and women reached financial equality?" and "How do American women fare globally?"

3. *Read.* For increased comprehension, students should focus on each third-level question as they read the material that answers it.
4. *Recite.* Students should answer each third-level heading immediately after reading the relevant material. If they are unable to answer a question, they should examine the material until they find the answer. Also, at the end of every major topic is a final recitation check called *Feedback.* (See sample *Feedback* questions on the following page.) If students cannot answer one of these self-test questions, they should note the correct answer given under the questions and look back at the text material to find out why this is the right answer. The recitation dimension of the SQ3R method will prevent them from deluding themselves into believing that they understand material when, in fact, they do not. This process is designed to replace surface recognition with a more thorough comprehension.
5. *Review.* After completing a chapter, students should once again briefly answer the questions posed in the third-level headings. Even better, they should have another person ask them the questions. They can complete their review of a chapter by using the *Review Guide* that appears at the end of each chapter.

Critical Thinking

Critical thinking—questioning commonly held assumptions—is traced by intellectual historians to fifth-century-B.C. Greece, particularly to the Athenians (Brinton 1963). Full-fledged Western interest in critical analysis did not appear until the eighteenth century, the period known as the Enlightenment (Gay 1966). Respect for critical and reasoned analysis, rather than judgments based on emotion, has been a key element of the Western world ever since.

Critical thinking is crucial for today's college students. First, the tradition of liberal education is the tradition of critical thought (Pelikan 1992). Second, as the nature of work continues to move from physical labor to cerebral activities, the facility for critical reasoning becomes an increasingly valuable asset on the job. Third, it doesn't take a rocket scientist to figure out the need for critical thought among all American voters; behavior of our political leaders makes this point daily. Finally, but not least, critical thinking is vital in personal contexts such as family life, decision making outside of work, and personal enrichment.

Critical thinking is incorporated in this edition in three ways:

1. **Critical thinking within the SQ3R method.** The promotion of critical thinking lies at the heart of the "question" step of the SQ3R

F E E D B A C K

1. The _____ is the set of mind that enables individuals to see the relationship between events in their personal lives and events in their society.

2. Which of the following is not one of the uses of the sociological imagination?
___ a. seeing the interplay of self and society
___ b. creating new aspects of culture not thought of by others
___ c. questioning aspects of social life most people take for granted
___ d. understanding the social forces that shape daily life

3. DuBois focused only on the American race question. T or F?

4. Match the three feminist theoretical frameworks with the following words and phrases.
___ a. liberal feminism   (1) patriarchy
___ b. radical feminism   (2) capitalism and patriarchy
___ c. socialist feminism   (3) equality of opportunity
                               (4) social and legal reform

Answers: 1. Sociological imagination; 2. B; 3. F; 4a. (3), 4b. (1), 4c. (2)

method. This point can be made by contrasting two approaches to learning: the *informational* approach and the *critical-thinking* approach. The objective in the informational approach is knowledge acquisition, important when one is exposed to new material. Following are some sample SQ3R informational questions.

- Do extremely isolated children develop human characteristics?
- To what extent do Americans exercise power through the ballot box?
- Do American workers like their jobs?

Although gaining information is essential, it is only the beginning. Critical thinking emphasizes further interaction with knowledge as it is being acquired. Following are some sample critical-thinking SQ3R questions:

- If prisons do not rehabilitate, what are some alternatives?
- Is the negative image of the poor in America accurate?
- What is required to prove the existence of the power elite in American society?

2. **Critical-thinking questions.** Critical-thinking questions are liberally interspersed throughout each chapter. Questions follow the *Doing Research* and *Sociology and You* features. A critical-thinking question is also included in each table, figure, and map. A set of four to six critical-thinking questions appears in the *Review Guide* at the end of each chapter. Consequently, a critical-thinking opportunity is offered at least every few pages. These wide-ranging questions encourage students to think critically and creatively about the ideas within a chapter. Sometimes students will apply these ideas to a particular aspect of

society. At other times students will use sociological ideas to analyze and understand events and experiences in their own lives.

3. **Critical feedback.** Each *Doing Research* closes with a series of critical-thinking questions under the heading, "Thinking About the Research," designed to help students better understand the piece of sociological research involved and to probe below the surface. For example, these questions are posed following the description of Stanley Milgram's study of group pressure and conformity:

- Discuss the ethical implications of Milgram's experiment.
- If the researcher had not been present as an authority figure during the experiment to approve the use of all shock levels, do you think group pressure would have been as effective? Explain.

4. **The sociological imagination.** It is easy to fall into a pattern of nonreflection about prevailing ideas that are passed from generation to generation. The feature *Using the Sociological Imagination*, described earlier, opens each chapter with a question designed to challenge some aspect of a social myth.

Review Guide

Each Review Guide begins with a chapter summary, followed by the chapter learning objectives. Next is a concept review of approximately 50 percent of the concepts introduced in the chapter. Students can test their grasp of key concepts by matching concepts with definitions. Several critical-thinking questions follow the concept review. These broad questions provide practice for essay tests. A set of multiple-choice questions then acts as a mini self-test. The feedback review consists of a sample of questions taken directly from the Feedback questions throughout the chapter. In most chapters, a graphic review feature tests understanding of a particular table or figure in the chapter. An answer key closes each Review Guide.

Supplements for the Updated Tenth Edition

Supplements for the Instructor

Instructor's Edition of *Sociology: The Essentials.* An Instructor's Edition (IE) of this text containing several features useful to instructors is available. The IE contains a walk-through of the several themes and many features of *Sociology* along with a complete listing of available bundles for this text. To obtain a copy of the Instructor's Edition, contact your Cengage Sales Representative.

Instructor's Resource Manual with Test Bank. This manual offers the instructor brief chapter outlines, student learning objectives, detailed lecture outlines, and teaching suggestions to facilitate in-class discussion. American Sociological Association (ASA) recommendations are noted for each chapter and matched to the A-heads of the detailed outlines to help instructors streamline their teaching methods with the American Sociological Association. Innovative class activities and Internet activities are included for each chapter. The test bank has 180 new questions, which are noted, and each multiple-choice and true/false item has the question type (factual, applied, or conceptual) and corresponding learning objective indicated.

ExamView® Computerized Testing. Create, deliver, and customize tests and study guides (both print and online) in minutes with this easy-to-use assessment and tutorial system. ExamView offers both a Quick Test Wizard and an Online Test Wizard that guide you step-by-step through the process of creating tests. The test appears on screen exactly as it will print or display online. Using ExamView's complete word-processing capabilities, you can enter an unlimited number of new questions or edit existing questions included with ExamView. Includes all new Tenth Edition test bank questions.

PowerLecture™ with ExamView®: A 1-stop Microsoft® PowerPoint® Tool. PowerLecture instructor resources are a collection of book-specific lecture and class tools on either CD or DVD. The fastest and easiest way to build powerful, customized media-rich lectures, PowerLecture assets include chapter-specific PowerPoint presentations, ExamView test files, images, animations and video, instructor manuals, test banks, useful Web links and more. PowerLecture media-teaching tools are an effective way to enhance the educational experience.

Wadsworth's Introduction to Sociology 2007 Transparency Masters. A set of black-and-white transparency masters consisting of tables and figures from Wadsworth's introductory sociology texts is available to help prepare lecture presentations. Free to qualified adopters.

Videos. Adopters of *Sociology* have several different video options available with the text.

Wadsworth's Lecture Launchers for Introductory Sociology. An exclusive offering jointly created by Wadsworth/Cengage Learning and Dallas TeleLearning, this video contains a collection of video highlights taken from the "Exploring Society: An Introduction to Sociology" Telecourse (formerly "The Sociological Imagination"). Each three- to six-minute-long video segment has been especially chosen to enhance and enliven class lectures and discussion of 20 key topics covered in any introductory sociology text. Accompanying the video is a brief written description of each clip, along with suggested discussion questions to help effectively incorporate the material into the classroom.

ABC® Videos (Introduction to Sociology Volumes I–IV). ABC Videos feature short, high-interest clips from current news events as well as historic raw footage going back 40 years. Perfect for discussion starters or to enrich your lectures and spark interest in the material in the text, these brief videos provide students with a new lens through which to view the past and present, one that will greatly enhance their knowledge and understanding of significant events and open up to them new dimensions in learning. Clips are drawn from such programs as *World News Tonight*, *Good Morning America*, *This Week*, *PrimeTime Live*, *20/20*, and *Nightline*, as well as numerous ABC News specials and material from the Associated Press Television News and British Movietone News collections.

Readers

***Classic Readings in Sociology*, Fourth Edition (edited by Eve L. Howard).** This series of classic articles written by key sociologists will complement any introductory sociology textbook. The reader serves as a touchstone for students: original works that teach the fundamental ideas of sociology.

***Understanding Society: An Introductory Reader*, Third Edition (edited by Margaret Andersen, University of Delaware, Kim Logio, St. Joseph's University, and Howard Taylor, Princeton University).** This reader includes articles with a variety of styles and perspectives, with a balance of the classic and contemporary. The editors selected readings that students will find accessible, yet intriguing, and have maximized the instructional value of each selection by prefacing each with an introduction and following each with discussion questions. The articles center on the following five themes: classical sociological theory, contemporary research, diversity, globalization, and application of the sociological perspective.

***Sociological Odyssey: Contemporary Readings in Introductory Sociology,* Third Edition (edited by Patricia A. Adler, University of Colorado and Peter Adler, University of Denver, Colorado).** This reader contains highly engaging, current articles that bring hot sociological topics to life for readers. Students enjoy the Adlers' balanced selection of timely readings, which cover issues and areas of interest to the general collegian, including male college cheerleaders, exploitation of immigrant laborers, teenage girls and cell phone usage, cyber romantic relationships, and cheating. In-depth introductions, explanation of theory, and discussion questions before each reading help guide students through the material.

***Sociological Footprints: Introductory Readings in Sociology,* Eleventh Edition (edited by Leonard Cargan, Wright State University and Jeanne H. Ballantine, Wright State University).** This anthology of 76 sociological studies provides a link between theoretical sociology and everyday life. *Sociological Footprints* offers classical, contemporary, popular, and multicultural articles in each chapter to show students a wide range of perspectives. Articles address today's most important social issues that are relevant to student's lives, social construction of sex, divorce trends, and American health care crisis. Each article is preceded by a short introductory essay, a series of guideline questions, and a list of glossary terms found in the article and therefore establish the student's reading of the article into a solid sociological framework.

Online Resources

Shepard, *Sociology,* Tenth Edition Student Companion Website. *www.cengage.com/sociology/shepard/* This website, available to all students, provides useful learning resources for each chapter of the book:

- Practice quizzes
- Glossary
- Flash cards
- Crossword puzzles
- Web links
- Video exercises
- Learning modules
- Virtual Explorations
- InfoTrac® College Edition with InfoMarks® exercises
- MicroCase exercises
- Map the Stats exercises
- Animations

WebTutor™ ToolBox for WebCT® or Blackboard®. Preloaded with content and available free via access code when packaged with this text, WebTutor ToolBox pairs all the content of this text's rich Student Companion Website with all the sophisticated course management functionality of a WebCT or Blackboard product. You can assign materials (including online quizzes) and have the results flow automatically to your gradebook. ToolBox is ready to use as soon as you log on—or, you can customize its preloaded content by uploading images and other resources, adding Web links, or creating your own practice materials. Students have access only to student resources on the Student Companion Website. Instructors can enter a pincode for access to password-protected Instructor Resources.

InfoTrac® College Edition with InfoMarks®. With each purchase of a new copy of the text comes a free four-month pincode to InfoTrac College Edition, the online library that gives students anytime, anywhere access to reliable resources. This fully searchable database offers 20 years' worth of full-text articles from almost 5,000 diverse sources, such as academic journals, newsletters, and up-to-the-minute periodicals including *Time, Newsweek, Science, Forbes,* and *USA Today.* This incredible depth and breadth of material—available 24 hours a day from any computer with Internet access—makes conducting research so easy, your students will want to use it to enhance their work in every course. Through InfoTrac's InfoWrite, students now also have instant access to critical-thinking and paper writing tools. Both adopters and their students receive unlimited access for four months.

Acknowledgments

No page of this edition is without the influence of my wife, Kay Shepard. The extent of her contributions make trying to be specific futile. That we both know is enough.

Several colleagues have provided thoughtful and helpful reviews for the Tenth Edition. Many thanks to the following individuals: Jeremy Cook, Oklahoma State University; Ron Stewart, Buffalo State College; Keith Kerr, Quinnipiac University; and Larry Rosenberg, Millersville University.

In addition to those individuals, the following colleagues have provided critiques of previous editions: G. William Anderson, Robert Anwyl, Paul J. Baker, Melvin W. Barber, Michael G. Bisciglia, Jerry Bode, Patricia Bradley, Ruth Murray Brown, Brent Bruton, Victor Burke, Bruce Bylund, David Caddell, Albert Chabot, Reba Chaisson, Stephen Childs, William T. Chute, Carolie Coffey, Kenneth Colburn Jr., William F. Coston, Jerome Crane, Ray Darville, Betty J. Daughenbaugh, Alline DeVore, Mary Van DeWalker, Mary Donahy, Susan B. Donohue, M. Gilbert Dunn, Lois Easterday,

Mark G. Eckel, Irving Elan, Ralph England, K. Peter Etzkorn, Mark Evers, Susan Farrell, Joseph Faulkner, Kevin M. Fitzpatrick, John W. Fox, Jesse Frankel, Larry Frye, Pamela Gaiter, Roberta Goldberg, Michael Granata, Ramona Grimes, James W. Grimm, Laura Gruntmeir, Justine Gueno, Rebecca Guy, Penelope J. Hanke, Thomas Harlach, Cynthia Hawkins, Kenneth E. Hinze, Amy Holzgang, Carla Howery, Emmit Hunt, David H. Kessel, Gary Kiger, James A. Kitchens, Joseph Kotarba, Irving Krauss, Mark LaGory, Raymond P. LeBlanc, Jerry M. Lewis, Jieli Li, Roger Little, Richard Loper, Roy Lotz, Scott Magnuson-Martinson, Judy Maynard, Elaine McDuff, R. Lee McNair, Doris Miga, Alfredo Montalvo, Edward V. Morse, Charles Mulford, Bill Mullin, Daniel F. O'Connor, Jon Olson, Charles Osborn, Thomas R. Panko, Margaret Poloma, Roksana Rahman, Carol Axtell Ray, Deborah Richey, Ellen Rosengarten, William Roy, Josephine A. Ruggiero, Steven Schada, Paul M. Sharp, Edward Silva, James Skellinger, LaFleur Small, Robert P. Snow, Mary Steward, Ron Stewart, Robert F. Szafran, David Terry, Ralph Thomlinson, Marcella Thompson, Charles M. Tolbert, Deidre Tyler, David Waller, Carrol W. Waymon, Michael E. Weissbuch, Dorether M. Welch, Lois West, Carol S. Wharton, Douglas L. White, Paul Wozniak, David Zaret, and Wayne Zatopek.

The Sociological Perspective

© Larry Harwood/Cengage Learning

LEARNING OBJECTIVES

- Illustrate the unique sociological perspective from both the micro and macro levels of analysis.
- Describe three uses of the sociological perspective.
- Distinguish sociology from other social sciences.
- Outline the contributions of the major pioneers of sociology.
- Summarize the development of sociology in the United States.
- Identify the three major theoretical perspectives in sociology today.
- Discuss feminist theory.
- Compare two recent developments in symbolic interactionism.

USING THE SOCIOLOGICAL IMAGINATION

Why do people commit suicide? Answers immediately come to mind: prolonged illness, loss of a lover, public disgrace, depression, heavy financial loss. Each of these explanations assumes personal crisis as the sole motivation. Yet, suicide is affected not only by personal trauma but also by social forces. Specifically, sociologist Emile Durkheim revealed that suicide rate varies with social characteristics. Highly socially integrated people—married, females, Catholics—exhibit lower rates of suicide. More socially isolated persons—unmarried, males, Protestants—show higher suicide rates. After 100 years, research continues to support Durkheim's findings and conclusions.

Durkheim is one of the pioneers of sociology who will be profiled in this chapter. Before turning to these pioneers, however, we will discuss the unique sociological perspective that Durkheim identified and developed (R. A. Jones 2005).

The Sociological Perspective

The Importance of Perspective

Why begin your study of sociology with a discussion of perspective? Perspective matters. Babies are usually brighter and better looking to their parents than to others. Newlyweds nearly always find their spouses infinitely more attractive than do their friends.

We all interpret what is happening around us through our own perspective—our own particular point of view. It can hardly be any other way because we normally do not even realize the extent to which our beliefs, attitudes, and views of reality are determined by our perspective. Sometimes, though, when our outlook is challenged, we may be jarred into realizing how much we take it for granted (see Worldview 1.1).

Perspective is reflected in a Native American's reply to white men in eighteenth-century Virginia. The Virginia colonists had offered to "properly educate" some young Indian boys at the College of William and Mary in Williamsburg. Here is the Native American's response:

WORLDVIEW 1.1

A World Turned Upside Down

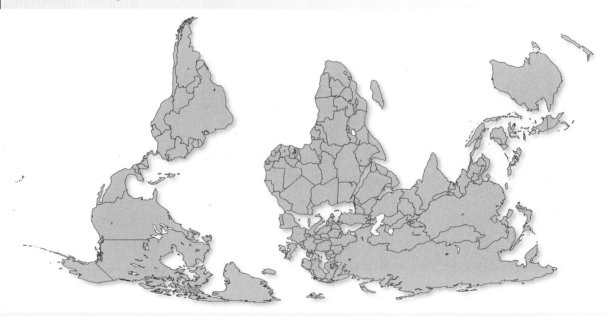

Without turning this map upside down, locate the United States. Do you find it more difficult than usual? Do you find this view of the world disorienting? Because you are so used to the conventional representation of the Earth, you may reject this worldview. So it is with any perspective.

In this book, you will be asked to abandon the typical American psychological perspective in favor of the sociological perspective.

1. How would you feel if this map were universally substituted for the one you

know? Describe your feelings, and explain why you feel this way.

2. What does your reaction to this map tell you about the power of the perspective you bring to a situation?

We know that you highly esteem the kind of learning taught in . . . [your] colleges. . . . But you, who are wise, must know that different nations have different conceptions of things; and you will not therefore take it amiss, if our ideas of this kind of education happen not to be the same with yours. We have had some experience of it; several of our young people were formerly brought up at the colleges of the northern provinces; they were instructed in all your sciences; but, when they came back to us, they were bad runners, ignorant of every means of living in the woods, unable to bear either cold or hunger, knew neither how to build a cabin, take a deer, nor kill an enemy, spoke our language imperfectly, were therefore neither fit for hunters, warriors, nor councelors; they were totally good for nothing.

We are however not the less obligated by your kind offer, though we decline accepting it; and, to show our grateful sense of it, if the gentlemen of Virginia will send us a dozen of their sons, we will take care of their education, instruct them in all we know, and make men of them. (Anonymous)

To the colonists' surprise, the benefits of a William and Mary education were not desired by the tribal elders. In fact, from the Native American perspective, the gentlemen of Virginia would be much better off to leave William and Mary and pursue an education in the tribe. Because of their own perspective, no William and Mary students transferred.

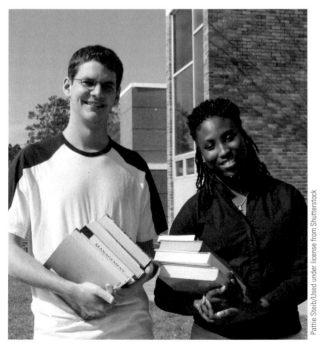

The decision to attend college is not merely a choice made by an isolated individual. Social relationships, with teachers, parents, peers, and others, influence a person's decision to become a college student. Sociologists study these patterned social relationships referred to as social structures.

Sociology has a unique perspective. The first segment of this chapter is devoted to this perspective—a focus on groups, not individuals, as an object of investigation.

Defining Sociology

At this point, how would you define sociology? As a novice to the field, you may at first view sociology as the study of human groups. This inclination has the advantage of tapping ideas you already possess. As you progress, however, you will acquire a more precise understanding of **sociology** as the scientific study of social structure.

In the meantime, your understanding of the concept of social structure will be understandably vague. For now it is sufficient for you to become familiar with two key aspects of the sociological perspective: (1) the interaction between social structure and the individual and (2) the idea of two levels of analysis.

Social Structure and the Individual: A Two-Way Street

What is unique about the perspective of sociology? Many of the leaders of the American Revolution reflect the essential characteristic of the sociological perspective. For example, both George Washington and John Adams, in accounting for the successful start of the American political experiment, pointed to social factors beyond their control. For Washington, the success of the revolution was due to the existence of new political ideas provided by the Enlightenment and the blessing of vast natural resources (Ellis 2007).

Similarly, sociologists do not concentrate on the idiosyncratic behavior of individuals, but on the patterns of behavior individuals share with others in their group or society. The starting point for sociology is the predictability and recurrence of social behavior. Sociologists are interested in the patterns of social relationships referred to as **social structures.** The average person tends to explain human behavior in individualistic or personal terms: a young man goes to war to prove his patriotism; a woman divorces her husband to develop her potential; a college student commits suicide to escape depression. Sociologists attempt to explain these same events without relying on the personal motivations of individuals. They look for social rather than personal explanations when they examine war, divorce, or suicide: young men go to war because they have been taught by their society to be patriotic; more women divorce because of the social trend toward sexual equality; college students commit suicide because of pervasive societal expectations of academic performance. Sociologists do not speak of a young man, a married woman, or a college student. They concentrate on categories of people—young men, married women, college students.

Pattie Steib/Used under license from Shutterstock

Can you think of an example of "patterned" social behavior? As you know, college students in a classroom do not all behave exactly alike. Some attempt to write down everything their professors say, some just listen to the lecture, some use tape recorders, and some tune in and out. Yet, if you visit almost any college or university, you will find patterned relationships. Professors lecture, students remain in their seats; professors give examinations, students take them. Although the individual characteristics of students and professors may vary from school to school, students and professors relate in similar patterned ways. It is the *recurrent* patterned interaction of people and the social structures created by such interaction that capture the attention of sociologists.

How does group behavior differ from individual behavior? Sociologists assume that social relationships are not solely determined by the particular characteristics of the individuals involved. Emile Durkheim (1966; originally published in 1895), a pioneering nineteenth-century sociologist, argued that we do not attempt to explain bronze in terms of its component parts (lead, copper, and tin). Instead, we consider bronze an alloy, a unique metal produced by the synthesis of several distinct metals. Even the consistency of bronze is not predictable from its components: bronze is hard, whereas lead, copper, and tin are soft and malleable. Durkheim reasoned that if a combination of certain metals produces a unique metal, some similar process might happen in groups of people. Indeed, people's behavior within a group setting cannot be predicted from the characteristics of individual group members. Something new is created when individuals come together as a collective. For example, after winning the Super Bowl, some exuberant and rowdy Denver Bronco fans tipped over trash cans and cars, spilled newspaper boxes, broke windows in buildings, tore down street signs, and started bonfires—behavior they would not have exhibited as lone individuals. When the World Trade Organization met in Seattle, many protestors peacefully demonstrated their opposition to its global economic effects, while others broke shop windows and looted stores. In 2005 civil unrest in French cities, involving for the most part French Arab rioters, destroyed more than 1,000 vehicles and many buildings. When Hurricane Katrina devastated New Orleans, terror from crime, both real and imagined, delayed evacuation of the dead and injured, caused some police officers to quit, and grounded helicopters.

On the other hand, the massacre of students at Virginia Tech led to unprecedented cooperation with its archrival, the University of Virginia. The historically bitter rivalry between the two schools was temporarily laid aside when the two university bands played in a first-ever joint concert the night before the 2007 football game. Cooperation extended beyond athletics. Many University of Virginia students and community members wore the Virginia Tech colors at a candlelight vigil. The University of Virginia Student Council established a memorial fund in conjunction with the Virginia Tech Alumni Association. Similarly, when Brazil, an ethnically diverse and often conflicted country, won the men's soccer World Cup, nearly 50,000 Brazilians peacefully and joyfully cheered in São Paulo.

Group members generally tend to act like each other even when their personal preferences are not the same as their group's. Some fraternity and sorority members, for example, drink alcoholic beverages even though, as individuals, they would prefer not to. Similarly, because of peer pressure, members of campus religious groups may not drink alcohol despite their preference for it. A study of helping behavior illustrates the power of social structure. Social psychologist Robert Levine (2003) found the best predictor of people's willingness to aid others in trouble to be population density. Helping behavior declines as population density increases. However, when individuals from large cities are in less densely populated places, they are as helpful as the residents. Due to the high degree of conformity within a group, similarities in social behavior exist despite extensive differences among individual members.

How is conformity related to group behavior? We live in groups ranging in size from a family to an entire society, and they all encourage conformity—often, conformity promoted by social forces that individuals do not create and cannot control. Because of this accommodation, people who belong to similar groups tend to think, feel, and behave in similar ways. American, Russian, and Chinese citizens, for instance, have distinctive eating habits, types of dress, religious beliefs, and attitudes toward family life. Groups of teenagers tend to listen to the same music, dress alike, and follow similar dating customs.

Why do people conform? Conformity within a group occurs in part because most of its members believe that their group's ways of thinking, feeling, and behaving are the best; they have been successfully taught to value their group's ways. And as already noted, some group members tend to comply even against their personal inclinations. Whether because members value their group's ways or because they yield to social pressures of the moment, the behavior of a group cannot be predicted simply from knowledge about the individual members. A group is not equal to the sum of its parts any more than is bronze.

Why is the existence of conformity important to sociology? Because conformity within societies exists, sociologists can attempt to understand, explain, and predict the often invisible social processes that permit successive generations to live predictable and orderly lives without each generation creating its own new guidelines for social living. Because a particular generation is spared this trouble, its members usually fail to ask why things are the way they are, or why things are changing. Sociologists, in contrast, constantly wrestle with these questions.

But don't people, in turn, affect society? Yes. If our discussion of sociology to this point causes you to believe that sociology is interested only in the effects of social structures on individuals, it has been misleading. It is necessary, however, to begin with this emphasis because without some considerable degree of conformity, there can be no group life. Still, this necessary nonindividualistic emphasis should not obscure the effect that individuals can have on social structures. Individuals are active, thinking beings who do not always follow social scripts. The interplay between individuals and their social structures is a two-way street: people are affected by social structures, and people change their social structures. Consequently, you should not conclude that all human behavior in groups is determined by preexisting social structures. For example, historian Joyce Appleby (2001) documents how the first generation born after the American Revolution fashioned a national culture based on an interpretation of their parents' creation of a new nation.

TABLE 1.1	
WHAT SOCIOLOGISTS SEE AND HOW THEY STUDY SUICIDE: Comparing Protestants and Catholics	
SOCIOLOGISTS FOCUS . . .	**SOCIOLOGISTS MIGHT INVESTIGATE . . .**
on the group, not the individual.	distinct characteristics of both Protestants and Catholics that affect their differential suicide rates.
on patterns of social behavior.	data to determine if Protestants and Catholics actually do consistently commit suicide at different rates.
on social forces that encourage conformity.	the ways that greater social integration among Catholics promotes a lower rate of suicide compared to more socially isolated Protestants.
on the effects people have on social structure.	means that Protestants may use to lower their suicide rate, such as the promotion of social involvement through church activities.

In summary, the sociological perspective focuses on the group, examines patterns of behavior, isolates patterns of conformity, and recognizes the effects of people on society. Table 1.1 illustrates these unique aspects of the sociological perspective using suicide rates among Catholics and Protestants.

Levels of Analysis: Microsociology and Macrosociology

An important distinction is implied in our discussion of the sociological perspective. A careful reading reveals the interest of sociology in both the interaction of people "within" social structures (*microsociology*) and the "intersection" of social structures per se (*macrosociology*).

Macrosociology and microsociology should be thought of as levels of analysis, because they determine which aspects of social structure a sociologist wishes to analyze. Macro means large; micro means small. Macro research focuses on groups as a whole. Microanalysis investigates the relationships within groups. Macro views from a distance, micro from close range (Helle and Eisentadt 1985a, 1985b; Scheff 1990; Hawley 1992).

Suppose for the moment that social structures are directly observable as tangible objects and that we are cruising at 20,000 feet at the controls of Sociologist One. On a clear day at this altitude we would be able to observe definite social structures, just as we can observe the lay of the land from an airplane. At 20,000 feet we will see no movement in the form of people interacting. We are at the macro level. Suppose we wish to get a closer look and start descending. As we descend, we will begin to see people interacting. When we focus on these relationships, we are at the micro level. Formal definitions of microsociology and macrosociology will make more sense against this general background.

Microsociology is concerned with the study of people as they interact in daily life. Consider the practice of knife fighting among street gangs. At the micro level, a sociologist attempts to explain participation in gang knife fighting based on the social relationships involved. For example, gang leaders, to validate their right to leadership, may feel that they must either fight members of their own gang who challenge their position or fight leaders of other gangs.

Macrosociology focuses on groups without regard to the interaction of the people involved. Some sociologists examine entire societies (Tilly 1978, 1986; Wallerstein 1979, 1984; Skocpol 1985; Lenski 1988; Nolan and Lenski 2005). We will use the term macrosociology in referring to the study of societies as a whole as well as to the relationships between social structures within societies. For example, sociologists might study the patterned relationships between the defense industry and the

1. Sociology is the scientific study of _____.
2. _____ explanations of group behavior are inadequate because human activities are influenced by social forces that individuals have not created and cannot control.
3. Microsociology focuses on relationships between social structures without reference to the interaction of the people involved. T or F?

Answers: 1. social structure; 2. Individualistic; 3. F

federal government, or the effect of the economy on the stability of the family.

The macro and micro levels of analysis are complementary; their combined use tells us more about social behavior than either one alone (Alexander et al. 1987; Sawyer 2001). To understand gang warfare, a microsociologist would want to know about the social relationships involved—the relationships between gang leaders and followers or between gang members and police on the beat. To supplement this understanding of gangs obtained at the micro level, the macro level can be used to examine the larger social structures of which gangs are a part. A macrosociologist might look for the aspects of a society or social structure that produce the poverty promoting delinquency in the first place—such as lack of education and joblessness.

Uses of the Sociological Perspective

Why study sociology? The sociological perspective offers you three personal benefits. First, it enables you to develop the *sociological imagination*. Second, sociological theory and research can be applied to important public issues. Third, the study of sociology can sharpen skills useful in many occupations.

The Sociological Imagination

How does the self interplay with society? In his famous essay "On Liberty," nineteenth-century philosopher John Stuart Mill wrote:

> It will probably be conceded that it is desirable [that] people should exercise their understandings, and that an intelligent following of custom, or even occasionally an intelligent deviation from custom, is better than blind and simple mechanical adhesion to it. (Mill 1993:68–69; originally published in 1859)

Knowing how social forces affect our lives can help prevent us from being prisoners of those forces. C. Wright Mills called this personal use of sociology the **sociological imagination**—the set of mind that enables individuals to see the relationship between events in their personal lives and events in their society. Events affecting us as individuals, Mills pointed out, are closely related to the ebb and flow of society (Mills and Etzioni 1999). Decisions, both minor and momentous, are not isolated, individual matters. Historically, for example, American society has shown a bias against childless marriages and only children. Couples without children have been considered selfish, and only children have often been labeled "spoiled" (Benokraitis 2007). These values date back to a time when there was a societal need for large families because of the high infant mortality rate and the need for labor on family farms. Only now, as the need for large families is disappearing, are we reading of the benefits of one-child families—to the child, to the family, and to society. The sociological imagination enables us to understand the effects of such social forces on our lives. With this understanding, we are in a stronger position to make autonomous decisions rather than merely conform (Game and Metcalfe 1996; Peck and Hollingsworth 1996; K. T. Erikson 1997).

This broadened social awareness permits us to read the newspaper with a more complete understanding of the implications of social events. Instead of interpreting an editorial opposing welfare as merely a selfish expression, we might see the letter as a reflection of the importance Americans place on independence and self-help (A. M. Lee 1990; Straus 2002).

Ashley Montagu has coined the term psychosclerosis to describe a mental condition that deprives people of the ability to accept new ideas. What are the symptoms of psychosclerosis, or "hardening of the mind"?

> Well, a certain lack of ability to see 360 degrees, to see all the way behind, sideways, as well as forward, to pick up the small and subliminal signals. To be wedded to fixed, stereotyped ideas that this is the right way and the only way. (Montagu 1977:50)

Because the sociological imagination expands our horizons, it may be considered an antidote for psychosclerosis.

How will the debunking theme stimulate your sociological imagination? Despite the availability of accurate information and explanations, people tend to cling to myths and false ideas about social life—impressions that are passed from generation to generation. Plato's "Allegory of the Cave" comes to mind. In *The Republic*, Plato describes a cave in which people have been chained, from childhood, so they cannot move. Forced to look only straight ahead, these prisoners cannot see behind them the blazing fire and a raised walkway in front of the fire. Men walk along the wall carrying all types of objects. The shadows on the wall in front of the prisoners, made by the walking men and the fire, are the only things the prisoners have ever seen. Their interpretations of these shadows constitute reality to them. Sociology attempts to replace common misconceptions about social life (shadows on the cave wall) with accurate information and explanations (Ruggiero 2001).

Because the task of sociology is to reveal the nature of human social behavior, it often leads us to question what we usually take for granted. What people take to be unassailable truth may, under scientific examination, be proven false. The sociological imagination, then, replaces common misconceptions about social life with accurate information and explanations.

How is the sociological imagination intellectually liberating? Like all liberal arts courses—anthropology, history, literature, and philosophy—sociology encourages intellectual liberation (Bierstedt 1963; Brouilette 1985). You can, for example, learn that the sacrificial murder of infants has been a normal and legitimate aspect of some societies and that those who participate in this practice are not evil and despicable. Among the Winnebago Indians of Wisconsin, for example, it was socially acceptable for parents to kill children who became too great a liability (Radin 1953). Among the Toda of southern India, female infants were frequently killed at birth; if twins of different sexes were born, the female was always killed (Murdock 1935). The Winnebago and Toda, like all humans, had been taught the ways of their group. Through the sociological imagination, you can become aware of the rationale for infanticide practiced by a society under extreme population pressure and living on the margin of subsistence.

Sociology provides a window to the social world that lies outside our own. It allows us to see the many social forces that shape our own lives. Sociology thus complements rather than replaces other ways of viewing human behavior. The more we seek to apply a range of different perspectives in attempting to interpret social forces and their meanings, the greater will be our understanding of our own behavior as well as the behavior of others. With such understanding comes the potential for greater personal freedom from social pressures. Consider divorce, for example. Social research reveals many factors in modern society that promote marital failure. Divorce is more likely to occur among couples who share particular characteristics: low levels

The sociological imagination allows us to see that joining a gang is a social act providing some young men and women with a sense of security and belonging they haven't found elsewhere.

Jack Star/Getty Images

SNAPSHOT OF AMERICA 1.1

Health Insurance

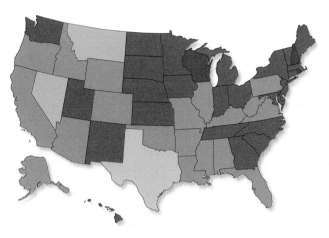

**Percentage of Children Under
18 Who Are Uninsured**

- 4.9% and below
- 5.0%–9.9%
- 10%–14.9%
- 15%–19.9%
- 20% and above

Source: U.S. Census Bureau, *Statistical Abstact of the United States*: 2004–2005

Americans tend to believe that all children have the right to recieve good medical care. Research has shown, however, that American children do not have equal access to medical care. One reason for this is a lack of health insurance. This map shows, by state, the percentage of children under age eighteen who do not have health insurance.

Interpret the Map

1. Which states have the fewest uninsured children? Which state has the most?
2. How does your state rate on health insurance coverage for children?
3. How might using your sociological imagination help you to explain why children in the United States have unequal access to medical care?

---

of education, early marriages, premarital pregnancies, or living arrangements with parents. As predictive as these characteristics are, they cannot be used to force couples not to marry. However, awareness of the inevitable pressures of such marriages may convince couples under these circumstances either not to marry, or at

least to anticipate and prepare more effectively for such problems. The successful application of the sociological imagination enables us to share Somerset Maugham's insight that "tradition is a guide and not a jailer."

Applied Sociology

Do sociologists have a moral responsibility to speak against and attempt to change aspects of social life they believe to be wrong? From the time it first appeared on the American academic scene in the late 1800s, sociology has steadily attempted to move from its origins as a social problem-solving discipline to a nonsocially involved science. During the intervening years, disagreement has periodically surfaced on the compatibility of these two viewpoints (A. M. Lee 1978; Weinstein 2000; Hamilton and Thompson 2002). Those arguing against the involvement of social scientists in the eradication of social ills view science as "value neutral"; that is, scientific research is supposed to be directed at discovering what actually exists, with no room for personal value judgments as to what *ought* to exist. The idea of value neutrality has dominated sociological thought for a long time. And although the idea of science as a value-neutral enterprise remains strong among sociologists, those favoring the interjection of standards of good and bad into scientific activity are on the increase (J. Q. Wilson 1993; Bickman and Rog 1997). This issue has gained considerable prominence

Humanist sociologists apply sociological theories and research findings to problems such as those that occupied Nobel Peace Prize winner Jane Addams—the urban poor in general and the plight of poor working women in particular.

in the form of **humanist sociology,** which places human needs and goals at the center (A. M. Lee 1978; Scimecca 1987; Giddens 1987), and **liberation sociology,** whose objective is to replace human oppression with greater democracy and social justice (Feagin and Vera 2001).

Does sociological research influence public policies and programs? Yes. Social scientists, for instance, contributed to the 1954 Supreme Court decision outlawing separate but equal schools for African Americans and whites. This is one of the earliest examples of the U.S. court system accepting social science research as a basis for legal decisions. Subsequent research has explored a variety of related topics, including the effects of school quality on student performance, the impact of social environment on measured IQ, the influence of school desegregation on the performance of African American and white students, and the effects of busing when desegregating schools. Referring to the application of sociology to public policy, William Julius Wilson, in his 1990 presidential address to the American Sociological Society, offered these closing words:

> *With the reemergence of poverty on the nation's public agenda, researchers have to recognize that they have the political and social responsibility as social scientists to ensure that their findings and theories are interpreted accurately by those in the public who use their ideas. They also have the intellectual responsibility to do more than simply react to trends or currents of public thinking. They have to provide intellectual leadership with arguments based on systematic research and theoretical analyses that confront ideologically driven and short-sighted public views.* (W. J. Wilson 1991:12)

Some sociologists advocate intervention beyond policy-related research. Following a long-neglected lead (Wirth 1931), clinical sociologists are now using sociological theories, principles, and research to promote social

Job Opportunities in Sociology

In general, all employers are interested in four types of skills regardless of what specific career path you choose. These skills are:

- The ability to work with others.
- The ability to write and speak well.
- The ability to solve problems.
- The ability to analyze information.

Because computers have revolutionized the office, for example, information analysis skills are becoming much more important to managers in all types of organizations. The increasing complexity of work demands greater critical thinking and problem solving skills. Knowledge is of limited use if you can't convey what you know to others.

The study of sociology helps students develop these general skills, so it is a solid base for many career paths. For sociology majors, the following list of possibilities is only the beginning—many other paths are open to you.

- Social services—in rehabilitation, case management, group work with youth or the elderly, recreation, or administration.
- Community work—in fund-raising for social service organizations, nonprofits, child-care or community development agencies, or environmental groups.
- Corrections—in probation, parole, or other criminal justice work.
- Business—in advertising, marketing and consumer research, insurance, real estate, personnel work, training, or sales.
- College settings—in admissions, alumni relations, or placement offices.
- Health services—in family planning, substance abuse, rehabilitation counseling, health planning, hospital admissions, and insurance companies.
- Publishing, journalism, and public relations—in writing, research, and editing.
- Government services—in federal, state, and local government jobs in such areas as transportation, housing, agriculture, and labor.
- Teaching—in elementary and secondary schools, in conjunction with appropriate teacher certification; also in universities, with research opportunities.

Think About It
1. Which of the career paths listed is most interesting to you? What is it about this area that you find interesting?
2. Evaluate your current strengths and weaknesses in the four primary skill areas.

Source: Adapted from *Careers in Sociology*, **American Sociological Association, 2006.**

change. Clinical sociologists provide help for individuals or serve as agents for change in organizations, communities, and even entire societies (Black 1984; Erickson and Simon 1998; Du Bois and Wright 2000; Koppel 2002; R. H. Hall and Tolbert 2004). Clinical sociologists may work as marriage and family therapists, might design intervention programs to reduce juvenile delinquency, or may redesign the social environment of cancer patients.

Sociology and Occupational Skills

How will the study of sociology contribute to the development of work skills? Most employers are interested in four types of skills: the ability to work well with others, the ability to write and speak fluently, the ability to solve problems, and the ability to analyze information (see Snapshot 1.1). Because computers have revolutionized the office, information analysis skills are becoming much more important to managers in all types of organizations. In addition, the increasing complexity of work demands greater critical analysis and problem-solving skills. The levels of each of these skills can be improved through the broad liberal arts foundation of sociology (Billson and Huber 1993; D. C. Miller 1994; Stephens 2002, American Sociological Association 2007).

What about more specific preparation for employment? In addition to general skills, specific sociology subfields offer preparation for fairly specialized jobs. Consider these examples:

- Training in race relations is an asset for working in human resources (personnel) departments, hospitals, or day-care centers.
- Background in urban sociology can be put to good use in urban planning, law enforcement, and social work.
- Courses focusing on gender and race serve as valuable background for work in community planning, arbitration, and sexual harassment cases.
- Training in criminology is sought by agencies dealing with criminal justice, probation, and juvenile delinquency.

- Courses in social psychology are valuable for sales, marketing, and advertising, as well as for counseling.

These jobs only scratch the surface; students of sociology are prepared to pursue many other careers. Consider this selected list: manager, executive, college placement officer, community planner, employment counselor, foreign service worker, environmental specialist, guidance counselor, health planner, journalist, labor relations specialist, marketing researcher, public

Engraved by Finden from a Miniature by Miss M. Gillies.

Harriet Martineau (1802–1876) emphasized sociology as a science and introduced feminism. Her profound deafness prevented her from earning a living as a teacher so she became an author.

Mary Evans Picture Library/The Everett Collection

FEEDBACK

1. The _____ is the set of mind that enables individuals to see the relationship between events in their personal lives and events in their society.

2. Which of the following is *not* one of the uses of the sociological imagination?
   a. seeing the interplay of self and society
   b. creating new aspects of culture not thought of by others
   c. questioning aspects of social life most people take for granted
   d. understanding the social forces that shape daily life

3. _____ sociology places human needs and goals at its center.

Answers: 1. sociological imagination; 2. b; 3. Humanist

relations supervisor, research analyst, writer, and editor (American Sociological Association, 2007).

The Social Sciences

Your initial understanding of sociology includes its unique focus and the uses of the sociological perspective. You can further clarify your grasp of sociology by distinguishing it from the other social sciences.

It was not until the nineteenth century that Auguste Comte coined the term sociology. Comte's ideas about sociology were embedded in his ideas about psychology, economics, and political theory. He viewed sociology as the one comprehensive social science.

Gradually, scholars developed the various social sciences to the point that they could be legitimately distinguished. Nevertheless, the social sciences have enough in common to overlap with one another at many points.

Anthropology

Because anthropology investigates culture, it is the social science most closely related to sociology. Historically, anthropologists have concentrated on the study of "primitive" or nonliterate societies, whereas sociologists have focused on the modern, complex, industrial societies of Western civilization. Because anthropologists are interested primarily in small nonindustrial societies, they tend to study entire societies. Sociologists, on the other hand, investigate more limited aspects of modern societies, such as political revolutions or the status of women vis-à-vis men. Today some anthropologists have extended their studies to include research in modern, complex societies, examining, for instance, cultural characteristics of neighborhoods and communities.

Psychology

Psychology focuses on the development and function of mental-emotional processes in human beings. Whereas sociology and anthropology concentrate on human groups, psychology takes as its domain the indi-vidual. Sociology and psychology have some common interest in an aspect of behavioral science called social psychology, a field that focuses on the interactions between individuals and groups. A social psychologist, for example, is interested in the ways in which groups promote conformity among their individual members.

Economics

Economics is the study of the production, distribution, and consumption of goods and services. It deals with such matters as the impact of interest rates on the flow of money and the effects of taxation on consumption. Economics uses highly developed mathematical models to predict changes in economic indicators. Although these economic models are not always correct, they are sophisticated, and they give economists greater predictive power than that of most other social scientists. Sociology and economics merge in an area of study called economic sociology, which concentrates on the interrelationships between the economic and noneconomic aspects of social life. An economic sociologist, for example, would be interested in the relationship between the extent of industrialization in a particular society and the degree of workers' involvement in union activity.

Political Science

Political science studies the organization, administration, history, and theory of government. Political scientists are concerned, for example, with voting patterns and political party participation. In recent years, political science has moved beyond its earlier preoccupation with the workings of government toward the study of political behavior in its broadest sense. This trend has inevitably bound closer together the interests of political scientists and political sociologists. Both political scientists and political sociologists share an interest in the social interaction that occurs within the political institution as well as the interaction between the political institution and other institutions, including economic and educational institutions. Both disciplines conduct research on the location and use of power,

F
E
E
D
B
A
C
K

Match the following fields of study and aspects of family life.

____ a. anthropology	(1) distribution of income and the family
____ b. sociology	(2) effects of slavery on family stability
____ c. history	(3) relationship between voting and the family
____ d. political science	(4) effects of early marriage on divorce
____ e. economics	(5) link between early childhood emotional support and familial retardation
____ f. psychology	(6) universality of the nuclear family

Answers: a. (6); b. (4); c. (2); d. (3); e. (1); f. (5)

the process of political socialization, the functioning of special interest groups, and the workings of political protest.

History

History examines past events in human societies. As a field of study, history is sharply divided into two camps (Landes and Tilly 1991). Humanistic historians believe that history cannot use the methods of the social sciences. They view any attempts at scientific generalization as the simplification of complex, unique human experiences rooted in specific times and places. Description, they argue, is the only legitimate function of the historian. Social scientific historians, in contrast, assume the existence of uniform, recurring patterns of human behavior that transcend time and place. The role of the historian, they contend, is to identify and verify these patterns. In recent years, the relationship between sociology and history has become more mutually beneficial. Sociology is currently more open to the use of historians' work in their attempts to document recurring patterns and provide explanations of historical reality; and more historians are now committed to the collection and analysis of primary sources with an eye to establishing recurrent and universal historical patterns.

Social Sciences and the Family

Here, by way of illustration, are distinctive social science approaches to the specific study of the family:

- *Sociologists.* Some sociologists investigating the family focus on the social causes of divorce. Factors thought to contribute to higher divorce rates in American society include early marriage, poor mate selection, economic prosperity, increased participation of women in the labor force, and the weakening of the stigma formerly associated with divorce (see Chapter 12, "Family").
- *Anthropologists.* An interest of some anthropological researchers is whether the nuclear family of married biological parents and their legitimate offspring exists universally in human societies. This topic can be studied by examining documented evidence on as many societies as possible and by attempting to find one or more societies where the nuclear family is not the norm (Leach 1982).
- *Psychologists.* Some psychologists are interested in studying familial retardation—cases of mental retardation that tend to appear in poor families with no known organic basis. Familial retardation is traced largely to an impoverished environment, including inferior nutrition and medical care and the relative absence of early emotional support and intellectual stimulation (Coon and Mitterer 2008).

- *Economists.* Economists use the family as the unit of analysis for the distribution of income. For example, currently, the bottom 60 percent of American families receive less than one-third of the nation's total income (Baumol and Blinder 2007).
- *Political scientists.* The relationship between voting and the family is a topic of research for political scientists. Political scientists have found, for example, that although voting on election day is done by private ballot, voting is primarily a group phenomenon. The family is the most important influence in forming political party identification and determining the voting behavior of its members. Political attitudes and behaviors are shaped, often unintentionally, by family members—who also exist in similar economic, religious, social class, and geographical environments (T. E. Patterson 2006).
- *Historians.* The relationship between slavery and the family has been the subject of research by historians. Using slavery in the United States as a case study, a controversial study by historian Henry Gutman (1983) presents evidence from twenty-one urban and rural communities in the South between 1865 and 1880 to challenge the long-standing assumption that slavery destroys the family structure. For example, Gutman concluded that most African American households during the slavery era had a father and mother present. In short, African American slave families tended to be stable and intact.

Founders of Sociology

European Origins

Sociology is a relatively new science, emerging as a distinct area of study in the late nineteenth century (L. J. Ray 1999; J. H. Turner 2002; J. H. Turner et al. 2006). In fact, it is only during the past half century that sociology has taken its place as a legitimate social science. The origins of sociology, however, can be seen in the work of eighteenth-century philosopher Adam Smith. A man of the Scottish Enlightenment, Smith believed that humans possess a natural affection for others. This led him to seek hidden forces of social harmony to match unseen physical forces like gravity and magnetism. In Smith's desire to do for the understanding of social life what Newton had done for the physical world lay the seeds of modern social science (G. S. Wood 2003).

Who wrote *The Wealth of Nations*, and what is the book's relevance to capitalism? Upon first meeting Scotsman Adam Smith (1723–1790), a female French novelist described him as harsh in speech, big in the teeth, and "ugly as the devil." Realizing that outward beauty is only skin deep, she later wrote a friend: "I wish that the devil would carry off all of our men of letters, all of our philosophers, and bring Mr. Smith to me" (Rae 1965:211–212). Famous for his absentmindedness, Smith was revered by other Enlightenment scholars for his intellectual gifts and practical nature. Smith began as a moral philosopher at the University of Glasgow, spent years as a tutor to the stepson of one of the men whose tax policies provoked the American Revolution, and during his last years held the position of Commissioner of Customs for Scotland (Muller 1993).

The Smith influence is most often associated with economics. This is small wonder; Smith's most famous book, *The Wealth of Nations* (Smith and Cannan 2000; originally published in 1776), is considered foundational to capitalism. Even sociologists themselves, though, often fail to place him among the founders of their field, despite the reflection of his ideas by the founders of sociology in the nineteenth century.

What were Smith's contributions to sociology? Smith made two major contributions to the development of sociology. His first achievement was to attribute much of human behavior to the influence of society. People learn honesty, for example, through participation in the family, church, community, and government. The roots of the human personality lie in our association with others.

In addition to emphasizing the sociological perspective, Smith laid the foundation for a sociological theory known as symbolic interactionism (discussed later in this chapter). He wanted to explain the development of the ability to follow society's rules. This ability, reasoned Smith, develops from the human need for the approval of others (Herman 2001). The concepts Smith used to express this idea in his second book, *The Theory of Moral Sentiments* (1976; originally published in 1759), inspired later sociologists to explore the relationship between human personality and society.

Smith thought like a sociologist and used a sociological perspective in his moral philosophy and economic analysis. The term sociology, however, had not yet been coined, and the field had not been established. Sociology was not formally introduced until the nineteenth century, some forty years after the death of Adam Smith.

What was occurring in the nineteenth century that sparked the emergence of sociology? Nineteenth-century Europe was torn by social change and controversy. The old social order of the Middle Ages—based on social position, land ownership, church, kinship, community, and autocratic political leadership—was crumbling under the social and economic influences of the Industrial and French Revolutions. The move from farm to factory resulted in the loss of a stabilizing community.

This social disharmony led Auguste Comte, Harriet Martineau, and other intellectuals to grapple with the restoration of community, social order, and predictability (Nisbet 1966; Mazlish 1993). Thus, in an important sense, the emergence of sociology as a distinct area of study was a conservative reaction to the social chaos of the nineteenth century. The central ideas of the major pioneers of sociology, all of whom were European, will shed further light on the origins of the field (J. A. Hughes, Martin, and Sharrock 1995; Camic 1998).

How do you think Auguste Comte's personality might have contributed to his breakup with his mentor? Auguste Comte (1798–1857) was a Frenchman born to a government bureaucrat. A small child, often ill, Comte proved early on to be an excellent student, although he had difficulty balancing his genuine interest in school and his rebellious and stubborn nature. In fact, his protest against the examination procedures at the Ecole Polytechnique (of MIT stature) led to his dismissal. A year later, Comte became secretary to the famous philosopher Henri Saint-Simon. Most of Comte's significant ideas were reflected in the writings he and Saint-Simon did together. Differences between Comte and his mentor slowly eroded their relationship, which finally ended in an argument over whose name was to appear on a publication.

What were Comte's major ideas? Comte is recognized as the father of sociology. The improvement of society was his main concern. If societies were to advance, Comte believed, social behavior had to be studied scientifically. Because no science of society existed, Comte attempted to create one himself and coined the term *sociology* to describe this science. Sociology, he asserted, should rely on **positivism**—the use of scientific observation and experimentation in the study of social behavior. When Comte wrote that sociology should rely on positivism, he meant that sociology should be a science based on knowledge of which we can be "positive," or sure. He also distinguished between **social statics,** the study of social stability and order, and **social dynamics,** the study of social change. This distinction between social stability and social change remains at the center of modern sociology.

As a result of his bitter breakup with Saint-Simon, Comte was almost prevented from writing his masterwork, *Positive Philosophy*. (It was in this work that Comte elucidated his principle of "cerebral hygiene." To prevent others from polluting his thoughts, Comte stopped reading.) After parting ways with Saint-Simon, Comte never found another well-paid position. He was forced to rely on intermittent activities—tutoring and testing in mathematics, for example. It was only after a few of his followers invited him to prepare some private lectures that Comte presented his ideas to the world.

Comte died before very many beyond his private circle came to appreciate his work. His belief that sociology could apply scientific procedures and promote social progress was later recognized and widely adopted by European scholars. Many nineteenth-century scholars came to embrace Comte's ideas; some did not.

What motivated Harriet Martineau to begin a writing career? Harriet Martineau (1802–1876) was an Englishwoman born into a solidly middle-class home. Her writing career, which included fiction as well as sociology, began in 1825 after the Martineau family textile mill was lost to a business depression (Webb 1960; Fletcher 1974; Pichanik 1980). Without family income, and with few immediate prospects for marriage following a broken engagement, Martineau was forced to seek a dependable source of income to support herself. She became a popular writer of celebrity stature, whose work outsold that of Charles Dickens.

What did Martineau contribute to the early development of sociology? Martineau, an important figure in the founding of sociology, is best known for her translation of Comte's *Positive Philosophy*. Done with Comte's approval, Martineau's translation remains the most readable. Despite being deaf, she also made original contributions in the areas of research methods, political economy, and feminist theory (Lengermann and Niebrugge-Brantley 1998; Hill and Hoecker-Drysdale 2001; Turner et al. 2006).

Martineau's book *How to Observe Manners and Morals* (1838) was the first methodology book in sociology. In it she emphasized several crucial research principles: the use of a theoretical framework to guide social observation, the development of predetermined questions in gathering information, objectivity, and representative sampling (Lipset 1962).

Martineau followed these research methods while conducting a comparative study of European and American society. Although not as well-known as Alexis de Tocqueville's *Democracy in America*, which was published at about the same time (2000; originally published in 1835), Martineau's *Society in America* (1837) remains her most widely read book today. Martineau compared America favorably with England, but she believed that America had not lived up to its promise of democracy and freedom for all its people. Slavery and the domination of women, she wrote, revealed a significant gap between American ideals and social practices (Terry 1983).

In *Society in America,* Martineau established herself as a pioneering feminist theorist who, like other early feminists, saw a link between slavery and the oppression of women. Consequently, she was a strong and outspoken supporter for the emancipation of both women and slaves.

Through her penetrating analysis of the subordination of American women, Martineau stands as an important envoy of contemporary feminist theory. Doors to the economic institution, she observed, were closed to women. Martineau thus identified economic dependency as the linchpin of female subordination. The oppression of women, in short, was due to sociological forces; subordination of women was part of the structure of society.

Why did Herbert Spencer not attend Cambridge University, his father's alma mater? Herbert Spencer (1820–1903), the sole survivor of nine children, was born to an English schoolteacher. Due to continual ill health, Spencer was taught exclusively by his father and uncle, mostly in mathematics and the natural sciences. Because of his poor background in Latin, Greek, English, and history, Spencer did not feel qualified to enter Cambridge University, his father's alma mater. His subsequent career became a mixture of engineering, drafting, inventing, journalism, and writing.

How did Spencer view society? To explain social stability, Spencer offered an analogy based on biology. Like a human body, a society is composed of interrelated parts working together to promote its well-being and survival. People have brains, stomachs, nervous systems, and limbs. Societies have economies, religions, governments, and families. Just as the eyes and the heart make essential contributions to the functioning of the human body, legal and educational institutions are crucial for a society's functioning.

Why did Spencer oppose social reform? Spencer introduced a theory of social change called social Darwinism, based on Charles Darwin's theory of evolution. Spencer believed that evolutionary social change led to progress, provided that people did not interfere. If left alone, natural social selection would ensure the survival of the fittest. On these grounds, Spencer opposed social reform. The poor deserve to be poor, the rich to be rich. Society profits from allowing individuals to find their own social class level without outside help or hindrance. To interfere with poverty—or the result of any other natural process—is to harm society.

When Spencer visited America in 1882, he was warmly greeted, particularly by captains of industry. After all, his ideas of noninterference provided a moral justification for their economic practices, whether respectable or ruthless. Later, public support for government intervention increased, and Spencer's ideas began to slip out of fashion. He reportedly died with a sense of having failed.

Spencer's theories have crumbled over time. But he brought to the table for the science of sociology a discussion of how societies are structured.

What were Karl Marx's basic commitments, and why, at one point, did he deny being a Marxist? The most influential of the nineteenth-century intellectuals was the German scholar Karl Marx (1818–1883). Although he did not consider himself a sociologist (his doctorate was in philosophy), his ideas have had a significant influence in shaping the field. Marx was deeply troubled by the long hours, low pay, and harsh working conditions in the capitalist system of his day. Preferring social activism to the abstractness of philosophy, Marx was guided by his conviction that social scientists should seek to change the world rather than merely observe it. In part through long association with his friend and collaborator Friedrich Engels, the son of an industrialist, Marx's commitment to democracy and humanism was channeled toward a concern for the poverty and inequality suffered by the working class. Forced by political pressures, Marx moved from Germany to France and finally settled in London, where he devoted the remaining decades of his life to a systematic analysis of capitalism.

Unfortunately, in the minds of many today, Marxism is equated with communism. Although some of his writings were later used as a basis for communism, it is likely that he would be as discouraged with the practice of communism in the twentieth century as he was with the utopian communists of his time. In fact, at one point Marx was so unhappy with others' erroneous interpretations of his work that he disavowed being a Marxist himself.

What is the legacy of Marx? Herbert Spencer and Karl Marx had different conceptions of society and social change. Spencer depicted society as a set of interrelated parts that promoted its own welfare. Marx described society as a set of conflicting groups with different values and interests; the selfishness and ruthlessness of capitalists harmed society. Spencer saw progress coming only from noninterference with natural, evolutionary processes. Marx disagreed.

Marx did, however, believe in an unfolding, evolutionary pattern of social change. He envisioned a linear progression of modes of production from primitive communism through slavery, feudalism, capitalism, and communism. However, he was also convinced that the transformation from capitalism to a classless society could be accelerated through planned revolution. Although his political objective in interpreting capitalism was to hasten its fall through revolution, he believed that capitalism would eventually self-destruct anyway because of its inherent contradictions.

Although recognizing the presence of several social classes in nineteenth-century industrial society—farmers, servants, factory workers, craftspeople, owners of small businesses, moneyed capitalists—Marx predicted that ultimately all industrial societies would contain only two social classes: the **bourgeoisie,** those who owned the means for producing wealth in industrial society, and the **proletariat,** those who labored for the bourgeoisie at subsistence wages. For Marx, the key to the unfolding of history was **class conflict**—conflict between those controlling the means for producing wealth and those laboring for them. Just as slaves rebelled against slave owners and peasants revolted against the landed aristocracy, wage workers would overtake the capitalists. Out of this conflict would emerge a classless society—one without exploitation of the powerless by the powerful.

Which capitalist institution did Marx consider primary? According to the principle of **economic determinism** (an idea often associated with Marx), the nature of a society is based on the society's economy. A society's economic structure determines its other systems: legal, religious, cultural, and political. Marx himself did not use the term economic determinism. The term was applied to his ideas by others—no doubt a consequence of his concentration on the economic sphere in capitalist society. Some interpreters mistakenly assume that because Marx perceived the economic institution as primary in capitalist society, he believed that all societies operated according to the same principles. Actually, Marx recognized that even in capitalist society

Karl Marx (1818–1883) was a social scientist who underscored the importance of conflict in social change. Parts of his writings were later used a basis for Communism.

economic and noneconomic institutions affect each other. Not only that, but he wrote that sometimes the economy "conditions" rather than "determines" the historical evolution of capitalist society.

Who was the first French sociologist, and what were some of his concerns? Emile Durkheim (1858–1917) was the son of a French rabbi. Durkheim originally intended to follow this family rabbinical tradition, but later experiences led him to become an agnostic, although he retained an intellectual interest in religion throughout his life. In fact, one of Durkheim's major concerns was social and moral order, an emphasis undoubtedly related to his upbringing in the home of a rabbi:

> From Judaic family training and an intimate environment Durkheim gained a deep and permanent concern for universal moral law and the problems of ethics, a concern that was not combined with any indulgent sense of humor. Indeed, he was eminently without humor and somewhat "heavy-handed." (Simpson 1963:1)

Durkheim was a brilliant student even during his early school years. In college, he was so intensely studious that his schoolmates nicknamed him "the metaphysician." He eventually became the first French academic sociologist.

What were Durkheim's foremost contributions? According to Durkheim, social order exists because of a broad consensus among members of a society. This consensus is especially characteristic of preindustrial, nonliterate, simple societies based on **mechanical solidarity**—social unity that comes from a consensus of values and beliefs, strong social pressures for conformity, and dependence on tradition and family. Witnessing the social upheaval brought on by industrial and democratic revolutions, Durkheim attempted to describe social order in complex, industrial society. In this modern, more complicated society, he contended, social order is based on **organic solidarity**—social unity based on a complex of highly specialized roles. These specialized roles render members of a society dependent on one another for goods and services.

Although Comte, Martineau, and other early sociologists emphasized the need to make sociology scientific, they did not have the necessary research tools. They did, however, influence later sociologists to develop more scientific methods: to replace armchair speculation with careful observation, to engage in the collection and classification of data, and to use data for formulating and testing social theories. Durkheim was one of the most prominent of these later sociologists. He first introduced the use of statistical techniques in his groundbreaking research on suicide. (See Using the

Sociological Imagination, at the beginning of this chapter, and Doing Research, later in the chapter.) In that study, Durkheim demonstrated that suicide involves more than an individual process. By revealing that suicide rates vary according to group characteristics—the suicide rate is lower among Catholics than Protestants and lower among married than single persons—Durkheim documented that human social behavior must be explained by social factors in addition to just psychological ones.

Who wrote *The Protestant Ethic and the Spirit of Capitalism*? Max Weber (1864–1920) was the eldest son of a well-to-do German lawyer and politician. Weber's father was a man rooted in the pleasures of this world. His mother, in stark contrast, was a strongly devout Calvinist who rejected the pleasure seeking of her husband. While Weber's father was preoccupied with life on earth, Weber's mother concentrated on her future salvation. Weber was affected psychologically by the conflicting values of his parents. Even though he suffered a complete mental breakdown, he eventually recovered to do some of his best work. His mother's influence is clearly reflected in his work, especially in his regard for the sociology of religion and in one of his most famous books, *The Protestant Ethic and the Spirit of Capitalism* (Ritzer 2007).

How did Weber contribute to the development of sociology? As a university professor trained in law and economics, Weber wrote on a wide variety of topics, including the relationship between capitalism and Protestantism, the nature of power, the development and nature of bureaucracy, the religions of the world, and the nature of social classes. Thanks to the quality of his work and the diversity of his interests, Weber is credited with being the single most important influence on the development of sociological theory.

Unlike Durkheim, Weber did not favor studying human beings as though they were physical things. Weber's approach was a subjective one: humans act on the basis of their own understanding of a situation. Consequently, sociologists must discover the personal meanings, values, beliefs, and attitudes that people bring to social situations. Understanding the subjective intentions of human social behavior could be accomplished through what Weber called the method of **verstehen**—understanding social behavior by putting oneself in the place of others. This empathetic perspective was apparently rooted in Weber's personal nature. According to Weber's wife, he "showed throughout his life an extraordinary appreciation for the problems other people faced and for the shades of mood and meaning that characterized their outlook on life" (Bendix 1962:8).

Max Weber's (1864–1920) model of a bureaucracy reflected greatly increased efficiency in business and government. Today, however, bureaucratic is often used as a synonym for unimaginative, plodding, or despotic.

Because Weber and Marx were not intellectual contemporaries, and because Weber disagreed with Marx on many points, it is often said that Weber was debating with Marx's ghost (Wiley 1987). It was out of honest differences on issues, not disrespect, that Weber opposed some of Marx's ideas. Although Marx was, at times, openly political in advocating his ideas, Weber stressed objectivity. He strongly counseled sociologists to conduct **value-free research**—research in which personal biases are not allowed to affect the research process and its outcome. Marx saw religion as retarding social change, whereas Weber believed that religion could promote change. Also, whereas Marx emphasized the role of the economy in social stratification, Weber's analysis involved several dimensions.

One should not conclude from these differences, however, that Weber's approach to sociology was antagonistic to Marx's. Actually, Weber was the most prominent of the second-generation German scholars who were concerned with power and conflict in society. Although there are significant differences in detail, both Marx and Weber took the problems of capitalism as their unifying theme (R. Collins 1994).

One of Weber's most notable contributions is the identification of rationalization as a key influence in the transition from preindustrial society. **Rationalization**—the use of knowledge, reason, planning, and objectivity—in industrial society marked a change from the tradition, emotion, superstition, and personal relationships of preindustrial society. With rationalization, for example, businesses were to be run on proven economic principles rather than custom. One of the most familiar effects of rationalization is the expansion of bureaucracy as the dominant type of organization. (See Doing Research, p. 25.) According to Weber, rationalization, with its emphasis on social control, would ultimately imprison us all in an "iron cage," robbing us of our freedom, autonomy, and individuality.

Sociology in America

Although the early development of sociology occurred in Europe, the maturation of sociology has taken place primarily in the United States. Because sociology has become a science largely through the efforts of American sociologists, it is not surprising that worldwide, most sociologists are American. Sociological writings in English are used by sociologists throughout the world, reflecting the tremendous global influence American sociologists have had since World War II. But, of course, American sociologists have been heavily influenced by the European originators of sociology. This is partly because early American sociologists had to rely on the writings of Smith, Comte, Martineau, Spencer, Marx, Durkheim, Weber, and others. Where would they have turned but to the existing body of literature published by the Europeans? There is, however, another reason for the strong European influence. Just as sociology in Europe was born during a time of rapid, disruptive change, sociology emerged in the United States during the period of rapid urbanization and industrialization following the Civil War. It is little wonder that Lester Ward (1841–1913), the founder of American sociology, picked up on Comte's conviction that sociology could promote social progress. Like Comte, Ward believed that industrial, urban society could be improved through sociological analysis.

Historically, how did American sociology develop? From its founding in 1892 to World War II, the first department of sociology at the University of Chicago stood at the forefront of American sociology. Housed at the University of Chicago have been a number of exceptional minds. George Herbert Mead, John Dewey, William I. Thomas, and Dorthony Swane Thomas pioneered the study of human nature and personality. Urban social problems such as prostitution, slums, and crime were studied extensively by Robert E. Park and Ernest Burgess. The study of social and cultural change was championed by William F. Ogburn. On the whole, the early Chicago School became closely linked with the idea of social reform. Emphasis on seeking solutions to problems of race, crime, and poverty continued at the University of Chicago after World War II with scholars such as Howard Becker, Joseph Gusfield, Herbert Blumer, David Riesman, and Erving Goffman (Abbott 1999).

Edith Abbott, Sophinista Breckenridge, Marion Talbott, and other female sociology researchers collaborated with the men at the Chicago School. They contributed to many studies of such urban problems as housing, child rearing, education, and poverty. Some of these women were professors in the Department of Sociology; others worked as university administrators (Deegan 1991, 2000). Many more women, having been denied university appointments, contributed to the development of social work rather than academic sociology.

After World War II, eastern universities such as Harvard and Columbia, midwestern universities such as Wisconsin and Michigan, and western universities such as Stanford and the University of California at Berkeley emerged as leading institutions fostering a diversity of sociological emphases. Sociologists such as George A. Lundberg underscored the application of the research methods of physical scientists; they wanted to make sociological research more sophisticated and scientific. Lundberg was known as a neo-positivist for his revival of Comte's scientific approach to sociology. A sociologist at Harvard, Talcott Parsons, returned to the development of general theories of society à la Weber, Durkheim, and other European scholars. Parsons was one of the leading proponents of functionalism (discussed later), which is one of several dominant theoretical perspectives in sociology today. Robert K. Merton, a Columbia University sociologist and a student of Parsons, advocated applying more specific theories, rather than general ones. Merton also stressed the importance of empirical research. By the 1970s, the number of practicing sociologists had grown dramatically. As a result, talented and influential sociologists were no longer located in only a few elite colleges and universities.

In their quest toward developing general sociological theories and establishing sociology as a science, sociologists between World War II and the 1960s had nearly lost sight of the idea that sociology could help solve social problems. Social reform emerged again during the turbulent 1960s and remains an important part of sociology today. "Humanist" sociologists, who owe much to sociologist C. Wright Mills, believe that sociology cannot be ethically neutral about important social issues. Rather, they argue, sociologists have an inherent obligation to question social arrangements and work for social transformation (A. M. Lee 1978; Scimecca 1987; Giddens 1987, 2000).

Two Early Contributors

Two early contributors who are often neglected in the history of American sociology are Jane Addams and W. E. B. Du Bois, close friends and coworkers. Although neither was a researcher or scientist, both were concerned about applying sociology to solve social problems.

How did Jane Addams contribute to American sociology? The best known and most influential of the early female social reformers in the United States was Jane Addams (1860–1935). Although her mother died when she was two years old, Addams's father provided a loving and comfortable home for her and her eight brothers and sisters. In addition to the physical comforts he provided, her father set an example with his love of learning.

Addams was an excellent student. Her earlier education emphasized practical knowledge and the improvement of "the organizations of human society." She attended The Women's Medical College of Philadelphia, but was compelled to drop out because of illness (Addams 1981; originally published in 1910).

During her childhood, Addams was exposed to many episodes of government corruption and to business practices that enriched owners and harmed workers. This was the beginning of her social conscience. While on one of her European trips, she saw the work being done to help the poor in London. With this example of social action and with her earlier exposure in the United States to corruption and exploitation, Addams began her life's work seeking social justice.

She cofounded Hull House in Chicago's slums. There, people who needed help—immigrants, the sick, the poor, the aged—could find some relief. Addams focused on the problems caused by the imbalance of power among the social classes. She invited sociologists from the University of Chicago to Hull House to witness firsthand the effects of the exploitation of the lower class. In addition to her work with the underclass, Addams was active in the women's suffrage and peace movements (Deegan 2000).

As a result of her tireless work for social reform, Addams was awarded the Nobel Peace Prize in 1931—the only sociologist to receive this honor. The irony is that Addams herself suffered a sort of class discrimination. She was not considered a sociologist during her lifetime, in part because she did not teach at a university. She was considered a social worker (a less prestigious career) because she was a woman and because she worked directly with the poor.

How did W. E. B. Du Bois contribute to an understanding of black communities? Together with other early African American sociologists, W. E. B. Du Bois (1868–1963), an African American educator and social activist, influenced the early development of sociology in the United States (Young and Deskins 2001). Although Du Bois's father was part white and part black (known then as a mulatto) and his mother was a direct descendant of a freedman Dutch slave, Du Bois was considered black by his associates. He attended an integrated high school in his native Great Barrington, Massachusetts, and was the first black to receive a diploma there. Du Bois earned a doctorate degree from Harvard University in 1895 and taught at a number of predominantly black universities during his career.

Du Bois learned firsthand about racial discrimination and segregation when he attended Fisk University in Nashville, Tennessee. From this experience and from teaching in rural, all-black schools around Nashville, Du Bois decided to fight what was then called "the Negro problem"—a racist policy based on the assumption that blacks were an inferior race. He did this by scientifically studying the social structure of black communities, first in Philadelphia and later in other places. In collaboration with Isabel Eaton, Du Bois published *The Philadelphia Negro,* a classic work on urban African Americans.

Besides contributing to the understanding of black communities, Du Bois worked for civil rights. He was the only black member of the Board of Directors of the National Association for the Advancement of Colored People (NAACP) when it was founded in 1910, and he was the editor of its journal *Crisis* for the next twenty-four years.

Du Bois's concern for his race did not stop at the borders of the United States. He was also active in the Pan African movement, which was concerned with the rights of all African descendants, no matter where

W. E. B. DuBois (1868–1963) focused on the question of race inside and outside the United States.

The Art Archive/Culver Pictures/Picture Desk

they lived. While continuing to document the experience and contributions of African people throughout the world, Du Bois died in the African country of Ghana, at the age of ninety-five (D. Lewis 1993, 2000).

F E E D B A C K

1. _____ is the idea that sociology should use observation and experimentation in the study of social life.
2. Which of the following sociologists contended that social change leads to progress, provided that people do not interfere?
   a. Durkheim   b. Weber   c. Martineau   d. Spencer   e. Marx   f. Du Bois
3. According to the principle of _____ a society's economic system molds the society's legal system, religion, art, literature, political structure, and other social arrangements.
4. According to Durkheim, the United States is a social order based on _____ solidarity.
5. According to Weber, the method of _____ involves an attempt to understand the behavior of others by putting oneself mentally in their place.
6. Du Bois focused only on the American race question. T or F?

Answers: 1. Positivism; 2. d; 3. economic determinism; 4. organic; 5. verstehen; 6. F

Theoretical Perspectives

Theory and Perspective

Do scientific theories compete with each other? It is normal in science for competing, even conflicting, theories to exist simultaneously. This may be because of insufficient evidence to determine which theory is accurate. Or it may be that different theories explain different aspects of reality. This is true even for the "hard" sciences, such as modern physics. Einstein's theory of general relativity, for example, was inconsistent with the widely accepted Big Bang theory of the origin of the universe. And Einstein was deeply bothered by quantum theory, whose laws have become the foundation of modern developments in such fields as chemistry and molecular biology (Hawking 1998). Just as theories in physics oppose one another, several major theoretical perspectives compete in sociology.

How does perspective affect perception? In *Beyond Good and Evil*, nineteenth-century philosopher Friedrich Nietzsche wrote:

> *It is more comfortable for our eye to react to a particular object by producing again an image it has often produced before than by retaining what is new and different in an impression.* (Nietzsche 1987:97; originally published in 1886)

As implied at the beginning of this chapter, a person's perspective draws attention to some things but blinds him or her to other possibilities. This fundamental power of perspective is graphically illustrated in two drawings that psychologists often use to illustrate the concept of perception. One, shown in Figure 1.1, is a picture of an old woman—or is it?

If you stare at the old woman long enough, she becomes a beautiful young woman with a feather boa around her neck. The other drawing (not shown in Figure 1.1) is a picture of a rabbit. If you stare at the rabbit long enough, a duck appears. If you continue to look, both drawings return to their original form. You cannot, however, see the old woman and the young woman or the rabbit and the duck at the same time.

Which image is real, the old woman or the young woman? The rabbit or the duck? It depends on your focus; your perspective influences what you see. So it is with any perspective. One perspective emphasizes certain aspects of an event; another perspective accents different aspects of the same event. When a perspective highlights certain parts of something, it necessarily places other parts in the background.

Sociology has three overarching theoretical perspectives: functionalism, conflict theory, and symbolic interactionism. Each of these perspectives provides a different slant on human social behavior (see Table 1.2). The exclusive use of any one of them prevents our seeing other aspects of social behavior, just as one cannot see the old woman and the young woman at the same time. All three perspectives together, however, allow us to see most of the important dimensions of social life that are of interest in sociology. These three perspectives can be placed within the context of macrosociology or microsociology. Functionalism and conflict theory are best viewed from the macro level, symbolic interactionism from the micro level.

FIGURE 1.1

The Importance of Perspectives

View I

View II

View III

Our perception of things is heavily influenced by our perspective. Decide whether you want to look at View I or View II (cover the other view). After examining View I or View II, look at View III. If you decided to look at View I, you should see the old woman in View III. If you decided to look at View II, you should see the young woman in View III. Can you see both? Try it on your friends.

TABLE 1.2

FOCUS ON THEORETICAL PERSPECTIVES:
Assumptions of Functionalism, the Conflict Perspective, and Symbolic Interactionism

This table contains the major assumptions of the three major theoretical perspectives. Although the assumptions of functionalism and conflict theory are direct opposites, each focuses on the nature of society. Symbolic interactionism uniquely addresses interpersonal issues in sociology.

Functionalism	The Conflict Perspective	Symbolic Interactionism
1. A society is a relatively integrated whole.	1. A society experiences inconsistency and conflict everywhere.	1. People's interpretations of symbols are based on the meanings they learn from others.
2. A society tends to seek relative stability.	2. A society is continually subjected to change.	2. People base their interaction on their interpretations of symbols.
3. Most aspects of a society contribute to the society's well-being and survival.	3. A society involves the constraint and coercion of some members by others.	3. Because symbols permit people to have internal conversations, they can gear their interaction to the behavior they think others expect of them and they expect of others.
4. A society rests on the consensus of its members.		

Source: The assumptions of functionalism and the conflict perspective are based on the work of Ralf Dahrendorf, "Toward a Theory of Social Conflict," *Journal of Conflict Resolution* 2 (June 1958):174. Adapted by permission of Sage Publications, Inc., © 1958.

Macrosociology: Functionalism

Functionalism emphasizes the contributions (functions) performed by each part of a society. For example, family, economy, and religion are all parts of a society. The family contributes to society by providing for the reproduction and care of its new members. The economy contributes by dealing with the production, distribution, and consumption of goods and services. Religion contributes by emphasizing beliefs and practices related to sacred things. Functionalism focuses on social integration, stability, order, and cooperation.

Who were the originators? Among the major pioneers of this perspective was, of course, Herbert Spencer, who compared societies to living organisms. Although sociologists no longer use the organism metaphor, they still view society as a system of interrelated parts. Emile Durkheim, a contemporary of Spencer's, greatly contributed to the development of this theoretical perspective. Two of the strongest supporters of functionalism have been American sociologists Talcott Parsons and Robert Merton.

What are the assumptions of functionalism? First, functionalists see the parts of a society as an integrated whole. A change in one part of a society leads to changes in other parts. A major change in the economy, for example, may change the family. This is precisely what happened as a result of the Industrial Revolution. Before the Industrial Revolution, when most people made their living by farming, the need for a large farm labor force was fulfilled by having many children. The

need for a large family declined dramatically as industrialization substituted machines for manual labor. Smaller families, then, became the norm.

Functionalists realize that societies are not perfectly integrated. Although the actual degree of integration varies, a certain degree of integration is necessary for the survival of the society.

Second, functionalists assume that societies tend to return to a state of stability after some upheaval has occurred. A society may change over time, but functionalists believe that it will return to a stable state by incorporating change in such a way that society will be similar to what it was before any change occurred.

The idea that a society changes yet maintains most of its original structure over time is referred to as **dynamic equilibrium**—a constantly changing balance among the parts. The student unrest on college and university campuses during the late 1960s is an illustration of dynamic equilibrium. The activities of student protestors did create some changes. The public no longer accepts all American wars as legitimate, universities are now more responsive to students' needs and goals, and the public is more aware of the importance of environmental protection. These changes, however, have not revolutionized American society. They have been absorbed into it, leaving it only somewhat different from the way it was before the student unrest. In fact, many of the student "radicals" are now part of the middle-class society they once rejected. Similarly, the widespread concern over corporate misbehavior that erupted in 2002 as a result of the Enron, WorldCom, and Arthur Andersen scandals failed to bring about any significant changes in American society.

Third, most aspects of a society exist to promote the society's survival and welfare. It is for this reason that all complex societies have economies, families, governments, and religions. If these elements did not contribute to a society's well-being and survival, they would disappear.

According to Robert Merton (1996), **manifest functions** are intended and recognized at the time; **latent functions** are unintended and unrecognized until later. In 1696, Scotland's Parliament established free public schools in all parishes. These schools were to enable the young to read Holy Scripture. Unintended consequences included the establishment of education as a way of life for middle-class Scots and the intellectual flowering of the Scottish Enlightenment in the eighteenth century (Herman 2001). More recently, the automakers' attempt to make it more difficult to hotwire expensive cars had the desired effect of reducing car theft. An unintended consequence, however, was the emergence of a new and more dangerous crime, carjacking (Mann 2002). One of the manifest functions of secondary schools, for example, is to teach math skills. A latent (and positive) function of schools is the development of close friendships.

Not all elements of a society make a positive contribution. Those that have negative consequences result in **dysfunction.** Bureaucratic rules, for example, often result in rigidity and impersonality. If a secretary in the sociology department treats you like a "number" rather than an individual, you don't like the inflexibility and impersonality.

Finally, according to functionalism, there is—among most members of a society—a consensus on values. Most Americans, for example, agree on the desirability of democracy, success, and equal opportunity. This consensus of values, say the functionalists, accounts for the high degree of cooperation found in any society.

There are legitimate criticisms of the functionalist perspective: It tends to legitimatize the status quo, and it neglects social change. Conflict theory, in contrast, takes social change as a focal point.

Macrosociology: Conflict Theory

Conflict theory emphasizes conflict, competition, change, and constraint within a society (Schellenberg 1982; Giddens 1979, 1987). Although this theoretical perspective was not very popular among most American sociologists until the 1960s, its roots go back as far as those of functionalism.

Who were the originators? Karl Marx, Max Weber, and Georg Simmel (1858–1918) are three German sociologists whose work underlies the conflict perspective. According to Marx, class conflict is inevitable in capi-

talist economies. Modern conflict advocates such as C. Wright Mills, Ralf Dahrendorf, and Randall Collins do not limit themselves to class conflict. They broaden Marx's insights to include conflict between any segments of a society, a point Weber had made much earlier (R. Collins 1994). For example, conflict exists between Republicans and Democrats, unions and management, industrialists and environmentalists.

How are the assumptions of conflict theory often the reverse of the functionalist perspective? Functionalism emphasizes a basic agreement on values within a society, concentrating on the ways people cooperate to reach common goals. Conflict theory, in contrast, focuses on the disagreements between various groups in a society or between societies. Groups and societies compete and conflict as they attempt to preserve and promote their own special values and interests.

How do conflict advocates view power? Because advocates of conflict theory see social living as a contest, their central question is, Who gets what? Their answer: those with the most **power**—the ability to control the behavior of others, even against their will—get the largest share of whatever is considered valuable in a society. Those with the most power have the most wealth, the most prestige, and the most privileges. Because some segments have more power than others, they are able to constrain the less powerful.

The wealthy and powerful, for example, can often manipulate income tax laws (Ehrenreich 1994; Barlett and Steele 2002; Huffington 2004; Johnston 2005, 2007). In 1998, U.S. oil and pipeline companies were required to pay only 5.7 percent of their domestic profits to the U.S. Treasury even though the official tax rate was 35 percent. Actually, half of the oil and pipeline firms surveyed were entitled to tax rebates (McIntyre 2000). According to a recent study by Citizens for Tax Justice, 250 of America's biggest corporations were taxed at only 20.1 percent in 1998. Moreover, forty-one of these companies received tax rebates for at least one year from 1996 to 1998. That is, they made additional money after filing their tax forms on the profits they reported to the Internal Revenue Service (IRS). For example, just before the accounting scandals erupted in 2002, Enron had paid no income taxes in four of the previous five years. Instead, the corporation was eligible for $382 million in tax refunds (Johnston 2002; Behr 2003). Tax evasion or avoidance is not limited to corporations. Millions of Americans pay less than their fair share of taxes annually, including more than 1,000 American taxpayers with incomes in excess of $200,000 who, in a given year, pay no federal income tax whatsoever (Ehrenreich 1994; Lewis and Allison 2002). The lowest 20 percent family income category in the United States gained 4.5 percent in after-tax income between

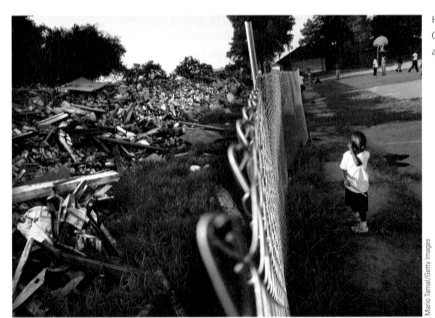

How does this photo of a post-Katrina New Orleans neighborhood emphasize the conflict approach to studying society?

Mario Tama/Getty Images

1979 and 2002, while the top 5 percent gained 111 percent (see Table 1.3). By 2010 the wealthiest 1 percent of Americans will receive 53 percent of President Bush's 2001 individual tax cuts, totaling nearly $500 billion (Citizens for Tax Justice 2002, 2007).

Other examples of wealthy and powerful interests getting their way are numerous. In 1985, Congress passed legislation continuing a federal sugar subsidy that placed an average of $250,000 in the pockets of some 12,000 American sugar growers. As a result, American consumers had to pay among the highest prices in the world for their sugar: a quantity of sugar costing 85 cents in Washington, D.C., sold for half that amount in Ottawa, Canada. According to the evidence, members of the House of Representatives were likely to

vote for the bill in direct proportion to the amount of campaign contributions they received from the sugar lobby (Birnbaum and Murray 1988; Stern 1988). The 1999 Ticket to Work and Work Incentives Improvement Act for the disabled contains a provision unrelated to disabilities that saves corporations $1.5 billion in taxes every two years (Barlett and Steele 2002). And President Bush's proposed budget for fiscal 2003 introduced new corporate farm subsidies that are projected to total $70 billion over the next decade (Gleckman 2002).

How does conflict theory explain social change? Because many conflicting groups exist and the balance of power among these groups may shift, conflict theory assumes that social change is continual. For example, the

TABLE 1.3

Average Income After Tax, 1979–2002 (in constant 2002 dollars, adjusted for changes in family size)

Family Income Category	After-Tax Incomes		
	1979	2002	% Change
Lowest 20%	13,200	13,600	4.5%
Second 20%	26,000	29,900	12%
Middle 20%	38,000	43,700	15%
Fourth 20%	49,800	61,700	24%
Highest 20%	87,700	130,000	48%
Top 5%	208,900	631,700	111%

As this table reveals, people in certain income categories profited more between 1979 and 2002 than others. Relate these data to the conflict perspective.

Sources: Congressional Budget Office, Effective Federal Tax Rates: 1979–2002, March 2005; Cited in Isaac Shapiro. "New IRS Data Show Income Inequality is Again on the Rise." Washington, DC: Center on Budget and Policy Priorities, 2005.

women's movement is attempting to change the balance of power between men and women. As this movement progresses, we see larger numbers of women in occupations once limited to men. More women are either making or influencing decisions in business, politics, medicine, and law. Gender relations are changing in other ways as well. More women are choosing to remain single, to marry later in life, and to have fewer children—or none. According to the conflict perspective, these changes are the result of increasing power among women. This brings up feminist theory.

What is feminist theory? **Feminist theory,** a branch of conflict theory, links the lives of women (and men) to the structure of gender relationships within society. Three frameworks within the broad umbrella of feminist theory can be isolated: liberal feminism, radical feminism, and socialist feminism (Freedman 2001; Andersen 2005). Each of these frameworks attempts to explain the relationship between gender, social class, and minority status.

Advocates of *liberal feminism* focus on equal opportunity for women and heightened public awareness of women's rights. Operating within the existing structure of society, liberal feminists look to legal and social change to rectify the abridgment of women's rights. If women are to receive the same treatment accorded men, long-standing laws, social policies, and social structures must be altered. Equal pay for equal work, abortion rights, and civil rights are among the issues uniting the liberal feminist perspective.

Liberal feminism attributes the subordination of women to unequal rights and to the ways in which males and females are taught by society to view themselves. *Radical feminists* trace the oppression of women to the fact that societies are dominated by men. A patriarchal society is controlled by males who use their power, prestige, and economic advantage to dominate females, including their sexuality. Male-controlled institutions—family, religion, economy, education, government—define women as subordinate to men and are used to ensure the perpetuation of female subordination. Radical feminists are especially interested in male supremacy and the ways men gain and maintain the power to control all social institutions (Crow 1998).

What does socialist feminism add to radical feminism? *Socialist feminism,* which is actually more radical than radical feminism, sees capitalism as a source of female oppression. This perspective on capitalism, however, goes beyond the Marxist interpretation. Socialist feminists contend that the problem of female oppression extends beyond the fact that under capitalist societies women are the property of men. Because female oppression exists in societies without a capitalist economic base, socialist feminists reason that more than capital-

ism is at work in female oppression. That something else is patriarchy. Patriarchy is the means for male domination of females in preindustrial and authoritarian societies (Khouri 2005). In capitalist society, the power relations of the class structure combine with the force of patriarchy to create and maintain male oppression of women.

Despite some significant differences in theoretical perspectives, feminists share at least two common themes. First, they believe that sociology carries a bias from years of theory and research shared by white middle-class males from western Europe and North America. Second, they believe that the nature of gender and gender relationships is sociological (rather than psychological), embedded as they are in social structures.

Which is the better perspective, functionalism or conflict theory? There is no "better" theoretical perspective. Each perspective emphasizes certain aspects of social life. The advantages of one perspective highlight the disadvantages of the other. Functionalism explains much of the consensus, stability, and cooperation within a society. Conflict theory explains much of the constraint, conflict, and change.

Because each perspective has captured an essential facet of society's nature, their combination, or synthesis, is a reasonable next step. Some attempts to combine functionalism and conflict theory have already been made (Dahrendorf 1958b; van den Berghe 1963, 1978). One of the most promising is the attempt to specify the conditions under which conflict and cooperation occur. Gerhard Lenski (1984; Nolan and Lenski 2005) contends that people cooperate—even share the fruits of their labors—when scarcity threatens their survival. But conflict, competition, and constraint are likely to occur when there is more than enough for everyone. Thus, as a society moves from a subsistence economy to an affluent one, conflict, competition, and constraint increase. And the more a deprived group questions the legitimacy of its condition, the more likely it is to conflict with privileged groups (Coser 1998), as illustrated by the civil rights and women's movements.

Microsociology: Symbolic Interactionism

Who were the originators of symbolic interactionism, and what were they thinking? Both functionalism and conflict theory deal with large social units and broad social processes—the state, the economy, evolution, and class conflict. At the close of the nineteenth century, some sociologists began to introduce a new focus on human social behavior. Instead of being preoccupied with larger structures per se (such as society and social class), they began to recognize the importance of people's interactions within social structures. Max Weber and Georg Simmel were among the earliest contributors

Emile Durkheim—The Study of Suicide

Emile Durkheim, the first person to be formally recognized as a sociologist and the most scientific of the pioneers, conducted a study that stands as a research model for sociologists today. His investigation of suicide was, in fact, the first sociological study to use statistics. In *Suicide* (1964; originally published in 1897), Durkheim documented his contention that some aspects of social behavior—even something as allegedly individualistic as suicide—can be explained without reference to individuals.

Like all of Durkheim's work, his study of suicide is best considered within the context of his concern for social integration (R. Collins 1994; Pickering and Walford 2000). Durkheim wanted to see if suicide rates within a social entity (e.g., a group, organization, or society) are related to the degree to which individuals are socially involved (integrated and regulated). In his study, Durkheim described three types of suicide: egoistic, altruistic, and anomic. He hypothesized that *egoistic suicide* increases when individuals do not have sufficient social ties. Because single (never married) adults, for example, are not heavily involved with family life, they are more likely to commit suicide than are married adults. On the other hand, he predicted *altruistic suicide* as more likely to occur when social integration is extremely strong. The Al Qaeda agents who slammed jetliners into the World Trade Center and the Pentagon in 2001 are one example, as are suicide bombers. Altruistic suicide need not be this extreme, of course. Military personnel who lay down their lives for their country are another illustration.

Durkheim forecasted his third type of suicide—*anomic suicide*—to increase when existing social ties are broken. For example, suicide rates increase during economic depressions. People suddenly without jobs, or without hope of finding any, are more prone to kill themselves. Suicide may also increase during periods of prosperity. People may loosen their social ties by taking new jobs, moving to new communities, or finding new mates.

Using precollected data from government population reports of several countries (much of it from the French government statistical office), Durkheim found strong support for his predictions. Suicide rates were, in fact, higher among unmarried than married people and among military personnel than civilians. They were also higher among people involved in nationwide economic crises.

Durkheim's primary interest, however, was not in the empirical (observable) indicators he used, such as suicide rates among military personnel, married people, and so forth. Rather, Durkheim used the results of his study to support several of his broader contentions: (1) social behavior can be explained by social rather than psychological factors; (2) suicide is affected by the degree of integration and regulation within social entities; and (3) because society can be studied scientifically, sociology is worthy of recognition in the academic world (Ritzer 2007). Durkheim was successful on all three counts. If Auguste Comte told us that sociology *could* be a science, Durkheim showed us *how* it could be a science.

Evaluate the Research

1. Do you believe that Durkheim's study of suicide supported his idea that much of social behavior cannot be explained psychologically? Why or why not?
2. Which approach do you think Durkheim followed in his study of suicide: functionalist, conflict, or symbolic interactionist? Support your choice by relating his study to the assumptions of the perspective you chose.

---

to the theoretical perspective of interactionism, which holds that groups exist only because their members influence one another's behavior. Later sociologists— namely, Charles Horton Cooley, George Herbert Mead, W. I. Thomas, Erving Goffman, Harold Garfinkel, and Herbert Blumer—expanded the insight of interactionism. They created symbolic interactionism, the most influential approach to interactionism.

What is symbolic interactionism? Symbolic interactionism targets a less abstract view of human social behavior than either functionalism or conflict theory. This perspective focuses on the actual interaction between people (Reynolds and Herman-Kinney 2003).

Basic to the symbolic interactionist perspective is the concept of a symbol. A **symbol** is something chosen to represent something else. It may be an object, a word, a gesture, a facial expression, or a sound. A symbol is something that is observable, something concrete; but it may represent something that is not observable, something abstract. A frown can be a symbol of either disapproval or concentration. Your school mascot, like the Hokie Bird of Virginia Tech, is a symbol of your college or university:

According to symbolic interactionism, social life can be likened to a theatrical performance. Don't we convey as much about ourselves in the way we dress as do the actors in this high school play?

Chris Ware/The Image Works

One cannot tell by looking at an X in an algebraic equation what it stands for; one cannot ascertain with the ears alone the symbolic value of the phonetic compound si; one cannot tell merely by weighing a pig how much gold he will exchange for; one cannot tell from the wave length of a color whether it stands for courage or cowardice, "stop" or "go." (L. A. White 1969:26)

More seriously, because African Americans were aware of past lynchings carried out by whites, blacks in 2007 took the sudden appearance of nooses across America as a symbol of hate (Fears 2007).

The meaning of a symbol is not determined by its own physical characteristics. Those who create and use the symbol assign a meaning to it. In 2007 presidential hopeful Mike Huckabee ran a televised political ad claiming to put politics aside momentarily to celebrate the birth of Christ. In the background of the talking head ad was a window pane, which appeared to form the shape of the Christian cross, thus using an object (window pane) to symbolize a central religious symbol. If people in a group do not share the same meaning for a given symbol, confusion results. For example, if some people interpret the red light of a traffic signal to mean "go," while others see the green light to mean "stop," chaos will result.

The importance of shared symbols is reflected in the formal definition of **symbolic interactionism:** the theoretical perspective that focuses on interaction among people—interaction based on mutually understood symbols.

What are the basic assumptions of symbolic interactionism? Herbert Blumer (1969a, 1969b), who coined the term *symbolic interactionism,* outlined three assumptions central to this perspective (see, again, Table 1.2). First, according to symbolic interactionism, we learn the meaning of a symbol from the way we see others reacting to it. For example, American musicians in Latin America soon learn that when audience members whistle at the end of a performance, they are unhappy; their whistling is a symbol of disapproval, as is the symbol of booing in the United States.

Second, once we learn the meanings of these symbols, we base our behavior (interactions) on them. Now that the musicians have learned that whistling symbolizes a negative response, they will, if the crowd begins whistling, decide against an encore. This is opposite the response they would have in the United States, where whistling has a very different meaning.

Finally, symbols are central to interaction in yet another way. We use the meanings of symbols to imagine how others will respond to our behavior. Through this capability, we can have "internal conversations" with ourselves. These internal conversations enable us to visualize how others will respond to us before we act. This is crucial because we guide our interactions with people according to the behavior we think others expect of us and we expect of others. Meanwhile, these others are also having internal conversations. The interaction (acting on each other) that follows is therefore symbolic interaction.

What approach has Erving Goffman added to the recent developments in symbolic interactionism? In an attempt to understand human interaction, Erving Goffman introduced **dramaturgy,** an approach that depicts human interaction as theatrical performance (Goffman 1961a, 1963, 1974, 1979, 1983; C. Lemert and Branaman 1997). Like actors on a stage, people (the performers) present themselves—by their dress, gestures, tone of voice—in such a way as to enhance their performance and create in others a favorable evaluation. Goffman labels this effort **presentation of self**. Through this kind of "impression management," we attempt to create a favorable evaluation of ourselves in the minds of others.

Think about your own impression management. Most students come to college concerned about acceptance by new peers. Consequently, you make decisions (consciously or not) about how you want others to view you. Choices in clothing are made—some students wear blue jeans and T-shirts; others choose a preppie look. Posters and magazines may be prominently displayed to present the desired image. Personal information such as athletic, musical, or sexual prowess will be appropriately leaked.

Performances, according to Goffman's theater analogy, may have a front and a back stage. College students may behave one way with dates and let down when alone with other campus classmates. Students may conduct themselves maturely when their parents are visiting campus, only to revert to old habits with their roommates. Behavior changes as it is managed in various settings.

Goffman's theoretical approach has stimulated the research of other sociologists. Dramaturgical symbolic interactionists have documented diverse arenas of American drama and ritual: social life, including singles bars; work, including nurses; athletics, including college football; politics, including effects of corruption; and religion, including the public's view of their social world (Deegan and Stein 1978; T. R. Young 1990; Hochschild 2003).

Although the feminist perspective grew most directly out of conflict theory, feminist sociologists also use the micro perspective. They have done very important research within the tradition of symbolic interactionism (West and Zimmerman 1987; Gardner 1994).

What are the limitations of symbolic interactionism? By virtue of being a microsociological theoretical perspective, symbolic interactionism sometimes fails to take the larger social picture into account. Social interaction in everyday life is sometimes affected by societal forces beyond the control (or even awareness) of individuals. Anthony Giddens (1979, 2000), in his *structuration theory*, attempts to overcome this potential limitation. Although Giddens recognizes that social structures influence the individual—a central point of both functionalism and conflict theory—he emphasizes the effects individuals have on social structures. He looks at individual recognition of social forces and their subsequent actions based on this recognition. Despite the micro limitations, symbolic interactionism is one of the most influential theoretical perspectives in sociology today (Fine 1993).

Each theoretical perspective seems lacking in some aspect. How, then, do these perspectives contribute to our study of sociology? Because the three theoretical perspectives have some contradictory assumptions and each is incomplete in itself, you may be somewhat confused. Where do perspectives in sociology now stand? The answer is easy: these three perspectives tend to complement one another because they focus on different aspects of social life.

Each of the perspectives could be used, for example, to better understand such deviant behavior as crime, juvenile delinquency, and mental illness. Functionalism emphasizes integration, stability, and consensus, so it accents the negative consequences of deviance. It underscores the cost of crime, including the money spent on police departments and prisons, violence in the streets, and the threat to public safety in inner cities and suburbs. The conflict perspective views deviance quite differently. Because various groups compete to influence the passage and enforcement of legislation serving their interests, many laws are created and enforced by those with political power. Consequently, those with the least influence—such as prostitutes and drug addicts—are the most likely to be classified as deviants. Interactionism focuses on the relationships within a group of deviants. It looks, for example, at the processes by which members of a drug-using group influence others to join them. It studies the ways drugs are bought from pushers, the methods for raising money to pay the pushers, and the ways people who use drugs interpret the user's activities.

The unique contribution of each of the three perspectives can be illustrated by examining another aspect of group life. American women currently earn about 76 cents for every dollar earned by men. An earning gap exists even between men and women in the same occupations with the same educational qualifications. How would functionalists, conflict advocates, and symbolic interactionists view this piece of social reality? Functionalists claim that income inequality between men and women is a dysfunction, or negative consequence, of the broader social condition of sexual inequality. Conflict advocates attribute income inequality to men's power in controlling women's income. Symbolic interactionists would point out that income inequality has existed for a very long time. It is only in recent years that many Americans have come to define income inequality as an unfair condition in need of eradication. In other words, income inequality now has a different symbolic meaning than it has had in the past.

Because each of these perspectives emphasizes different aspects of social life, one perspective cannot be said to be inherently superior to the others. Except for sheer personal preference, the old witchlike lady is no worse than the beautiful young woman, the rabbit no

28 SOCIOLOGY

FEEDBACK

1. A _____ perspective is a set of assumptions accepted as true by its advocates.
2. Indicate whether the following statements represent functionalism (F), conflict theory (C), or symbolic interactionism (S).
   ____ a. Societies are in dynamic equilibrium.
   ____ b. Power is one of the most important elements in social life.
   ____ c. Religion helps hold a society together morally.
   ____ d. Symbols are crucial to social life.
   ____ e. A change in the economy leads to a change in the family structure.
   ____ f. People conduct themselves according to their subjective interpretations of reality.
   ____ g. Many elements of a society exist to benefit the powerful.
   ____ h. Different segments of a society compete to achieve their own self-interests rather than cooperate to benefit others.
   ____ i. Social life should be understood from the viewpoint of the individuals involved.
   ____ j. Most members of a society agree that democracy is desirable.
   ____ k. Social change is constantly occurring.
   ____ l. Conflict is harmful and disruptive to society.
3. According to the _____ approach, humans use impression management to control the attitudes and responses of others toward them.
4. Match the three feminist theoretical frameworks with the following words and phrases.
   ____ a. liberal feminism   (1) patriarchy
   ____ b. radical feminism   (2) capitalism and patriarchy
   ____ c. socialist feminism   (3) equality of opportunity
                                   (4) social and legal reform

Answers: 1. theoretical; 2. a. (F), b. (C), c. (F), d. (S), e. (F), f. (S), g. (C), h. (C), i. (S), j. (F), k. (F), l. (F); 3. dramaturgical; 4. a. (3), (4), b. (1), c. (2)

better than the duck. By the same token, no theoretical perspective is innately superior to another. Each of the three major perspectives has its own worth, because each tells us something different about group life. The remaining chapters will illustrate each perspective in the context of the chapter topic.

SUMMARY

REVIEW GUIDE

1. Sociology is the scientific study of social structure. It maintains a group rather than an individual focus. It emphasizes the patterned and recurrent social relationships between group members and uses social factors to explain human social behavior. Macrosociology and microsociology are levels of analysis crucial to understanding the sociological perspective.
2. Sociology benefits both the individual and the public. First, through the sociological imagination, individuals can better understand the relationship between what is happening in their personal lives and the social events occurring in their society. The sociological imagination promotes the questioning of conventional, and often misleading, ways of thinking, and it provides a vision of social life that extends far beyond the often narrow confines of one's limited personal experience. Second, sociological research contributes to public policies and programs. Third, sociology enhances the development of occupational skills.

3. It was not until the nineteenth century that the process of differentiating the various social sciences seriously began. Five of the major social sciences—anthropology, psychology, economics, political science, and history—are clearly distinguishable from sociology. Although each is a separate field in its own right, each shares some areas of interest with sociology.
4. Sociology is rather young. It was born out of the social upheaval created by the French and Industrial Revolutions. In an attempt to understand the social chaos of their time, early sociologists emphasized social stability and social change. Sociology received its start primarily from the writings of European scholars Adam Smith, Auguste Comte, Harriet Martineau, Herbert Spencer, Karl Marx, Emile Durkheim, and Max Weber.
5. Auguste Comte, the generally acknowledged father of sociology, believed that society could advance only if studied scientifically. Harriet Martineau contributed to research methods, political theory, and feminism.

According to Herbert Spencer, social progress occurs only if people do not interfere with natural processes. Karl Marx argued that history unfolds according to the outcome of conflict between social classes. In capitalist societies, the conflict is between the ruling bourgeoisie and the ruled proletariat.

6. Emile Durkheim shared with the earlier pioneers a concern for social order. Two of his major contributions were the nonpsychological explanation of social life and the introduction of statistical techniques in social research. One of Max Weber's major contributions was also methodological. His method of *verstehen* assumed an understanding of human social behavior based on mentally putting oneself in the place of others. He also explored the process of rationalization as it existed in the transition from traditional to industrial society.

7. American sociology has been heavily influenced by the early European sociologists, in part because it, too, was born during a time of social upheaval (following the Civil War). From its founding in the 1800s to World War II, the hotbed of American sociology was the University of Chicago. After World War II, sociology departments in the East and Midwest rose to prominence.

8. A theoretical perspective is a set of assumptions considered true by its advocates. Functionalism and conflict theory are the province of macrosociology. Symbolic interactionism is part of microsociology.

9. According to functionalism, a society is an integrated whole, seeks a dynamic equilibrium, is composed of elements promoting its well-being, and is based on the consensus of its members. The conflict perspective contradicts these assumptions. A society experiences conflict at all times, is constantly changing, and rests on the domination of some of its members by other members.

10. Feminist theory, a form of conflict theory, can be divided into three frameworks—liberal feminism, radical feminism, and socialist feminism. Although there are important differences among these theoretical frameworks, they each link the lives of women and men to the structure of gender relationships within society.

11. According to symbolic interactionism, people's interpretations of symbols are based on the meanings they learn from others. People base their interaction on their interpretations of symbols. Because symbols permit people to have internal conversations, they can gear their interaction to the behavior they think others expect of them and they expect of others.

12. No single sociological perspective is all-encompassing. Each perspective makes a unique contribution; combined, these perspectives shed considerable light on social life.

LEARNING OBJECTIVES REVIEW

- Illustrate the unique sociological perspective from both the micro and macro levels of analysis.
- Describe three uses of the sociological perspective.
- Distinguish sociology from other social sciences.
- Outline the contributions of the major pioneers of sociology.
- Summarize the development of sociology in the United States.
- Identify the three major theoretical perspectives in sociology today.
- Discuss feminist theory.
- Compare two recent developments in symbolic interactionism.

CONCEPT REVIEW

____ a. economic determinism
____ b. mechanical solidarity
____ c. positivism
____ d. dynamic equilibrium
____ e. social structure

____ f. bourgeoisie
____ g. macrosociology
____ h. sociology
____ i. *verstehen*
____ j. symbol

____ k. latent function
____ l. conflict theory
____ m. presentation of self
____ n. social dynamics
____ o. theoretical perspective

1. a set of assumptions accepted as true by its advocates
2. the theoretical perspective that emphasizes conflict, competition, change, and constraint within a society
3. an unintended and unrecognized consequence of some element of a society
4. the ways people attempt to create a favorable evaluation of themselves in the minds of others
5. the study of social change

6. the method of understanding the behavior of others by putting oneself mentally in another's place
7. patterned, recurring social relationships
8. the scientific study of social structure
9. the use of observation, experimentation, and other methods of the physical sciences in the study of social life
10. something that stands for or represents something else
11. the idea that the nature of a society is based on the society's economy

12. social unity based on a consensus of values and norms, strong social pressure for conformity, and dependence on tradition and family

13. the assumption by functionalists that a society both changes and maintains most of its original structure over time

14. members of industrial society who own the means for producing wealth

15. the level of analysis that focuses on relationships between social structures without reference to the interaction of the people involved

CRITICAL-THINKING QUESTIONS

1. Think of a recent time you conformed. Were you responding to group pressure? Explain.

   _____

   _____

2. Apply the sociological imagination to your college choice. Identify the social forces (for example, family, friends, school, and the media) that you believe had the most influence on your decision.

   _____

   _____

3. Discuss the possible benefits of sociology to your future personal and work life.

   _____

   _____

4. Sociology, psychology, anthropology, economics, political science, and history are all considered social sciences. What do they have in common that justifies this grouping? Explain.

   _____

   _____

5. Max Weber introduced the concept of *verstehen*. How would you use this approach to social research if you wanted to investigate the importance of money to your peers?

   _____

   _____

6. Think of an aspect of human social behavior (e.g., college sports or fraternities and sororities) that you would like to know more about. Which of the three theoretical perspectives would you use? Explain your choice.

   _____

   _____

7. Think about some social change (e.g., increased drug use or later age at first marriage) that you think is particularly significant. Is this change better explained by conflict theory or functionalism?

   _____

   _____

8. Could any of your behavior over the past week be described by the dramaturgical perspective? Explain, using personal examples.

   _____

   _____

MULTIPLE-CHOICE QUESTIONS

1. **According to C. Wright Mills, the _____ is the set of mind that allows individuals to see the relationship between events in their personal lives and events in their society.**
   a. sociological awareness
   b. macrosociology
   c. sociological imagination
   d. microsociology
   e. moral imperative

2. **Shepard defines sociology as**
   a. the scientific study of socialism and other political-economic systems.
   b. the opposite of psychology.
   c. the scientific arm of social work.
   d. the study of how culture affects the personality.
   e. the scientific study of social structure.

3. **Microsociology focuses on**
   a. class conflicts.
   b. relationships among social structures.
   c. the way individuals interact with each other in social relationships.
   d. material very similar to psychology.
   e. emotional-mental development.

4. **Macrosociology focuses on**
   a. relationships between individuals within groups.
   b. political systems.
   c. various systems of producing, distributing, and consuming goods and services.
   d. how individuals interpret and react to their own culture.
   e. relationships between social structures without concern for the interaction of the individuals involved.

5. The sociological perspective that places human needs and goals as the primary concern of sociology is referred to as
   a. the sociological imagination.
   b. qualitative sociology.
   c. the ministerial perspective.
   d. humanist sociology.
   e. pure sociology.

6. Positivism is defined as
   a. the approach that should be used in dealing with the soft sciences.
   b. a religion that started in the early 1900s in France.
   c. a mental attitude designed to ensure success in life.
   d. the use of observation and experimentation in research.
   e. the balance of power between groups.

7. Karl Marx saw class conflict as
   a. one of the major economic consequences of World War I.
   b. the product of insecurities of the Cold War.
   c. one of the social and economic consequences of the social chaos of nineteenth-century Europe.
   d. conflict between those controlling the means for producing wealth and those laboring for them.
   e. conflict among those controlling the means for producing wealth.

8. Karl Marx's political objective was to
   a. end the British Monarchy.
   b. obtain a seat in the British Parliament.
   c. explain the workings of capitalism to hasten its fall through revolution.
   d. promote the use of religion as the opium of the masses.
   e. identify the distinctive characteristics of monogamy to enhance family stability.

9. According to Emile Durkheim, the social unity based on a complex of highly specialized roles that make members of a society dependent on one another is
   a. mechanical solidarity.
   b. organic solidarity.
   c. positivism.
   d. value freedom
   e. dialectic materialism.

10. The theorist who wrote *The Protestant Ethic and the Spirit of Capitalism* was
    a. Emile Durkheim.
    b. Alexis de Tocqueville.
    c. Max Weber.
    d. Lester Ward.
    e. Robert Merton.

11. Functionalism is defined as
    a. the theoretical perspective that emphasizes the contributions made by each part of society.
    b. the type of society characterized by weak family ties, competition, and impersonal social relationships.
    c. the form of control by which authority is split evenly between husband and wife.
    d. a system in which strategic industries are owned and operated by the state.
    e. a sense of identification with the goals and interests of the members of one's social class.

12. _____ emphasizes conflict, competition, change, and constraint within a society.
    a. The functionalist perspective
    b. Critical theory
    c. The symbolic interactionist perspective
    d. Social revolutionist theory
    e. Conflict theory

13. The concept of patriarchy is most closely associated with
    a. liberal feminism.
    b. socialist feminism.
    c. conservative feminism.
    d. Marxist feminism.
    e. radical feminism.

14. Impression management is a central concept of which of the following theoretical perspectives?
    a. functionalism
    b. conflict theory
    c. symbolic interactionism
    d. critical theory
    e. exchange theory

FEEDBACK REVIEW

True-False
1. Microsociology focuses on the relationships between social structures without reference to the interaction of the people involved. T or F?
2. W. E. B. Du Bois focused only on the American race question. T or F?

Fill in the Blank
3. _____ explanations of group behavior are inadequate because human activities are influenced by social forces that individuals have not created and cannot control.
4. According to Durkheim, the United States is a social order based on _____ solidarity.

5. The _____ is the set of mind that enables individuals to see the relationship between events in their personal lives and events in their society.

6. A _____ perspective is a set of assumptions accepted as true by its advocates.

Matching

7. Match the following fields of study and aspects of family life.

_____ a. anthropology
_____ b. sociology
_____ c. history
_____ d. political science
_____ e. economics
_____ f. psychology

(1) distribution of income and the family
(2) effects of slavery on family stability
(3) relationship between voting and the family
(4) effects of early marriage on divorce
(5) link between early childhood emotional support and familial retardation
(6) universality of the nuclear family

8. Indicate whether the following statements represent functionalism (F), conflict theory (C), or symbolic interactionism (S).

_____ a. Religion helps hold a society together morally.
_____ b. Symbols are crucial to social life.
_____ c. A change in the economy leads to a change in the family structure.
_____ d. Many elements of a society exist to benefit the powerful.
_____ e. Social life should be understood from the viewpoint of the individuals involved.
_____ f. Social change is constantly occurring.

9. Match the three feminist theoretical frameworks with the words or phrases.

_____ a. liberal feminism
_____ b. radical feminism
_____ c. socialist feminism

(1) patriarchy
(2) capitalism and patriarchy
(3) equality of opportunity

Multiple Choice

10. **Which of the following is not one of the uses of the sociological imagination?**

a. seeing the interplay of self and society
b. capacity for creating new aspects of culture not thought of by others
c. ability to question aspects of social life most people take for granted
d. capability of understanding the social forces that shape daily life

11. **Which of the following sociologists contended that social change leads to progress, provided that people do not interfere?**

a. Durkheim
b. Weber
c. Martineau
d. Spencer
e. Marx
f. Du Bois

GRAPHIC REVIEW

Table 1.3 contains data on average income (after taxes) in America—by income category. Answering the following questions will test your understanding of this table.

1. What is the most important generalization you can make from the data?

_____
_____
_____

2. How would conflict theorists interpret the data?

_____
_____
_____

3. Would functionalists agree with the interpretation of conflict theorists? Why or why not?

_____
_____
_____

ANSWER KEY

Concept Review	Multiple Choice	Feedback Review	
a. 11	1. c	1. F	c. F
b. 12	2. e	2. F	d. C
c. 9	3. c	3. Individualistic	e. S
d. 13	4. e	4. organic	f. C
e. 7	5. d	5. sociological	9. a. 3, 4
f. 14	6. d	imagination	b. 1
g. 15	7. d	6. theoretical	c. 2
h. 8	8. c	7. a. 6	10. b
i. 6	9. b	b. 4	11. d
j. 10	10. c	c. 2	
k. 3	11. a	d. 3	
l. 2	12. e	e. 1	
m. 4	13. e	f. 5	
n. 5	14. c	8. a. F	
o. 1	15. a	b. S	

2

Social Research

Spencer Grant/PhotoEdit

USING THE SOCIOLOGICAL IMAGINATION

Does lower church attendance result in a rise in juvenile delinquency? The finding of a statistical link between church attendance and delinquency (delinquency increasing as church attendance decreases) meets the test of common sense. We can easily speculate on why this would be the case. An observed relationship between these two events, however, does not necessarily mean that one causes the other. In fact, delinquency increases as church attendance decreases because of a third factor—age. Age is related to both delinquency and church attendance. Older adolescents not only go to church less often, but are more likely to be delinquents. The apparent relationship between church attendance and delinquency, then, is actually produced by a third factor—age—that affects both of the original two factors.

Mistaken ideas can survive when people rely on sources of knowledge not grounded in the use of reason and not stimulated by the search for truth. A major benefit of sociological research lies in its replacement of false beliefs with more accurate knowledge. Before we turn to the logic of scientific research and the methods of sociological research, it will be helpful to consider some sources of knowledge.

Sources of Knowledge

Nonscientific Sources of Knowledge

We all have a problem in assessing the truth of information we encounter. That problem is called "motivated reasoning": the tendency to accept less critical information that is consistent with ideas we already have and to reject information that contradicts our present beliefs. Recent research on the brain helps explain why this happens. When confronted with information we consider neutral, brain activity is centered in areas that control higher reasoning. Exposure to political information (about which we usually have strong feelings) triggers activity in areas of the brain involving emotion (McArdle 2008). Given the power of motivated reasoning in filtering information, we need to be aware of nonscientific sources of knowledge that tend to short-circuit reasoning power, leaving us prey to erroneous thinking. The rest of this chapter explores how science promotes the use of reasoning rather than emotion in evaluating information.

How do we know what we think we know? Four major nonscientific sources of knowledge are intuition, common sense, authority, and tradition. *Intuition* is quick and ready insight that is not based on rational thought. To intuit is to have the feeling of immediately understanding something because of insight from an unknown inner source. For example, the decision against dating a particular person because "it feels wrong" is a decision based on intuition.

Common sense refers to opinions that are widely held because they seem so obviously correct. The problem with commonsense ideas is that they are often wrong. Some people take as common sense, for example, that property values will decline when African Americans move into a white middle-class neighborhood. Yet, the research doesn't support this idea. As philosopher Alasdair MacIntyre writes, "Because common sense is never more than an inherited amalgam of past clarities and past confusions, the defenders of common sense are unlikely to enlighten us" (MacIntyre 1998:117).

An *authority* is someone who is supposed to have special knowledge that we do not have. A king believed to be ruling by divine right is an example of an authority. Reliance on authority is often appropriate. It is more reasonable to accept a doctor's diagnosis of an illness than to rely on information from a neighbor whose friend had the same symptoms (although even a single doctor's diagnosis should not be accepted uncritically). In other instances, however, authority can obscure the truth. Astrologers who advise people to guide their lives by the stars are an example of a misleading authority. Advocates of science and supporters of authority are currently debating an approach to teaching science: evolution versus intelligent design.

The fourth major nonscientific source of knowledge is *tradition*. Despite evidence to the contrary, it is traditional to believe that an only child will be self-centered and socially inept. In fact, to avoid these alleged personality traits, most Americans still wish to have two or more children (Sifford 1989). And barriers to equal opportunity for women persist in industrial societies despite evidence that traditional negative ideas about the capabilities of women are fallacious.

Nonscientific sources of knowledge often provide false or misleading information (Adler and Clark 2003), sometimes leading to completely opposite conclusions. One person's intuition tells him to buy oil stocks, whereas another person's intuition tells her to avoid all energy stocks. One person's commonsense conclusion may deem the feminist movement as harmful to the family, while it seems perfectly obvious to someone else that feminism promotes family values.

Intuition, common sense, authority, and tradition engender motivated reasoning because conceptions they generate are emotion-laden. Information from these sources of knowledge come as pronouncements

When forming opinions about global warming, Americans are increasingly relying on scientific sources. The protestors in this picture are concerned about the harm that will be done to the planet by relying on nonscientific sources regarding global warming.

Jewel Samad/Getty Images

that attempt to deny any opportunity for falsification. Without choice there is only room for emotional thinking.

A Model for Doing Research

In an effort to obtain accurate knowledge, social scientists use a research model known as the *scientific method*. This model follows several distinct steps: identifying a problem, reviewing the literature, formulating hypotheses, developing a research design, collecting data, analyzing data, and stating findings and conclusions (Schutt 2006).

Identifying the Problem

Researchers begin by choosing an object or topic for investigation. A research question may be chosen for a variety of reasons—because it interests the researcher, addresses a social problem, tests a major theory, or responds to a government agency's needs.

Reviewing the Literature

Once the object of study has been identified, the researcher must examine the literature for relevant theories and previous research findings. For example, a sociologist investigating suicide will probably develop an approach that makes use of the classic study of suicide by Emile Durkheim as well as other studies on the topic.

Formulating Hypotheses

From a careful examination of relevant theory and previous research findings, a sociologist can state one or more **hypotheses**—tentative, testable statements of relationships among variables. These variables must be defined precisely enough to be measurable. One hypothesis might be, "the longer couples are married, the less likely they are to divorce." The independent variable (length of marriage) and the dependent variable (divorce) must be defined and measured. Scientists measure variables through the use of **operational definitions**—definitions of abstract concepts in terms of simpler, observable procedures. Divorce could be defined operationally as the legal termination of marriage. If so, the measurement of divorce would be qualitative—the couple is either legally married or not. Length of marriage could be measured quantitatively—the number of years a couple has been legally married.

Developing a Research Design

A research design describes the procedures the researcher will follow for collecting and analyzing data. Will the study be a survey or a case study? If it is a survey, will data be collected from a cross section of an entire population, as with the Harris and Gallup polls, or will a sample be selected from only one city? Will simple percentages or more sophisticated statistical methods be used? These and many other questions must be answered while the research design is being developed.

Prominent among these question is the determination of the population and sample to be used in the study. A **population** consists of all those people with the characteristics a researcher wants to study. A population can be all college sophomores in the United States, all former drug addicts living in Connecticut, or all current inmates of the Ohio State Penitentiary. Most populations are too large and inaccessible to collect information on all members. For this reason, a sample is drawn. A **sample** is a limited number of cases drawn from the larger population. The sample must be selected carefully if it is to have the same basic characteristics as the population. If a sample is not representative of the population from which it is drawn, the survey findings cannot be used to make generalizations about the entire population (Winship and Mare 1992). The U.S. Census Bureau regularly uses sample surveys in its highly accurate work. The Gallup and Harris polls are also recognized as reliable sample indicators of national trends and public opinion.

Collecting Data

There are three basic ways of gathering data in sociological research: asking people questions, observing behavior, and analyzing existing materials and records. Sociologists studying the harmony in interracial marriages could question couples about their communication skills and compatibility. They could locate an organization with a large number of interracially married couples and observe couples' behavior. Or they could compare the divorce rate among interracially married couples with the divorce rate of the population as a whole.

Analyzing Data

Once the data are collected and classified, they can be analyzed to determine whether or not the hypotheses are supported. This is not as easy or automatic as it sounds, because results are not always obvious. Because the same data can be interpreted in several ways, judg-

ments have to be made. Guarding against personal biases is especially important in this phase of research.

Stating Findings and Conclusions

After analyzing the data, a researcher is ready to state the conclusions of the study. It is during this phase that the methods are described and the hypotheses are formally accepted, rejected, or modified. The conclusions of the study are then related to the theory and research findings on which the hypotheses are based, and directions for further research are suggested. Depending on the findings, the original theory itself may have to be altered. Whether the statement of conclusions appears in a scientific journal, a book, or online, it includes a description of the methods used. By making the research procedures public, scientists make it possible for others to either duplicate the research, conduct a slightly different study, or proceed in a very different direction.

Using the Research Model

Realistically, do sociologists follow these steps? Some sociologists believe that this research model is too rigid to capture spontaneous, subjective, and changeable social behavior. They may prefer to discover their findings without preconceived ideas of outcome, without hypothesis-biased observations, and without an inflexible research design. Even though most sociologists do follow the model, they do not do so mechanically. They may conduct exploratory studies prior to stating

---

F E E D B A C K

Listed here are the steps in the research model. Following are some concrete examples related to the sociability of the only child. Indicate the appropriate example for each step number.

_____ Step 1: identifying the problem

_____ Step 2: reviewing the literature

_____ Step 3: formulating hypotheses

_____ Step 4: developing a research design

_____ Step 5: collecting data

_____ Step 6: analyzing data

_____ Step 7: stating findings and conclusions

    a. The researcher reads past theory and research on the sociability of only children.

    b. From previous research and existing theory, the researcher states that only children appear to be more intelligent than children with siblings.

    c. The researcher collects data on only children from a high school in a large city.

    d. The researcher writes a report giving evidence that only children are more intelligent than children with brothers or sisters.

    e. The researcher decides to study the intelligence level of only children.

    f. The researcher classifies and processes the data collected to test a hypothesis.

    g. The researcher decides on the data needed to test a hypothesis, the methods for data collection, and the techniques for data analysis.

Answers: Step 1: e; Step 2: a; Step 3: b; Step 4: g; Step 5: c; Step 6: f; Step 7: d

SNAPSHOT OF AMERICA 2.1

Suicide Rates for the Year 2002

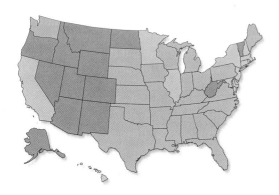

Deaths per 100,000 Populatuion

- Above average: 14.4 or more
- Average: 10.5 to 14.3
- Below average: 10.4 or fewer

Source: *National Vital Statstics Reports,* from the Centers for Disease Control and Prevention, 2005.

SNAPSHOT OF AMERICA 2.2

Population Density

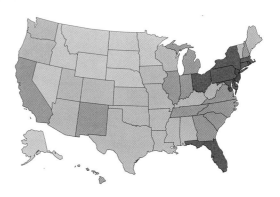

People per Square Mile

- 1,000 or above
- 260–999
- 130–259
- 26–129
- 25 or below

Source: U.S. Census Bureau, *Statistical Abstact of the United States: 2004–2005*

Interpret the Map

1. Emile Durkheims's study of suicide suggested that one factor in the suicide rate is the degree to which the individual has group ties. One indication of social ties is population density. Based on Snapshot 2.2, where would you expect to find the highest suicide rates in the United States? Does Snapshot 2.1 agree with your predictions?

2. Durkheim noted that "psychological explanations are insuffient when analyzing social behavior." In your own words, tell what this statement means.

3. Which perspective do you think Durkheim followed in his study of suicide: the functionalist, the conflict, or the symbolic interactionist? Give reasons for your answer.

4. Using what you have learned from Durkheim's research, formulate a hypothesis about mass suicide.

---

hypotheses or before developing research designs. Or they may alter their hypotheses and research designs as their investigations proceed.

This does not mean that these steps can be ignored in conducting sociological research. Most sociologists more or less go through this process anyway. And all researchers, even those who say they do not follow this procedure, have the model in mind as they do their work. Moreover, research reports are evaluated with this method in mind. If researchers seriously violate the research process, their findings and conclusions may not be viewed as credible.

Science as a Source of Knowledge

Science is considered a superior source of knowledge. In part, this is because science is based on the principles of objectivity and verifiability.

What is objectivity? According to the principle of **objectivity,** scientists are expected to prevent their personal biases from influencing the interpretation of their results. A male, antifeminist biologist investigating aptitudes, for example, is supposed to guard against any unwarranted tendency to conclude that males make better scientists than females. Researchers must interpret their data solely on the basis of merit; the outcome they personally prefer is irrelevant. This is what Max Weber (1918–1946) meant by value-free research (see Chapter 1).

Can scientists really be objective? Sometimes, scientists unintentionally let their personal biases influence their work. For example, in the 1950s, pioneering sex researcher Alfred Kinsey revolutionized popular thinking about sex in America. He has been accused of being both a homosexual and a masochist, characteristics said to unduly influence his research. James H. Jones (1997) makes this charge and presents evidence that Kinsey was a man with an ideological agenda whose research methods undermine his claim to objectivity. Kinsey's generalizations about the American population were based, according to Jones, on data gathered largely from volunteers, including disproportionate numbers of male prostitutes, gays, and prison inmates.

Historian Howard Zinn (2003, 2005) emphasizes the impossibility of being completely neutral with respect to the outcomes of research. Because researchers bring their own interests to work, they must decide what to report and what to omit from a wide array of possible theories and "facts." This process of choice, whether consciously or unconsciously, reflects personal preferences.

When investigating the cause of a murder, modern police have to be guided by the principle of multiple causation.

methods. Verifiability is important because it exposes scientific work to critical analysis, retesting, and revision by colleagues. If researchers repeating a study produce results at odds with the original study, the original findings will be questioned. Under these circumstances, erroneous theories, findings, and conclusions will not survive (Begley 1997).

Causation and the Logic of Science

The Nature of Causation

What is causation? Scientists assume that an event occurs for a reason. According to the concept of **causation,** events occur in predictable, nonrandom ways, and one event leads to another. Why does this book remain stationary rather than rise slowly off your desk, go past your eyes, and rest against the ceiling? Why does a ball thrown into the air return to the ground? Why do the planets stay in orbit around the sun?

More than 2,000 years ago, Aristotle claimed that heavier objects fall faster than lighter ones. In the late 1500s, Galileo contended that all objects fall with the same acceleration (change of speed) unless slowed down by air resistance or some other force. But it was not until the late 1600s that Sir Isaac Newton developed the theory of gravity. We now know that objects fall because the Earth has a gravitational attraction for objects near it. The planets remain in orbit around the sun because of the gravitational force created by the sun.

Because scientists assume causation, one of their main goals is to discover cause-and-effect relationships. They attempt to discover the factors—there is usually more than one—that cause events to happen.

Why multiple causation? Leo Rosten, noted author and political scientist, once wrote, "If an explanation relies on a single cause, it is surely wrong." Events in the physical or social world are generally too complex to be explained by any single factor. For this reason, scientists are guided by the principle of **multiple causation,** which states that an event occurs as a result

How can subjectivity be reduced? Scientists, then, cannot possibly be completely objective. But if subjectivity in research cannot be eliminated, it can be reduced. According to Zinn, the best approximation of objectivity exists when researchers strive for the truth and follow specific safeguards: continually reexamine their thinking, permit contradicting evidence to alter their view, and make public any evidence that runs counter to their view of the truth. If researchers are aware of their biases, they can consciously take them into account. They can be more careful in designing research instruments, selecting samples, choosing statistical techniques, and interpreting results. According to Swedish economist Gunnar Myrdal (1983), personal recognition of biases is insufficient; public exposure of them is essential. Personal values, Myrdal contends, should be explicitly stated, so that those who read a research report can be aware of the author's biases.

What is verifiability? **Verifiability** means that a study can be repeated by other scientists. This is possible because scientists report in detail on their research

---

FEEDBACK

1. Intuition is quick and ready insight based on rational thought. T or F?
2. The major problem with nonscientific sources of knowledge is that such sources often provide erroneous information. T or F?
3. Define objectivity and verifiability as used in science.
4. According to Gunnar Myrdal, it is enough that scientists themselves recognize their biases. T or F?

Answers: 1. F; 2. T; 3. Objectivity exists when an effort is made to prevent personal biases from distorting research. Verifiability means that any given piece of scientific research done by one scientist can be duplicated by other scientists.; 4. F of several factors operating in combination. What, for example, causes crime? Cesare Lombroso, a nineteenth-century Italian criminologist, believed that the predisposition to commit crimes was inherited and that criminals could be identified by certain physical traits (large jaws, receding foreheads). Modern criminologists reject Lombroso's (or anyone else's) one-factor explanation of crime. They cite numerous factors that contribute to crime, including drugs; excessive materialism; peer pressure; hopeless poverty in slums; and overly lax, overly strict, or erratic child-rearing practices. Each of these factors is a variable.

Causation and Variables

How, in a research project, are variables related to causation? A **variable** is a characteristic (such as age, education, or social class) that is subject to change. It occurs in different (varying) degrees. Some materials have greater density than others; some people have higher incomes than others; the average literacy rate is higher in developed countries than in developing ones. Each of these is a **quantitative variable,** a variable that can be measured and given a numerical value. Because differences can be measured numerically, individuals, groups, objects, or events can be pinpointed at some specific point along a continuum. In contrast, a **qualitative variable** consists of variation in kind rather than in number. It denotes differences in distinguishing characteristics of a category rather than differences in numerical degree. Sex, marital status, and group membership are three qualitative variables often used by sociologists. This is an either/or or yes/no variable: people are either male or female; they are either single or married; they are either Greek or they are not (see Table 2.1).

Once scientists identify the variables (qualitative or quantitative) to investigate, they define these variables as either independent or dependent. Variables that cause something to occur are **independent variables.** Variables in which a change (or effect) can be observed are **dependent variables.** Marital infidelity is an independent variable (although, of course, not the only one) that may cause the dependent variable of divorce. The independent variable of poverty is one of several independent variables that can produce a change in the dependent variable of hunger. Whether a variable is dependent

---

TABLE 2.1

Forming Quantitative and Qualitative Variables

Please indicate whether you strongly agree, agree, disagree, or strongly disagree with each of the following statements:

	Strongly Agree	Agree	Disagree	Strongly Disagree
a. Most schoolteachers really know what they are talking about.	1	2	3	4
b. To get ahead in life, you have to get a good education.	1	2	3	4
c. My parents encouraged me to get a good education.	1	2	3	4
d. Children should not be allowed to quit school at sixteen.	1	2	3	4
e. Too little emphasis is put on education these days.	1	2	3	4
f. My parents think that going to college is essential.	1	2	3	4

These questions are designed to measure the value people place on education. The total score of this measure can range from 6 for those strongly agreeing with all six statements (highest value on education) to 24 for those strongly disagreeing with all six statements (lowest value on education). Like any quantitative variable, most actual scores will fall somewhere between these two extremes.

In your own words, describe your views on the importance of education.

_____

_____

_____

Answers to this question could be used by a sociologist to classify people into one of two categories: those who place a high value on education or those who place a low value on education.

WORLDVIEW 2.1

The Wired World

This map shows the number of telephone mainlines per thousand people for 2002. This data is an example of a quantitative variable.

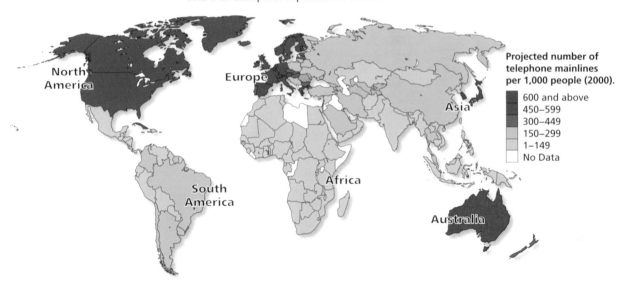

Projected number of telephone mainlines per 1,000 people (2000).

- 600 and above
- 450–599
- 300–449
- 150–299
- 1–149
- No Data

Source: United Nations Development Programme, 2004.

Interpret the Map

1. Explain why the data in this map constitutes a quantitative variable.

2. What would need to be done with the data to make it a qualitative variable?

3. If you were using telephone mainlines per 1,000 people in research, would it likely be a dependent or independent variable?

or independent varies with the context. Hunger may be a dependent variable in a study of poverty; it may be an independent variable in a study of crime.

An **intervening variable** influences the relationship between an independent variable and a dependent variable. A government support program, for example, may intervene between poverty and hunger. With a strong safety net, very poor parents and their children may experience no more hunger than working-class families. Poverty is the cause of hunger, but it does not have to be if there is government intervention in the form of income and food. The poor *without* a safety net will experience more hunger; the poor *with* a safety net will not.

What is a correlation? After identifying which independent and dependent variables they will study, sociologists attempt to find out how the variables are related. They describe these relationships in terms of correlations. A **correlation** exists when a change in one variable is associated with a change (either positively or negatively) in the other.

A correlation may be positive or negative. A **positive correlation** exists if both the independent vari-

able and the dependent variable change in the same direction. A positive correlation exists if we find that grades (dependent variable) improve as study time (independent variable) increases (see Figure 2.1). In a **negative correlation,** the variables change in opposite directions. For example, an increase in the independent variable is linked to a decrease in the dependent variable. A negative correlation exists if we find that grades (dependent variable) go down as time spent videogaming (independent variable) increases.

What are the criteria for establishing a causal relationship? Establishing causation is much more complicated than establishing a correlation between two variables. In a causal relationship, one variable actually causes the other to occur. Three standards are commonly used for establishing causality (Lazarsfeld 1955; Hirschi and Selvin 1973). These standards can be illustrated in the mistaken assumption that lower church attendance causes higher juvenile delinquency. (See Using the Sociological Imagination at the beginning of the chapter.)

1. *Two variables must be correlated.* Some researchers found that juvenile delinquency increases as church attendance declines (R. Stark, Kent,

FIGURE 2.1

Positive and Negative Correlations

Examine the graphs below. Choose two examples of your own to illustrate a positive and a negative correlation. Briefly explain why you would expect the type of correlation in each case.

In a positive correlation, increases in one variable are associated with increases in the other. Grades (dependent variable) improve with time spent studying (independent variable).

In a negative correlation, increases in one variable are associated with decreases in the other. Grades (dependent variable) decrease as time spent watching television (independent variable) increases.

and Doyle 1982). Does this negative correlation mean that lower church attendance causes higher delinquency? To answer this question, the second criterion of causality must be met.

2. *All possible contaminating factors must be taken into account.* A major problem in establishing causality lies in the control of all relevant variables. Holding contaminating factors constant (ruling out their influence) is one of the greatest challenges in science.

Although all cause-and-effect relationships involve a correlation, the existence of a correlation does not necessarily indicate a causal relationship. Just because two events vary together (are correlated) does not mean that one causes the other. Two totally unrelated variables may have a high correlation. In fact, the correlation between lower church attendance and delinquency is known as a **spurious correlation**—an apparent relationship between two variables, which is actually produced by a third variable that affects both of the original two variables.

The negative relationship between church attendance and delinquency occurs because age is related to both church attendance and delinquency (older adolescents attend church less frequently, and older adolescents are also more likely to be delinquents). Thus, before we can predict that a causal relationship exists between church attendance and delinquency, we need to take into account all variables relevant to the rela-

tionship. In this instance, the age variable reveals that the relationship between church attendance and delinquency is not a causal one. Church attendance is the contaminating factor here.

3. *A change in the independent variable must occur before a change in the dependent variable can occur.* The cause must occur before the effect. Does lack of church attendance precede delinquency, or vice versa? Logically, either one can precede the other, or they can occur simultaneously. Thus, even if the original correlation between church attendance and delinquency is maintained after holding age and other possible contaminating factors constant, causality between these two variables still cannot be established. Why? Because it cannot be determined which occurred first.

The Experiment as a Model

Describing an experiment is an excellent way to illustrate causation. Although used infrequently by sociologists, experiments provide insight into the nature of all scientific research because they are grounded in the concept of causation.

An **experiment** takes place in a laboratory in an attempt to eliminate all possible contaminating influences. By ruling out extraneous factors, a researcher can determine the effects (if any) of an independent variable on a dependent variable. According to the logic of experiments, if the dependent variable changes when the experimental (independent) variable is introduced but does not change when it is absent, the change must have been caused by the independent variable.

The basic ingredients of an experiment are a *pretest,* a *posttest,* an *experimental variable,* an *experimental group,* and a *control group.* Suppose a researcher wants to study, experimentally, the effects of providing information on drug use (the experimental, or independent, variable) on student attitudes toward drug use (the dependent variable). After selecting a class of eighth graders, the researcher could first measure the teenagers' attitudes toward drug use (pretest). Then, at a later time, a film demonstrating the harmful effects of drugs might be shown to the class. After the movie, the students could again be questioned about their attitudes toward drug use (posttest). Any changes in their attitudes toward drug use that took place between the pretest and the posttest could be attributed to the experimental variable. Such a conclusion might be wrong, however, because the change could have been due to factors other than the experimental variable: A student in the school might have died from an overdose, a rock star might have publicly endorsed drug use, or a pusher might have begun selling drugs to the students.

The conventional method for controlling the influence of contaminating variables is to select a control group as

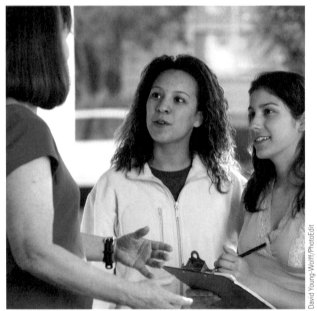

David Young-Wolff/PhotoEdit

The survey is the most widely used research method for collecting data in sociology. Surveys are best conducted in person, although use of the telephone is now widespread. An advantage of the survey is that it permits the gathering of information about a large number of people.

Assuming that the members of each group had similar characteristics and that their experiences between the pretest and the posttest were the same, any difference in attitudes toward drug use between the two groups could reasonably be attributed to the students' exposure or lack of exposure to the film.

How can experimental and control groups be made comparable? The standard way to make experimental and control groups comparable—except for exposure to the experimental variable—is through matching or randomization. In **matching,** participants in an experiment are matched in pairs according to all factors thought to affect the relationship being investigated. Members of each pair are then assigned to one group or the other. In **randomization,** which is preferable to matching, subjects are assigned to the experimental or control group on a random (chance) basis. Assignment to one group or the other can be determined by flipping a coin or by having subjects draw numbers from a container. Whether matching or randomization is used, the goal is the same—to form experimental and control groups that are alike with respect to all relevant characteristics except the experimental variable. If this requirement has been met, any significant change in the experimental group, as compared to the control group, can be attributed with considerable confidence to the experimental variable. That is, a causal link will have been established between the independent and dependent variables.

well as an experimental group. In the preceding example, half of the eighth-grade class could have been assigned to the **experimental group,** the group exposed to the experimental variable, and half to the **control group,** the group not exposed to the experimental variable.

F E E D B A C K

1. Match the following concepts and statements.

_____ a. causation   _____ f. independent variable

_____ b. multiple causation   _____ g. dependent variable

_____ c. variable   _____ h. correlation

_____ d. quantitative variable   _____ i. spurious correlation

_____ e. qualitative variable

(1) something that occurs in varying degrees

(2) the variable in which a change or effect is observed

(3) a change in one variable associated with a change in another variable

(4) the idea that an event occurs as a result of several factors operating in combination

(5) a factor that causes something to happen

(6) the idea that the occurrence of one event leads to the occurrence of another event

(7) a factor consisting of nonnumerical categories

(8) when a relationship between two variables is actually the result of a third variable

(9) a variable consisting of numerical units

2. An _____ attempts to eliminate all possible contaminating influences on the variables being studied.

3. The group in an experiment that is not exposed to the experimental variable is the _____ group.

4 Experimental and control groups are made comparable in all respects except for exposure to the experimental variable through _____ or _____.

Answers: 1. a. (6), b. (4), c. (1), d. (9), e. (7), f. (5), g. (2), h. (3), i. (8); 2. experiment; 3. control; 4. matching, randomization

BASIC STATISTICAL MEASURES

Statistics are methods used for tabulating, analyzing, and presenting quantitative data (Levin and Fox 2002). Sociologists, like all scientists, use statistical measures. The statistics you will encounter in this textbook and in the sources you are likely to read later, such as *The Wall Street Journal, Time, Newsweek,* and *The Economist*, are easily comprehended. Among the most basic statistical measures are averages (modes, means, medians) and correlations.

An *average* is a single number representing the distribution of several figures: Suppose the following annual salary figures were those of nine highly paid Major League Baseball players:

  $3,300,000 (catcher)
  $4,500,000 (first base)
  $3,600,000 (second base)
  $4,900,000 (starting
    pitcher)
  $3,600,000 (third base)
  $5,300,000 (left field)
  $4,200,000 (center field)
  $6,100,000 (right field)
  $4,300,000 (shortstop)

There are three kinds of averages that can make these numerical values more meaningful (see Figure 2.2). Each gives a different picture.

The *mode*—in this case $3,600,000—is the number that occurs most frequently.

The mode is appropriate only when the objective is to indicate the most popular number. If a researcher were to rely on the mode alone, though, in a report of these Major League Baseball salaries, readers would be misled, because no mention is made of the wide range of salaries ($3,300,000 to $6,100,000).

The *mean* is the measure closest to the everyday meaning of the term *average*. It lies somewhere in the middle of a range. The mean of the salary figures listed earlier—$4,422,222—is calculated by adding all the salaries together ($39,800,000) and dividing by the number of salaries (9). The mean, unlike the mode, takes all of the figures into account,

but it is distorted by the extreme figure of $6,100,000. Although one player earns $6,100,000, most players make considerably less—the highest-paid player earns nearly twice as much as the lowest-paid player. The mean is distorted when there are extreme values at either the high or low end of a scale; it is more accurate when extremes are not widely separated.

The *median* is the number that divides a series of values in half; that is, half of the values lie above it, half below. In this example, the median is $4,300,000—half of the salaries are above $4,300,000, half are below. What is the advantage here? The median is not distorted by extremes.

FIGURE 2.2

Mode, Mean, and Median of Salaries

Suppose that you were negotiating your salary with a Major League Baseball team. Also suppose that your agent presented these average salaries as currently accurate with the news that management would give you only one chance to declare your bottom-line salary. Which type of average would you choose? Explain why.

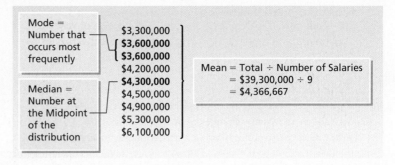

Mode =
Number that
occurs most
frequently

Median =
Number at
the Midpoint
of the
distribution

  $3,300,000
  $3,600,000
  $3,600,000
  $4,200,000
  $4,300,000
  $4,500,000
  $4,900,000
  $5,300,000
  $6,100,000

Mean = Total ÷ Number of Salaries
    = $39,300,000 ÷ 9
    = $4,366,667

Quantitative Research Methods

Because sociologists find it difficult to create experimental and control groups, they tend to rely more on other research methods, classified as either quantitative or qualitative. Quantitative research uses numerical data. Qualitative research rests on descriptive data.

Quantitative research methods include survey research and precollected data. About 90 percent of the research published in major sociological journals is based on surveys, so this approach is discussed first.

Survey Research

The survey, in which people are asked to answer a series of questions, is the most widely used research method among sociologists. It is ideal for studying large numbers of people.

How are effective surveys conducted? In survey research, care must be taken in selecting respondents and in formulating the questions to be asked (T. R. Black 2001). Researchers describe the people surveyed based on populations and samples.

How are representative samples selected? A **random sample** is selected on the basis of chance so that each member of a population has an equal opportunity of being selected. A random sample can be selected by assigning each member of the population a number and then drawing numbers from a container after they have been thoroughly scrambled. An easier and more practical method, particularly with large samples, uses a table of random numbers in which numbers appear without pattern. After each member of the population has been assigned a number, the researcher begins with any number in the table and goes down the list until enough subjects have been selected.

If greater precision is desired, a **stratified random sample** can be drawn. With this method, the population is divided into categories such as sex, race, or age. Subjects are then selected randomly from each category. In this way, the proportion of persons in a given category reflects the population at large.

How is information gathered in surveys? In surveys, information is obtained through either a questionnaire or an interview. A **questionnaire** is a written set of questions that survey participants answer by themselves. In an **interview,** a trained interviewer asks questions and records answers. Questionnaires or interviews may be composed of either closed-ended or open-ended questions (May 2001).

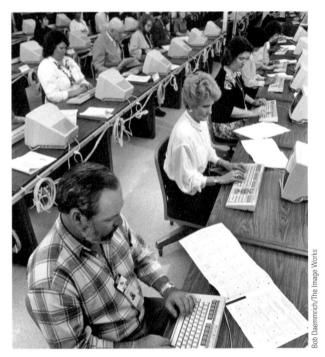

These employees of the U. S. Census Bureau are entering data from one of the government agency's many surveys. The results of these surveys, considered to be representative of the United States population, are widely utilized for decision making by private individuals, business organizations, and political leaders.

Closed-ended questions are those for which a limited, predetermined set of answers is possible—multiple-choice questions, for example. Because participants must choose from rigidly predetermined responses, closed-ended questions sometimes fail to elicit the participants' underlying attitudes and opinions. On the positive side, closed-ended questions make answers easier to quantify and compare. **Open-ended questions** ask the respondent to answer in his or her

FEEDBACK

1. Match the following terms and statements:
   ____ a. population
   ____ b. representative sample
   ____ c. random sample
   ____ d. sample
   ____ e. survey
   (1) selected on the basis of chance so that each member of a population has an equal opportunity of being selected
   (2) all those people with the characteristics the researcher wants to study within the context of a particular research question
   (3) a limited number of cases drawn from the larger population
   (4) a sample that has basically the same relevant characteristics as the population
   (5) the research method in which people are asked to answer a series of questions
2. Use of company records would be an example of using _____ data.

Answers: 1. a. (2), b. (4), c. (1), d. (3), e. (5); 2. precollected

READING TABLES AND GRAPHS

Tables and graphs present information concisely. They are chock-full of fascinating findings, which you can easily decode. Just follow the steps outlined here. The steps are keyed to Table 2.2 and Figure 2.3.

1. Begin by reading the title of the table or graph carefully; it will tell you what information is being presented. Table 2.2 shows median annual incomes in the United States by sex, race, and education.

2. Find out the source of the information. You will want to know whether the source is reliable, whether its techniques for gathering and presenting data are sound. The figures originated from the U.S. Bureau of the Census, a highly trusted source.

3. Read any notes accompanying the table or graph. Not all tables and graphs have notes, but if notes are present, they offer further information about the data. The notes in Table 2.2 and in Figure 2.3 explain that all the data refer to the total money income of full-time and part-time workers, ages 25 and over, in a March 2003 survey.

4. Examine any footnotes. Footnotes in Table 2.2 and Figure 2.3 indicate that the data are categorized by the highest grade actually completed. Although you may have assumed this, "years of schooling" could have referred to the total number of years in school, regardless of the grade level attained.

5. Look at the headings across the top and down the left-hand side of the table or graph. To observe any pattern in the data, it is usually necessary to keep both types of headings in mind. Table 2.2 and Figure 2.3 show the median annual income of African American and white males and females for several levels of education.

TABLE 2.2

Median Annual Income by Sex, Race, and Education

Examine the following data on median annual income by sex, race, and education. State the conclusions you can make from these data. Which findings most surprise you?

Demographic Group	Overall Median Income	Median earnings in dollars as compared with education attainment*				
		Less than HS	HS or GED	Associate's Degree	Bachelor's Degree	Master's Degree
White males	$34,176	$18,447	$30,174	$39,239	$51,160	$62,460
African American males	$26,807	$15,727	$25,095	$35,945	$41,664	$51,146
White females	$22,639	$11,508	$19,483	$26,313	$32,003	$42,447
African American females	$21,548	$12,268	$18,036	$25,068	$35,107	$41,553

*Based on highest grade completed.

Note: These figures include the total money income of full-time and part-time workers, ages 25 and over, as of March 2003.

Source: U.S. Bureau of the Census

own words. Answers to open-ended questions, however, are not easy to quantify, and the interviewer may inadvertently change the meaning of the respondent's answers by rephrasing them.

What are the advantages and disadvantages of closed-ended survey research? Surveys based on closed-ended questions have the advantage of precision and comparability of responses. They permit the use of statistical techniques, a feature they have in common with experiments. Statistical techniques can be used because of still other advantages in survey research. For example,

surveys permit the collection of large samples, which in turn permit more detailed analysis; surveys include a large number of variables; and variables in surveys can be quantified.

The survey research method has several disadvantages, however (Schuman 2002). (1) Due to the large samples usually involved, surveys are expensive. (2) Because survey questions are predetermined, interviewers cannot always include important unanticipated information, although they are encouraged to write such information in the margin or on the back of the interview form. (3) The response rate—particularly

6. Find out what units are being used. Data can be expressed in percentages, hundreds, thousands, millions, billions, means, and so forth. In Table 2.2 and Figure 2.3, the units are dollars and years of schooling.

7. Check for trends in the data. For tables, look down the columns (vertically) and across the rows (horizontally) for the highest figures, lowest figures, trends, irregularities, and sudden shifts. If you read Table 2.2 vertically, you can see how income varies by race and sex within each level of education. If you read the table horizontally, you can see how income varies with educational attainment for white males, African American males, white females, and African American females. A major advantage of presenting these data as a graph (Figure 2.3) is that the sudden shifts, trends, irregularities, and extremes are easier to spot than they are in tables.

8. Draw conclusions from your own observations. Again, look at Table 2.2 and Figure 2.3. Although income tends to rise with educational level for both African Americans and whites, it increases much less for African American men and for women of both races than for white men. At each level of schooling, African American men earn less than white men. In fact, white male high school dropouts have incomes only $6,548 below African American male high school graduates; white male high school graduates earn only $5,771 less than African American males with some college but no degree. White women and African American women both appear to improve their earning power through high school and college education.

FIGURE 2.3

Median Annual Income by Gender, Race, and Education

Clearly, this table documents the income advantage that white males in the United States have over white females and African Americans of both sexes. Explain how this situation can be used to challenge the existence of a true meritocracy.

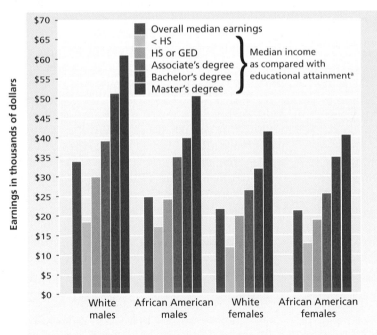

Legend:
- Overall median earnings
- < HS
- HS or GED
- Associate's degree
- Bachelor's degree
- Master's degree
} Median income as compared with educational attainment[a]

Note: These figures include the total money income of full-time and part-time workers, ages 25 and over, as of March 2003.

[a] In terms of highest degree completed

Source: U.S. Bureau of the Census

for mailed questionnaires—is often low. A respectable return rate is about 50 percent, although researchers make an effort to obtain a return rate of 80 percent or higher. Even in interviews, some people are not available and some refuse to answer the questions. Because nonresponses can make the sample unrepresentative, surveys may be biased. (4) The phrasing of survey questions may also introduce bias. For example, negatively phrased questions are more likely to receive a no answer than neutrally phrased questions. It is better to ask, "Are you in favor of abortion?" than "You aren't in favor of abortion, are you?" Respondents also interpret the same question differently. If asked about the extent of their drug use, some respondents may include alcohol in their answers; others may not. (5) Surveys cannot probe deeply into the context of the social behavior being studied; they draw specific bits of information from respondents, but they cannot capture the total social situation. (6) Survey researchers must be on guard for the *Hawthorne effect,* when unintentional behavior on the part of researchers influences the results they obtain from those they are studying (Roethlisberger and Dickson 1964; originally published in 1939). As researchers and survey participants interact, participants detect cues regarding

John Birdsall/The Image Works

Some aspects of social life, such as neighborhood youth gangs, are best studied in their natural setting. Why do you think this is so?

what the researchers are trying to find. The participants, depending on the circumstances, may attempt to please the researcher, frustrate the researcher's goals, or give "socially acceptable" answers.

Precollected Data

The use of previously collected information is a well-respected method of obtaining data. This is known as **secondary analysis.** In fact, the first sociologist to use statistics in a sociological study—Emile Durkheim—relied on precollected data (see Doing Research).

What are the major types of precollected data? The sources for precollected data are as varied as government reports, company records, voting records, prison records, and reports of research done by other social scientists. The U.S. Census Bureau is one of the most important sources of precollected data for sociologists. The Census Bureau collects information on the total population every ten years and conducts countless specific surveys each year. Consequently, detailed information exists on such topics as income, education, race, sex, age, marital status, occupation, death rates, and birth rates. Other government agencies also collect information. The U.S. Department of Labor regularly collects information on the nation's income and unemployment levels across a variety of jobs. The U.S. Department of Commerce issues monthly reports on various aspects of the economy.

What are the advantages and disadvantages of precollected data? Precollected data provide sociologists with inexpensive, quality information. Existing sources of information also permit the study of a topic over a long period of time. With census data, for example, we can begin in 1960, with the onset of the War on Poverty,

and trace the changes in the relative income levels of African Americans and whites. Also, because the data have been collected by others, the researcher cannot influence the answers.

Because the information was collected for different purposes, there is also a downside to precollected data. The existing information may not exactly suit the current researcher's needs. Also, the people who originally collected the data may have been biased. Finally, sometimes precollected data are simply too old to be valid.

Qualitative Research Methods

Some aspects of social reality can be revealed best through qualitative (nonquantitative) research methods. Most qualitative research methods include field research and the subjective approach (Neuman 2005; Lincoln and Denzin 2005; Strauss and Corbin 2007).

Field Research

Field research investigates aspects of social life that cannot be measured quantitatively and that are best understood within a natural setting. The world of prostitution, the inner workings of a Mafia family, and events during a riot are all topics best studied by field research.

The most popular approach to field research is the **case study**—a thorough investigation of a single group, incident, or community. Case studies are accomplished primarily through intensive observation, information obtained from informants, and informal interviews. Newspaper files, formal interviews, official records, and surveys can be used to supplement these techniques.

This method assumes that the findings in one case can be generalized to similar situations. The conclusions of a study on drug use in Chicago, for example, should apply to other large cities as well. It is the researcher's responsibility to indicate factors that are unique and that would not apply to other situations. Researchers conducting case studies often use the technique of participant observation.

What is participant observation? In **participant observation,** a researcher becomes a member of the group being studied. A researcher may join a group with or without informing its members that he or she is a sociologist. A compelling account of covert participant observation appears in *Black Like Me,* a book written by John Howard Griffin (1961). Griffin, a white journalist, dyed his skin to study the life of African Americans in the South. Although he had previously visited the South as a white man, his experiences while posing as an African American were quite different.

Sociologists sometimes reveal themselves as participant researchers. Elliot Liebow studied two dozen disadvantaged African American men who hung around a particular corner in Washington, D.C. His study illustrates the open approach to participant observation. Even though he was a white outsider, Liebow was allowed to participate in the daily activities of the men: "The people I was observing knew that I was observing them, yet they allowed me to participate in their activities and take part in their lives to a degree that continues to surprise me" (Liebow 2003:253).

What are the advantages and disadvantages of field studies? Field studies can produce a depth and breadth of understanding unattainable with quantitative research methods. Uniquely, they can reveal insights into a social situation from the experiences of the people involved. Adaptability is another advantage. Even after a field study has begun, it is easily altered when new insights or oversights are discovered. (This is in contrast to survey research, where changes are not practical once research is in progress.) In addition, field studies are valuable where survey research would be either impossible or biased—as in a study of skid-row derelicts or organized crime. (Imagine studying a delinquent gang with a set of questions about their activities.) Because of these advantages, field studies may produce insights and explanations not likely to be unearthed through quantitative research.

Disadvantages do exist, however. The findings from one case may not be generalizable to similar situations. One mental hospital or community may be quite unlike any other mental hospital or community. If possible bias of the survey research sample is a major problem, so is the potential bias of the field researcher. In the absence of more precise measuring devices, the researcher often has to rely on personal judgment and interpretation. Because of personal blind spots or emotional attachment, the researcher may not accurately interpret his or her observations. Moreover, the lack of objectivity and standardized research procedures makes it difficult for another researcher to duplicate a field study. Because of these disadvantages, many sociologists regard the results of field studies as insights that must be investigated further with more precise quantitative methods (see Table 2.3).

The Subjective Approach

The subjective approach to research has a long and honorable place in sociology. Recall from Chapter 1 Max Weber's method of *verstehen*, in which the subjective intentions of people are discovered by imagining oneself in their place. The subjective approach, then, studies an aspect of human social behavior by ascertaining the interpretations of the participants themselves. A prominent example of the **subjective approach** is ethno-

methodology, a development in microsociology that attempts to uncover taken-for-granted social routines.

How does ethnomethodology work? **Ethnomethodology** is the study of processes people develop and use in understanding the routine behaviors expected of themselves and others in everyday life. Ethnomethodologists assume that people share the meanings that underlie much of their everyday behavior. Through observing others and through a process of trial and error in social situations, people develop a sense of appropriate behavior. This understanding prevents them from making silly or serious social errors and saves them from having to continually decide the fitting behavior for particular situations. Predictable, patterned behavior is a result of this process (Livingston 1987; Atkinson 1988; Hilbert 1990; Pollner 1991).

How can ethnomethodologists discover what is going on in the minds of individuals as they construct a mental sense of social reality? Because they are not mind readers, ethnomethodologists have had to be inventive. Harold Garfinkel (1984) is a prominent advocate of ethnomethodology. He believes the best course to understanding people's construction of social reality is to deprive them momentarily of their mental maps of daily routines. If people are deprived of their definitions of expected behaviors, they reconstruct a coherent picture of social reality. Ethnomethodologists can learn by observing this process of reconstruction.

Garfinkel writes of situations that his students have created. Here the researchers can observe what people do when deprived of their taken-for-granted social routines. The following passage describes a situation in which an experimenter (E) is attempting to deprive a subject (S) of his sense of expected routine by asking for more detailed information than is normally required in everyday situations. In the context of watching television, the experimenter first asks, "How are you tired? Physically, mentally, or just bored?"

S: *I don't know, I guess physically, mainly.*
E: *You mean that your muscles ache or your bones?*
S: *I guess so. Don't be so technical.*
(*After more watching*)
S: *All these old movies have the same kind of old iron bedstead in them.*
E: *What do you mean? Do you mean all old movies, or some of them, or just the ones you have seen?*
S: *What's the matter with you? You know what I mean.*
E: *I wish you would be more specific.*
S: *You know what I mean! Drop dead!* (Garfinkel 1984:43)

The researcher continues this type of conversation until the subject is disoriented and can no longer respond within a previously developed frame of reference. The researcher can then observe the subject's

TABLE 2.3

Research Methods: Advantages and Disadvantages

A summary of the advantages and disadvantages of several basic research methods is presented here. Suppose that you wished to do a study of the relationship between dating and self-esteem in high school. From the table, select the method that you think is best suited to your research. Explain the advantages and disadvantages of this method to your research.

Research Method	Definition	Advantages	Disadvantages
Quantitative Methods Survey research	People answer a series of questions, usually predetermined.	• Precision and comparability of answers. • Use of statistical techniques. • Information on large numbers of people. • Detailed analysis.	• Expensive due to large numbers. • Low response rate. • Phrasing of questions introduces bias in favor of certain answers. • Researchers' behavior can affect answers given.
Secondary analysis	Information gathered by one researcher is used by another researcher for a different purpose.	• Inexpensive. • Can study a topic over a long period of time. • Researchers' influence on subjects avoided.	• Information collected for a different reason may not suit another researcher's needs. • Original researcher may already have introduced biases. • Information may be outdated.
Experiment	Occurs in a laboratory setting with a minimum of contaminating influences.	• Can be replicated with precision. • Variables can be manipulated. • Can be relatively inexpensive. • Permits the establishment of causation (rather than just correlation).	• Laboratory environment is artificial. • Not suited to most sociological research. • Number of variables studied is limited.
Qualitative Methods Case study	Thoroughly investigates a small group, incident, or community.	• Provides depth of understanding from group members' viewpoint. • Unexpected discoveries and new insights can be incorporated into the research. • Permits the study of social behavior not feasible with quantitative methods.	• Difficult to generalize findings from one group to another group. • Presence of researcher can influence results. • Hard to duplicate. • Takes lots of time. • Difficult to be accepted as a group member (in case of participant observation).

FEEDBACK

1. Field studies are best suited for situations in which _____ measurement cannot be used.
2. A _____ is a thorough investigation of a small group, an incident, or a community.
3. In _____, a researcher becomes a member of the group being studied.
4. According to the _____ approach, some aspect of social structure is best studied through an attempt to ascertain the interpretations of the participants themselves.
5. _____ is the study of the processes people develop and use in understanding the routine behaviors expected of themselves and others in everyday life.

Answers: 1. quantitative; 2. case study; 3. participant observation; 4. subjective; 5. Ethnomethodology

Marianne Bertrand and Sendhil Mullainathan—What's in a Name?

Social researchers infrequently use the experiment as a method of study. And when they do use the experimental method, it is usually a field or "natural" experiment—an experiment done outside the laboratory in the context of normal social life (Babbie 2007). Marianne Bertrand and Sendhil Mullainathan (2002) conducted a field experiment on racial discrimination in the labor market.

Bertrand and Mullainathan began their research with the knowledge that racial inequality exists in the U.S. labor market. This chapter has documented the presence of this inequality. Using race as their experimental variable, the authors asked two research questions: (1) do employers actively discriminate against African Americans in hiring by rejecting them in favor of whites who are only equally qualified, and (2) does improving the credentials of African

Americans compared with whites overcome this discrimination? With their field experiment design, Bertrand and Mullainathan sent résumés in response to a series of Chicago and Boston help-wanted ads and used the rate of callback for interviews as the measure of success for each résumé. To manipulate perception of race, the researchers randomly assigned half of the résumés very white-sounding names such as Emily Walsh or Brendan Baker. The other half of the résumés were randomly given very African American–sounding names like Lakisha Washington or Jamal Jones. The researchers responded to each employment ad by sending two higher-quality résumés (one with a white-sounding name, the other with an African American–sounding name) and two lower-quality résumés (one with a white-sounding name, one with an African American–sounding name). They responded to more than 1,300 ads for a wide variety of white-collar jobs, sending out almost 5,000 résumés.

Even though race was assigned randomly, callback rates for African Americans and whites were quite different. Applicants with white names received one callback for every ten or so résumés. Fifteen callbacks were required for résumés with African American names. In other words, white names attracted 50

percent more callbacks than African American names. Because applicants' names were randomly assigned, the researchers could safety attribute the differential callback rate to racial discrimination.

Race, they found, also influenced the likelihood of higher-quality résumés being rewarded with invitations to interviews. Whites with better résumés received 30 percent more invitations to interviews than whites with lower-quality résumés. In contrast, African American applicants did not benefit from appearing better qualified. That is, African Americans saw little benefit from upgrading their credentials.

Bertrand and Mullainathan conclude that discrimination is an important reason for poor African American performance in the U.S. labor market. African Americans get fewer interview opportunities and are not highly rewarded for improving their observable qualifications. As long as these conditions exist, racial inequality will persist.

Evaluate the Research
1. Discuss the implications of these findings for African American incentives.
2. Relate this research to the tendency to blame the victim.
3. Identify some public policy implications of these findings.

creation of a new definition for expected or "normal" social interaction.

Ethics in Social Research

The Issue of Ethics

Research is a distinctly human activity. Although there are principles for conducting research, such as objectivity and verifiability, scientists sometimes forgo them. At times, the ethics of research is not honored by research-

ers (Greenberg 2003; K. Chang 2002; Goodstein 2002; Spotts 2002).

Unfortunately, there is a long list of examples of ethical lapses in medical research. During the Nuremberg trials, twenty Nazi doctors were convicted of conducting sadistic experiments on concentration camp inmates. From 1932 to 1972, the Public Health Service of the U.S. government deliberately did not treat approximately 400 syphilitic African American sharecroppers and day laborers, so that biomedical researchers could study the full evolution of the disease (J. M. Jones 1993; Reverby, 2000). Federal investigators in the United States have

TABLE 2.4

FOCUS ON THEORETICAL PERSPECTIVES:
Investigating School Violence and School Funding

This table illustrates the research method a sociologist of a particular theoretical persuasion would *most likely* choose to investigate school violence and school funding. Any of the three sociologists, of course, could use *any* of the three research methods.

Theoretical Perspective	Research Method	Approach to the Research Question
Functionalism	Survey	• A questionnaire on violence in high school is sent to a national, random sample of principals. The survey examines a possible relationship between incidence of school violence and level of school funding.
Conflict theory	Case study	• A particular high school with low funding is studied with respect to a relationship between school violence and school funding. Researchers interview administrators, teachers, and students.
Symbolic interactionism	Participant observation	• Concealing her identity, a researcher takes a temporary job at a high school with low funding. She attempts to covertly observe a possible link between school violence and school funding.

documented more than ten years of fraud in some of the most important breast cancer research ever conducted (Crewdson 1994). In a Cincinnati hospital, from 1960 to 1972, approximately 100 women and men with cancer were subjected to experimental radiation over their entire bodies in U.S. military–funded research on the effects of nuclear war. Twenty-one died within a month of exposure and nearly all of them died within months of this "therapy" (Stephens 2002). More recently, the worldwide leading researcher of human cloning, South Korean Woo Suk Hwang, revealed that most—and perhaps all—of his stem cell cultures were fake.

Plagiarism is an ethical lapse that can occur in any area of research. For example, in 2001 three prominent historians (Stephen Ambrose, Doris Kearns Goodwin, and Joseph Ellis) faced charges of including in their writings the work of others without citation.

Several eminent social scientists have been criticized for conducting research that many scientists consider unethical. In each case, subjects were placed in stressful situations without being informed of the true nature of the experiment (Milgram 1963, 1965, 1974; Zimbardo, Anderson, and Kabat 1981). These and other studies have created great interest in a code of ethics. There is, in fact, a formal code of ethics for professional sociologists (American Sociological Association 1997).

A Code of Ethics in Sociological Research

The formal code of ethics for sociologists covers important areas beyond research, including relationships with students, employees, and employers. In broad terms, the code is concerned with maximizing the benefits of sociology to society and minimizing the harm that sociological work might create. Of importance in the present context are the research-related aspects of the code.

Sociologists are committed to objectivity; the highest technical research standards; accurate reporting of their methods and findings; and protection of the rights, privacy, integrity, dignity, and autonomy of the subjects of their research. Because the first several topics have already been covered in this chapter, the focus in this section is on the rights, privacy, integrity, dignity, and autonomy of participants in sociological research.

Occasionally, adherence to the code is documented. Mario Brajuha, a graduate student at a major American university, kept detailed field notes while engaging in a participant observation study of restaurant work (Brajuha and Hallowell 1986). Because of suspected arson after a fire at the restaurant where he was employed as a waiter, his field notes became the object of interest by the police, the district attorney, the courts, and some suspects. By refusing to reveal the contents of his field notes, Brajuha protected the rights of those individuals described in his notes. He did so in the face of a subpoena; threats of imprisonment; and the specter of personal harm to himself, his wife, and his children. The case was finally dropped after two difficult years.

Although lapses in ethics are rare in sociological research, much can be learned from such cases. Laud Humphreys (1979) studied homosexual activities in men's public bathrooms ("tearooms"). By acting as a lookout to warn the men of approaching police officers, he observed their activities closely. As participants left the tearooms, Humphreys recorded their license plate numbers to obtain their addresses for personal interviews. After waiting a year, Humphreys falsely presented himself as a survey researcher to obtain additional information.

Did Humphreys violate the code of ethics as a covert participant observer? Yes, Humphreys violated the privacy of these people. Most did not want their sexual activities known, and Humphreys did not give them the opportunity to refuse to participate in the study. Humphreys also deceived the men by misrepresenting himself in both the tearooms and their homes. Finally, by recording his observations, Humphreys placed these people in jeopardy of public exposure, arrest, or loss of employment. (Actually, because of his precautions, none of the subjects were injured as a result of his research. In fact, to protect their identities, Humphreys even allowed himself to be arrested.)

There are more recent examples of ethics breaches in research. In 2006, Dr. Claudia Henschke published a study concluding that 80 percent of lung cancer deaths could be avoided with the widespread use of CT (computerized tomography) scans. This news first surprised the medical community and then disappointed it as the members subsequently learned that the cigarette maker Liggett funded the Henschke research (G. Harris 2008).

Unethical decisions are made by groups as well as by individual researchers acting alone. In 2004, the National Cholesterol Education Program at the U.S. National Heart, Lung, and Blood Institute (NLHBI) urged millions of Americans to take statin drugs. This is ethically problematic because medical researchers had been questioning for years the effectiveness of statins in lowering cholesterol. Moreover, eight of the nine NLHBI panel members had earlier received research financing from one or more of the companies making statins (Mahar 2008).

Do ethical concerns make research harder? Yes, but it is the researcher's responsibility to decide when a particular action crosses an ethical line—a decision not always

Should You Believe Everything You Read?

It is sometimes said that we are living in the "age of instant information." One unfortunate side effect is the tendency for studies and research results to be reported in the media without background or explanation. There are, however, some easy steps you can follow that will make you a savvy consumer in the information marketplace.

BE SKEPTICAL. Be suspicious of what you read. The media sound-bite treatment tends to sensationalize and distort information. For example, the media may report that $500,000 was spent to find out that love keeps families together. In fact, this may have been only one small part of a larger research project. Moreover, chances are the media have oversimplified even this part of the researcher's conclusions.

CONSIDER THE SOURCE OF INFORMATION. The credibility of a study may be affected by who paid for the results. For example, you should know whether a study on the relationship between cancer and tobacco has been sponsored by the tobacco industry or by the American Cancer Society. Suppose that representatives of tobacco companies denied the existence of any research linking throat and mouth cancer with snuff dipping. Further suppose that an independent medical researcher concluded that putting a "pinch between your cheek and gum" has led to cancer in humans. The self-interest of the tobacco companies taints their objectivity.

At the very least, you want to know the source of information before making a judgment. This caution is especially relevant to Internet research.

Because Internet information varies widely in its accuracy and reliability, sources must be evaluated with great care.

DO NOT MISTAKE CORRELATION FOR CAUSATION. Remember that a correlation between two variables does not mean that one caused the other. At one time, the percentage of Americans who smoked was increasing at the same time that life expectancy was increasing. Did this mean that smoking caused people to live longer? Actually, a third factor—improved health care—accounted for the change. Do not assume that two events are related causally just because they occur together.

Think About It
Bring to class an article reporting on a study. These can often be found in news magazines. Be prepared to explain how these three safeguards apply to the article's report.

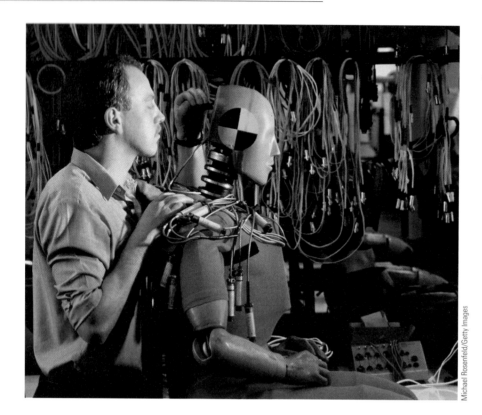

It is generally thought that using human corpses in automobile crash tests is unethical. Ethical standards also apply to potentially harmful social research utilizing live humans.

Michael Rosenfeld/Getty Images

easy to make, because moral lines are often blurred. Moreover, the researcher must balance the interests of those being studied against the need for accurate, updated data.

Although Kai Erikson is one of the most sensitive and outspoken critics of disguised observation, he has defended it on grounds that it is—on occasion—the only way to obtain relevant information:

Some of the richest material in the social sciences has been gathered by sociologists who were true participants in the group under study but who did not announce to other

1. Situations a, b, and c describe three research situations involving possible ethical violations (Babbie 2007). Match each situation with the appropriate aspect of the social science code of ethics for research on human subjects.
   (1) concern for participants' privacy
   (2) avoidance of deception
   (3) obligation not to harm participants
   a. After a field study of deviant behavior during a riot, law enforcement officials demand that the researcher identify those people who were observed looting. Rather than risk arrest as an accomplice after the fact, the researcher complies.
   b. A research questionnaire is circulated among students as part of their university registration packet. Although students are not told they must complete the questionnaire, the hope is that they will believe they must, thus ensuring a higher completion rate.
   c. Researchers obtain a list of gay men they wish to study. They contact them with the explanation that each has been selected "at random" from among the general population to take a sampling of "public opinion."
2. Match the concepts on the left side with the definitions on the right side.
   ____ a. reliability   (1) when a measurement technique yields consistent results on repeated applications
   ____ b. validity   (2) the duplication of the same study to ascertain its accuracy
   ____ c. replication   (3) when a measurement technique actually measures what it is designed to measure

Answers: 1. a. (3), b. (2), c. (1); 2. a. (1), b. (3), c. (2)

members that they were employing this opportunity to collect research data. . . . It would be absurd, then, to insist as a point of ethics that sociologists should always introduce themselves as investigators everywhere they go and should inform every person who figures in their thinking exactly what their research is about. (K. T. Erikson 1967:368)

Balance is the key to the issue of ethics. At the very least, subjects—whether in experiments, surveys, or field studies—should be protected from social, financial, psychological, or legal damage (M. Hunt 1999).

A Final Note

Reliability, Validity, and Replication

Researchers can be guided by all the important research considerations we have discussed in this chapter and still not conduct a good study. They can be mindful of objectivity, sensitive to the criteria of causation, and careful in the selection of the most appropriate method (survey, precollected data, field study). Still, they may fail to produce knowledge superior to that yielded by intuition, common sense, authority, or tradition.

What else must a researcher do? Sociologists must pay careful attention to the quality of measurement (Babbie 2007). Consequently, they must emphasize reliability and validity in the creation and evaluation of the measuring devices they use for the variables they wish to investigate.

What is meant by reliability? A measurement technique must yield consistent results on repeated applications—a requirement called **reliability.** Reliability is tested by repeated administration of a measurement technique, such as a questionnaire, to the same subjects to ascertain whether the same results occur each time. Suppose a researcher, after deciding to study satisfaction with day care among parents, designs a questionnaire. If, on repeated applications of the questionnaire to a sample of parents, the level of satisfaction remains consistent, then confidence in the reliability of the measurement device rises. Should, on the other hand, the level of satisfaction—from one administration of the questionnaire to the next—vary over a period of time, then we would doubt that satisfaction with child care is actually being measured.

The problem of reliability is also an issue in qualitative research. Suppose that our field researcher is also interested in satisfaction with day care among the children. If the level of satisfaction among the children seems different each day to the researcher, then doubt is raised about the reliability of the measurement technique being used.

Even when a measurement technique is reliable, it still may not produce scientifically sound results. This is because a measurement technique must be not only reliable, but valid.

What is validity? **Validity** exists when a measurement technique actually measures what it is designed to measure. Thus, a technique intended to measure parental satisfaction with day care may yield consistent results on repeated applications to a sample of parents, but not really be measuring satisfaction at all. The measurement device might be tapping parental need to view day care positively (to mask guilt feelings about permitting someone else to be the care provider during working hours). Children at a day-care center may appear satisfied to the visiting researcher because they are neglected during the day and welcome his or her attention, or because the day-care provider has coached them to appear satisfied. A measurement technique, in short, may be consistently measuring something very different from what it purports to measure.

How does replication contribute to the self-corrective nature of research? **Replication**—the duplication of the same study to ascertain its accuracy—is closely linked to both reliability and validity in that reliability and validity problems unknown to original researchers are likely to be revealed as subsequent social scientists repeat the research. It is partially through replication that scientific knowledge accumulates and changes over time.

A major goal of scientific research is to generate knowledge that is more reliable than can be obtained from such nonscientific sources as intuition, common sense, authority, and tradition. Through efforts to be objective and to make their research subject to replication by others, researchers attempt to portray reality as accurately as possible. The methods of research presented in this chapter are the specific tools sociologists use to create knowledge of social life that is as accurate as possible.

However, empirical results obtained through the use of research methods are not the final goal of science. As Gerhard Lenski has stated, "Science is more than method: its ultimate aim is the development of a body of 'verified' general theory" (Lenski 1988:163). For this reason, there is constant interaction between sociological theory and research methods. Theory is used to develop hypotheses capable of being supported or falsified through testing. These results, in turn, may support existing theory, alter it, or lead to its ultimate rejection and the creation of a new theory. Lenski points out that divorced from research methods, "sociological theory has more in common with seminary instruction in theology and biblical studies" (1988:165) than it does with the natural sciences model that sociology is emulating. Theory is trustworthy and useful only to the extent that it has been tested and found to be valid.

SUMMARY

1. People tend to get information from such nonscientific sources as intuition, common sense, authority, and tradition. Generally speaking, these sources are inadequate for obtaining accurate knowledge about social life. The advantage of scientific knowledge is grounded in its objectivity and verifiability.

2. A research model involves several distinct steps: identifying the problem, reviewing the literature, formulating hypotheses, developing a research design, collecting data, analyzing data, and stating findings and conclusions. These steps form a model for scientific research.

3. Objectivity is the aim of social scientists. Subjectivity can be minimized if researchers make themselves aware of their biases and make their biases public when presenting their findings. Through verification of past research, erroneous theories, findings, and conclusions can be exposed to other scientists.

4. The concept of causation—the idea that the occurrence of one event leads to the occurrence of another event—is central to science. All events have causes, and scientists attempt to discover the factors causing those events.

5. Three criteria must be met before a cause-and-effect relationship can be said to exist. First, two variables must be correlated. That is, change in the independent variable (the causal factor) must be associated with a change in the dependent variable (effect). Second, the correlation must not be spurious, that is, due to the effects of a third variable. Third, it must be shown that the independent variable always occurs before the dependent variable. Scientists think in terms of multiple causation because events are usually caused by several factors, not simply by a single factor.

6. The experiment is a good example of a research method that is based on the idea of causation. Sociologists, however, do not generally use the experimental research method because it is often impossible to control the relevant social variables.

7. Two major quantitative research methods are the survey and precollected data. Surveys can draw on large samples, are quantitative, include many variables, are relatively precise, and permit the comparison of responses. Researchers must take care to collect representative samples for their surveys. Precollected data permit high-quality research at reasonably low cost and reveal changes in variables over an extended period of time.

8. Field studies are best used when some aspect of social structure cannot be measured quantitatively, when interaction should be observed in a natural setting, and when in-depth analysis is needed. The case study is the popular approach to field research. Some sociologists have adopted a subjective approach in which emphasis is on ascertaining the subjective interpretations of the participants themselves.

9. Researchers have an ethical obligation to protect participants' privacy and to avoid deceiving or harming participants. Preserving the rights of subjects is sometimes weighed against the value of the knowledge to be gained. While usually harmless, these compromises sometimes place the subjects in jeopardy.

10. Demands on researchers go even further. They must design measurement devices that give consistent results each time they are used (reliability), and measurement techniques must actually measure what they intend (validity).

LEARNING OBJECTIVES REVIEW

- Identify major nonscientific sources of knowledge; then explain why science is a superior source of knowledge.
- Discuss cause-and-effect concepts, and apply the concept of causation to the logic of science.
- Differentiate the major quantitative research methods used by sociologists.
- Describe the major qualitative research methods used by sociologists.
- Explain the steps sociologists use to guide their research.
- Describe the role of ethics in research.
- State the importance of reliability, validity, and replication in social research.

CONCEPT REVIEW

Match the following concepts with the definitions below them

_____ a. participant observation
_____ b. experiment
_____ c. verifiability
_____ d. subjective approach
_____ e. experimental group

_____ f. independent variable
_____ g. objectivity
_____ h. correlation
_____ i. population
_____ j. sample

_____ k. field research
_____ l. case study
_____ m. survey
_____ n. replication

1. the group in an experiment exposed to the experimental variable
2. a statistical measure in which a change in one variable is associated with a change in another variable
3. a research approach for studying aspects of social life that cannot be measured quantitatively and that are best understood within a natural setting
4. a thorough, recorded investigation of a small group, incident, or community
5. all those people with the characteristics a researcher wants to study within the context of a particular research question
6. the principle of science stating that scientists are expected to prevent their personal biases from influencing their results and their interpretation of the results

7. a variable that causes something to happen
8. the type of field research technique in which a researcher becomes a member of the group being studied
9. a principle of science by which any given piece of research can be duplicated (replicated) by other scientists
10. a research method in which people are asked to answer a series of questions
11. the duplication of the same study to ascertain its accuracy
12. a limited number of cases drawn from a population
13. laboratory research that attempts to eliminate all possible contaminating influences on the variables being studied
14. a research method in which the aim is to understand some aspect of social reality through the study of the subjective interpretations of the participants themselves

REVIEW GUIDE

CRITICAL-THINKING QUESTIONS

1. Suppose a noncollege friend insists that you are wasting your time in college because the experience gained from the "university of hard knocks" is all a person needs to know the truth. What arguments would you use to defend science as a better source of knowledge?

2. Suppose your sociology professor reports on his recent study showing that men are generally better business managers than women. If you are concerned about a possible lack of objectivity on his part, what questions could you ask him to boost your confidence in his results?

3. The experiment is the research model for investigating causal relationships. What is there about the nature of causation and the design of experiments that supports this claim?

4. Do you think that a selected sample of 3,000 individuals could yield an accurate picture of the leisure habits of Americans? Why or why not?

5. Pretend that you are a sociologist studying the relationship between the receipt of welfare payments and commitment to working. Describe the research method you would use, and show why it is the most appropriate to this topic.

MULTIPLE-CHOICE QUESTIONS

1. The concept of intuition refers to
   a. quick and ready insight that is not based on rational thought.
   b. opinions that are widely held because they seem so obviously correct.
   c. someone who is supposed to have special knowledge that we do not have.
   d. the fourth major nonscientific source of knowledge.
   e. a variable that causes something to happen.

2. **According to your text, causation can be asserted when**
   a. going from particular instances to general principles.
   b. there are only a limited number of cases taken from society.
   c. events occur in a predictable, nonrandom way, and one event leads to another.
   d. people develop and use routine behaviors expected of themselves and others in everyday life.
   e. a change in one variable is often accompanied by a change in another variable.

3. **Several factors have been shown to influence crime rates in poor neighborhoods. This illustrates the principle of**
   a. the poverty/crime hypothesis.
   b. multiple causation.
   c. verifiability.
   d. criminology.
   e. variance.

4. **A variable that causes something else to occur is called**
   a. a dependent variable.
   b. a correlation variable.
   c. a causation variable.
   d. an independent variable.
   e. a qualitative variable.

5. **The term *correlation* is defined as**
   a. a change in one variable associated with a change in the other.
   b. an apparent relationship between two variables, which is actually produced by a third variable that affects both of the original two variables.
   c. an event that occurs as a result of several factors operating in combination.
   d. something that occurs in different degrees among individuals, groups, objects, and events.
   e. a research method in which people are asked to answer a series of questions.

6. **All of the following are criteria for establishing a causal relationship *except*:**
   a. All possible contaminating factors must be controlled.
   b. A relationship representing a spurious relationship must exist.

   c. The independent variable must occur before the dependent variable.
   d. Two variables must be correlated.

7. **All of the following statements about experiments are true *except*:**
   a. A description of the experiment provides an excellent means to illustrate causation.
   b. A controlled experiment provides insight into the nature of all scientific research.
   c. Controlled experiments take place in a laboratory.
   d. The basic idea of the controlled experiment is to rule out the effects of extraneous factors to see the effects of an independent variable on a dependent variable.
   e. Controlled experiments do not need a control group because of the controlled atmosphere the laboratory provides.

8. **The experimental group is exposed to the experimental variable; the group that is not exposed to the experimental variable is the**
   a. natural group.
   b. experiential group.
   c. control group.
   d. dependent group.
   e. independent group.

9. **A standard way of making experimental and control groups comparable in all respects, except for exposure to the experimental variable, is through**
   a. qualifying or quantifying.
   b. matching or randomizing.
   c. pretesting or posttesting.
   d. testing or retesting.
   e. verification or replication.

10. **A written set of questions that survey participants are asked to fill out by themselves is called**
    a. a survey.
    b. an interview.
    c. a questionnaire.
    d. a survey research.
    e. an independent variable.

11. **Use of data from the U.S. Bureau of the Census is an example of**
    a. primary analysis.
    b. population sampling.
    c. the Hawthorne effect.
    d. secondary analysis.
    e. a case study.

12. **The type of research used for studying aspects of social life that cannot be measured quantitatively and that are best understood in a natural setting is called**
    a. field research.
    b. survey research.

c. participant observation.

d. analysis of precollected data.

e. content analysis.

13. **Ethnomethodologists assume that**

a. the subjective approach relies too much on intuition.

b. the behavior of people is random.

c. the underlying factor explaining human behavior is ethnicity.

d. questionnaires need to be tightly structured.

e. people share the meanings underlying much of their everyday behavior.

14. **A tentative, testable statement of a relationship among variables is called**

a. a hypothesis.

b. an operational definition.

c. a formal argument.

d. a correlation.

e. a conclusion.

15. **In Laud Humphreys's study of homosexual activities occurring in men's public bathrooms ("tearooms"), what ethical standard did he violate?**

a. He studied homosexuals.

b. He acted as a participant observer.

c. He violated the privacy of the participants.

d. He used the research to become famous.

e. He did not violate any ethical standards.

REVIEW GUIDE

FEEDBACK REVIEW

True-False

1. The major problem with nonscientific sources of knowledge is that such sources often provide erroneous information. T or F?

2. According to Gunnar Myrdal, it is enough that scientists themselves recognize their biases. T or F?

Fill in the Blank

3. The group in an experiment that is not exposed to the experimental variable is the _____ group.

4. Field studies are best suited for situations in which _____ measurement cannot be used.

5. In _____, a researcher becomes a member of the group being studied.

6. An _____ attempts to eliminate all possible contaminating influences on the variables being studied.

7. Use of company records would be an example of using _____ data.

8. According to the _____ approach, some aspects of social structure are best studied through an attempt to ascertain the interpretations of the participants themselves.

Matching

9. Listed here are the steps in the research model. Following are some concrete examples related to the sociability of the only child. Indicate the appropriate example for each step number.

____ Step 1: identifying the problem

____ Step 2: reviewing the literature

____ Step 3: formulating hypotheses

____ Step 4: developing a research design

____ Step 5: collecting data

____ Step 6: analyzing data

____ Step 7: stating findings and conclusions

a. Read past theory and research on the sociability of only children.

b. From previous research and existing theory, a researcher states that only children appear to be more intelligent than children with siblings.

c. A researcher collects data on only children from a high school in a large city.

d. A researcher writes a report giving evidence that only children are more intelligent than children with brothers or sisters.

e. A researcher decides to study the intelligence level of only children.

f. A researcher classifies and processes the data collected in order to test a hypothesis.

g. A researcher decides on the data needed to test a hypothesis, the methods for data collection, and the techniques for data analysis.

10. Situations a, b, and c describe three research situations involving possible ethical violations (Babbie 2007). Match each situation with the appropriate aspect of the social science Code of Ethics for research on human subjects.

____ a. After a field study of deviant behavior during a riot, law enforcement officials demand that the researcher identify those people who were observed looting. Rather than risk arrest as an accomplice after the fact, the researcher complies.

_____ b. A research questionnaire is circulated among students as part of their university registration packet. Although students are not told they must complete the questionnaire, the hope is that they will believe they must, thus ensuring a higher completion rate.

_____ c. Researchers obtain a list of right-wing radicals they wish to study. They contact the radicals with the explanation that each has been selected "at random," from among the general population, to take a sampling of "public opinion."

(1) concern for participants' privacy

(2) avoidance of deception

(3) obligation not to harm participants

Table 2.2 displays the median annual income in the United States by sex, race, and education. Demonstrate your understanding of the information in this table by answering the following questions.

GRAPHIC REVIEW

1. What does this table tell us about the relationship among sex, race, and education in the United States?

_____

_____

_____

2. Identify the demographic group that enjoys the greatest economic benefits of education.

_____

_____

_____

3. Identify the demographic group that benefits the least, economically, from higher levels of education.

_____

_____

_____

ANSWER KEY

Concept Review

a. 8
b. 13
c. 9
d. 14
e. 1
f. 7
g. 6
h. 2
i. 5
j. 12
k. 3
l. 4
m. 10
n. 11

Multiple Choice

1. a
2. c
3. b
4. d
5. a
6. b
7. e
8. c
9. b
10. c
11. d
12. a
13. e
14. a
15. c

Feedback Review

1. T
2. F
3. control
4. quantitative
5. participant observation
6. experiment
7. precollected
8. subjective
9. Step 1: e
   Step 2: a
   Step 3: b
   Step 4: g
   Step 5: c
   Step 6: f
   Step 7: d
10. a. 3
    b. 2
    c. 1

3

Culture

Louis-Laurent Grandadam/Getty Images

LEARNING OBJECTIVES

- Identify the three major dimensions of culture.
- Describe and illustrate the interplay between language and culture.
- Discuss cultural diversity and its promotion within a society.
- Describe and illustrate the relationship between cultural diversity and ethnocentrism.

- Outline the advantages and disadvantages of ethnocentrism, and discuss the role of cultural relativism in combating ethnocentrism.
- Explain the existence of cultural similarities that are shared around the world.
- Explain the relationship between culture and heredity.

USING THE SOCIOLOGICAL IMAGINATION

Does the existence of cultural diversity rule out cultural similarities? In other words, what, if anything, could the United States have in common with Afghanistan? What could we Americans have in common with a society that does not approve of women appearing in public with uncovered faces? What do we share with the citizens of Bhutan, who believe that gods assume human form to direct normal day-to-day affairs? How do we relate to the culture of Ingushetia, a republic inside Russia, where it is legal for men to have several wives—all at the same time?

Many Americans, in enthusiastic endorsement of multiculturalism, ignore the similarities among cultures. In reality, however, despite the surface differences, social scientists document a wide variety of "cultural universals" shared by all cultures. Like the United States, for instance, Iraq (or Bhutan or Nigeria) has families, schools, houses of worship, economics, and governments. Later in this chapter we elaborate on this evidence. First, however, we examine the nature of culture and society.

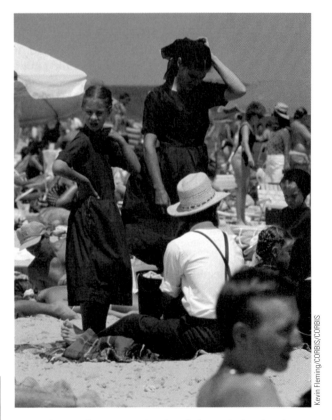

Kevin Fleming/CORBIS/CORBIS

Coming from a different culture than that of other sunbathers doesn't prevent this Amish family from enjoying a day at the beach.

Dimensions of Culture

Culture and Society

What is the difference between culture and society? **Culture** is a people's way of life that is passed on from generation to generation. It consists of physical objects as well as patterns of thinking, feeling, and behaving. On the material side, American culture includes such physical objects as skyscrapers, fast-food restaurants, videogames, and cars. On the nonmaterial side, American culture includes various beliefs, rules, customs, a family system, and a capitalist economy. Although culture and society are tightly interwoven and cannot exist without each other, they are not identical. A **society** is a group of people living in a defined territory and participating in a common culture. Culture is that society's total way of life.

Why is culture important? Culture underlies human social behavior. What people do and don't do, what they like and dislike, what they believe and don't believe, and what they value and discount are all based, in large part, on culture. Culture provides the blueprints people in a society use to guide their relationships with others.

It is from culture that teenage girls are stimulated to compete for positions on the women's basketball team, and it is from culture that teenage boys come to believe that "pumping iron" is a gateway to popularity.

Human social behavior, then, is based on culture. And because culture is not innate, human behavior must be learned. The concepts presented in this chapter help substantiate this point.

If you wanted to describe and analyze a culture, what would you look for? How could you begin to classify the components of a people's way of life? The sociological classification system consists of three major dimensions of culture: the normative (standards for behavior), the cognitive (knowledge and beliefs), and the material (tangible objects). An elaboration of each dimension will help you better understand the nature of culture.

The Normative Dimension

The normative dimension of culture, which consists of the standards for appropriate behavior of a group or society, is heavily tied into functionalism with its emphasis on social integration, stability, and consensus. The most important aspects of the normative dimension are norms, sanctions, and values.

What are norms? **Norms** are rules defining appropriate and inappropriate behavior (Hechter and Opp 2005). A Hindi peasant in India can be found lying dead of starvation beside perfectly healthy and edible cattle, because in India cows are sacred. A Far Eastern woman may have her head severed for going to bed with a man before marriage. Roman emperors sent relatives to small isolated islands for life for disgracing the family. Each of these instances is a result of cultural norms—rules that specify ways of behaving for specific situations. Norms help explain why people in a society or group behave similarly in similar circumstances.

William Graham Sumner (1906), an early sociologist, wrote perceptively about norms. Anything, he stated, can be considered appropriate when norms exist that approve of it. This is because once norms are learned, members of a society use them to guide their social behavior. Norms are so ingrained that they guide our social behavior without our awareness. In fact, we may not even be consciously aware of a norm until it has been violated. For instance, we do not think about standing in line for concert tickets as a norm until someone attempts to step in front of us. Then it immediately registers. Taking one's turn in line is expected behavior, and cutting in line is violating that norm.

Norms vary widely from society to society. While visiting the Persian King Gelese in the nineteenth century, Sir Richard Francis Burton witnessed the "Ku to Man" ceremony performed in his honor. The dramatic climax of this ceremony was the ritual execution of twenty-three resigned and terrified slaves (Kruschwitz and Roberts 1987). Norms also change within the same society. Premarital sexual intercourse among teenage American women has increased dramatically since the 1950s (see Figure 3.1). Interestingly, the rate of premarital intercourse among teenagers has declined since the 1990s, indicating that sexual norms may be changing once again (Risman and Schwartz 2002; "U.S. Teen Sexual Activity" 2005).

Norms range from relatively minor rules, such as applause after a performance, to extremely important ones, such as laws against stealing. Sumner identified three basic types of norms: folkways, mores, and laws. These three types of norms vary in their importance within a society, and their violation is tolerated to different degrees.

What are folkways? Rules that cover customary ways of thinking, feeling, and behaving but lack moral overtones are called **folkways.** For example, whether one uses a cell phone in a restaurant is not a moral issue; it qualifies as a folkway. Folkways in the United States include shaking hands when introduced; opening doors for older persons; and, if you are a male, removing your hat when you enter a house.

Because folkways are not considered vital to the welfare of the group, disapproval of violators is not very great. Those who consistently violate folkways—say, by persistently talking loudly in quiet places, wearing shorts with a suit coat and tie, or wearing a different-colored sock on each foot—may appear odd. We may avoid these people, but we do not consider them wicked or immoral.

Some folkways are more important than others, and the social reaction to their violation is more intense. Failure to offer a woman a seat on a crowded bus draws little reaction, but obnoxious behavior at a party after excessive drinking may bring a severe negative response from others (see Table 3.1).

What are mores? **Mores** (pronounced MOR-ays; singular, *mos*) are norms of great moral significance. They are thought to be vital to the well-being of a society. Conformity to mores elicits significant social approval; violation of this type of norm evokes strong disapproval. For example, Americans subscribe to a long-standing mos requiring able-bodied men to work for a living. Able-bodied men who do not work are stigmatized (Waxman 1983).

FIGURE 3.1

Premarital Sexual Experience Among Teenage Women in the United States

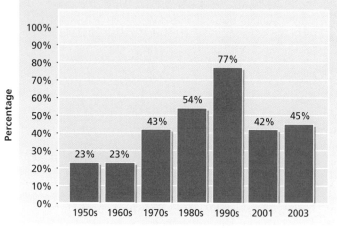

Note: Studies in each decade covered a two-year period.
Source: Based on Sandra L. Hofferth, John R. Kahn, and Wendy Baldwin, "Premarital Sexual Activity Among U.S. Teenage Women over the Past Three Decades," *Family Planning Perspectives* 19 (1987):49; Planned Parenthood Federation of America, Inc. "Planned Parenthood Fact Sheet: Sexuality Education in the U.S"; www.ppfa.org/ppfa/sex-ed.html/ (June, 1995); and "Assessing Health Risk Behaviors Among Young People," U.S. Department of Health and Human Services, Centers for Disease Control and Prevention, 2002; Kaiser Family Foundation, January, 2005.

INTERNET LINK   The Internet has an extensive list of sources that pertain to premarital sexual activity. Several of these provide useful data for the sociologist researching in this field. To begin, go to Campaign for Our Children at www.cfoc.org/.

TABLE 3.1

Do's and Taboos Around the World

Bulgaria	• A nod means no, and a shake of the head from side to side means yes.
Great Britain	• Never touch the Queen, not even to gently guide her, and don't offer to shake her hand unless she extends her hand to you first.
Germany	• Shaking hands while your other hand is in your pocket is considered impolite.
Greece	• Unlike the British, the Greeks do not respect lines, or queues, in public places.
Italy	• Italians consider it unfeminine for a woman to pour wine.
Russia	• Whistling at public gatherings is a sign of disagreement and disapproval.
Turkey	• It is considered rude to cross your arms over your chest or to put your hands in your pockets while talking to someone.
Oman	• It is an insult to sit in such a way as to face your host with the soles of your shoes showing. Do not place your feet on a desk, table, or chair.
Iran	• Shaking hands with a child shows respect for his parents.
Saudi Arabia	• At political events, it is customary for men to greet elders and dignitaries by kissing their right front shoulder.
The People's Republic of China	• Some pushing and shoving in stores or when boarding public transportation is common and not considered rude.
India	• Since the head is considered a sacred part of the body in India, you should not pat children on the head or touch an older person's head.
Japan	• Women should avoid wearing high heels in order not to risk towering over Japanese counterparts.
South Korea	• Koreans, especially women, cover their mouths when laughing to avoid impolitely showing the inside of their mouths.

Source: Roger E. Axtell, *Do's and Taboos Around the World*, revised and expanded ed. (New York: Wiley, 2007).

Although conformity to folkways is generally a matter of personal choice, conformity to mores is a social requirement. As in the case of folkways, some mores are more vital to a society than others. Failure to stand while the national anthem is being played is not as serious a violation of American mores as using loud profanity during a church service. Choosing, unnecessarily, to live on welfare is a more serious breach of appropriate conduct than either.

Some mores are more serious than others. A **taboo** is a mos so important that its violation is considered repugnant. Although definitions of incest vary from society to society, the incest taboo is generally regarded as the only taboo existing in all societies. There are, however, taboos other than the prohibition against having sexual relations with relatives. Hindus in India have a taboo forbidding the killing of cows. The mother-in-law taboo in some societies prohibits or severely restricts social contact between a husband and his wife's mother. While modern Saudi Arabia has lifted the taboo on women traveling abroad without a male companion, women are still forbidden to drive (Fattah 2007). Taboos in Western societies prohibit transvestism and bestiality (having sexual relations with animals). As Christie Davies (1982) points out, taboos such as those against transvestism and bestiality cannot have a psychological or biological origin, because some societies do not have such taboos. The world generally shares a taboo against the use of

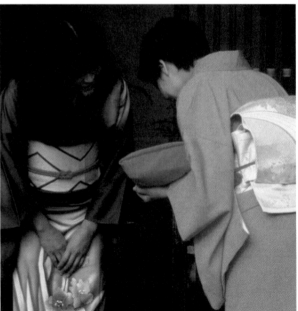

In Japanese culture an emphasis on politeness has enabled people to learn to live harmoniously in close quarters.

Wayne Eastep/Getty Images

nuclear weapons. And it has been taboo in the United States to torture political prisoners.

How do laws differ from mores? **Laws,** the third type of norm, are norms that are formally defined and enforced by officials. Folkways and mores emerge slowly and are often unconsciously created, but laws are consciously created and enforced.

The Taliban militia in Afghanistan created a strict set of gender-apartheid laws. Women were not permitted to work outside the home, and females were banned from schools. Women could go outside only if accompanied by a male relative, women had to hide themselves behind painted windows, and women could not be examined by male physicians.

Mores are an important source of laws, as was the case in many of the Taliban laws. At one time in human history, the norm against murder was not written down. But as civilization advanced, the norm prohibiting murder became formally defined and was enforced by public officials. But not all mores become laws. For example, it is not against the law to cheat on a college examination (although one can be expelled or otherwise punished). Nor have all laws been mores at one time. Fines for overtime parking and laws against littering have never been mores.

Laws often remain on the books for a long time after the mores of a society have changed (Koon, Powell, and Schumaker 2002). It is illegal in Minnesota to hang male and female undergarments on the same clothesline; card playing on trains is prohibited in New York; elephants in Natchez, Mississippi, cannot legally drink beer; and in Portland, Oregon, it is against the law to roller-skate in public bathrooms. Mores and laws sometimes overlap. Although private citizens may show strong disapproval of a father who fails to support his children, public officials can do little unless the mother is willing to start legal action.

Because norms must be learned and accepted by individuals, conformity to them is not automatic. For this reason, groups must have some means for teaching norms and encouraging conformity to them. They do this, in part, through sanctions.

How are norms enforced? Sanctions are rewards and punishments used to encourage conformity to norms. They can be formal or informal. **Formal sanctions** are sanctions that may be given only by officially designated persons, such as judges and college professors. Formal sanctions range widely in their severity. From the Middle Ages to the Protestant Reformation, it was an unpardonable sin for lenders to charge interest on money. (This practice, called *usury,* was condemned in the Bible.) This civil crime was punishable on the third offense by public humiliation and social and economic ruin (Cockburn 1994). A Saudi Arabian girl gang-raped by seven men in 2007 was given 200 lashes and six months in prison for being in a car with a male who was not a relative. A British schoolteacher in 2007 was jailed for two weeks in Sudan after her class of seven-year-olds named a teddy bear Muhammad in the class-

room. Both women received pardons after international outrage.

Formal sanctions may be less extreme. Courts across the United States have handed down punishments involving public shaming. For example, some courts have required child molesters to place, in front of their residences, signs describing their crimes (Nasser 1996).

Formal sanctions can also be positive. Soldiers earn the Congressional Medal of Honor for heroism, and some students receive an A for academic performance.

Informal sanctions are sanctions that can be applied by most members of a group. They, too, can be positive or negative. Informal sanctions include thanking someone for pushing your car out of a snowbank to glaring harshly at someone who is cheating on a test. In school, the label "fag" is used against perceived male effeminacy to reinforce norms of masculinity (Pascoe 2007).

Sanctions are not used randomly or without reason. Specific sanctions are associated with specific norms. Teenagers who violate their parents' curfew are not supposed to be beaten or locked in a closet. Convicted murderers are not invited to the White House for special celebrations. Instead, a society or group develops appropriate sanctions for following or failing to follow specific norms (A. V. Horwitz 1990).

Sanctions, most often, do not have to be used. After we reach a certain age, most of us conform without the threat of sanctions. We may conform because we believe that the behavior expected of us is appropriate, because we wish to avoid guilty feelings, or because we fear social disapproval. In other words, if we have been properly socialized, we will mentally sanction ourselves before breaking a norm.

The severity of the sanction for a deviant act varies from one society to another and from one time to another. A convicted thief in the United States might be sentenced to only a few months in jail. A convicted thief in North Yemen might have to pick up his amputated hand and salute the judge with it. During the 1960s, possession of an ounce of marijuana could result in several years in prison. Today, possession of a small amount of marijuana in most states is a misdemeanor, and in several states it is equivalent to a minor traffic violation. In 2007, the U.S. Sentencing Commission voted to reduce retroactively prison sentences for crack-cocaine offenses.

Sometimes, informal sanctions are illegally imposed. A soccer sportscaster was shot in the knees in Avellino, Italy, by irate fans. The rash of "dowry deaths" in India, not too long ago, is also an example of illegal sanctioning. According to police statistics, almost 400 women were set on fire in New Delhi. Although some of these deaths were accidental and some were suicides, it is suspected that many of them were murders. Some of the mothers-in-law, upset over the small size of their son's wife's dowry, poured gasoline over their daughters-in-law and threw matches on them. A Jordanian father killed his daughter recently upon learning of her pre-marital relations with her new husband. The dishonor she brought to the family could not be excused despite her pregnancy (Khouri 2005).

So far, we have discussed norms and sanctions as aspects of normative culture. Although norms and sanctions are relatively specific and concrete, the next major component of the normative dimension of culture—values—is rather broad and abstract. Values are much more general than norms.

What are values? **Values** are broad cultural principles that most people in a society consider desirable. Values are so general that they do not specify precise ways of thinking, feeling, and behaving. Thus, different societies or different groups within the same society can have quite different norms based on the same value. For instance, the value of freedom has been expressed differently in America and in the former Soviet Union. In the Soviet Union, as Robin Williams (1970) notes, freedom was expressed in the right to such things as employment, medical care, and education. Americans have different norms based on the value of freedom— the right to free speech and assembly, the right to engage in private enterprise, the right to a representative government, and the right to change where they live and work. In addition, patriotism for some citi-

zens might involve death for their country. For others, refusing to fight in a war they consider immoral might be patriotic. Identical values do not result in identical norms (Hurley 2000).

Why are values important? Values are important because they have a tremendous influence on human social behavior—mainly because norms are based on them. A society that values democracy will have norms ensuring personal freedom; a society that values humanitarianism will have norms providing for its most unfortunate members; a society that values hard work will have norms against laziness. Values are also important because they are so general that they tend to permeate most aspects of daily life. In America, for example, the value of freedom affects more than a person's political life. It affects such diverse areas as relationships in the family, treatment within the legal system, the operation of organizations, and the choice of religious affiliation.

What are the basic values in American society? Any attempt to state the basic values of American society is risky. Because America has so many diverse groups and because it is constantly changing, any one set of values is unlikely to receive unanimous support (Etzioni 2001). Despite these problems, Robin Williams (1970) has done an excellent job of outlining fifteen major values guiding the daily lives of most Americans. Whether or not his classification is complete, it provides a picture of the major values influencing Americans. These values include achievement and success, activity and work, humanitarianism, efficiency and practicality, progress, material comfort, equality, freedom, democracy, individuality, science and rationality, external conformity, group (racial, ethnic, religious) superiority, morality, and patriotism. Following is a sample of the values identified by Williams:

- *Achievement and success.* Americans emphasize achievement, especially in the world of work. Success is supposed to be based on effort and competition, and it is viewed as a reward for performance. Wealth is a symbol of success and personal worth, and quantity often represents quality.
- *Activity and work.* We tend to stress action over inaction in almost all instances. For most Americans, continuous and regular work is an end in itself, advancement is to be based on merit rather than favoritism, and all citizens are supposed to have the opportunity to perform at their best.
- *Efficiency and practicality.* Americans pride themselves on getting things done by the most rational means. We search for better, faster ways of doing things, praise good workmanship, judge work

KEVORK DJANSEZIAN/AP Photos

The U. S. women's soccer team celebrates its gold medal victory over Brazil in the 2004 Olympics. What cultural values do these women demonstrate?

performance by its practical consequences, and depend on science and technology.

- *Equality.* Americans advocate equality for all citizens. In interpersonal activities, we tend to treat one another as equals, to defend everyone's legal rights, and to favor equal opportunity for everyone.
- *Democracy.* Americans emphasize that all citizens are entitled to equal rights. All Americans should have equal rights and equal opportunity under the law, and power should not be concentrated in the hands of the elite.
- *Group superiority.* Despite their concern for equality of opportunity, Americans tend to place greater value on people of their own race, ethnic group, social class, or religious group.

These values, clearly, are interrelated. The values of achievement, success, activity, and work are related to the values of efficiency and practicality. Equally obvious is the conflict between some values: Americans value group superiority while stressing equality and democracy.

Are these values prominent in American society today? Williams identified fifteen major American values approximately thirty years ago. Although these values have remained remarkably stable over the years, some have changed somewhat. For example, today there is less emphasis on group superiority than in the past, as can be seen in the apparent decline of openly racist attitudes and behavior (R. Farley 1998; Rochon 1998). However, it is usually norms and behavior, rather than underlying values, that change radically. In other words, it is probably because of civil rights laws that many Americans are now less likely to make overt racist statements and discriminate against minority members. They simply behave in a way that is consistent with changed norms. So, as yet, racism remains part of the fabric of American culture.

The norms related to hard work and activity have also changed in recent years. According to Lionel Lewis (1982), many Americans now work as hard at their leisure (e.g., long-distance running and mountain climbing) as they do at their jobs.

Williams's analysis of major American values remains basically sound today, but it is not surprising that some, quite validly, believe that his list is incomplete. George and Louise Spindler (1983), for example, wish to add optimism, honesty, and sociability to the list of major American values.

Although functionalism is the basis of the normative dimension of culture, symbolic interactionism is related to the cognitive and material dimensions. Be alert to these theoretical relationships.

The bride, groom, and guests march in a traditional wedding ceremony in Indonesia. What values are reflected in this event?

The Cognitive Dimension

Do beliefs matter? **Cognition** is the process of thinking, knowing, or processing information. The cognitive dimension of culture, then, refers to its complex of ideas and knowledge. The most important aspect of the cognitive dimension of culture is **beliefs**—ideas concerning the nature of reality. Actually, beliefs are influential whether they are true or false. The Romans believed that Caesar Augustus was a god; the Tanala, a hill tribe of Madagascar, believed that the souls of their kings passed into snakes; and many Germans believed that pictures of Hitler on their walls would prevent the walls from crumbling during bombing raids. In 1953, Edmund Hillary, a New Zealander, became the first man to climb to the top of Mount Everest (29,035 feet). When he descended, he was greeted with hatred by the Sherpas of Nepal, who "believed that Buddhist gods resided up there, in the clouds, and they did not want to accept that the first human foot there had not been one of their own" (Hoffer 1999:122). On the American frontier, it was believed that a dog howling in the distance foretold a death in one's family, that anyone who brought a shovel into a cabin would carry out a coffin, and that a bird on one's windowsill or inside one's home meant sorrow in the future. While each of these beliefs is false, other beliefs—such as the belief that the human eye can distinguish more than 7 million colors or the belief that no life exists on Mars (Begley 1994; Lemonick 1994)—are factually supported by evidence.

Regardless of their truth, beliefs are important, because people tend to base their behavior on what they believe. Beliefs are also important because people are very attached to them, sometimes in ways that seem bizarre to outsiders. After surrendering their property and families following a siege, members of a Chinese village fought to the death rather than cut off their ponytails (Lilla 2007).

The Material Dimension

How do a society's physical objects reflect its culture? **Material culture** consists of the concrete, tangible objects within a culture—automobiles, basketballs, chairs, highways, birth-control pills, art, jeans. Artifacts, or physical objects, have no meaning or use apart from the meanings people give them. Consider newspaper and pepper as physical objects. Of course, each of these two things separately has some meaning for you, but can you think of a use for them in combination? Some midwives have used pepper and newspaper in a process known as nettling. An elderly medical doctor tells the story of his first encounter with nettling:

The ink of my medical license was hardly dry, and as I was soon to find out, my ears would not be dry for some time. I had never delivered a baby on my own and faced my maiden voyage with some fear.

Upon entering Mrs. Williamson's house, I found a local midwife and several neighbors busily at work preparing for the delivery. My fear caused me to move rather slowly and my happiness over my reprieve prompted me to tell the women that they were doing just fine and to proceed without my services.

Having gotten myself off the hook, I watched the ladies with a fascination that soon turned to horror.

At the height of Mrs. Williamson's labor pains, one of the neighbors rolled a piece of newspaper into a funnel shape. Holding the bottom end of the cone she poured a liberal amount of pepper into it. Her next move was to insert the sharp end of the cone into Mrs. Williamson's nose. With the cone in its "proper" place, the neighbor inhaled deeply and blew the pepper from the cone into the inner recesses of Mrs. Williamson's nose—if not her mind.

Suddenly alert, Mrs. Williamson's eyes widened as her senses rebelled against the pepper. With a mighty sneeze, I was introduced to nettling. The violence of that sneeze reverberated through her body to force the baby from her womb in a skittering flight across the bed. An appropriately positioned assistant fielded the baby in midflight and only minor details of Orville's rite of birth remained.

Before this doctor was introduced to nettling, this particular combination of newspaper and pepper had no meaning for him. And until nettling was devised, the combination was without meaning for anyone, even though the separate physical objects existed as part of the culture.

Physical objects do not have the same meanings and uses in all societies. Although it is conventional to use a 747 jet for traveling, it is possible that a 747 downed in a remote jungle region of the world could be used as a place of worship, a storage bin, or a home. In the United States, out-of-service buses, trains, and trolley cars have been converted to restaurants.

Clearly, the cultural meaning of a physical object is not determined by the physical characteristics of the object. People use newspaper and pepper during childbirth or make a temple of a downed 747 jet because of cognitive and normative definitions. The meanings of physical objects are based on the beliefs, norms, and values people hold with regard to them. This is readily apparent when new meanings of a physical object are considered. At one time, only pianos and organs were

---

TABLE 3.2

FOCUS ON THEORETICAL PERSPECTIVES: Culture

This table illustrates the unique slant that each of the three major theoretical perspectives takes on culture. Select a different aspect of culture for each perspective and make up your own table.

Theoretical Perspective	Aspect of Culture	Example
Functionalism	• The norm requiring students to listen to their teachers on request	• This norm promotes better education for everyone because of a lack of disorder.
Conflict theory	• Drug use by a member of a drug subculture	• Because of a lack of power, he or she is more likely to go to jail than an upper-class user.
Symbolic interactionism	• Confederate flag on the back window of a truck	• To the driver, the flag may be a symbol of cultural heritage; to an African American, it may represent slavery.

used in church services; guitars, drums, and trumpets were not "holy" enough to accompany a choir. Yet, many churches today use these "worldly" instruments regularly in their worship activities. The instruments have not changed, but the cultural meanings placed on them have.

Ideal and Real Culture

There is sometimes a gap between cultural guidelines and actual behavior. This gap is captured in the concepts of ideal and real culture. **Ideal culture** refers to cultural guidelines publicly embraced by members of a society; these are the guidelines we *claim* to accept. **Real culture** refers to *actual* behavior patterns. Sometimes subterranean real cultural patterns are publicly denied because they conflict with the ideal culture. For example, one aspect of America's ideal culture is honesty. Yet, in real culture, some taxpayers annually violate both the letter and spirit of existing tax laws, some businesspeople engage in dishonest business practices, some students cheat on exams, and some college athletes do the "high $500" handshake (during which a team booster leaves illegal money in their palms). These are not isolated instances. These real cultural patterns are often

passed from generation to generation. Keep in mind that we are not referring to cases of individual deviance, such as people who murder, rape, and rob. These types of antisocial behavior violate even real culture.

Does the fact that we sometimes ignore cultural guidelines make ideal culture meaningless? Absolutely not. In an imperfect world, ideal culture provides high standards. These ideals are targets that most people attempt to reach most of the time. Otherwise, social chaos would prevail. Ideal culture also permits the detection of deviant behavior. Individuals who deviate too far from the ideal cultural pattern are sanctioned. This helps preserve the ideal culture.

Culture as a Tool Kit

According to the dominant view, culture guides behavior by providing the values or ends toward which behavior is directed. Ann Swidler, a contemporary critic of the culture-as-a-way-of-life approach, thinks that culture should be viewed as a " 'tool kit' of symbols, stories, rituals, and worldviews, which people may use in varying configurations to solve different kinds of problems" (Swidler 1986:273). In this view, culture provides

FEEDBACK

1. _____ consists of all the material objects as well as the patterns of thinking, feeling, and behaving that are passed from generation to generation among members of a society.
2. A _____ is composed of a people living within defined territorial borders who share a common culture.
3. _____ are rules defining appropriate and inappropriate ways of behaving.
4. _____ are rewards and punishments used to encourage desired behavior.
5. Indicate whether the following are formal sanctions (F) or informal sanctions (I).
   ____ a. A mother spanks her child.
   ____ b. A professor fails a student for cheating on an exam.
   ____ c. A jury sentences a person to life in prison for espionage.
   ____ d. A husband separates from his wife after she has an affair.
6. _____ are broad cultural principles embodying ideas about what most people in a society consider desirable.
7. _____ are ideas concerning the nature of reality.
8. Indicate whether each of the following best reflects a belief (B), folkway (F), mos (M), law (L), or value (V).
   ____ a. conception that God exists
   ____ b. norm against cursing aloud in church
   ____ c. norm encouraging the eating of three meals daily
   ____ d. idea of progress
   ____ e. norm against burning a national flag
   ____ f. norm encouraging sleeping in a bed
   ____ g. norm prohibiting murder
   ____ h. norm against overtime parking
   ____ i. idea that the Earth's orbit is elliptical
   ____ j. idea of freedom
9. _____ culture consists of the concrete, tangible objects within a culture.
10. _____ culture refers to aspects of culture publicly embraced by members of a society.

Answers: 1. Culture; 2. society; 3. Norms; 4. Sanctions; 5. a. (I), b. (F), c. (F), d. (I); 6. Values; 7. Beliefs; 8. a. (B), b. (M), c. (F), d. (V), e. (M), f. (F), g. (L), h. (L), i. (B), j. (V); 9. Material; 10. Ideal a range of choices to be applied in defining and solving the problems of living. Because the content of any culture is not fully consistent, the "tools" that are chosen vary among individuals and situations. This view of culture is gaining visibility. Its eventual prominence in the sociological approach to culture remains to be seen (Hays 1994).

Language and Culture

Culture is the social heritage of humans. This heritage is altered by each generation and must be learned by new members of society. Both the creation and the transmission of culture depend heavily on the human capacity to develop and use symbols, the most significant of which make up language. The following discussion relies on concepts central to symbolic interactionism. Symbolic interactionism, you will recall, emphasizes our use of symbols in subjective interpretations of human social behavior.

Symbols, Language, and Culture

Once again, what are symbols? **Symbols**—things that stand for, or represent, something else—can range from physical objects to words, sounds, smells, and tastes. As you saw in Chapter 1, however, the meaning of a symbol is not dictated by its characteristics. There is nothing intrinsically good about the sound created by applause. In the United States, for example, applause warms the heart of an entertainer, politician, or professor, but in Latin America it symbolizes disapproval. Of course, once meaning has been assigned to applause and we learn to associate it with approval or disapproval, it seems as though the appropriate meaning is determined by the applause itself. In the Islamic world, the required head scarf for women is a symbol of modesty. Americans see it as a symbol of male oppression of women. A more recently defined symbol, a red ribbon insignia, iden tifies AIDS supporters (Peyser 1993). Tiger Woods is worth $100 million to Nike as a symbol for the firm's new golf division. The baseball that Barry Bonds hit for his record-breaking seventy-third home run in 2001 is estimated to be a $1 million symbol. Because it symbolized to the Arab world a takeover, the American flag first draped over the head of a Saddam Hussein statute in 2003 was quickly replaced by an Iraqi flag.

In Lewis Carroll's *Through the Looking Glass,* Humpty Dumpty says to Alice with some finality, "When I use a word, it means just what I choose it to mean—neither more nor less." So it is with symbols. Of course, the same word can symbolize very different things. Shortly after September 11, 2001, American teenagers appropriated newly current words for their own purposes.

Ground-zero bedrooms are a total mess; mean teachers are terrorists; a disciplined student experienced a jihad; and out-of-style clothes are burqas (Wax 2002). Even the symbol "nine-eleven" was expanded from its conventional meanings as a numerical value and an emergency phone number to include the 2001 terrorist attacks in New York, Washington, and Pennsylvania.

Symbolic culture is not limited to vocal, written, or material symbols. **Gestures,** whether in the form of facial expression, body movement, or posture, also carry culturally defined and shared symbolic meanings. Most Americans know that shaking one's fist at another driver, forming a circle with one's forefinger and thumb, and extending one's thumb in the direction of traffic while walking along a highway each convey a commonly held meaning to other Americans. Sometimes, though not frequently, gestures may change within a culture. For example, it used to be that men publicly holding hands could only suggest a homosexual relationship. When NFL players hold hands in the huddle on national television today, it symbolizes team solidarity. And different gestures transmit different meanings in various cultures. When Americans hold a thumb upright, either strong approval or the desire for a ride is assumed. The same gesture in Nigeria is taken as an insult (Axtell 2007).

What is the relationship between language and culture? Language frees humans from the limits of time and place. It allows us to create culture. The Wright brothers' victory over gravity did not come just from their own solitary efforts. They constructed their airplane according to aerodynamic principles known for some time before their first successful flight. They could read, discuss, and recombine existing ideas and technology.

Equipped with the symbols of language, humans can transmit their experiences, ideas, and knowledge to others. Children can be taught things without any actual experience on their part. Although it may take some time and repetition, children can be taught the dangers of fire and heights without being burned or toppling from stairs. This process of social learning, of course, applies to other cultural patterns as well, such as exhibiting patriotism, consuming food, or staying awake in class.

The Sapir-Whorf Hypothesis

According to Edward Sapir (1929) and Benjamin Whorf (1956), language is our guide to reality; our view of the world depends on the particular language we have learned. Our perception of reality is at the mercy of the words and grammatical rules of our language. And because our perceptions are different, our worlds are different. This is known as the **hypothesis of linguistic relativity.**

TABLE 3.3

Symbols Created for Internet Communication

The symbols in the left column were created to stand for the symbols in the right column. Why are both columns called symbols?

:-)	Happy
;-)	Winking
:-(	Sad
:-D	Laughing
:-O	Surprised
:-@	Screaming
:-P	Tongue in cheek
R	Are
UOK?	Are you OK?
B	Be
B4	Before
BBL	Be back later
BRB	Be right back
BTW	By the way
XLNT	Excellent
4	For, Four
GR8	Great
L8R	Later
LOL	Laughing out loud
LUV	Love
NO1	No one
OIC	Oh, I see
PLS	Please
PCM	Please call me
C	See
THX	Thanks
TTUL	Talk to you later
2	Too, to, two
2DAY	Today
2MORO	Tomorrow
WAN2	Want to
?	What
U	You
YR	Your

Subsequent research challenges the extreme assertion that language entirely determines thinking. Studies do reveal, however, that language significantly shapes thought (Boroditsky 2001; Prinker 2007; Kenneally 2008; Lupyan 2008). This is important in examining the relationship between language and perception.

What can vocabulary reveal about a culture? When something is important to a society, its language will contain many words to describe that entity. The Agta of the Philippines have thirty-one different verbs meaning "to fish" (M. Harris 1990). The importance of time in American culture is reflected in many words—*era, moment, interim, recurrent, century, afternoon, semester, eternal, annual, meanwhile, regularly,* and *semester,* just to name a few. When something is unimportant to people, they may not even have a word for it. When Christian missionaries first went to Asia, they were dismayed because the Chinese language contained no word for—and therefore no concept of—sin. Other missionaries were no less dismayed to learn that Africans and Polynesians had no way to express the idea of a single, all-powerful God. The Tzeltal language in southern Mexico recognizes five basic colors—white, black, red, green, and yellow. The language of the Jalé of New Guinea has terms only for black and white (Berlin and Kay 1969).

Does the hypothesis of linguistic relativity mean people are forever prisoners of their language? As far as we know, all people have the genetic capacity to learn any language. Thus, although the principle of linguistic relativity states that a person's view of the world is colored by his or her language, it does not state that people are forever trapped by their language. Exposure to another language can alter a person's perception of the world.

People can begin to view the world differently as they learn a new language. Most people, however, confine themselves to their native language. Consequently, they tend not to change their view of the world. Language, then, so necessary for the creation and transmission of culture, is also constricting.

What other factors help to shape our perception of reality? Application of the Sapir-Whorf hypothesis of linguistic relativity is almost limitless; it extends far beyond vocabulary into areas as diverse as visual perception, audio perception, and ideas about privacy and space. The Japanese use paper walls as sound barriers and are not bothered by noise from adjacent rooms. Americans staying at a hotel in Japan may complain that they are being bombarded with noise because Westerners have not been conditioned to screen out sound (E. T. Hall 1990). At the other extreme, Germans feel very protective about their privacy. German executives, for example, have a closed-door policy. Problems arise, then, in subsidiaries of American firms in Germany because American executives leave their doors open. The German preference for physical barriers to protect their privacy is reflected in the prevalence of double doors in German hotels and other public and private buildings (E. T. Hall 1976, 1979).

1. _____ are things that stand for, or represent, something else.
2. Because of _____, culture can be created and transmitted.
3. According to the hypothesis of linguistic relativity, words and the structure of a language cause people to live in a distinct world. T or F?
4. The language we learn determines forever how we see the world. T or F?

Answers: 1. Symbols; 2. language; 3. T; 4. F

Cultural Diversity and Similarity

Inherent in the discussion thus far is the relationship between sociology's major theoretical perspectives and other central cultural concepts. Symbolic interactionism is intimately connected with the role of language in culture and with the cognitive and material dimensions of culture. The assumptions of functionalism underlie the normative dimension of culture. As you view cultural diversity and similarity, you will again be aware of the appropriate theories. Conflict theory, with its emphasis on inconsistency, conflict, and change, undergirds any discussion of cultural diversity, whereas functionalism is the basis for a sociological analysis of cultural similarity.

Cultural Diversity

In ancient Rome, adultery was an accepted practice among the upper classes, and divorce was granted upon agreement of the couple. In modern Italy, adultery is frowned on, and divorce has only recently been legalized. Under Saudi Arabian religious laws, adulterers are stoned to death in public. Some American married couples advertise their "swinging" behavior in newspapers and magazines. Eskimos used to offer their wives as sexual favors to visitors—even strangers—and felt offended if the offer was refused. These examples reflect the almost endless cultural variations in the world. Not only do different societies have different beliefs, norms, values, and sanctions, but various groups within the same society have their own cultural patterns. Even in a simple society, cultural diversity makes it impossible for all members to participate in all aspects of culture. In modern societies, cultural diversity is staggering. Given the biological similarity among humans, this diversity must be explained by nongenetic factors.

How is cultural diversity promoted? Cultural diversity exists in all societies in part due to the presence of social categories. A **social category** is a group of persons who share a social characteristic such as age, sex, or religion. Members of social categories are expected to participate in aspects of culture unique to them. A particular age, gender, or religion, for example, often has distinctive activities associated with it.

Cultural diversity also stems from the existence of groups—known as *subcultures* and *countercultures*—that are somewhat different from the larger culture. Although these groups do participate in the dominant culture—they may speak the language, work at regular jobs, eat and dress like most others, and attend recognized houses of worship—they also think, feel, and behave in ways that set them apart. Subcultures and countercultures are more prevalent in large, complex societies.

What is a subculture? A **subculture** is a group that is part of the dominant culture but differs from it in some important respects. By tradition, Americans like to see themselves as part of a large, single culture. This view is fueled by the long-standing conception of the United States as a "melting pot" that absorbs peoples from around the world into "American" culture. This viewpoint minimizes the presence of cultural diversity in the United States. Since the 1960s, there has been a movement toward an emphasis on the cultural uniqueness of subgroups. This movement, known as **multiculturalism,** accents the viewpoints, experiences, and contributions of minorities (women as well as ethnic and racial minorities). In fact, many scholars are concerned that the current emphasis on cultural differences is unduly fragmenting American society (Schlesinger 1998).

Sociologists have routinely studied subcultures, both before and after the rise of multiculturalism. The residents of southern Appalachia are an illustration of early sociological research on subcultures. According to some sociologists in the early 1960s, these people subscribe to ways of thinking, feeling, and behaving quite different from those of middle-class America. Middle-class Americans believe in progress, planning ahead, status seeking, education, social participation, and freedom to determine one's destiny. In contrast, Jack Weller (1980) contended, southern Appalachians are fatalistic, present oriented, unambitious, and nonparticipative. Although they are different from middle-class Americans, Weller notes that southern Appalachians are not opposed to the dominant culture. Rather, Weller argues, the southern Appalachian subculture constitutes a cultural adaptation to the reality of living a long-standing deprived and frustrating existence.

These skinheads are giving the Nazi salute during their training outside Moscow. Are they part of a subculture or a counterculture?

Another example of a subculture in the United States is the Old Order Amish, a group that has consciously attempted to retain its unique cultural identity ever since entering the United States in the early 1700s. The highly religious 100,000-strong Amish, who live primarily in the Midwest, exert tremendous efforts to maintain physical and cultural separation from the dominant culture. You may not have seen any Amish in person, but you probably have seen them depicted in the movies or on television as very traditionally dressed farmers traveling by horse and buggy. The Amish are also culturally conservative in their view of women (subservient to men), birth control (prohibited), education (send their children to Amish schools), and technology (only recently beginning to welcome machines, electricity, and indoor plumbing) (Janofsky 1997; Zellner and Kephart 2000).

The Amish generally reject living and participating in the dominant culture, but members of another group—devotees of the Grateful Dead band—both live and participate in the dominant culture. "Deadheads," as they call themselves, subscribe to a subculture constructed around the band. They share a unique language, mode of dress (featuring tie-dyed clothing), and liberal attitudes toward drug use. Deadheads tend to have mainstream jobs and lives. For them, participation in the subculture revolves around hearing the band when it is close enough to attend. Many Deadheads, however, take off for months to follow the band on tour around the country (R. G. Adams 1995).

A less extreme subculture is that of San Francisco's Chinatown. Early Chinese immigrants brought much of their native culture with them to America and have attempted to retain it by passing it from generation to generation. Although Chinese residents of Chinatown have been greatly affected by American culture, they have retained many cultural patterns of their own, such as language and family structure.

Other subcultures in America are those formed by athletes, actors, and surfers, as well as by people in mental hospitals, convents, and universities. (For other excellent examples, see Fine 1996; Redhead 1997; Meuller 1999; Zellner and Kephart 2000.)

What is a counterculture? A **counterculture** is a subculture that is *deliberately* and *consciously* opposed to certain central aspects of the dominant culture. A counterculture can be understood only within the context of its underlying opposition to the dominant culture (see Doing Research). Countercultures openly defy the norms, values, and beliefs of the dominant culture. Rebelling against the dominant culture is central to their members. For example, a counterculture was formed on American college and university campuses in the 1960s. What Edward Suchman (1968) called the "hang-loose" ethic rejected such central aspects of American culture as material success, achievement, hard work, efficiency, authority, premarital chastity, and the nuclear family. A more recent counterculture, the militia movement, is based on political opposition. Members of the militia counterculture congregate in isolated compounds, sharing a way of life fashioned to resist the authority of the federal government. Fully armed, members of the militia organize their lives around a future revolution against their oppressor (Stern 1996; Ferber 1998). Many Americans view the 1995 Oklahoma City federal building bombing that killed 168 people as either part of the militia movement or an outgrowth of it. The skinheads, a neo-Nazi counterculture founded in the United States at the beginning of the 1980s, is a counterculture based on racial, ethnic, and sexual orientation intolerance. This intolerance appears in violent acts against gay men and lesbians, ethnic minorities (particularly Jews), and blacks. Skinheads draw their members from among young, white, working-class males. Prime marks of membership are a shaved head and a quasi-military outfit. Skinheads see themselves as patriotic defenders of a belief in racial superiority and "sexual correctness" who have a right to use violence against their targets (Wooden 1995).

Prison counterculture surfaced at the 1999 trial of John King. King was convicted of the gruesome truck-dragging murder of James Byrd Jr. During an earlier prison stretch, King had become a member of a prison gang named the Confederate Knights of America's Texas Rebel Soldiers. Dedicated to white supremacy, this prison counterculture advocates violence in many forms. For example, the norm "blood in, blood out" involves the entrance requirement of a violent act and the threat of losing one's own blood for getting "out," or leaving the group (Galloway 1999). Certain types of delinquent gangs, motorcycle gangs, drug groups, and religious groups may also form countercultures (Cushman 1995; Roszak 1995; Zellner 1995; Zellner and Kephart 2000; Heineman 2002).

How do cultural differences affect attitudes toward others? Once members of a society learn their culture, they tend to become strongly committed to it. In fact, they tend to become so committed that they cannot conceive of any other way of life. If members of other groups seem different, they may be considered strange. Cultural differences may even become signs of alleged inferiority. This tendency to judge others in relation to one's own cultural standards is referred to as **ethnocentrism**.

People who spend most of their lives with others culturally similar to themselves—who hardly ever deal with people different from themselves—will almost inevitably use their own cultural standards to judge others. The ethnocentric eye may see those who are different as inferior, ignorant, crazy, or immoral. We are shocked by a people such as the Ik, who willingly abandon their chil-

dren at three years of age (Turnbull 1994). And it is hard to understand the Yanomamö women of Venezuela and Brazil who measure the affection of their husbands by the number of beatings they are given (Chagnon 1997).

Less exotic examples of ethnocentrism are plentiful—extreme nationalism being one. Members of all societies tend to offer themselves as exemplary models. The Olympic Games are much more than an arena for young men and women to engage in healthy and exuberant competition; they are also an expression of ethnocentrism. Influential political and nationalistic undercurrents run through the Olympics. A country's final ranking in this athletic competition for gold, silver, and bronze medals is frequently taken as a reflection of the country's worth and status on the world stage.

Ethnocentrism also exists within societies. Regional rivalries in the United States are a source of many humorous stories, but these jokes reflect an underlying ethnocentrism. Boston is said by some (mostly Bostonians) to be the hub of the universe. Some inhabitants of the Eastern seaboard believe that only a wasteland separates them from the Pacific Ocean, while some more enlightened Easterners are willing to concede the addition of California in describing all that is good in the United States. Of course, this long-standing ethnocentric viewpoint has been subject to attack by those in Middle America, who have always felt that virtue

SNAPSHOT OF AMERICA 3.1

Immigration to the United States

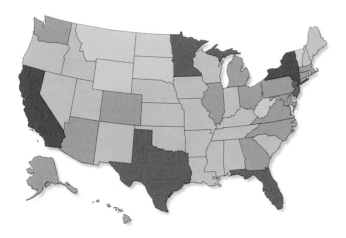

Number of Immigrants

- 100,000 or above
- 50,000–99,999
- 10,000–49,999
- 1,000–9,999
- 999 or below

Source: Bureau of U.S. Citizenship and Immigration Services: *2004 Yearbook of Immigration Statistics.*

The United States has long been the destination of immigrants from around the world. These immigrants have contributed to the nation's rich cultural diversity in a wide variety of ways. This map shows the number of immigrants who came to the United States in 2004 and the states in which they settled.

Interpret the Map

1. Which states had the most immigrants in 2004? Which states had the fewest?
2. About how many immigrants settled in your state in 2004?
3. Why would it be correct to say that the United States is a nation of immigrants?

Ben Curtis/AP Photo

These victims of genocide in West Darfur, Sudan illustrate the disastrous consequences of extreme ethnocentrism.

resides with them. And finally, the members of country clubs, churches, and schools all over America typically feel that their particular ways of living should be adopted by others.

Are there advantages of ethnocentrism? Up to this point, ethnocentrism has been portrayed as either politically incorrect or ridiculous. It may be both. But to some degree it is inevitable. And its inevitability is rooted partly in the advantages it offers to social life. Imagine the expense and effort required to create the integration, high morale, and loyalty that ethnocentrism provides a society or group. Few things draw people closer together than shared loyalty or conviction that what they are doing is right and superior vis-à-vis the actions and beliefs of others. Such commitment makes people feel good about themselves and their fellow group members. Fires of nationalism and patriotism are not fanned by declaring that your nation is slightly below average! Finally, ethnocentrism promotes social stability because people who are convinced that truth and beauty are theirs seldom entertain the need for change.

People are ethnocentric not simply because they recognize the benefits of ethnocentricity for social life. No. Ethnocentrism would be pervasive even if it did not offer these advantages. We are taught the rightness of our culture from a variety of sources (home, church, peers, schools, mass media), so it is only natural that this "rightness" becomes the yardstick for evaluating other cultures and how their members think, feel, and behave. Judgments of all kinds are nearly always

alloyed with socially derived convictions regarding right and wrong. Ethnocentrism, then, is a predictable by-product of the transmission of culture.

What are the disadvantages of ethnocentrism? A price may be paid for the integration, morale, loyalty, and stability that ethnocentrism provides. Extreme ethnocentrism has ill effects within societies as well as between them. On the intrasocietal side, extreme ethnocentrism may create such a high degree of integration and stability that innovation is hampered. Societies whose members are too firmly convinced of their righteousness may choke off exploration for new solutions to persistent problems. Further, they may reject, without examination, solutions that might be gleaned from other societies. And if heightened ethnocentrism impedes both the internal creation of ideas and the external influence for solutions to long-standing problems, it may create an even greater disadvantage in facing the challenge of new problems and unfamiliar circumstances.

On the intersocietal front, social and political conflict is nearly always a result of extreme ethnocentrism. Global peace and welfare become secondary goals in a world in which ethnocentrism intensifies intersocietal conflicts based on power struggles for economic superiority and/or military supremacy.

We have already seen some of the negative consequences of ethnocentrism for a society. Ethnocentrism has harmful effects on individuals as well. **Culture shock**—the psychological and social stress we may experience when confronted with a radically different cultural environment—is one such negative consequence. Even cultural anthropologists, who are trained professionals, may experience culture shock. Napoleon Chagnon gives this account of his reaction to his first contact with the Yanomamö:

My heart began to pound as we approached the village and heard the buzz of activity within the circular compound. Mr. Barker commented that he was anxious to see if any changes had taken place while he was away and wondered how many of them had died during his absence. I felt into my back pocket to make sure that my notebook was still there and felt personally more secure when I touched it.

The entrance to the village was covered over with brush and dry palm leaves. We pushed them aside to expose the low opening to the village. The excitement of meeting my first Yanomamö was almost unbearable as I duck-waddled through the low passage into the village clearing.

I looked up and gasped when I saw a dozen burly, naked, sweaty, hideous men staring at us down the shafts of their drawn arrows! Immense wads of green tobacco were stuck between their lower teeth and lips making them look even more hideous, and strands of dark-green slime dropped or

Jacquelynne Eccles— Teenagers in a Cultural Bind

Adolescence is often marked by drama and difficulty. Jacquelynne Eccles and colleagues (1993) investigated the experience of American teenagers entering a Midwestern junior high school and discovered that some teenage troubles are more than hormonal—they are cultural as well. The relationships among seventh graders, teachers, and parents are embedded in the normative and cognitive aspects of culture.

Eccles conducted a two-year study of students from twelve school districts in middle-class Michigan communities. She studied 1,500 early adolescents moving from sixth grade in an elementary school to seventh grade in a junior high school. Students filled out questionnaires at school in two consecutive years—the sixth and seventh grades. This procedure permitted Eccles to document changes the teenagers experienced after the first year of their transition.

The findings were not encouraging. The relationships between students and teachers deteriorated

over the year, a deterioration Eccles found related to the culture of junior high school. At the very time when young adolescents especially needed supportive relationships outside the home, personal and positive relationships with teachers fell victim to cultural and organizational changes in junior high school. There was increased grouping based on academic achievement and increased public evaluation comparing students with one another. This heightened emphasis on student ranking comes just when young adolescents are at the peak of concern about their status relative to their peers. In addition, in the junior high culture, the students experienced less opportunity to participate in classroom decision making. The resulting diminished school motivation and lessened academic self-confidence is as predictable as it is unfortunate. The culturally defined educational structure of junior high works directly against the fact that adolescents develop better in emotionally supportive environments.

Eccles's news was no better on the home front. In fact, the results in the family paralleled those reported for the school. Excessive parental control over teenagers went up at the same time school motivation and self-esteem of the junior high students went down.

As a check on these general findings, Eccles compared students in more culturally supportive schools and families with those in less supportive

ones. In both the school and the family settings, she found more positive results in culturally supportive environments. Students who were able to participate in school and family decision making showed higher levels of academic motivation and self-esteem than their peers with less opportunity to participate.

The solution, Eccles concludes, exists in a cultural change within both the schools and the family. Schools and the family need to develop balanced cultural expectations of young adolescents based on their developmental needs. At this age, young people have a growing need for autonomy. Neither cracking down on them nor relinquishing control strikes the proper balance. The task is for the family and the school to provide "an environment that changes in the right way and at the right pace" (Eccles et al. 1993:99).

Evaluate the Research

1. Do you recall your first year in junior high school? Analyze your experience based on the cultural definitions that existed in your school and home at the time.
2. Do adolescent students belong to a subculture? A counterculture? Explain.
3. Which of the three theoretical perspectives is most helpful in understanding the social relationships Eccles describes? Apply this perspective to explain her findings.

---

hung from their nostrils—strands so long that they clung to their pectoral muscles or drizzled down their chins. We arrived at the village while the men were blowing a hallucinogenic drug up their noses. One of the side effects of the drug is a runny nose. The mucus is always saturated with the green powder and they usually let it run freely from their nostrils. . . . My next discovery was that there were a dozen or so vicious, underfed dogs snapping at my legs, circling me as if I were to be their next meal. I just stood there holding my notebook, helpless and pathetic. Then the stench of the decaying vegetation and filth hit me and I almost got

sick. I was horrified. What kind of welcome was this for the person who came here to live with you and learn your way of life, to become friends with you? They put their weapons down when they recognized Barker and returned to their chanting, keeping a nervous eye on the village entrances. (Chagnon 1997:11–12)

If trained anthropologists can experience culture shock, it is easy to understand why immigrants are stunned by cultural practices foreign to their own. As a U.S. native, you might be surprised at the shocks,

WORLDVIEW 3.1

Patterns of Tourism

Although people often want to observe and experience cultures different from their own, exposure to cultural diversity can be uncomfortable. Most international tourist travel occurs among countries sharing common cultural traditions and languages.

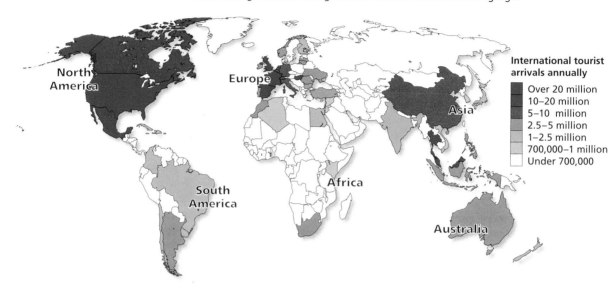

International tourist arrivals annually

- Over 20 million
- 10–20 million
- 5–10 million
- 2.5–5 million
- 1–2.5 million
- 700,000–1 million
- Under 700,000

Adapted from the *Student Atlas*: DK Publishing, Inc.

Interpret the Map

1. Identify the world regions that receive the highest and lowest number of tourists.

2. Are there any reasons to believe that these travel patterns might change in the near future? If so, what factors might bring about this change?

---

large and small, that internationals face in this country. When one newcomer was asked by a checkout clerk if he wished paper or plastic, he got very confused and upset. How could he choose paper or plastic when he thought payment required money? More seriously, imagine the difficulties of dating if you came from a culture in which interaction with the opposite sex among the young is limited to relatives (Pipher and Cohen 2002). You will probably better understand culture shock if you take an overseas job assignment (Wagster 1993).

Cultural Relativism

Ethnocentrism can injure others when we attempt to impose our ways of thinking, feeling, and behaving on them. Despite good intentions, outsiders have, in many instances, harmed members of other cultures.

What can be done to reduce the negative personal effects of ethnocentrism? Awareness of ethnocentrism and the harm it can cause is a necessary first step to its control. An important second step lies in a perspective known as **cultural relativism.** According to this perspective, values, norms, beliefs, and attitudes are not in themselves correct or incorrect, desirable or undesirable; they simply exist within the total cul-

tural framework of a people and should be evaluated in relation to their place within the larger cultural context of which they are a part rather than according to some alleged universal standard that applies across all cultures. Cultural relativism gives us a unique window through which to observe cultural variations.

Offering one's mate for sexual activity with an overnight guest is not allowed in most societies. Hans Ruesch, in his novel *Top of the World* (1959), however, reveals a society in which this practice was not only acceptable but expected. In traditional Eskimo society, it was a serious personal affront to the husband if a guest refused to "laugh" (have sexual intercourse) with his wife.

In the following passage from Ruesch's novel, an Eskimo named Ernenek tries to explain how his rage caused him to accidentally kill a guest who had refused to have sexual relations with his wife Asiak. Ernenek and Asiak are talking with a potential ally in the matter, a white man whose life Ernenek has saved:

"You have saved my life, Ernenek," the white man said, "and I wish to straighten things out so that you need no longer fear my companions. But you will have to stand before a judge. I will help you explain things."

"You are very kind," said Ernenek, happy.

"You said the fellow you killed provoked you?"

"So it was."

"He insulted Asiak?"

"Terribly."

"Presumably he was killed as you tried to defend her from his advances?"

Ernenek and Asiak looked at each other and burst out laughing.

"It wasn't so at all," Asiak said at last.

"Here's how it was," said Ernenek. "He kept snubbing all our offers although he was our guest. He scorned even the oldest meat we had."

"You see, Ernenek, many of us white men are not fond of old meat."

"But the worms were fresh!" said Asiak.

"It happens, Asiak, that we are used to foods of a quite different kind."

"So we noticed," Ernenek went on, "and that's why, hoping to offer him at last a thing he might relish, somebody proposed him Asiak to laugh with."

"Let a woman explain," Asiak broke in. "A woman washed her hair to make it smooth, rubbed tallow into it, greased her face with blubber and scraped herself clean with the knife, to be polite."

"Yes," cried Ernenek, rising. "She had purposely groomed herself! And what did the white man do? He turned his back to her! That was too much! Should a man let his wife be so insulted? So somebody grabbed the scoundrel by his miserable little shoulders and beat him a few times against the wall—not in order to kill him, just wanting to crack his head a little. It was unfortunate it cracked a lot."

"Ernenek has done the same to other men," Asiak put in helpfully, "but it was always the wall that went to pieces first."

The white man winced, "Our judges would show no understanding for such an explanation. Offering your wife to other men!"

"Why not? The men like it and Asiak says it's good for her. It makes her eyes sparkle and her cheeks glow."

Cultural Relativism

Different behaviors, traditions, and expectations can often result in misunderstandings between people of different cultures. As just noted, learning to look at things from a point of view different from your own, and not making value judgments based on your beliefs and norms, is called cultural relativism. Having mutual respect and understanding for other cultures is sometimes more effective than modern technology and money in producing change and goodwill between nations.

Cultural relativism is illustrated in the true story of a young Peace Corps volunteer who was sent to a remote village to help build a well. The stream that was near the village was used for everything from watering goats to bathing to washing clothes to cooking and drinking. It was obvious that clean drinking water would benefit the vil-

lage and improve health. Armed with plans, equipment, and budget and schedule, the hopeful volunteer arrived ready to begin.

At first, the village people were not very willing to help. After several weeks of lonely effort the volunteer met with the council to ask why nobody was helping her with this urgent project. "A well would be nice," the people agreed, "but what we really need is a good soccer field where we can play without getting hurt on the stones and uneven ground." So the volunteer agreed that some of the money and equipment could be used to build a good soccer field first.

After several weeks of effort, the soccer field was complete and a village soccer team was formed. Now work was able to start on the well, but once again the villagers seemed reluctant to help. Another council meeting was held, and the volunteer was told, "Ah yes, the well would be nice, but what we really need is a bridge across the stream so other villages can easily come to play soccer on our field." Since she couldn't dig the well alone, the volunteer agreed that some more time and money would be used to build a bridge. Unfortunately, the bridge proved to be more difficult than expected, and by the time it was com-

plete, the budget and schedule were both used up.

The volunteer went back to the capital, disappointed and resentful that she had not been able to improve the village. Some weeks later, she was invited back by the villagers for a festival to celebrate the success of the soccer tournament they had arranged. When she arrived she was astonished to find a new well in the very center of the village. She asked the village elders for an explanation.

"The soccer tournament is important to us," she was told, "because it gives us pride and importance and gives us a reason to meet with the people of the other villages. We really never wanted a well."

"Then why did you build it?" she asked.

"We didn't build it because we wanted it," was the answer. "We built it because YOU wanted it."

Think About It

1. What assumptions did the volunteer make about the needs of the villagers? What were the actual needs? Who was more right about what the villagers needed? Why?

2. Describe a recent time when you made assumptions that turned out to be culturally based.

"Don't you people borrow other men's wives?" Asiak inquired.

"Never mind that! It isn't fitting, that's all."

"Refusing isn't fitting for a man!" Ernenek said indignantly. (Ruesch 1959:87–88)

Now apply cultural relativism. How did this norm fit with other aspects of Eskimo culture? Under conditions at that time, all Eskimo possessions were handmade, difficult to replace, subject to hard use, and easily destroyed. Within this cultural context, lending wives made sense, as Ernenek explains:

Anybody would much rather lend out his wife than something else. Lend out your sled and you'll get it back cracked, lend out your saw and some teeth will be missing, lend out your dogs and they'll come home crawling, tired—but no matter how often you lend out your wife she'll always stay like new. (Ruesch 1959:88)

Cultural relativism does not require our accepting other cultural ways as our own, nor does it require engagement in alien cultural practices. But because it is impossible, as a cultural relativist, to view aspects of another person's culture ethnocentrically, cultural relativism removes barriers between ourselves and those who are culturally different; it is essentially an antidote to ethnocentrism.

Cultural relativism can help us adjust more easily when we are meeting new people or entering new situations. For example, your adjustment to a college roommate from another country, another region of the United States, or even another social class will be smoother if you attempt to understand your roommate from the viewpoint of his or her own cultural background. And if significant cultural differences exist between you and

your spouse, cultural relativism can enhance your marital relationship. It is also useful at work, where you will need to cooperate with people from other cultural environments. And given the growing importance of America's international business ties, someday you may well be practicing cultural relativism with people from Japan, Germany, Egypt, or Saudi Arabia.

Cultural Similarity

The world today seems to display an overwhelming diversity of social and cultural behaviors: in the late 1990s, the president of Ingushetia, a republic inside Russia, signed a bill into law making it legal for a man to have multiple wives. German husbands and wives wear their wedding band on their right hand. In Bulgaria, a nod means no and a shake of the head means yes. Upon entering a home in India, you may be adorned with a garland of flowers—which you should remove immediately as a sign of humility. It is not polite to accept food in Iran until your host has offered it to you a number of times. Despite these surface differences, sociologists and anthropologists have identified many behaviors that are shared by all cultures. All societies have families, schools, houses of worship, economies, governments, and systems of prestige.

A recent poll compared Arabs (eight countries) and Americans with regard to the values they consider most important to be taught to their children (Zogby 2002). As Table 3.4 reveals, Arabs and Americans generally subscribe to common values. The same values (self-respect, good health and hygiene, responsibility) are among the top three for both Arabs and Americans. In fact, convergence is very high on five of the top six values. The biggest differences are on a couple of values lower on both

---

TABLE 3.4

A Comparison of Values Among Arabs and Americans

This table contains a comparison of the ranking of values to be taught to children among Arabs and Americans. Do the data support the existence of cultural diversity or cultural similarity? Explain.

Value	Arab Rank	American Rank
Self-respect	1	2
Good health and hygiene	2	3
Responsibility	3	1
Respect for elders	4	5
Achieve a better life	5	8
Self-reliance	6	4
Religious faith	7	12
Serious work habits	8	6
Obedience	9	11
Creativity	10	9
Tolerance of others	11	10
Respect for authority	12	7

Source: James J. Zogby, *What Arabs Think: Values, Beliefs, and Concerns* (Utica, NY: Zogby International, 2002), 94.

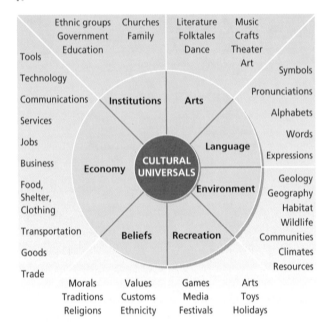

FIGURE 3.2

Cultural Universals

hospitality, housing, incest taboos, inheritance rules, joking, language, law, medicine, marriage, mourning, music, obstetrics, property rights, religious rituals, sexual restrictions, status differences, and toolmaking. When each of these universals is examined more closely, the similarity among cultures becomes even more apparent (see Figure 3.2).

How are cultural universals expressed? Just because societies share a cultural universal does not mean that they express it in the same way. Usually, cultures develop different ways of demonstrating the same universal trait. These may be thought of as **cultural particulars.** Traditionally, in the United States (although it has changed in recent years), the division of labor was based on gender roles. Women performed the domestic tasks and men worked outside the home. The Manus of New Guinea, on the other hand, place the man completely in charge of child rearing. Among the Mbuti pygmies, the Lovedu of Africa, and the Navajo and Iroquois Indians, men and women share equally in domestic and economic tasks (Little 1975).

Why do cultural universals exist? The biological similarity shared by all humans helps to account for the presence of cultural universals. If a society is to survive, children must be born and cared for and some type of family structure must exist. Groups that have deliberately eliminated the family—such as the Shakers—have disappeared. The existence of two sexes limits the type of marital relationships that can exist: one male and one female (monogamy), several females and one male (polygyny), one female and several males (polyandry), several males and females (group marriage), and two people of the same sex (homosexual marriage). Because people become ill, there must be some sort of medical

lists. Americans place greater emphasis on teaching their children respect for authority. Arabs give the teaching of religious faith a more prominent place. For both, shared values focus on personal and family concerns rather than on external matters such as political issues.

George Murdock (1945) identified about seventy **cultural universals,** general cultural traits thought to exist in all known cultures. These universals included athletic sports, cooking, courtship, division of labor, education, etiquette, funeral rites, family, government,

**F
E
E
D
B
A
C
K**

1. Indicate which of the following are social categories (SC), subcultures (S), or countercultures (C).

   _____ a. Chinatown in New York City   _____ d. females

   _____ b. motorcycle gang   _____ e. revolutionary political group

   _____ c. Catholics   _____ f. the superrich

2. The tendency to judge other societies or groups according to one's own cultural standards is known as _____.
3. The psychological and social stress one may experience when confronted with a radically different cultural environment is called _____.
4. _____ is the idea that any given aspect of a particular culture should be evaluated in terms of its place within the larger cultural context of which it is a part rather than in terms of some alleged universal standard that applies across all cultures.
5. _____ are general cultural traits thought to exist in all known cultures.
6. Which of the following is not one of the reasons that cultural universals exist?

   a. physical environment
   b. biological similarity of humans
   c. cultural predetermination
   d. problems of maintaining social life

Answers: 1. a. (S), b. (C), c. (SC), d. (SC), e. (C), f. (SC); 2. ethnocentrism; 3. culture shock; 4. Cultural relativism; 5. Cultural universals; 6. c care. Because people die, there must be funeral rites, mourning, and inheritance rules. Because babies are born, some form of obstetrics is required. Because food is necessary, cooking must exist. So goes the list of human biological similarities and their influence on culture.

A second source of cultural universals is the physical environment. For example, the awesomeness of nature and people's inability to explain physical phenomena—such as the eclipse of the sun or the creation of the universe—helped to encourage the development of religion. Because humans cannot survive in extreme climates without artificial protection, some form of housing must be created. Conflicts often occur over natural territorial borders such as rivers and mountains, so some provision must be made for maintaining order within as well as between societies.

Finally, cultural universals exist because societies face many of the same problems. If a society is to survive, certain social provisions must be made. New members must be socialized; goods and services must be produced and distributed; means of dealing with the supernatural must be devised; tasks must be assigned; and work must be accomplished.

Culture, Society, and Heredity

Culture and Heredity

What is the relationship between culture and biology? **Instincts** are genetically inherited, complex patterns of behavior that always appear among members of a particular species under appropriate environmental conditions. Although nonhuman animals are heavily hardwired for action through instincts, human infants cannot go very far on the basis of their genetic heritage alone. They lack immediate and automatic solutions to the problems they face. And later, without instincts to determine the type of shelter to build, the type of food to eat, the time of year to have children, or the type of mating pattern to follow, humans are forced to create and learn their own ways of thinking, feeling, and behaving. That is, humans are forced to rely on the culture they have created—even for such basic needs as eating and reproducing.

If humans were controlled by instincts, they would all behave in the same way. If, for example, women had an instinct for mothering, then all women would want children, and all women would love and protect their children. In fact, some women do not want to have children, and some women who give birth abuse or abandon, or kill, their children (Hrdy 2000).

Is genetic heritage without influence on human behavior? Although humans lack instincts, they are nevertheless affected by their genetic nature. Research on twins reared together and apart has examined the relative influence of heredity and environment on personality traits. It is estimated that about 50 percent of the diversity in measured personality characteristics is attributable to genetic heritage (Tellegen et al. 1993; Messner 1997).

In addition, humans have reflexes—simple, biologically inherited, automatic reactions to physical stimuli. A human baby, for example, cries when pinched; its eyes blink in response to a foreign particle; and the pupils of its eyes contract in bright light. Humans also have biologically inherited drives, or impulses, to reduce discomfort. They want to eat, drink, sleep, associate with others, and have sexual relations.

Genetically inherited personality traits, reflexes, and drives, however, do not "determine" human social behavior. Instead, culture channels the expression of these biological characteristics (Good et al. 1994). Most boys in traditional American culture, for example, are taught not to cry in response to pain, whereas boys in Jewish and Italian cultures learn to pay more attention to physical discomfort and express it more obviously (Zborowski 1952, 1969). To cite another example, humans have inherited the capacity for love. Awareness of this fact, however, does not allow us to predict the precise ways that different groups of people will express the ability to love. The people of one culture may believe that being one of several husbands is the most natural form of marriage, whereas members of another culture may endorse monogamous marriage.

Some social and biological scientists are challenging the traditional sociological perspective on human nature. Their divergent perspective is called *sociobiology*.

Sociobiology

Sociobiology is the study of the biological basis of human behavior. Sociobiologists argue that like physical characteristics, human social behavior is shaped through the evolutionary process. Thus, sociobiology is the application of Darwinian natural selection and modern genetics to human social behavior (Pinker 2003).

How do sociobiologists view human behavior? According to Darwin's theory of evolution, all organisms evolve through the process of natural selection. Those organisms best suited to an environment survive and reproduce themselves, and the rest perish. If human behavior is based on millions of years of evolution, say the sociobiologists, then most human behavior is basically self-protective (Dawkins 1990). Those behaviors that improve our chances for survival are genetically retained and reproduced. Parental affection and care, friendship, sexual reproduction, and the education of children must therefore be biologically based because they contribute to the survival of the human species.

Kenneth Sponsler/Used under license from Shutterstock

Studies of identical twins show that about half of your personality traits are inherited.

Some sociobiologists even contend that there may be specific genes for such things as aggression, religion, homosexuality, group loyalty or altruism, the creation of hierarchies, and the incest taboo.

Sociobiologists believe that the sharp line usually drawn between human and nonhuman animals is inappropriate. DNA testing shows that humans and apes (chimpanzees, gorillas, orangutans, and gibbons) are more closely related than either are to monkeys. Moreover, the chimpanzees' closest relative is not the gorilla but humans; we differ in only 1.6 percent of our DNA (Diamond 1993, 1999).

Nonhuman animals learn and transmit their knowledge—as when a group of chimpanzees spontaneously begin to use long sticks to ferret ants from an anthill for a meal (De Waal 1999; Whiten et al. 1999). Many nonhuman animals, assert sociobiologists, exhibit intelligence of a kind formerly thought to be unique to humans, including use of language (Begley 1993; Linden 1993a; Fagot, Wasserman, and Young 2001; Hauser 2001). For the first time among wild animals, chimps have been observed deliberately, consistently, and repeatedly using a tool as a weapon against one of their own kind (Linden 2002a, 2002b; Mercader et al. 2007). These findings tend to support pioneer sociobiologist Edward Wilson's (1978, 1986) contention that the study of human behavior must begin with the genetic heritage of humans.

Contemporary evolutionary social scientists report some aspects of social behavior they believe are genetically based. Women, it is reported, look for different characteristics in men they prefer to date and marry than

men value in women (Buss, Malamuth, and Winstead 1998). And stepfathers are more likely to abuse their stepchildren than biological fathers their own offspring (Daly and Wilson 1997).

What do the critics of sociobiology say? Some critics of sociobiology fear that this perspective will be used as a justification to label specific races inferior; others fear that it will be used to argue for the superiority of the male (Andersen 2005). Still other critics fear that the sociobiological perspective will be used to bolster the assumption of innate human selfishness. Marxists object to sociobiology on the grounds that it justifies the existence of Western capitalism.

Critics join in common objection to a genetic explanation of human social behavior. There is, they contend, too much social diversity around the world for human behavior to be explained on strictly biological grounds (Parker and Easton 1998). The human brain and the unique human capacity for using language and creating social life have allowed humans to overcome any contribution to behavior that might come from their genes. Humans, liberated from the confines of their genes by a large, well-developed cerebral cortex—which permits, among other things, abstract thinking—create and transmit a dazzling array of ways for thinking, feeling, and behaving. Nonhuman species, which either have no cerebral cortex or have an undeveloped one, behave as they do because of a strict genetic code. Birds do not walk south for the winter, salmon do not fly upstream, and lions do not prefer ferns to fresh meat. Critics believe that sociobiology should not sidetrack sociologists from their efforts to understand and explain human behavior from the sociological perspective. They agree with the nineteenth-century English philosopher John Stuart Mill: "Of all the vulgar modes of escaping from consideration of the effect of social and moral influence upon the human mind, the most vulgar is that of attributing the diversities of conduct and character to inherent natural differences" (Zinn 2003:33).

Since the early days of sociobiology, some common ground has emerged between sociologists and sociobiologists. Anthropologist Marvin Harris (1990) agrees that there is some biological continuity between human and nonhuman animals and that there is a basic human nature. Evolutionary psychologists and some sociologists now contend that human biology and the human capacity for creating a nearly infinite variety of ways for thinking, feeling, and behaving are two sides of the same coin. The genetic heritage of humans, they argue, shapes and limits human nature and social life. Consequently, they contend, sociologists should not overlook the complex relationship between genetic heritage and the human capacity for creating social life (Lopreato 1990; Weingart et al. 1997; Konner 1999; Pinker 2002; Ridley 2003).

1. A genetically inherited, complex pattern of behavior that always appears among members of a particular species under appropriate environmental conditions is a(n)

   a. reflex.

   b. instinct.

   c. drive.

   d. need.

2. Scientists agree that humans have instincts for self-preservation, motherhood, and war. T or F?

3. Indicate which of the following are drives (D), which are reflexes (R), which are instincts (I), and which are human creations (H).

   ____ a. eye blinking in dust storm

   ____ b. need for sleep

   ____ c. reaction to a loud noise

   ____ d. socialism

   ____ e. sex

   ____ f. racial inequality

4. According to sociology, culture totally determines the nature of human behavior. T or F?

5. _____ is the study of the biological basis of human behavior.

6. Sociologists now agree that genetic heritage plays no part in the shaping and limiting of social life. T or F?

Answers: 1. b; 2. F; 3. a. (R), b. (D), c. (R), d. (H), e. (D), f. (H); 4. F; 5. Sociobiology; 6. F

SUMMARY

1. Culture consists of all the material objects as well as the patterns of thinking, feeling, and behaving passed from generation to generation within a society. A society is a group of people living within defined territorial borders who share a culture.

2. The three broad categories of culture are the normative, cognitive, and material dimensions. The normative dimension of culture is composed of norms, sanctions, and values. Norms are rules defining appropriate and inappropriate ways of behaving. There are several types of norms. Folkways are not considered vital to a group and may be violated without significant consequences. When mores—norms considered essential—are violated, social disapproval is strong. Some norms become laws. Sanctions—positive and negative, formal and informal—are used to encourage conformity to norms. Values are broad cultural principles defining the desirable. Norms and values are not the same thing; the same value can be expressed through radically different norms. In complex societies, some values conflict.

3. Language is an important aspect of the cognitive dimension of culture. Beliefs, another important aspect of the cognitive dimension, are ideas concerning the nature of reality. Whether or not they are actually true, beliefs have a great influence on the members of a society.

4. Material culture is composed of the concrete, tangible aspects of a culture. Aspects of material culture—desks, trucks, cups, money—have no inherent meanings. Material objects have meanings only when people assign meanings to them.

5. Cultural guidelines and actual behavior do not always match. Sociologists distinguish between the cultural guidelines a society claims to accept (ideal culture) and the behavior patterns actually practiced (real culture).

6. Because humans are capable of creating and communicating arbitrary symbols such as language, they have the ability to create and transmit culture. Gestures, such as facial expressions or body movements, also carry culturally defined and shared symbolic meanings. Language is also important because it organizes people's view of reality. According to the hypothesis of linguistic relativity, people actually live in different worlds because their languages make them aware of different aspects of their environment. People are not forever trapped by their language, but they are limited by it unless they learn to see the world from another linguistic viewpoint.

7. Wide cultural variation exists from one society to another as well as within a single society. Because all humans are basically the same biologically, cultural diversity must be explained by nongenetic factors. Cultural diversity within

societies is promoted by the existence of social categories, subcultures, and countercultures.

8. Ethnocentrism—judgment of another group in relation to one's own cultural standards—is a major consequence of cultural diversity. The consequences of ethnocentrism are both positive and negative. The principle of cultural relativism—the idea that any given aspect of a culture should be evaluated according to its place within the larger cultural context of which it is a part—helps to combat ethnocentrism.

9. Some cultural universals are found in all societies. Their expression in specific practices, however, varies widely from one society to another. Cultural universals exist because of the biological similarity of humans, common limitations of the physical environment, and the common problems of sustaining social life.

10. Contrary to popular belief, humans do not have instincts. Most behavior among nonhuman animals is instinctual, but human behavior is not the sole product of genetic heritage. Human behavior is learned. Even genetically inherited reflexes and drives do not determine how humans will behave, because people are heavily influenced by culture. Although culture does not determine human nature, it does significantly condition it. Sociobiologists are now arguing for recognition of the role of biology in human behavior.

LEARNING OBJECTIVES REVIEW

- Identify the three major dimensions of culture.
- Describe and illustrate the interplay between language and culture.
- Discuss cultural diversity and its promotion within a society.
- Describe and illustrate the relationship between cultural diversity and ethnocentrism.
- Outline the advantages and disadvantages of ethnocentrism and discuss the role of cultural relativism in combating ethnocentrism.
- Explain the existence of cultural similarities that are shared around the world.
- Explain the relationship between culture and heredity.

CONCEPT REVIEW

Match the following concepts with the definitions listed below them.

_____ a. cultural universals
_____ b. sociobiology
_____ c. sanctions
_____ d. informal sanctions
_____ e. real culture
_____ f. cognition

_____ g. society
_____ h. cultural particulars
_____ i. laws
_____ j. taboos
_____ k. ethnocentrism
_____ l. mores

_____ m. subculture
_____ n. formal sanctions
_____ o. culture shock
_____ p. gestures

1. the study of the biological basis of human behavior
2. general cultural traits thought to exist in all known cultures
3. subterranean patterns for thinking, feeling, and behaving
4. rewards and punishments that may be applied by most members of a group
5. norms that are formally defined and enforced by designated persons
6. norms so strong that their violation is thought to be punishable by the group or society or even by some supernatural force
7. the widely varying, often distinctive ways societies demonstrate cultural universals
8. the psychological and social stress one may experience when confronted with a radically different cultural environment

9. the process of human thinking
10. norms of great moral significance, thought to be vital to the well-being of a society
11. rewards and punishments used to encourage desired behavior
12. a group of people who live within defined territorial borders and who share a common culture
13. rewards and punishments that may be given only by officially designated persons
14. the tendency to judge others in terms of one's own cultural standards
15. a group that is part of the dominant culture but differs from it in some important respects
16. facial expressions, posture, and body movements that carry culturally defined and shared symbolic meanings

CRITICAL-THINKING QUESTIONS

1. Distinguish the normative, cognitive, and material aspects of culture. Cite illustrations to show an understanding of the differences.

2. The ability to use language is cited by sociologists as a characteristic separating humans from nonhuman animals. Discuss the relationship between language and culture. Provide examples.

3. Functionalists and conflict theorists would be expected to have different views of countercultures. Identify these differences, and defend the position you prefer in relation to the theoretical perspective you support.

4. Discuss the relationship between cultural relativism and ethnocentrism. Give one or more examples to make the connection.

5. Sometimes you hear it said that people are just "naturally" selfish. Do you agree or disagree that humans are capable only of pursuing their own self-interest? Defend your position.

MULTIPLE-CHOICE QUESTIONS

1. **In your text, *culture* is defined as**
   a. a number of persons who share a social goal.
   b. rules defining appropriate and inappropriate ways of behaving for specific situations.
   c. the material objects as well as the patterns for thinking, feeling, and behaving that are passed from generation to generation among members of a society.
   d. the idea that any given aspect of a particular culture should be evaluated in relation to its place within the larger cultural context of which it is a part.
   e. people who live within defined territorial borders and who share common ties.

2. **This chapter distinguishes among three major dimensions of culture: the normative, the material, and the**
   a. cognitive.
   b. ecological.
   c. ethical.
   d. genetic.
   e. noncognitive.

3. _____ **are rules defining appropriate and inappropriate ways of behaving for specific situations.**
   a. Folkways
   b. Values
   c. Mores
   d. Norms
   e. Laws

4. _____ **are applied only by officially designated persons, such as judges, executioners, or college professors.**
   a. Sanctions
   b. Formal sanctions

   c. Laws
   d. Mores
   e. Statutes

5. **The example, in the text, of a doctor's introduction to nettling demonstrates that**
   a. material aspects of culture have the same meanings and uses in all cultures.
   b. it is conventional to use a 747 jet for traveling.
   c. separate physical objects existing as part of a culture may be combined to have meaning for some people and no meaning for others.
   d. newspaper and pepper are an ineffective way of delivering babies.
   e. laws regulating midwifery are unnecessary.

6. **Which of the following statements is *not* consistent with the "tool kit" view of culture?**
   a. Culture provides a range of choices to be applied in the definition and solution of the problems of living.
   b. Culture is a means for "fixing" social problems.
   c. Culture should be viewed as a group of symbols, stories, rituals, and worldviews.
   d. Because the content of any culture is not fully consistent, the symbols, stories, rituals, and worldviews that are chosen vary among individuals and situations.

7. **The Sapir-Whorf hypothesis holds that**
   a. people are forever prisoners of their language.
   b. our view of the world depends on the particular language we have learned.
   c. Germans are less concerned about privacy than are Americans.

d. exposure to another language can cause a person to become marginalized in their own culture.

e. functionalism is more plausible than conflict theory.

8. **Societies experience cultural diversity in part because of**

a. biological variation.

b. social policy.

c. social categories.

d. social determinism.

e. social disaggregation.

9. **In your text, a *counterculture* is referred to as**

a. a subculture that is deliberately and consciously opposed to certain central aspects of the dominant culture.

b. the material objects as well as the patterns for thinking, feeling, and behaving that are passed from generation to generation among members of a society.

c. general cultural traits thought to exist in all known cultures.

d. aspects of culture publicly embraced by members of a society.

e. a group that is part of the dominant culture but differs from it in some important respects.

10. **Negative consequences of extreme ethnocentrism include all of the following *except***

a. culture shock.

b. intrasocietal fragmentation.

c. intersocial conflict.

d. resistance to innovation.

e. feelings of cultural relativism.

11. **The idea that a given aspect of a particular culture should be evaluated in relation to its place within the larger cultural context of which it is a part, rather than according to some alleged universal standard that applies across all cultures, is known as**

a. cultural universalism.

b. ethnocentrism.

c. ecological adaptation.

d. cultural relativism.

e. pragmatism.

12. _____ **are general traits thought to exist in all cultures.**

a. Cultural particulars

b. Cultural universals

c. Ecological adaptations

d. Material adaptations

e. Cultural adaptations

13. **Simple, biologically inherited, automatic reactions to physical stimuli are known as**

a. reflexes.

b. impulses.

c. instincts.

d. drives.

e. needs.

14. _____ **is the study of the biological basis of human behavior.**

a. Anthropology

b. Ethnomethodology

c. Sociolinguistics

d. Sociobiology

e. Social psychoogy

FEEDBACK REVIEW

True-False

1. According to the hypothesis of linguistic relativity, words and the structure of a language cause people to live in a distinct world. T or F?

2. Sociologists now agree that genetic heritage plays no part in the shaping and limiting of social life. T or F?

Fill in the Blank

3. _____ consists of all the material objects as well as the patterns for thinking, feeling, and behaving that are passed from generation to generation among members of a society.

4. Because of _____, culture can be created and transmitted.

5. The tendency to judge other societies or groups according to one's own cultural standards is known as _____.

6. _____ is the idea that any given aspect of a particular culture should be evaluated in relation to its place within the larger cultural context of which it is a part rather than according to some alleged universal standard that applies across all cultures.

7. _____ are broad cultural principles embodying ideas about what most people in a society consider to be desirable.

Matching

8. Indicate whether these statements best reflect a belief (B), folkway (F), mos (M), law (L), or value (V).

____ a. conception that God exists

____ b. norm against cursing aloud in church

____ c. norm encouraging the eating of three meals daily

____ d. idea of progress

____ e. norm against burning a national flag

____ f. norm encouraging sleeping in a bed

____ g. norm prohibiting murder

____ h. norm against overtime parking

____ i. idea that the Earth's orbit is elliptical

____ j. idea of freedom

9. Indicate whether the following are formal sanctions (F) or informal sanctions (I).

_____ a. A mother spanks her child.

_____ b. A professor fails a student for cheating on an exam.

_____ c. A jury sentences a person to life in prison for espionage.

_____ d. A husband separates from his wife after she has an affair.

10. Indicate which of the following are social categories (SC), subcultures (S), or countercultures (C).

_____ a. Chinatown in New York City

_____ b. motorcycle gang

_____ c. Catholics

_____ d. females

_____ e. revolutionary political group

_____ f. the superrich

GRAPHIC REVIEW

The trend toward increased premarital sexual experience among teenage females in the United States is graphically illustrated in Figure 3.1. The following questions ask you to relate this trend to the normative dimension of culture.

1. During the 1950s, in the United States, what type of norm did the prohibition against premarital sex among teenage females represent?

_____

_____

_____

2. What type of norm regarding premarital sex among teenage females exists in America today?

_____

_____

_____

3. Compare the types of formal and informal sanctions used to discourage premarital sexual activity among American female teenagers during the 1950s and today.

_____

_____

_____

ANSWER KEY

Concept Review

a. 2
b. 1
c. 11
d. 4
e. 3
f. 9
g. 12
h. 7
i. 5
j. 6
k. 14
l. 10
m. 15
n. 13
o. 8
p. 16

Multiple Choice

1. c
2. a
3. d
4. b
5. c
6. b
7. b
8. c
9. a
10. e
11. d
12. b
13. a
14. d

Feedback Review

1. T
2. F
3. Culture
4. language
5. ethnocentrism
6. Cultural relativism
7. Values
8. a. B
   b. M
   c. F
   d. V
   e. M
   f. F
   g. L
   h. L
   i. B
   j. V

9. a. I
   b. F
   c. F
   d. I
10. a. S
   b. C
   c. SC
   d. SC
   e. C
   f. SC

Socialization over the Life Course

4

Ron Fehling/Masterfile

LEARNING OBJECTIVES

- Discuss the contribution of socialization to the process of human development.
- Describe the contribution of symbolic interactionism to our understanding of socialization, including the concepts of the self, the looking-glass self, significant others, and role taking.
- Compare and contrast the life course theories of Freud, Erikson, and Piaget.

- Distinguish among the concepts of desocialization, resocialization, and anticipatory socialization.
- Better understand the socialization process of young people.
- Describe the stages of adult development.
- Discuss the unique demands of socialization encountered in late adulthood.

- Compare and contrast the application of functionalism and conflict theory to the socializing effects of mass media.

Does violence on television lead to real-life violent acts? Until fairly recently, Americans strongly disagreed about the spillover effect of television violence. Moreover, social scientists hesitated to claim a causal link between television violence and actual violence. No longer. After hundreds of studies, researchers now find common ground in citing the connection between televised aggression and personal aggressiveness.

This link between televised violence and real violence is another example of a culturally transmitted human social behavior. As humans learn the culture around them, they adopt certain patterns of behavior. This learning process begins at birth and continues into old age. It is called *socialization*.

The first section of this chapter documents the importance of socialization for human development. A discussion of theoretical perspectives, with an emphasis on symbolic interactionism, then follows. The next two sections portray socialization as it unfolds throughout the life cycle. The chapter closes with an examination of mass media in light of functionalism and conflict theory.

Social Participation or Social Deprivation

According to an enduring popular belief, human nature is spoiled by civilization. Society corrupts. Jean-Jacques Rousseau's "noble savage" has been a frequent theme in romantic novels. Underlying this idea is the belief in a fundamental human nature that exists prior to social contact. Whether this is true or not, the nature of humans is shaped by socialization. Human beings at birth are helpless and without knowledge of their society's ways of thinking, feeling, and behaving. If a human infant is to learn how to participate in social life, much cultural learning has to take place.

Nearly all the human social behavior we consider natural and normal is learned. How, for example, is one supposed to behave at a cocktail party? Certainly, an individual is expected to have a drink. But what will the drink be—white wine or sparkling water? Asking seriously for a Bloody Mary with fresh cow's blood would result in signs of disgust and social disapproval. Yet, in some African tribes, a glass of cow's blood would be relished (Douglas 1979). Children in some Palestinian ref-

ugee camps aspire to becoming suicide bombers just as other children wish to be policemen or athletes. These statements from children will seem shocking to most people:

"Everyone here [residents of a refugee camp near Bethlehem] *wants to be a Shahid* [martyr], *everyone has a friend who died and went to heaven," says Salem, a genial 10-year-old. "Everything I could wish for is in paradise," says Abdullah, who is 13. "I would have green gardens and fresh fruit trees and we would have freedom. Not like here."* (Miller and Penniman 2002:49)

One source of these aspirations is the family. Palestinian mothers have been observed celebrating their children's success in blowing up other children. Their infants are sometimes outfitted as suicide bombers (Will 2002). Preference for cow's blood or admiration for a suicide martyr are, like nearly all aspects of social life, acquired through **socialization**—the process of learning to participate in group life through the acquisition of culture.

Learning about the countless aspects of social life begins at birth and continues throughout life. For example, infants in most American homes are taught to eat certain foods, to sleep at certain times, and to smile at certain facial expressions. But socialization is not limited to the early years; it is a lifelong process enabling people to fit into all kinds of social groups. Socialization must occur if freshmen are to adjust to their new environment, if plebes are to survive at West Point, if presidents of the United States are to govern, and if nursing-home residents are to adapt to their unfamiliar setting. Would-be executives who prefer warm cow's blood to dry white wine will have to keep their tastes a secret or abandon their occupational aspirations.

How is socialization related to personality? Although socialization is a lifelong process, the most important learning occurs early in life. Childhood cases of extreme social isolation reveal that without early prolonged and intensive social contact, children do not learn such basics as walking, talking, and loving. Without socialization, a human infant cannot develop a **personality**—the relatively organized complex of attitudes, beliefs, values, and behaviors associated with an individual.

The Importance of Socialization

How can the effects of socialization be assessed? An accurate assessment of socialization would require an experiment comparing a control group of normally socialized infants with an experimental group of socially isolated infants. Assuming that these two

Monkeys fail to develop psychologically and socially when deprived of their mothers. Can we generalize from monkeys to human children?

When agitated by a mechanical toy bear or a rubber snake, these infant monkeys consistently ran to their cloth mothers for security and protection.

Apparently, infant monkeys need intimacy, warmth, physical contact, and comfort. Indeed, Harlow has shown that infant monkeys raised in isolation become distressed, apathetic, withdrawn, hostile adult animals. They never exhibit normal sexual patterns; and as mothers they either reject or ignore any babies they may have, sometimes even abusing them physically.

Can we generalize from monkeys to humans? Generalizing research findings from nonhumans to humans is risky, because people are not monkeys. Nevertheless, many experts on human development believe that a human infant's emotional need for affection, intimacy, and warmth—like Harlow's monkeys—is as important as the infant's physiological need for food, water, and physical protection. Babies denied close human contact usually have difficulty forming emotional ties with others. Touching, holding, stroking, and communicating appear to be essential to normal human development. Although Harlow's findings with monkeys cannot be applied directly to humans, similar findings on human children have given them considerable credibility. According to a Lawrence Casler study (1965), for example, the developmental growth rate of institutionalized children can be improved by only twenty minutes of extra touching a day. Cases of socially isolated children provide additional support.

Social Isolation among Humans

There is dramatic evidence that children deprived of social contact do not develop all of the characteristics associated with being human. Prominent among this evidence are cases of children who have been deprived—both socially and emotionally. Fortunately, documentation includes observations of the changes that occurred when such unfortunate children were placed in an environment designed to socialize them. We will look at the histories of three children—Anna, Isabelle, and Genie—who were socially and emotionally neglected or abused (K. Davis 1947; Curtiss 1977; Pines 1981).

Who was Anna? Anna was the second child born to her unmarried mother. They lived with Anna's grandfather, who was incensed by the latest evidence that his daughter did not measure up to his strict moral code. Anna and her mother were forced to move out of the house. After several months of moving from one place to another, repeated failures at adoption, and no further alternatives, Anna and her mother returned to the grandfather's house. Anna's mother so feared that the sight of the child would anger her father that she

groups of children were biologically the same, the differences between them at the end of the experimental period could be attributed to socialization. For obvious reasons, there is no such evidence on human infants. There has, however, been some experimental research with monkeys and some nonexperimental evidence from studies of socially isolated children.

How do monkeys react to social isolation? Experiments by Harry Harlow (Harlow and Zimmerman 1959; Harlow and Harlow 1962; Harlow 1967; Blum 2002) have shown the negative effects of social isolation among rhesus monkeys. In one experiment, infant monkeys were separated from their mothers at birth and then exposed to two artificial mothers—wire dummies of the same approximate size and shape as real adult monkeys. One of the surrogate (substitute) mothers had an exposed wire body; the other was covered with soft terry cloth. Free to choose between them, the infant monkeys consistently spent more time with the soft, warm mother. Even when the exposed wire surrogate became the only source of food, the terry cloth mother remained their preference. Clearly, closeness and comfort were more important to these monkeys than food.

confined Anna to an attic-like room on the second floor of their farmhouse. Anna was kept alive on milk alone until she was discovered at the age of five. Barely alive, she was extremely emaciated and undernourished. Her legs were skeleton-like, her stomach bloated. Apparently, Anna had seldom been moved from one position to another, and her clothes and bed were filthy. Positive emotional attention was unfamiliar to her. When she was found, Anna could exhibit no signs of intelligence, nor could she walk or talk.

During the first year and a half after being found, Anna was in a county home for children. Among other things, she learned to walk, understand simple commands, eat by herself, tend to personal neatness somewhat, and recall people she had seen. But her speech was that of a one-year-old. Anna was then transferred to a school for retarded children, where she made some further progress. But still, at the age of seven, her mental age was only nineteen months, and her social maturity was that of a two-year-old. A year later she could bounce and catch a ball, participate as a follower in group activities, eat normally (although with a spoon only), attend to her toilet needs, and dress herself (except for handling buttons and snaps). Significantly, she had acquired the speech level of a two-year-old. By the time of her death at age ten, Anna had made some additional progress. She could carry out instructions, identify a few colors, build with blocks, wash her hands, brush her teeth, and try to help other children. Her developing capacity for emotional attachment was reflected in the love she had developed for a doll.

Who was Isabelle? Nine months after Anna was found, Isabelle was discovered. She, too, was the child of a single mother and was kept in isolation for fear of social disapproval. Isabelle's mother, a deaf-mute since the age of two, stayed with the child in a dark room, secluded from the rest of the family. When found at the age of six and a half, Isabelle was physically ill from an inadequate diet and lack of sunshine. Her legs were so bowed that when she stood, the soles of her shoes rested against each other, and her walk was a skittering movement. Unable to talk except for a strange croaking sound, Isabelle communicated with her mother by means of gestures. Like an animal in the wild, she reacted with fear and hostility to strangers, especially men. Some of her actions were those of a six-month-old infant.

Isabelle's first IQ score was near the zero point, and her social maturity was at the level of a two-and-a-half-year-old. Despite the belief that Isabelle was feebleminded, an intensive program of rehabilitation was begun. After a slow start, Isabelle progressed through the usual stages of learning and development, much like any normal child progressing from ages one to six. Although the pace was faster than normal, the stages

of development were in their proper order. It took her only two years to acquire the skills mastered by a normal six-year-old. By the time she was eight and a half, Isabelle was on an educational par with children her age. By outward appearances, she was an intelligent, happy, energetic child. At age fourteen, she participated in all the school activities normal for other children in her grade. To Isabelle's good fortune, she, unlike Anna, benefited from intensive instruction at the hands of trained professionals. Her ability to progress may also have been due to the presence of her mother during the period of isolation.

Who was Genie? A more recent case did not end as happily as Isabelle's. Genie, from the time she was nearly two, had been kept isolated in a locked room by her father. When she was found, at the age of thirteen, much of her behavior was subhuman. Because Genie's father severely punished her for making any vocal sounds whatsoever, she was completely silent. In fact, she neither sobbed when she cried nor spoke when in a fit of rage. Because she had hardly ever worn clothing, Genie did not notice changes in temperature. Never having been given solid food, she could not chew. Because she had spent her entire life strapped in a potty chair, Genie could not stand erect, straighten her arms or legs, or run. Her social behavior was primitive. She blew her nose on whatever was handy or into the air when nothing was available. Without asking, she would take from people things that attracted her attention.

Attempts to socialize Genie over a four-year period were not successful. At the end of efforts to rehabilitate her, Genie could not read, could speak only in short phrases, and had just begun to control some of her feelings and behavior. Genie paid a high price—her full development as a human being—for the isolation, abuse, and lack of human warmth she experienced.

Anna, Isabelle, and Genie are the most fully documented cases of socially isolated children, but other cases continue to surface. For example, six-year-old Betty Topper was kept chained to a bedpost (by a two-foot-long chain and a harness normally used on horses) since she was a year old in a home littered with trash and debris. Early reports state that when found, thirty-pound Betty was covered with her own waste and could only make moaning sounds (McNary 1999). Reports of her social development should be available in the future.

What other disruptions in social contact retard human development? The implication from case studies such as Anna, Isabelle, and Genie is unmistakable: the personal and social development associated with being human is acquired through intensive and prolonged social contact with others. But cases of extreme social isolation are not the only evidence for this generalization. Children can be affected adversely when the degree of contact with others is limited or when emotional attachments are not formed (Blum 2002).

In pioneering research, René Spitz (1946a, 1946b) compared the infants in an orphanage with those in a women's prison nursery. After two years, some of the children in the orphanage were retarded, and all were psychologically and socially underdeveloped for their age. By the age of four, a third of them had died. No such problems were observed among the prison nursery infants. Not one of them died during this period, even though the physical environment was not as clean as that of the orphanage. Spitz traced the difference between the two groups of infants to the emotional, physical, and mental stimulation offered in each setting. In the prison nursery, the infants' mothers were with them for the first year of their lives. The mothers were not present in the foundling home, and the infant-nurse ratio there was seven or eight to one. They were isolated from the other children as well. Also, in contrast to the infants in the prison nursery, the infants in the orphanage lacked the stimulation normally provided by toys. Social isolation, then, need not be extreme—as it was for Anna, Isabelle, and Genie—to cripple social and personality development.

Further research supports Spitz's conclusions. William Goldfarb (1945), for example, investigated a group of children reared in an institutionalized infant home. He compared these children, who were in the infant home from shortly after birth to three years of age, with children who had spent nearly all their lives in foster homes. He found that the institutionalized children suffered personality defects from their early years of deprivation that persisted even after they had been placed in foster homes. Compared with the children who had known only foster homes, the institutionalized children had lower IQ scores, were more emotionally and socially immature, had more difficulty caring about others, and were more passive and apathetic.

Lytt Gardner (1972) attributed a condition known as *deprivation dwarfism* to emotional deprivation. He conducted an intensive study of six "thin dwarfs"—children who were underweight and short for their age and who had retarded skeletal growth. Gardner found that these children had come from families missing the normal emotional attachment between parents and children. When such children were removed from their hostile family environments, they began to grow physically; on returning home, their growth was stunted once again. According to Gardner, deprivation dwarfism is a concrete example—an experiment of nature, so to speak—demonstrating "the delicacy, complexity and crucial importance of infant-parent interactions" (Gardner 1972).

In another study, some infants adopted from a Lebanese orphanage were compared with children who remained in the orphanage. Initially, all of these children existed in stark cribs, lying on their backs in a barren room. Human touch was limited to diaper changing. After one year, the children's intellectual, motor, and social development was on a six-month-old level. Those children left in the orphanage continued to be underdeveloped, whereas the adopted infants made many developmental strides. In another orphanage study, some preschool-age children (average IQ 64) were placed in an institution for retarded adults, where each was "adopted" by an older woman and received much attention from patients and staff members. Children who remained in the orphanage (or another orphanage) lost an average of 20 IQ points over three years. An average gain of 28 points was experienced among the "adopted" orphans (Ornstein and Ehrlich 2000). A more recent study also documents the relationship between early childhood social contact and IQ (Carey 2007). In this study, children raised in orphanages scored an average IQ eight points below children placed in foster homes in their early years. Children reared from birth by their own biological parents scored, on average, 28 points above children from foster homes.

The process of socialization, then, permits us to develop the basic characteristics we associate with being human. It is also through socialization that we learn culture and learn how to participate in society.

FEEDBACK

1. _____ is the process through which people learn to participate in group life through the acquisition of culture.
2. According to sociologists, no fundamental human nature exists prior to social contact. T or F?
3. Thanks to recent breakthroughs, research findings on the need of infant monkeys for warmth and affection can easily be applied to humans. T or F?
4. The cases of Anna, Isabelle, and Genie indicate that the personal and social development associated with being human is acquired through intensive and prolonged social contact with others. T or F?
5. Social isolation need not be extreme to damage social and personality development. T or F?

Answers: 1. Socialization; 2. T; 3. F; 4. T; 5. T

Theoretical Perspectives and Socialization

Functionalism, Conflict Theory, and Symbolic Interactionism

How does each theoretical perspective view the socialization process? Each of the three major theoretical perspectives sheds light on the processes of socialization, although the symbolic interactionist perspective facilitates a more complete view than the other two. The contribution of functionalism is more implied than explicit. That is, the very concept of socialization is based on the idea that people fit into groups by virtue of learning the culture of their society. People take their place in society by learning and accepting what society expects of them. The process of socialization assumes continuity and stability—as does functionalism. In fact, if it were otherwise, the existence of society would not be possible.

The conflict perspective views socialization (the learning of roles and acceptance of statuses) as a way of perpetuating the status quo. When people are socialized to accept their family's social class, for example, they help perpetuate the existing class structure. People are socialized into accepting their social fate before they have enough self-awareness to realize what is happening. Once social class socialization has taken place, it is very difficult to overcome. Consequently, socialization maintains the social, political, and economic advantages of the higher social classes. People who do not challenge their lot in life can never mount a revolution against the class structure.

The contribution of symbolic interactionism to our understanding of the socialization process is more precise than either functionalism or conflict theory. Symbolic interactionism helps us appreciate the subtleties of socialization: the development of the self-concept, the role of symbols and language in interpreting the social environment, the process of learning and assuming roles, and the social antecedents of human nature (Hewitt 2002).

Symbolic Interactionism and Socialization

As indicated in Chapter 1, Adam Smith, in the eighteenth century, laid the foundation for symbolic interactionism. This passage from Smith's *The Theory of Moral Sentiments* suggests the nature of symbolic interactionism:

> When I endeavor to examine my own conduct . . . and either to approve or condemn it . . . , I divide myself, as it were, into two persons. . . . The first is the spectator, whose sentiments with regard to my own conduct I endeavor to enter into, by placing myself in his situation, and by

considering how it would appear to me, when seen from that particular point of view. The second is the agent, the person whom I properly call myself, and of whose conduct. . . . I was endeavoring to form some opinion. The first is the judge; the second the person judged of. (Smith 1976/1759:113)

Charles Horton Cooley (1864–1929) and George Herbert Mead (1863–1931) were the originators of symbolic interactionism. At the turn of the twentieth century they challenged the belief, prominent in their day, that human nature is biologically determined. To them, like Adam Smith, human nature is a social product (Herman 2001).

Symbolic interactionism, when applied to socialization, involves a number of key concepts, including the self-concept, the looking-glass self, significant others, role taking (the imitation stage, the play stage, the game stage), and the generalized other.

What is the looking-glass self? From watching his own children at play, Cooley formulated insights about the development of the **self-concept**—an image of oneself as an entity separate from other people—that still stand today. Cooley and Schubert (1998) noted the many ways that children interpret the reactions of others toward them. From such insights, children learn to judge themselves in relation to how they imagine others will react to them. Thus, others serve as mirrors for the development of the self.

Cooley called this way of learning the **looking-glass self**—a self-concept based on our perception of others' judgments of us. We use others as mirrors reflecting back our imagined reactions of them to us. According to Cooley, the looking-glass self is the product of a three-stage process that is constantly taking place. First, we imagine how we appear to others. Next, we imagine the reaction of others to our imagined appearance. Finally, we evaluate ourselves according to how we imagine others have judged us. The result of this process is a positive or negative self-evaluation. Of course, this is not a conscious process, and the three stages occur in rapid succession in any given instance.

Suppose that you have a new professor you want to impress. To do so, you prepare especially well for the next day's class. As you participate in the class discussion the next day, you have an image of your performance (stage 1). After finishing your comments, you think your professor is disappointed (stage 2). Because you wished your professor to be impressed, you feel bad about yourself (stage 3).

How can the looking glass be distorted? Because the looking-glass self comes from our imagination, the mirrors we use may be distorted. The looking glass may not accurately reflect others' opinions of us. In

Others serve as mirrors for the development of our self-concepts. The looking-glass self is based on our perceptions of others' judgments of us. As these children play, they are building their self-concepts through imagining the judgments of the other.

Maxim Slugin/Used under license from Shutterstock

the preceding example, your professor may have been so impressed that he could not show his genuine reaction. If so, you misread her lack of expression and silence. Parents, to take another example, may punish a child for some deed that, in fact, endears the child to them. A child sent to his room for using obscene language may not know how hard it was for his parents to conceal their amusement. An anorexic teenager may continue to threaten her health by not eating because she believes that others see her as fat.

Although we may misinterpret others' perceptions of us, this does not diminish the effectiveness of the looking-glass process. Even though we incorrectly believe that a professor dislikes us, the consequences to us are just as real as if our interpretation were true. E. L. Quarantelli and Joseph Cooper (1966) found the self-concept of dental students to be nearer to the evaluations they *thought* their instructors had given them than to the evaluation their instructors had actually given them. Despite the possibility of distortion in the looking glass, the relationship between self and others is well established.

Can we go beyond the looking glass? Although not denying that our self-evaluations are heavily influenced by our perception of others' evaluations of us, Viktor Gecas and Michael Schwalbe (1983) think that other factors also affect self-concept. They suggest going beyond the looking glass to consider other sources of influence on the development of self-concept.

According to Gecas and Schwalbe, the looking-glass self-orientation depicts human beings as oversocialized,

passive conformists. That is, self-concepts are assumed to be based solely on the real or imagined opinions of others. (This is ironic, because Cooley and other symbolic interactionists view individuals as active and creative.) To correct this passive view of human beings, Gecas and Schwalbe developed the argument that individuals' self-concepts are also derived from personal judgments regarding how effectively they control their environments. Individuals with power on the job, for example, tend to have higher self-esteem than powerless individuals (Kanter 1993). True, part of the response is due to social recognition (or lack of it) from others; part of this effect, however, is attributable to the presence or absence of control over one's fate. A feeling of self-efficacy, in short, contributes to self-esteem.

Are all people in our looking glass equally significant? One's self-concept is not equally influenced by everyone. As George Herbert Mead pointed out, some people are more important to us than others (Mead 1934). Those whose judgments are most important to our self-concept are called **significant others.** For a child, significant others are likely to include mother, father, grandparents, teachers, and playmates. Teenagers place heavy reliance on their peers. The variety of significant others is greater for adults, ranging from spouses, parents, and friends to ministers and employers.

What is role taking? Because humans have language and the capacity for thinking, we can carry on silent conversations. That is, we can think something to ourselves and respond internally to it. This facility is crucial for anticipating the behavior of others. Through internal conversation, we can imagine the thoughts, emotions, and behavior of others in all social situations. This ability enables us to engage in **role taking**—the process that allows us to take the viewpoint of another individual and then respond to ourselves from that imagined viewpoint.

Role taking is a cognitive process that permits us to play out scenes in our mind and anticipate what others will say or do. Thus, we avoid a trial-and-error method, which would be necessary if we could not mentally anticipate the behavior of others. If, for example, you wanted to ask your employer for a raise, and if you could not mentally put yourself in your boss's place, you would have no idea of the objections that she might raise. But by role taking her reaction, you can be ready to justify your request.

How does the ability for role taking develop? According to Mead, the ability for role taking is the product of a three-stage process: the imitation stage, the play stage, and the game stage. In the **imitation stage,** which begins around age one and a half to two years, the child imitates (without understanding) the physical and verbal behavior of a significant other. This is the first step in developing the capacity for role taking. At about the age of three or four, a young child can be seen playing at being mother, father, police officer, teacher, or astronaut. This play involves acting and thinking as a child imagines another person would. This is what Mead called the **play stage**—the stage during which children take on roles of others one at a time.

The third phase in the development of role taking Mead labeled the **game stage,** the stage in which children learn to engage in more sophisticated role taking. After a few years in the play stage, children are able to consider the roles of several people simultaneously.

Games they play involve several participants, and there are specific rules designed to ensure that the behavior of the participants fits together. All participants in a game must know what they are supposed to do and what is expected of others in the game. Imagine the confusion in a baseball game if young first-base players have not yet mastered the idea that the ball hit to a teammate infielder will usually be thrown to them. In the play stage, a child may pretend to be a first-base player one moment and pretend to be a base runner the next. In the game stage, however, first-base players who drop their glove and run to second base when an opposition player hits the ball will not remain in the game for very long. It is during the game stage that children learn to gear their behavior to the norms of the group.

When do we start acting out of principle? During the game stage, a child's self-concept, attitudes, beliefs, and values gradually come to depend less on individuals and more on generalized referents. Being an honest person is no longer merely a matter of pleasing significant others such as one's mother, father, or minister. Rather, it begins to seem wrong *in principle* to be dishonest. As this change takes place, a **generalized other**—an integrated conception of the norms, values, and beliefs of one's community or society—emerges.

What is the self? According to Mead, the self is composed of two analytically separable parts: the "me" and the "I." The **"me"** is the part of the self formed through socialization. Because it is socially derived, the "me" accounts for predictability and conformity. Yet, much human behavior is spontaneous and unpredictable. An angry brother may, for example, spontaneously and unaccountably yell hurtful words at a sister he loves. Afterward, his reaction may be, "I don't know what came over me. How could I do that to a person I care

---

TABLE 4.1

FOCUS ON THEORETICAL PERSPECTIVES: Socialization

Each theoretical perspective has a unique view of the socialization process. This table identifies these views and illustrates the unique interpretation of each view with respect to the influence of the mass media on the socialization process.

Theoretical Perspective	View of Socialization	How the Media Influences Socialization
Functionalism	• Stresses how socialization contributes to a stable society	• Network television programs encourage social integration by exposing the entire society to shared beliefs, values, and norms.
Conflict theory	• Views socialization as a way for the powerful to keep things the same	• Newspaper owners and editors exercise power by setting the political agenda for a community.
Symbolic interactionism	• Holds that socialization is the major determinant of human nature	• Through words and pictures, children's books expose the young to the meaning of love, manners, and motherhood.

for?" To account for this spontaneous and unpredictable part of the self, Mead wrote of another dimension of the self—the "**I.**"

The "I" doesn't operate just in extreme situations of rage or excitement but interacts constantly with the "me" as we conduct ourselves in social situations. According to Mead, the first reaction of the self comes from the "I"; but before we act, the initial impulse is directed in socially acceptable channels by the "me." Thus, the "I" normally takes the "me" into account before acting. However, the uniqueness and unpredictability of much human behavior demonstrates that the "me" does not always control the innovative, unpredictable dimension of the self.

Psychology and Life Course Theories

Because socialization involves our entire life course and because the development of human personality is part of that life course, we will veer temporarily from sociology and present several psychologically oriented life course theories. The theories range from the psychoanalytic theory of Sigmund Freud to the psychosocial perspective of Erik Erikson and the cognitive perspective of Jean Piaget.

A Psychoanalytic Perspective

Aside from his theories of the unconscious, Sigmund Freud's (1856–1939) greatest contribution was his work regarding the influence of early childhood experiences on personality development. Experiences within the family during the first few years of life, Freud contended, largely shape future psychological and social functioning.

What is the composition of the human personality? According to Freud, the personality has three parts: the id, the ego, and the superego. The *id* is made up of biologically inherited urges, impulses, and desires. It is selfish, irrational, impulsive, antisocial, and unconscious. The id operates on the pleasure principle—the principle of having whatever feels good. Newborn infants are said to be totally controlled by the id; they want their every desire fulfilled without delay. Very early in life, members of society—usually parents—interfere with the pleasure of infants. Infants gradually learn to wait until it is time to eat, to control their bowels and bladder, and to hold their temper. In many ways, they must face the fact that others are not around merely to satisfy their impulses.

To cope with this denial of pleasure, children begin to develop an *ego*—the conscious, rational part of the personality that thinks, plans, and decides. The ego is ruled by the reality principle, which allows us to delay action until a time when the gratification of our desires is more likely. Thus, the ego mediates between the biological, unconscious impulses of the id and the denying social environment. But the ego is not itself sufficient to control the id. Nor need it be.

At about four or five years of age, the *superego*—roughly the conscience—begins to develop. It contains all the "right" and "wrong" ideas that we have learned from those close to us, particularly our parents. By incorporating their ideals into our personality, we develop what may be thought of as an internal parent. Parents are no longer the only source of punishment for wrongdoing; we punish ourselves through guilt feelings. At the same time, we feel good about ourselves when we live up to the standards contained in our superego. Through this internal monitoring system, we learn to channel our behavior in socially acceptable ways and to repress socially undesirable thoughts and actions.

What is the relationship among the id, the ego, and the superego? Freud did not see the id, ego, and superego as separate regions of the brain or as little people battling and negotiating inside our heads. Rather, he saw them as separate, interacting, and conflicting processes within the mind. The id demands satisfaction; the superego prohibits it. The ego supplies rational information in this conflict; it attempts to gain satisfaction within the limits set by the superego and the social environment. The following example, although an oversimplification, should clarify the relationship between these three parts of the personality:

> *Let's say you are sexually attracted to an acquaintance. The id clamors for immediate satisfaction of its sexual desires, but is opposed by the superego (which finds the very thought of sexual behavior shocking). The id says, "Go for it!" The superego icily replies, "Never even think that again!" And what does the ego say? The ego says, "I have a plan!"* (Coon and Mitterer 2008:369)

If the ego has difficulty controlling the id, then some sexually related experience—sexual intercourse, masturbation, even rape—may follow. Should the superego be dominant, sexual desires may be sublimated into other activities, such as working, dancing, studying, or stamp collecting. If the ego is not overwhelmed by the id or superego, it will encourage more socially acceptable behavior, such as dating or marriage.

Psychosocial Development

Psychoanalyst Erik Erikson (1902–1994) extended the human development research of his mentor, Sigmund Freud. Like his teacher, Erikson (1973) emphasized the

1. In its approach to socialization, _____ emphasizes social interaction based on symbols.
2. According to the looking-glass process, our _____ is based on how we think others judge the way we look and act.
3. Those individuals whose judgments of us are the most important to our self-concept are _____.
4. _____ is the process of mentally assuming the viewpoint of another individual and then responding to oneself from that imagined viewpoint.
5. According to George Herbert Mead, children learn to take on the roles of individuals, one at a time, during the _____ stage.
6. In Mead's theory of the self, the _____ is the unsocialized side, and the _____ is the socialized side.

Answers: 1. symbolic interactionism; 2. self-concept; 3. significant others; 4. Role taking; 5. play; 6. "I", "me"

role of the ego as the mediator between the individual and society and wrote of "crises" accompanying each stage in human development. Erikson, however, differed from Freud in his underlying assumptions and in his specific ideas on the relationship between self and socialization. To Freud, the human personality is almost totally determined in early childhood. Erikson, in contrast, believed that personality may change at any time in life. Whereas Freud attributed most human behavior to universal instincts, Erikson emphasized cultural variations. Although Freud contended that parents are the most influential forces in personality development, Erikson considered the impact of peers, friends, and spouses to be significant as well.

According to Erikson, all individuals pass through a series of eight developmental stages from infancy to old age. Each of these developmental stages involves a psychosocial crisis or developmental task (see Table 4.2). Personal identity is firmly established or impaired, depending on how successfully an individual handles each turning point. The effects of successfully or unsuccessfully meeting these crises are cumulative: successful management of a crisis at an earlier stage increases the chances of mastering later developmental tasks; those who fall behind in one stage will have increasingly greater difficulty in the following stages. Those individuals whose egos adequately meet the psychosocial crisis at each developmental stage are the most mature and the happiest and have the most stable personal identities.

Several additional points about Erikson's eight stages of psychosocial development should be kept in mind. First, Erikson does not use the word *crisis* to mean a threat or catastrophe. Rather, it is a turning point, a crucial period during which growth or maladjustment may occur, depending on how the crisis is handled. Second, although the way a crisis in one developmental period is handled affects the management of later crises, no crisis is necessarily solved permanently. Thus, a sense of trust developed early in life may, because of social experiences, turn to distrust later; and a sense of self-doubt

may, through relationships with others, later change to self-confidence. Third, as already implied, each crisis must be solved via interaction with others, and characteristics developed at earlier stages of development (trust, autonomy, industry, and so forth) require the support of others if they are to be maintained. Fourth, each crisis is solved within a particular social setting—family group, school, peer group, neighborhood, or workplace.

Cognitive Development

Jean Piaget (1896–1980) was concerned primarily with the development of intelligence, or cognitive abilities—thinking, knowing, perceiving, judging, and reasoning. According to Piaget (1981; Piaget and Inhelder 1973; Piaget and Valsiner 1999), children gradually develop cognitive abilities through interaction with their social setting—through the process of socialization. Piaget contended that children are not merely passive recipients of social stimuli; they are actively engaged in interpreting their environment as they attempt to adjust to it. Cognitive ability, in Piaget's theory, advances in stages. Learning to solve problems in one stage must precede problem solving at later stages. Piaget has isolated four stages of cognitive development: the sensorimotor stage, the preoperational stage, the concrete operational stage, and the formal operational stage.

What cognitive development occurs in the sensorimotor stage? The *sensorimotor stage,* where the basis for thought is laid, begins at birth and lasts until the age of eighteen months to two years. It is called the sensorimotor stage precisely because most of its activities are associated with learning to coordinate body movements with information obtained through the senses—touching, hearing, seeing, feeling. Only gradually do children come to realize that they are an object separate from other things. Initially, they do not realize that they can cause things to happen. A baby shaking bells on its crib does not realize who has caused the sound.

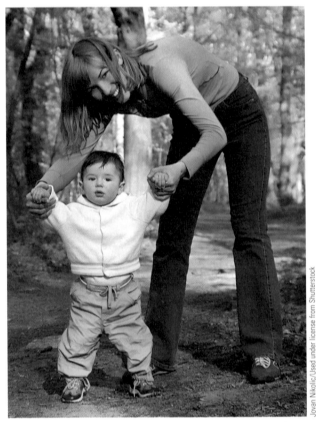

During Piaget's sensorimotor stage of development, children are learning to coordinate their bodily movements. Learning to walk is one way children learn to use their senses—touching, hearing, seeing, and feeling.

One of the most significant developments during this stage is the development of a sense of "object permanence"—a sense that objects exist even when they cannot be seen. Once this level of thought is reached, a child will, for example, pursue a ball that has rolled under a bed. Later a child can anticipate the reappearance of an object. Instead of looking into the entrance of a tunnel after an electric train has entered, a child will watch for the train's reappearance at the other end of the tunnel. During this stage, children come to see their world as an understandable and predictable place.

How do children think during the preoperational stage? Between the ages of two and seven, during the *preoperational stage,* children learn to think symbolically and to use language. Initially, they have difficulty distinguishing between what they call something (a symbol) and the object itself; an object and the symbol used to represent it are the same to them. If you throw away a milk carton that has been used as a child's toy truck, you will have a difficult time convincing the child that it was only a milk carton.

Another dominant characteristic of children during the preoperational stage is self-centeredness. Children's extreme egocentrism is clearly reflected in their inability to see things from others' points of view. Many children believe that the sun and the moon follow them when they are taking a walk. In one type of experiment, children have been placed opposite adults with a two-sided mirror between them. When asked what the adults see in their side of the mirror, the children answer "me!"

In the preoperational stage, thoughts and operations are not reversible. Consider this conversation researcher John Phillips (1975) had with a four-year-old boy:

> *"Do you have a brother?"*
> *"Yes."*
> *"What's his name?"*

TABLE 4.2

Erik Erikson's Stages of Psychosocial Development

Examine Erik Erikson's eight stages of psychosocial development. Do any or all of these stages ring true in your life? Explain.

Theoretical Perspective	Crisis	Favorable Outcome
1. First year of life	• Trust versus mistrust	• Faith in the environment and in others
2. Ages two to three	• Autonomy versus shame and doubt	• Feelings of self-control and adequacy
3. Ages four to five	• Initiative versus guilt	• Ability to begin one's own activities
4. Ages six to twelve	• Industry versus inferiority	• Confidence in productive skills; learning how to work
5. Adolescence (ages twelve to eighteen)	• Identity versus role confusion	• Integrated image of oneself as a unique person
6. Early adulthood (ages eighteen to thirty-five)	• Intimacy versus isolation	• Ability to form bonds of love and friendship with others
7. Middle adulthood (ages thirty-five to sixty)	• Generativity versus stagnation	• Concern for family, society, and future generations
8. Late adulthood (over age sixty)	• Integrity versus despair	• Sense of dignity and fulfillment; willingness to face death

"Jim."

"Does Jim have a brother?"

"No."

What about the stage of concrete operations? Increasing abstractness dominates the stage of *concrete operations,* which spans the ages of seven to eleven. A child begins to think logically about time, quantity, and space; handle arithmetic operations; and place things logically in categories. In addition, children can reverse thoughts and operations during this stage. The reversibility of thought allows an older child to recognize that A-B-C is the reverse of C-B-A and to recognize that if he has a brother, then his brother also has one. A certain degree of abstract thinking is reflected in the children's ability to imagine themselves in the place of others and to gear their behavior to others.

Despite the movement toward abstract thinking that occurs during the stage of concrete operations, the child still has difficulty if concrete objects are not involved. Logical thinking is tied to action, to the observation or manipulation of concrete symbols and objects. For example, children at this stage of development can master the concept of the conservation of quantity by watching someone pour liquids into different-sized glasses. In the preoperational stage, if children see water poured from a slender glass into a wide glass, they assume that the wider glass has less water in it because the level of the water appears to be lower. During the stage of concrete operations, however, children come to realize that the amount of water is the same in both glasses. They could not grasp this idea without an understanding of concrete operations. But it is only in the next stage that completely abstract thought becomes possible.

What is unique about thinking in the stage of formal operations? A fundamental change in cognitive ability begins to occur some time after the age of eleven. Children learn to think without the aid of concrete objects and manipulations; they can begin to think in terms of abstract ideas and principles. Gradually, the adolescent can reason about hypothetical matters. For example, if you ask younger children, "What would life be like if people had wings?" they would say something like "That's silly" or "People don't have wings." But during the stage of formal operations, children become willing to speculate on such matters. Children also learn during the *formal operations* stage to consider relationships that are logical, even if they are ridiculous. If younger children are asked to complete the sentence "If day is dark, night is _____," they will argue that day is not dark. Older children will answer that night is light. It is during this stage, then, that the capacity for adult thinking is developed.

F E E D B A C K

1. Listed here are the three major parts of the personality according to Freud. Match these parts with the examples.
   ____ a. id
   ____ b. ego
   ____ c. superego
   (1) A student decides not to cheat on an examination because the professor seems to be watching the class closely.
   (2) A student does not cheat on an examination because she feels too guilty about it.
   (3) A man kills his girlfriend in a fit of jealous rage.
2. The major role of the _____ in Freud's theory of the personality is to mediate between innate impulses and the conscience.
3. Whereas Freud believed that the human personality is almost totally determined in early childhood, Erikson held that personality can change _____.
4. According to Erikson, the successful management of a psychosocial crisis at an earlier stage increases the chances of mastering later crises. This means that the effects of success or lack of success at earlier crisis points are _____.
5. Match Jean Piaget's stages of cognitive growth with the statements below them.
   ____ a. sensorimotor stage
   ____ b. concrete operational stage
   ____ c. formal operational stage
   ____ d. preoperational stage
   (1) A child thinks without the aid of concrete objects and manipulations.
   (2) A child recognizes that 3 + 2 is the same as 2 + 3.
   (3) A child learns that a cat that runs under a couch still exists.
   (4) A child believes that a box is a spaceship.
6. The ability to imagine oneself in the place of others develops during the stage of _____.

Answers: 1. a. (3), b. (1), c. (2); 2. ego; 3. at any time; 4. cumulative; 5. a. (3), b. (2), c. (1), d. (4); 6. concrete operations

Socialization and the Life Course

Desocialization, Resocialization, and Anticipatory Socialization

Over our life course, old ways of life must be abandoned, new ways adopted, and preparation made for transitions from one period of life to another. *Desocialization, resocialization,* and *anticipatory socialization,* concepts that describe these processes, are associated with symbolic interactionism (Lachman et al. 1994; Lachman and James 1997). Erving Goffman's work on "total institutions" provides an excellent, albeit extreme, illustration of these concepts.

Goffman believes that people are not always free to manage their own lives but are under the direction of others. In *Asylums,* Goffman (1961b) writes about places such as mental hospitals and prisons as **total institutions**—places in which residents are separated from the rest of society. These residents are controlled and manipulated by those in charge, the purpose being to change the residents. The first step toward change is **desocialization**—the process of relinquishing old norms, values, attitudes, and behaviors. In extreme situations, such as mental hospitals and prisons, desocialization involves an attempt to obliterate the resident's old self-concept. Along with the self-concept go the norms, values, attitudes, and behaviors associated with the personal identity the resident had upon entering the institution. This is accomplished in many ways. Replacing personal possessions with standard-issue items promotes sameness among the residents. It deprives them of the personal effects (long hair, hair brushes, baseball caps, T-shirts) that identify them as unique individuals. Serial numbers and the loss of privacy also contribute to the breakdown of past identity.

Once the self-concept has been fractured, **resocialization**—the process of learning to adopt new norms, values, attitudes, and behaviors—can begin. The staff, using an elaborate system of rewards and punishments, attempts to instill a new self-concept in the residents. Rewards for conformity to these new identities include extra food or special periods of privacy. Punishments for nonconformity involve loss of special privileges, physical punishment, or physical isolation.

The concepts of desocialization and resocialization were developed by Goffman to analyze social processes in extreme situations. They still apply to other social settings, including basic training in the U.S. Marine Corps and plebe (freshman) year at the United States Military Academy. In much less extreme form, these concepts illuminate changes in our normal life course. Desocialization and resocialization occur as a child makes the transition into the adolescent subculture, when young adults begin their occupational careers, and as the elderly move into retirement or widowhood.

Can socialization begin prior to joining a group? **Anticipatory socialization**—the process of preparing oneself for learning new norms, values, attitudes, and behaviors—does not generally occur in the extreme social settings represented by total institutions. This is because anticipatory socialization involves voluntary change. Teenagers, because they want to resemble those their own age, may willingly abandon many of the norms, values, attitudes, and behaviors learned previously. Consequently, preteens begin very early to observe the ways of teenagers, their new **reference group**—a group used to evaluate oneself and from which to acquire attitudes, values, beliefs, and norms. University seniors, normally seen on campus only in jeans and oversized sweatshirts, suddenly, as graduation nears, are wearing tailored suits and more serious expressions. Initiating their transition into the business world, they are talking with friends who have graduated as well as with company recruiters. By anticipating the new environment they are about to enter, graduating seniors are, in effect, preparing themselves for the resocialization they know awaits them (Atchley 2003).

Are these life stages culturally universal? Stages of the life course as we know them were unknown in previous societies. A "midlife" crisis could not have been imagined in preindustrial society, because that is when people died. Even as late as 1900, life expectancy in the United

This soldier in training is now part of a total institution. This drill sergeant is tearing down his previous self-concept so that a new one appropriate to combat can be developed.

SHANNON STAPLETON/CORBIS

States was just over forty-nine years (Atchley 2001). The concept of childhood did not exist in medieval society. The problems of adolescence are a relatively recent development.

Among the Truks of Micronesia, the specter of death looms at forty.

> In Truk, they rely basically on bread, fruit and fish. In order to get fish a young man has to be strong and agile; he has to be a good paddler of a canoe, a good navigator to go out on the reef and gather fish. When the Trukese reaches about forty years of age, his strength begins to decline; he also does not climb trees as well as he used to. And when his strength begins to wane, he begins to feel that his life is ebbing away and he begins to prepare himself for death. (Kübler-Ross 1986:28)

Thus, the stages in the life cycle are not defined strictly by biology or psychology. The process encompasses social and cultural factors as well (Kübler-Ross 1997a, 1997b).

Socialization of the Young

What are the major effects of the family? The child's first exposure to the world occurs within the family. Being dependent and highly impressionable, the child is virtually defenseless during the first few years of life. Through close interaction with a small number of people—none of whom the child has selected—the child learns to think and speak; internalizes norms, beliefs, and values; forms some basic attitudes; develops a capacity (or incapacity) for intimate and personal relationships; and acquires a self-image. By the time the child develops some independence and judgment, much of the socializing work of the family has been accomplished. Development does not end at age five, but a foundation for later development has been firmly established (Santrock 2002).

The impact of the family reaches far beyond its direct effects on the personal and social development of the child. Our family's social class significantly affects how others treat us and what we think of ourselves. Our family of birth largely determines our place in society. Jean Evans offers an illustration of this in the case of Johnny Rocco, a twenty-year-old living in a city slum:

> Johnny hadn't been running the streets long when the knowledge was borne in on him that being a Rocco made him "something special"; the reputation of the notorious Roccos, known to neighbors, schools, police, and welfare agencies as "chiselers, thieves, and trouble-makers" preceded him. The cop on the beat, Johnny says, always had some cynical smart crack to make. . . . Certain children were not permit-

SNAPSHOT OF AMERICA 4.1

Rates of Imprisonment

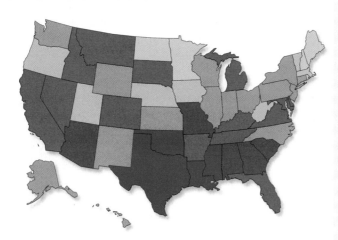

Rates of Imprisonment

- 600 or more
- 500–599
- 400–499
- 300–399
- 200–299
- 199 or less

Source: U.S. Department of Justice, Bureau of Justice Statistics, 2005.

The U.S. has one of the highest rates of imprisonment in the industrialized world—over four times that of any Western European country. Justice officials worry that some prisons function as "schools for crime." If prisons do first desocialize and then resocialize inmates toward criminal identity, then the U.S. prison system is unintentionally increasing the criminal portion of the population. This map shows the number of prisoners with sentences of more than one year per 100,000 U.S. residents.

Interpret the Map

1. Where does your state rank in terms of imprisonment rate? Can you relate the extent of imprisonment in your state to the nature of the socialization that occurs in your state?

2. Do the states adjoining your state have imprisonment rates that are similar or dissimilar to your state?

In addition to teaching academic subjects, schools offer a "hidden curriculum" – informal and unofficial aspects of culture taught to prepare children for life in the larger society. The children in this photograph are learning the values of order and waiting in turn.

Kevin Radford/Superstock

ted to play with him. Wherever he went—on the streets, in the neighborhood, settlement house, at the welfare agency's penny milk station, at school, where other Roccos had been before him—he recognized himself by a gesture, an oblique remark, a wrong laugh. (J. Evans 1954:11)

The effects of social class usually extend into adulthood. For example, in a major study of occupational attainment, Peter Blau and Otis Duncan make this observation about the relationship between one's family and subsequent work life:

> *The family into which a man is born exerts a profound influence on his career, because his occupational life is conditioned by his education, and his education depends to a considerable extent on his family.* (Blau and Duncan 1967:330)

What about school? In school, children are under the care and supervision of adults who are not relatives. The first year of school involves a transition from an environment saturated with personal relationships to an impersonal environment. Rewards and punishments are based on performance rather than affection. Although a mother may cherish any picture that her child creates, a teacher evaluates her students by more objective standards and informs those who are not meeting these standards. Slowly, children are taught to be less dependent emotionally on their parents. In addition, the school links children to the broader society. It creates feelings of loyalty and allegiance to something beyond the children's families.

The socialization process in school involves more than reading, writing, and arithmetic. Underlying the formal goals of the school is the **hidden cur-**

riculum—the subterranean informal and unofficial aspects of culture that children are taught as preparation for life in the larger society. The hidden curriculum teaches children discipline, order, cooperation, and conformity—characteristics thought to be needed for success in modern bureaucratic society, whether the child becomes a doctor, college president, secretary, assembly-line worker, or professional athlete.

Is socialization in the public school always functional? Not according to educational critic John Holt (1995). Life in schools, for example, is run by the clock. Whether or not a student understands something he or she has been working on, and whether or not the child is ready to switch to a different subject, a bell signals that all children must move to the next scheduled event. Getting through a predetermined set of activities within a given time period often becomes more important than learning. And there are rules and regulations, Holt says, to cover almost all activities—how to dress, how to wear one's hair, which side of the hall to walk on, when to speak in class. Teachers reward children with praise and acceptance when they recite the "right" answers, behave "properly," or exhibit "desirable" attitudes.

How do peer groups contribute to socialization? The **peer group**—composed of individuals of roughly the same age and interests—is the only agency of socialization that is not controlled primarily by adults. Children usually belong to several peer groups. A child may belong to groups composed of neighborhood children, schoolmates, service club members, and religiously affiliated friends.

The peer group provides children with experiences they are unlikely to capture elsewhere (Adler and Adler

1998). In the family and at school, children are subordinated to adults, but through the peer group, young people have an opportunity to engage in give-and-take relationships. Children experience conflict, competition, and cooperation. The peer group also gives children experience in self-direction. Children can begin to make their own decisions; experiment with new ways of thinking, feeling, and behaving; and engage in activities that involve self-expression.

The peer group further promotes independence from adults by introducing the child to a social milieu that is often in conflict with the adult world. Children learn to be different from their parents in ways that contribute to the development of self-sufficiency.

The capacity for intimacy is enhanced because the peer group provides an opportunity for children to develop close ties with friends outside the family, including members of the opposite sex. While children are making close friends with a few individuals, they are also learning to get along with large numbers of people, many of whom are quite different from themselves. This helps develop the social flexibility needed in a mobile, rapidly changing society.

The rise of formal education has contributed immensely to the emergence of a peer world that is not only separate from adults but also beyond their control. Children are isolated from adult society by being set apart, in school, for most of their preadult lives.

Keiko Ikeda—High School Reunions

If you asked most Americans to talk about their experience at a recent high school reunion, what would they say? "It was great seeing old friends." "I was curious about how things turned out for people I loved and hated as a teenager." "I plan to get together with some old friends in the near future." High school reunions are generally thought to be playful amusement, a time to recapture fond memories of youth.

One researcher wished to investigate the meaning of high school reunions. Keiko Ikeda (1998) studied eight reunions in the American Midwest. He observed these reunions armed with a camera, tape recorder, and notebook. After each reunion, he conducted in-depth, life-story interviews with samples of participants.

Like all in-depth observational studies, the results are too complex and varied for comprehensive reporting. One aspect of Ikeda's study,

however, reveals the cultural nature of high school reunions. He compared several reunions of one high school—tenth, fifteenth, twentieth, thirtieth, fortieth, and fiftieth. Ikeda focused on the relative emphasis on the past and the present. As you can see from the following passage, the past becomes more important as age increases:

In the earlier reunions (the tenth and fifteenth years), a concern with relative status and a sense of competitiveness is expressed, often blatantly, through award-giving ceremonies. . . . The hall was decorated in the school colors, and images of the high school mascot were present, but beyond this no high school memorabilia were displayed. The music, too, was current, and not the rock 'n' roll of the late sixties and early seventies.

The twentieth-year reunion of the Class of '62 is typical of a transitional phase in which elements from the past begin to assume an important role. The past is expressed in high school memorabilia . . . in . . . films and slides taken during high school, and in . . . high school anecdotes that are playfully interwoven throughout the ceremonial events.

In the thirtieth-year reunion of the Class of '52, the past firmly occupied center stage. A carefully crafted, chronological narrative of the senior year, entitled "The Way

We Were," was read, in which major class activities were recalled month by month. . . .

In the fiftieth-year reunion, we find a dramatic disappearance of all ritual activities. According to the president of the Class of '32, his class had held reunions every ten years since graduation, and in earlier ceremonies they had given awards, but this time, "none of the folks in the reunion committee felt like doing that kind of thing." It seemed that attendees at the fiftieth-year reunion, for the most part, had risen above concerns of past and present and were content to celebrate together the simple fact that they all still had the vigor to attend a reunion.

Source: Keiko Ikeda, *A Room Full of Mirrors* (Stanford, CA: Stanford University Press, 1998), 143–145.

Evaluate the Research

1. Ask several adults to describe the activities at one or more high school reunions they have attended. Compare their descriptions with Ikeda's findings.

2. Suppose you had a class assignment to study an upcoming reunion at your school. Select a research question you would want to ask. Identify the research methods you would use.

WORLDVIEW 4.1

Availability of Television   The mass media play a key role in the socialization process. Since nearly every U.S. home has at least one television (the majority have more), this medium is one of the most influential in the United States. This map shows that ownership of televisions varies widely around the world.

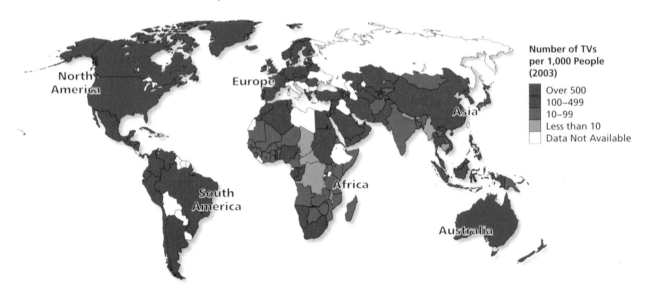

Number of TVs per 1,000 People (2003)
- Over 500
- 100–499
- 10–99
- Less than 10
- Data Not Available

Source: The World Bank, *2005 World Development Indicators*.

Interpret the Map

1. What geographical factor(s) might contribute to the density of TV households in South America?

2. Do you think the attitudes of members of societies with more televisions are influenced more by government advertising than members of societies with fewer televisions? Why or why not?

Because they are separated from the adult world for such a long time, young people are forced to depend on one another for much of their social life (*The State of Our Nation's Youth* 2001).

Who has more influence on young people—friends or family? An important factor contributing to the dependence of adolescents on each other is the distribution of the population in advanced industrial societies. Most Americans now live in either urban or suburban areas. In addition to the prevalence of the dual-employed family in American society, both parents may commute many miles to work. If so, they spend much of their time away from home. Consequently, once children reach the upper levels of grade school, they may be spending more time with their peers than they do with their parents.

Urie Bronfenbrenner (1970) notes that peer groups fill the vacuum created in the lives of children who receive an insufficient amount of attention from their parents. And according to Judith Harris (1998), peers are more important than parents in socializing children. Even though most sociologists do not agree with this author's extreme conclusion, they do believe that

the peer group is having a growing effect on social development.

What role do the mass media play in socialization? Mass media are means of communication designed to reach the general population. It is not technology per se that makes these media "mass." Newspapers existed in the American colonies, but they were limited in circulation and aimed at more local audiences (DeFleur and Ball-Rokeach 1982). Technologies become mass media only when their audiences include any average person who wishes exposure to them. Sociologists agree that the mass media are a powerful socializing agency. A primary function of the mass media is informing children about their culture (Riesman 2001).

From the mass media—television, radio, newspapers, magazines, movies, books, the Internet, tapes, and disks—children learn the behavior expected of individuals in certain social statuses. Although these popular images may be distorted (detective and police work are not as exciting and glamorous as depicted) or stereotyped, the media display role models for children to imitate (Cavender and Fishman 1998). The media often present characters in such simple, one-sided forms that it is easy to recognize behavior suitable for men, women, heroes, and villains. Learning these role models helps to integrate the young into society:

The mass media, by their content alone, teach many of the ways of the society. This is evident in the behavior we take for granted—the duties of the detective, waitress, or sheriff; the functions of the hospital, advertising agency, and police court; behavior in hotel, airplane, or cruise ship; the language of the prison, army, or courtroom; the relationship between nurses and doctors or secretaries and their bosses. Such settings and relationships are portrayed time and again in films, television shows, and comic strips; and all "teach"—however misleadingly—norms, status positions, and institutional functions. (Elkin and Handel 2001:189)

Mass media also offer children ideas (sometimes real, sometimes idealized) about values in their society: achievement and success, activity and work, democracy and equality. Mass media are also purveyors of consumerism and violence (Bagdikian 2004).

Is there a relationship between television violence and real-life violence? The only activity that American children engage in more than watching television is sleeping. Ninety-nine percent of all households in the United States have at least one television set; the average number of sets in a household is 2.24; 66 percent have three or more television sets. The average U.S. household watches 6 hours and 47 minutes daily. At 65, the average American will have spent nine years in front of a television. Any medium with which people spend so much time must contribute significantly to the socialization process.

Consider the relationship between violence on television and real-life violence. By the time an average American child completes elementary school, he will have seen 8,000 murders. At age eighteen, that same average child will have witnessed 200,000 violent acts. As noted in "Using the Sociological Imagination" at the beginning of this chapter, social scientists have, in the past, been reluctant to recognize a causal connection between media violence and real-life violence. However, after hundreds of studies involving more than 10,000 children, most now conclude that watching aggressive behavior on television significantly increases aggression (Strasburger 1995; W. Dudley, 1999; Anderson and Bushman 2002; Carter and Weaver 2003).

Researchers now generally agree that at times the effect of television is direct, concrete, and dramatic. A nine-year-old girl, for example, was gang-raped with a beer bottle by three teenage girls and a boy three days after they had seen a similar incident on a television movie. At least twenty-nine Americans have shot themselves while imitating the Russian roulette scene from *The Deer Hunter*. A two-year-old girl died when her older

At the end of elementary school it is estimated that the average American child will have viewed 200,000 violent acts. What is the relationship between media violence and real-life violence?

brother, age five, set the house on fire with matches while imitating behavior he had seen on a television show. A sixteen-year-old murdered his mother to kick off an intended killing spree modeled by the *Scream* movies. Would-be copycats followed the massacre at Virginia Tech, just on the basis of the news reports. Television's effects, of course, are usually more hidden, subtle, and long term:

[N]ot every child who watches a lot of violence or plays a lot of violent games will grow up to be violent. Other forces must converge, as they did [at Columbine]. . . . But just as every cigarette increases the chance that someday you will get lung cancer, every exposure to violence increases the chances that some day a child will behave more violently than they would otherwise. (To Establish Justice, 1999: vi)

Extreme pressure from a variety of sources has caused the major television networks to institute ratings for their programs (Farhi 1997). Most critics of violence on television have seen this self-policing move as an evasion tactic by an industry under public attack. Among other things, the critics strongly recommend the "V chip," which a parent can preset to block any program rated V-for-violent. More will be said about the effects of the mass media in the socialization process in the closing section of this chapter.

Are there other agencies of socialization during childhood and adolescence? The family, school, peer group, and mass media are the major agencies of socialization during childhood and adolescence, but they are not the only ones. Although religion does not have the same degree of influence in all societies, it can affect the moral outlook of young people, even in secular societies such as the United States. Athletic and academic teams teach young children to compete, cooperate, follow rules, make friends, and handle disagreeable peer relationships. Other potentially significant agencies of socialization are youth organizations such as the Cub Scouts, Girl Scouts, and YMCA.

Do these agencies of socialization ever conflict? Yes. In complex societies, conflict between agencies of socialization is inevitable. Some families work at cross-purposes with the schools; the church and the peer group may place conflicting demands on adolescents; parents may believe that the mass media are undermining the values they have taught their children. Conflict may even occur within a single agency of socialization.

One of a child's parents may emphasize materialism, whereas the other parent minimizes the importance of possessions. Little League baseball coaches may talk of fair play but encourage their players to win at all costs.

Early and Middle Adulthood Socialization

While early socialization lays an important foundation, the socialization process is not over at age eighteen. Although indispensable, childhood and adolescent socialization does not completely prepare us to meet the demands of adult life. If adults are to participate successfully in society, they must continue to undergo socialization (Lachman et al. 1994; Lachman and James 1997; Santrock 2002). And with the extension of the life span, the elderly confront new situations and roles (Clausen 1986; Hogan and Astone 1986; Atchley 2003; Hillier and Barrow 2006).

There are several models of adult developmental stages (van Gennep 1960; E. H. Erikson 1982; Neugarten 1968; R. Gould 1975; Levinson 1978, 1986). Despite differences in detail, it is possible to paint a general portrait of adult socialization.

What are the stages of development in adulthood? *Early adulthood* begins toward the end of the teen years and extends into the late thirties. (See Figure 4.1 for a graphic presentation of one model of the stages of development in adulthood.) This period involves a

---

FIGURE 4.1

Developmental Periods in the Eras of Early and Middle Adulthood

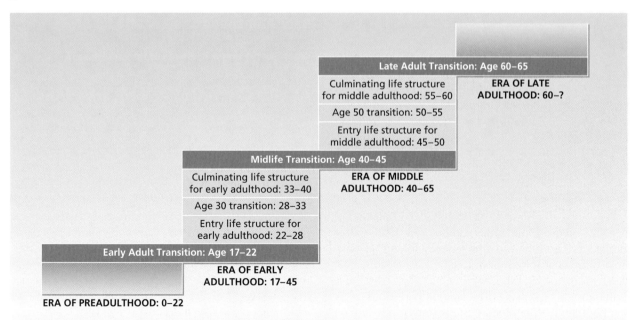

Source: From *The Seasons of a Man's Life* by Daniel J. Levinson et al. Copyright © 1978 by Daniel J. Levinson. Reprinted by permission of Alfred A. Knopf, a division of Random House, Inc.

move beyond adolescence and a preliminary step into adulthood; it ends when the individual has made a life within the adult world. Young people forge a temporary link between themselves and the adult world that involves, among other things, choosing an initial occupation and establishing a new family through marriage. Toward the end of the twenties and early thirties, some of the earlier provisional choices may be reevaluated. Marriages may be in jeopardy, divorce may be a possibility, and extramarital affairs are not uncommon. Between ages thirty and forty, a settling-down period tends to occur, during which adults, more conscious of their mortality, attempt to make a place in life for themselves and their families. Central to this period is an interest in achieving occupational success, contributing to society, and establishing a solid family life.

Next comes the transition from early to *middle adulthood*. This is the period when new questions about one's place in the world may arise. Adults may wonder what they have done with their lives and what they wish to do with the rest of their working years. Sometime during the forties, choices are made either to continue the path already taken or to establish a somewhat different life. In the case of the latter, a number of things may occur, including such drastic steps as an occupational change, a divorce, an extramarital affair, or a move to a new community. Typically, more subtle changes take place. For example, family life improves or deteriorates, work becomes more important, or thoughts turn toward retirement. Between the mid-forties and retirement age, adults complete the tasks of middle adulthood. Acceptance of one's level of occupational achievement occurs. Emphasis on success is replaced by concern with personal relationships and the "small pleasures of life," including children and grandchildren.

Does this model apply to both men and women? Research supporting the preceding model is primarily based on the male experience. Because insufficient research has been done on females, it remains to be seen how well these developmental stages fit women. Preliminary work, however, indicates that the model is less applicable to women than to men. Women have unique socialization experiences (Rosenfeld and Stark 1987).

What are some of the socialization experiences unique to women? In early adulthood, women may experience conflict between the ideal wife-mother model they learned as children and the modern roles they may now value. Men often reinforce the traditional female domestic roles by expecting their mates, who have demanding jobs outside the home, to perform most of the household work (Pendleton, Poloma, and Garland 1980; Skinner 1983; Hochschild 1997; Hochschild and Machung 2003). Many women now experience added

pressure in a society that expects more of them than it does of men.

For women who spend most of their adult lives as full-time wives and mothers, middle adulthood presents some special problems. Those women who have based their identity and self-worth on their home, husbands, and children may experience a void when grown children don't require as much attention and husbands at the peak of their careers are preoccupied with work. Also, economically dependent women who experience the death of a husband or a divorce may experience multiple shocks—declining level of living, breaking into the job market, returning to school.

Although signs of early aging—such as wrinkles, weight gains, and loss of body shape—come to men as well as women, it is females who feel the greatest negative repercussions. This is in large part due to the double standard of aging. Many of the characteristics associated with masculinity—aggressiveness, competitiveness, ambitiousness, and decisiveness—are heightened by physical maturity. In contrast, traditionally held images of femininity are diminished in midlife.

The current generation of more highly educated and occupationally successful young women will be in a much stronger position to handle the social and economic difficulties of middle adulthood. Actually, some evidence indicates that women in middle adulthood are faring better than the popular image suggests, despite the continuing pay gap between men and women (Isaacs 2007b). The loss of family responsibilities is opening new doors of opportunity for many women at the midpoint of their lives (Kendall 2008). Many women in midlife are initially entering the labor force, resuming earlier careers, or pursuing higher education.

Late Adulthood Socialization

What are the major demands of late adulthood? Length of life is closely linked with the economic base of a society. A study of the Bushmen, a hunting and gathering people, shows that 40 percent of them die before the age of fifteen. Even a century ago, the life expectancy of most people did not exceed thirty-five to forty years. In modern societies, adults can now expect to live beyond seventy. This longer life span exposes aging people to demands unique in human history. The major challenge in American society during *late adulthood* is the withdrawal from participation in certain major aspects of social life (Mitford 1998; Cavanaugh and Blanchard-Fields 2001; Benokraitis 2007). Roles are lost because statuses are lost. Two of the most important lost statuses are those of worker and spouse. One cannot act like a worker if no coworker or bosses are present; one cannot act like a wife if no husband exists.

Older people are expected to retire from work. There are many cues: workmates begin to encourage retire-

An increasingly significant proportion of Americans live into retirement years. While not all older people adjust well to separation from meaningful work roles, this couple appears to be adapting nicely to the role requirements of late adulthood.

ment; a high level of job performance is no longer expected; deserved promotions are given to younger people; rewards for work diminish; duties are reduced; demotions occur; retraining programs are not made available. Moreover, retirement is considered by most people to be the right thing to do.

Although older workers are expected to retire, they receive little preparation for life without work (Seale 1998). They are simply cut off at an arbitrary age—to be creative and active or to stagnate and be passive. They are forced to assume what has been called a "roleless status" (Shanas 1980; Hooyman and Kiyak 2007). The transition from work to retirement may carry negative consequences for social identity and self-esteem:

> *An individual may intellectually apprehend the processes of compulsory retirement policies and yet be emotionally unable to accept them as beyond control. If the older worker incorporates the negative societal stereotypes, a gradual shift in self-image will occur, one that is difficult to overcome in the retirement years. When the negative appraisal summarized in "We see you as a bumbling old fool" becomes internalized to "I am a bumbling old fool," a downward spiral is set in motion that is difficult to break.* (Hendricks and Hendricks 1987:332)

Most older people must also face the loss of a spouse. There is little or no preparation for a single life before the death of one's husband or wife, and very few single people sixty-five and older remarry, particularly women. There is, however, intense socialization immediately following the death of a wife or husband. Those who have lost their spouses are expected to be as active as they were before the death occurred. Those who resume normal functioning and establish their independence are praised, whereas those who withdraw and mourn too long are criticized. However, many members of the community—doctors, social workers, friends, ministers, relatives—are available during the mourning period to help the widowed person learn to establish an independent life. This is particularly important in a society where the elderly frequently live alone rather than with relatives. Like retirement, there is scant cultural definition of roles for a widowed person. Widowhood is another roleless status.

Why, then, do aging adults tend to be well adjusted? Gerontologists have long puzzled over the consistent research finding that a large majority of aging adults adjust relatively well in a youth-oriented society—even one that pushes them aside. To solve this puzzle, noted gerontologist Robert Atchley (2003) studied 1,271 residents of a small Midwest town of 25,000 over a twenty-year period. In the process, he developed **continuity theory** to explain how people adjust psychologically and socially to the aging process. Continuity theory presumes that most aging people maintain consistency with their past lives and use their life experiences to intentionally continue developing in self-determined channels. Because change and continuity are not mutually exclusive, Atchley contends, details can change as long as basic life themes and patterns are maintained. Moreover, adaptation via continuity can vary from one area of a person's life to another. Atchley found that the highest degree of continuity occurs in people's values, beliefs, and ideas, followed by continuity in social relationships. Not surprisingly, given their changes in status (retirement, widowhood, health), aging people tend to be less successful in maintaining continuity in activities.

There are, in fact, some signs of continuity. For example, about three quarters of baby boomers intend to continue working after retirement, building lives around both work and leisure (Freedman 2007). And in 2007 the U.S. Census Bureau reported that Americans over fifty years of age remarry at a rate similar to the population as a whole (55.8 percent for males; 43.5 percent for females).

How do older people face death? Americans are not very well socialized when it comes to handling death (Moller 1995a, 1995b). Death is simply not a subject very much discussed in the United States. Americans do not even die. They "pass on," "go to their just reward," "go to a better world," "leave us," "buy the farm," or just "go." In fact, sixty-six euphemisms for dying have been identified (DeSpelder and Strickland 1991). Americans do not like to talk about death, particularly to the dying. Relatives often attempt to keep the dying person from knowing that he or she is dying, and dying people sometimes keep their own prognosis secret (Atchley 2001; Leming and Dickinson 2002). Consequently, dying Americans do not know how to fulfill the status of a dying person. The status of a dying person is, like retirement and widowhood, a status almost devoid of roles. Dying Americans face conflicting advice. One person may advise a dying person to accept death after a long and fruitful life. Another person may offer Dylan Thomas's advice:

> *Do not go gentle into that good night,*
> *Old age should burn and rave at close of day;*
> *Rage, rage against the dying of the light.*
> (D. Jones 1971:207)

Death is handled differently in other cultures (Kübler-Ross 1997a; Seale 1998). Among Alaskan Indians, the dying participate in the timing of and planning for death; they play a very active role in their own dying and death. According to Lois Grace, a deacon of the Kawaiahao Church in Hawaii, the Hawaiian people believe that death is a constant companion, whether they are hunting, partying, or staying at home. When someone dies, a big luau among friends and relatives follows a wailing procession. Children are prominently included in the funeral so that they can learn early that death is a part of life. Among the Japanese, death is not something to fear. A dead person "goes to the pure land, the other shore, a place often described as beautifully decorated pools and silver, gold and *lapis lazuli*" (Kübler-Ross 1986:30). Memorial services held by ministers are more for the family than for the deceased.

The American aversion to the topic of death has, until lately, prevented us from gaining much knowledge about death and dying. Partly because of the gradual aging of the American population, death and dying are topics of greater interest to researchers. Consequently, we are beginning to learn how dying people face death. We know, for example, that people are more accepting of death if they are permitted to talk openly and freely about it (Martocchio 1985; Sanders 1992). Those in **hospices**—organizations designed to provide support for the dying and their families—appear to be better socialized in part because they can share their feelings with others who are dying (Mor et al. 1988; DeSpelder and Strickland 1991).

Elisabeth Kübler-Ross (1997b) has identified five stages of the terminally ill. In the first stage, *denial and isolation,* there is a refusal to accept the fact that one is

going to die; some individuals even attempt to convince themselves that they have been suddenly cured. In the second stage, *anger* is the dominant emotion. After the dying person has accepted the fact that this is actually happening to him or her, the reaction becomes "Why me?" Patients in this stage may express anger in various ways. The third stage involves *bargaining*. Most bargains are made with God and involve asking for more time. Postponement is the basis of bargaining: "If only I could see my son married" or "If only I could see my grandchild born." The fourth stage, *depression,* begins when the terminally ill patient can no longer deny the reality of dying. The predominant emotion during this stage is a deep sense of loss for oneself and for one's family. In the *acceptance,* or last, phase, the dying person is able to think about death with a degree of quiet expectation. Although this is not a happy stage, it is, as one patient expressed it to Kübler-Ross, a time for "the final rest before the long journey."

Whether these five stages apply to cultures other than the United States has yet to be determined. And as some sociologists have pointed out, even for American society, these stages are not necessarily followed in all cases. People may move back and forth among the stages and may exhibit characteristics of more than one stage at the same time. It is dangerous to view these stages in the dying process as an inevitable progression.

The Sociology of the Life Course

Some qualifications to this description of early, middle, and late adulthood need to be noted. First, as you have already seen, the model of early and middle adulthood is based primarily on samples of males. Second, although this model may describe a general pattern, not all people go through each of these stages. Some individuals handle developmental tasks in unique ways, and some skip periods of growth and change. Third, the ages given for each period are only approximate; individuals may enter stages at different ages. Although these qualifications do nothing to diminish the value of examining adulthood from a developmental perspective, they do alert us to the existence of sociological factors that differentially affect the socialization experiences of both women and men. Thus, complete reliance on this model for predicting what will happen to people and how they will behave masks variations traceable to sociological factors that need to be discussed (Rossi 1974; George 1993). The model omits, for example, important social influences of gender, social class, race, and ethnicity. For purposes of illustration, let's consider the effects of gender and social class on retirement and widowhood, the two major late adulthood adjustments.

How does social class influence retirement? **Social class** refers to a segment of a population whose members have a relatively similar share of society's desirable things and who share attitudes, values, norms, and an identifiable lifestyle. (Social class is explored in detail in Chapter 8, "Social Stratification.") Two desirable things associated with social class are economic resources and occupational status. Each of these sociological factors affects the experience of retirement.

Financial security is one of the most important determinants of satisfaction with life among retirees (Atchley and Miller 1983; Calasanti 1988). Occupational status also ranks high as a predictor of satisfaction with retirement. First, retirees who held higher-status jobs are generally more financially secure than those with lower-status jobs. Second, higher-status occupations have other characteristics promoting better adjustment to retirement. For example, higher-status occupations involve more interpersonal contact and greater participation in a variety of structured activities. The social skills and the more positive attitudes toward group activities can be transferred into more satisfying non-work retirement activities (Kohn and Schooler 1983). One positive consequence is less social isolation and loneliness.

How do gender differences affect widowhood? Widowhood negatively influences the quality of life of both elderly men and elderly women. Gender, however, appears to exercise differential effects. The major sociological factors associated with gender are the degree of financial security (gender and social class interact) and the extent of social involvement.

Adjustment to widowhood appears to be related to financial resources (Brubaker 1991). And older widows are usually at a disadvantage financially, especially if they have been economically dependent on their husband and are not covered by their husband's pension plan (Choi 1992). The situation is even worse for economically dependent women who are under sixty years of age, because they do not qualify for Social Security benefits. Even if they qualify for Social Security benefits, they live well below the poverty line (Watterson 1990). Thus, poverty or near poverty is the fate of a substantial proportion of widows in the United States (S. Thomas 1994).

Although women are at a disadvantage in adjusting to widowhood financially, they benefit from their greater capabilities for emotional expression and for attaining and maintaining a network of social support involving friends, relatives, and other widows. Widowers, on the other hand, are more socially isolated, have more difficulty displaying grief, express greater loneliness, and recover more slowly from the loss of their spouse (T. B. Anderson 1984; J. K. Burgess 1988; Morgan 1989).

Socialization and the Mass Media: Functionalist and Conflict Theories

Because the mass media are playing an increasingly important role in the socialization process in American society, additional coverage of them is appropriate. This coverage involves application of the functionalist and conflict theories to the mass media. Socialization through the family, schools, and religion are covered later in separate chapters.

Functions of the Mass Media

A major problem in identifying the functions of the mass media is separating their unique functions from those of other institutions that use the mass media for their own purposes. Although isolating the societal

FEEDBACK

1. The informal and unofficial things that are taught to children in school to prepare them for life in the larger society are called the _____.
2. The _____ provides children with needed experiences they are unlikely to obtain within the family.
3. Because there is almost no cultural definition of what retired and widowed persons should do, they are in what has been called a _____ status.
4. Match the following stages of dying with the accompanying statements.
   ____ a. denial and isolation   ____ c. bargaining   ____ e. acceptance
   ____ b. anger   ____ d. depression
   (1) "Why me?"
   (2) "What will Martha do without me?"
   (3) "I would be fine if I could just see Julia graduate."
   (4) "I am going to be fine, you'll see; just leave me alone for a while."
   (5) "My marriage has been a good one. There is little left for me to do."

Answers: 1. hidden curriculum; 2. peer group; 3. roleless; 4. a. (4), b. (1), c. (3), d. (2), e. (5)

functions of the mass media is difficult, some successful attempts have been made (Lasswell 1948; C. R. Wright 1960, 1974; McQuail 1987).

What are the functions of the mass media? The media have consequences for societies, groups, and individuals. We focus here only on societal consequences:

1. *The mass media provide valuable information regarding events inside and outside a society.* Warnings of natural disasters or military invasions are delivered by the mass media. The media also provide data regarding political candidates, stock market trends, or consumer products. In addition, by publicizing information about deviant behavior, the media enhance social control.

2. *The mass media promote social continuity and integration.* The mass media expose an entire population to the society's dominant beliefs, values, and norms. The media socialize the young, continue its process of socialization into adulthood, and transmit new cultural developments to the population.

3. *The mass media provide entertainment.* By providing amusement, diversion, and relaxation, the media contribute to the reduction of social tension. People are better able to perform their social functions, and society is more stable.

4. *The mass media explain and interpret meanings of events and information.* This process tends to support established cultural forms and to enhance the building of consensus among members of a society. Obviously, the media are involved in the process of socialization as they attempt to translate meanings of events and information for the population.

5. *The mass media can help mobilize society.* The mass media can galvanize a society for war, motivate great numbers of people to join a humanitarian movement, or move citizens to protest government policies.

The functions just outlined are positive ones. Functional analysis, as you know, also involves the identification of negative consequences, or dysfunctions.

What are some dysfunctions of the mass media? The media may foster panic while delivering information; increase social conformity; legitimate the status quo and impede social change while promoting social continuity and integration; divert the public from serious long-term social issues through a constant menu of trivial entertainment and short-term emphasis; shape public views through editorializing as they "interpret" the meanings of events and information; and create violence via public mobilization (Rosenwein 2000).

Conflict Theory and the Media

As you have come to expect, conflict theory underscores more fundamental drawbacks to the mass media. In this section we examine the mass media from the conflict perspective, beginning with the neo-Marxian model and concluding with the power elite model.

What is the Marxian view of the media in America? Marxists consider the media to be monopolized by the ruling class for the purposes of earning a profit and bolstering their power. In the process, workers are exploited by being paid less than they deserve, consumers are overcharged, and the ruling class receives excessive profits. They view the media as necessary to disseminate the **ideology** of the ruling class (i.e., the set of ideas they use to justify and defend their interests and actions) and to reduce the impact of competing ideas. By controlling the marketplace of ideas, the ruling class can dilute the development of class consciousness and subsequent political action on the part of the workers. The media, then, are seen as a tool of manipulation by which the ruling class maintains its power.

There is a non-Marxian theory that also sees a few elite individuals in control of American society and, of

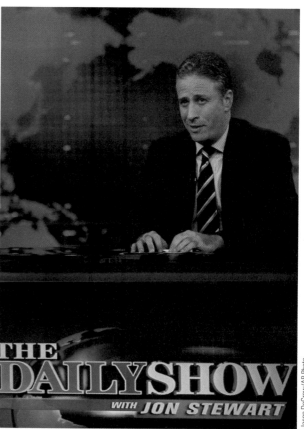

The mass media serve several functions in modern American society. How does the *Daily Show* with Jon Stewart comply with these functions?

MySpace and Personal Identity

MySpace is now competitive with Facebook, ranking as the third most popular website in the United States. It is estimated that MySpace has more than 300 million accounts, adding more than 200,000 registrants daily. Since its launching in 1999, MySpace has been a presence for today's college students, and promises to be even more influential in the future. Consequently, its place in the socialization of America's youth merits examination.

As just discussed, Erik Erikson depicted personality development as the result of working through a series of successive crises throughout life. If successful in meeting these crises, a teenager learns trust, self-control, self-confidence, and independence. At adolescence, the development of personal identity moves from the family to the outside world. By age eleven or twelve,

a preteen spends about 40 percent of her time with friends. Teens set out to find their place with peers.

It is true: it is important that teens present themselves to others in face-to-face interaction. However, MySpace can be an important vehicle for identity development because it provides a unique place to present oneself to others: here is what I look like; these are the things I like and dislike; this is what I value. Because writing requires a fuller picture of one's identity, MySpace is a unique venue for a more detailed and comprehensive self-portrait to present to peers.

A MySpace Web page encourages teenagers to more freely give and receive feedback on personal attributes. They learn to evaluate the appropriateness of attitudes, beliefs, values, and ways of acting by seeing how others react to them and their friends. Communication via MySpace provides what seems to be a safe place to try out things one might not risk in person. This guarantees feedback often missed in face-to-face interaction.

Another use of MySpace in identity formation is the creation of a shadow Web page, a Web page under a fake name. A shadow Web page may contain information about an individual that is considered off-limits to friends. Rosen suggests that a shadow website may be an aid in handling Erikson's adolescent crisis of identity versus role confusion:

Consider, for example, the teen who is struggling with her sexuality, political views, or any other topic that might cause distress or embarrassment among her friends if posted on her [public] MySpace page. The shadow page allows free expression . . . and the possibility of experimentation without embarrassment or consequences. (Rosen 2007:79)

While MySpace can serve these functions, it can also have dysfunctions. Cyber bullying and sexual abuse are just two examples.

Think About It

1. Does your experience approximate the description just given? Explain. What about your friends?
2. Do you think that the dysfunctions associated with MySpace outweigh the benefits? Develop your answer.

course, in control of the mass media. This is the theory of the **power elite**—a unified coalition of top military, corporate, and government leaders (the executive branch in particular). The power elite theory is covered in greater depth in Chapter 14 ("Political and Economic Institutions"). For now, we will concentrate on this theory as it applies to the mass media.

Do members of the power elite control the mass media? Thomas Dye (2001) explores three lines of indirect evidence: the organizational concentration of the media, the agenda-setting power of the media, and the media's ability to socialize the population.

The mass media are increasingly becoming concentrated in fewer corporate hands. Likening the mass

media to a near totalitarian state, Ben Bagdikian (2004) documents the decline, since 1983, in the number of corporations controlling most of America's print publications, television, radio, and movies—from fifty to five. Despite the growth of cable television, American television is still controlled primarily by three major networks: the American Broadcasting Companies (ABC), Columbia Broadcasting System (CBS), and the National Broadcasting Company (NBC). This control is still intact despite the recent competition of the Fox network, a relatively new network making some bold moves to become a big player. Most of America's local television stations must affiliate with one of these four networks because they cannot afford to produce all of their own programming. Each of the networks is permitted to own local stations, a combination that covers more than one-third of households in the United States. Some ground has been lost to cable television—more than 69 percent of all households in the United States are cable subscribers. The four networks still hold just more than 40 percent of all television viewing. Moreover, all four networks are now part of larger corporations, a move that greatly magnifies their power. ABC, for example, is now part of The Walt Disney Company, a corporation that owns, in addition to its famous movie studio and theme parks, the ABC television network; eight television stations; six radio networks; nineteen radio stations; four daily and thirty-five weekly newspapers; numerous magazine and book publications; and such cable television holdings as ESPN (an all-sports network), the Arts and Entertainment Network, and Lifetime Cable Network (*Directory of Corporate Affiliations 2007*).

Of the four networks, Fox is the only one located outside the United States. News Corp. Ltd, an Australian-based corporation, in 2000 also owned Fox Television Stations, HarperCollins Publishers, *TV Guide,* the *Boston Herald,* News American Publishing Company, News Ltd. of Australia, and Twentieth Century-Fox Film Corporation (*Directory of Corporate Affiliations* 2007). News Corp. Ltd.'s principal activities are the printing and publishing of newspapers, magazines, and books; commercial printing; television broadcasting; film production and distribution; and motion picture studio operations.

Newspapers and radio stations receive most of their national and international news from two wire services: the Associated Press (AP) and United Press International (UPI). Newspaper ownership is increasingly becoming monopolized as major newspaper chains absorb local papers. The New York Times Company owns not only *The New York Times* but also many other daily newspapers along with radio and television stations. Consider the concentration of power created by the merger of TimeWarner and AOL:

When AOL took over TimeWarner, it also took over: Warner Brothers Pictures, Morgan Creek, New Regency, Warner Brothers Animation, a partial stake in Savoy Pictures, Little Brown & Co., Bullfinch, Back Bay, Time-Life Books, Oxmoor House, Sunset Books, Warner Books, the Book-of-the-Month Club, Warner/Chappel Music, Atlantic Records, Warner Audio Books, Elektra, Warner Brothers Records, Time-Life Music, Columbia House, a 40 percent stake in Seattle's Sub-Pop records, Time *magazine, Fortune, Life, Sports Illustrated, Vibe, People, Entertainment Weekly, Money, In Style, Martha Stewart Living, Sunset, Asia Week, Parenting, Weight Watchers, Cooking Light, DC Comics, 49 percent of the Six Flags theme parks, Movie World and Warner Brothers parks, HBO, Cinemax, Warner Brothers Television, partial ownership of Comedy Central, E!, Black Entertainment*

Television, Court TV, the Sega channel, the Home Shopping Network, Turner Broadcasting, the Atlanta Braves and Atlanta Hawks, World Championship Wrestling, Hanna-Barbera Cartoons, New Line Cinema, Fine Line Cinema, Turner Classic Movies, Turner Pictures, Castle Rock productions, CNN, CNN Headline News, CNN International, CNN/SI, CNNN Airport Network, CNNfi, CNN radio, TNT, WTBS, and the Cartoon Network. (Alterman 2003:22–23).

Finally, as of 2008, Rupert Murdock's News Corp. media empire now extends to thirty-five television stations, Fox News Radio, movie studio 20th Century Fox, *The New York Post,* and *The Wall Street Journal.*

Similar lists can be made for each of the other four media giants (*Directory of Corporate Affiliations* 2007). In 1996, the federal government loosened regulations on radio station ownership. Scooping up hundreds of local stations, six chains broadcasted to 42 percent of the national radio audience (Fonda 2003). The Federal Communications Commission (FCC) in 2003 attempted but failed to loosen media ownership restrictions, eliminating the long-standing ban on the cross-ownership of newspapers, television stations, and radio stations in the same cities. The commission ruling, rejected by a federal appeals court, would also have permitted television broadcast networks to own more stations at the local and national level, up to 45 percent of the national audience (Ahrens 2003). Continuing its effort to loosen the law, the FCC voted in 2007 to permit broadcasters in the nation's twenty largest markets to also own a newspaper.

The ability of the media to set the political agenda is also underscored by advocates of the power elite model. Through agenda setting, the media select for the public which issues are important for consideration (R. H. Turner and Killian 1987). The media can create, publicize, and dramatize an issue to the point that it becomes a topic for debate among both political leaders and the public. Or the media can downplay an issue or an event so that it disappears from public view. Although agenda setting is not accomplished only in the television arena, television is the most powerful force: "TV is the Great Legitimator. TV confers reality. Nothing happens in America, practically everyone seems to agree, until it happens on television" (W. A. Henry 1981:134). Not only can the media attempt to tell us *what* to think, as the functionalists suggest; they are also said to be extremely successful in determining what we think *about.*

Conflict theorists also underscore the media's role in socializing the population. It is through television (and increasingly the Internet) that most Americans stay in touch with affairs in this country as well as events in the rest of the world. Because television is the most widely shared experience in the United States, it has the domi-

F
E
E
D
B
A
C
K

1. The _____ are those means of communication that reach large audiences without any personal interaction between the senders and the receivers of messages.
2. Which of the following was *not* discussed in the text as a function of the mass media?
   a. provision of information
   b. creation of consensus
   c. mobilization of society
   d. promotion of social discontinuity and disintegration
3. Which of the following was *not* presented as indirect evidence for the power elite model of the mass media?
   a. foreign infiltration into the media industry
   b. organizational concentration of the media industry
   c. agenda setting
   d. socialization of the population

Answers: 1. mass media; 2. d; 3. a

nant role in portraying what the media consider to be the most important beliefs, norms, attitudes, and values for people to hold. Through entertainment programming, news coverage, and advertising, Americans are constantly being exposed to a particular view of the most desirable and appropriate ways to think, feel, and behave.

The foregoing analysis does not prove that the mass media primarily serve the interests of the elite. It does, however, suggest the potential for the media to be doing so. Conflict theorists, of course, contend that this potential is being realized.

SUMMARY

1. Socialization, the process of learning to participate in group life through the acquisition of culture, is one of the most important social processes in human society. Without it, we would not be able to participate in group life and would not develop many of the characteristics we associate with being human. Evidence of the importance of socialization has been shown in studies of isolated monkeys and children. Socially isolated primates—including humans—do not develop as we would expect.

2. Several major theories emphasize the importance of socialization for human development. Although functionalism and conflict theory bear on the socialization process, the most fully developed sociological approach to socialization comes from symbolic interactionism. Key concepts include the self-concept; the looking-glass self; significant others; role taking; the imitation, play, and game stages; and the generalized other.

3. Several major life course theories emphasize the importance of socialization for human development from a psychological viewpoint. Sigmund Freud accentuated the interaction between the biological nature of human beings and the social environment. Erik Erikson, a student of Freud's, described a series of developmental stages that occur from infancy to old age. Each of these developmental stages is accompanied by a psychosocial crisis, or developmental task. In Erikson's view, socialization and

personality development are not confined to childhood; they are lifelong processes. According to Jean Piaget, the ability to think, know, and reason develops through interaction with others. If we are to mature cognitively, Piaget contends, each of us must pass through four identifiable stages in the proper developmental sequence.

4. Symbolic interactionism contributes to our understanding that socialization is a lifelong process. The concepts of total institutions, desocialization, resocialization, anticipatory socialization, and the reference group are particularly important in this regard.

5. During childhood and adolescence, the family, school, peer group, and mass media are the major agencies of socialization. The family makes a tremendous impression on the child both because it is the first agency of socialization and because the child is dependent and highly impressionable.

6. The school is generally the first agency of socialization that is controlled by nonrelatives. In school, children are exposed to new standards of performance applied to everyone, are encouraged to develop loyalties beyond their own families, and are prepared in a number of ways for adult life. In addition, by exposing children to such skills as reading and writing and such subject matter as mathematics and English, schools train children to be disciplined, orderly, cooperative, and conforming.

R
E
V
I
E
W

G
U
I
D
E

7. The peer group is the first agency of socialization that is not controlled by adults and consequently provides young people with experiences they cannot easily obtain elsewhere. In peer groups, young people learn to deal with others as equals, gain experience in self-direction, establish some degree of independence from their parents and other adults, develop close relationships with friends of their own choosing, and learn to associate with various types of people.

8. Specific effects of the mass media are difficult to document. Still, we know that television, radio, newspapers, and other mass media play a significant part in the socialization process by introducing children to various aspects of their culture.

9. Socialization in infancy, childhood, and adolescence makes a profound contribution to personality development and one's ability to participate in social life. There is, however, an increasing recognition that socialization continues throughout life, even to old age. There is much interest now in viewing adulthood from a developmental perspective.

10. The extension of life expectancy through medical science has created problems of adjustment unparalleled in human history. At an arbitrary age, most older workers are deprived of a meaningful work life. They are forced to retire whether or not they are still productive. Just as there is little preparation for retirement, there is hardly any prior socialization for losing a husband or wife. Like retirement, widowhood is a roleless status. Despite these changes, aging adults tend to be psychologically well adjusted. Unlike people in some other cultures, Americans are not well socialized to face the prospect of dying. Despite this relative lack of socialization, Americans tend to go through several stages in the process of dying.

11. The mass media encompass those means of communication that reach large audiences without any personal interaction between the senders and the receivers of messages. The major forms of mass communication are television, radio, newspapers, magazines, movies, books, tapes, disks, and now the Internet.

12. Although the functions of the media are intertwined with other institutions, it is possible to isolate their major functions. The media provide information, promote social continuity and integration, supply entertainment, explain and interpret events and information, and can mobilize the society when necessary. Some negative consequences, or dysfunctions, are also associated with the mass media.

13. Both the neo-Marxist and power elite views have been applied to the mass media. Proponents of the power elite model point to concentration of power in the media, the agenda-setting power of the media, and the media's ability to socialize the population as at least indirect evidence of a ruling class in America.

LEARNING OBJECTIVES REVIEW

- Discuss the contribution of socialization to the process of human development.
- Describe the contribution of symbolic interactionism to our understanding of socialization, including the concepts of the self, the looking-glass self, significant others, and role taking.
- Compare and contrast the life course theories of Freud, Erikson, and Piaget.
- Distinguish among the concepts of desocialization, resocialization, and anticipatory socialization.
- Better understand the socialization process of young people.
- Describe the stages of adult development.
- Discuss the unique demands of socialization encountered in late adulthood.
- Compare and contrast the application of functionalism and conflict theory to the socializing effects of mass media.

CONCEPT REVIEW

Match the following concepts with the definitions listed below them.

____ a. significant others
____ b. ideology
____ c. looking-glass self
____ d. personality

____ e. total institution
____ f. anticipatory socialization
____ g. resocialization
____ h. hospices

____ i. social class
____ j. peer group
____ k. "I"
____ l. reference group

1. the process of preparing oneself for learning new norms, values, attitudes, and behaviors
2. organizations designed to provide support for the dying and their families
3. the spontaneous and unpredictable part of the self
4. the set of ideas used to justify and defend the interests of those in power in a society
5. one's self-concept based on perceptions of others' judgments

6. a group composed of individuals of roughly the same age and interests

7. a relatively organized complex of attitudes, beliefs, values, and behaviors associated with an individual

8. a group we use to evaluate ourselves and to acquire attitudes, beliefs, values, and norms

9. the process of learning to adopt new norms, values, attitudes, and behaviors

10. those persons whose judgments are the most important to an individual's self-concept

11. a segment of the population whose members have a relatively similar share of the desirable things and who share attitudes, values, norms, and an identifiable lifestyle

12. a place in which residents separated from the rest of society are controlled and manipulated by those in charge

CRITICAL-THINKING QUESTIONS

1. Defend the proposition that human nature is more a matter of nurture than nature. Cite evidence in your argument.

2. What do you think is the greatest contribution that symbolic interactionism has made to our understanding of the socialization process? Why?

3. You are now a college student. Have you undergone (or are you currently undergoing) desocialization, resocialization, or anticipatory socialization? Provide examples from your own experience.

4. Evaluate the statement that socialization ends after childhood. Be specific in your line of argument.

5. Discuss the pros and cons of the mass media from the viewpoint of sociologists. Use specifics in developing your thoughts.

MULTIPLE-CHOICE QUESTIONS

1. _____ is the process through which people learn to participate in group life through the acquisition of their culture.
   a. Imprinting
   b. Behavior modification
   c. Socialization
   d. Imitation
   e. Acculturation

2. The relatively organized system of attitudes, beliefs, values, and behaviors associated with an individual is known as
   a. personality.
   b. internalization.
   c. humanization.
   d. socialization.
   e. dualism.

3. René Spitz's study of infants in an orphanage and infants in a women's prison nursery found that
   a. the infants in both the orphanage and the prison nursery developed normally.
   b. social isolation need not be extreme to damage social and personality development.
   c. social isolation must be extreme to damage social and personality development.

   d. infants in both the orphanage and the women's prison died at early ages.
   e. children in prison nurseries are worse off than children in orphanages.

4. _____ allows for the greatest understanding of the socialization process.
   a. Functionalism
   b. Conflict theory
   c. Symbolic interactionism
   d. Sociobiology
   e. Game theory

5. The _____ is based on our perceptions of others' judgments of us.
   a. significant other self
   b. alternative self
   c. distorted self-concept
   d. other-directed self
   e. looking-glass self

6. The process of mentally assuming the viewpoint of another individual and then responding to oneself from that imagined viewpoint is referred to as
   a. role reversal.
   b. play acting.
   c. gaming.
   d. role taking.
   e. role distancing.

7. According to George Herbert Mead, children learn to take on the roles of other individuals, one at a time, during the _____ stage.
   a. game
   b. imitation
   c. work
   d. play
   e. oedipal

8. According to Freud, the id, ego, and superego should be thought of as
   a. separate regions of the brain.
   b. separate, interacting, and conflicting processes within the mind.
   c. a mass of impulses, urges, and desires.
   d. multiple personalities.
   e. little people negotiating inside our heads.

9. According to Erikson, all individuals pass through a series of eight developmental stages from infancy to old age, each of which involves a _____ crisis or _____ task.
   a. psychosocial; developmental
   b. sociobiological; cognitive
   c. developmental; psychopathic
   d. psychosocial; cognitive
   e. developmental; cognitive

10. Jean Piaget was concerned primarily with the development of _____, which is the ability to think, know, perceive, judge, and reason.
    a. psychosocial development
    b. cognitive ability
    c. self-concept
    d. socialization
    e. role taking

11. _____ are most likely to participate in anticipatory socialization.
    a. New prison inmates
    b. Teenagers
    c. Office workers
    d. New mental patients
    e. Cab drivers

12. A group composed of individuals of roughly the same age and interests is known as a
    a. subculture.
    b. culture cohort.
    c. peer group.
    d. resocialization group.
    e. society.

13. Because there is almost no cultural definition of what retired and widowed persons should do, they occupy what has been called a(n) _____ status.
    a. undefinable
    b. casual
    c. master
    d. limbo
    e. roleless

14. According to the Marxian view, the media in America are
    a. a threat to social conformity.
    b. the "mouthpiece" of organized labor.
    c. a means of production.
    d. a tool of manipulation by which the ruling class maintains its power.
    e. a threat to the proletarian revolution.

15. The power elite is composed of
    a. top university, corporate, and government elite.
    b. those in direct control of the mass media.
    c. top military, corporate, and government leaders.
    d. top religious, university, and corporate leaders.
    e. those who place controlling others over economic success.

FEEDBACK REVIEW

True-False

1. According to sociologists, no fundamental human nature exists prior to social contact. T or F?
2. Thanks to recent breakthroughs, research findings on the need of infant monkeys for warmth and affection can easily be applied to humans. T or F?

Fill in the Blank

3. In its approach to socialization, _____ emphasizes social interaction based on symbols.
4. According to the looking-glass process, our _____ is based on how we think others judge the way we look and act.
5. Whereas Freud believed that the human personality is almost totally determined in early childhood, Erikson held that personality can change _____.
6. According to Erikson, the successful management of a psychosocial crisis at an earlier stage increases the chances for mastering later crises. This means that the effects of success or lack of success at earlier crisis points are _____.
7. The informal and unofficial things that are taught to children in school to prepare them for life in the larger society are called the _____.
8. Because there is almost no cultural definition of what retired and widowed persons should do, they are in what has been called a _____ status.

Matching

9. Match Jean Piaget's stages of cognitive growth with the statements.
   ____ a. sensorimotor stage
   ____ b. concrete operational stage
   ____ c. formal operational stage
   ____ d. preoperational stage

(1) A child thinks without the aid of concrete objects and manipulations.

(2) A child recognizes that 3 + 2 is the same as 2 + 3.

(3) A child learns that a cat that runs under a couch still exists.

(4) A child believes that a box is a spaceship.

10. Match the following stages of dying with the accompanying statements.

_____ a. denial and isolation

_____ b. anger

_____ c. bargaining

_____ d. depression

_____ e. acceptance

(1) "Why me?"

(2) "What will Martha do without me?"

(3) "I would be fine if I could just see Julia graduate."

(4) "I am going to be fine, you'll see; just leave me alone for a while."

(5) "My marriage has been a good one. There is little left for me to do."

GRAPHIC REVIEW

Snapshot of America 4.1 shows the rate of imprisonment by state.

1. Why was a rate used as the measure rather than the total number of inmates?

_____

_____

_____

2. Why would a high number like 100,000 U.S. residents be used in the calculation?

_____

_____

_____

ANSWER KEY

Concept Review	Multiple Choice	Feedback Review
a. 11	1. c	1. T
b. 5	2. a	2. F
c. 6	3. b	3. symbolic interactionism
d. 8	4. c	4. self-concept
e. 13	5. e	5. at any time
f. 1	6. d	6. cumulative
g. 10	7. d	7. hidden curriculum
h. 3	8. b	8. roleless
i. 12	9. a	9. a. 3
j. 7	10. b	b. 2
k. 4	11. b	c. 1
l. 9	12. c	d. 4
	13. e	10. a. 4
	14. d	b. 1
	15. c	c. 3
		d. 2
		e. 5

5

Social Structure and Society

Anwar Hussein/Getty Images

Are humans genetically selfish? Many would say that, by their very nature, people are basically selfish. Evidence indicates otherwise. If humans, as a species, shared the trait of selfishness, we would not expect to find a society populated by cooperative members. Yet, the economic relationships in hunting and gathering societies are grounded in the practice of sharing. Lacking a conception of private property or ownership, members of hunting and gathering societies view even thrift as an indication of selfishness. If genetically programmed to be selfish, these people could not exist without a sense of personal ownership. They would be incapable of such communal concern.

Members of hunting and gathering societies, like members of most groups, know what is expected of them and what they can expect from others, and they engage in the same basic social patterns time after time. All groups have patterned and predictable social relationships that are passed from generation to generation. These patterned social relationships are referred to as *social structure*, which is the basic subject matter of sociology.

Social Structure and Status

Your understanding of sociology as the study of social structure will jell in this chapter. You learned in Chapter 3 that culture shapes human social behavior—that in the absence of biological preprogramming, culture has to provide the raw material for thinking, feeling, and behaving. Without culture, humans would have no blueprint for social living. This chapter will lead you through a set of sociological concepts that demonstrate the flow between culture and social structure.

In *As You Like It*, William Shakespeare wrote:

All the world's a stage. And all the men and women merely players; They have their exits and their entrances; And one man in his time plays many parts.

All members of a group have parts they are expected to play. Students are expected to attend class, listen to the instructor, fulfill class requirements, and take examinations. Professors are expected to teach class, make assignments, and grade examinations. On any American college campus, you will find similar relationships between students and faculty members. Here, each student–faculty interaction is orderly and predictable. If, however, you found yourself in a class where students shouted the professor down, the custodian presented lessons, and the students did all the teaching, you would miss the presence of structure. Without the order and predictability you expected, you would become apprehensive about the educational process in general and about relating to those unfamiliar professors. To fit in, you would need some awareness of the underlying social structure.

Fortunately, we are usually spared such confusion because groups are relatively predictable. Upon entering a new group, we bring some knowledge of the ways people normally relate to one another. In other words, in our minds, we carry a social map for various group situations. We have mental images—however unconscious and hazy—of the group structure in which we want to participate. This awareness permits us to engage in patterned social relationships within groups without personal embarrassment or social disruption. This underlying pattern of social relationships is called **social structure**.

The mental maps of social structure are not part of our genetic heritage; they must be learned from others. In the process, we learn statuses and roles—major elements of social structure.

What do sociologists mean by status? People may refer to themselves as students, doctors, welders, secretaries, mothers, or sons. Each of these labels refers to a **status**—a position a person occupies within a social structure. Status helps us define who and what we are in relation to others within the same social structure. Some social statuses are acquired at birth. For example, a newly born female instantly becomes a child and a daughter. From then on, she assumes an increasingly larger number and variety of statuses.

How do people acquire social statuses? There are two basic types of social statuses—ascribed and achieved statuses. An **ascribed status** is neither earned nor chosen; it is assigned to us. At birth, an infant is either a male or a female. Except in instances of sex-change operations, we cannot select our gender. Age is another prominent example of an ascribed social status. In some societies, religion and social class are ascribed by the family of birth. If you were born into a lower-class home in India, for example, you would not be permitted to rise to a higher social class.

An **achieved status** is earned or chosen because people have some degree of control and choice. In most modern societies, for example, an individual can decide to become a spouse or a parent. Occupations are also achieved statuses in modern societies; people have latitude to choose their work. Plumber, electrician, sales representative, nurse, executive, lawyer, and doctor are achieved statuses.

Sociologists are interested in the relationships among social statuses. A sociologist investigating college athletics, for example, may focus on the status of the athlete in relation to the statuses of teammate, coach, parent, and professor (see Figure 5.1).

What is a status set? In addition to these relationships, an athlete will occupy various other statuses that may be totally unrelated to that of athletics. A **status set** is all of the statuses that an individual occupies at any particular time. One athlete may be an architect major, an artist, and a church choir member; another may be a service club leader, a motivational speaker, and a mother. Each of these statuses is part of another network of statuses. Assume, for example, that in addition to being an athlete, an individual is also a part-time jazz musician. In this status, she would interact with the statuses of nightclub owner, dancer, and fellow musician, among others.

Are all of a person's statuses equal? Some statuses are more important than others. **Master statuses** are important because they influence most other aspects of a person's life. Master statuses may be achieved or ascribed. In industrial societies, occupations—achieved for the most part—are master statuses. Our occupation strongly influences such matters as where we live, how well we live, and how long we live. "Criminal" can be an achieved master status because its effects permeate the rest of that person's life.

Eileen M. Collins, commander of the 2005 space shuttle Discovery, was the first woman to command a U. S. space shuttle. Her master status of astronaut is achieved; her master status of woman is ascribed.

FIGURE 5.1

The Interrelationships of Social Statuses

Social statuses cannot exist in isolation. Without other statuses (parent, administrator, referee), the status of college athlete would have no meaning. All statuses, then, are interrelated with other statuses.

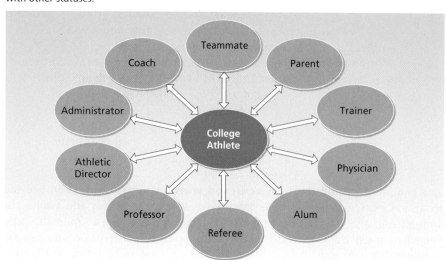

1. _____ refers to the patterned relationships among individuals and groups.
2. A _____ is a position a person occupies within a group.
3. Match the type of status with the examples that best illustrate it.
   ____ a. ascribed status
   ____ b. achieved status
   ____ c. master status
   ____ d. status set
   (1) basketball coach, mother, author, daughter, professor
   (2) occupation, gender, race, ethnicity
   (3) president of the United States
   (4) sex

Answers: 1. Social structure; 2. status; 3. a. (4), b. (3), c. (2), d. (1)

Ascribed master statuses are no less important to a person's life than achieved master statuses. A person who acquires an immunity deficiency through a blood transfusion in a hospital becomes an AIDS victim who will likely suffer prejudice and discrimination in employment, housing, and social relationships. The physically handicapped often have similar experiences.

Age, gender, race, and ethnicity are also ascribed master statuses because they significantly affect the likelihood of achieving other social statuses. Age categories, gender, and various racial and ethnic minorities do not normally constitute social classes because all social classes contain males, females, young people, retired persons, racial categories, and ethnic people. But these statuses are especially influential in social class placement because they affect other social statuses that a person may occupy. When will the United States have a female president? Would you let a nineteen-year-old or a ninety-year-old handle your case in court—or remove your appendix?

Social statuses are similar to the parts performers play on the stage. Prostitute, pimp, police officer, and judge are parts that individuals play in real life. The behavior of an individual depends largely on the part, or status, that individual holds. This is because parts in real life are based on culturally defined roles.

Social Structure and Roles

Roles

What are roles? **Roles** are the culturally defined rights and obligations attached to a status; they indicate the behavior expected of an individual holding that particular status. Any status carries with it a variety of roles. For example, the roles of a modern doctor include keeping abreast of new medical developments, scheduling office appointments, diagnosing illnesses, and prescribing treatment.

Roles can be thought of as the glue that holds a network of social statuses together; the roles of one status are matched with the roles of other statuses through rights and obligations. **Rights** inform individuals of behavior they can expect from others. **Obligations** inform individuals of the behavior others expect from them. The rights of one status correspond to the obligations of another. Doctors, for example, are obligated to diagnose their patients' illnesses. Correspondingly, patients have the right to expect a diagnosis. Patients have an obligation to keep their office appointments, and doctors have a right to expect that they will do so.

To continue the stage metaphor, roles are the script that indicates to the actors (status holders) what beliefs, feelings, and actions are expected of them (see Table 5.1). Just as a playwright or screenwriter specifies the content of a performer's part, culture determines the parts played in real life. Mothers, for instance, follow dissimilar maternal "scripts" in different cultures. Most American mothers tend to emphasize independence more than most German mothers do.

As noted in Chapter 1, Erving Goffman, with his dramaturgical approach, deserves the most credit for developing the analogy between the stage and social life. In a sense, Goffman's life's work was devoted to the distinction between culturally prescribed roles and the fulfillment of those roles.

Role Performance and Social Interaction

How is role performance tied to social interaction? Statuses and roles provide the basis for group life. But roles are activated only when people in statuses engage in role performance. This occurs through the process of social interaction.

Role performance is the actual conduct, or behavior, involved in activating a role. Although some role performance can occur in isolation, as when a student studies alone for an examination, most of it involves social interaction. **Social interaction** is the process of two or more persons influencing each other's behavior. For example, before two boys begin to fight, they have probably gone through a process of insult, counter-insult, and challenge. Fortunately, most social interaction is not as negative and violent, but the same process of influence and reaction is involved (Turner 2002c).

Not all social interaction follows expected role performance. Spectators at a professional golf tournament are not supposed to cheer bad shots or boo good ones. When some American fans at the 1999 Ryder Cup competition did both, they insulted European players and fans.

If statuses are analogous to the parts of a play and roles are the script, then social interaction represents the way actors respond to cues given by other actors, and role performance is the actual performance. Table 5.1 illustrates this metaphor.

Relationship between Culture and Social Structure

How are culture and social structure related? Figure 5.2 outlines the links that connect culture and social structure. Starting at the top of this figure, the first major bond between culture and social structure is the concept of *role*—culturally defined rights and obligations attached to a social status. The second link is a *status*—a position that a person occupies within a group. It is through the roles attached to each status that culture enters the picture. Yet people do not always follow roles exactly. The manner in which roles are actually carried out *is role performance*, the third link in the conceptual chain. Role performance occurs through *social interaction*. This is the fourth link between culture and social structure. Social interaction based on roles is observable as patterned relationships, which constitute social structure. In turn, existing social structure affects the creation of—and changes in—culture. (Note the two-headed arrow in Figure 5.2.)

TABLE 5.1

The Stage Analogy

This table draws an analogy between rehearsed behavior on the stage and real social behavior. In your own words, describe the parallel between the sociological concepts on the right and the stage terms on the left.

Stage	Social Life
Parts	Statuses
Script (lines)	Roles
Cues	Social interaction
Actual performances	Role performance

Are all these terms necessary? Concepts are as necessary for sociology as they are for physics. Both sociologists and physicists deal with abstractions, with unobservable phenomena whose existence must be assumed from the observation of things that can be seen. Before 1970, physicists had never directly observed individual atoms. Instead, they used indirect methods, such as photographs of X rays beamed at a small crystal of rock salt, to infer the existence of atoms. Just as physicists have used indirect methods, sociologists use the observation of patterned relationships (something concrete) to infer the existence of social structure (something abstract). If the concepts and relationships outlined in Figure 5.2 were not known, it would be as difficult for sociologists to establish the existence of social structure as it would have been for the ancient Greeks to document the existence of atoms. Concepts and their relationships are the necessary building blocks of any area of scientific study.

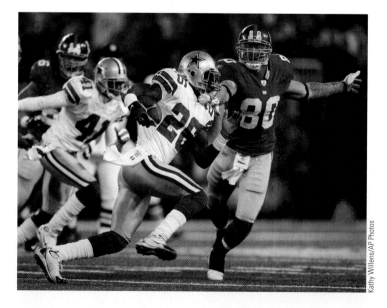

Each of these Dallas Cowboys and New York Giants professional football players have particular roles and statuses on their teams.

Kathy Willens/AP Photos

FIGURE 5.2

The Links Between Culture and Social Structure

Sociologists concentrate on the study of social structure. They have developed a set of concepts and a conception of their relationship in order to understand the basic nature of social structure.

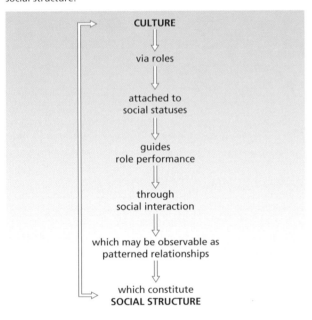

How does social interaction differ from playacting? Although the stage metaphor is a valid one, there is danger in taking it literally. First, "delivery of the lines" in life is not the conscious process practiced by stage performers. Most real-life role performance occurs without much forethought because we act in ways that have been unconsciously adopted through observation, imitation, and socialization.

Second, there is more of a discrepancy between roles and role performance in real life than between a stage script and its performance. Although performers may sometimes ad-lib, change lines to suit themselves, or introduce a little "business," overall they adhere to the script. Differences between a role and role performance in real life are neither as easy to detect nor as easy to control as departures from a script.

Keep in mind, however, that the range of acceptable responses is not limitless. Only certain responses are considered culturally legitimate. It is not an appropriate response for the professor to bodily eject the student from her office, and the student would be foolish to pound the professor's desk in protest.

Role Conflict and Role Strain

The existence of statuses and roles permits social life to be predictable, orderly, and recurrent. At the same time,

SNAPSHOT OF AMERICA 5.1

Unemployment Rates

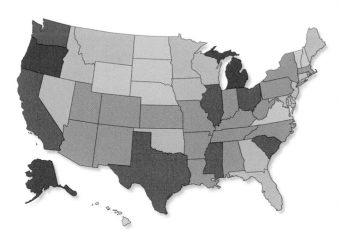

Unemployment Rate (2004)

- 7% or above
- 6.0%–6.9%
- 5.0%–5.9%
- 4.0%–4.9%
- 3.9% or below

Source: U.S. Department of Labor, Bureau of Labor Statistics, 2005.

For many Americans, occupation serves as their primary master status. Americans tend to define themselves and be defined by what they do for a living. Sometimes, however, that master status is disrupted by unemployment. Unemployment is the situation that occurs when a person does not have a job but is actively seeking one. Unemployment prevents people from fulfilling their role expectations as earner, which can also affect the performance of spousal and parental roles. This map shows the percentage of each state's civilian labor force, aged 16 and older, that was unemployed in 2004.

Interpret the Map

1. Which states had the highest unemployment rates in 2004? Which states had the lowest rates?
2. About what percentage of the civilian labor force in your state was unemployed in 2004?
3. How can unemployment affect the performance of spousal and parental roles?

Philip Zimbardo—Adopting Statuses in a Simulated Prison

Did the behavior of American soldiers at Abu Ghraib prison in Iraq surprise you? Well, it didn't surprise Philip Zimbardo, professor emeritus at Stanford University. He has seen "good" people do "bad" things before—in a college psychology laboratory! Zimbardo designed an experiment to observe the behavior of people without record of crime or violence in a mock "prison." He was amazed at the rapidity with which statuses were adopted and roles fulfilled by "liberal" college students randomly assigned to play prisoners and guards. This experiment reveals the ease with which people can be socialized to statuses and roles when they find themselves in a social structure that warrants it. Zimbardo's own words describe the design and results of this experiment (Zimbardo 2008).

In an attempt to understand just what it means . . . to be a prisoner or a prison guard, Craig Haney, Curt Banks, Dave Jaffe and I created our own prison. We carefully screened over 70 volunteers who answered an ad in a Palo Alto city newspaper and ended up with about two dozen young men who were selected to be part of this study. They were mature, emotionally stable, normal, intelligent college students from middle-class homes. . . . They appeared to represent the cream of the crop of this generation. None had any criminal record. . . .

Half were arbitrarily designated as prisoners by a flip of a coin, the others as guards. These were the roles they were to play in our simulated prison. The guards . . . made up their own formal rules for maintaining law, order and respect, and were generally free to improvise new ones during their eight-hour, three-man shifts. The prisoners were unexpectedly picked up at their homes by a city policeman in a squad car, searched, handcuffed, fingerprinted, booked at the Palo Alto station house and taken blindfolded to our jail. There they were stripped, deloused, put into a uniform, given a number and put into a cell with two other prisoners where they expected to live for the next two weeks. . . .

At the end of only six days we had to close down our mock prison because what we saw was frightening. It was no longer apparent to most of the subjects (or to us) where reality ended and their roles began. The majority had indeed become prisoners or guards, no longer able to clearly differentiate between role playing and self. There were dramatic changes in virtually every aspect of their behavior, thinking, and feeling. . . . We were horrified because we saw some boys (guards) treat others as if they were despicable animals, taking pleasure in cruelty, while other boys (prisoners) became servile, dehumanized robots who thought only of escape, of their own individual survival and of their mounting hatred for the guards. We had to release three prisoners in the first four days because they had such acute situational traumatic reactions as hys-terical crying, confusion in thinking, and severe depression. Others begged to be paroled, and all but three were willing to forfeit all the money they had earned [$15 per day] if they could be paroled. By then (the fifth day) they had been so programmed to think of themselves as prisoners that when their request for parole was denied they returned docilely to their cells. . . .

About a third of the guards became tyrannical in their arbitrary use of power, in enjoying their control over other people. They were corrupted by the power of their roles and became quite inventive in their techniques of breaking the spirit of the prisoners and making them feel they were worthless. . . . By the end of the week the experiment had become a reality.

Source: Published by permission of Transaction, Inc., from *Society*, vol. 9, no. 6. Copyright © 1972 by Transaction, Inc.

Evaluate the Research

1. If you were asked to discuss Zimbardo's experiment within the context of one of the three major theoretical perspectives, which would you choose? Why?

2. One of Zimbardo's conclusions, not stated in the preceding account, is that the brutal behavior found in real-life prisons is not due to the antisocial characteristics or personality defects of guards and prisoners. Can you argue, sociologically, that he is right in this conclusion? How?

3. There was some controversy over the ethics of this experiment. Do you think this experiment could be carried out today under the sociological code of ethics discussed in Chapter 2? Why or why not?

each person holds many statuses, and each status involves many roles. This diversity invites conflict and strain.

What is the difference between role conflict and role strain?
Role conflict occurs when the performance of a role in one status clashes with the performance of a role in another status. College students who hold the statuses of student and employee often find it difficult to balance study and work demands.

Because groups and subcultures have unique role prescriptions, they also exhibit unique role conflicts. As young, affluent, successful individuals, athletes attract

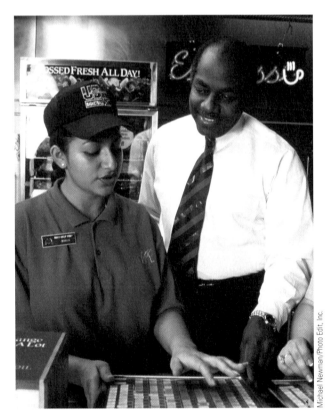

The process of role selection prior to behavior occurs in nearly all instances of social interaction. The range of acceptable responses does have limits. This is especially so in jobs at fast food restaurants where almost all behavior is rigidly programmed.

a large number of "friends." Because the athletes pay all the bills, these friends often evolve into employees paid to do certain functions for the athletes. Role conflict for athletes inevitably flows from the mixing of business and personal relationships.

Role strain occurs when some of the roles of a single status clash. College basketball coaches, for example, have to recruit for the next season while attempting to win games in the current season. University professors teach classes, spend time with students, and publish their research findings. Each of these roles is time consuming, and the fulfillment of one role may interfere with the performance of others. If expectations for you as a college student pressure you to perform well academically, join a fraternity or sorority, pursue a sport, date, and participate in other college activities, you will probably experience some degree of role strain.

How can role conflict and role strain be managed? Role conflict and strain may lead to discomfort and confusion. To reduce anxiety and have smoother relationships with others, we often solve role dilemmas by setting priorities. When roles clash, we decide which role is more important to us and act accordingly. A student who misses work frequently because of school requirements can eliminate the role conflict completely by quitting work and assigning priority to school. If she remains in both statuses, she can reduce work hours or cut down on extracurricular school activities. A university professor may have to decide to emphasize either teaching or research. By ranking incompatible roles based on their importance, we reduce role conflict and strain.

We can also segregate roles. This is especially effective for reducing the negative effects of conflicting roles. An organized crime member may reduce role conflict by segregating his criminal activities from his role as a loving father. A college coach experiencing the role strain associated with coaching and recruiting simultaneously can decide to give priority to one of those roles. He may, for example, let his assistant coach do most of the recruiting until the season ends.

F E E D B A C K

1. _____ are culturally defined rights and obligations attached to statuses.
2. The rights of one social status correspond to the _____ of another one.
3. _____ occurs when the performance of a role in one status clashes with the performance of a role in another status.
4. _____ occurs when the roles of a single status are inconsistent.
5. Match the following concepts with the examples listed beside them.
   ____ a. role   (1) A mother is expected to take care of her children.
   ____ b. role performance   (2) A husband and wife discuss disciplining one of their children.
   ____ c. social interaction   (3) A university president hands out diplomas at a graduation ceremony.
6. Which of the following is *not* one of the differences between the stage and social life?
   a. There is considerably more discrepancy between roles and role performance in social life than between a stage script and its performance.
   b. Unlike the stage, there are no cues and responses in real life.
   c. Role performance in real life is not the conscious process that actors go through on the stage.
   d. In social life, the cues and responses are not as programmed and predictable as on the stage.

Answers: 1. Roles; 2. obligations; 3. Role conflict; 4. Role strain; 5. a. (1), b. (3), c. (2); 6. b

Reducing Conflict in Two-Career Families

Families with two working adults have special strains. While in 1960, less than 20 percent of married women with young children worked outside of the home, by 2004, the figure was about 65 percent. This increase has resulted in added role conflict for women. In a two-career family, the woman is more likely to suffer from conflict because she is still generally expected to balance her traditional homemaker roles with her career roles. The women are not the only ones who suffer, however. The effects of this conflict are felt by husbands and children as well. Because you will likely be faced with the stress associated with dual-career families, you would be wise to learn now some techniques for reducing role conflict.

1. **Focus on the positive.** Conflict can be reduced when couples define their situation positively. If both partners are working from choice rather than necessity, it can be helpful to begin each day remembering some of the reasons why they first made the choice for both to work. These reasons might include additional income or personal satisfaction.

2. **Put family needs first.** Role conflict can be most effectively managed when family roles are placed ahead of working roles. When a baby-sitter fails to show up, when a child is sick, or when a parent–teacher conference is called, one of the parents can place these demands above work-related demands. Placing a higher priority on family needs will help keep the family support structure intact.

3. **Assume one role at a time.** Conflict can be reduced if a person focuses on only one role at a time. Leaving job-related problems at work and family issues at home is often difficult but is very effective in reducing role conflict.

4. **Find the compromise balance.** Although many men take active roles in child care today, in order to meet family obligations, women still make the most compromises in their careers. With the increasing number of women in better paying professional careers, we should expect more equality in career compromises between husbands and wives.

Think About It

Identify three ways that you believe would help reduce role conflict in a dual-career family. Provide specific examples not given in the text.

Can conformity to roles ever be complete? No. Complete conformity to all roles is impossible. This poses no problem as long as role performance occurs within accepted limits. Professors at research-oriented universities may be permitted to emphasize teaching over research. Coaches may accent fair play, character building, and scholarship. But professors at research universities who do too little publishing or coaches who win too few games usually will not be rewarded for very long. At some point, they will have exceeded the acceptable limits for deviation from expected role performance.

Theoretical Perspectives and Social Structure

If you stop to think about it for a moment, you will realize that the concepts of role, social status, and patterned relationships reflect the concerns of functionalism with stability, order, and consensus (see Table 5.2). The concepts of role behavior and social interaction have the strong flavor of symbolic interactionism. Also present in the study of social structure, but not as apparent, are the ideas of conflict theory. To illustrate conflict, consider the struggles between street gangs and law enforcement officials and the clashes between teenagers and their parents. To illustrate change, consider women's more prominent role in sports and the economy, and ponder the effects of emerging globalization on your career choices.

Society

A society is the largest and most self-sufficient social structure in existence. As you may recall from Chapter 3, a **society** is composed of a people living within defined territorial borders, sharing a common culture. Theoretically, a society is independent of all outsiders. It contains enough smaller social structures—fam-

TABLE 5.2

FOCUS ON THEORETICAL PERSPECTIVES: Illustrating Social Structure Concepts

This table illustrates how each theoretical perspective might view a major concept. The concepts could be switched to any other theoretical perspective and illustrated from that perspective. Associate each concept with a different theoretical perspective and provide your own example.

Theoretical Perspective	Social Structure Concept	Example
Functionalism	Role	• Social integration is promoted by culturally defined rights and obligations honored by group members.
Conflict theory	Ascribed master status	• Ascribed master statuses such as gender and race empower some to subjugate others.
Symbolic interactionism	Social interaction	• Roles are carried out by individuals based on the symbols and meanings they share.

ily, economy, government, religion—to meet most of the needs of its members. As you will see, preindustrial societies met these criteria; they truly could be independent and self-sufficient. But modern societies, although capable of caring for most members' needs, must have political, military, economic, cultural, and technological ties with other societies. In fact, as we move toward a global society, the fate of all people becomes linked.

Anthropologists and sociologists have classified societies in various ways. Because all societies must fulfill the need for food, one classification system (the one we use in this chapter) is based on the way subsistence is addressed. This classification system is based on the evolution from the four types of preindustrial societies (hunting and gathering, horticultural, pastoral, and agricultural societies) to industrial societies and finally to postindustrial societies (Haviland 2007). A society's solution to the subsistence problem is clearly reflected in its culture and social structures (see Table 5.3).

Preindustrial Societies

What is unique about a hunting and gathering society? The **hunting and gathering society** survives by hunting animals and gathering edible foods such as wild fruits and vegetables. Th is is the oldest solution to the subsistence problem. In fact, it was only about 9,000 years ago that other methods of solving the subsistence problem emerged.

Although not all hunting and gathering societies are exactly alike, they do share some basic features (Beals, Hoijer, and Beals 1977; Service 1979). Hunting and gathering societies are usually nomadic; that is, they must move from place to place as the food supply and seasons change. Because nomads must carry all their possessions with them, they accumulate few material

goods. Hunting and gathering societies also tend to be very small—usually fewer than fifty people—with members scattered over a wide area. Because the family is the only institution in hunting and gathering societies, it tends to all the needs of its members. On a day-to-day basis, association is typically limited to the immediate family. In fact, these societies are organized either as self-sufficient family groups or as loose combinations of families or bands. Thus, hunting and gathering societies are tied together by kinship. Most members are related by blood or marriage, although marriage is usually limited to those outside the family or band.

Cooperation or competition? Which is more valued in preindustrial society? Economic relationships within hunting and gathering societies are based on cooperation—members sharing their surplus with other members. Most members of this type of society are kin, and sharing takes place without the expectation of reciprocation. Members of hunting and gathering societies seem simply to give things to one another. In fact, the scarcer something is, the more freely it is shared. Generosity and hospitality are valued; thrift is considered a reflection of selfishness. Because the obligation to share goods is one of the most binding aspects of their culture, members of hunting and gathering societies have little or no conception of private property or ownership.

Without a sense of private ownership and with few possessions for anyone to own, hunting and gathering societies have no social classes, no rich or poor. These societies also lack status differences based on political authority because they have no political institutions; there is no one to organize and control activities. According to E. Adamson Hoebel, when the Inuits wanted to settle disputes, they had dueling songs. The people involved in the dispute prepared and sang songs to express their sides of the issue. Their families,

TABLE 5.3

Comparison of Major Types of Society

Carefully examine this comparison of major types of society. Choose one type of society and explain why the nature of its culture and social structure fits with its means of subsistence and technological base.

Type of Society	Time of Origin	Means of Subsistence	Technological Base	Culture and Social Structure
Hunting and Gathering	First type of society to emerge among humans	Hunting animals and gathering edible foods in nature	Simple handmade tools	• Small nomadic bands based on kinship and cooperation; common ownership of property; scant division of labor along sex and age lines within families
Horticultural	About 9,000 years ago	Domesticating plants	Slightly more advanced handmade tools (digging sticks, hoes, spades)	• Less nomadic living in more permanent settlements; bands organized around families; more conflict among bands; neglible division of labor based on sex and age
Pastoral	About 9.000 years ago	Domesticating animals (goats, sheep, cattle)	Handmade horticultural tools; meat-cutting tools; knowledge in such areas as grazing land, carrying capacity, breeding, size and composition of herds, weather, water supply	• Live in long-term villages; some trade emerges; women remain at home while males attend herds; greater economic surplus leads to more complex division of labor; greater competition over surplus
Agricultural	About 5,000–6,000 years ago	Permanent land cultivation and harnessing of animal energy	Plow and animal energy	• Increased productivity releases people from land; more complex division of labor branching outside economic roles; establishment of smaller cities; emergence of separate political, economic, and religious institutions; state replaces family as prime mover; appearance of distinct social classes; emergence of trade and money economy
Industrial	About 250 years ago	Application of science and technology to production	Power-driven machines	• Economy shifts from subsistence to an open market due to greater food surplus; growth of large cities; importance of kinship declines; education moves out of homes and increases in importance; women become less subordinate to husbands; influence of religion less pervasive; social institutions become more complex, specialized, and independent; social relationships more impersonal
Postindustrial	Around 1970	Development of service industries	Intellectual	• Greater social instability in the form of divorce, crime, and rapid social change; less social and cultural consensus; reduced gender inequality; individualism increases; urban population moves out of large cities

as choruses, accompanied them. Those listening to the duel applauded their choice for the victor (Hoebel 1983).

The division of labor in hunting and gathering societies is limited to the sex and age distinctions found in most families. Men and women are assigned separate tasks, and certain tasks are given to the old, the young, and young adults. This scant division of labor exists because there are no institutions beyond the family. Also, more leisure time is available in hunting and gathering societies than in any other. Today, few true hunting and gathering societies remain other than the Khoi-San (Bushmen) in southern Africa, the Kaska Indians in Canada, and the Yanomamö of Brazil.

What is a horticultural society? According to archaeological evidence, **horticultural societies** solved the subsistence problem primarily through the domestication of plants. This type of society came into being about 9,000 years ago. The transition from hunting and gathering to horticultural societies was not abrupt. It occurred over several centuries. In fact, some hunters and gatherers in the Middle East were harvesting wild grains with stone sickles 10,000 years ago, and early horticultural societies also hunted and gathered (Nolan and Lenski 2005).

The shift from hunting and gathering to horticulture led to more permanent settlements. Subsistence no longer required people to move frequently in order to take advantage of available food. Instead, people could work a land area for extended periods of time before moving on to more fertile soil. This relative stability in simple horticultural societies—and particularly in the permanent settlements in the Middle East and southeastern Europe—permitted the growth of large societies with greater population densities.

The family is even more basic to social life in horticultural societies than in hunting and gathering societies. In hunting and gathering societies, the survival of the band usually has uppermost priority. In horticultural societies, primary emphasis is on providing for household members. This is because producing food in horticultural societies can be accomplished through the labor of family members. With the labor necessary for survival, households can be more self-sufficient and independent: "The family is to the . . . [horticultural]

economy as the manor to medieval European economy, or the industrial corporation to modern capitalism: each of these is the central production-institution of its time" (Sahlins 1968:75).

While hunters and gatherers seldom engage in interband raids, there is considerable intervillage conflict in horticultural societies. Much of the fighting, however, resembles modern contact sports more than warfare. Satisfaction is often gained more from defeating the enemy in games than from killing him. Although conflict does sometimes lead to battle and death, religious beliefs and rituals keep slaughter within bounds (Plog and Bates 1990). Still, Napoleon Chagnon (1997) estimates that at least 24 percent of all male deaths among the Yanomamö (South American preindustrial people) are due to warfare—a figure consistent with those for other tribes that constantly feud with their neighbors.

How is the subsistence problem addressed in pastoral societies? Most horticultural societies keep domesticated animals such as pigs and chickens. They do not, however, depend economically on the products of these animals the way *pastoralists* (herders) do. In **pastoral societies**, food is obtained primarily by raising and taking care of domesticated animals. For the most part, these are gregarious (herd) animals such as cattle, camels, goats, and sheep, all of which provide both milk and meat. Because food grains are needed to supplement the food obtained from animals, pastoralists must also either cultivate some land or trade with people who do (Nanda and Warms 2006; Peoples and Bailey 2008).

There is more physical mobility in pastoral societies than in those based more fully on the cultivation of land, although the degree of mobility may vary greatly. Permanent (or at least long-term) villages can be

Judging from this photograph, to what type society do these Thailand rice farmers belong?

Digitalpress/Dreamstime LLC

This farmer lives in an agricultural society. How does his society differ technologically from a horticultural society?

maintained if, as seasons change, herd animals are simply moved to different pastures within a given area. In such societies, the women remain at home while the males take the herds to different pastures. A permanent village, however, cannot exist if, as the weather changes, an entire population moves with the herd (nomads).

Because both horticultural and pastoral societies can produce a surplus of food, they usher in important social changes unknown in hunting and gathering societies. With a surplus food supply, some members of the community are free to create a more complex division of labor. People can become political and religious leaders or makers of surplus goods such as pottery, spears, and clothing. Because nonedible goods are produced, an incentive to trade with other peoples emerges.

The creation of a surplus also permits the development of social inequality, although it is limited by the size and dependability of the surplus. Even a relatively small surplus, however, means that some families (or villages or clans) have more wealth than others. Competition over resources can take the form of warfare, which in turn leads to the emergence of political leaders or possibly the creation of slavery.

What are the characteristics of an agricultural society? The transition from horticultural to **agricultural society** was made possible largely through the invention of the plow (Nolan and Lenski 2005). The longer the horticulturalists worked the soil with their digging sticks, hoes, and spades, the greater their disadvantage. With these crude implements, horticulturalists could not control the weeds, and they could not dig deeply enough to reach the soil's nutrients, which recede farther below the root line each year of cultivation. The plow, which appeared about 5,000 to 6,000 years ago, did not solve these problems completely, but it was effective enough to permit the permanent cultivation

of land. The plow not only controlled the weeds but also turned the weeds into fertilizer by burying them under the soil. By digging more deeply into the ground, the plow was able to reach nutrient-rich dirt that had sunk below root level. The result was more productivity—more food per unit of land.

Moreover, the plow permitted a shift from human to animal energy. Only humans could use a hoe or spade, but oxen could pull a plow. Using draft animals also increased productivity because larger areas could be cultivated with fewer people. As a result, more people were released from the land. This, then, freed those people to engage in noneconomic activities such as formal education, concerts, and political rallies. This led, in turn, to other significant changes: the introduction of occupations not directly tied to farming (such as politician, blacksmith, and haberdasher); the establishment of cities; the development of more complex economic specialization; and the emergence of separate political, economic, and religious institutions.

Although family ties remained important, the state replaced the kinship group as the guiding force for agricultural societies. Advanced agricultural societies were monarchical, headed by a king or an emperor who desired as much control as possible. Distinct social classes appeared for the first time. Wealth and power were based on land ownership, which was controlled by the governing elite. While the elite enjoyed the benefits of the economic surplus, the peasants did most of the work.

Urban merchants held less prestige than the governing class. They were better off than the peasants; but they, too, worked hard for their living. An economy, based on trade and a monetary system, began to emerge as an identifiable institution. The agricultural economy developed two basic sectors: a rural, agricultural sector and an urban, commercial sector. Institutional specialization was also reflected in the increasing separation of the political and religious elite. Although rulers were believed to be divinely chosen, few of them doubled as religious leaders.

Industrial Societies

The Industrial Revolution was "a kind of discontinuous leap in human history; a leap as important as that which had lifted the first . . . [horticultural] settlements above the earlier hunting communities" (Milberg and Heilbroner 2007:102). The Industrial Revolution created **industrial society**—a society whose subsistence is based primarily on the application of science and technology to the production of goods and services.

What happens to societies as new technologies appear? Alberto Martinelli and Neil Smelser (1990) have identified some basic structural changes that occur in societies shifting from an agricultural to an industrial economic

Agricultural Employment   As societies move from the preindustrial to the postindustrial stage, fewer people are required to raise food to feed the population. This map shows the percentage of each country's population invoved in the production of agricultural products.

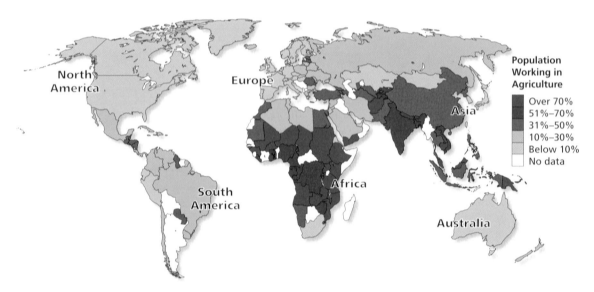

Source: U.S. Central Intelligence Agency, *The World Factbook, 2005.*

Interpret the Map

1. After examining this map, what generalizations about types of societies around the world would you make? Explain.

2. Which countries do you think could be ready to move from one type of society to another? Be specific about countries and types of societies.

3. What parts of the world are the least likely to change in the near future? Explain your answer.

---

base. Industrialism brings with it a change away from simple, traditional technology (plows, hammers, harnesses) toward the application of scientific knowledge. With this change comes the discovery and adoption of new, complex technological devices. Early examples of industrial technology include the steam engine and the use of electrical power in manufacturing. More recent technological developments include nuclear energy, aerospace-related inventions, the computer, and nanotechnology. These technological devices are used in factories and in agriculture.

In industrial societies, the trend is away from human and animal power toward power-driven machines. These machines are operated by wage earners who produce goods for sale on the market. With this mechanization, the shift is away from subsistence farming toward the selling of agricultural goods on the open market. Farmers are able to produce a surplus sufficient to support themselves and many others. This food surplus supports urbanization, the movement of people from farms and villages to gigantic urban concentrations.

Thus, industrial society is also an urban society, with vast concentrations of people located in and around many urban centers. In fact, about half of the world's population now lives in cities. And the world's urban population is expected to grow by 72 percent (to almost 5 billion). By 2030, half of the world's population will, for the first time, live in cities (*2007 World Population Data Sheet* 2007; United Nations 2007).

The concept of structural differentiation also helps to describe industrial society. **Structural differentiation** occurs when a single social structure divides into two or more social structures. Under the new circumstances, these new separate social structures operate more efficiently than the one alone could. Under the domestic form of production, for example, the family performs a variety of economic roles. In the change from domestic to factory production, economic activities are moved from the family to the factory. Because economic and familial functions are now separated into two institutions, these activities become increasingly segregated.

Other kinds of changes occur through structural differentiation. The education of the young moves from the home to the school. Because an industrial society requires an educated labor force, education is extended from the elite to the masses. Kinship declines in importance as the immediate family begins to separate socially and physically from the extended family. Personal choice and love replace arranged marriages. Women, through their entrance into the workforce, become less subordinate to their husbands. Religion no longer pervades all aspects of social life as various institutions—political, economic, familial, educational—become more and more independent of religious influence. Individual mobility increases dramatically, and social class is based more on occupational achievement than on the social class of one's parents.

Because the United States has been an industrial society for so long, its characteristics are taken as a given. The effects of industrialization are easier to observe in societies currently moving from an agricultural to an industrial economic base. For example, Vietnam and Malaysia, two countries currently undergoing industrialization, are experiencing these kinds of changes at the beginning of the twenty-first century (Phu 1998; S. Singh 1998).

How do preindustrial and industrial societies differ? Ferdinand Tönnies, Emile Durkheim, and Robert Redfield were three early sociologists who, each with his own unique approach, compared preindustrial and industrial societies. Contemporary sociologists generally agree that each of these three men were successful at isolating central features differentiating the two types of societies.

Ferdinand Tönnies (1957; originally published in 1887), an early German sociologist, distinguished between **gemeinschaft** (German for "community") and **gesellschaft** (German for "society"). The former—closely approximating preindustrial society—is based on tradition, kinship, and intimate social relationships. The latter—representing industrial society—is characterized by weak family ties, competition, and less personal social relationships.

Shortly after Tönnies formulated his distinction, Emile Durkheim (1997; originally published in 1893) made a similar observation. He distinguished the two types of societies by the nature of their social solidarity—the degree to which a society is unified.

Social solidarity, Durkheim contended, is predicated on a society's division of labor. In societies in which the division of labor is simple—in which most people are doing the same type of work—**mechanical solidarity** is the foundation for social unity. A society based on mechanical solidarity achieves social unity through a consensus of beliefs, values, and norms; strong social pressures for conformity; and dependence on tradition and family. In this type of society, which is best observed in small, nonliterate societies, people tend to behave, think, and feel in much the same ways; to place the group above the individual; and to emphasize tradition and family.

Modern industrial societies, in contrast, involve a very complex and differentiated division of labor. Because people perform very specialized jobs, they are dependent on others and must cooperate with them. Whereas farmers in a simple society are largely self-sufficient, members of an industrial society depend on a variety of people to fulfill their needs—barbers, bakers, manufacturers, and other suppliers of services. This modern industrial society is based on **organic solidarity**. It achieves social unity through a complex of specialized statuses that makes members of a society interdependent. Dependence and need for cooperation replace the homogeneity of beliefs, values, and norms characteristic of simpler societies.

The term *organic solidarity* is based on an analogy with biological organisms. If a biological organism composed of highly specialized parts is to survive, its parts must work together. Similarly, the parts of a society based on organic solidarity must cooperate if the society is to survive. In modern societies, individualism tends to replace strict group conformity; people do not share all ways of thinking, feeling, and behaving; and the importance of kinship declines.

In this same vein, anthropologist Robert Redfield (1941) distinguished between a folk society and an urban society. A **folk society** rests on tradition, cultural and social consensus, family, personal ties, little division of labor, and an emphasis on the sacred. In an **urban society**, social relationships are impersonal and contractual; the importance of the family declines; cultural and social consensus is diminished; economic specialization becomes even more complex; and secular concerns outweigh sacred ones.

Postindustrial Society

Some societies, including the United States, have passed beyond industrial society into **postindustrial society**. The postindustrial economic base is grounded more in service industries than in manufacturing industries and relies on expertise in production, consumption, and government (Kumar 2004).

What are the major features of postindustrial society? Daniel Bell (1999) identifies five major features of postindustrial society:

1. *For the first time, the majority of the labor force is employed in services rather than in agriculture or manufacturing.* Today, the United States is the only country in the world in which more than half of all employment is in services—trade,

finance, transportation, health, recreation, research, and government. In fact, more than 75 percent of all employed workers in the United States are in service jobs.

2. *White-collar employment replaces much blue-collar work.* The shift to a service economy in postindustrial society—with an emphasis on office work—leads to a preponderance of white-collar workers. They outnumbered blue-collar workers in the United States for the first time in the mid-1950s, and the gap is still increasing. The most rapid growth among white-collar workers has been in professional and technical employment.

3. *Theoretical knowledge is the key organizing feature.* In postindustrial society, knowledge is used for the creation of innovations as well as for the formulation of government policy (computers permit economic forecasting so that various economic theories can be tested). As theoretical knowledge becomes more important, so do educational and research institutions.

4. *Through new means of technological forecasting, society can plan and control technological change.* In industrial society, technological change is uncontrolled. That is, the effects of a technological innovation are not assessed before its introduction. The pesticide DDT was introduced as a benefit to agriculture before its harm to living things had been determined. The internal combustion engine contributes to our affluence and economic growth but contaminates the environment. Technology is more intensely scrutinized in postindustrial society. In the race to provide the automobile industry a lighter and cheaper metal, the aluminum industry claimed that a one-ton increase of aluminum in place of steel would reduce carbon dioxide pollution by twenty tons over the life of an average car. This claim was successfully challenged by a scientific study showing that it would take thirty-two to thirty-eight years of driving aluminum-intensive vehicles to offset the pollution released by the production of aluminum required to build those vehicles (Chamberlain 1999). Technological assessment in postindustrial society will permit us to consider the effects—good and bad—of an innovation before it is introduced.

5. *Intellectual technology dominates human affairs.* A new intellectual technology dominates in postindustrial society, much as production technology has dominated industrial society for the past 150 years. With "intellectual technology," mathematical-type problem-solving rules can, in many cases, replace human judgment. With modern computers, it is possible to consider, simultaneously, a large number of interacting variables. This capability allows us to manage the large-scale organizations that prevail in postindustrial society. Intellectual technology enables complex organizations—including government at national, state, and local levels—to set rational goals and to identify the means for reaching them.

The problem of subsistence ushered in industrial society primarily through the application of science and technology to the production of goods and services. Assembly line workers are common in industrial society.

John A. Rizzo/Getty Images

Occupational knowledge in an industrial society, like the knowledge needed by these science technicians, cannot be learned in the home. What does this mean for education in an industrial society?

Laurence Gough/Used under license from Shutterstock.

These are some of the most general features of postindustrial society. Considerable detail on the nature of politics and work in postindustrial society is presented in Chapter 14, "Political and Economic Institutions."

What effect has the postindustrial transformation had on the stability of society? According to Francis Fukuyama (1999, 2000), the United States and other developed nations have undergone a thirty-year economic transition equivalent in magnitude to the Industrial Revolution. This economic transition, he asserts, has brought with it social instability.

Fukuyama identifies several signposts of "the great disruption" in a number of deteriorating social conditions, starting in the mid-1960s:

> *Crime and social disorder began to rise, making inner-city areas of the wealthiest societies on earth almost uninhabitable. The decline of kinship as a social institution, which has been going on for more than 200 years, accelerated sharply in the second half of the twentieth century. Marriages and births declined and divorce soared; and one out of every three children in the United States and more than half of all children in Scandinavia were born out of wedlock. Finally, trust and confidence in institutions went into a forty-year decline. (Fukuyama 1999:55)*

Does Fukuyama expect social instability to lessen or increase? Currently, Fukuyama sees indications of a return to social stability as the great disruption runs its course. The establishment of new social norms, he believes, is reflected in the slowing down of increases in divorce, crime, distrust, and illegitimacy. In the 1990s, Fukuyama notes, many developed societies have even seen a reversal of these rates; crime, divorce, illegitimacy, and distrust have actually declined:

> *This is particularly true in the United States, where levels of crime are down a good 15 percent from their peaks in the early 1990s. Divorce rates peaked in the early 1980s, and births to single mothers appear to have stopped increasing. Welfare rolls have diminished almost as dramatically as crime rates, in response both to the 1996 welfare-reform measures and to opportunities provided by a nearly full-employment economy in the 1990s. Levels of trust in both institutions and individuals have also recovered significantly since the early 1990s. (Fukuyama 1999:80)*

Fukuyama finds the roots of a return to social order in two places. For one thing, human nature leads us away from social disorder toward social order. What's more, human reason permits us to create solutions to problems of social disorder. Thus, humans are geared by nature and by reason away from Durkheim's condition of anomie (normlessness): "The situation of normlessness . . . is intensely uncomfortable for us, and we will seek to create new rules to replace the old ones that have been undercut" (Fukuyama 1999:76). Because culture can be changed, it can be used to create new social structures better adapted to changing social and economic circumstances.

F E E D B A C K

1. A _____ is the largest and most self-sufficient group in existence.
2. The _____ society is the oldest known solution to the subsistence problem.
3. The _____ society solved the subsistence problem primarily through the domestication of plants.
4. The transition from horticultural to agricultural society was made possible largely through the invention of the _____.
5. _____ occurs when one social structure divides into two or more social structures that operate more successfully separately than the one alone would under the new circumstances.
6. A society based on _____ solidarity achieves social unity through a complex of highly specialized roles that makes members of a society dependent on one another.
7. Which of the following is *not* one of the major features of postindustrial society?
   a. emphasis on theoretical knowledge
   b. employment of the majority of the labor force in service industries
   c. reliance on intellectual technology
   d. increased dependence on skilled blue-collar workers
   e. shift toward the employment of white-collar workers

Answers: 1. society; 2. hunting and gathering; 3. horticultural; 4. plow; 5. Structural differentiation; 6. organic; 7. d

SUMMARY

1. Relationships among individuals are patterned. This underlying pattern of social relationships is called social structure. Social scientists have developed the concepts of status, role, role performance, and social interaction to explain the existence of social structure.

2. A status is a position that a person occupies within a social structure. Individuals occupying interrelated statuses usually behave toward one another in orderly and predictable ways. Some statuses are ascribed, or assigned, to people; other statuses are achieved, or earned. Still other statuses (master statuses) are so significant that they affect most other aspects of a person's life.

3. Roles—the glue that binds a network of social statuses—are the culturally defined rights and obligations attached to social statuses. Rights inform one person of the behavior that can be expected from another person; obligations inform individuals of the behavior that others expect from them. The rights of one status correspond to the obligations of another status.

4. Role performance occurs when roles are put into action. Role performance takes place through social interaction. Role conflict occurs when the performance of a role in one status clashes with the performance of a role in another status. Role strain occurs when the roles of a single position are inconsistent. Role conflict and role strain, which may lead to discomfort and confusion, can be reduced by setting priorities and segregating roles. Because of role conflict and role strain, an individual's conformity to all roles is impossible. This does not cause a problem so long as roles are carried out within certain tolerance limits.

5. A society is composed of a people living within defined territorial borders, sharing a common culture. The way a society solves the problem of subsistence heavily influences the society's culture and social structures. Historically, societies have become larger and more complex as the means for solving subsistence problems have improved. The major types of societies are hunting and gathering, horticultural, pastoral, agricultural, industrial, and postindustrial.

6. Although both the horticultural and agricultural revolutions brought significant change, the greatest transformation came with the Industrial Revolution. The leap from preindustrial to industrial society was so great that many scholars have been able to isolate the basic characteristics distinguishing these two types of societies.

7. Postindustrial society is on the rise. This type of society has a predominately white-collar labor force with the following features: concentration in service industries; base of technical knowledge; reliance on technical forecasting; and dependence on computer technology for organizing and making decisions.

R E V I E W   G U I D E

LEARNING OBJECTIVES REVIEW

- Explain what sociologists mean by social structure.
- Distinguish and illustrate the basic building blocks of social structure.
- Define the concept of society.
- Describe the means of subsistence practiced by the major types of preindustrial societies.
- Discuss the characteristics of industrial society.
- Compare and contrast preindustrial, industrial, and postindustrial societies.

CONCEPT REVIEW

Match the following concepts with the definitions listed below them.

____ a. society
____ b. urban society
____ c. industrial society
____ d. organic solidarity
____ e. achieved status

____ f. rights
____ g. hunting and gathering society
____ h. social structure
____ i. role strain
____ j. master status

____ k. structural differentiation
____ l. gesellschaft
____ m. folk society
____ n. role performance

1. a slot within a social structure occupied because of an individual's efforts
2. a society that solves the subsistence problem through hunting animals and gathering edible fruits and vegetables
3. roles informing individuals of the behavior that can be expected from others
4. when one social structure divides into two or more social structures that operate more successfully separately than the one alone would under the new circumstances
5. a society based on tradition, cultural and social consensus, family, personal ties, little division of labor, and an emphasis on the sacred
6. the actual conduct involved in putting roles into action
7. occurs when the roles of a single status are inconsistent with one another
8. patterned recurring social relationships among individuals and groups
9. people who live within defined territorial borders, sharing a common culture
10. a status that affects most other aspects of a person's life
11. Tönnies's term for the type of society characterized by weak family ties, competition, and impersonal social relationships
12. a society whose subsistence is based primarily on the application of science and technology to the production of goods and services
13. social unity based on a complex of highly specialized statuses that makes members of a society dependent on one another
14. a society in which social relationships are impersonal and contractual, kinship is deemphasized, cultural and social consensus are not complete, the division of labor is complex, and secular concerns outweigh sacred ones

CRITICAL-THINKING QUESTIONS

1. Suppose that a college friend of yours wants to know the meaning of the term *social structure*. Use the stage analogy to develop an understandable answer. Use examples.

2. Have you experienced role conflict or role strain lately? If so, describe the situation. If not, explain why you have been immune to role conflict and role strain, making clear the meaning of the concepts.

3. Discuss some of the basic distinguishing features of preindustrial and industrial societies. Illustrate your answer.

4. In what ways will your work life be different in a postindustrial society than in an industrial society?

MULTIPLE-CHOICE QUESTIONS

1. _____ is defined as patterned, recurring social relationships.
   a. Informal structure
   b. Social structure
   c. Constructive structure
   d. Anomic structure
   e. Symbolic interaction

2. The type of status that is neither earned nor chosen but is assigned to us is called
   a. achieved status.
   b. ascribed status.
   c. acquired status.
   d. assigned status.
   e. symbolic status.

3. The president of a corporation is an example of a(n)
   a. achieved status.
   b. ascribed status.
   c. acquired status
   d. assigned status.
   e. symbolic status.

4. _____ are culturally defined rights and obligations attached to statuses, indicating the behavior expected of individuals holding them.
   a. Norms
   b. Values
   c. Mores
   d. Roles
   e. Beliefs

5. Rights inform individuals of behavior they can expect from others; _____ inform individuals of the behavior others can expect from them.
   a. exchange principles
   b. roles
   c. statuses
   d. cues
   e. obligations

6. _____ occurs when the performance of a role in one status clashes with the performance of a role in another status.
   a. Role conflict
   b. Role strain
   c. Role distancing
   d. Role conformity
   e. Role deformity

7. By ranking incompatible roles based on their importance to us, we can reduce role _____ and _____.
   a. conflict; strain
   b. conflict; behavior
   c. conflict; ambiguity
   d. strain; ambiguity
   e. ambiguity; distancing

8. A(n) _____ is composed of a people living within defined territorial borders, sharing a common culture.
   a. society
   b. institution
   c. community
   d. social group
   e. formal organization

9. The _____ society solved the subsistence problem through the application of science and technology to the production of goods and services.
   a. hunting and gathering
   b. horticultural
   c. agricultural
   d. industrial
   e. postindustrial

10. Emile Durkheim used the term _____ to describe social unity that is achieved through a consensus of beliefs, values, and norms; strong social pressure for conformity; and a dependence on tradition and family.
    a. organic solidarity
    b. mechanical solidarity
    c. gesellschaft society
    d. gemeinschaft society
    e. traditional society

FEEDBACK REVIEW

Fill in the Blank

1. _____ refers to the patterned relationships among individuals and groups.

2. _____ are culturally defined rights and obligations attached to statuses.

3. The rights of one social status correspond to the _____ of another one.

4. _____ occurs when the roles of a single status are inconsistent.

5. A _____ is the largest and most self-sufficient group in existence.

6. The transition from horticultural to agricultural society was made possible largely through the invention of the _____.

7. A society based on _____ solidarity achieves social unity through a complex of highly specialized roles that makes members of a society dependent on one another.

Multiple Choice

8. **Which of the following is *not* one of the differences between the stage and social life?**
   a. There is considerably more discrepancy between roles and role performance in social life than between a stage script and its performance.
   b. Unlike the stage, there are no cues and responses in real life.
   c. Role performance in real life is not the conscious process that actors go through on the stage.
   d. In social life, the cues and responses are not as programmed and predictable as on the stage.

9. **Which of the following is *not* one of the major features of postindustrial society?**
   a. emphasis on theoretical knowledge
   b. employment of the majority of the labor force in service industries
   c. reliance on intellectual technology
   d. increased dependence on skilled blue-collar workers
   e. shift toward the employment of white-collar workers

Matching

10. Match the type of status with the examples that best illustrate it.
    ____ a. ascribed status
    ____ b. achieved status
    ____ c. master status
    ____ d. status set
    ____ e. role
    ____ f. role performance
    ____ g. social interaction
    (1) basketball coach, mother, author, daughter, professor
    (2) occupation, gender, race, ethnicity
    (3) president of the United States
    (4) sex
    (5) a husband and wife discuss disciplining one of their children
    (6) a university president hands out diplomas at a graduation ceremony
    (7) a mother is expected to take care of her children

ANSWER KEY

Concept Review
a. 10
b. 15
c. 2
d. 13
e. 14
f. 1
g. 4
h. 3
i. 9
j. 8
k. 11
l. 5
m. 12
n. 6
o. 7

Multiple Choice
1. b
2. b
3. a
4. d
5. e
6. a
7. a
8. a
9. d
10. b

Feedback Review
1. Social structure
2. Roles
3. obligations
4. Role strain
5. society
6. plow
7. organic
8. b
9. d
10. a. 4
    b. 3
    c. 2
    d. 1
    e. 7
    f. 6
    g. 5

Groups and Organizations

J. Scott Applewhite/AP Photos

OUTLINE

6

LEARNING OBJECTIVES

- Define the concept of the group, and differentiate it from social categories and social aggregates.
- Outline the basic characteristics of primary and secondary groups.
- Describe the five major types of social interaction.

- Discuss the nature of group processes: task accomplishment, leadership styles, and decision making.
- Define the concept of formal organization, and identify the major characteristics of bureaucracy.
- Discuss the advantages and disadvantages of bureaucracy.

- Distinguish between formal and informal organization.
- Describe the iron law of oligarchy, and demonstrate its importance with examples.
- Discuss prejudice and discrimination in organizations.

Most people assume that conflict should be avoided because it is disruptive and interferes with group effectiveness. Conflict does have some negative consequences. Not as well known, however, are some social benefits of conflict. Giving full consideration to the different opinions in a group, for example, can prevent a group from being a victim of the self-deceptive belief that the majority is right. In other words, openness to disagreement and conflict can prevent groupthink.

The space shuttle *Challenger* is an excellent example of a disastrous decision that was based on conformity to group ideas. The *Challenger* was launched from Kennedy Space Center on January 28, 1986. Just over a minute after the launch, the *Challenger* disintegrated, and in the instant of explosion, took the lives of all seven astronauts on board.

As is true for all space mission personnel, the *Challenger* team was composed of a number of specialists. The engineers on the team had recommended against takeoff that morning because crucial parts had never been tested at so low a temperature. As victims of groupthink, the NASA leaders screened out this opposition; they acted as if the engineers lacked the intelligence and courage to make the "right" decision. The leaders downgraded the opinions of the engineers and convinced most of the group that the decision, except for the engineers, was unanimous. By avoiding consideration of a dissenting view, the majority lost the shuttle's passengers and harmed NASA's long-term objectives (Moorhead, Ference, and Neck 1991).

Concept of the Group

Americans are socialized to think in individualistic terms. In fact, one of our country's values, as noted in Chapter 4 ("Socialization over the Life Course"), is individuality. This individualistic orientation is often a stumbling block to those first introduced to the concept of the group. Perhaps it will ease the transition into thinking about groups (rather than individuals) if we begin by focusing on what is thought of as an individualistic activity—but in fact is a group endeavor. Art is a likely candidate for this purpose.

Think of painting and you are likely to have images of Rembrandt, Renoir, or Picasso. The art of dance will call to mind Nureyev, Baryshnikov, or Savion Glover. It

may be Gershwin, Sondheim, or Webber who comes to the fore when the subject of musical achievement on Broadway is the topic. In short, when we think of art, we think of individual artists. Without denigrating the talent, hard work, and achievements of individual artists whose works are precious to us, Howard Becker wants to underscore the fact that even such an allegedly individualistic pursuit as art "involves the joint activity of a number, often a large number of people" (Becker 1982:1).

Becker begins his book *Art Worlds* with this story told by Anthony Trollope, the famous nineteenth-century English novelist:

It was my practice to be at my table every morning at 5:30 a.m.; and it was also my practice to allow myself no mercy. An old groom, whose business it was to call me, and to whom I paid £5 a year extra for the duty, allowed himself no mercy. During all those years at Waltham Cross he was never once late with the coffee which it was his duty to bring me. I do not know that I ought not to feel that I owe more to him than to any one else for the success I have had. By beginning at that hour I could complete my literary work before I dressed for breakfast. (Becker 1982:1)

Trollope's groom is one member of the complex network of social relationships that permitted this artist to accomplish his work. Moreover, the activities of both Trollope and his groom had to mesh with those of many other people (printers, publishers, and editors, to mention only a few) if the literary works of this Victorian writer were ever to become available to the public. Some of these relationships were personal and face-to-face; others were not. This chapter elaborates on the variety of group relationships and the contexts in which they appear throughout social life.

A **group** is composed of people who are in contact with one another; share some ways of thinking, feeling, and behaving; take one another's behavior into account; and have one or more interests or goals in common. Because of their characteristics, groups play an important role in the lives of their participants as well as influencing the societies in which they exist.

Types of Groups

Categories, Aggregates, and Groups

A group should be distinguished from a **social category**—people who share a social characteristic. A taxpayer, a woman, and a college graduate each belong to a social category. A group is also sometimes confused with a **social aggregate**—people who happen to be at the same place at the same time, such as students waiting in line for concert tickets. Although neither categories nor aggregates are groups, some of their members may form groups. Victims of a disaster (an aggregate)—such as residents of a town devastated by a tornado—may work together to cope with the emergency. Citizens of a state (a social category) may band together in an organized tax revolt. These people may form a group if they begin to interact regularly; share ways of thinking, feeling, and behaving; take one another's behavior into account; and have some common goals.

Primary Groups

Primary and secondary groups are two principal types of groups (see Table 6.1). At the extremes, the characteristics of these two types of groups—and the relationships that occur within them—are opposites. But most groups sit at different points along a continuum from primary to secondary.

Why is a primary group described as tightly integrated? A **primary group** is composed of people who are emotionally close, know one another well, and seek one another's company. The members of a primary group have a "we" feeling and enjoy being together. These groups are characterized by relationships that are intimate, personal, caring, and fulfilling.

Charles Horton Cooley, one of the founders of symbolic interactionism, coined the term *primary group*. Primary groups are the most important setting for socialization. The family, neighborhood, and childhood play groups, Cooley observed, are primary because they are the first groups an infant experiences. People, of course, participate in primary groups throughout life. Close friends in high school and college, neighbors who keep an eye on one another's children, and friends who meet weekly for golf are examples of primary groups.

What conditions promote the development of primary groups? A number of conditions favor the develop-

People in primary groups are emotionally close, know each other well, enjoy each other's company, and share a feeling of "we-ness." The family is the most important primary group to which we belong.

ment of primary groups and **primary relationships**. Although primary relationships may occur in their absence, the likelihood of having a primary relationship is increased if these conditions exist.

- *Small group size.* It is difficult for the members of large groups to spend enough time together to develop close emotional ties; the chances of knowing everyone fairly well are far greater in small groups. Thus, members of a bridge foursome are more likely to develop primary relationships than are members of a metropolitan chamber of commerce, even when both groups meet regularly.
- *Face-to-face contact.* People have maintained primary relationships despite separation by war, prison, or residential changes and have even established long-distance primary relationships through e-mail or telephone conversations, but primary

TABLE 6.1

Characteristics of Primary and Secondary Groups

You belong to both primary and secondary groups. Select one of each, and describe your experience with the functions it is supposed to perform.

	Nature or Relationship	Function	Examples
Primary Group	• Close and continuous social interaction	• Provides emotional support • Contributes to socialization process • Promotes conformity and social control	• Family • Soldiers in combat • Street gang
Secondary Group	• Impersonal social interaction	• Aids achievement of group goals	• Aerobics class • NFL football team • Telecommunications law firm

relationships occur more easily when interaction is face-to-face. People who can see each other and who can experience such nonverbal communication as facial expressions, tone of voice, and touch are much more likely to develop close ties.

- *Continuous contact.* The probability of developing a primary relationship also increases with prolonged contact. Intimacy rarely develops in a short period of time. In spite of reported love at first sight, most of us require repeated social contact for development of a primary relationship.

- *Proper social environment.* The development of primary relationships is also affected by the social environment in which interaction occurs. If individuals are expected to relate to one another strictly on the basis of status or role, primary relationships are unlikely to develop. Total personalities are usually not considered, and personal concern for one another does not seem appropriate. Lawyers who see judges in court face-to-face, in the presence of a small number of people, and over a long period of time are nevertheless unlikely to develop primary relationships with them. Forming primary relationships when unequal statuses are involved is always difficult. This is the reason primary relationships do not usually develop between students and professors, bosses and employees, or judges and lawyers.

What are the functions of primary groups? First, primary groups provide emotional support through caring, personal, and intimate relationships. During World War II, the German army refused to disintegrate despite years of being outnumbered, undersupplied, and outfought. These conditions should have led to desertion and surrender, but they did not. According to Edward Shils and Morris Janowitz (1948), strong primary relationships within German combat units accounted for the Germans' stability and resilience against overwhelming odds. Although social cohesion developed among American soldiers during the Korean conflict, it was not the same as during World War II. Because of special combat conditions and a personnel rotation system, "the basic unit of cohesion was a two-man relationship rather than one that followed squad or platoon boundaries" (Janowitz 1974:109).

Second, primary groups contribute to the socialization process. For infants, the family is the primary group that provides the emotional support necessary for the development of both an integrated personality and a sense of self. The family also conveys information about culture so that children learn how to participate in social life. In addition, primary groups promote adult socialization as individuals adjust to new and changing social environments—as they enter college, take new jobs, change social classes, marry, and retire. Primary groups make these adjustments less painful because membership in them helps people fit into new social situations. Frequent interaction with people who demonstrate genuine concern for others aids adjustment to a new social setting.

Finally, primary groups promote conformity and contribute to social control. The stability and perpetuation of a society depend on the members' acceptance of the society's norms and values. Unless most of the members of a society support the norms and values, that society cannot continue to exist in its present form. Primary groups teach new members to accept norms and values and provide group pressure to promote conformity to them. William F. Whyte's (1993) study of an Italian slum gang illustrates social control within primary groups. Whyte found that rank within this gang determined how well the members performed in sports. One member of the gang, a former semiprofessional baseball player, was one of the gang's best players. Yet, after losing status in the gang, he could no longer play very well. According to Whyte, the man indicated that he would play baseball better when gang members were not involved. Whyte also reported that bowling scores corresponded with status in the gang—the higher the rank, the higher the score. If a lower-ranked member began to bowl better than those above him, the others used verbal remarks—"You're bowling over your head" or "How lucky can you get?"—to remind him that he was stepping out of line.

Secondary Groups

Why is a secondary group not an end in itself? A **secondary group** exists to accomplish a specific purpose. Unlike a primary group, a secondary group is impersonal and goal oriented; it involves only a segment of its members' lives. Work groups, volunteer groups formed during disasters, and environmental organizations are all examples of secondary groups. The relationships of the group members are **secondary relationships**—impersonal interactions involving only limited parts of their personality. Interactions between salespersons and customers, employers and workers, and dentists and patients are secondary relationships.

Secondary relationships are not necessarily unpleasant. But the purpose of the group is to accomplish a task, not to enrich friendships. In fact, if friendship becomes more important than the task, a secondary group may become ineffective. If the members of a college basketball team become more interested in the emotional relationships among themselves or with their coach than in playing their best basketball, their play on the court could suffer.

Do secondary groups ever include primary relationships? Although primary relationships are more likely to occur

These Goth teenagers gathered in a Swedish park are clearly a group. Are they an in-group or an out-group?

Ladi Kim/Alamy Limited

in primary groups and secondary relationships in secondary groups, there are a number of exceptions. Many secondary groups are settings for primary relationships. Members of a work group may relate to one another in a manner that is personal, demonstrate genuine concern for one another as total personalities, and have relationships that are fulfilling in themselves. Similarly, members of a primary group occasionally engage in secondary interaction. One family member may, for example, lend money to another member of the family at a given interest rate with a specific repayment schedule.

Reference Groups

How do we use groups as a point of reference? We use certain groups to evaluate ourselves and to acquire attitudes, values, beliefs, and norms. Groups used in this way are called **reference groups**. Reference groups may include families, teachers, college classmates, student government leaders, college Greek organizations, rock groups, or professional football squads.

Reference groups influence self-esteem and behavior. Say, for instance, that because of the imagined reaction by reference groups (such as parents and teachers), a student is motivated to study. The resulting grade point average creates a sense of accomplishment and boosted self-esteem.

You may have a reference group without being a member; you may only aspire to be a member. For example, you need not be a member of a ballet troupe to view dancers as a reference group. You need only assess yourself in terms of their standards and subscribe to their beliefs, values, and norms. Junior high school

girls, for example, may imitate high school girls' leadership style or athletic interests. Junior high school boys may copy high school boys' taste in clothing and music.

Reference groups need not be positive. Observing the behavior of a group you dislike may reinforce a preference for contrary ways of acting, feeling, and behaving. For example, a violent gang may provide a blueprint for behavior to avoid.

In-Groups and Out-Groups

How is social life an interplay between in-groups and out-groups? In-groups and out-groups can exist only in relation to each other; one is the flip side of the other. An **in-group** encourages intense identification and loyalty. The level of identification and loyalty is sufficiently compelling that members tend to exclude others. An **out-group** is a group toward which in-group members feel opposition and competition. It is from membership in these groups that people divide into "we" and "they."

Where are these groups found? In-groups and out-groups develop in virtually every social arena—for example, college Greek organizations, athletic teams, political parties, racially or ethnically divided neighborhoods, and countries at war. Because of different tastes, perspectives, and values, everyone prefers some groups over others. And by virtue of being in some groups, you are out of others. As a college student, you can easily identify many of the in-groups and out-groups in your social landscape. For example, jocks and cheerleaders are in-groups for some, out-groups for others. Phi Beta Kappa is an in-group for its members and an out-group for students uninterested in academics. At Columbine High School, students knew about a number of cliques beyond the publicized jocks and cheerleaders—preppies, stoners, gangbangers, skaters, nerds. Different schools have variations of these in- and out-groups. At Austin High School in Texas, "kickers" wear oversize belt buckles, large hats, and cowboy boots. California schools have surfer cliques (A. Cohen 1999). A "we" versus "they" mentality can even develop between generations. Baby boomers (Americans born between 1945 and 1964), for instance, are accused of bashing Generation X (America's twenty-somethings) as unambitious complainers who want the good life handed to them (Giles 1994).

The formation of in-groups and out-groups depends on the establishment and protection of group boundaries. If nothing distinguishes "us" from "them," then there can be no "ins" and "outs." Group boundaries remind in-group members that movement over the line places them on the outside. To the outsiders, group boundaries form an entrance barrier (Hogg and Abrams 2001).

How are group boundaries maintained? Maintaining something as important as group boundaries requires commitment from the group members. Unfortunately, this may involve clashes with outsiders. Skinheads (neo-Nazis) periodically attack Jews or destroy their property. Members of an urban gang may injure or kill a member of another gang who has entered their territory. Boundaries are also maintained through group symbols. Symbols may be as benign as a right-to-life bumper sticker or as intentionally threatening as having one's head shaved. Groups also may adopt certain words and rituals (handshake, high five) to demarcate themselves. These are easily recognized at a social club initiation ceremony.

As this description of group boundaries suggests, in-group and out-group relationships can have serious consequences. Let's consider some of these repercussions.

What is the sociological importance of in-groups and out-groups? Membership in an in-group has both positive and negative aspects. Those on the inside may benefit from heightened self-esteem and a sense of social identity. However, in-group members may foster an inflated sense of their own worth while promoting an unrealistically negative view of others. This distortion of reality, which in-groups often deliberately create, is self-deluding. If such self-delusion becomes central to the identity of in-group members, psychological and social development is retarded. Moreover, some members may be personally damaged on realizing the false basis for their self-esteem.

Harm is not confined to in-group members. In fact, out-group members experience many more negative effects. The increase in self-esteem enjoyed by in-group members can come at the expense of outsiders. Even more serious is the physical damage that may be done by extremist groups such as neo-Nazis or certain religious factions willing to kill others in the name of their deity (O'Brien and Palmer 1993). Historically, racial or ethnic majorities have viewed minorities as outsiders—effectively denying them everything from a sense of worth to equal opportunity (Schaefer 2007). Males have been accused of using a similar pattern of dominance against females (Andersen 2005).

In-group/out-group relationships, of course, generally do not take such extreme forms. Spirited competition among sororities and friendly athletic rivalries are more typical expressions of these groups. The relative harmlessness of most cases, however, should not mask the potentially serious consequences of in-groups and out-groups, consequences related to conflict, power, and self-respect.

Social Networks

What are social networks? As individuals and as members of primary and secondary groups, we interact with many people. All of a person's social relationships constitute his or her **social network**—a web of social relationships that joins a person to other people and groups. The typical American knows between 500 and 2,500 people, and many of them know each other (Milgram 1967). This total set of social relationships—this social network—includes family members, work colleagues, classmates, physicians, church members, close friends, car mechanics, and store clerks. Social networks tie us to hundreds or thousands of people within our community, throughout the country, and even around the world (Doreian and Stokman 1997).

The Internet is expanding the amount of interaction and the flow of information within networks. Before the Internet, for example, environmental activists across the United States had to depend on slower, more cumbersome means of communication, such as the print media, the telephone, and letter writing. With the Internet, environmental organizations can supply almost unlimited information to anyone with access to the Web. For example, volunteers can now easily recruit others to join them in addressing the Chesapeake Bay's environmental problems. Protests in various regions of the country can be organized very quickly, and feedback among network members can be instantaneous.

This increased ease, speed, and frequency of social contact can promote a sense of membership in a particular network. Whereas opponents to gun control were once largely unaware of each other, they may now feel part of a nationwide social network.

Are social networks groups? Although a person's social network includes groups, it is not a group per se. A social network lacks the boundaries of a group, and it does not necessarily involve close or continuous interaction. Because many of the relationships are sporadic and indirect, not all participants experience a feeling of membership. Although a social network does not preclude close ties—like those of college alums who maintain contact via telephone, computers, and reunions—most ties in a social network are secondary in nature.

How strong are the ties in a social network? Social relationships within a network involve both strong and weak ties (Granovetter 1973; L. C. Freeman 1992). Strong ties exist in primary relationships, such as those with one's parents or spouse. These ties are emotionally close and intimate and are based on a genuine concern for the other person(s). Weak ties, most often found in secondary relationships, include salespersons, distant relatives, and most coworkers. These ties are more impersonal and goal oriented (Mizruchi and Stearns 2001).

The number and type of strong ties are heavily influenced by level of education. More highly educated people have a larger number of strong ties, a greater variety of strong ties, and more strong ties outside the family. Not unexpectedly, older people have fewer strong ties; and outside the home, urban dwellers have more strong ties than do rural residents. Although women have as many strong ties as men, more of these ties are within the family (C. S. Fischer 1982; Marsden 1987).

What are the functions of social networks? Social networks serve several important functions. First, networks provide a sense of belonging (Putnam 2001). Second, they furnish support in the form of help and advice. Finally, networks are a useful tool for those entering the labor market. Getting to know relevant people is often useful to one's career (Putnam 1996; Petersen, Saporta, and Seidel 2000).

The use of networks to locate new places of employment deserves special attention because of gender differences. Although the social networks of American females and males are equally extensive, men are more enmeshed in "good-old-boy networks" containing many higher-status contacts (Lin, Ensel, and Vaughn 1981). In the past, women tended to have more ties with family members and fewer higher-status ties. This situation is changing, however, as gender inequality diminishes in the United States and as more women are seeing networks as a pathway to higher status and more powerful positions.

FEEDBACK

1. A _____ is composed of several people who are in contact with one another; share some ways of thinking, feeling, and behaving; take one another's behavior into account; and have one or more interests in common.

2. A _____ group is composed of people who are emotionally close, who know one another well, who seek one another's company because they enjoy being together, and who have a "we" feeling.

3. Listed here are some examples of primary and secondary relationships. Indicate which examples are most likely to be primary relationships (P) and which are most likely to be secondary relationships (S).
   ____ a. a marine recruit and his drill instructor at boot camp
   ____ b. a married couple
   ____ c. a manager and his professional baseball team
   ____ d. professors and students
   ____ e. car salespersons and their prospective customers

4. Which of the following is *not* a condition that promotes the development of primary groups?
   ____ a. small group size
   ____ b. face-to-face contact
   ____ c. continuous contact
   ____ d. interaction on the basis of status or role

5. Match the following terms and statements:
   ____ a. group   (1) skinheads
   ____ b. in-group   (2) spectators at a Fourth of July fireworks display
   ____ c. social aggregate   (3) orchid growers
   ____ d. social category   (4) teenage subculture
   ____ e. reference group   (5) members of a regular Saturday night poker game

6. A _____ is a web of social relationships that joins a person directly to other people and groups and, through those individuals and groups, indirectly to additional parties.

Answers: 1. group; 2. primary; 3. a. (S), b. (P), c. (S), d. (S), e. (S); 4. d.; 5. a. (5), b. (1), c. (2), d. (3), e. (4); 6. social network

Nancy Palmeiri/AP Photos

The level of conflict is high between Bentley and Winona State in a 2008 NCAA Division II semifinal game. Is this instance of conflict disruptive to society?

Social Interaction in Groups

Social interaction is crucial to groups. Robert Nisbet (1970) describes five types of social interaction basic to group life: cooperation, conflict, social exchange, coercion, and conformity.

Cooperation

What is cooperation, and when are people most likely to cooperate? **Cooperation** is a form of interaction in which individuals or groups combine their efforts to reach some common goal. Cooperation usually occurs when reaching a goal depends on united resources and efforts. The survivors of a plane crash in a snow-covered mountain range must cooperate to survive. Victims of floods, mudslides, tornadoes, droughts, or famines must help one another to survive their crisis. This is not to say that cooperation exists only during emergencies. Children who agree to a set of rules for a game, couples who agree to share household duties, students who

march in support of a community project, and groups who organize to prevent a major power-line installation are cooperating. Indeed, without some degree of cooperation, social life could not exist.

Conflict

What is conflict, and are all consequences negative? Individuals or groups who work together to obtain certain benefits are cooperating. Those who work against one another for a larger share of the rewards are in **conflict**. In conflict, defeating the opponent is considered essential. In fact, defeating the opponent may become more important than achieving the goal and may bring more joy than winning the prize.

As pointed out earlier in "Using the Sociological Imagination," conflict is usually considered a disruptive form of interaction—one to be minimized or eliminated. A cooperative, peaceful society is assumed to be better than one in conflict. Although conflict does have some undesirable consequences, it can also be socially beneficial.

According to Georg Simmel (1858–1918), a major benefit of conflict is the promotion of cooperation and unity *within* opposing groups. Although conflict may lead to the separation of groups, it can strengthen relationships within groups. Thus, the Revolutionary War drew American colonists together even though it brought them into conflict with the British. Similarly, a labor union becomes more integrated during the process of collective bargaining. And police officers who attempt to stop a husband from physically abusing his wife may find themselves under attack by both mates. According to historian Stephen Oates (1977), the "outside meddling" of Stephen Douglas and other Democrats united the Illinois Republican party and secured the U.S. Senate nomination for Abraham Lincoln. The threat of an impeachment trial of President Clinton united the Democratic party during the 1998 congressional elections. And the events of 9/11 unified Americans with increased levels of patriotism.

Another positive consequence of conflict is the rejuvenation of norms and values. Equality of opportunity, freedom, and democracy have always been important values to Americans. But until the nonviolent civil rights demonstrations, the urban violence in large northern cities, and the abuse experienced by civil rights supporters, as well as the poverty, inequality, and lack of opportunity among America's minorities were largely invisible to white society. Once these problems were forcibly called to the public's attention, actions were taken to reaffirm these basic values. Among these actions were Lyndon Johnson's War on Poverty, the Civil Rights Act of 1964, the Equal Employment Opportunity Act of 1972, and affirmative action programs (Fineman 1992).

Conflict may also be beneficial when it challenges norms, beliefs, and values. Domestic opposition to the Vietnam War in the 1960s—which tragically involved unparalleled conflict and violence on college and university campuses—ended the military draft, brought into question the legitimacy of political and military intervention in other countries, enlarged the American value of patriotism to include opposition to certain wars, and encouraged higher education to be more responsive to students' needs. These events did alter some norms, beliefs, and values—a lesson not lost on African Americans during the 1992 riots in Los Angeles.

A study by Douglas Maynard (1985) concludes that conflict can have benefits even for young children. In a study of first-grade children, Maynard observed a sense of social structure emerging from disputes and arguments. Authority and friendship patterns were also the product of social conflict.

Social Exchange

What is the nature of a social exchange relationship? In *The Nicomachean Ethics*, Aristotle observed:

> *All men, or most men, wish what is noble but choose what is profitable; and while it is noble to render a service not with an eye to receiving one in return, it is profitable to receive one. One ought, therefore, if one can, to return the equivalent of services received, and to do so willingly.*

In this passage, Aristotle touches on **social exchange**, a type of social interaction in which one person voluntarily does something for another, expecting a reward in return. If one person helps a friend paint her room, expecting that some equivalent help will be returned, the relationship is one of exchange. In an exchange relationship, it is the benefit derived rather than the relationship itself that is central. An individual who does something for someone else obligates that person. This obligation can be repaid only by a return favor. Thus, the basis of an exchange relationship is reciprocity, the idea that you should do for others as they have done for you (P. M. Blau 1986).

Exchange relationships are not always exploitative; many exchange relationships are based on trust, gratitude, and affection. You may help someone move out of a dormitory into an apartment because he has been friendly to you. You may expect some response—help when you move, or maybe just an expression of appreciation. We usually do anticipate some response even though we may be unaware of it.

What is the difference between cooperation and exchange? Although both cooperation and social exchange involve working together, there is a significant difference between these two types of interaction. In cooperation, individuals or groups work together to achieve a shared goal. Reaching this goal, however, may or may not benefit those who are cooperating. And although individuals or groups may profit from cooperating, that is not their main objective. For example, a group may work to build and maintain an adequate supply of blood for a local blood bank without thought of personal benefit. If, however, the primary objective of the group is to guarantee blood availability for its own members, then it has an exchange relationship with the blood bank. In cooperation, the question is, How can we reach our goal? In exchange relationships, the implied question is, What's in it for me?

Coercion

In what way is coercion an involuntary interaction? In **coercion,** an individual or group compels others to behave

These campaign volunteers are celebrating after Hillary Clinton won a Nevada presidential precinct caucus. Is this an example of cooperation or exchange?

in certain ways. The central element is domination. In a sense, coercion is an unequal exchange: because of superior power, someone can get something from someone else without repaying. Domination may occur through physical force, such as imprisonment, torture, or death. But coercion is generally expressed more subtly through social pressure—ridicule, rejection, ostracism, withdrawal of love, denial of recognition. Coercion is a part of many social relationships. Parents may coerce children with a curfew; employers may coerce employees with dismissal; professors may coerce students to study via the threat of a test. Obviously, conflict theory underlies this type of social interaction.

Conformity

Do most people conform to group pressures? Conformity is behavior matching group expectations. When we conform, we adapt our behavior to fit the behavior of those around us. While some people are more likely to conform than others are, most people conform to the expectations of some group most of the time. Social life—with all its uniformity, predictability, and orderliness—simply could not exist without this type of social interaction. Without conformity, there could be no churches, families, universities, or governments; without conformity, there could be no culture or social structures (see "Doing Research").

The Business of Groups

Nineteenth-century social philosopher Georg Simmel was one of the first to recognize the importance of groups for society. But, he argued, group members face a dilemma. Belonging to a group requires that members submit to social control, losing some personal freedom in the process. If these people remain outside the group, however, they lose such possible benefits of membership as economic security and social acceptance (Simmel 1964; originally published in 1908). For this reason, most groups encounter both cooperation and conflict. Because these pressures can interfere with the smooth functioning of the groups, social scientists have devoted considerable attention to task accomplishment within groups, various leadership styles for groups, and group decision making.

What is the nature of task accomplishment within groups? Most groups are designed to accomplish certain tasks, which can range from planning leisure activities to manufacturing a product. At the same time, members of groups have self-images and desires that do not necessarily mesh with the goals of the group as a whole. Beginning with Robert Bales (1950), social scientists have tried to understand the ways in which task accomplishment within groups could expedite both the achievement of the group's goals and the needs of individual group members.

Bales developed a scheme, called *interaction process analysis*, to classify group interactions into several categories. He identified twelve separate categories, which can be understood in relation to two basic problems that, according to Bales, must be solved by any social system. First, *instrumental problems* must be solved. Instrumental problems are those that are directly related to achieving the goals of the group. A group trying to elect a political candidate must solve problems regarding fund-raising, advertising, arrangements for speeches and rallies, charges from the opposing candidate, and so on. Second, *social-emotional problems* must receive attention. These are the problems involving individual satisfaction, disagreements, and other related matters that inevitably arise whenever individuals try to coordinate their activities.

A number of laboratory studies conducted by Bales and his associates have further indicated that both

---

TABLE 6.2

Illustrating Types of Social Interaction

A type of social interaction is illustrated from the viewpoint of a particular theoretical perspective. Each interaction can be viewed from either of the other two perspectives. Associate a type of social interaction with a different theoretical perspective, and make up your own example.

Theoretical Perspective	Type of Social Interaction	Example
Functionalism	Conformity	• Team integration is promoted when basketball players accept their roles on the floor.
Conflict theory	Coersion	• Conflict in prisons is kept in check by the superior power of the guards.
Symbolic interactionism	Social exchange	• Two college students may work together in preparing for an exam because passing a course carries the same symbolic value for each of them.

instrumental and social-emotional processes are important, as Stephen Wilson notes:

> . . . *if the group is going to do its job, (1) task-oriented interaction must be high; (2) more answers must be given than questions asked; (3) there must be more interaction contributing to the solidarity of the group (positive social-emotional) than detracting from solidarity (negative social-emotional). In other words, there must be, as Bales's theory suggests, some degree of balance to keep the individuals together as a group.* (S. Wilson 1978:66–67)

What is the nature of leadership in groups? It was once believed that leaders are born and that people have to possess certain traits to be effective as leaders. It is now known that no leadership trait is absolutely necessary for effective leadership, but Ralph Stogdill has suggested that successful leaders are likely to have the following combination of traits:

> *The leader is characterized by a strong drive for responsibility and task completion, vigor and persistence in pursuit of goals, venturesomeness and originality in problem solving, drive to exercise initiative in social situations, self-confidence and sense of personal identity, willingness to accept consequences of decision and action, readiness to absorb interpersonal stress, willingness to tolerate frustration and delay, ability to influence other persons' behavior, and capacity to structure social interaction systems to the purpose at hand.* (Stogdill 1974:81)

Such traits are likely to increase the effectiveness of leaders only if they are translated into behavior, which is the reason appropriate leadership behavior has been the subject of considerable research. Studies by Rensis Likert (1961, 1967) led him to emphasize the importance of supportive behavior, in which leaders try to provide direction in a way that maintains others' sense of personal worth and importance. Similarly, Ralph White and Ronald Lippitt (1960) found that democratic leaders (who try to build consensus about decisions) are more effective than authoritarian ones (who give orders without considering the preferences of group members). Both, however, were found to be more effective than laissez-faire leaders (who make little or no attempt to organize group activities). Subsequent studies suggest that appropriate leadership behavior depends on the situation presented by the group to be led. Fred Fiedler (1967), for example, has argued that leaders should be more authoritarian if they face a situation in which they have very high or very low amounts of control. In intermediate situations, leaders should be more democratic and concerned with the quality of human relationships.

If Fiedler's theory is correct, the popular leader of an infantry platoon (high control) or the disliked supervisor of an ineffective work crew (low control) should be more authoritarian than the well-liked director of a research team facing an unstructured task (intermediate control). Fiedler's theory has been criticized on several grounds, but it is an indication of the increasing tendency for social scientists to understand leadership in its situational context (Ashour 1973; Schriesheim and Kerr 1977).

Can groups make better decisions than individuals? Groups are used in many settings for the solution of a variety of problems. In some situations, group decisions can be shown to be superior to the decisions that would be made by individuals acting alone. For example, researchers have found that groups are useful for solving intellectual tasks because they allow people to hear several points of view and provide a setting in which the merits of various ideas can be discussed (H. H. Kelley and Thibaut 1969).

However, group pressure can also harm the quality of decisions by leading to excessive conformity and loss of individuality. The Spanish writer José Rodó has written that "every society to which you remain bound robs you of a part of your essence and replaces it with a speck of the gigantic personality which is its own." Actually, the same thing could be said of each group to which an individual is strongly committed. For the sake of group membership, we may relinquish part of our individuality and independence.

These business people are in the process of making a decision. Name one possible positive consequence and one negative consequence of group decision making.

The tendency to conform to group pressure has been dramatically illustrated in a classic experiment by Solomon Asch (1955) in which many participants publicly denied their own senses in order not to deviate from majority opinion. Asch assembled numerous small groups of male college students, ostensibly for experiments in visual judgment. He asked the groups to compare the length of lines printed on two cards (see Figure 6.1). One card showed a single vertical line, which was to be used as a standard for comparison with three lines of different lengths on a second card. Each participant was asked to identify the line on the second card that matched in length one of the lines on the first card. In each group, all but one of the subjects had been instructed by Asch to choose a line that obviously did not match. The naive subject—the only member of each group unaware of the real nature of the experiment—was forced either to select the line he actually thought matched the standard line or to yield to the unanimous opinion of the group. In earlier tests of individuals in isolation, Asch had found that the error rate in matching the lines was only 1 percent. Under group pressure, however, the naive subjects went along with the majority's wrong opinion over one-third of the time. If this large proportion of naive subjects yielded to group pressure in a group of strangers, it is not difficult to imagine the conformity in groups to which people are emotionally committed (Myers 2001).

How does "groupthink" discourage alternative points of view? Irving Janis (2006) has argued that many group decisions are likely to be the product of **groupthink**, a situation in which pressures toward uniformity discourage members of the group from expressing their reservations about group decisions. During the Kennedy administration, for example, a decision to launch the ill-fated Bay of Pigs invasion against Cuba was made only because several top advisers to the president failed

FIGURE 6.1

Cards for Asch's Experiments

Which of the lines on card A matches the line on card B? You must be thinking, "What a no-brainer." You may be surprised to learn that in a group setting, many people associated the other two lines in card A with the line on card B. Read about Asch's experiment in the text.

Card A   Card B

to reveal their reservations about the plan. Another example of groupthink is said to be found in the George W. Bush administration's buildup to the invasion of Iraq in 2003. The small number of designers of this plan are accused of selecting only intelligence that could lead to allegations of the presence of weapons of mass destruction and Al Qaeda in Saddam Hussein's Iraq to convince those who did not share their convictions that war was necessary (Heilbrunn 2008; Kaplan 2008; Weisberg 2008). Such hesitancy to destroy the illusion of "assumed consensus" often leads to group decisions that have not been subjected to critical evaluation. Research indicates that groupthink can be avoided only when leaders or group members make a conscious effort to see that all group members participate actively and that points of disagreement and conflict are tolerated.

We have been viewing groups as relatively small social units. Now, we turn to larger social units known as for-

F
E
E
D
B
A
C
K

1. Match the following types of social interaction with the examples beside them:
   ____ a. cooperation   (1) paid blood donors
   ____ b. conflict   (2) students doing what a professor assigns
   ____ c. social exchange   (3) Saddam Hussein invading Kuwait
   ____ d. coercion   (4) flood victims helping each other
   ____ e. conformity   (5) employees forced to work overtime or be fired
2. According to Robert Bales, interaction within groups must be designed to solve both _____ problems and _____ problems.
3. Social scientists agree that leaders must have certain traits if they are to be effective. T or F?
4. Solomon Asch's experiment demonstrates the positive consequences of group pressure. T or F?
5. A situation in which pressures toward uniformity discourage members of a group from expressing their reservations about group decisions is called _____.

Answers: 1. a. (4), b. (3), c. (1), d. (5), e. (2); 2. instrumental, social-emotional; 3. F; 4. F; 5. groupthink

These impatient bank customers are clearly experiencing the effects of formal organizational structure. Do you think most organizations are bureaucratic in nature?

Randy Faris/Corbis Super RF/Alamy Limited

mal organizations. It will become evident that groups are very much a part of any formal organization.

Formal Organizations

Even in the early part of this century, Charles Horton Cooley (1902) foresaw a shift from primary groups toward secondary groups. Until the 1930s, most Americans lived on farms or in small towns and villages. Nearly all of their daily lives were spent in primary groups—families, neighborhoods, and churches. As industrialization and urbanization advanced, however, Americans became more enmeshed in secondary groups. Born in hospitals, educated in large schools, employed by huge corporations, regulated by government agencies, cared for in nursing homes, and buried by funeral establishments, Americans—like members of other industrialized societies—find themselves within secondary groups known as *formal organizations*. This major change accompanies the transition from preindustrial to industrial society. The trend is continuing. Citizens of modern societies have become highly dependent on large organizations—both business and nonbusiness. Understanding the nature of organizations is essential to comprehending industrial and postindustrial society (Blau and Meyer 1987; Pfeffer 1997).

Theoretical Perspectives and Formal Organizations

Although it is not always explicit, functionalism and conflict theory are intimately involved in the sociological study of formal organizations. Bureaucracy, for example, with its emphasis on structure, stability, rules, and hierarchy, is based squarely on functionalism. When sociologists examine the dynamics of organizations, they don't abandon functionalism, but conflict theory does play a predominant role.

The Structure of Formal Organizations

A **formal organization** is deliberately created to achieve one or more goals. Examples of formal organizations are high schools, colleges, corporations, government agencies, and hospitals.

How are formal organizations and bureaucracies related? Most formal organizations today are **bureaucracies**—formal organizations based on rationality and efficiency. Although bureaucracies are popularly thought of as monuments to inefficiency, they have proved to be an efficient form of organization for industrial society. Despite his deep concern about its negative consequences, Max Weber, who first analyzed the nature of bureaucracy, wrote of its efficiency.

What are the major characteristics of bureaucracy? From a study of history and an observation of events of his day, Max Weber sketched the basic organizational principles that were being developed to promote rational and efficient administration. Weber identified the following characteristics of a bureaucracy:

- *The organization has a division of labor based on the principle of specialization.* In a bureaucracy, each person is responsible for certain functions or tasks. (See Figure 6.2 for an organizational chart outlining the division of labor in a modern university.) A specialized division of labor improves organizational performance because an individual can become an expert in a limited area of organizational activity.
- *The organization has a hierarchy of authority.* To clarify the concept of authority, it is necessary to distinguish it from power. Power is to the ability to control the behavior of others, even against their will. Authority is the exercise of legitimate power—power that produces compliance because those subjected to it believe that obedience is the proper

Stanley Milgram— Group Pressure and Conformity

Can a group move a normally nonviolent person to physically and severely punish a victim? As incredible as this may sound, researcher Stanley Milgram (1964) revealed this as a possibility.

As noted in the text, Solomon Asch (1955) demonstrated that group pressure can influence people to such an extent that they will deny their true sense perceptions. Specifically, they can be pressured to match up uneven lines—lines they originally perceived as different lengths. Milgram wanted to know if group pressure could have the same effect on behavior. Can group pressure cause people to treat others in ways they otherwise would not?

To test this question, Milgram could have chosen a behavior relatively easy to induce, such as sharing food with a stranger or doing minor damage to someone else's property. Instead, Milgram asked research participants to administer increasingly higher electric shocks to people who appeared to be in pain. And these research participants were people just like you and me!

Milgram placed eighty males in an experimental situation in which a team of three individuals (one naive subject who did not know the nature of the experiment and two "confederates" who did know the nature of the experiment) tested a fourth person (another confederate in on the experiment) on a learning task. Each mistake by the fourth party brought an apparent electric shock from the

naive subject. The two confederate "testers" (participants cooperating with Milgram) on each team suggested increasingly higher shock levels for successive mistakes made by the "learner" (the confederate in the electric chair). The researcher recorded the degree to which the team's naive member resisted or went along with group pressure to increase the voltage levels.

After explaining to the naive tester that the experiment was investigating the effects of punishment (electric shocks) on memory, the researcher strapped the learner into an electric-chair apparatus in full view of the other three team members. These three were seated in front of a large shock generator whose switches were labeled from 15 to 450 volts in 15-volt increments. The lowest-level group of switches was labeled "Slight Shock"; the highest-level group of switches read "Danger: Severe Shock."

The naive subject controlled the maximum shock that could be administered. He could keep the shock level at 15 volts throughout or increase the shock level in line with the group's recommendation. It was his decision.

The learner in the electric chair not only answered questions; as the shocks became stronger, he grunted, protested, and finally demanded that the experiment stop. Each of these complaints was actually prerecorded and played in coordination with a given shock level. No discomfort was indicated until a 75-volt shock was administered. At 120 volts, the learner shouted that the shocks were becoming painful. Groans followed a 135-volt shock, and at 150 volts, the learner cried out to be released from the experiment, complaining of a heart condition. At 180 volts, the learner cried out, "I can't stand the pain," and 285 volts was followed by an agonized scream. From 300 to 450 volts, the victim continuously shouted desperately that he would no longer answer questions.

Milgram first needed to know how far the naive subject would go in administering shocks *without* group influence. To accomplish this, an identical experiment was run minus the two confederates in the punishing group. Response to group pressure was measured by the difference in the naive subject's behavior in the two situations.

Group pressure heavily affected the level of shock administered by the naive subject. That is, the average shock level in the three-person situation was significantly higher than in the one-person situation. Perhaps most interesting are the results on *maximum* shock levels. When alone, only two naive subjects went beyond the point where the learner first vehemently protested. Under group pressure, twenty-seven of the naive subjects went beyond this point. Nineteen of the naive subjects went above 255 volts ("Intense Shock"), and ten went into the group of voltages labeled "Danger: Severe Shock." Seven even reached 450 volts (the highest shock level possible).

The research by Milgram and Asch reveals the power of group pressure to create conformity in thought and behavior. Although conformity must occur to a degree for social structure and society to exist, many scholars worry about the extent to which social pressure can determine how humans think and act.

Evaluate the Research

1. Discuss the ethical implications of Milgram's experiment. (You may want to refer to Chapter 2 for pointers about ethics in social research.)

2. If the researcher had not been present as an authority figure during the experiment to approve the use of all shock levels, do you think group pressure would have been as effective? Explain.

response. Bureaucratic organizations tend to have a pyramidal shape. The greatest amount of authority is concentrated in the few positions at the top, with decreasing amounts in the expanding number of lower positions (see Figure 6.2). Organizational effectiveness and efficiency are enhanced by a hierarchy that can coordinate the many statuses involved in a highly specialized organization.

- *Organizational affairs are based on a system of rules and procedures.* Rules and procedures guide the performance of work and provide a framework for decision making. They stabilize the organization because they coordinate activities and cover most situations.
- *Members of the organization maintain written records of their organizational activities.* These records of work and activities are kept in files. This organizational "memory" is essential to smooth functioning, stability, and continuity.
- *Statuses in the organization—especially managerial ones—are considered full-time jobs.* Bureaucratic statuses are not considered avocational. This increases the commitment of the organization's members and demands their full attention.
- *Relationships within the organization are impersonal, devoid of favoritism.* For example, personnel are selected on the basis of technical and professional qualifications and promoted on the basis of merit. Similarly, relationships with clients are conducted without regard to personal considerations. The norms prohibiting the influence of favoritism and personal relationships in organizational activities are intended to ensure equal treatment for those who work in the organization as well as for those who are served by it.
- *Employees of bureaucratic organizations do not own their positions.* Bureaucratic positions cannot be sold or inherited. This allows organizations to fill positions with the most qualified people. Those in bureaucratic positions are held responsible for their use of equipment and facilities.

For Weber, bureaucratic organization involved a set of strategies that would promote order in human relationships. Weber believed that the achievement of order would be increasingly difficult in societies undergoing industrialization. In preindustrial societies, activities in one aspect of life had been closely related to activities in every other aspect. As societies modernized, people's activities became less closely related, and traditional systems of norms and values began to weaken. As Kenneth Thompson indicates, Weber saw bureaucracy—with its hierarchy of authority, system of rules and procedures, and inherent safeguards—as a way to maintain control in an increasingly complex society:

The spread of the bureaucratic form of organization to all spheres was part of a general process of rationalization in

---

FIGURE 6.2

University Organization Chart

Each organizational position and department within a bureaucracy has certain tasks associated with it. The connecting lines indicate who reports to whom and who has authority over whom.

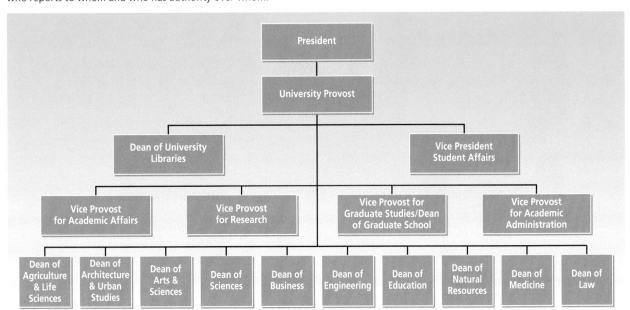

modern society. The reason why it would inevitably spread was because its characteristics of precision, continuity, discipline, strictness and reliability make it technically the most satisfactory form of organization for those who sought to exercise organizational control. (K. Thompson 1980:10–11)

Do modern organizations practice these characteristics of bureaucracy? In his formulation of bureaucratic characteristics, Weber used the **ideal-type method**. This method isolates, to the point of exaggeration, the most basic characteristics of some social entity. It involves the construction of a *pure* type. Because real life is seldom as pure as an abstract description, social entities do not have all the characteristics of an ideal type to the maximum degree. Universities, for example, are less bureaucratic than business firms, in part because universities employ large numbers of professionals whose quality of work is adversely affected by a bureaucratic environment (Litwak 1961).

Practically speaking, then, the characteristics of bureaucracy are best thought of as a series of continua; various organizations emphasize each characteristic of bureaucracy to different degrees. In fact, some units within an organization may be more bureaucratic than others (R. H. Hall 1962). In a university, for example, the president's office is much more bureaucratic than an academic department of philosophy or biology. The more closely an organization is based on the characteristics identified by Weber, the more bureaucratic it is.

Advantages of Bureaucracy

What are the advantages of bureaucracy? Citing the advantages of bureaucracy, Weber wrote:

The decisive reason for the advance of bureaucratic organization has always been its purely technical superiority over any other form of organization. The fully developed bureaucratic mechanism compares with other organizations exactly as does the machine with the nonmechanical modes of production. (Gerth and Mills 1958:214)

Weber's model of bureaucracy was formulated in response to the inadequacies of earlier organization forms. In monarchies, for example, positions were appointed on the basis of favoritism and politics. Administrators had little or no training for the work they were doing. Because these officials were wealthy, administration was only an avocation. And because high positions were politically based, power could be lost overnight. Moreover, when a king or queen was replaced, the entire leadership structure changed drastically.

As Weber correctly saw, earlier types of organizational structure were out of step with the rise of the capitalist market economy, which required steadiness, precision, continuity, speed, efficiency, and minimum cost—advantages bureaucracy could offer. **Rationalism**—the solution of problems on the basis of logic, data, and planning rather than tradition and superstition—was

SNAPSHOT OF AMERICA 6.1

Membership in Labor Unions

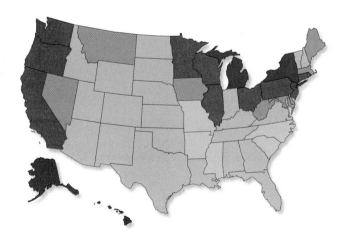

Union Membership Rates (2004)

- 20% or above
- 15.0%–19.9%
- 10.0%–14.9%
- 5.0%–9.9%
- 4.9% or below

Source: U.S. Department of Labor, Bureau of Labor Statistics, 2005.

Labor unions, organizations created by workers to improve working conditions, are bureaucracies. They have a division of labor, a hierarchy of authority, a system of rules and procedures, written records, and promotion based on merit within the organization. This map shows the percentage of employed workers in each state who are members of labor unions.

Interpret the Map

1. Which states had the highest union membership rates in 2004? Which states had the lowest rates?
2. Why do you think it is that not all workers are members of labor unions?

on the rise, and the characteristics of bureaucracy were consistent with this sweeping trend.

As strange as it might sound, bureaucracy is designed to protect individuals. Without rules, procedures, and the norm of impersonal treatment (which everyone hates), decision making would be arbitrary. It might sound great, for example, to abolish final exams; but then grading might not be as objective. This, of course, is not to say that most professors in a nonbureaucratic setting would be abusive or that favoritism does not occur in bureaucratic organizations. Nevertheless, formally established rules and procedures and the norm of impersonality ensure an important degree of equal treatment.

A famous account of the benefits of specialization is found in Adam Smith's *The Wealth of Nations*. Published in 1776, this book attempts to answer the question: What is the source of wealth in a society? For centuries, the mercantilists had found wealth in the gold and silver in the royal treasury. Smith, expressing the new economic thought of the French physiocrats, gives a different answer: the wealth of nations resides in the productive power of labor, power that is best promoted by the division of labor. Smith illustrates with a description of a pin factory, citing the benefits that come when "one man draws out the wire, another straightens it, a third cuts it, a fourth points it, a fifth grinds it at the top for receiving the head" (A. Smith 1965:4/1776). By this process, Smith notes, ten men working in the factory can make 48,000 pins daily, whereas a man working alone could make only 20 or so pins. Although recognizing pin making as a "trifling" example, Smith emphasizes that the division of labor is readily applicable to any type of "art and manufacture."

Disadvantages of Bureaucracy

What are the disadvantages of bureaucracy? Although Weber praised bureaucracy as the most efficient type of organization, he was well aware of undesirable side effects. He expressed concern that the formal rationality of bureaucracy would lead to socially undesirable outcomes. For one thing, Weber noted that bureaucracy represented "the concentration of the material means of management in the hands of the master" (Gerth and Mills 1958:221). This leads to concerns about who controls bureaucratic organizations. In the words of Charles Perrow, one of the foremost contemporary scholars of organizations:

At present, without huge, disruptive, and perilous changes, we cannot survive without large organizations. Organizations mobilize social resources for ends that are often essential and even desirable. But organizations also concentrate those resources in the hands of a few who are prone to use them for ends we do not approve of, for ends we may not even be aware of, and, more frightening still, for ends we are led to accept because we are not in a posi-

tion to conceive alternative ones. The investigation of these fearful possibilities has too long been left to writers, journalists, and radical political leaders. It is time that organizational theorists began to use their expertise to uncover the true nature of bureaucracy. This will require a better understanding both of the virtues of bureaucracy and its largely unexplored dangers. (Perrow 1986:6)

Weber himself raised the specter of what he called the *iron cage of rationality*. He feared the likelihood of rationality spreading to all aspects of social life, creating a dehumanizing social environment and entrapping everyone. George Ritzer (1998, 2004, 2006) refers to the "McDonaldization of society" as a metaphor to express Weber's iron cage of rationality. McDonald's, Ritzer argues, is the epitome of rationalism in organizations (see "Doing Research", Chapter 1). On the dehumanizing effects of extreme rationality, Ritzer writes:

. . . fast-food restaurants are . . . often dehumanizing settings in which to eat or work. Customers lining up for a burger or waiting in the drive-through line, and workers preparing the food often feel as though they are part of an assembly line. Hardly amenable to eating, assembly lines have been shown to be inhuman settings in which to work. (Ritzer 2004:17)

It is the McDonaldization of everything from lube jobs to medicine that Ritzer believes threatens modern society with Weber's iron-cage nightmare. If the nightmare is still there when we awaken, then the rational has become irrational (Alfino, Caputo, and Wynyard 1998).

Although Weber was deeply concerned about the societal implications of bureaucracy, others have focused on bureaucracy's internal inefficiencies (Hummel 1987; Leibenstein 1987). The famous Parkinson's law, named after C. Northcote Parkinson (1993), states that work always expands to fill the time available for its completion. According to this principle, there is often little relationship between the amount of work to be done and the size of the staff assigned to do it. Thus, even when there is more staff than needed to do a given amount of work, those assigned to do the work will always find plenty to do, and their supervisors may even feel that additional help is needed. Parkinson's law is obviously intended to portray wastefulness in bureaucracies.

It is important to note that Parkinson's law applies to the bureaucratic form. The blame for wastefulness in bureaucracy lies not with lazy individuals but with the way this type of organization grows. According to Parkinson, the phenomenon he describes exists because bureaucrats want to multiply the number of people under them in the hierarchy and because bureaucrats spend much of their time creating work for one another.

In addition, although promotion on the basis of merit is supposed to decrease favoritism and increase efficiency, it may also have some drawbacks. Those selected for promotion in an organization are, if the

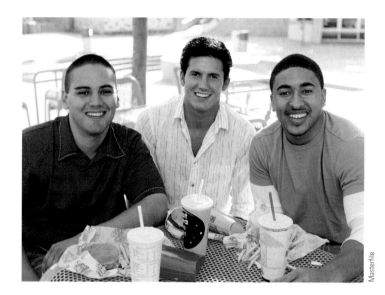

Eating fast food seems like a harmless activity (fat content aside). However, many sociologists think the rationality and standardization undergirding fast-food industry promote dehumanization. Do you?

Masterfile

system is based on merit, those who are doing their present jobs well. As long as they continue to do well in their new jobs, these people are likely to continue to be promoted. The problem, according to Laurence Peter and Raymond Hull (1996), is that people are promoted until they find themselves in jobs that are beyond their capabilities. Thus, according to the Peter Principle, employees in a bureaucratic hierarchy tend to rise to their level of incompetence. Organization members who have reached their level of incompetence shift their concern from performing competent work to sustaining organization values. They also begin to evaluate subordinates on the basis of promptness, courtesy, and cooperativeness rather than competence.

In *The Organization Man*, William H. Whyte (1972; Whyte and Nocera 2002) maintains that organizations have become so influential that they shape our nature. Organizations, Whyte argues, reward people who are good team players and who can sell their personalities and promote the organization at the same time. "Organization men and women" sell themselves, adjust to different social environments, remain flexible, remain uncommitted to any set of values, and adjust their personalities to the people and groups around them.

These popular ideas about bureaucracy have gained wide acceptance, even though scientific evidence supporting them does not exist, and even though Parkinson's law and the Peter Principle were both satirically written. It is not that these ideas are necessarily wrong; we simply cannot evaluate their validity at this time because they defy scientific study. There are, however, other negative consequences of bureaucracy that are more widely recognized by social scientists. Among these are goal displacement and trained incapacity.

What is goal displacement? When organizational rules and regulations become more important than organizational goals, **goal displacement** occurs (Merton 1968).

Professors who emphasize grades more than learning are practicing goal displacement. Social workers who become more concerned with eligibility requirements than with delivering services are engaging in goal displacement.

What is trained incapacity? According to Thorstein Veblen (1933), **trained incapacity** exists when previous training prevents someone from adapting to new situations. Trained incapacity involves tunnel vision, inflexibility, and inability to change. Let me offer an illustration based on personal experience. At one point during my graduate training in sociology, the federal government was giving me financial support. With a child on the way, I felt it was necessary to earn some additional money. The chairperson of the sociology department permitted me to teach one course, providing the university administration agreed that the rules would permit receiving money from the federal government and the university at the same time. I posed this question to the appropriate university official and watched a perplexed bureaucrat fumble through the code of regulations for a pat answer, which simply would not present itself. After minutes of obvious discomfort, he found what he was looking for. With the relief of a winner at Russian roulette, this man said excitedly, "Here it is!" Although the new regulation could have been interpreted otherwise, this administrator chose the interpretation that bolstered his original inclination to deny permission. I later learned that earlier policy had prohibited persons on grants from obtaining additional financial support from the university. This bureaucrat had become so accustomed to this way of doing things that he could not interpret the new ruling in any way other than through the old organizational viewpoint that had become so indelibly his own. By past training and experience, he had developed an incapacity to meet the new organizational realities.

More seriously, the crew of the USS *Vincennes*, on July 3, 1988, mistook an Iranian passenger airliner for an Iranian F-14 jet fighter heading their way. Instead of destroying a menacing military plane, the *Vincennes* crew killed all of the nearly 300 civilian passengers on a regularly scheduled flight over the Persian Gulf. This tragic accident occurred because the *Vincennes* was created to operate on the open sea, not on a small body of water with all sorts of air and water vehicles constantly coming into its radar sights. Instead of identifying the passenger airliner for what it was, the crew acted on the basis of its training and did what it thought was necessary for self-protection. The crew's training resulted in a tragedy, essentially preventing the crew from adapting to the new environment. The crew's incapacity stemmed from the nature of its training.

Do alternatives to bureaucracy exist? Weber developed his theory of bureaucracy for the organizations that were in the changing society in which he lived. During the 1960s, some sociologists began to see bureaucracy as an outmoded organizational form. Warren Bennis (1965), for example, contended that bureaucracy was no longer able to deal with the individual goals of employees and the increasing rates of social change. He therefore predicted that bureaucracies would be replaced by **organic-adaptive systems**—organizations based on rapid response to change rather than on the continuing implementation of established administrative principles:

> *The key word will be "temporary"; there will be adaptive, rapidly changing temporary systems. These will be "task forces" organized around problems-to-be-solved. The problems will be solved by groups of relative strangers who represent a set of diverse professional skills. The group . . . will evolve in response to the problem rather than to programmed role expectations. The "executive" thus becomes coordinator or "linking pin" between various task forces. . . . People will be differentiated not vertically, according to rank and role, but flexibly and functionally according to skill and professional training.* (Bennis 1965:34–35)

Bureaucracy, of course, has not disappeared, and Bennis (1979) has qualified some of his earlier predictions. Increasing numbers of organization researchers would agree, however, that bureaucracy is best suited for relatively stable, predictable situations. In situations characterized by rapid change and uncertainty, other forms of organization are likely to emerge (Mintzberg 1979). Such attempts to relate the structure of organizations to the situations they face have been labeled the *contingency* approach to organization theory (Burns and Stalker 1966; P. R. Lawrence and Lorsch 1967; Shepard and Hougland 1978). Although the contingency approach has been criticized on some grounds (Pennings 1975; Pfeffer 1978), it is considered a useful corrective to the traditional assumption that there exists "one best way" to organize.

A strong advocate of alternatives to the bureaucratic form of organization, Joyce Rothschild (1979, 1986) has studied cooperative organizations—organizations intended to be on the other end of the spectrum. Cooperative organizations, owned and managed by their employees, are characterized by full membership participation, minimization of rules, promotion of primary social relationships, hiring and promotion based on intimate ties and shared values, elimination of status differences, and a deemphasis on job specialization.

Although some analysts contend that nonbureaucratic forms of organizations are best suited to new business technologies and the future environmental context (Drucker 1988; T. Peters 1988; Mazaar 1999), most scholars agree that bureaucracy will remain the dominant organizational form. Bureaucracies, it is contended, persist in part because they work best with the technologies and environments characteristic of the modern era. This does not mean, however, that some organizations cannot be nonbureaucratically structured. It is just that such organizations will be few in number. Nonbureaucratic organizational principles are most useful as a supplement to bureaucratic organization, as in research laboratories or in organizational work teams (S. P. Robbins 1990).

What are some of the differences among formal organizations? Sociologists, such as Amitai Etzioni (1975), have attempted to take differences in formal organizations into account by classifying organizations into categories. Some organizations, such as churches, political parties, universities, and service clubs, are joined because people want to join; they are joined because of personal interest or emotional commitment. Because members are free to join or not, these are voluntary organizations. Etzioni refers to them as *normative organizations* because shared understandings provide an important basis for coordination of members' activities. Organizations that people are forced to join are *coercive organizations*. Examples include prisons, custodial mental hospitals, concentration camps, forced-labor camps, elementary schools, and military systems using the draft. Other organizations are *utilitarian organizations*; people join them because of the benefits they derive from membership. Work organizations, including factories, offices, and professional firms, are examples of utilitarian organizations. Utilitarian organizations fall somewhere between the first two types. People are not forced to work (it is not a coercive organization), and yet they must work to be self-supportive (it is not a voluntary organization). Thus, utilitarian organizations are neither entirely voluntary nor entirely coercive; they have some elements of both.

Whereas Etzioni classified organizations according to the motivation for membership, Peter Blau and Richard

F E E D B A C K

1. A formal organization is _____
   _____ .
2. The ability to control the behavior of others even against their will is _____ .
3. Which of the following has *not* been seen as an advantage of bureaucracy?
   ____ a. its avoidance of the use of inappropriate criteria in hiring employees
   ____ b. its use of rules to provide definite guidelines for behavior within the organization
   ____ c. its success in hiding the true nature of authority relationships
   ____ d. its encouragement of administrative competence in managers
4. Social workers who become more concerned with eligibility requirements than with delivering services to their low-income clients are engaging in _____ .

Answers: 1. a group that has been deliberately and consciously created for the achievement of one or more specific goals; 2. power; 3. c; 4. goal displacement

Scott (1982) have classified organizations according to the recipients receiving the most benefit. In *mutual-benefit organizations* (political parties, professional associations, labor unions, social clubs), the prime beneficiary is the membership. These organizations exist to promote the interests of those who belong to them. *Business organizations*, on the other hand, are expected to serve the interests of their owners. Because these organizations are profit oriented, they emphasize maximum gain at the least cost. The prime beneficiary of a *service organization* is the organization's clients. The goal of social work agencies, schools, and hospitals is to aid those people who qualify for their services. *Commonwealth organizations* (the military, the Department of State, police and fire departments, the Environmental Protection Agency) are intended to serve the general public.

Dynamics of Formal Organizations

The discussion of organizations thus far has described the formal framework on which organizations are based. But organizations also have dynamic processes. These processes unfold from the structure of the formal organization and lead to the modification of that structure (R. H. Hall and Tolbert 2004). This section is concerned with the informal structure within formal organizations as well as the tendency toward a concentration of power at the top.

Informal Organization

It is not necessary to observe any formal organization very long before realizing that it contains patterns of interaction and relationships that are not anticipated by the formal organization chart. Bureaucracies are designed to act as secondary groups. Relationships among organization members are supposed to be gov-

erned by the hierarchy of authority, by rules and procedures, and by impersonality. But, as anyone who has worked in a bureaucratic organization knows, there are primary relationships as well. Primary relationships emerge as part of the **informal organization**—a group (within a formal organization) in which personal relationships are guided by unofficial norms, rituals, and sentiments that are not part of the formal organization. Based on common interests and personal relationships, informal groups usually form spontaneously.

When was informal organization first recognized? The existence of informal organization within bureaucracies was first documented in the mid-1920s, when a group of Harvard researchers were studying the Hawthorne plant of the Western Electric Company in Chicago. In a study of fourteen male operators in the Bank Wiring Observation Room, F. J. Roethlisberger and William Dickson (1964; originally published in 1939) observed that work activities and job relationships were based on norms and social sanctions of that particular group of male operators. Group norms prohibited "rate busting" (doing too much work), "chiseling" (doing too little work), and "squealing" (telling group secrets to supervisors). Conformity to these norms was maintained through ridicule, sarcasm, criticism, hostility, and "binging" (hitting deviants on the arm). The group also had its own unique stratification structure, with soldermen at the bottom, wiremen in the middle, and inspectors at the top.

The Need for Informal Organization

Why does informal organization exist? Informal groups spontaneously develop to meet needs that are ignored by the formal organization. Whereas modern organizations are designed to be impersonal, informal groups offer personal affection, support, and humor. In addition to humanizing the formal organization, informal organization is a means of protection. Informal organization encourages conformity, and the resulting soli-

WORLDVIEW 6.1

Human Development

Each year, the United Nations ranks the countries of the world in terms of their human development. The human development index (HDI) is based on each country's life expectancy, educational attainment, and adjusted real income. This map shows how the world's countries ranked on the HDI in 2004.

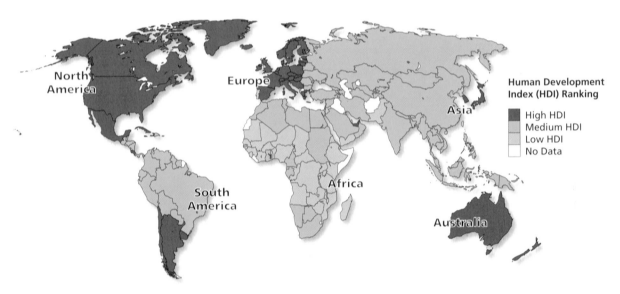

Source: United Nations Development Programme, *Human Development Report 2004*.

Interpret the Map

Support or refute the following statement: "To achieve a high ranking on the HDI, a country must have well-established and stable bureaucracies in the areas of health care, education, and the economy."

---

darity protects group members from maltreatment by those outside the group.

How does informal organization affect the formal organization? Early studies, starting with the Bank Wiring Observation Room, concluded that informal organization resulted in behavior counterproductive to the organization's goals. In the Hawthorne study, for example, workers enforced the informal norm of producing an average of two units a day. Workers who turned out more than two units per day were forced by the group to stop exceeding the informal production quota.

Later research has revealed beneficial consequences as well. Among carpenters and bricklayers, Raymond Van Zelst (1952) found, for example, that establishing work groups on the basis of friendship patterns increased workers' job satisfaction; decreased friction and anxiety among work partners; and created a friendly, cooperative atmosphere—all of which contributed to lower turnover and higher productivity.

Joseph Bensman and Israel Gerver (1963) described the use of an illegal tool in an airplane factory. This tool, called a tap, is used to thread nuts so they will fit with wing

bolts that are slightly out of alignment. Otherwise, the defect would have to be reported. The tap is outlawed—its use can lead to dismissal—because a tapped nut does not have proper holding power under flight conditions. However, use of the tap was tolerated by management as long as it was handled discreetly. In fact, without its use, the productivity of the plant would have suffered. Thus, the use of the tap was controlled informally by supervisors, inspectors, and workers themselves. In short, to avoid detection by Air Force inspectors, supervisors and workers evolved an informal organization complete with norms and sanctions for the systematic use of an illegal device. The formal organization considered itself to have benefited from this informal subterfuge. (Of course, the same could not be said for the passengers who traveled in the planes.)

The Role of Power in Formal Organizations

The word *organization* itself implies power. A hierarchy of authority is an expression of power distribution—in decreasing amounts from top to bottom. And organizations are political in more subtle ways. Every organization has people whose interests and goals conflict. Many preferences of top executives are not shared by wage workers; the outlooks of sales managers are often at odds with the views of production managers. Because a rational choice between all these conflicting views often requires more information than decision makers can assemble, decision making is usually political. Those with the most power usually get their way (Cyert and March 1963; H. A. Simon 1976; Pfeffer 1981).

Achievement of organizational goals requires the exercise of power. Although the employment of a power advantage is beneficial in organizations, power may be monopolized by individuals for their own limited purposes. This tendency is articulated in the *iron law of oligarchy* (Michels 1949; originally published in 1911).

How does the pyramidal shape of bureaucratic organizations promote the iron law of oligarchy? According to the **iron law of oligarchy** formulated by German sociologist Robert Michels, power tends to become increasingly concentrated in the hands of a few members of any organization. Michels observed that even in organizations intended to be democratic, leaders eventually gain control and other members become virtually powerless. In fact, he concluded that this increased concentration of power occurs because those in power want to remain in power:

> It is organization which gives birth to the dominion of the elected over the electors, of the mandatories over the mandators, of the delegates over the delegators. Who says organization, says oligarchy. (Michels 1949:401)

Politics in the United States is a prime example of Michels's principle. Voter apathy is reflected in low voter turnout for elections at all levels of government. Because so few are involved in the political process, those who are politically active gain an inordinate amount of power. The same thing occurs in organizations such as labor unions. Membership passivity over an extended period of time permits those at the top to consolidate power. Leaders are able to stay in power by building a loyal staff, controlling communications, and marshalling resources (money, jobs, favors).

Why does organization lead to oligarchy? According to Michels, three organizational factors encourage oligarchy. First, organizations need a hierarchy of authority to delegate decision making. Agreeing with Weber, Michels contends that the organizational problems of any large group require the development of a bureaucracy. Even in highly democratic organizations, such as the socialist parties of Europe, it is necessary for the masses to delegate much of the decision making to a few leaders. Delegation means creating a hierarchy of authority.

Second, the advantages held by leaders at the top of the hierarchy allow them to consolidate their power.

They are able to monopolize power because of their social and political skills, and once in power, they use these skills to strengthen their positions. Moreover, merely by doing their jobs, leaders of the organization gain special knowledge not available to the membership. They can create a staff that is loyal to them; they can control the channels of communication; they can use organizational resources to increase their power.

Third, oligarchy in organizations is also encouraged by the characteristics of the followers. The membership tends to defer to the greater skills possessed by their leaders. The nonelites believe that their leaders are more articulate and are doing a better job than they could do. In addition, the masses are passive, even indifferent; they do not want to spend a lot of time in organizational activities. The general membership may, in fact, feel grateful to those at the top for the leadership they are providing.

What are the consequences of oligarchy? Democratic organizations—the type of organization in which oligarchy presents the greatest problem—suffer from this tendency toward oligarchy, contended Michels, as those in power come to care more about perpetuating their power than about organizational goals. Oligarchic leaders become excessively careful and conservative as the maintenance of their positions becomes more important to them than considering the interests of the membership. As a result, the desires of the membership may be sacrificed.

Must oligarchy always develop? Research done since Michels published his theory has tended to support the iron law of oligarchy, the operation of which has been documented in a variety of organizations, including political parties, legislatures, churches, the American Legion, the American Sociological Association, the American Medical Association, and labor unions.

There is, however, opposition to oligarchy in organizations; otherwise, all organizations would be oligarchic. According to Alvin Gouldner, if there is an iron law of oligarchy, there must also be an iron law of democracy: "If oligarchical waves repeatedly wash away the bridges of democracy, this eternal recurrence can happen only because men doggedly rebuilt them after each inundation" (Gouldner 1955:506). Sentiment from the masses is continuously bubbling up to depose oligarchic leaders in democratic organizations. The survival of democracy in the face of oligarchy can be seen in the democratic removal of entrenched labor leaders and politicians. Although the odds against turning out entrenched leaders are high, it happens often enough to remind us that oligarchy does not completely destroy democracy. Moreover, if they are to remain in power, oligarchic leaders have to pay attention to the desires of their followers. Although mere responsiveness to the membership may not be the ideal form of democracy, its presence indicates that even oligarchic leaders often do not have a totally free hand to do as they please (Sayles and Strauss 1967; Voss and Sherman 2000).

It is also possible for members of democratic organizations to erect barriers to the development of oligarchy. A classic example is the typesetters' International Typographical Union (ITU). According to Seymour Lipset, Martin Trow, and James Coleman (1956), the ITU has been able to avoid oligarchic leadership because of its internal two-party system. Every two years, each party offers a complete slate of candidates for all offices. Because the two parties are of equal strength, the turnover in office compares favorably with the turnover in national political elections.

Other research concludes that concentration of power does not inevitably lead to an abandonment of the organization's goals for the selfish benefit of the leaders. Research on the policies of churches toward civil rights and on other controversial issues has led James Wood to note that "many organizations exist precisely to foster values that are precarious when left to individuals" (J. R. Wood 1981:101). Wood found that such values—including those that would favor the enactment of controversial policies regarding civil rights—can most effectively be implemented when leaders have a strong organizational base of support. Contrary to Michels's predictions, leaders in such situations often enacted policies that resulted in controversy, a loss of membership, and a loss of professional positions within the organization. Given a choice between the leaders' personal benefits and the maintenance of the organization's values, the leaders often chose to maintain the organization's values.

Joyce Rothschild and J. Allen Whitt (1987) also question the alleged inevitable connection between organization and oligarchy. They challenge the iron law of oligarchy through an investigation of cooperatives— organizations owned and operated by the workers themselves. Although not all cooperatives prosper or even survive, Rothschild and Whitt maintain that the very existence and persistence of cooperatives contradict the assumption that organization leads to oligarchy.

Privacy in the Workplace

In *The Unwanted Gaze*, Jeffrey Rosen (2000) attempts to document the destruction of privacy in the United States. He examines the legal, technological, and social changes eroding control over personal information. Because even deleted e-mail, for example, can be retrieved by others, a record of private self-disclosures exists permanently. Private preferences, tastes, and tendencies are revealed on every website we hit. During the impeachment of President Clinton, prosecutors legally gained access to Monica Lewinsky's bookstore purchases and unsent love letters on her own computer.

If Monica Lewinsky's loss of privacy is shocking to you, consider workplace vulnerability. Probably more than three-fourths of the *Fortune* 1,000 companies now have employee-monitoring software, and many of these firms keep their monitoring practices secret. More and more corporations are firing employees who violate prohibitions against the personal use or abuse of telecommunications technology. E-mailing a friend or coworker a heated threat to file a discrimination suit against the company for failing to promote you could result in termination even if you later deleted it (Faltermayer 2000). Business ethicist Laura Hartman offers this sobering picture:

A multitude of basic and inexpensive computer monitoring products allows managers to track web use, to observe downloaded files, to filter sites, to restrict your access to certain sites, and to know how much time you have spent on various sites. These include products such as WebSense, Net Access Manager, WebTrack and Internet Watchdog.

One particular firm, SpyShop.com, claims to service one-third of the Fortune 500 firms. This firm sells items such as a truth-telling device that links to a telephone. You are told that you can interview a job candidate on the phone and the device identifies those who lie. Another firm, Omnitracks, sells a satellite that fastens to the top or inside of a truck. The product allows trucking firms to locate trucks at all times. If a driver veers off the highway to get flowers for her or his partner on Valentine's Day, the firm will know what happened.

SpyZone.com sells an executive investigator kit that includes the truth phone I mentioned earlier, as well as a pocket recording pen. Other outlets sell pinhole lens camera pens, microphones that fit in your pocket. The motto of one firm is "In God we trust. All others we monitor." That firm offers a beeper buster: a computer program that monitors calls placed to beepers within a certain vicinity. A screen on your computer will show you all the numbers so that you can determine whether the individual is being distracted during working hours. (Hartman 2000:12)

Think About It

To what extent do employees have a right to privacy in the workplace? Analyze workplace privacy from a functionalist, conflict, and symbolic interactionist perspective.

Keith Brofsky/Getty Images

Women do not have a voice among these high-level business executives. Why do you think societal gender inequality is still pretty consistently reproduced in organizations?

Race, Ethnicity, Gender, and Social Class in Organizations

By the nature of organizations, power is concentrated in relatively few positions at the top. Mirroring the broader society, top positions in the hierarchy of organizations are held by white men from higher social classes. Consequently, minorities, women, and lower-class members are subject to the same patterns of prejudice and discrimination they experience in their nonwork lives (D. T. Miller 1999). The term *glass ceiling* has been used to reflect the fact that so few women and minorities get promoted to the more powerful (higher) positions.

How does the treatment of minorities mirror society? One of the most graphic reflections of inequality in organizations is found in the secretly taped meetings of Texaco's upper managers in 1996. The racial discrimination suit filed by Texaco's African American employees was settled in their favor for $115 million shortly after the content of these tapes hit the mass media. Texaco could hardly deny reparations for company racism when it became public knowledge that one top executive used the "N" word and another joked that "black jelly beans were stuck to the bottom of the bag" (Solomon 1996).

Much more systematic evidence of inequality exists. Even among employees with identical education, white men are disproportionately promoted over African Americans, Latinos, and Native Americans. It works the same way with dismissal: African American employees in the federal government (the environment most favorable to minorities) are discharged at a rate double that of whites (Zwerling and Silver 1992; DeWitt 1995). These and other studies provide convincing evidence of prejudice and discrimination against minorities in organizations (Feagin 1991; Cose 1995; see Chapter 9).

Is societal gender inequality reproduced in organizations? Yes it is. Like racial and ethnic minorities, women are less likely to be promoted than white men, even when they have the same level of education. For one thing, women are judged by higher criteria than men. Women, too, are more likely to be let go than men (White and Althauser 1984; Aldrich 1999; H. Smith 1999).

Sociologist Rosabeth Moss Kanter (1993) conducted a classic study of gender inequality in organizations. She found that women are usually treated as tokens, facing the charge that they owe their positions to just being women. The visibility of women due to their small number puts them under the glare of a spotlight. Stress is created by the fear of making mistakes easily seen by others in the organization. Women are more likely to be socially isolated, denied access to informal networks, and without powerful mentors. This isolation, in turn, reduces their chances for advancement in the hierarchy.

Because men overwhelmingly outnumber women in the upper ranks of organizations, Kanter found that even successful women find themselves in a minority status. Successful women in male peer groups become symbols of what-women-can-do. On the one hand, this makes them highly visible in an environment in which success depends on visibility. On the other hand, they sometimes feel the loneliness of being different from the other group members, having to sacrifice their privacy and anonymity. Because of their visibility, the few women in the upper ranks also feel extra pressure to avoid making mistakes. Even those women who have experienced little or no discrimination feel that they have to work harder than the average man to be successful.

Another indicator of gender inequality is **sexual harassment**—the use of one's superior power in making unwelcome sexual advances. Sexual harassment has been documented in nearly all types of organizations, including the military, the government, corporations, factories, and universities (Lott 1994). More will be said about this topic in Chapter 10, which more fully explores gender inequality in American society.

How is social class perpetuated in organizations? Organizations perpetuate social class levels through the placement of working-class people in low-paying, temporary, and dead-end jobs. Members of the upper and middle classes enter higher-level jobs with higher salaries and wages and greater opportunities for advancement. And this cannot be explained by social class differences in education. Persons from upper-class families tend to receive more favorable treatment than do lower-class persons with comparable educational credentials (Blau and Meyer 1987). Educational qualifications are often overshadowed by stereotypes associated with class differences in speech, dress, and behavior. Employees from lower-class backgrounds are more likely to be denied promotion, to receive lower pay raises, and to be dismissed than their higher-class coworkers who display proper "manners" and dress more stylishly.

Racial, ethnic, gender, and social class inequalities are mirrored in organizations. Not only do organizations reflect societal inequalities—they perpetuate them.

Organizations and Their Environments

The **organizational environment** consists of all the forces outside an organization that exert an actual or potential influence on the organization. An organization's environment includes the general technological, economic, political, cultural, and related conditions in which the organization operates (Hawley 1968; Meyer and Scott 1985; Hannan and Freeman 1989; W. R. Scott 1992; DiMaggio 2001). Interest in environments has increased in part because of the growing realization that organizations can survive only to the extent that they can generate enough support among environmental actors to continue receiving needed resources (in the form of money, new members, or whatever). It is equally true, however, that many organizations have sufficient resources to exert significant influence on their environments (Florida and Kenney 1991). Whether they are influencing outside actors or are the recipients of influence, organizations are frequently involved in interorganizational relationships.

What is the nature of interorganizational relationships? An **interorganizational relationship** is a pattern of interaction among authorized representatives of two or more formally independent organizations. Interorganizational relationships may be entered for many reasons, but most researchers would agree that a need for resources possessed by other organizations and an awareness of the availability of those resources in the other organizations contribute to the formation of interorganizational relationships (S. Levine and White 1961; Cook 1977; Hougland and Sutton 1978; Alter and Hage 1992).

Although organizations become involved in interorganizational relationships because outside resources are needed, all parties are concerned that the interests of their own organizations be protected. As Figure 6.3 shows, although the police must interact with fourteen other organizations in the area of problem youth, only three of these relationships are regularly cooperative. Six of the relationships, including those with important organizations such as the juvenile court and the detention center, are regularly marked by conflict (R. H. Hall and Tolbert 2004).

FIGURE 6.3

Patterns of Cooperation and Conflict in Interorganizational Relationships Involving the Police

This figure illustrates the nature of interorganizational relationships. Even when the police need the help of other organizations to reach their goals, only three of their fourteen interorganizational relationships are cooperative.

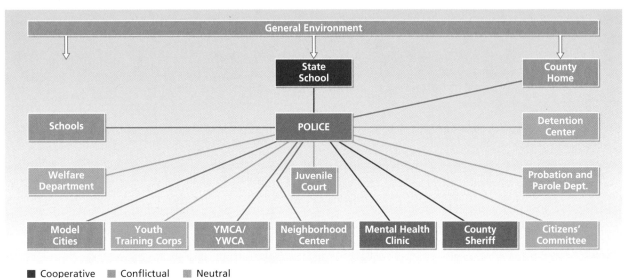

■ Cooperative ■ Conflictual ■ Neutral

Source: Richard H. Hall, *Organizations*, 7th ed. (Englewood Cliffs, NJ: Prentice-Hall, 2001), 225. Reprinted by permission.

**F
E
E
D
B
A
C
K**

1. A group in which personal relationships are guided by rules, rituals, sentiments, and traditions not provided for by the formal organization is known as a(n) _____ organization.
2. The effects of informal organization on the formal organization are almost always negative. T or F?
3. The theory known as the _____ emphasizes the danger of power becoming concentrated in the hands of a few members of any organized group.
4. Cooperative patterns tend to dominate in interorganizational analysis. T or F?

Answers: 1. informal; 2. F; 3. iron law of oligarchy; 4. F

SUMMARY

**R
E
V
I
E
W

G
U
I
D
E**

1. Group members interact, share culture, consider one another's behavior, and have some common interests and goals. Groups are sometimes confused with social categories and social aggregates.

2. A primary group is composed of individuals who know one another well, interact informally, and like being together. A secondary group is goal oriented, and its members interact without personal or emotional involvement. Likewise, a primary relationship is intimate and personal, whereas a secondary relationship is impersonal.

3. Primary groups are more likely to develop when a relatively small number of people are involved, when interaction is face-to-face, when people are in contact over a long period of time, and when the social environment is appropriate. Primary groups are formed in part because of the emotional support and sense of belonging they provide members. This type of group serves society by aiding in the socialization process and by helping to maintain social control.

4. We use reference groups to evaluate ourselves and to acquire attitudes, values, beliefs, and norms. Reference groups may be those we merely aspire to join, or those we use as negative role models.

5. Particularly intense feelings of identification and loyalty envelop in-groups. The feelings involve opposition, antagonism, and competition toward out-groups. Efforts must be made to maintain group boundaries, the consequences of which have significant social implications.

6. A social network (not a group per se) links a person with a wide variety of individuals and groups. The social relationships that are formed may be "strong" (parents and children) or "weak" (distant relatives, members of a college class). Social networks provide a sense of belonging and social support and help in the job market. More women are now using networking, once the sole province of men, for career advancement.

7. Five types of social interaction are basic to group life. In cooperation, individuals or groups combine their efforts to reach some common goal. When individuals or groups work against one another to obtain a larger share of the valuables, they are in conflict. In conflict, defeating one's adversary is thought to be necessary. Social exchange is a form of interaction in which one person voluntarily does something for another person with the expectation that a reward will follow. Coercion exists when one person or group is forced to behave according to another person's or group's demands. When a person behaves as others expect, conformity occurs.

8. Most groups experience both cooperation and conflict. This is partly because groups satisfy some personal needs, such as social acceptance, while thwarting others, such as personal freedom. Social scientists have been very interested in the ways in which interaction within groups can achieve group goals while fulfilling the needs of individual group members.

9. Moving beyond the trait theory of leadership, social scientists have focused on appropriate leadership behavior. More recently, studies of leadership suggest that appropriate leadership behavior depends on the situation a leader faces.

10. In some situations, group decision making has been shown to be superior to individual decision making. Intellectual tasks, for example, are better handled through groups. On the other hand, group pressure may damage the quality of decision making because of the excessive conformity of group members. Many group decisions are reached in situations in which group members are discouraged from expressing their reservations. At the extreme, this is called groupthink.

11. A formal organization—the epitome of a secondary group—is deliberately created to achieve some goal. There are many types of formal organizations in modern soci-

ety, but most share one feature—they are bureaucratic. According to Max Weber, bureaucracy is the most efficient type of formal organization. Using the ideal-type method, Weber outlined the characteristics that give bureaucracy its technical superiority over other types of organizations.

12. Although Weber saw bureaucracy as an efficient organizational form, he was well aware of its negative side. When rules and regulations become more important than organizational goals (as they often do), those who are supposed to be served by the bureaucracy suffer. Adherence to rules and regulations may also make bureaucrats overly conservative; such people learn their jobs so well that they cannot adapt to changing times and circumstances. Such problems have led researchers to examine alternatives to bureaucracy. Although most social scientists agree that bureaucracy is appropriate for many situations, contingency theorists have suggested that other types of organizations may fit particular circumstances.

13. The existence of primary groups and primary relationships within formal organizations has received considerable attention. These informal groups can either help or hinder the achievement of formal organizational goals, depending on the circumstances.

14. The exercise of power in organizations is receiving increasing attention. Power can benefit organizations by facilitating change and adapting to changing conditions, but power can also be used for the benefit of a small, unrepresentative group within the organization. The applicability of the iron law of oligarchy to modern organizations is a continuing source of debate among sociologists.

15. Minorities, women, and lower social classes suffer from power differentials in organizations. For this reason, organizations serve to perpetuate the inequality that exists in the broader society.

16. Organizations exist within an environment. They depend on factors within that environment for necessary resources, and they also attempt to exert influence on portions of the environment. Both of these situations encourage involvement in interorganizational relationships, which allow organizations to adapt to environmental pressures but involve conflict as well as cooperation.

LEARNING OBJECTIVES REVIEW

- Define the concept of the group, and differentiate it from social categories and social aggregates.
- Outline the basic characteristics of primary and secondary groups.
- Describe the five major types of social interaction.
- Discuss the nature of group processes: task accomplishment, leadership styles, and decision making.
- Define the concept of formal organization, and identify the major characteristics of bureaucracy.
- Discuss the advantages and disadvantages of bureaucracy.
- Distinguish between formal and informal organization.
- Describe the iron law of oligarchy, and demonstrate its importance with examples.
- Discuss prejudice and discrimination in organizations.

CONCEPT REVIEW

Match the following concepts with the definitions listed below them.

_____ a. social exchange
_____ b. rationalism
_____ c. ideal-type method
_____ d. bureaucracy
_____ e. out-group

_____ f. iron law of oligarchy
_____ g. organizational environment
_____ h. group
_____ i. conflict
_____ j. secondary group

_____ k. formal organization
_____ l. primary relationship
_____ m. social category
_____ n. conformity
_____ o. in-group

1. a type of formal organization based on rationality and efficiency
2. occurs when individuals or groups work against one another to obtain a larger share of the limited valuables in a society
3. a type of social interaction in which an individual behaves toward others in ways expected by the group
4. a social structure deliberately created for the achievement of one or more specific goals

5. several people who are in contact with one another; share some ways of thinking, feeling, and behaving; take one another's behavior into account; and have one or more interests or goals in common
6. a method that involves isolating the most basic characteristics of some social entity
7. a group toward which one feels intense identification and loyalty

8. the principle stating that power tends to become concentrated in the hands of a few members of any social structure

9. all the forces outside the organization that exert an actual or potential influence on the organization

10. a group toward which one feels opposition, antagonism, and competition

11. a relationship that is intimate, personal, based on a genuine concern for another's total personality, and fulfilling in itself

12. the solution of problems on the basis of logic, data, and planning rather than tradition and superstition

13. a group that is impersonal and task oriented and involves only a segment of the lives and personalities of its members

14. a number of persons who share a social characteristic

15. when one person voluntarily does something for another, expecting a reward in return

CRITICAL-THINKING QUESTIONS

1. Identify one of your primary groups. Discuss the three functions of primary groups using your own experience to illustrate.

2. Think of a club you belong to on campus or a club you joined as a high school student. Was this a primary or secondary group? Support your answer with the use of sociological concepts and actual experiences.

3. Recall an in-group of which you are (or were) a member. Discuss the consequences this group had for others. Defend the right of this group to exist on sociological grounds.

4. Think of an example of groupthink at your school. Analyze this situation in relation to positive or negative consequences.

5. Most Americans think of bureaucracies in a very negative way. Develop an argument you could use the next time someone endorses the prevailing view of bureaucracy in your presence.

6. Analyze your school as a bureaucracy. Give an example of the following characteristics of bureaucracy: (a) a system of rules and procedures; (b) impersonality and impartiality. Discuss a positive and negative consequence of each characteristic.

MULTIPLE-CHOICE QUESTIONS

1. **Several people who are in contact with one another; share some ways of thinking, feeling, and behaving; take one another's behavior into account; and have one or more interests or goals in common are known as a**
   a. commune.
   b. clique.
   c. group.
   d. social category.
   e. social aggregate.

2. **Which of the following are considered the two principal types of groups?**
   a. primary and small groups
   b. primary and reference groups
   c. secondary and small groups
   d. secondary and corporate groups
   e. primary and secondary groups

3. **A group is sometimes confused with a_____, a number of people who happen to be in the same place at the same time.**
   a. social collectivity

   b. social category
   c. social class
   d. social aggregate
   e. mob

4. **A web of social relationships that joins a person directly to other people and groups and, through those individuals and groups, indirectly to additional parties is known as a**
   a. social network.
   b. relationship web.
   c. social structure.
   d. safety net.
   e. social complex.

5. **Robert Nisbet has described five types of social interaction basic to group life. Which of the following is _not_ one of them?**
   a. cooperation
   b. conflict
   c. coercion
   d. consensus
   e. conformity

6. Which of the following statements has been found to be most accurate concerning group decisions?
   a. Individual decisions are always better than group decisions.
   b. Group pressure has little effect on group decisions.
   c. Group decisions can be superior to individual decisions.
   d. Mistakes made by groups are harder to correct.
   e. Group decisions are always better than individual decisions.

7. The type of formal organization based on rationality and efficiency is known as
   a. a commune.
   b. an ecclesia.
   c. a hierarchy.
   d. a profession.
   e. a bureaucracy.

8. According to the text, _____ is the ability to control the behavior of others even against their will.
   a. authority
   b. prestige
   c. privilege
   d. ostracism
   e. power

9. The solution of problems on the basis of logic, data, and planning rather than on the basis of tradition and superstition is known as
   a. groupthink.
   b. rationalism.
   c. mythology.
   d. existentialism.
   e. jurisprudence.

10. A judge who is more concerned with following legal procedures than reaching just decisions is engaging in
    a. contingency thinking.
    b. categorical thinking.
    c. goal displacement.
    d. coercion.
    e. trained incapacity.

11. The part of a formal organization in which personal relationships are guided by rules, rituals, and sentiments not provided for by the formal organization itself is known as
    a. the enforcement of mores zone.
    b. the application of legitimate authority network.
    c. informal organization.
    d. the Peter Principle unit.
    e. the contingency group.

12. _____ states that power tends to become concentrated in the hands of a few members of any social structure.
    a. The iron law of oligarchy
    b. The Peter Principle
    c. Parkinson's law
    d. The law of demographic transition
    e. The iron law of responsibility

13. _____ consists of all the forces outside the organization that exert an actual or potential influence on the organization.
    a. Organizational environment
    b. Bureaucratic environment
    c. Organic-adaptive environment
    d. Organizational context
    e. External organization

FEEDBACK REVIEW

True-False

1. Social scientists agree that leaders must have certain traits if they are to be effective. T or F?
2. The effects of informal organization on the formal organization are almost always negative. T or F?

Fill in the Blank

3. A _____ is a web of social relationships that joins a person directly to other people and groups and, through those individuals and groups, indirectly to additional parties.
4. Social workers who become more concerned with eligibility requirements than with delivering services to their low-income clients are engaging in _____.

Multiple Choice

5. Which of the following is not a condition that promotes the development of primary groups?
   a. small group size
   b. face-to-face contact
   c. continuous contact
   d. interaction on the basis of status or role

6. Which of the following has *not* been seen as an advantage of bureaucracy?
   a. its avoidance of the use of inappropriate criteria in hiring employees
   b. its use of rules to provide definite guidelines for behavior within the organization
   c. its success in hiding the true nature of authority relationships
   d. its encouragement of administrative competence in managers

Matching

7. Listed here are some examples of primary and secondary relationships. Indicate which examples are most likely to be primary relationships (P) and which are most likely to be secondary relationships (S).
    ____ a. a marine recruit and his drill instructor at boot camp
    ____ b. a married couple
    ____ c. a manager and his professional baseball team
    ____ d. professors and students
    ____ e. car salespersons and their prospective customers

8. Match the following terms and statements.
    ____ a. group
    ____ b. in-group
    ____ c. social aggregate
    ____ d. social category
    ____ e. reference group
    (1) skinheads
    (2) spectators at a Fourth of July fireworks display
    (3) orchid growers
    (4) teenage subculture
    (5) members of a regular Saturday night poker game

GRAPHIC REVIEW

Table 6.3 displays the sales, assets, and profits of corporations by size of the organization. Answering the following questions will check your understanding of this information as it relates to the material in this chapter.

1. Choose one characteristic of bureaucracy to show that this type of organization makes possible the development of extremely large corporations.

    _____

    _____

    _____

2. Relate the concentration of power in the top 100 companies to the iron law of oligarchy.

    _____

    _____

    _____

ANSWER KEY

Concept Review	Multiple Choice	Feedback Review
a. 15	1. c	1. F
b. 12	2. e	2. F
c. 6	3. d	3. T
d. 1	4. a	4. social network
e. 10	5. d	5. goal displacement
f. 8	6. c	6. d
g. 9	7. e	7. c
h. 5	8. e	8. a. S
i. 2	9. b	b. P
j. 13	10. c	c. S
k. 4	11. c	d. S
l. 11	12. a	e. S
m. 14	13. a	9. a. 5
n. 3		b. 1
o. 7		c. 2
		d. 3
		e. 4

Deviance and Social Control

7

Susan Walsh/AP Photos

LEARNING OBJECTIVES

- Define deviance, and explain its relative nature.
- Define social control, and identify the major types of social control.
- Describe the biological and psychological explanations of deviance.
- Discuss the positive and negative consequences of deviance.
- Differentiate the major functional theories of deviance.
- Compare and contrast cultural transmission theory and labeling theory.
- Discuss the conflict theory view of deviance.
- Distinguish the four approaches to crime control.

How often is rape committed by a stranger? Unfortunately, many young people today know someone who has been raped. When women find themselves outdoors and alone, they are often fearful of possible surprise attacks by dangerous strangers. This is a genuine concern, and women are smart to be cautious; yet, more than 80 percent of reported rapes in the United States are committed by individuals who are known to their victims. Experts use the terms *date rape* and *acquaintance rape* to describe this reality.

As date or acquaintance rape illustrates, deviant behavior may involve perversion. Deviance, however, is not limited to "nuts, sluts, and perverts" (Liazos 1972). **Deviance** is *any* behavior that departs from societal or group norms. It can range from criminal behavior (recognized by almost all members of a society as deviant) to wearing heavy makeup (considered deviant by some religious groups).

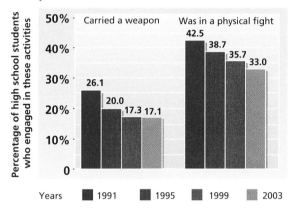

Two Types of Youth Deviance

What does the graph indicate about the trend in youth violence?

Percentage of high school students who engaged in these activities

Carried a weapon: 26.1, 20.0, 17.3, 17.1
Was in a physical fight: 42.5, 38.7, 35.7, 33.0

Years: 1991, 1995, 1999, 2003

Source: Centers for Disease Control and Prevention, "Healthy Youth!" (2004).

Deviance and Social Control

The Nature of Deviance

Without conformity and predictability in human behavior, society could not exist. Order and stability are the cornerstones of social life. All the same, we know that social life includes people who breach norms. Some people violate norms by robbing banks, assaulting others, or committing murder. Some incidents of less severe deviance have been high profile because they involved prominent sports figures whose behavior was captured on national television. Dennis Rodman, the flamboyant NBA player, cross-dressed, had tattoos all over his body, and regularly changed the color of his hair (yellow, red, blue, green, purple and orange, multicolored, and so forth). During one season, Rodman head-butted a referee and deliberately kicked a photographer in the groin. Even more dramatic was the behavior of Mike Tyson, former heavyweight boxing champion, who twice bit the ear of an opponent in a world championship fight (Starr and Samuels 1997). Figure 7.1 illustrates the frequency of two types of juvenile deviance.

Most behavior tends to fall within a normal range of social expectations. *Deviance*, behavior outside this range, may be either *positive* (overconforming) or *negative* (underconforming).

What is the distinction between positive and negative deviance? Negative deviance involves behavior that underconforms to accepted norms. Negative deviants reject the norms, misinterpret the norms, or are unaware of the norms. This is the kind of behavior popularly associated with the idea of deviance. There is, however, another type of deviance. **Positive deviance** encompasses behavior that overconforms to social expectations (Heckert 2003). Positive deviants conform to norms in an unbalanced way. Underlying this type of deviance is an idealization of group norms that leads people to extremes of perfectionism. In its own way, positive deviance can be as disruptive and hard to manage as negative deviance. Few people find a society characterized by slavish and unchallenged conformity any more desirable than a disorderly and unpredictable society.

This more complete picture of the nature of deviance is illustrated in Figure 7.2. Think about the norms related to personal appearance in American society. For example, the mass media are constantly telling young people that thin is good and fat is bad. Negative deviants will miss the mark on the obese side. Positive deviants may push themselves to the point of anorexia. Most young people will weigh somewhere between these two extremes.

How is identifying deviance a challenge? Deviance is a matter of social definition. In a diverse society like the United States, it is often difficult to agree on what does and does not constitute deviance. Jerry Simmons asked a sample of respondents to list categories or types of persons they considered deviant:

FIGURE 7.2

The Distribution of Deviance Relative to the Norm of Leanness

Most people conform to norms within a socially acceptable range. Only at the extremes of this normal range of behavior does deviance appear. We are accustomed to talking about negative deviance, behavior that involves underconformity. We are much less aware of positive deviance, or overconforming behavior. Leanness as the desirable standard for physical appearance is illustrated in this figure. Think of another example that exists among college students.

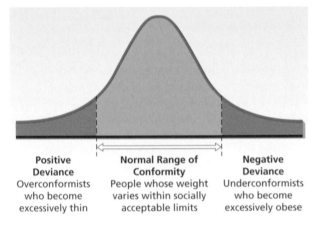

Positive Deviance	Normal Range of Conformity	Negative Deviance
Overconformists who become excessively thin	People whose weight varies within socially acceptable limits	Underconformists who become excessively obese

Source: Adapted from Jay J. Coakley, *Sport in Society*, 6th ed. (Boston, WBC/McGraw-Hill, 2006).

The sheer range of responses predictably included homosexuals, prostitutes, drug addicts, radicals, and criminals. But it also included liars, career women, Democrats, reckless drivers, atheists, Christians, suburbanites, the retired, young folks, card players, bearded men, artists, pacifists, priests, prudes, hippies, straights, girls who wear makeup, the President of the United States, conservatives, integrationists, executives, divorcees, perverts, motorcycle gangs, smart-alec students, know-it-all professors, modern people, and Americans. (Simmons 1969:3)

To this list, Leslie Lampert would add obese people. For a week she wore a "fat suit," adding 150 pounds to her normal body weight, to experience firsthand what it feels like to be an overweight woman in American society. According to Lampert, American "society not only hates fat people, it feels entitled to participate in a prejudice that at many levels parallels racism and religious bigotry" (Lampert 1993:154).

Why do sociologists limit their study of deviance? Because minor instances of behavior that some might consider deviant occur frequently in modern societies, sociologists generally reserve the term *deviance* for violations of *significant* social norms. Significant norms are those that are highly important either to most members of a society or to the members with the most power. For sociologists, then, a **deviant** is a person who violates one or more of society's most highly valued norms.

Public reactions to deviants are usually negative and involve attempts to change or control the deviant behavior. Sociologists, however, do not make such value judgments. To them, deviant behavior is neither good nor bad; it is simply behavior that violates social norms and draws a negative response from others.

Conditions Affecting Definitions of Deviance

Identifying deviance is clearly a relative matter. Insight into the relativity of deviance began early in the history of sociology. According to William Graham Sumner, norms determine whether behavior is deviant or normal (Sumner 1906). Because norms vary from group to group, society to society, and time to time, the behavior considered to be deviant varies. Emile Durkheim contended that no behavior is deviant in itself, because deviance is a matter of social definition.

According to Sumner, Durkheim, and many other sociologists, the deviance label depends on three circumstances: the social status and power of the individuals involved, the social context in which the behavior occurs, and the historical period in which the behavior takes place.

Actresses Mary-Kate Olsen and Ashley Olsen are famous anorexics. How would you explain to your parents that their behavior is "positive"?

William Chambliss— Saints and Roughnecks

William Chambliss (1973), over a two-year period, observed the behavior of two teenage gangs at Hanibal High School. In addition to gang activity, Chambliss documented the responses of parents, teachers, and police to the delinquent behavior.

The Saints—eight children of white, stable, upper-middle-class parents—were pre-college, good students, and active in school affairs. The Roughnecks—six young men from lower-class white families—were poor students.

Both gangs engaged in similar amounts of delinquency. Yet, the Saints were slated for bright futures, while the Roughnecks were on track to dead-end lives. That was the picture held by parents, teachers, and law enforcement officials. And that was the way events actually turned out.

Why? The answer makes Chambliss's study sociologically fascinating. Deviance is relative to social class; it may be *primary* or *secondary*.

Hanibal townspeople simply never recognized the Saints' behavior as delinquent. When caught stealing or driving drunk, the Saints were seen as good boys "playing pranks" or "sowing wild oats." How else, parents, teachers, and law enforcement officials asked themselves, could you explain such behavior from well-mannered, well-dressed teenagers with nice cars and money in their pockets?

No such benefit of the doubt was extended to Roughnecks caught in the same acts. When these badly dressed, rude, not-so-rich boys were caught by the police stealing property or drinking excessively, they were seen as heading for trouble.

The Roughnecks engaged in secondary deviance—their delinquency evolved into a lifestyle and permeated their personal identities. Others responded to them as deviants, and they gradually incorporated this labeling into their self-definitions. The more they saw themselves as deviants, the more time they spent in deviant activities. Two of the boys dropped out of high school: one sentenced to thirty years for murder, the other sentenced for life on a murder conviction. Another Roughneck finished high school and become a gambler. There was scant information on a fourth Roughneck: he was "driving a truck up north." Two went to college on athletic scholarships and turned their lives around.

In contrast, the Saints engaged in primary deviance. Because, claimed Chambliss, they were not labeled, the Saints' acts of deviance did not come to dominate their lifestyle or self-image. In fact, seven of the eight Saints attended college right after high school. Five of these graduated in four years, one in six years, and the last in seven years. Four of the college graduates went to graduate school. Among these four were a lawyer, a doctor, and a PhD student.

Evaluate the Research

1. From your understanding of Chambliss's study, is deviance socially created? Explain.
2. Which of the three major theoretical perspectives best explains Chambliss's findings? Support your choice.

How do social status and power influence the definition of deviance? Researcher William Chambliss (1973) demonstrated the importance of social status and power in defining deviance. As a participant observer, Chambliss watched two teenage gangs for two years at Hanibal High School. While at Hanibal High, Chambliss observed the school and nonschool activities of two gangs: the Saints, a gang of eight upper-middle-class white boys, and the Roughnecks, a gang of six lower-class white students. Although the Roughnecks and the Saints engaged in many of the same types of deviant behavior, the Roughnecks were judged by community members as worthless kids who were poor students, heavy drinkers, and violent individuals. Why, Chambliss asked, "did the community, the school, and the police react to the Saints as though they were good, upstanding, nondelinquent youths with bright futures but react to the Roughnecks as though they were tough, young criminals, who were headed for trouble?" (Chambliss 1973:28). The different treatment of these two gangs seemed to be traceable to their social class backgrounds. Lower-class people are more likely to be labeled deviants than are middle- or upper-class individuals, even when they behave in similar ways. (See Doing Research for more detail.)

Of course, high status does not always protect those who engage in deviant behavior. After discovery of some falsification of his résumé in 2001, newly hired Notre Dame football coach, George O'Leary, had his job offer withdrawn (Kornheiser 2001). Noted historian Joseph Ellis was forced to take a year's leave of absence from his university teaching position following disclosure that, in his classes, he made false claims of personal military service. Although neither O'Leary nor Ellis committed prosecutable offenses, their deviant behavior was punished by official sanctions and personal embarrassment.

Charges of child sexual abuse by Catholic priests have been very costly to the Catholic church. In addition to a damaged reputation resulting from a storm of highly negative media coverage, the Catholic church is faced with declining parishioner financial contributions and payments of some $100 million in settlements agreements.

How do definitions of deviance vary with social context? Behavior can be considered conforming or deviant only within a particular social context; behavior considered deviant in one society may be acceptable and even encouraged in others. Suppose a group of people believe that gods live in rocks and trees and that these spirits are capable of granting them favors: power over their enemies, the ability to heal the sick, and the foresight to predict the future. Among the traditional Alaskan Eskimo, such people would be awarded high social status and deference. But in a highly industrialized society, members of such a group would be considered strange.

Deviance is also relative from group to group within the same society. Conformity to the norms of one group may earn intense disapproval from another. Teenagers, for example, are usually torn between conforming to the demands of their peers and meeting the expectations of their parents.

The possibility of being considered deviant from some group's point of view is likely in a society as complex as modern America. For instance, deviance can even vary within a geographical area. In Florida, it is illegal to play casino games on land. Three miles offshore in international waters, it is acceptable to play slot machines, cards, and other games of chance.

How do definitions of deviance vary over time? Societal approaches to cigarettes and drugs illustrate labeling variations over time. Gerald Markle and Ronald Troyer (1988), who studied the changing definition of cigarette smoking in American society, found that definitions of smoking have varied considerably. The definition of smoking was quite negative during the early 1900s, when smoking was banned by fourteen states. It became positive during the 1950s when advertising began to depict smoking as glamorous and sophisticated. Today, smoking is prohibited in places as diverse as restaurants, government buildings, corporate offices, airplanes, elevators, physicians' offices, and university buildings. By 2008, fourteen states had statewide smoking bans covering bars (whether they serve food or not), casinos, restaurants, and workplaces. Twenty states had less comprehensive statewide

Definitions of deviance change from time to time. Although same-sex marriage is still viewed by many Americans as a form of deviance, prohibitions have loosened to the point that many gays, such as this California couple in 2008, are willing to legally acknowledge their living relationship.

Marcio Jose Sanchez/AP Photos

bans. Only sixteen states had no statewide smoking prohibitions.

Like the United States, Europe has traditionally taken a strong stance against drug use. There is now a movement toward liberalization. While only a few countries favor legalizing hard drugs, several are less militantly opposed. As part of this shift from a war on drugs to an informal truce, some countries are tolerating marijuana use, and even the user of heroin or other hard drugs is less likely to be criminalized. Germany now authorizes local governments to permit citizen possession of drugs for "personal use." In the Netherlands, drug use is now considered a public health issue rather than a criminal matter. Portugal has decriminalized the use of all previously banned narcotics from marijuana to crack cocaine. Even though the 2002 Nevada ballot initiative to legalize possession of up to three ounces of marijuana failed, its very presence reflects growing support for the legalization of marijuana in the United States (Hawkins 2002; J. Stein 2002). Since 1996, twelve states have legalized medical marijuana use. Actually, one-third of Americans now favor legalizing marijuana, up from 12 percent in 1969 (Carroll 2005).

There are even signs of some loosening of laws on crack cocaine. Extremely harsh penalties for possession of crack cocaine were enacted during the 1980s' "tough on crime" era. For example, crack offenders spend more time in prison than people convicted of manslaughter. One mother was condemned to 24 years in jail at the age

1. _____ is behavior that violates norms.
2. A major problem that sociologists have in defining deviance is its _____.
3. _____ is the means for promoting conformity to a group's or society's rules.

Answers: 1. Deviance; 2. relativity; 3. Social control

of 72 for refusing to testify against her son. Possession of 100 times more powder cocaine is required for the same sentence involving crack offenders. In 2007, the U.S. Supreme Court voted 7–2 to unbind federal district judges from the prior commission guidelines that created the large discrepancy in punishments given to crack and powder cocaine offenders. A day later, the U.S. Sentencing Commission voted unanimously to give incarcerated crack cocaine offenders an opportunity to reduce their sentences to a level comparable to powder cocaine offenders (Fears 2007).

Forms of Social Control

Human behavior is predictable most of the time. We generally know what to expect from others and what others expect of us. You know by now: this is no accident. Just as laws of physics help us predict the actions of the physical world around us, social and cultural mechanisms promote order, stability, and predictability in social life. We feel confident that drivers will not drive on the wrong side of the road, waiters will not pour wine over our heads, and strangers will not expect us to accept them as overnight guests. Without **social control**—means for promoting conformity to norms—social life would be capricious, even chaotic. There are two broad types of social control: internal control, which lies within the individual, and external control, which exists outside the individual.

What is the distinction between internal and external social control? Internal social control is self-imposed and is acquired during the socialization process. For example, most people most of the time refrain from stealing, not just because they fear arrest or lack the opportunity to steal, but because they consider theft to be wrong. The norm against stealing has become a part of the individual. This is known as the *internalization* of social norms.

Unfortunately for society, the internalization of norms is not complete in all societies or groups. If it were, we would have no deviance at all. Because the process of socialization does not ensure the conformity of all people all of the time, external social control must also be present.

External social control is based on sanctions—rewards and punishments—designed to encourage desired behavior. Positive sanctions, such as awards, salary increases, and smiles of approval, are used to encourage conformity. Negative sanctions, such as fines, imprisonment, and ridicule, are intended to stop socially unacceptable behavior. Sanctions may be formal or informal. Ridicule and gossip are examples of informal sanctions; imprisonment, capital punishment, exile, and honorary titles are formal sanctions.

Biological and Psychological Explanations of Deviance

As explained in Chapter 3, sociologists, despite the rise of sociobiology, remain skeptical of biological explanations of human behavior. The search for some biological foundation for deviance, however, has a long history and is still occurring.

Search for a Biological Basis for Deviance

What are some biological explanations of deviance? One of the earliest proponents of biological causes of deviance was Cesare Lombroso (1918), an Italian physician who saw a relationship between physical traits and crime. Lombroso believed that criminals were accidents of nature, throwbacks to an earlier stage of human evolutionary development. American psychologist William Sheldon (1949), a direct intellectual heir of Lombroso, identified three basic types of body shape: endomorphs (soft and round), mesomorphs (muscular and hard), and ectomorphs (lean and fragile). Sheldon concluded that criminals were more likely to be mesomorphs.

The theories of Lombroso and Sheldon are not generally considered valid today, but the search for a connection between physical characteristics and deviant behavior continues. Contemporary research in this area is much more sophisticated methodologically than the work of Lombroso and Sheldon (Fuller and Thompson 1978; Rowe 1983, 1986; Mednick, Gabrielli, and Hutchings 1984; Brunner et al. 1993). The most influential contemporary attempt to establish a link between biology and criminal behavior appears in a book entitled *Crime and Human Nature*, by James Q. Wilson and Richard Herrnstein (1985). The authors argue that Lombroso was thinking in the right direction. Although criminals are not born, the authors contend that many people are born with "con-

Because of its hurtful nature, gossip can be a very effective informal sanction.

stitutional factors"—personality, intelligence, anatomy—that predispose them to serious criminal behavior. Wilson and Herrnstein point to research on twins and adopted children. Criminality is said to be more common among identical twins than among fraternal twins. This is taken to be important because identical twins are the same genetically, whereas fraternal twins are only as genetically similar as brothers and sisters. Sons of chronic offenders raised in noncriminal homes are reported to have a three times greater probability of following a life of crime than sons of nonoffenders who have been adopted by law-abiding parents. Wilson and Herrnstein are careful to point out that they are referring only to genetic predispositions toward criminal behavior. People with these biological predispositions do not necessarily become criminals; these predispositions may be offset by socialization in an emotionally supportive environment, such as the family.

How do sociologists evaluate biological explanations of deviance? Sociologists generally have not placed much stock in biological explanations of deviance. There are five main reasons for this.

1. Biological theories ignore the fact that deviance is more widely distributed throughout society than are the hereditary and other physical abnormalities that are supposedly the causes of deviance.
2. Biological theories almost totally discount the influence of social, economic, and cultural fac-

tors. These theories fail to explain how a person can have a genetic predisposition for something that is relative.
3. Early biological theories of deviance were based on methodologically weak research.
4. There are ideological problems and controversial implications inherent in the biological approach. Forced sterilization of "defectives" and related eugenic strategies for reducing crime have created a storm of protests.
5. Finally, biological factors are more often invoked to explain the deviance of armed robbers, murderers, and heroin addicts than, say, the crimes of corporate executives, government officials, and other high-status persons (Empey 1982).

Personality as the Source of Deviance

What common focus do psychological theories of deviance share? As varied as they are, psychological explanations of deviance all locate the origin of criminality in the individual personality (J. Hagan 1994). This is to be expected because psychology, as discussed in Chapter 1, concentrates on the development and functioning of mental-emotional processes in human beings. More specifically, psychological theories of deviance take for granted the existence of a "criminal personality": a pathological personality with measurable characteristics that distinguish criminals from noncriminals. Pathological personality traits may be either inherited (and therefore related to biological explanations of deviance) or socially acquired (Raine 1993).

Although Sigmund Freud did not attempt to explore the origins of deviance, his biologically based, instinctual approach to human personality lends itself to such exploration. Psychologists who have adapted Freud's theories to the study of deviance believe that deviance is due to unconscious personality conflicts created in infancy and early childhood. For example, criminality is said to be the result of biologically based desires beyond the control of a defective ego and superego. Other psychological explanations of crime are based on the identification of specific personality traits. Delinquents are said to be more aggressive, hostile, rebellious, or extroverted than nondelinquents (Glueck and Glueck 1950; Eysenck 1977). Some advocates of the psychological search for the criminal personality contend that criminals are born rather than made, an approach closely akin to biological explanations of deviance (Yochelson and Samenow 1994).

What is the sociological critique of psychological theories of deviance? First, psychological theories, like biological ones, often ignore social, economic, and cultural factors shown by sociological research to affect the likelihood, frequency, and types of deviance. Second, psychological

AP Photos

How did the Reverend King's philosophy of nonviolence, which, however, encouraged the breaking of laws, benefit American society?

theories focus on deviance such as murder, rape, and drug addiction with relatively little to say about such deviance as white-collar crime. Third, psychological theories tend to view deviance as the result of physical or psychiatric defects rather than as actions considered deviant by social and legal definitions. Fourth, psychological theories of deviance cannot explain why deviant behavior is engaged in by individuals not classifiable as pathological personalities. Finally, psychological theories, with their emphasis on pathology—deviants are "psychopaths" or "antisocial" personalities—suggest eugenic solutions to the crime problem that are unacceptable to some segments of society. For example, because more deviance occurs among certain racial and ethnic groups, birth control might be seen as a solution to the crime problem that is the most consistent with psychological theories.

This brings us to sociological explanations of deviance. The specific theories of deviance will now be discussed within the frameworks of functionalism, symbolic interactionism, and conflict theory.

Functionalism and Deviance

Functionalism sheds light on deviance, both in a general sense and via the strain theory of Robert Merton. We will examine both of these contributions as well as that of control theory.

Costs and Benefits of Deviance

Functionalism attempts to explain both the negative and the positive effects of social forces on society. It is easy to think of deviance as having negative social con-

---

FEEDBACK

1. Recent research has established a definite causal relationship between certain human biological characteristics and deviant behavior. T or F?
2. According to James Q. Wilson and Richard Herrnstein,
    a. studies of adopted children cannot help in understanding possible biological antecedents of criminal behavior.
    b. intelligence is not correlated with criminality.
    c. emotionally supportive families cannot help prevent criminal behavior.
    d. constitutional factors predispose individuals to criminality.
3. Which of the following is not one of the psychological approaches to deviance?
    a. extroversion
    b. race
    c. unconscious conflicts
    d. IQ
4. Psychological theories of deviance still have not developed the breadth to be able to explain white-collar crime. T or F?

Answers: 1. F; 2. d; 3. b; 4.T

sequences. It is harder to imagine benefits of deviance, but functionalism tells us that there are some.

What are the negative effects of deviance on society? Widespread violation of norms threatens the foundation of social life. What would happen if bus drivers decided to create their own routes each morning, if local television stations aired what they preferred rather than what had been scheduled, if parents took care of their children's needs only when it was convenient? The result of such nonconformity is social disorder from which follows unpredictability, tension, and conflict.

Deviance erodes trust. If bus drivers do not follow the planned route, if television stations constantly violate their schedules, if parents support their children only sporadically, trust will be undermined. A society characterized by widespread suspicion and distrust cannot function smoothly. Take, for instance, the effects of increased rape in the United States. As pointed out in "Using the Sociological Imagination," rape overwhelmingly occurs between people who know each other. Almost one-fourth of college women report being victims of rape. Consequently, increasing female distrust often creates a barrier to normal dating relationships.

Unpunished deviance encourages others to be nonconformists. If bus drivers regularly pass people waiting for the bus, those people might bombard the buses with rocks. If television stations offer random programming, customers may damage their facilities in protest. If parents neglect their children, more teenagers may turn to delinquency. Deviance stimulates more deviance.

Deviant behavior is also expensive. For one thing, it diverts resources, both human and monetary. Police may have to spend their time dealing with wayward bus drivers, irate passengers, and mobs of annoyed television viewers rather than performing more serious duties. And if delinquency skyrockets, the criminal justice system will be even more jammed than it already is.

How does deviance benefit society? Despite its negative effects, deviance can sometimes benefit society. Emile Durkheim observed that deviance clarifies norms. By exercising social control, society defines, adjusts, and reaffirms norms; it defends its values. When parents are taken to court or lose their children because of neglect, for example, society's expectations are demonstrated to other parents and children.

Deviance can be a temporary safety valve. Teenagers listen to music, watch television programs, and wear clothes that adults may view as deviating from expected behavior. Involvement in this relatively minor deviant behavior may relieve some of the pressure teenagers feel from the social demands of the adult world (parents, teachers, clergy).

Deviance increases unity within a society or group. When deviance reminds people of something they

value, it strengthens their commitment to that value. The exposure of government spies selling secrets to an enemy intensifies feelings of patriotism. Learning of parents who abuse their children may encourage other parents to rededicate themselves to proper child care.

Deviance promotes needed social change. Suffragettes who took to the streets in the early 1900s scandalized the nation but helped bring women the right to vote. Prison riots in the past have led to the reform of inhumane conditions. As deplorable as it sounds, the discovery of sexual molestation of young children has led to an improvement of day-care facilities. Looting and other deviant behavior in inner cities has promoted some government efforts to revitalize urban American.

Strain Theory

According to Durkheim (1964; originally published in 1893), **anomie** is a social condition in which norms are weak, conflicting, or absent. In the absence of shared norms, individuals are uncertain as to how they should think and act, and societies become disorganized. Robert Merton (1968) has adapted this concept of anomie to deviant behavior. He calls it strain theory.

According to Merton's **strain theory**, deviance is most likely to occur when there is a discrepancy between a culturally prescribed goal (e.g., economic success) and a legitimate means of obtaining it (education). The resulting strain causes some people to engage in deviant behavior. Merton asserts that culture determines the things people should want (goals) and the legitimate ways (means) of obtaining these things. In American society, an important goal is success and the acquisition of the material possessions that usually accompany it. Although everyone is taught to value material success, contends Merton, some are denied access to the legitimate means for achieving it (Passas and Agnew 1997).

How do people respond to strain? Merton suggests five general modes of adaptation. By way of illustration, Merton uses the American goal of success (see Table 7.1). One possible response is *conformity*—pursuing culturally approved goals through legitimate means. Here, people accept the goal and the means to achieve it. The very wealthy come to mind. But this category also includes poor people who continue to work hard in conventional jobs in the hope of improving life for themselves and their children. By definition, conformity is not deviant behavior. All of the other four responses are considered deviant, however.

In *innovation*, the individual accepts the goal of success but adopts illegitimate means for achieving it. Teenagers may steal an expensive automobile. Others may use prostitution, robbery, drug dealing, or other lucrative criminal behavior to be successful. Innovation is the most widespread and obvious type of deviant response.

TABLE 7.1

Conforming and Deviant Responses in Merton's Strain Theory

Describe the deviant behavior of some people you know using two of Merton's deviant responses.

Success as a Culturally Approved Goal	Hard Work as the Socially Accepted Way to Succeed	Conformity Response	Deviant Responses	Examples
Accepts goal of success	• Accepts hard work as the apppropriate way to succeed	Conformity—works hard to succeed		Business executive
Accepts goal of success	• Rejects hard work as the appropriate way to succeed		**Innovation**—finds illegal ways to succeed	Criminal
Rejects goal of success	• Accepts hard work as the appropriate way to succeed		**Ritualism**—acts as if he wants to succeed but does not exert much effort	Unmotivated teacher
Rejects goal of success	• Rejects hard work as the appropriate way to succeed		**Retreatism**—drops out of the race for success	Skid-row alcoholic
Rejects goal of success	• Rejects hard work as the appropriate way to succeed		**Rebellion**—substitutes new way to achieve new goal	Militia group member

Source: Adapted from Robert K. Merton, *Social Theory and Social Structure*, rev. ed. (New York: Free Press, 1968).

In *ritualism*, the individual rejects the goal (success) but maintains the legitimate means. People go through the motions without really believing in the process. One example is the bureaucrat who continues to go about the daily routines of work while abandoning any idea of moving up the ladder. A college student may come to believe that college is without meaning, that the whole experience is a disappointment, that her goal of extreme economic success is not going to be realized. Yet, because of family expectations or lack of alternatives, she may continue to go through the motions needed to graduate.

Retreatism is a deviant response in which both the legitimate means and the approved goals are rejected. Skid-row alcoholics, drug addicts, and bag ladies have dropped out. Retreatists fail to be successful by either legitimate or illegitimate means; they do not even seek success.

In *rebellion*, people reject both success and the approved means for achieving it. At the same time, they substitute a new set of goals and means. Some militia group members in the United States illustrate this response. They may live in near isolation while pursuing their goal of changing society. They may use such means as creating their own currency, deliberately violating gun laws, and threatening (or engaging in) violent behavior against law enforcement officers.

How has strain theory been applied? Strain theory has been used most extensively in the study of juvenile delinquency. Albert Cohen (1977) uses the theory to explain the prevalence of gang delinquency among lower-class youth. According to Cohen, lower-class youth are poorly equipped to succeed in the middle-class world of school, which rewards verbal skills, academic success, neatness, and the ability to delay gratification. To protect their self-esteem, the youth reject these rules of success and self-worth and create their own status standards—standards they are able to achieve. Being tough, destructive, and daring is a vital part of this status system.

According to Richard Cloward and Lloyd Ohlin (1998), deviance does not always result from the strain created by a discrepancy between cultural goals and socially approved means for reaching these goals. Cloward and Ohlin refine strain theory, emphasizing that deviant behavior is not an automatic response; like any other type of behavior, deviance must be learned. And learning must take place through observation of others. Thus, deviance is more likely to occur as observation of others increases.

The Cloward and Ohlin concept helps explain why deviance occurs more in some parts of a society than in others. In middle- and upper-class homes, young people are encouraged to value education—the most important socially approved means for success—and are exposed constantly to people who are, through legitimate means, successful. Poor young people, on the other hand, can more easily model their behavior after pimps, drug dealers, and other types of criminals who have illegitimately acquired material symbols of success—money, clothes, cars, and women. And when juvenile delinquents are sent to reform schools or prisons, they increase their opportunity to learn criminal behavior because they associate with well-practiced veterans. Cloward and Ohlin's use of strain theory overlaps with the symbolic interactionist approach to deviance, discussed in the next section.

How can strain theory be evaluated? Merton's strain theory has had great staying power, in part because of its applicability to juvenile delinquency and crime. Another plus is its emphasis on social structure rather than on individuals.

Strain theory has some shortcomings. Because it is in the functionalist tradition, it assumes a consensus in values; it assumes that everyone values success and defines success in economic terms. Also, strain theory does not seem to explain an individual's preference for one mode of adaptation over another. Why, for instance, do some people choose to conform rather than to rebel? Finally, although strain theory helps explain crime and delinquency, it offers no help in explaining other types of deviance, such as mental illness or drug abuse.

Control Theory

Travis Hirschi's (1972) control theory is also based on Durkheim's concept of deviance. According to **control theory,** conformity to social norms depends on a strong bond between individuals and society. If that bond is weak or is broken for some reason—if anomie is present—deviance occurs. In control theory, then, social bonds *control* the behavior of people; it is the social bond that prevents deviance from occurring. Individuals conform because they fear that deviance will harm the relationships they have with others. They don't want to "lose face" with family members, friends, or classmates.

What are the basic elements of social bonds? According to Hirschi, the social bond has four intertwined dimensions:

1. *Attachment.* The likelihood of conformity varies with the strength of the ties one has with parents, friends, and institutions such as schools and churches. The stronger the attachment, the more likelihood of conformity.
2. *Commitment.* The greater one's commitment to legitimate social goals such as educational attainment and occupational success, the more likely one is to conform. The commitment of adults exceeds that of teenagers; and the commitment of individuals who believe their hard work will be rewarded is greater than the commitment of those who think they cannot compete within the system.
3. *Involvement.* Participation in legitimate social activities increases the probability of conformity, both because it positively focuses time and energy and because it encourages contact with others whose good opinion one values.
4. *Belief.* Subscription to norms and values of society promotes conformity. Respect for the rules of social life strengthens one against the temptations of deviance.

In short, when social bonds are weak, the chances for deviance increase. Individuals who lack attachment, commitment, involvement, and belief have little incentive to follow the rules of society.

FEEDBACK

1. Which of the following is *not* one of the benefits of deviance for society?
   a. It decreases suspicion and mistrust.
   b. It promotes social change.
   c. It increases social unity among members of a society.
   d. It provides a safety valve.
   e. It promotes clarification of norms.
2. According to _____ theory, deviance is most likely to occur when there is a discrepancy between culturally prescribed goals and the socially approved means of obtaining them.
3. A college professor who simply goes through the motions of teaching classes without any thought of success is an example of which of the following responses in strain theory?
   a. rebellion
   b. conformity
   c. ritualism
   d. innovation
   e. retreatism
4. The higher commitment to societal values and norms among adults than teenagers is consistent with the principles of _____ theory.

Answers: 1. a; 2. strain; 3. c; 4. control

How has control theory been received by other social scientists? Generally speaking, empirical support for Hirschi's control theory has been relatively strong. Several studies demonstrate the theory's power to explain both nondelinquent and deviant behavior (Bernard 1987; Rosenbaum 1987; Sampson and Laub 1990).

Scientific support weakened, though, when Hirschi attempted to broaden his control theory of delinquency into a more general theory encompassing all types of crime (Gottfredson and Hirschi 1990; Hirschi and Gottfredson 1990). Critics are skeptical that a specific theory of conformity and delinquency can explain crime in all its varieties, ranging from insider trading to murder. The existence of weak social bonds may satisfactorily account for pimps, drug dealers, and carjackers. But it is much more difficult for control theory to explain socially well-integrated individuals who embezzle corporate funds, use public monies for private purposes, or dump toxic waste into rivers (J. Hagan 1994; McCaghy et al. 2007). In fairness, expansion of control theory needs further research.

Symbolic Interactionism and Deviance

According to theories based on symbolic interaction, deviance is learned through socialization. Concepts such as the looking-glass self, significant others, primary group, reference group, and the generalized other, discussed in Chapter 4, underlie both cultural transmission theory and labeling theory.

Cultural Transmission Theory

What is cultural transmission theory? Cultural transmission theory emerged from the "Chicago School" during the 1920s and 1930s. According to **cultural transmission theory**, deviance is part of a subculture; it is transmitted through socialization. Clifford Shaw and Henry McKay (1929) observed that delinquency rates stay high in certain neighborhoods, even though the ethnic composition changes over the years. According to Shaw and McKay, delinquency in these neighborhoods persists because it is transmitted through play groups and gangs. As new ethnic groups enter the neighborhoods, they learn delinquent behavior from the neighborhood residents (Warr 2002).

Edwin Sutherland and Donald Cressey (1992) helped to extend cultural transmission theory to deviance that occurs from generation to generation and from one ethnic group to another. According to **differential association theory**, deviant behavior—like religious choice, political affiliation, or sport preference—is learned, principally, in primary groups. The more individuals are exposed to people who break the law, the more likely they are to become criminals.

At least three other factors impinge on the likelihood of deviance through differential association. First, an individual's vulnerability is affected by the ratio of exposure to deviants and nondeviants. Second, a person is more likely to become a deviant if his or her significant others are deviants. Finally, the earlier a person is exposed to deviants, the more likely that person will become a deviant.

How do sociologists evaluate cultural transmission theory? Cultural transmission theory shows that conformity in one setting may be deviant in another. Beyond this, cultural transmission theory explains why individuals placed in prison may become even more committed to deviance; in prison one is exposed almost exclusively to criminals. **Recidivism**—a return to crime after prison—makes sense in the light of cultural transmission theory because ex-convicts normally return to the significant others, primary groups, and reference groups through which they originally learned to be deviants.

Cultural transmission theory leaves unanswered questions. It cannot account for the reason some individuals become deviant without exposure to great numbers of people who advocate breaking the law. Middle-class delinquents are an illustration. Conversely, cultural transmission theory cannot say why most people reared in crime-ridden environments do not become deviants. Cultural transmission theory is limited in scope because it deals only with crime. Finally, although cultural transmission theory explains how deviance is learned, it fails to explain why some behaviors are considered deviant whereas others are not. An explanation of this relativity of deviance is a major strength of labeling theory.

Labeling Theory

Is deviance defined by the act or the label? Although strain theory and cultural transmission theory help us understand why deviance occurs, neither explains the relativity of deviance. **Labeling theory** attempts to fill this gap. In this theory, no act is inherently deviant; deviance is a matter of social definition. According to labeling theory, deviance exists when some members of a group or society label others as deviant. Howard Becker, a pioneer of labeling theory, states the heart of this perspective:

Social groups create deviance by making the rules whose infraction constitutes deviance, and by applying these rules to particular people and labeling them as outsiders. From this point of view, deviance is not a quality of the act the person commits, but rather a consequence of the application by others of rules and sanctions to an "offender." The deviant is one to whom that label has successfully been applied;

deviant behavior is behavior that people so label. (Becker 1991:9)

Labeling theory explains, for example, why unmarried pregnant teen girls are more severely ostracized than the biological fathers. Even today, the males are seldom considered deviant because cultural ideas about their sexual responsibility are different. And, of course, it is easier to stigmatize women because advanced pregnancy is undeniably visible. Labeling theory also explains why a middle-class youth who steals a car may go unpunished for "borrowing" the vehicle, whereas a lower-class youth goes to court for stealing. Too often, lower-class youth are "expected" to be criminals while middle-class youth are not.

Does deviant behavior necessarily carry a label? Edwin Lemert's (2000) distinction between primary and secondary deviance helps clarify the labeling process. In cases of **primary deviance**, a person engages only in isolated acts of deviance. For example, when college students are asked to respond to a checklist of unlawful activities, most admit to having violated one or more norms. Yet, the vast majority of college students have never been arrested, convicted, or labeled as criminals. Certainly, those who break the law for the first time do not consider themselves criminals. If their deviance stops at this point, they have engaged in primary deviance; deviance is not a part of their lifestyle or self-concept. Juveniles, likewise, may commit a few delinquent acts without becoming committed to a delinquent career or regarding themselves as delinquents.

Secondary deviance, on the other hand, refers to deviance as a lifestyle and personal identity. A secondary deviant is a person whose life and identity are organized around deviance. In this case, the deviant status engulfs individuals and becomes a status that overshadows all other statuses. Individuals identify themselves primarily as deviants and organize their behavior largely in relation to deviant roles. Other people label them as deviant as well and respond to them accordingly. When this occurs, these individuals usually begin to spend most of their time committing acts of deviance; deviance becomes a way of life, a career (Kelly 1996). Secondary deviance is reflected in the words of Carolyn Hamilton-Ballard, known as Bubbles to her fellow gang members in Los Angeles:

> *Because of my size, I was automatically labeled a bully-type person. . . . I mean, people saw that Bloods jacket and since everybody thought I was crazy, I started acting crazy. At first it was an act, but then it became me. After being the target for drive-bys and going through different things, that became my life-style. I started retaliating back and I got more involved.* (Johnson and Johnson 1994:209)

What are the consequences of labeling? Erving Goffman illuminates negative effects of labeling when he wrote about **stigma**—an undesirable characteristic or label used by others to deny the deviant full social acceptance (Falk 2001). Stigmatic labels—such as jailbird, crazy, fat, homeless, and cripple—spoil an individual's entire social identity (Heatherton et al. 2000; Hudak and Kihn 2001; Link and Phelan 2001). One attribute is used to discredit the individual's entire worth. The words of a forty-three-year-old bricklayer, who was unemployed during the Depression, illustrate this point:

> *How hard and humiliating it is to bear the name of an unemployed man. When I go out, I cast down my eyes because I feel myself wholly inferior. When I go along the street, it seems to me that I can't be compared with an average citizen, that everybody is pointing at me with his finger. I instinctively avoid meeting anyone. Former acquaintances and friends of better times are no longer so cordial. They greet me indifferently when we meet. They no longer offer me a cigarette and their eyes seem to say, "You are not worth it, you don't work."* (Goffman 1963:17)

Obviously, it is the labeling of an attribute rather than the attribute itself that causes psychological pain and suffering. What matters, for example, is the manner in which an attribute is acquired. Someone with a severe speech impediment caused by a war injury in Iraq is not likely to be labeled by those who know the origin of the impediment.

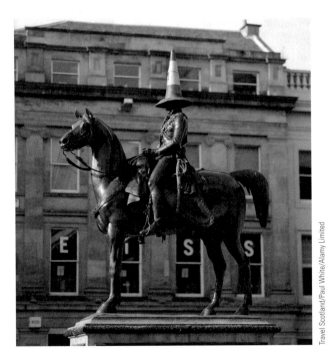

Travel Scotland/Paul White/Alamy Limited

Some young people decorated this Duke of Wellington statue in Glasgow, Scotland with a traffic cone. Were they engaged in primary or secondary deviance?

The application of stigmas varies by time and place. Today's active membership in a political party in the United States is considered a sign of good citizenship. To the generation of George Washington and Thomas Jefferson being an advocate of a political party was a major stigma that made one unacceptable for national office. Even campaigning for office was a sign of unworthiness (Ellis 2007).

Does labeling cause deviance? According to labeling theory, labeling is all that is required for an act to be deviant. Deviance, however, cannot be defined solely based on the reactions of others. If an act is not deviant unless it is labeled, there can be no such thing as undetected deviance.

Deviance cannot be defined without some reference to norms, but this does not mean that labeling processes are unimportant. Future behavior may be affected by the way others respond to an individual's deviance. In this sense, labeling produces deviants. To assign a person the label of delinquent, drug addict, or prostitute is a form of social penalty that may lead to rejection and exclusion from conventional groups and consequently may evoke further deviant acts. Labeling leads to deviance when it produces secondary deviance.

Recall the study of the Roughnecks and the Saints, two teenage high school gangs. According to Chambliss (1973), labeling the Roughnecks as deviants promoted additional deviance. Labeling the Saints as upstanding young men promoted conformity to most conventional norms (see "Doing Research").

How do sociologists assess the value of labeling theory? Labeling theory has contributed several insights. First, it has established that deviance is a matter of social definition, that deviance is relative. Second, labeling theory highlights, through concepts such as secondary deviance and stigma, the acquisition of a deviant self-concept that is likely to lead to a career of deviance. Third, labeling theory applies to a greater variety of deviance than the other theories. It can be applied to mental illness and prostitution as easily as to drug dealing and homosexuality.

Labeling theory also has its limitations. It does not explain habitual deviants who have never been detected and labeled, nor does it deal with isolated acts of deviance (primary deviance). It is also possible that labeling may halt rather than create deviance, as in the case of a man who frequents pornographic movie houses until he is suddenly discovered and labeled by his friends. Also, with its accent on relativity, labeling theory tends to overlook the fact that some acts, such as murder and incest, are banned in virtually all societies. Finally, labeling theory tends to generate sympathy for deviants, who are depicted as innocent victims of labeling.

Labeling theory implies that those who react against deviants are the culprits. Lawmakers do not cause people to commit criminal acts. It seems inappropriate to say that a police officer who arrests a person is somehow at fault. If the criminal law were abolished, acts that result in injuries would not disappear. In other words, a "mugging by any other name hurts just as much" (Nettler 1984).

Mental Illness and Labeling Theory

It is especially interesting to examine mental illness from the labeling perspective, because mental disorders, being a part of the medical field, are assumed to have an objective basis like cancer or arteriosclerosis. Labeling theory offers a unique way of viewing mental illness.

How does labeling theory approach mental illness? By virtue of its foundation in symbolic interactionism, labeling theory views mental illness as the result of social interaction in which others respond to us and we imagine what those responses mean. Mental illness is considered a matter of social definition (Collier 1993).

According to psychiatrist Thomas Szasz (1986), mental illness, as defined by the medical community, is a myth. Szasz sees the behaviors associated with mental disorder as adaptations to interaction-based stresses threatening to overwhelm an individual. In response to "problems of living," people (primarily unconsciously) impersonate a sick person to obtain help from others.

Thomas Scheff (1974, 1999) has formulated a theory of mental illness (based on the labeling perspective) in which most psychiatric symptoms of mental illness are seen as violations of social norms. Because of temporary pressures, contends Scheff, anyone might exhibit behaviors that violate norms—withdrawing socially, becoming overly suspicious, muttering to oneself. This primary deviance is usually considered transitory and is ignored by others. If this rule breaking is exaggerated and distorted by others, it can lead to a label—mentally ill. If the process of labeling goes far enough, mental illness may become a career path; it may reach the stage of secondary deviance in which it becomes part of a person's self-concept. Thus, a person may be successfully labeled mentally ill by a social process when, in fact, the original deviance may have been only temporary acts to handle temporary pressures.

Scheff's critics present data showing that the prime factor determining the negative reaction of others to mental patients is the patients' behavior, not the label such people have acquired (D. Phillips 1963, 1964; Link and Cullen 1983). More recent research concludes that labels do matter and that labeling theory cannot be dismissed as a framework for understanding responses to the mentally ill (Link 1987; Link et al. 1987).

F E E D B A C K

1. Which of the following did Edwin Sutherland mean by differential association?
   a. Crime is more likely to occur among individuals who have been exposed more to unfavorable attitudes toward the law than to favorable ones.
   b. People become criminals through association with criminals.
   c. Crime is not transmitted culturally.
   d. Crime comes from differential conflict.
2. _____ theory is the only sociological theory that takes into account the relativity of deviance.
3. _____ refers to deviance as a lifestyle and personal identity.
4. _____ is an undesirable characteristic used by others to deny the deviant full social acceptance.
5. According to labeling theory, mental illness is a matter of _____.

Answers: 1. a; 2. Labeling; 3. Secondary deviance; 4. Stigma; 5. social definition

Conflict Theory and Deviance

Deviance in Capitalist Society

Of the several conflict perspectives of deviance, Marxist researchers have forged an influential line of theory and research (Quinney 1980; Chambliss and Seidman 1982; Swaaningen 1997; Arrigo 1999). Focusing on class conflict, Marxist criminologists see deviance as a product of the exploitative nature of the ruling class. Deviance is behavior that the rich and powerful see as threatening to their interests. Consequently, the rich and powerful determine which acts are deviant and to what extent deviants should be punished.

According to the Marxian interpretation, the rich and powerful use the law to maintain their position. They can do this, argue the Marxists, because a society's beliefs, values, and norms (including, of course, laws) exist to protect the interests of the ruling class.

By what means do capitalism and deviance become linked? Steven Spitzer (1980) theorized some basic ways in which the culture of a capitalist society supports its economic system. First, critics of capitalism are considered deviants because their beliefs challenge the economic, political, and social basis of capitalism. Second, because capitalism requires a willing workforce, those who will not work are considered deviants. Third, those who threaten private property, especially that belonging to the rich, are prime targets for punishment. Fourth, because of society's need for respect of authority, people who show a lack of respect for authority—agitators on the job, people who stage nonviolent demonstrations against established practices—are treated as deviants. Fifth, certain activities and characteristics are encouraged or discouraged, depending on their congruence

with the requirements of the economic system. Athletics are approved because they foster competition, achievement, teamwork, and winning. Nonathletic males are wimps (Eder 1995; Adler and Adler 1998).

The Marxian perspective is not the only way to understand deviance from the viewpoint of conflict theory. The relationship between minorities and the judicial system is another way; white-collar crime is yet another (J. W. Coleman 1987, 1989; G. S. Green 1997; F. E. Hagan 2007).

Race, Ethnicity, and Crime

What is the relationship among race, ethnicity, and crime? Advocates of conflict theory point to the differential treatment minorities receive in the American criminal justice system (see Figure 7.3). They cite statistics showing that, at all points in the criminal justice process, African Americans and Latinos are dealt with more harshly than whites—from arrest through indictment, conviction, sentencing, and parole (J. H. Skolnick 1998; Schaefer 2007). Even when the criminal offense is the same, African Americans and Latinos are more likely than whites to be convicted, and they serve more time in prison than whites (Huizinga and Elliot 1987; Bridges and Crutchfield 1988; Klein, Turner, and Petersilia 1988). African Americans constitute 40 percent of U.S. prison inmates and 40 percent of those with death sentences (Sourcebook of Criminal Justice Statistics 2003). Although African Americans account for only 13 percent of the total population in the United States, more than 43 percent of inmates under the death penalty are African American.

Consider further illustrations of the differential experience of African Americans. During the four years following 1977, when the death penalty was legalized again in the United States, state attorneys were twice as likely to ask for the death penalty when an African

Even when they have committed the same crimes, African Americans and Latinos are more severely punished than whites.

American was found guilty of murdering a white person than when an African American murdered an African American or when a white killed a white (McManus 1985). Although nearly one-half of all homicide victims in the United States are African American, more than 80 percent of prisoners on death row are there for murdering whites (Death Penalty Information Center 2008). Prosecutors are less likely to seek the death penalty when an African American has been killed, and juries and judges are less prone to impose the death penalty in cases involving African American victims. Finally, white murder defendants are more likely to work out plea bargains than are African American defendants (Pratt 1998).

Why do minorities have a higher probability of being arrested, convicted, and severely punished? Conflict theory suggests several reasons for differences in the way minorities and whites are treated in the criminal justice system. For one thing, conflict theorists point to the fact that minorities generally do not have the economic resources to buy good legal services. Thus, the outcomes of their trials are not likely to be as favorable to them.

Sociologists who follow the conflict perspective believe that crimes against whites tend to be punished more severely than crimes against minorities because society sees minority interests as less important than the interests of whites. **Victim discounting** reduces the seriousness of crimes directed against members of lower social classes (Gibbons 1985). According to the logic behind victim discounting, if the victim is less valuable, the crime is less serious, and the penalty is less severe.

White-Collar Crime

What is white-collar crime? According to Edwin Sutherland (1940, 1983), the originator of the term, **white-collar crime** is any crime committed by respectable and high-status people in the course of their occupations. Or, as one researcher put it, lower-status people commit crimes of the streets; higher-status people engage in "crimes of the suites" (Nader and Green 1972; D. R. Simon 2007). Officially, the term *white-collar crime* is reserved for economic crimes such as price fixing, insider trading, illegal rebates, embezzlement, bribery of a corporate customer, manufacture of hazardous products, toxic pollution, and tax evasion (Geis, Meier, and Salinger 1995; J. W. Coleman 1997; Calavita, Pontell, and Tillman 1999; Weisburd and Waring 2001).

U.S. citizens were dismayed in 2002 when it was discovered that many large corporations, including Enron and WorldCom, had illegally underreported their earnings in order to inflate their stock value. Moreover, we were shocked to learn, these corporations were aided in this deception by some accounting firms—such as Arthur Andersen—that were supposed to audit them. And all the while, brokers with national investment houses such as Merrill Lynch were encouraging investors to purchase stock they actually believed to be worthless in order to protect their firms' lucrative investment banking contracts with some of these same large corporations. Due to the subsequent collapse of stock values, thousands of workers lost their jobs (and their retirement nest eggs), and individual stockholders and pension funds lost billions.

How can white-collar crime be viewed from the conflict perspective? Advocates of conflict theory point out that white-collar crime is extremely harmful to society. Its perpetrators, however, are treated more leniently than other criminals because of their class position (Henry and Lanier 2001).

The costs of white-collar crime are higher than is generally thought. According to the U.S. Department of Justice, white-collar crime costs in excess of $200 billion annually—an amount eighteen times greater than the costs of street crime. Illegal working environments (e.g., factories that expose workers to toxic chemicals) account for about one-third of all work-related deaths in the United States. Five times more Americans are killed each year from illegal job conditions than are murdered on the streets. None of these figures, of course, reflect the costs to society of a demoralized citizenry bombarded by news of criminal acts by corporate, political, and religious leaders (D. R. Simon 2007).

Despite this social harm, much more tolerance is shown white-collar criminals than criminals in the lower classes. Penalties are both tougher and more likely to be imposed for crimes that are committed by lower-class people than for those committed by people in higher social classes. Drug law violators (more likely to be from the lower class), for example, are treated more harshly

FIGURE 7.3

Who's Behind Bars?

According to data analyzed for this report, as of Jan. 1, 2008 more than **1 in every 100 adults is behind bars**. For the most part, though incarceration is heavily concentrated among men, racial and ethnic minorities, and 20- and 30-year olds. Among men the highest rate is with black males aged 20-34. Among women it's with black females aged 35-39. How would you explain the figures below?

MEN

White men ages 18 or older **1 in 106**

All men ages 18 or older **1 in 54**

Hispanic men ages 18 or older **1 in 36**

Black men ages 18 or older **1 in 15**

Black men ages 20-34 **1 in 9**

WOMEN

White women ages 35-39 **1 in 355**

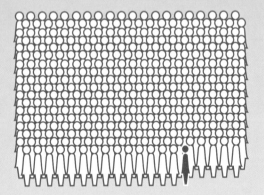

Hispanic women ages 35-39 **1 in 297**

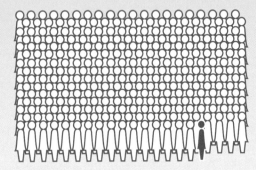

All women ages 35-39 **1 in 265**

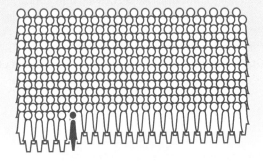

Black women ages 35-39 **1 in 100**

Source: Jenifer Warren, "One in 100: Behind Bars in America." Washington, DC: Pew Center on the States and the Public Safety Performance Project, 200 Based on analysis of "Prison and Jail Inmates at Midyear 2006," published June 2007 by the U.S. Department of Justice, Bureau of Justice Statistics. All demographic statistics, with exception of "1 in every 100 adults" are midyear 2006, not 2008 figures.

than embezzlers. Although fewer than 400 embezzlers are in prison, thousands of drug law violators are incarcerated. In federal court, where most white-collar cases are tried, probation is granted to 40 percent of antitrust violators, 61 percent of fraud defendants, and 70 percent of embezzlers. In general, convicted white-collar criminals are less likely to be imprisoned; and if they are, these white-collar criminals receive shorter sentences. Moreover, when sentenced, white-collar criminals are more likely to be placed in a facility with extra amenities (e.g., tennis courts or private rooms), which critics have dubbed "Club Fed" (Gest 1985; U.S. Department of Justice 1987; Reiman 2000).

In light of the conflict perspective, it is interesting to examine the legal fate of top executives at corporations such as Enron, WorldCom, and Tyco. With so many corporations involved and with the magnitude of the personal and economic consequences, the legal response was expected to be unusually harsh. These corporate officials (and their lax boards of directors) have already been rebuked by the public; and in 2002 the federal government enacted the stiffest legal penalties for financial fraud since the Great Depression in the 1930s (VandeHei and Hilzenrath 2002). Some criminal prosecutions by the federal government were successful. Former Tyco executives, Dennis Kozlowski and Mark Swartz were sentenced to eight to twenty-five years for corporate looting. Bernard Ebbers, former CEO of WorldCom, was sentenced to twenty-five years for an $11 billion accounting fraud. The former CEO of Adelphia Communications, John Rigas, received a fifteen-year prison sentence for financial irregularities. Scott Sullivan, former chief financial officer of WorldCom, got a five-year prison term for financial fraud. Citicorp agreed to pay $2 billion to investors for helping in the Enron accounting scandal. Ken Lay (who subsequently died) and Jeffrey Skilling, former top executives of Enron, were convicted in 2006.

In 2004, nationally recognized celebrity and business person, Martha Stewart, was sentenced to five months in prison and five months home confinement for obstructing an investigation into illegal stock trading. Many Americans consider her case an example of the leniency typically shown to white-collar criminals.

However, within a year after these scandals broke, penalties experienced by the perpetrators were relatively light. To avoid further investigations by New York State Attorney General Eliot Spitzer, major brokerage firms forged a financial settlement in 2002; but the amount of settlement was a fraction of either the money they made for themselves or the money they lost for their trusting investors. While this earlier settlement with Spitzer did not prevent the Securities and Exchange Commission (SEC) from fining the ten top investment firms in 2003, the $1.4 billion penalty pales beside the money the firms earned. Citigroup, for example, earned $43 billion between 2000 and 2002, but paid only a $400 million fine for its transgressions during the 1990s (Fogarty and Iwata 2003; Shell 2003).

TABLE 7.2

FOCUS ON THEORETICAL PERSPECTIVES: Illustrating Deviance

This table provides illustrations of deviance using concepts associated with a particular theoretical perspective. Construct some examples of your own.

Theoretical Perspective	Sociological Concept	Examples of Deviance
Functionalism	Anomie	• Delinquent gangs sell drugs because they do not agree with the norms and values against drug use advocated in the larger society.
Conflict theory	White-collar crime	• A convicted Wall Street stock manipulator (a more powerful member of society) may spend less time in prison than a factory worker (a less powerful member of society) found guilty of the same crime.
Symbolic interactionism	Labeling	• Some high school students reject dating because they have been consistently treated and described as "not cool."

1. Marxist criminologists see deviance as a product of the exploitative nature of the _____.
2. Which of the following is *not* one of the basic ways the culture of a capitalist society supports the society's economic system?
   a. People whose beliefs clash with capitalism are labeled deviants.
   b. Capitalism requires a willing workforce.
   c. Innovation is rewarded.
   d. People who fail to show respect for authority are likely to be considered deviant.
3. The process of reducing the seriousness of crimes based on the lower social value placed on the victims is called _____.
4. _____ is any crime committed by respectable and high-status people in the course of their occupations.

Answers: 1. ruling class; 2. c; 3. victim discounting; 4. White-collar crime

What are the strengths and weaknesses of the conflict approach to deviance? Of course, not all definitions of deviance can be attributed to the exploitation of the weak by the powerful. Laws against murder and rape are meant to protect both the rich and the poor. Moreover, some laws—consumer and environmental laws—exist despite the opposition of big business. Most proponents of the conflict perspective would not conclude that deviance exists only because of cultural and social arrangements supportive of the ruling class. Nevertheless, the conflict approach to deviance, like labeling theory, underscores the relativity of deviance. Perhaps the greatest strength of conflict theory is its ability to link deviance to social inequality and power differentials.

Crime in the United States

Crime is the most important type of deviance in modern society. Most Americans think of **crime**—acts in violation of the law—as including a narrow range of behavior. On the contrary, in the United States, more than 2,800 acts are considered federal crimes, and many more acts violate state and local statutes.

If a society is to prevent and control crime, that society must be aware of just how much crime is being committed—and what type of crime. The major source of America's crime statistics is the FBI's *Uniform Crime Reports* (U.S. Department of Justice 2007a). These official statistics are gathered from police departments across the country. Reports are submitted voluntarily by law enforcement agencies.

How is crime distributed in the United States? Data on arrests are collected for major categories of crimes (crime index offenses): murder, forcible rape, robbery, aggravated assault, burglary, larceny-theft, and motor vehicle theft (see Table 7.3). In 2006, the number of

index crimes known to the police totaled 11,401,313. Violent crime—murder, forcible rape, aggravated assault, and robbery—made up 12 percent of the

The job of the forensic scientist is to examine evidence—fingerprints, DNA, handwriting, firearms – for indications that a crime has occurred. Here Eric Szmanda, one of the stars of the television series CSI, is pretending to be a forensic scientist.

CBS-TV/THE KOBAL COLLECTION/The Picture Desk

TABLE 7.3

Crimes in the United States

In this table you will find the rates for the major crime categories tracked annually by the FBI. Note the rate changes that occurred between 1997 and 2006. Pretending that you are a reporter for your school newspaper, use these data to summarize crime trends in the United States.

| Types of Crime | 2006 | | 1997–2006 | |
	Number of Crimes	Crime Rate (per 100,000 inhabitants)	Percent Change in Number of Crimes	Percent Change in Crime Rate
Violent crime	1,417,745	473.5	−13.3	−22.5
Murder	17.034	5.7	−6.4	−16.3
Forcible rape	92,445	30.9	−3.8	−14.0
Robbery	447,403	149.4	−10.3	−19.7
Aggravated assault	860,853	289.5	−15.9	−24.8
Property crime	9,983,568	3,334.5	−13.6	−22.7
Burglary	2,183,746	729.4	−11.2	−20.6
Larceny-theft	6,607,013	2,206.8	−14.7	−23.7
Motor vehicle theft	1,192,809	398.4	−11.9	−21.2

Source: U.S. Department of Justice, Federal Bureau of Investigation, *Uniform Crime Reports*, 2007.

known crimes, and property crime—burglary, larceny-theft, and motor vehicle theft—accounted for about 88 percent. Figure 7.4 displays the percent distribution of crime in America. Figure 7.5 shows the *Uniform Crime*

FIGURE 7.4

Types of Crimes Americans Commit: FBI Crime Index Offenses, 2003 (Percent Distribution)

This figure shows the contribution each major type of crime makes to the total of U.S. crime. What surprises you most about these figures?

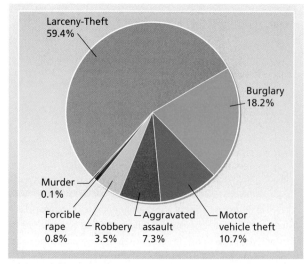

Larceny-Theft 59.4%
Burglary 18.2%
Murder 0.1%
Forcible rape 0.8%
Robbery 3.5%
Aggravated assault 7.3%
Motor vehicle theft 10.7%

Source: U.S. Department of Justice, Federal Bureau of Investigation, *Uniform Crime Reports,* 2003.

Reports estimate of the frequency at which crimes occur in the United States.

How much crime is there in the United States today? Crime increased sharply between the 1960s and the early 1990s. For example, the FBI index of violent crime increased from a big-city offense rate per 100,000 of 860 in 1969 to over 1,100 in 2001. The violent crime rate is considerably higher than in most other industrialized countries.

As Table 7.3 shows, however, both violent crime and property crime have declined since 1997. Because homicide receives the most publicity, it can be used to highlight this general, across-the-board reduction in crime. The murder rate in the United States has declined about 16 percent since 1997. This decline, however, has gained dramatic momentum only since the mid-1990s as the economy boomed and unemployment in the inner city diminished (Blumstein and Wallman 2006).

Measurement of Crime

How reliable are *Uniform Crime Reports*? Although the FBI statistics provide considerable information about crime, they have serious limitations. Some critics portray the FBI statistics as over representing the lower classes and minorities but undercounting the middle and upper classes. Amateur thefts and minor assaults are not as likely to be reported to the police as murders and auto thefts. Prostitutes and intoxicated persons are subject to arrest in public places, but they are relatively safe in private settings, where the police cannot enter without a warrant.

FIGURE 7.5

How Often Do Americans Commit Crime? FBI Crime Clock: 2007

The numbers below, of course, do not mean that one crime index offense actually occurs every 2.7 seconds. It does mean that when *all* crime index offenses for 2007 are divided by the total number of seconds in a year, there are enough of them to be spaced out every 2.7 seconds. That is frightening enough.

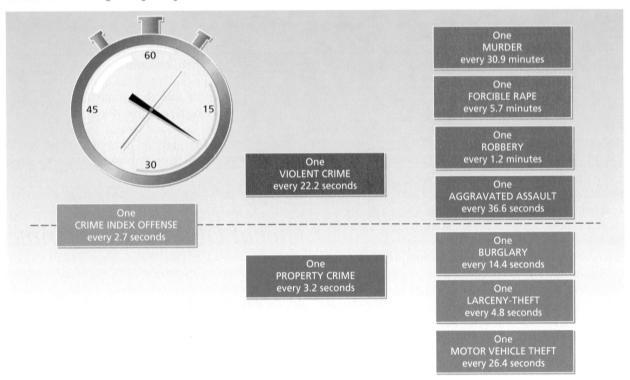

Source: U.S. Department of Justice, Federal Bureau of Investigation, *Uniform Crime Reports, 2007.*

INTERNET LINK The FBI publishes its *Uniform Crime Reports* annually. For additional information on the latest crime statistics for the United States, you can visit the FBI's home page at www.fbi.gov or you can access data from the U.S. Department of Justice at www.ojp.usdoj.gov

Also, about two-thirds of U.S. crimes are not reported to the police. In addition, crime reporting varies from jurisdiction to jurisdiction and from crime to crime. For instance, government officials, corporate executives, and other white-collar offenders are seldom included in official crime statistics. Finally, in such areas as unfair advertising and job discrimination, the laws are enforced through special administrative boards rather than the criminal justice system (Box 1981; K. N. Wright 1987; Ermann and Lundman 1996; J. W. Coleman 1997).

In response to these criticisms, the National Crime Victimization Survey (NCVS) was launched in the early 1970s. This survey is conducted semiannually for the Bureau of Justice Statistics by the U.S. Census Bureau. The bureau uses state-of-the-art survey research techniques (U.S. Department of Justice 2007). At points, the NCVS and the *Uniform Crime Reports* contradict each other. For example, the NCVS indicates that the FBI statistics significantly underestimate the total amount of crime victimization in the United States (U.S. Department of Justice 2007b).

Whom are we to believe? Both the *Uniform Crime Reports* and the National Crime Victimization Survey have their strengths. Because most people know relatively little about the technical definitions of crime, a major strength of the *Uniform Crime Reports* is its use of experienced police officers; they decide if an act has violated a statute. The NCVS has two attractive advantages. First, it helps to compensate for the massive official underreporting of crime. Second, the surveys are scientifically sound. At the very least, the NCVS is an increasingly important supplement to the FBI's official statistics. Together, these two sources provide a relatively complete account of the extent and nature of crime in the United States (K. N. Wright 1987; U.S. Department of Justice 2007).

1. According to the FBI's *Uniform Crime Reports*, crime in the United States has increased since 1989. T or F?
2. According to the National Crime Victimization Survey, does the *Uniform Crime Reports* underestimate or overestimate the amount of crime in America? _____

Answers: 1. F; 2. underestimate

Juvenile Crime

Juvenile crime refers to violations of the law committed by those less than eighteen years of age. Juvenile offenders are the third largest category of criminals in the United States. Teenage criminal activity includes theft, murder, rape, robbery, assault, and the sale of illegal substances. Juvenile delinquent behavior can extend beyond crime to deviance that only the young can commit (such as failing to attend school, fighting at school, and drinking and smoking under age).

What is the trend in juvenile crime? Violent juvenile crime reached its lowest level in a decade in 2001, a fall of about 30 percent since 1995 (U.S. Department of Justice 2002). Since 1993, the juvenile murder arrest rate has dropped by half, juvenile arrests for weapons violations have declined by a third, and the juvenile rape arrest rate has gone down by a fourth. There have also been fewer juvenile victims

of murder—down from almost 3,000 to about 2,000. Juvenile crime, in short, returned to the rates typical of the years prior to the crack epidemic of the late 1980s.

Several factors are said to account for this decline in juvenile crime. First, there has been a decline in the demand for crack cocaine. Second, remaining crack gangs that provided guns to juveniles have reached truces. Third, police have clamped down on illegal guns, and repeat violent juvenile offenders have been given stiffer sentences.

Global Differences in Crime

Crime rates vary within any society. Sociological factors such as gender, age, race, ethnicity, social class, geographical region, and neighborhood each contribute to these variations. Crime rates also vary from society to society. One way to examine cross-cultural differences

SNAPSHOT OF AMERICA 7.1

Violent Crime

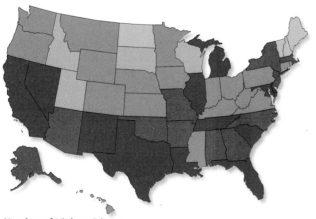

Number of Violent Crimes Reported per 100,000 Population (2003)

- 700 or above
- 550–699
- 400–549
- 250–399
- 150–249
- 149 or below

Source: Federal Bureau of Investigation, *Uniform Crime Reports*, 2003.

Although experiencing a recent decline, the U.S. has one of the highest violent crime rates of the major industrialized countries. In fact, the U.S. has the highest murder, rape, and robbery rates, and keeps pace in burglaries and auto thefts. This map indicates the number of violent crimes by state per 100,000 residents.

Interpret the Map

1. Create a graph showing how the violent crime rate in your state compares with the rates in other states.
2. Pose a question that relates to the relative ranking of your state with other states.
3. What sociological conclusion can you draw from this map?

Look Out for Identity Thieves!

One of the newest forms of deviance is "identity theft"—stealing credit information belonging to another person. The frequency of identity theft began skyrocketing around the turn of the present century (O'Brien 2000; Grabosky and Smith 2001; O'Harrow 2001). The results to victims can be devastating. Veronica, a California college student, cannot get a student loan because someone in Michigan is using her identity to open credit card accounts and obtain telephone service. In another case, Meredith rented a room in her house to another woman. After discovering Meredith's Social Security number, this boarder used it to fill out several credit card applications in Meredith's name and to run up more than $70,000 in charges. In yet another example, Graciela has been a victim of identity theft for more than ten years. A thief gained access to her Social Security number, birth cer-

tificate, and driver's license. With this information, the imposter has obtained credit cards, purchased furniture, bought cars, and obtained welfare. (All of these examples and more are available through the Privacy Rights Clearinghouse at www.privacyrights .org, 2008.)

Beth Givens of the Privacy Rights Clearinghouse (a nonprofit group for consumers' privacy rights) explains that identity theft can occur in many ways. A thief can steal a wallet or purse, get copies of credit card slips from trash, or steal someone's mail. There are also high-tech methods of identity theft. The most common method is through illegal access to the computers of credit-rating companies. These companies maintain credit reports that provide valuable information about a consumer—Social Security number, birth date, credit card numbers, address. Although credit-rating companies try to prevent high-tech identity theft, the very nature of their service makes this information accessible through computer terminals. This access is an open invitation to criminals.

The victims of identity theft obviously suffer great damage. Unless the thief is caught in the act, there seems to be little the police can do to stop this kind of crime. At the same time, many victims report that their victimization is compounded by abusive col-

lection agencies. Victims also have to spend time and money cleaning up the mess imposters have created for them—sometimes up to ten years or more. Victims are often scarred emotionally, feeling violated, hopeless, and angry.

The main goal of imposters in identity theft is to purchase items at no cost. In these cases, the victims still maintain their identities. But what if identity theft also involved actually losing one's identity?

What would happen if a person's identity were actually "stolen"? A movie—*The Net*—showed what it might be like to have your identity taken away. In this movie, a woman's entire identity is obliterated by thieves. By stealing the documents that would prove her identity and destroying all of her existing computer records, the criminals hijack her identity. Using her photograph and Social Security number, the villains in the movie create a whole new identity for her, including a new name, bad credit report, and criminal record. As the woman in the movie laments, "They knew everything about me. It was all on the Internet!"

Think About It
Which theoretical perspective would be most useful in analyzing identity theft? Explain your choice, and apply that perspective to identity theft.

---

in crime is to compare crime rates in the United States with those of other countries.

Global Crime Comparisons

How do the levels of crime in other countries compare with crime in the United States? Data on crime rates reveal some striking differences between the United States and other countries (Kalish 1988; MacKellar, Landis, and Yanagishita 1995; F. E. Hagan 2007). The following specific statements can be made:

- The U.S. homicide rate per 100,000 population is around 5.5; the rate of homicide in Europe is less than 2 per 100,000.

- The U.S. rape rate was 32 per 100,000, roughly six times higher than the average for Europe.
- U.S. rate for robbery was around 145 per 100,000 compared with the average European rate of less than 50 per 100,000.
- For crimes of theft and auto theft, the ratio of U.S. rates to average European rates was roughly 2 to 1.
- Burglary was the only crime examined for which the U.S. rate was less than double the average for European countries.
- U.S. crime rates were also higher than those of Canada, Australia, and New Zealand, but the differences were smaller than when compared with Europe. For burglary and auto theft, the rates were quite similar.

FEEDBACK

1. In which of the following categories of crime is the United States *not* the leader in the world?
   a. murder
   b. rape
   c. robbery
   d. burglary
2. Crime in Japan refutes a predicted positive relationship between modernization and crime. T or F?

Answers: 1. d.; 2. T.

Clearly, Americans watching the evening news do not have to fear being deceived by sensationalizing reporters telling them of the homicides, rapes, and robberies that seem to be occurring everywhere in the country. There *is* a lot of crime in the United States. Of the major industrialized countries, the United States has the highest murder, rape, and robbery rates and keeps pace, as well, in burglaries and auto thefts. Despite the rise in crime rates in Europe during the 1990s, crime remains a significantly greater problem in the United States than in any other country.

Are these cross-cultural data really to be trusted? You read earlier about the difficulties associated with gathering accurate U.S. crime data. So, it is not unreasonable to question the accuracy of data coming from many different cultures. There are problems associated with data recording, data collection, and varying definitions of crime from one country to another. Despite some notable dissenters (Balrig 1988; Gerber 1991), experts generally agree, though, that the differences in crime rates across societies are too great and too consistent to be denied credibility.

Crime and Modernization

According to one line of argument, increased crime is associated with economic development (Fenwick 1982). If this is the case, then the world's leading countries, economically speaking, will predictably have the highest levels of crime. This prediction, however, is contradicted by the relatively low rates for most types of crimes in such highly industrialized countries as Japan, Germany, and Switzerland.

Approaches to Crime Control

The **criminal justice system** comprises the institutions and processes responsible for enforcing criminal law. It includes police, courts, and a correctional system (Tonry 2000; Inciardi 2006; Siegel and Senna 2007; Siegal 2008). A criminal justice system may draw on four approaches to punishment—deterrence, retribution, incarceration, and rehabilitation—which may be used in varying combinations.

Deterrence

In 1994, Michael Fay, an American teenager convicted of car vandalism, was lashed four times with a four-foot rattan cane in a Singapore prison (M. Elliott 1994). The experience of being "caned" is said to be sufficiently bad that future crime is effectively discouraged. Singapore is by no means alone in the use of corporal punishment. Practices such as caning are designed to protect society through deterrence. The **deterrence** approach emphasizes intimidation, using the threat of punishment to discourage criminal actions. There is considerable debate on the effectiveness of deterrence (DiIulio and Piehl 1991).

Does punishment deter crime? Until the mid-1960s, social scientists generally rejected the idea that punishment deters crime. This rejection was based partly on earlier studies, which indicated that the death penalty did not deter murder any more effectively than life imprisonment. More recent research has led to a reevaluation of the relationship between punishment and crime. Although the evidence to support this point is inconclusive, social scientists may have discounted the deterrence doctrine prematurely (Greenberg and Kessler 1982; Logan 1982; Hook 1989; Yunker 2001; Liptak 2008a).

Investigation of the complicated relationship between punishment and crime is only in its infancy, but some research indicates that the threat of punishment does deter crime if potential lawbreakers know two things: that they are likely to get caught and that the punishment will be severe. In the United States, however, punishment for crime is usually not certain, swift, or severe. Consequently, punishment does not have the deterrent effect that it could have.

Capital punishment is a special case. Murder is usually committed not after a rational consideration of the act or its consequences but during an outburst of emotion. Under such irrational circumstances, one would not expect the fear of capital punishment to be a deterrent, and research shows that it is not. If the death penalty were a deterrent to murder, then a decline in its use should be followed by an increase in the murder rate. But research in many countries, including the United States, indicates that the murder rate remains constant or even drops following a decline in the use of the death penalty. Other research indicates that the use of capital punishment as a sanction is neither swift nor certain (Acker, Bohm, and Lanier 1998; Sarat 1998; Schonebaum 1998; M. E. Williams 2004).

How do Americans feel about capital punishment? Actually, American attitudes toward capital punishment have shifted dramatically over the years. In colonial America, murderers, robbers, arsonists, counterfeiters, and many other types of criminals were regularly hung in the public square before large crowds. By the 1780s, the death penalty was beginning to be a sign of social backwardness by many. By the Civil War, three northern states had abolished it and the other northern states reserved it for murderers. Later in the nineteenth century, the death penalty lost even more public favor. In the late 1960s, the death penalty was abolished, only to gain acceptance less than a decade later.

According to a 2007 Gallup poll, about 70 percent of the American population currently supports the death penalty for murder. However, support for the death penalty decreases to 47 percent when life imprisonment without parole is an alternative (Gallup 2007). Attitude toward the death penalty in the United States varies according to race and ethnicity. Seventy-three percent of whites favor the death penalty, compared with 55 percent of African Americans. This difference increases dramatically when life imprisonment without parole is possible. This variation in attitude toward the death penalty is not surprising, considering that African Americans are disproportionately sentenced to death row (U.S. Department of Justice, 2006). Although African Americans constitute about 13 percent of the U.S. population, they make up about 40 percent of death row inmates.

Between 1970 and 2000, the frequency of executions in the United States spiked (Banner 2003; Zimring 2004). Since 2000, however, the number of inmates sentenced to death and executed has dropped. And starting in early 2003, there were new rumblings of the death penalty debate. Illinois Governor George Ryan, formerly a proponent of the death penalty, had emptied his state's death row. After a three-year review of Illinois death row cases, Ryan had taken some unprecedented steps. He began by pardoning four death row inmates, three of whom were released from prison immediately. And for the remaining 167 inmates, Ryan announced a commutation of the death sentences to life without parole. Citing injustice, arbitrariness, and capriciousness, Ryan indicated that he would no longer "tinker

with the machinery of death" in a system "haunted by the demon of error" (Pierre and Lydersen 2003).

Despite the current scientific debate, about three-fourths of Americans are convinced that the death penalty is a deterrent to murder. And, of those Americans who favor the death penalty, more than three-fourths say that they would continue to favor it even if confronted with conclusive evidence that the death penalty does not act as a deterrent to murder and that it does not lower the murder rate. Rather than a need for deterrence, feelings of revenge and a desire for retribution, then, appear to contribute more to the support of capital punishment (Zimring 2004).

Increasing American uneasiness about the death penalty is reflected in politics and in the courts (S. Moore 2007). Partly in response to the release of death row inmates based on DNA evidence, Congress passed the Innocence Protection Act in 2004 (K. Johnson 2008). Twelve states have suspended the death penalty since 2000, and in 2006 two-thirds of the states executed no one. At the close of 2007, New Jersey became the first state in forty years to abolish the death penalty (Kim 2008). A 2008 Nebraska Supreme Court did rule the electric chair to be cruel and unusual punishment. In 2007, the U. S. Supreme Court established a de facto moratorium on the death penalty when it agreed to hear a case challenging the constitutionality of lethal injection (Greenhouse 20XX). The effects of the subsequent 2008 Supreme Court ruling that lethal injection as currently practiced is constitutional remains to be seen. This decision could have a mixed effect as some states resume executions while others do not (Liptak 2008a).

Retribution

What is the philosophy behind the retributive approach to criminal punishment? The public may demand **retribution**—that criminals pay compensation equal to their offenses against society. When an eye is taken, an eye must be returned. Although the law allows designated officials to exact retribution, it does not tolerate acts of personal vengeance. The legal system removes responsibility for the punishment of a criminal from the hands of individuals and places it in the hands of officials. A convicted criminal repays the state rather than the harmed individual(s). Retribution, in the name of justice, is a social rather than a personal matter. If an individual whose husband is a victim of homicide exacts the price of wrongdoing by shooting her husband's killer, she is also responsible to society.

The retributive philosophy is concerned with past rather than future crime. Like deterrence, the incarceration and rehabilitation approaches to punishment, to which we now turn, focus on preventing further crime by already existing offenders.

Incarceration

What is the justification for incarceration? Criminals who are not on the street cannot commit crimes. The **incarceration** approach removes criminals from society. In repressive societies, such as Nationalist China and the former Soviet Union, people may spend their entire lives in prison camps for crimes ranging from political opposition to murder. The incarceration philosophy, of course, may take less extreme forms. The United States, for example, is less severe, even though recent decades have seen an increasing emphasis on incarceration (Grapes 2000; Mauer 2006). Compared with a historical average of the 1925–1970 period, the current rate of imprisonment in the United States has increased five times. Moreover, the current incarceration rate is six to eight times the rates of Western European countries. With less than 5 percent of the world's population, the U.S. has almost a quarter of its prisoners (Garland 2001; Western and Pettit 2002; Aizenman 2007; Liptak 2008b).

What does the American public say? There are some signs of a decline in the "get-tough" approach. According to a recent study, American public opinion now favors crime prevention and treatment over severe punishment. In general, adults have shifted from several previously held convictions: deterrence through stricter sentencing, capital punishment for more crimes, and fewer paroles for convicted felons. By a 65 to 32 percent margin, they endorse prevention through job and vocational training, provision of family counseling, and more neighborhood activity centers for youth. In 1994, Americans were pretty evenly divided between addressing the underlying causes of crime or empha-sizing deterrence through strict punishment. Almost two-thirds of Americans now agree that crime reduction is best achieved through rehabilitation (*Changing Public Attitudes Toward the Criminal Justice System* 2002).

Rehabilitation

Rehabilitation is an approach to crime control that attempts to resocialize criminals. Rehabilitative programs have traditionally been introduced within the prisons, with programs aimed at giving prisoners both social and work skills that will help them adjust to society after release.

Do prisons rehabilitate criminals? Unfortunately, because 30 to 60 percent of those released from penal institutions will be back in prison two to five years later, it is difficult to believe that rehabilitation programs in the United States are working (Elikann 1996; Zamble and Quinsey 1997). This high rate of recidivism—past offenders returning to prison—can be attributed partly to characteristics of the offenders and the problems they face while in the criminal justice system (such as the stigma of being an ex-convict). The tendency toward recidivism may also be due to the inadequacies of prisons. Prisons are schools for crime: ex-convicts often leave prison more committed as criminals and with a higher likelihood of continued criminal involvement (Abramsky 2001).

Social scientists also recognize that it is difficult to change attitudes and behavior within the prison subculture. Although prison guards and officials hold the power, the informal rules of the prison subculture have

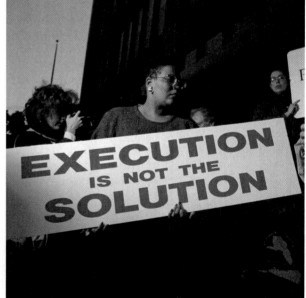

Capital punishment is one the most controversial issues in American society today.

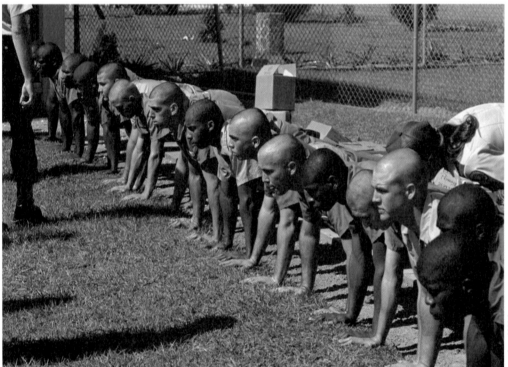

Juvenile boot camps are used as an alternative to confinement in juvenile corrections facilities. These boot camps focus on discipline and physical conditioning, and typically are restricted to nonviolent or first-time offenders.

Steve Starr/CORBIS

greater effect on prisoners' behavior. Conformity with the "inmate code" stresses loyalty among inmates as well as opposition to correctional authorities. This conformity is enforced verbally and physically. Consequently, involvement within prison criminal networks often increases in prisons, promoting a continued life of crime (J. Hagan 1993; Sherman 1993a).

There is yet another precipitating factor leading to continued criminal activity following imprisonment: the changing nature of employment in today's economy. Extended separation from a rapidly changing society leads to the obsolescence of employable job skills and to the deterioration of outside contacts who might be helpful in finding a job. Also, a released prisoner is likely to transfer the toughness reinforced in prison life to the workplace. This transfer of prison norms does not work because most jobs in the service economy require interpersonal skills (J. Hagan 1994).

If prisons do not rehabilitate, what are some alternatives? Several alternatives are being tried. One is the combination of prison and probation. A mixed, or split, sentence, known as *shock probation*, is designed to shock offenders into recognizing the realities of prison life. Prisoners must serve part of their sentence in an institution. The remainder of the sentence is then suspended, and the prisoner is placed on probation.

Another approach relies on community-based programs designed to reintroduce criminals into society. By getting convicts out of prison for at least part of the day, community-based programs help break the inmate code. At the same time, prisoners have a chance to

become part of society, participating in the community but under professional guidance and supervision.

A third strategy, *diversion*, is aimed at preventing, or greatly minimizing, the offender's penetration into the criminal justice system. Diversion involves a referral to a community-based treatment program rather than to a penal institution or a probationary program. Because offenders are handled outside the formal system of criminal law, the lawbreakers are not as likely to acquire stigmatizing labels and other liabilities (N. Morris and Tonry 1990; Lanier and Henry 2004).

Will any of these alternatives be successful? Most of the programs are relatively recent and have not yet been adequately evaluated. But as now being implemented, it appears that alternatives to prison offer little advantage over conventional probation programs. Still, defenders of alternative sanctions argue for their usefulness, if appropriately applied to those incarcerated who are a lower risk for recidivism (Tonry and Lynch 1996). And, given public concern about the skyrocketing costs of the criminal justice system, use of alternatives will probably increase with an eye to more effective application (Siegel and Senna 2007).

Crime Prevention

The approaches just described target current criminals. A different sentiment for crime prevention concentrates on steering potential offenders away from criminal activity in the first place, by providing treatment for addicts and mentoring and counseling for poor children; creating

1. Match the approaches to punishment on the left side with the examples on the right.

   ____ a. rehabilitation   (1) imprisonment without parole

   ____ b. deterrence   (2) the practice of caning

   ____ c. retribution   (3) employment and educational programs in prison

   ____ d. incarceration   (4) death penalty for murder

2. Prisons currently do an adequate job rehabilitating inmates. T or F?
3. It appears that social scientists have prematurely discounted the deterrence doctrine. T or F?
4. Research supports capital punishment as a deterrent to murder. T or F?

Answers: 1. a. (3), b. (2), c. (4), d. (1); 2. F; 3. T; 4. F

living-wage jobs in poor neighborhoods; improving public education for poor children, and funding after-school and late-night recreation programs. Given Americans' punitive philosophy of corrections, it seems unlikely that we will invest heavily in "precrime" programs.

Crime Control: Domestic and Global

Crime Control in the United States

A society can adopt various combinations of the four approaches to crime control. Singapore, for example, takes deterrence, retribution, and incarceration very seriously. The upside is that Singapore has safe streets and clean subways. The downside is a reduction in individual liberty and increased brutality toward prisoners.

The United States does not have a consistent commitment to any one of the major approaches. Public opinion and public policy are split between the less punitive approach of rehabilitation and the other three approaches.

Public and political concern about violent crime has led to new penal policies that reflect a turn from rehabilitation toward a more punitive attempt to control crime. This endeavor involves more emphasis on incarceration as a method of deterring crime and on retribution as a method of compensation. By way of illustration, many Americans support California's "three strikes" statute, which doubles the standard penalty for second-time serious or violent crimes and requires twenty-five years to life for three-time felons (W. Booth 1994; Morin 1994; Shichor and Sechrest 1996). Due to this more severe approach, the adult prison population more than quadrupled since 1980, standing at 2.2 million (Warren 2008).

Even recent rehabilitative initiatives have been harsher. Boot camps, for example, have been created for young nonviolent first offenders. The tough, military-style environment is intended to resocialize these young people by instilling self-discipline, commitment to work, and respect for the law (Katel 1994).

Although the latest national crime bill devoted some of its $30 billion to crime prevention programs (such as more police, youth programs, and the banning of nineteen semiautomatic assault-type weapons), it also has a strong punitive accent. This legislation vastly increases the number of federal crimes subject to the death penalty and imposes mandatory life sentences for repeat violent offenders. Funds are also available to states for new prisons.

A cross-cultural comparison in approaches to crime control will be informative. It reveals some striking differences between the United States and other countries.

Global Crime Control

All countries use incarceration. They vary significantly, however, in their emphasis on imprisonment (Christie 2000). Even western European countries, which share many cultural traditions, have varying imprisonment rates. England and Wales are near the top in inhabitants per 100,000 in prison. Germany, France, and Spain are near the middle, and Greece and the Netherlands are near the bottom. As stated earlier, the imprisonment rate in the United States is six to eight times that of any western European country. Actually, the United States is at the very top worldwide in the number of prison inmates. No other industrialized country is even in the world's top ten. For the first time, more than 1 in every 100 adults is in an American jail or prison (Warren 2008). As shown in Figure 7.6, the United States has more inmates than the 26 largest European inmate populations.

The United States is the only democracy to use the death penalty often. The European Union formally stated its opposition to the death penalty in 2007. By 2006, 90 countries had eliminated the death penalty. In 2007, the United Nations General Assembly called for a moratorium on executions with the intent of ultimately abolishing the death penalty (Lederer 2007).

As implied earlier, not all industrialized countries have moved in the more punitive direction. In Sweden, emphasis remains on rehabilitation, treatment, and job training. Among the Japanese, criminals are resocialized

FIGURE 7.6

International Comparisons

The U.S. inmate population compared to the 26 largest European inmate populations (years vary).

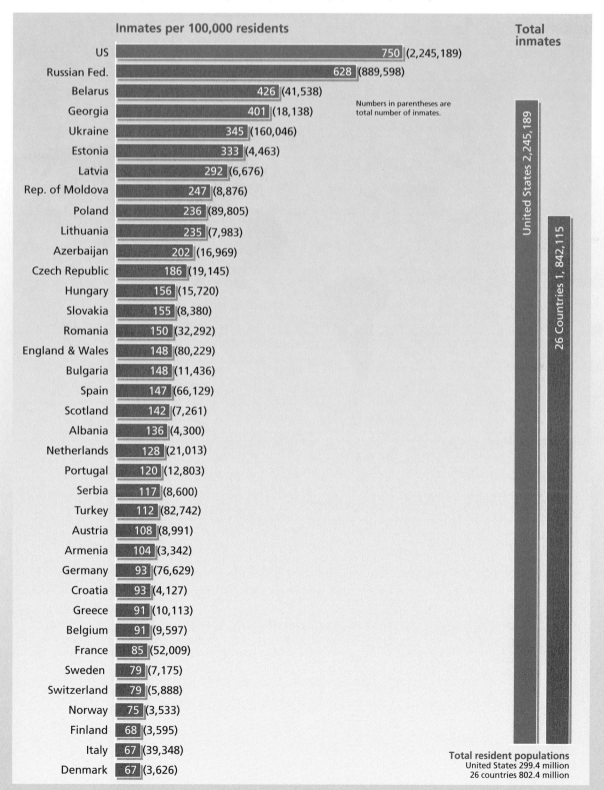

Inmates per 100,000 residents

Country	Rate	Total inmates
US	750	(2,245,189)
Russian Fed.	628	(889,598)
Belarus	426	(41,538)
Georgia	401	(18,138)
Ukraine	345	(160,046)
Estonia	333	(4,463)
Latvia	292	(6,676)
Rep. of Moldova	247	(8,876)
Poland	236	(89,805)
Lithuania	235	(7,983)
Azerbaijan	202	(16,969)
Czech Republic	186	(19,145)
Hungary	156	(15,720)
Slovakia	155	(8,380)
Romania	150	(32,292)
England & Wales	148	(80,229)
Bulgaria	148	(11,436)
Spain	147	(66,129)
Scotland	142	(7,261)
Albania	136	(4,300)
Netherlands	128	(21,013)
Portugal	120	(12,803)
Serbia	117	(8,600)
Turkey	112	(82,742)
Austria	108	(8,991)
Armenia	104	(3,342)
Germany	93	(76,629)
Croatia	93	(4,127)
Greece	91	(10,113)
Belgium	91	(9,597)
France	85	(52,009)
Sweden	79	(7,175)
Switzerland	79	(5,888)
Norway	75	(3,533)
Finland	68	(3,595)
Italy	67	(39,348)
Denmark	67	(3,626)

Numbers in parentheses are total number of inmates.

Total inmates

United States 2,245,189

26 Countries 1, 842,115

Total resident populations
United States 299.4 million
26 countries 802.4 million

Source: Jenifer Warren, "One in 100: Behind Bars in America." Washington, DC: Pew Center on the States and the Public Safety Performance Project, 2008, p.35. Based on International Centre for Prison Studies at King's College, London, "World Prison Brief." Data downloaded January 2008.
Note: Rates are for total number of residents, not just adults. Figures in this chart may not align with others due to differences in counting methods.

to recognize their moral shortcomings; moral responsibility is placed on offenders (Feeley and Simon 1992). This stands in stark contrast to the current warehousing approach in the United States, where the prison population has increased sixfold since 1970 (Western and Pettit 2002).

WORLDVIEW 7.1

Death Penalty Policy

Countries vary in their approach to the control of crime. The most extreme form of social control, the death penalty, is utilized in many countries, while some countries have abolished capital punishment completely. This map shows variations in national policy regarding the death penalty.

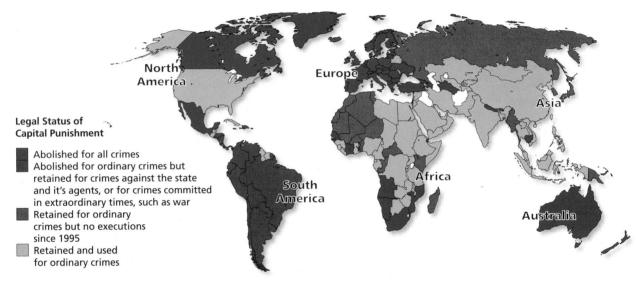

Legal Status of Capital Punishment

- Abolished for all crimes
- Abolished for ordinary crimes but retained for crimes against the state and it's agents, or for crimes committed in extraordinary times, such as war
- Retained for ordinary crimes but no executions since 1995
- Retained and used for ordinary crimes

Source: Amnesty International Online, 2005.

Interpret the Map

1. Do you notice any pattern in the use of the death penalty? Describe it.

2. What additional information would you need to determine if capital punishment is an effective deterrent to crime? Explain.

FEEDBACK

1. Toward which approach to crime control has the United States been moving?
   a. rehabilitation
   b. deterrence
   c. incarceration
   d. retribution

2. Of these western European countries, which has the highest rate of imprisonment?
   a. Spain
   b. England and Wales
   c. Germany
   d. Netherlands

3. The imprisonment rate in the United States is four times that of any western European country. T or F?

Answers: 1. c; 2. b; 3. T

SUMMARY

1. Although social life requires conformity and order, it also involves deviance, behavior that violates norms. Deviance is difficult to define because it is a relative matter. What is considered deviant depends on the characteristics of the individuals, the social context, and the historical era.

2. All societies need some means of social control to promote conformity to their norms. Internal social control, promoted through the socialization process, enables individuals to control their own behavior. External social control is applied by other people.

3. According to biological explanations of deviance, there is a link between physical characteristics and deviant behavior. Early on, Cesare Lombroso believed that criminals were throwbacks to an earlier stage of human evolutionary development. William Sheldon attributed crime to body shape. Despite later research efforts, there is still no convincing proof that genetic characteristics cause people to be deviant.

4. Psychological theories of deviance attribute deviant behavior to pathological personality traits. Sociologists remain as skeptical of psychological theories of deviance as they are of those rooted in biology.

5. According to functionalists, deviance has both negative and positive consequences for society. On the negative side, deviance encourages social disorder, erodes trust, encourages further nonconformity in others, and diverts resources from other social needs. On the positive side, deviance helps clarify norms, offers a safety valve, increases social unity, and brings about needed social change.

6. According to strain theory, deviance occurs because of a discrepancy between cultural goals and socially acceptable means of achieving those goals. Control theory attempts to explain conformity rather than deviance. Conformity, which of course excludes deviance, is based on the existence of a strong bond between individuals and society.

7. The symbolic interactionist perspective yields two theories of deviance. Cultural transmission theory contends that deviance is learned, just like any other aspect of culture. According to labeling theory, an act is deviant only if other people respond to it as if it were deviant. Isolated norm violation may have no serious consequences for individuals. If individuals are labeled as habitual devi-

ants, however, they may organize their lives and personal identities around deviance. Mental illness illustrates this theory.

8. The conflict perspective, when applied to the study of deviant behavior, emphasizes social inequality and power differentials. The most powerful members of a society are said to determine group norms and, consequently, the definition of deviant. Conflict theorists relate deviance to capitalism, pointing to the relationship between race, ethnicity, and crime.

9. The official crime statistics—the *Uniform Crime Reports* published annually by the FBI—are useful in providing an estimate of crime, but they underestimate national crime in America. The National Crime Victimization Survey (NCVS) was launched to overcome this limitation. There is some inconsistency in the findings of these two sources.

10. The United States has the greatest crime problem in the world. The gap between the United States and other countries remains substantial, despite the rising crime rates in Europe during the 1990s.

11. The four approaches to crime control are deterrence, retribution, incarceration, and rehabilitation. The United States is torn between the less punitive approach of rehabilitation and the harsher approaches of deterrence, retribution, and incarceration.

12. Increasingly, sociologists recognize that prisons are not successful at rehabilitating criminals. In the past, reform programs took place within prisons. In recent years, there has been an interest in rehabilitating criminals outside the prison system. During the 1970s, Americans began to take a harsher view of criminals. Even social scientists are beginning to believe that punishment can deter crime under certain circumstances, although there is no convincing evidence that capital punishment affects the murder rate.

13. Of late, the United States has increasingly moved toward incarceration as a solution to crime control. The imprisonment rate in the United States is currently six to eight times that of any Western European country and nearly double that of the former Soviet Union.

LEARNING OBJECTIVES REVIEW

- Define deviance, and explain its relative nature.
- Define social control, and identify the major types of social control.
- Describe the biological and psychological explanations of deviance.
- Discuss the positive and negative consequences of deviance.
- Differentiate the major functional theories of deviance.
- Compare and contrast cultural transmission theory and labeling theory.
- Discuss the conflict theory view of deviance.
- Distinguish the four approaches to crime control.

CONCEPT REVIEW

Match the following concepts with the definitions listed below them.

____ a. labeling theory   ____ e. social control   ____ i. deviance

____ b. differential association theory   ____ f. anomie   ____ j. recidivism

____ c. retribution   ____ g. white-collar crime

____ d. control theory   ____ h. cultural transmission theory

1. behavior that violates norms
2. a social condition in which norms are weak, conflicting, or absent
3. the conception that conformity is based on the existence of a strong bond between individuals and society
4. the theory that deviance is part of a subculture transmitted through socialization
5. the public demand of compensation from criminals equal to their offense against society
6. the theory that crime and delinquency are more likely to occur among individuals who have been exposed to more unfavorable attitudes toward the law than to favorable ones
7. the theory that deviance exists when some members of a group or society label others as deviants
8. means for promoting conformity to a group's or society's rules
9. a crime committed by respectable and high-status people in the course of their occupations
10. a return to crime after imprisonment

CRITICAL-THINKING QUESTIONS

1. Identify a time in your life when you were considered a deviant by others. Discuss the attempts at social control you experienced. Be specific as to the types of social control and their concrete application to you.

_____

2. Name some type of deviant behavior you would like yourself and/or others to engage in because you think society would benefit from it. State the sociological case for the social benefit of this type of deviance.

_____

3. Describe someone you know who falls into one of the four deviant responses identified by strain theory. Use specific characteristics of this person to show that he or she fits the response you selected and not the other three responses.

_____

_____

4. Everyone has observed someone who has been labeled deviant by some members of society. Discuss the consequences of this labeling for the person identified as a deviant.

_____

5. The text outlined several distinct approaches to crime control. Choose one approach and explain why you believe it has been successful or unsuccessful in relation to functionalism, conflict theory, or symbolic interactionism.

_____

_____

MULTIPLE-CHOICE QUESTIONS

1. **Behavior that violates norms is known as**
   a. deviance.
   b. anomie.
   c. stigma.
   d. recidivism.
   e. social control.

2. **Emile Durkheim contended that no behavior is deviant in itself, because deviance is a matter of social**
   a. pressure.
   b. biological determination.
   c. definition.
   d. control.
   e. indifference.

3. **The process of promoting conformity to a group's or society's rules is known as**
   a. coercion.
   b. influence.
   c. social cohesion.
   d. social control.
   e. groupthink.

4. **Imprisonment, capital punishment, medals, and honorary titles are examples of**
   a. informal sanctions.
   b. internal social control.
   c. formal sanctions.

d. stigmas.

e. differential rewards.

5. **According to Emile Durkheim, _____ is a social condition in which norms are weak, conflicting, or absent.**

a. chaos

b. ennui

c. enmity

d. melancholia

e. anomie

6. **The strain theory has been used most extensively in the study of**

a. white-collar crime.

b. scientific fraud.

c. alcoholism.

d. prison riots.

e. juvenile delinquency.

7. **Theories based on symbolic interaction say that deviance is the product of**

a. behavior modification.

b. cerebral mutations.

c. socialization.

d. trial and error.

e. psychological adaptation.

8. **Clifford Shaw and Henry McKay used _____ to explain the continuation of high delinquency rates in certain neighborhoods even though the ethnic composition changed over the years.**

a. cultural transmission theory

b. differential association theory

c. functionalism

d. labeling theory

e. deviant zone theory

9. **According to _____, deviance exists when some members of a group or society label others as deviants.**

a. labeling theory

b. functionalism

c. strain theory

d. cultural transmission theory

e. symbolic interactionism

10. **Isolated acts of norm violations are known as**

a. secondary deviance.

b. noncriminal activity.

c. learned deviance.

d. primary deviance.

e. independent deviance.

11. **Undesirable characteristics used to deny a deviant full social acceptance are known as**

a. social shields.

b. social brands.

c. social stigmas.

d. informal sanctions.

e. formal sanctions.

12. **Marxist criminologists view deviance as a product of**

a. socialization.

b. language acquisition.

c. differential association.

d. the harmful effects of religion.

e. class conflict.

13. **The official statistics from the FBI's *Uniform Crime Reports* are known to be biased mainly due to**

a. undercounting of the lower class.

b. overcounting of the middle and upper classes.

c. overrepresentation of the lower classes and minorities.

d. human error in compiling the statistics.

e. the upper class not being included due to their high social status.

14. **The _____ is comprised of the police, the courts, and the correctional system.**

a. justice complex

b. social legal charter

c. system of jurisprudence

d. legal elite

e. criminal justice system

15. **Which of the following is the most accurate statement regarding crime and punishment?**

a. Crime and punishment are not related in industrial societies.

b. Punishment deters crime if potential lawbreakers believe they are likely to be caught and severely punished.

c. Punishment deters crime if potential lawbreakers believe that the criminal justice system is fair.

d. Punishment deters crime only if children are taught that an eye for an eye is the best philosophy.

e. Potential lawbreakers are less likely to commit crimes if they have friends or relatives who have been punished by the criminal justice system.

FEEDBACK REVIEW

True-False

1. Recent research has established a definite causal relationship between certain human biological characteristics and deviant behavior. T or F?

2. It appears that social scientists have prematurely discounted the deterrence doctrine. T or F?

3. The imprisonment rate in the United States is six to eight times that of any Western European country. T or F?

Fill in the Blank

4. A major problem that sociologists have in defining deviance is its _____.

5. According to labeling theory, mental illness is a matter of _____.

6. The process of reducing the seriousness of crimes based on the lower social value placed on the victims is called _____.

Multiple Choice

7. **Which of the following is *not* one of the benefits of deviance for society?**
   a. It decreases suspicion and mistrust among members of a society.
   b. It promotes social change.
   c. It increases social unity.
   d. It provides a safety valve.
   e. It promotes clarification of norms.

8. **Which of the following did Edwin Sutherland mean by differential association?**
   a. Crime is more likely to occur among individuals who have been exposed more to unfavorable attitudes toward the law than to favorable ones.
   b. People become criminals through association with criminals.
   c. Crime is not transmitted culturally.
   d. Crime comes from differential conflict.

9. **In which of the following categories of crime is the United States *not* the leader in the world?**
   a. murder
   b. rape
   c. robbery
   d. burglary

Matching

10. Match the approaches to punishment with the examples below.
    ____ a. rehabilitation
    ____ b. deterrence
    ____ c. retribution
    ____ d. incarceration
    (1) imprisonment without parole
    (2) the practice of caning
    (3) employment and educational programs in prison
    (4) death penalty for murder

GRAPHIC REVIEW

Attitudes toward the death penalty for murder vary among Americans. The results of a Gallup poll, contained in Figure 7.4, clearly reflect a difference between blacks and whites. Answer these questions to explore the meaning of these results.

1. Do you think that the strong support for capital punishment among white Americans is due more to its deterrent effects or to the desire for revenge? Explain.

   _____

   _____

2. Why do you think African Americans are less supportive of the death penalty than white Americans?

   _____

   _____

   _____

ANSWER KEY

Review

a. 7
b. 6
c. 5
d. 3
e. 8
f. 2
g. 9
h. 4
i. 1
j. 10

Multiple Choice

1. a
2. c
3. d
4. c
5. e
6. e
7. c
8. a
9. a
10. d
11. c
12. e
13. c
14. e
15. b

Feedback Review

1. F
2. T
3. T
4. relativity
5. social definition
6. victim discounting
7. a
8. a
9. d
10. a. 3
    b. 2
    c. 4
    d. 1

Social Stratification

8

Ankya/Used under license from Shutterstock.

- Explain the relationship between social stratification and social class.
- Compare and contrast the three dimensions of stratification.
- State the major differences between the functionalist, conflict, and symbolic interactionist approaches to social stratification.

- Identify the distinguishing characteristics of the major social classes in America.
- Discuss the extent of poverty and the perceptions of poverty in the United States.
- Evaluate U.S. commitment to poverty programs.

- Outline some of the consequences of social stratification.
- Describe upward and downward social mobility in the United States.
- Discuss the major features of global stratification.

Is federal spending on social welfare programs excessive? A significant proportion of the U.S. population is unhappy with the amount of tax dollars being spent on the poor. Americans, however, seriously overestimate the proportion of the federal budget that is spent on welfare. Less money in real terms was spent on welfare recipients in 2002 than in 1997. Besides, these welfare expenditures account for less than 1 percent of total federal expenditures, a reduction of three-fourths since 1970, and less than one-fifth of 1 percent of the total gross national product (GNP) of the United States.

Negative attitudes toward welfare recipients are part of American culture, and these attitudes stigmatize the poor. Although all societies have social inequality, not all societies develop negative (and false) stereotypes of those at the bottom. We will return to this topic later in this chapter, but for now we turn to the dimensions on which social classes are created.

In George Orwell's Animal Farm, the animals overthrow their human master to form their own soon-to-be stratified society.

Dimensions of Stratification

Because of the human tendency to rank one another, inequality emerges. Soon some people are at the top of the hierarchy and others are at the bottom. This tendency is a recurring theme in literature. Dr. Seuss writes of the Sneetches, birds whose rank depends on whether or not they have a large star on their stomach. Star-bellied Sneetches have high status; plain-bellied Sneetches have low status. In the classic novel *Animal Farm,* George Orwell created a barnyard society where the pigs ultimately take over the previously classless animal society by altering one of their cardinal rules from "All animals are equal" to "All animals are equal—but some animals are more equal than others."

What is the difference between social stratification and social class? The fictional creatures mentioned in the previous paragraph are engaged in **social stratification**—the creation of layers (strata) of a population who possess unequal shares of scarce desirables, the most important of which are income, wealth, power, and prestige (R. Levine 2006). Each of the layers is a **social class**—a segment of the population whose members hold a relatively similar share of scarce desirables and who share values, norms, and an identifiable lifestyle. The number of social classes within a stratification struc-

ture varies. A stratification structure might include upper upper, middle upper, lower upper, upper middle, middle middle, lower middle, upper lower, middle lower, and lower lower classes. Or, like old coal-mining towns in America or like underdeveloped countries, there might be only an upper class and a lower class.

Karl Marx and Max Weber made the most significant early contributions to the study of social stratification. Marx demonstrated the importance of the economic foundations of social classes, while Weber accented the prestige and power aspects of stratification. Together, the work of these two great minds reveals to us that stratification is multidimensional (Gerth and Mills 1958; McCall 2001).

The Economic Dimension: Karl Marx

Although both Marx and Weber were concerned with inequality, their emphases were different. For Marx, the economic factor was an independent variable explaining the existence of social classes. Weber, on the other hand, viewed the economic dimension as a dependent variable. That is, Weber was more concerned with the economic consequences of stratification.

Marx identified several social classes in nineteenth-century industrial society—laborers, servants, factory workers, craftspeople, proprietors of small businesses, moneyed capitalists—but predicted that capitalist societies would ultimately be reduced to two social classes. Those who owned capital (the means of production)—the **bourgeoisie**—would be the rulers. Those who worked for wages—the **proletariat**—would be the ruled. The propertied capitalist class could exploit the labor of those without property. Even though the workers performed all the labor, they would be kept at a subsistence wage level, and the capitalists would enjoy the economic surplus created by that labor. In short, Marx predicted that because the capitalists owned the means of production (factories, land), they would both rule and exploit the working class. The working class would have nothing to sell but its labor.

Marx's analysis went deeper: Marx contended that all aspects of capitalist society—work, religion, government, law, morality—were economically conditioned. Consequently, the capitalists controlled all social institutions, which they used to their advantage. They could structure the legal system, educational system, and government to suit their own interests. For Marx, all of capitalist society was a superstructure resting on an economic foundation; the economy determined the nature of the society.

Whereas Marx foresaw only two social classes, Weber envisioned several social classes. Marx focused on the relationship to the means of production as the cause of social stratification; Weber examined the consequences of people's relationships to the economic institution. These consequences Weber termed **life chances**—the likelihood of securing the "good things in life," such as housing, education, good health, and food. The probability of possessing these desirables was directly related to economic resources, such as real estate, wages, inheritance, and investment profits.

By almost any measure—food, clothing, housing, or health—the standard of living has steadily improved in Europe and America. Still, there is considerable economic inequality, even within Western countries; and the United States is no exception. To understand the extent of economic inequality, it is necessary to distinguish income from wealth. **Income** is the amount of money received (within a given time period) by an individual or group. **Wealth** refers to all the economic resources possessed by an individual or group.

What is the extent of economic inequality in America? America was not supposed to have much economic inequality. As Richard Morin puts it:

> It was the great story of the American Century: the fusing after World War II, of the broadest and most prosperous middle class the world had ever seen. In its ticky-tack ramblers and backyard barbecues and unbridled sense of confidence lay a historic reversal—the steady closing of physical and economic gaps between America's rich and poor. (Morin 1994:14)

It has not turned out that way. Income inequality has been increasing for the last thirty years. Between 1980 and 1990, the share of America's wealth held by the top 1 percent of the population almost doubled, increas-

ing from 22 percent to 39 percent. Since then the top 1 percent has accumulated more than 70 percent of all earnings growth. In fact, the United States is now the most economically polarized and unequal of the major Western countries and continues to become even more unequal (K. Phillips 2002; Palast 2004; Sherman 2007). In 2006, more than 36 million Americans were living in poverty (U.S. Bureau of the Census 2007a). At the other extreme, there are nearly 3 million millionaires and some 800 billionaires (Collins and Yeskel 2005). And between 1978 and 2005, CEO pay increased from 35 times to 262 times the average worker's pay (Sawhill and Morton 2007). Economist Paul Samuelson describes income inequality in America this way:

> If we made an income pyramid out of building blocks, with each layer portraying $500 of income, the peak would be far higher than Mt. Everest, but most people would be within a few feet of the ground. (Samuelson and Nordhaus 2004:386)

The truth in Samuelson's statement is supported by government figures on the distribution of income. Annual income is currently more concentrated at the top than it has been since 1929 (Aron-Dine 2007; Juliano 2007; Sherman 2007). The share of after-tax income now going to the top 1 percent of American households is the highest ever recorded. Similarly, the shares of after-tax income going to the middle and bottom fifths of households are the smallest on record. Between 1979 and 2005, the richest 1 percent of households showed a gain of 228 percent in after-tax income ($745,000). On the other hand, the poorest fifth gained 6 percent ($900) and the middle fifth gained 21 percent ($8,700; see Figure 8.1). Consequently, in 2005 the richest 1 percent of households received 70 times as much in average after-tax income than the poorest one-fifth of households and 20 times as much as the middle one-fifth of households. And the pace of this widening gap between the haves and the have-nots is accelerating (Allen 2007; Johnston 2007b).

Although these income distribution figures reveal persistent economic inequality, they do not show the full nature of the concentration of wealth (what one owns) in America. Because the federal government regularly publishes data on national income, income comparisons are not difficult to calculate. Accurate data on wealth holdings are much harder to obtain. Still, some research provides a picture of the concentration of wealth in the United States. The top 1 percent of American households holds about 33 percent of the nation's wealth; the top 20 percent has 84 percent of the wealth; the bottom 80 percent of households owns 16 percent of the nation's wealth (see Figure 8.2). Since the early 2000s, movement toward the concentration of wealth has become even more dramatic (Collins and Yeskel 2005; D. A. Gilbert 2008).

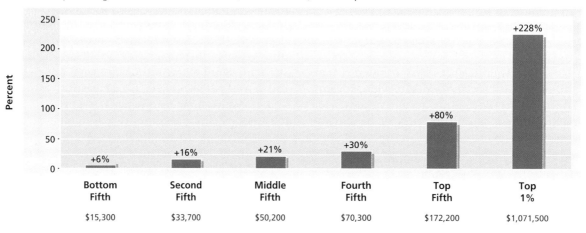

FIGURE 8.1

Change in Average Real After-Tax Income 1979-2005

To what percentages do the labels "lowest fifth," "middle fifth," and "top fifth" refer?

	Bottom Fifth	Second Fifth	Middle Fifth	Fourth Fifth	Top Fifth	Top 1%
	+6%	+16%	+21%	+30%	+80%	+228%

Average 2005 Income

| $15,300 | $33,700 | $50,200 | $70,300 | $172,200 | $1,071,500 |

Source: Arloc Sherman, "Income Inequality Hits Record Levels, New CBO Data Show." Center on Budget and Policy Priorities, 2007. http://www.cbpp.org/12-14-07inc.htm

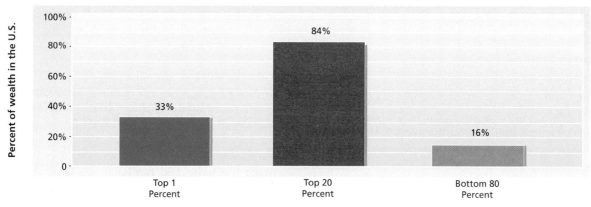

FIGURE 8.2

Shares of Wealth

Is this picture of the distribution of wealth different from what you would expect? Explain.

Top 1 Percent	Top 20 Percent	Bottom 80 Percent
33%	84%	16%

Percen of U.S. population

Source: Edward N. Wolff, "Changes in Household Wealth in the 1980s and 1990s in the U.S." (Working Paper No. 407). The Levy Economics Institute, May 2004.

What are the consequences of growing income inequality? Inferior health care, loss of leisure time, restricted diets, low-quality housing, and poor police protection are just a few of the disadvantages brought about by the growing disparity between rich and poor in the United States. But, this is the widest rich–poor gap since the Depression in 1929, and it has other less visible negative consequences. The growing economic disparity between the rich and the poor is lowering relative educational levels because college attendance goes up with income level. This, in turn, decreases the proportion of the workforce with college backgrounds, thereby retarding the growth of the skilled workforce. The quality of the workforce is further damaged because the increase of low-income students leads to poorer overall academic performance at all grade levels. The rate of job growth is affected by the income gap, and the rich are even more empowered to influence government for their own interests (Bernstein 1994; Zepezauer and Naiman 1998; Mishel, Schmitt, and Bernstein 2001; Palast 2004).

The Power Dimension: Max Weber

You will recall that **power** is the ability to control the behavior of others, even against their will. Individuals or groups who possess power are able to use it to enhance their own interests, often—but not necessarily—at the expense of society.

Why is power a separate dimension of stratification? According to Marx, those people in a society who own and control capital have the power. Sociologists agree that economic success heightens the chances of increased power. The wealthy, for example, have the power to influence tax laws in their favor. However, many sociologists, beginning with Weber, have argued that economic success and power do not necessarily overlap. This argument is based on several points. First, the fact that money can be used to exert power does not mean that it will be used in this way. Money is a resource that can be used to enhance power, but a deci-

sion must be made to use it for that purpose. For example, there are other families as wealthy as the Kennedys who have not used their economic resources to gain political power.

Second, money and ownership of the means of production are not the only resources that can be used as a basis for power. Expert knowledge can be used to expand power, too. For example, many lawyers convert their expertise into substantial amounts of political power. Eloquence as a speaker is another possible resource. And fame was a factor in 1952, when Albert Einstein was offered the presidency of Israel. He refused by saying, "I know a little about nature, and hardly anything about men."

Power is also attached to the social positions we hold. In organizations, elected officers have more power than rank-and-file members do. People in top executive positions in the mass media are powerful, even if they themselves do not have great wealth.

Finally, we can overcome a scarcity of resources if we have large numbers of people on our side or if

Cyber Bullying and Stratification

As noted, social stratification exists whenever scarce desirables are unequally distributed. In school, among peers, prestige is a most cherished and scarce valuable. Because prestige is such a scarce commodity in school, the charmed circle is small. Those excluded from valued cliques experience varying degrees of social anxiety and loss of self-esteem. And it is the most socially isolated young people that are at the greatest risk of being bullied.

Bullying involves intentional aggression promoted by differences in power or strength (Nansel et al. 2001). We are less interested here in bullying based on coercion (through physical strength) than bullying based on status differences. While bullying is age-old, *cyber bullying* has emerged only in the last

decade and obviously entails the use of technology. Cyber bullying may be direct (sending messages to someone else personally) or by proxy (enlisting the help of others, with or without their knowledge) (Aftab 2006).

Cyber bullying is perpetrated through a wide range of technological devices, including e-mails, instant messaging, and cell phone cameras. While traditional bullying is upclose and personal, cyber bullying is often anonymous and occurs among people who have never met. A physical bully can be avoided; a cyber bully is almost impossible to avoid.

Victims of cyber bullying suffer from social anxiety and low self-esteem, both associated with low social status in the youth culture (Kowalski, Limber, and Agaston 2008). Megan Meier is a case in point. After a period of name-calling and exclusion from valued school cliques, Megan craved emotional connection. Her mother thought that a MySpace page would be good for her sensitive, socially isolated daughter. But Megan's neediness only made her a highly vulnerable target for abuse. This thirteen-year-old girl became the victim of a group attack on MySpace, labeling her a liar, a whore, and worse. Megan fought back for a while,

but the avalanche of hate messages, and the variety of the sources sending them, led her to hang herself with a belt.

Megan's case has an added twist, illustrating that cyber bullying is not limited to the young. She was the victim of a hoax. Another neighborhood mother, after creating a shadow MySpace profile of a boy named "Josh Evans," had her daughter join her in convincing Megan that Josh liked her. A few weeks later, Josh sent Megan MySpace messages telling her that he hated her. Other girls, most not aware that Josh was fictitious, piled on with vicious insults. Megan's subsequent depression led to her suicide (Maag 2007).

Think About It
1. Which theory of stratification do you think best encompasses cyber bullying? Defend your choice.
2. Describe the research methods you think most appropriate for studying cyber bullying.
3. Do you have any personal knowledge of cyber bullying? Does the case you have in mind conform to the description of cyber bullying given here? Elaborate.

Popular celebrities such as Lindsay Lohan and Paris Hilton have considerable wealth. Their prestige rating is stronger in some circles than in others.

we are skillful at organizing our resources. Hitler, for example, was able to transform limited resources into a mass political movement. He gained absolute power by promising to deliver Germany from the economic destitution it suffered following World War I.

The Prestige Dimension: Max Weber

A third dimension of social stratification is **prestige**—recognition, respect, and admiration attached to social positions. Prestige is defined by one's culture and society (J. M. Berger et al. 1998; Webster and Hysom 1998). In the first place, favorable social evaluation is based on the norms and values within a group. Honor, admiration, respect, and deference are extended to dons within the Mafia, but despite America's fascination with the underworld, Mafia chiefs do not have high prestige outside their own circles. Second, prestige is voluntarily given, not claimed. Scientists cannot proclaim themselves Nobel prize winners; journalists cannot award themselves Pulitzer prizes; corporate executives cannot grant themselves honorary doctorates. Recognition must come from others. Third, those accorded similar levels of prestige share identifiable lifestyles. The offspring of upper-class families are more likely to attend private universities and Episcopalian churches; children from lower-class homes are less likely to attend college at all and tend to belong to fundamentalist religious groups. In fact, some sociologists view social classes as subcultures.

How is conspicuous consumption related to the prestige dimension? People with sufficient economic resources may use their wealth to enhance their prestige; they may consume goods and services to display their wealth to others (Frank 2000). Thorstein Veblen (1995; originally published in 1899) called this phenomenon **conspicuous consumption**. Examples are numerous and often astonishing: guests at Cornelius Vanderbilt's seventy-room, $10 million "summer cottage" at Newport could choose between salt water and freshwater in their bathrooms; guests preferring milk at Baron Alfred de Rothschild's mansion had their choice of milk from Jersey, Hereford, or Shorthorn cows (Wallechinsky and Wallace 1981). There are many more contemporary illustrations of status symbols as well. Millionaire Malcolm Forbes's son paid $156,450 for an almost 200-year-old bottle of wine believed to have belonged to Thomas Jefferson. Wall Street broker Jack Grubman was investigated for temporarily upgrading his rating of AT&T stock in a successful effort to enroll his children in the prestigious 92nd Street Y preschool, an important status symbol in New York City (McGinn and Peterson 2002). Companies are constantly creating products that promote conspicuous consumption. Chanel, for example, has introduced a line of diamond-studded handbags priced at $260,150.

How is prestige distributed? Prestige distribution coincides with a society's value system. People in statuses that are considered the most important—that are valued the most highly—have the most prestige. Because Americans value the acquisition of wealth and power, they tend to assign higher prestige to persons in positions rewarded with wealth and power. The most stable source of prestige in modern society is associated with occupations (Nakao, Hodge, and Treas 1990; Nakao and Treas 1990).

The prestige rankings of selected occupations in the United States are presented in Table 8.1. Obviously, white-collar occupations (doctors, ministers, school-teachers) tend to have higher prestige than blue-collar occupations (carpenters, plumbers, mechanics). As expected, physicians, lawyers, and dentists have high occupational prestige, whereas janitors, cab drivers, and garbage collectors have low prestige. You may find it somewhat surprising, however, that college professors have more prestige than bankers—a demonstration that even though wealth and power usually determine prestige, there are exceptions.

What is the basis for occupational prestige? According to Robert Hodge, Paul Siegel, and Peter Rossi (1964), all societies rely on comparable factors when determining occupational prestige. In fact, Robert Hodge, Donald Treiman, and Peter Rossi (1966) have shown that the ranking of occupations based on prestige is similar in advanced as well as underdeveloped societies. Occupational prestige scores vary according to the following factors: compensation; education, skills, and ability required; power associated with the occupation; the importance of the occupation to the society; and the nature of the work (mental, or white-collar, work versus manual, or blue-collar, work). More recently, occupational prestige has been shown to be affected by gender.

Why do different cultures evaluate occupations similarly? Donald Treiman (1977) attempted to answer this question through an examination of nearly 100 studies of occupational prestige in sixty countries. From these studies, which included countries of western and eastern Europe, North America, Africa, and South America, Treiman concluded that all societies must accomplish certain objectives. For example, domestic peace must be maintained, and a variety of goods—including food, housing, clothing, and transportation—must be produced and distributed. Culturally dissimilar societies develop similar occupational positions because of the functions that each must perform. This is clearly a functionalist explanation.

Not only do societies around the world develop similar occupational positions, but these positions tend to be arranged in occupational prestige hierarchies that are substantially similar throughout the world:

> In all societies, ranging from highly industrialized nations like the United States to peasant villages in up-country Thailand, the basic pattern of occupational evaluations is the same—

TABLE 8.1

Prestige Rankings of Selected Occupations in the United States

Occupation	Prestige Score
Surgeon	87
Lawyer	75
College professor	74
Engineer	71
School principal	69
Pharmacist	68
Registered nurse	66
Accountant	65
Professional athlete	65
Public grade school teacher	64
Banker	63
Druggist	63
Veterinarian	62
Police officer	61
Actor	60
Journalist	60
TV anchorman	60
Businessperson	60
Nursery school teacher	55
Social worker	52
Funeral director	49
Jazz musician	48
Mail carrier	47
Insurance agent	46
Secretary	46
Bank teller	43
Automobile dealer	43
Evangelist	41
Sales clerk	36
Bus driver	32
Waitress	29
Garbage collector	28
Janitor	22
Prostitute	14
Panhandler	11

Why do you think the highest-listed prestige score is 87? What occupations might rate a higher score?

Source: General Social Survey, National Opinion Research Center, 1996.

professional and higher managerial positions are most highly regarded, lower white-collar and skilled blue-collar jobs fall in the middle of the hierarchy, and unskilled service and laboring jobs are the least respected. (Treiman 1977:103)

This is the case, according to Treiman, because certain occupations in all societies have greater control over scarce resources than do other occupations.

1. _____ involves the creation of layers of people possessing unequal shares of scarce desirables.
2. Match the concepts with the examples.
  ____ a. life chances (1) the respect accorded doctors
  ____ b. power (2) a politician conforming to the interests of a lobby
  ____ c. prestige (3) living in a neighborhood with good police protection
  ____ d. conspicuous consumption (4) a lavish wedding
3. The top 1 percent of American households hold one-third of the total wealth. T or F?
4. The most stable source of prestige in modern societies is associated with _____.

Answers: 1. Social stratification; 2. a. (3), b. (2), c. (1), d. (4); 3. T; 4. occupations

Differential access to these resources—skill, authority, property—creates differential power. The power created by control over scarce resources is used to acquire special privilege. Power and privilege, in turn, are used to garner prestige. In this way, a single worldwide occupational prestige hierarchy is created. This is a conflict explanation.

Are the economic, power, and prestige dimensions related? Generally, there is a close relationship among these three dimensions of stratification. Those who are high on one dimension tend to be high on the other two. Power, particularly political power, is a decided economic advantage to its holders. For example, the wealthy pay for political favors either directly (as with money) or indirectly (as through inside information on lucrative business deals). Of the 535 members of the U.S. Congress, 170 have a net worth of at least $3.1 million. Ten members of President George W. Bush's first cabinet were millionaires (Edsall 2002b; Keller 2002). And former President George H. W. Bush and his son George W. Bush are by no means the only Americans who have used their wealth as springboards to the presidency of the United States.

Prestige and wealth are also closely intertwined. High prestige is heavily dependent on economic resources. The lifestyle usually associated with high prestige cannot be maintained without the economic clout to purchase such necessities as a nice home in the proper neighborhood or to send one's children to the best schools. But those with high prestige normally have no cause for concern, for with prestige come special economic rewards and unique business opportunities for generating wealth. Most of the wealth amassed by Tiger Woods, who is in billionaire range, did not come directly from prize money in professional golf tournaments. Because of his prestige, people want to buy any product he even implicitly endorses. Ranking behind only Oprah Winfrey in annual income (+$100,00 million), Tiger has multimillion-dollar contracts with corporate giants such as Nike, Buick, American Express, Accenture, Gillette, and PepsiCo.

Prestige and power often mix quite well. Prestige is attached to social positions requiring important decision making. And prestige may be converted into power. Ronald Reagan and Arnold Schwartzenegger are both actors who became governors of California. Reagan, of course, went on to become president.

Explanations of Stratification

Obviously, social class exerts a powerful influence on social life. The social class into which people are born and to which they rise or fall affects their existence as much as any other social factor. For this reason, if for no other, it is of benefit to understand why the process of stratification is a mainspring of society.

In most societies, inequality—in income, wealth, power, and prestige—seems to be a basic fact of life (Chris Tilly 1999). Even societies deliberately organized to eliminate inequality are not classless. Despite the former Soviet Union's attempt to create an unstratified society, great differences in rank and rewards developed. Leading members of the Communist Party, the dominant elite, were able to obtain privileges for themselves and their families. For example, Politburo members and national secretaries of the Communist Party drove black Zil limousines, hand-tooled cars worth more than $100,000 each, while the majority of Russian people had no chance of ever owning any kind of car. The Chinese Communist Party has also apparently become an entrenched, self-interested ruling class (Wortzel 1987).

Why is stratification a universal feature of social life? There are several explanations.

The Functionalist Theory of Stratification

According to the functionalist theory of stratification, inequality renders a service. The most qualified people fill the most important positions and perform their tasks competently; they are then rewarded for their efforts. The functionalist theory recognizes that inequality exists not only because certain jobs are more important than others but also because these jobs often involve special talent and training.

To encourage people to make the sacrifices (years of education and so forth) necessary to fill these jobs, society attaches special monetary rewards and prestige to these positions. For example, doctors make more money and have more prestige than bus drivers because a high level of skill is required in the medical profession and our society's need for highly qualified doctors is great. Once qualified people assume positions, however, they must be motivated to do their jobs competently. It is not happenstance, then, that greater rewards go to those who do their work better than others do. Thus, the more competent doctors, secretaries, and salespersons usually receive more of the "just desserts" than those who perform their tasks less well.

The functionalist theory has dominated sociological thinking for a long time, but it has been a very difficult theory to test. Although some studies support at least some aspects of the functionalist theory (Grandjean 1975; Cullen and Novick 1979), several weaknesses should be noted.

What are the weaknesses of the functionalist theory? Although the functionalist theory is elegant, concise, and consistent with traditional American beliefs about inequality, it has several failings (Tumin 1953). First, there are many people who have power, prestige, and wealth whose contributions to society do not seem very important. Many top athletes, film stars, and rock singers make much more money than the President of the United States. Why should scientists make almost no money for groundbreaking books reporting their research, whereas screenwriters regularly receive $500,000 plus per script?

Second, the functionalist theory ignores the barriers to competition faced by some members of society—the poor, women, the aged, African Americans, and various other ethnic groups. A wealth of talent exists among people in these categories, but much of it has never been tapped, because such people have not experienced equal opportunity to compete. They have not had the opportunity to become educated because they are too poor, the wrong color, or the wrong sex. And even if they are technically qualified, they may lose higher-level positions to less qualified individuals who happen to possess particular social characteristics.

Third, the functionalist theory overlooks the inheritance of social class level. In some societies, one's social class is determined at birth. Even in modern societies, in which the rate of mobility is supposed to be high, children tend to remain in or near the social class of their parents. This means that talented daughters and sons from lower-class homes cannot compete on an equal basis with the offspring of middle- and upper-class people. Consequently, many talented individuals—who might have preferred being surgeons, lawyers, or executives—are truck drivers, plumbers, and short-order cooks.

Finally, the functionalist theory of stratification has an ethnocentric bias. That is, it assumes that all people in all societies will be motivated to compete for a greater portion of the scarce desirables. In fact, in some societies, greater emphasis is placed on a more or less equitable distribution of wealth, power, and prestige than on individual competition for larger shares.

The Conflict Theory of Stratification

Melvin Tumin (1953) concluded that stratification exists not because it is necessary for the benefit of society but because it helps people holding the most power and economic resources to maintain the status quo. From his perspective, the functionalist theory is used as a justification for keeping some people at the top of the class structure and some at the bottom. This viewpoint is consistent with the conflict theory of stratification.

How does conflict theory account for inequality? According to the conflict theory of stratification, inequality exists because some people are willing to exploit others. The

According to conflict theory, inequality is a result of the elite exploiting the less powerful. The power of nineteenth-century industrialists is represented by financier J. P. Morgan.

The Art Archive/Culver Pictures/Picture Desk

elite hold a monopoly over the desirables, and they use their monopoly to dominate others. Stratification, from this perspective, is based on force rather than on consent. Those with wealth, power, and prestige are able to maintain, perhaps even increase, their share of desirables in a society (Perrucci and Wysong 2007).

This conflict theory of stratification is based on Marxian ideas regarding class conflict. For Marx, all of history has been a class struggle between the powerful and the powerless, the exploiters and the exploited. Capitalist society is the final stage of the class struggle. Although (according to conflict theorists) the capitalists are outnumbered, they are able to control the workers by creating a belief system that legitimates the status quo. Those who own the means of production are able to disseminate their ideas, beliefs, and values through the schools, the media, the churches, and the government. In this way, the dominant ideas of the elite become the dominant ideas of the ruled. Marx used the term **false consciousness** to refer to working-class acceptance of the dominant ideology (even though the ideology worked against working-class interests).

Because Marx believed that class conflict occurs in every historical epoch, he predicted that the working class would eventually shed its false consciousness—and its chains—for a revolution based on a true class consciousness. That is, it would realize its exploitation and reject the dominant belief system. After overthrowing the capitalists, the working class would create a socialist society in which the means of production and all property would be owned equally by the people. This would be a transitional stage between capitalist society and the classless society of communism. Ultimately, socialism would be replaced by communism, which would put an end to human misery and exploitation.

Marx's work was somewhat neglected by American sociologists for a time, although it has indirectly influenced the field through the negative reactions of many sociologists. Why has Marx been neglected? For one thing, many of his ideas—exploitation, revolution, classless society, communism—do not fit well with the American emphasis on capitalism, achievement, and upward mobility. In addition, many of Marx's predictions appear to have been wrong. Socialist revolutions have occurred in noncapitalist societies; communism did not produce classless societies; the middle class rather than the working class has expanded in capitalist societies.

Marxists do not consider that any of the preceding factors disprove Marx's theory. First, the so-called socialist and communist revolutions in precapitalist societies were essentially the result of a misinterpretation of Marx, who insisted that capitalism was an essential precondition that must occur before a communist society is possible. Second, Marx's theory defines the middle class as part of the working class. This is because the middle class also does not own the means of production, and its gradual expansion fulfills Marx's prediction of the increasing "proletarianization" of labor (i.e., more and more people, including small business owners, doctors, and other professionals, lose ownership of their labor power and come to work for someone else). Finally, according to Marxists, it is premature to assert that Marx was wrong simply because capitalism has not yet disappeared. Marx gave no timetable for the socialist revolution, and Marxist theorists see evidence of increasing contradictions within capitalism (Therborn 1982, 1984; E. O. Wright 1985, 2000).

Some sociologists have begun to look past the apparent miscalculations in Marx's theories to appreciate some of the assumptions on which they are based. They have substantial appreciation for his insight into the ways the elite use society's institutions, including the state, to achieve their own ends.

German sociologist Ralf Dahrendorf (1959) has advocated shifting from the Marxist emphasis on class conflict to conflict between groups, such as unions and corporations. Dahrendorf has attributed stratification more to the possession of power than to ownership of property. America's legal system, for example, is used by the wealthy for their benefit; education helps the elite perpetuate their power; the political system is skewed toward the interests of the powerful. Classes, then, for the followers of the conflict perspective, are created through the struggle for scarce resources (Domhoff 1996; Giddens 1987, 2002).

The Symbolic Interactionism Theory of Stratification

Why was the inequality between former Soviet political leaders and average Russian people tolerated by the masses? Why does the poorest one-fifth of Americans not revolt against the richest one-fifth of the population? Social stratification systems can persist over the long haul only if people believe in their legitimacy.

How is the legitimacy of a stratification structure incorporated into the minds of individuals? Social stratification persists only as long as its legitimacy is accepted. Symbolic interactionism helps us understand the legitimation process; it must occur through socialization. Symbols, in the form of language, explain to the young the existence of stratification structure and the reasons for people being located in particular strata. According to this perspective, American children are taught that one's place in the stratification structure is the product of talent and effort. Those "on top" have worked hard and have used their abilities, and those "at the bottom" have lacked the talent and motivation to succeed; therefore, it is not fair to challenge the system. In this way, people come to believe in the fairness of the existing system.

The legitimacy of a stratification structure penetrates even deeper. Legitimacy views are incorporated into an individual's self-concept.

How do ideas about inequality become part of the self-concept? Certain self-images are implied in the legitimization of stratification. Those at the bottom of the stratification structure suffer from lower self-esteem. How could it be otherwise when messages from all sides tell them of their inferiority? After all, as the symbolic interactionists document, self-esteem is based on our perceptions of others' opinions. The looking-glass process is at work (see Table 8.2).

Moreover, a person's sense of self-worth affects the amount of rewards he or she feels entitled to receive. Thus, people at the bottom of the heap tend not to challenge the stratification structure because of the conviction that they are undeserving. Those at the top blame the victims; the victims blame themselves.

The reverse is true for the higher classes. Those profiting most from the stratification structure tend to have higher self-esteem. This, in turn, fuels their conviction that the present arrangement is just. In short, people's self-concepts help preserve the status quo.

TABLE 8.2

FOCUS ON THEORETICAL PERSPECTIVES: Social Stratification

This table summarizes what issues of social stratification might be of interest to each of the major perspectives and predictions that they would make. Why would the symbolic interactionists be more likely than the functionalists to look at issues of self-esteem?

Theoretical Perspective	Research Topic	Expected Result
Functionalism	• Relationship between job performance and pay	• Pay levels increase with job performance.
Conflict theory	• Relationship between social class level and the likelihood of punishment for a crime	• The chances for prosecution decrease as social class level increases.
Symbolic interactionism	• Link between social class level and self-esteem	• Self-esteem is higher among the upper class than the lower class.

---

1. Match the theories of stratification with the examples.

   ____ a. functionalist theory

   ____ b. conflict theory

   ____ c. symbolic interactionism

   (1) Corporate executives make more money because they determine who gets what in their organizations.

   (2) Engineers make more money than butlers because of the education they possess.

   (3) Ghetto children tend to have low self-esteem.

2. Which of the following is *not* a criticism of the functionalist theory of stratification?

   a. There are people at the top whose jobs do not seem very important.

   b. It has a bias toward cultural relativism.

   c. It overlooks the barriers to competition.

   d. It glosses over the inheritance of social class level.

3. The conflict theory of stratification is based on Marx's ideas on class _____.

4. According to the symbolic interactionists, the deepest penetration of the beliefs supporting the stratification structure comes as a result of

   a. the "I"

   b. evolution

   c. conflict

   d. the self-concept

Answers: 1. a. (2), b. (1), c. (3); 2. b; 3. conflict; 4. d

Stratification in American Society

Class Consciousness

Americans have always had a sense of inequality, but they have never had a sophisticated understanding of **class consciousness**—a sense of identification with the goals and interests of the members of one's own social class (Centers 1949; Hodge and Treiman 1968; Jackman and Jackman 1983). In part because the American public has historically shown relatively little interest in class differences, American scholars began to investigate inequality rather late. It was not until the 1920s that sociologists began to systematically identify social classes in American communities. Since that time, however, the material available on this subject has been voluminous. Early efforts to study stratification were through intensive case studies of specific communities (W. L. Warner and Lunt 1941; Hollingshead 1949; Hollingshead and Redlich 1958; W. L. Warner 1960). Only relatively recently have attempts been made to describe the stratification structure of America as a whole (Rossides 1997; Chris Tilly 1999; Perrucci and Wysong 2007; D. A. Gilbert 2008).

Donna Eder—Who's Popular, Who's Not?

Can you remember what it was like to be in middle school? Donna Eder (1995) formed a research team to answer this question. Focusing on the rituals and daily speech routines of twelve- to fourteen-year-olds, Eder paints a portrait of teenage school culture.

At Woodview Middle School, Eder and her research team observed lunchtime interaction and attended extracurricular activities—talent shows, athletic practices, athletic contests, cheerleading tryouts, cheerleading practice, and choir and band practices and performances. After several months of observation, the research team conducted informal interviews with individuals and groups. To capture interaction for closer study, the researchers obtained student and parental permission for audio and video recordings. Our discussion is confined to Eder's findings about social stratification in middle school.

In the sixth grade, there were no elite groups. Woodview seventh and eighth graders, however, did not see each other as equals. This central fact was evident to the researchers after only a few days of observation. Popular seventh graders were divided along gender lines, with popular boys in one group and popular girls in another. By the eighth grade, the two groups intermingled. In both grades, popularity was based not on who was best liked, but on who was most visible in the school. Being visible meant that many others knew who you were and wanted to talk with you.

Status differences could arise during the seventh and eighth grades because cheerleading and team sports existed as a vehicle to high visibility. Realizing the source of their prestige, male athletes took every opportunity to display symbols of their team affiliation. Team uniforms, jerseys, and athletic shoes were among the most important items of dress. Bandages, casts, and crutches were worn with pride.

Girls could not use sports to gain visibility because female athletics were not valued by faculty, administrators, or students. But because boys' athletic events were the most important and best-attended school activities, girls used cheerleading to make themselves widely known. Besides appearing at basketball and football games, cheerleaders appeared in front of the entire student body at pep rallies and other school events.

Boys made fun of this high-status female activity by mockingly imitating cheers. One male coach joined the mockery by telling football players to either practice harder or he would get them cheerleading skirts. He then pretended to cheer in a falsetto voice.

Girls, on the other hand, regarded cheerleaders highly. Popular girls in the seventh and eighth grades either were cheerleaders or were friends of cheerleaders. Flaunting their status (just as the male athletes did), cheerleaders put on their uniforms as far ahead of games as possible and wore their cheerleading skirts for extracurricular school activities. Some girls even had T-shirts emblazoned with "cheerleader" on the back.

Evaluate the Research
1. Although it happened a while ago, middle school is something you experienced firsthand. Which of the three major theoretical perspectives do you think best explains the stratification structure described in this feature? Give reasons for your choice.
2. Was the stratification structure at your school similar to Woodview's? Explain why or why not, giving examples to support your conclusion.

Identification of Social Classes

Any attempt to describe the social class structure of American society is hazardous. Social classes are fluid, replete with exceptions, and abstract. Nevertheless, sociologists have identified the major features of the American class structure. According to Daniel Rossides (1997), the upper class comprises about 1 percent of the population, the upper middle class about 14 percent, the middle middle class about 30 percent, the working class about 30 percent, and the lower class about 25 percent. Despite the hazards involved, a brief description of each of these social classes follows. (Figure 8.3 presents a similar picture of the American class structure.)

Are members of the upper class all alike? Despite America's emphasis on the openness and fluidity of its class structure, its upper class contains only a fraction of its population (1 to 3 percent). Moreover, there is even stratification within the upper class. At the pinnacle rests the upper upper class, the 1 percent of the American population that we recognize as the aristocracy. Its members represent the "old" money amassed for generations by families whose names appear in the Social Register—Ford, Rockefeller, Vanderbilt, du Pont. Gaining entrance into this most elite of clubs is a long-run proposition because the basis of membership is blood rather than sweat and tears. Money must age over generations to have any chance of entering the orbit of the blue bloods.

FIGURE 8.3

American Class Structure

What does the shape of the American class structure indicate about stratification in the United States?

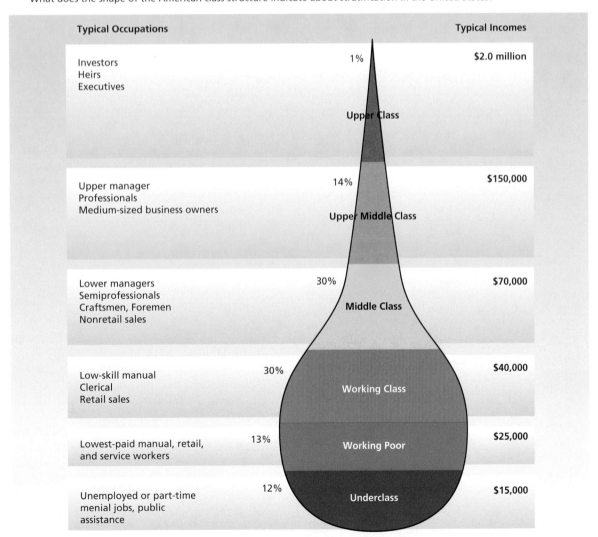

Typical Occupations		Typical Incomes
Investors, Heirs, Executives	1% — Upper Class	$2.0 million
Upper manager, Professionals, Medium-sized business owners	14% — Upper Middle Class	$150,000
Lower managers, Semiprofessionals, Craftsmen, Foremen, Nonretail sales	30% — Middle Class	$70,000
Low-skill manual, Clerical, Retail sales	30% — Working Class	$40,000
Lowest-paid manual, retail, and service workers	13% — Working Poor	$25,000
Unemployed or part-time menial jobs, public assistance	12% — Underclass	$15,000

Source: Adapted from Dennis Gilbert, *The American Class Structure*, 7th ed. (Thousand Oaks, CA: Pine Forge Press, 2008), p. 13.

Obviously, members of the upper upper class pay enormous attention to lineage, which they protect in several ways: sending their children to the best private secondary schools and universities, maintaining rather exclusive interaction with one another, and promoting a high rate of intramarriage.

The lower upper class, the "new rich," are in the upper class more often because of achievement rather than birth; membership is based more on earned income than inherited wealth. The lower upper class is composed of people with much more varied socioeconomic backgrounds. Many members of the lower upper class inherited some wealth, although usually not fortunes. Others may have amassed fortunes by being shrewd investors on the stock market, creating a chain of hardware stores, becoming a Metropolitan Opera star, or earning millions from athletic talent. Although they may actually be better off financially, people who have entered the lower upper class in their own lifetimes are generally barred from the exclusive social circles in which members of the upper upper class travel.

The upper class as a whole wields tremendous power over domestic and international decisions and over events affecting vast numbers of people (Dye 2001). If the United States has a power elite, this is it (see Chapter 14).

What is the composition of the middle class? The upper middle class (10 to 15 percent of the population) is composed of those who have been successful in business, the professions, politics, and the military. Basically, this class is made up of individuals and families who benefited from the tremendous corporate and professional expansion following World War II. Members of this class earn enough to live pretty well and to save some money. They are typically college educated and have high educational and occupational aspirations for their children. Although they do not exercise national

or international power, their influence is felt in their communities, where they tend to be active in voluntary and political organizations.

The middle middle class is large (30 to 35 percent of the population) and heterogeneous. Its members include small business owners; small farmers; small, independent professionals (small-town doctors and lawyers); semiprofessionals (clergy, teachers, nurses, firefighters, social workers, and police officers); middle-level managers; and sales and clerical workers. Their income level, which is at about the national average ($22,199 in 2000) does not permit them to live as well as the upper middle class. Typically, these people have only a high school education, although many may have had some college or may have completed college.

Members of the middle class are interested in civic affairs. Their participation in political activities is less than in the classes above them but higher than in either the working class or the lower class. The 1980s witnessed a dramatic decrease in the proportion of Americans that can be classified as middle class. Since 1980, the proportion of middle-income Americans has declined by 20 percent (Morin 1991; Wallich and Corcoran 1992).

What is unique about the working class? The working class (often referred to as the lower middle class) comprises about one-third of the population. Its members typically include either blue-collar workers who work with their hands—roofers, delivery truck drivers, and machine operators—or lower-level salespeople and clerical workers (Rubin 1994). The economic resources of its members are, on the whole, fewer than those of the classes above, although some of these workers (e.g., plumbers) may earn more than some middle-class people. Members of the working class experience considerable economic pressure because their incomes are below the national average, their employment is less stable, and they are generally without the hospital insurance and retirement benefits enjoyed by those in the middle class. The specter of unemployment or illness is both real and haunting. Outside of union activities, members of the working class have little opportunity to exercise power, and their participation in organizational life, including politics, is scant (Zweig 2000).

What groups of people fall into the lower class, and what is their most common shared characteristic? Lumped into the lower class (20 to 25 percent of the population) are the working poor, the frequently unemployed, the underpaid, and the permanently unemployed. The most common shared characteristic of the lower class is lack of value in the labor market. There are many routes into the lower class—birth, physical or mental disability, old age, loss of a marriage partner, lack of education or training, occupational failure, alcoholism. There are,

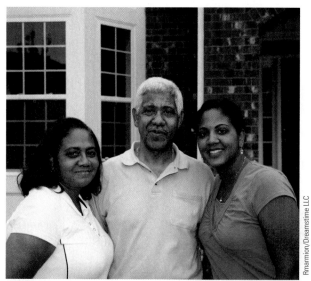

This family fulfills the American image of comfortable middle-class living.

Rmarmion/Dreamstime LLC

however, very few paths out (W. J. Wilson et al. 1988; M. B. Katz 1993; D. A. Gilbert 2008).

The **working poor** (about 13 percent of the population) consists of people employed in low-skill jobs with the lowest pay. Its members are typically the lowest-level clerical workers, manual workers (laborers), and service workers (fast-food servers). Lacking steady employment, the working poor do not earn enough to rise above the poverty line. The working poor tend not to belong to organizations or to participate in the political process (Barrington 2000; Munger 2000).

The **underclass** (about 12 percent of the population) is composed of people who are usually unemployed and who come from families with a history of unemployment. They either work in part-time menial jobs (unloading trucks, picking up litter) or are on public assistance. In addition to a lack of education and skills, many of the members of the underclass have other problems. Physical or mental disabilities are common, and many are single mothers with little or no income. We explore the underclass in the following section on poverty in the United States as well as in the next chapter on racial and ethnic minorities.

Poverty in America

Widespread poverty existed throughout the United States when Michael Harrington helped to make it a political issue in his book *The Other America* (1962). Forty years later, scholars continue to underscore the presence of poverty in affluent America (Newman 1999, 2007; Gilbert 2008). We turn now to the measurement of poverty.

Measuring Poverty

On an absolute scale, poverty embodies a lack of housing, food, medical care, and other necessities for maintaining life. **Absolute poverty** is the absence of enough money to secure life's necessities. On a relative scale, though, it is possible to have those things required to remain alive—and even to live in reasonably good physical health—and still be poor. **Relative poverty** is measured by comparing the economic condition of those at the bottom of a society with that of other strata. Because poverty is determined by the standards that exist within a society, a definition of poverty is not the same in India as in the United States.

How can relative poverty be measured? According to advocates of the relative measurement of poverty, the poverty threshold should be raised as the standard of living in a country rises. An increasing amount of spendable income does not make people feel less poor if the income levels of those around them are increasing at the same or a faster rate. A common measure of relative poverty is an income comparison between the lowest fifth of the population and the other four-fifths (refer to Figure 8.1). Such a measure means, of course, that a segment of the population will always be poor unless the total income distribution approaches equality.

How can absolute poverty be measured? Poverty in America has traditionally been measured in an absolute way. Absolute measurement of poverty involves drawing a "poverty line," an annual income level below which people are considered poor. In 2004, that figure was set at $19,484 for a family of four. The poor, as measured by the federal government standard, comprises more than 12 percent of the American population (see Figure 8.4). The following description of America's poor is based on this absolute measure of poverty.

Identifying the Poor

Minorities, people who live in female-headed households, children under eighteen years of age, elderly people, and people with disabilities make up the most disadvantaged groups living in the United States today.

How are race and ethnicity related to poverty? Nearly 44 percent of the poor in America today are non-Latino white (Mishel, Bernstein, and Allegretto 2006). The poverty rate for African Americans and Latinos is much higher than that for whites, however. The poverty rate for whites is 8.2 percent; for African Americans and Latinos more than 20 percent. African Americans and Latinos together account for only about one-fourth of the total population, but they make up nearly half of the poor population (see Figure 8.5).

FIGURE 8.4

Number of Poor and Poverty Rate: 1959–2006

This graph shows two types of information: (1) the number of poor in the total population and (2) the poverty rate as a percentage of the total population. Why is it often helpful to have related information plotted on the same graph?

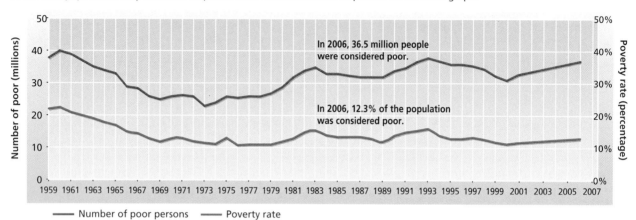

Number of poor persons ——— Poverty rate

Note: The data points represent the midpoints of the respective years. The latest recessionary period began in July 1990.
Sources: U.S. Bureau of the Census, *Poverty in the United States: 1995*, Current Population Reports, P60–185 (Washington, DC: U.S. Government Printing Office, 1996); U.S. Bureau of the Census, "Poverty 1997—Poverty Estimates by Selected Characteristics," *Current Population Survey, March 1998*, 1999a, at www.census.gov/hhes/poverty/poverty97/pv97est1.html; and U.S. Bureau of the Census, "Household Income at Record High. Poverty Declines in 1998, Census Bureau Reports." United States Department of Commerce News Release, September 30,1999; and U.S. Bureau of the Census, *Poverty in the United States: 2001*, Current Population Reports, P60-219 (Washington, DC: U.S. Government Printing Office, 2002), U.S. Bureau of the Census, "Income, Poverty, and Health Insurance Coverage in the United States: 2006") 2007. http://www.census.gov/prod/2007pubs/p60-233.pdf

FIGURE 8.5

The Distribution of Poverty in the United States

What are the most important conclusions you would reach from this figure?

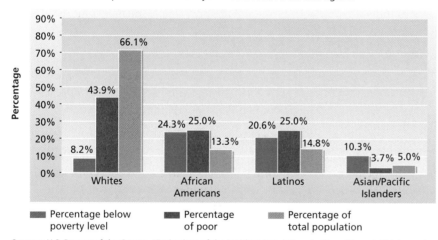

Percentage below poverty level   Percentage of poor   Percentage of total population

Sources: U.S. Bureau of the Census, "Projections of the Resident Populations by Race, Hispanic Origin, and Nativity: Middle Series, 2001 to 2005," at www.census.gov/population/projections/nation/summary/np-t5-b. pdf; and U.S. Bureau of the Census, *Poverty in the United States: 2001,* Current Population Reports, P60-219 (Washington, DC: U.S. Government Printing Office, 2002), U.S. Bureau of the Census, "Income, Poverty, and Health Insurance Coverage in the United States: 2006," 2007. http://www.census.gov/prod/2007pubs/p60-233

How are gender and age related to poverty? Another large segment of the poor population is made up of female-headed households. We can look at this issue in two different ways. We can look at all poor households as a group and determine what proportion of them is headed by females. When we do this, we find that about one-half of poor households are female headed. In contrast, when we look at nonpoor households, we find that only about 14 percent are headed by females. Another approach would be to look at all female-headed households as a group and determine what proportion of them is poor. We find that the poverty rate for these households is about 28 percent, compared with just greater than 10 percent for all families.

By either measure, then, households headed by females are poorer than those headed by males. A related factor is the poverty rate for children under eighteen years of age. The current rate for this is 17.4 percent—the highest rate for any age group in the United States (U.S. Bureau of the Census 2007a). The high poverty rates for women and children reflect a trend in U.S. society. Between 1960 and today, women and children make up a larger proportion of the poor. Sociologists refer to this trend as the **feminization of poverty** (Edin and Lein 1997; Edin and Kefalas 2006; Sidel 1998, 2006).

There are several reasons women have a higher risk of being poor. As we discuss in more detail in Chapter 10, women earn only 80 cents for every dollar earned by men. Women with children find it more difficult to find and keep regular, long-term employment. A lack of good child-care facilities adds to the likelihood that they will not be able to continue working.

Older Americans make up another large segment of the poor. Over 9 percent of people aged sixty-five or older live in poverty (U.S. Bureau of the Census 2007). Another large segment of the poor are people who have disabilities. This group is three times as likely to live in poverty as people who are able-bodied. Also, people

who have severe disabilities are more likely to live in poverty than are those who have less severe disabilities (Freedman, Martin, and Schoeni 2004).

This description should not leave the impression that the able-bodied poor in the United States do not work. About 20 million of the poor in America are over eighteen years of age. Of these, about 9 million worked either full-time or part-time in 2006. In other words, 45 percent of poor Americans are working. Fifty-five percent are ill, disabled, retired, at home or not working for family reasons, or looking for work. About 11 percent are in school, or not working for some unspecified reason (U.S. Bureau of the Census 2007a). At any particular point in time, about one-third of welfare mothers are working. Nearly all of the poor in America are either underage children, working people, or persons not working for legitimate reasons. This evidence contradicts the prevailing assumption that the poor are able-bodied individuals simply waiting for others to take care of them

Is this an accurate picture of America's poor? At the very least, this absolute measure of poverty is a reasonable approximation—reasonable, in part, because prior to 1963 the federal government had neither a definition nor a measure of poverty at all. Developed by Mollie Orshansky (1965) of the Social Security Administration, the federal government's absolute measure of poverty was based on an "economy" food plan: a plan designed

SNAPSHOT OF AMERICA **8.1**

Percentage of Population in Poverty

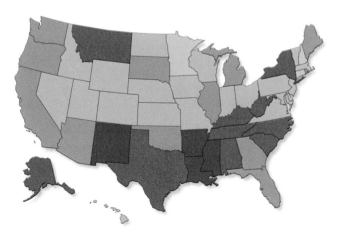

Although the U.S. economy is prosperous, some people are concerned that many have not benefited from this prosperity. In fact, many people still live in poverty. This map shows the average percentage of the poor by state for 2002 to 2004.

Interpret the Map

1. Can you make any generalization about poverty from this map?

2. If you were the governor of your state, what would your platform on poverty be? Be specific.

Percentage of
Population in Poverty

- 17.0% or above
- 14.0%–16.9%
- 11.0%–13.9%
- 8.0%–10.9%
- 7.9% or below

Source: U.S. Bureau of the Census, 2005.

Ewing Galloway/Jupiter Images

Poverty is greatest among American children under six years of age.

by the U.S. Department of Agriculture as nutritionally adequate for use in emergencies or, temporarily, when family funds are low. The poverty cutoffs (based on factors such as family size, gender of the head of the household, and place of residence) were based essentially on the amount of income left after allowing for an adequate diet at minimum cost.

Critics point to some shortcomings of this approach. This definition reflects inflation, but it ignores social and economic changes such as the feminization of poverty and the higher costs of child care, transportation, and health care. The federal poverty level also considers the cost of living to be uniform across all states (except Hawaii and Alaska): a family living in Manhattan in New York is assumed to spend the same amount on food, shelter, and clothing as a family in Manhattan, Kansas (Bhargava and Kuriansky 2002). According to Mark Rank (2003), the number of people ever in poverty is a better indicator of the extent of poverty than the number of people in poverty in any particular year. By age 75, he contends, 59 percent of Americans will have spent at least a year below the poverty line during adulthood, and 68 percent will have experienced near poverty.

Perceptions of Poverty

How do Americans view the poor? The traditional American view of the poor has roots extending back to ancient Greece. In the seventh or eighth century B.C.E., the Greek poet Hesiod expressed for his contemporaries this view about work and the poor:

> *Work! Work, and then Hunger will not be your companion, while fair-wreathed and sublime Demeter will favor you and fill your barn with her blessings. Hunger and the idling man are bosom friends. Both gods and mortals resent the lazy man, a man no more ambitious than the stingless drones that feed on the bees' labor in wasteful sloth. Let there be order and measure in your own work until your barns are filled with the season's harvest. Riches and flocks of sheep go to those who work. If you work, you will be dearer to immortals and mortals; they both loathe the indolent. (Hesiod 1983:74–75)*

There is another example of this approach to poverty: it is the adaptation of Charles Darwin's theory of biological evolution to the social world. According to Darwin, only the fittest plants and animals survive in a particular environment. Those plants and animals that survive carry physical characteristics that give them a competitive edge for survival. The process of natural selection ensures that the fittest survive. When applied to human society, this view is called *social Darwinism*. It contends that the fittest individuals survive as a result of social selection and that the inferior are eliminated by the same process (Strickland and Shetty 1998).

Although social scientists have long rejected social Darwinism, this philosophy has contributed to that part of America's basic value system that is variously labeled the work ethic, the achievement and success ideology, and the ideology of individualism (Gilens 1999). Individualism in America involves several central beliefs:

1. Each individual should work hard and strive to succeed in competition with others.
2. Those who work hard should be rewarded with success (seen as wealth, property, prestige, and power).
3. Because of widespread and equal opportunity, those who work hard will, in fact, be rewarded with success.
4. Economic failure is an individual's own fault and reveals lack of effort and other character defects. (Feagin 1975:91–92)

According to the ideology of individualism, those at the bottom are where they belong because they lack the ability, energy, and motivation to survive in a competitive social world. From this vantage point, it is the poor who are to blame for their condition; those who fail deserve to do so (Ryan 1976; O'Toole 1998; N. P. Barry 1999; Zinn 2003). In its most extreme form, the ideology of individualism contends that legislation to protect the poor is harmful because protective legislation interferes with the process of natural selection, thus slowing social evolution (Lux 1990).

To test whether the ideology of individualism continues to affect the American public's view of the poor, James Kluegel and Eliot Smith (1990) conducted a nationwide survey. Respondents were asked to judge the reasons usually cited to explain poverty in America. They found that individualistic explanations were the most popular. Lack of thrift, lack of effort, lack of ability, and loose morals and drunkenness—all individualistic causes—were considered the most important reasons for poverty. Much less emphasis was placed on structural explanations that locate the causes of poverty in social and economic factors, factors beyond the control of the poor (see Table 8.3). Although structural explanations for poverty have grown in importance in recent years, the individualistic interpretation remains dominant, particularly among affluent, higher-status, white, older Americans, those furthest from the situation being studied (W. J. Wilson 1997; Goode and Maskovsky 2001; "National Survey on Poverty in America" 2001; Draut 2002; Eitzen and Smith 2003).

Does race influence attitudes toward the poor? In a complicated way, it does. Americans do not reject welfare in principle; the poor who are not lazy and who try to find jobs deserve help, most say. It is the "undeserving" poor—those perceived as shiftless—who are the target of public scorn (R. C. Lieberman 1998; M. K. Brown 1999; Neubeck and Cazenave 2001). This is where race is thought to enter the mix. Due, in part, to the media's long-standing negative portrayal of the African American poor, welfare has become enshrouded in racial stereotypes. While Americans do not dislike welfare because African Americans benefit from it, many do believe that welfare benefits undeserving African Americans who prefer a government handout to work (Gilens 1999; Blank and Haskins 2001).

Americans, of course, do not condemn just the African American poor. While race complicates the issue, many Americans are opposed to welfare for *all* "undeserving" poor, regardless of their race and ethnicity.

How accurate is the negative image of the poor? Several studies challenge the view that the poor, whatever their race or ethnicity, are lazy, opposed to work, and prefer to live off others. In a recent study of the working poor in the inner city (Harlem), Katherine Newman (1999) documents the many African Americans and Latinos who are trying to form families, work steadily, and support themselves. If, Newman argues, Americans realized that most minorities are members of the hard-working

---

TABLE 8.3

Perceived Reasons for Poverty in the United States

Reasons for Poverty	Percentage Replying		
	Very Important	Somewhat Important	Not Important
Individualistic Explanations			
• Lack of thrift and proper money management by poor people	64%	30%	6%
• Lack of effort by the poor themselves	53	39	8
• Lack of ability and talent among poor people	53	35	11
• Loose morals and drunkenness	44	30	27
Structural Explanations			
• Low wages in some businesses and industries	40	47	14
• Failure of society to provide good schools for many Americans	46	29	26
• Prejudice and discrimination against blacks	31	44	25
• Failure of private industry to provide enough jobs	35	39	28
• Being taken advantage of by rich people	20	35	45

How would you have answered this survey? Explain your answers.

Source: Adapted with permission from James R. Kluegel and Eliot R. Smith, *Beliefs About Inequality: Americans' Views of What Ought to Be* (New York: Aldine de Gruyter, 1990), p. 79. Copyright © 1990 by James R. Kluegel and Eliot R. Smith.

poor, Americans would be less ready to label them as undeserving.

For years, studies of welfare mothers have indicated that about three-fourths prefer employment to staying at home. Most welfare mothers say they are not presently working because of their children, housework, bad health, or lack of skills (Poddel 1968; Warren and Berkowitz 1969). A national survey indicates a strong work ethic among the poor. A representative sample of poor Americans was asked, "Which do you think is more important in life: working hard and doing what is expected of you or doing the things that give you personal satisfaction and pleasure?" Seventy-eight percent subscribed to the idea of hard work, which was higher than for the nonpoor in the sample (65 percent). In addition, the poor favored the idea of workfare—requirement by the government that physically able poor people work before they can receive poverty benefits—to a somewhat greater degree than the nonpoor did (59 percent versus 51 percent). This evidence suggests that the poor believe in the same work-related values and attitudes as other Americans (I. A. Lewis and Schneider 1985). More recent studies report no difference in desire to work between the poor and the nonpoor. Whether male or female, black or white, young or old, poor people appear to share the middle-class identification of self-esteem with work and self-support. The aspirations of the poor match those of the nonpoor; both groups desire an adequate education and a home in a good neighborhood (W. J. Wilson 1997; Lichter and Crowley 2002).

In summary, many Americans have tended to blame the poor for their poverty, despite evidence that the poor are not the lazy, shiftless, willingly dependent people they are popularly thought to be. Both the attitudes and the behavior of the poor contradict the negative label often applied to them (Blee and Billings 1999; O'Connor 2002).

Responses to the Problem of Poverty

Before the mid-1960s, fighting poverty was not a major goal of the federal government (O'Connor 2002). Some programs, such as Social Security and Aid to Families with Dependent Children (AFDC), had been enacted during the Great Depression. These measures did not usually reach the lowest levels of needy citizens, however. Finally in 1964, President Lyndon Johnson marshaled the forces of the federal government to begin a War on Poverty.

What were the goals of the War on Poverty? The philosophy behind the War on Poverty was to help poor people help themselves (Jacoby 1997; Barry 1999; Patterson 2000). President Johnson's predecessor, John F. Kennedy, believed that if the chains of poverty were to be broken,

it had to be through self-improvement, not temporary relief. Accordingly, almost 60 percent of the first poverty budget was earmarked for youth opportunity programs and the work experience program (work and job training designed primarily for welfare recipients and unemployed fathers).

Hopes for positive results from the War on Poverty were high. However, not all of the programs were as successful as predicted. Indeed, some have come under severe criticism. These criticisms center around supposed widespread abuses and the fear that the system encourages people to become dependent upon the government longer than is necessary. "Fixing" the way social welfare should be provided and payments should be distributed has been the focus of many hot political debates.

Welfare Reform

Prior to 1996, the main source of welfare assistance for the needy was AFDC (Dervarics 1998). The most recent legislation on welfare reform, enacted in 1996, replaced AFDC with a program known as Temporary Assistance for Needy Families (TANF).

What is the nature of welfare reform? The Temporary Assistance for Needy Families (TANF) program limits the amount of time those able to work can receive welfare payments. The bill has three major elements: it reduces welfare spending, it increases state and local power to oversee welfare rules, and it adds new restrictions on welfare eligibility. For example, benefits to children of unwed teenage mothers are denied unless the mothers remain in school and live with an adult. Cash aid to able-bodied adults will be terminated if they fail to get a job after two years.

Has welfare reform worked? It is too early to make a final evaluation of this latest attempt at welfare reform. However, according to the Administration of Children and Families, the government agency that administers TANF, welfare rolls have decreased substantially since the new welfare-to-work program went into effect. In 1996, for example, more than 12 million people were receiving welfare. In 2003, the total number of TANF recipients was slightly less than 5 million. This number represents approximately 2 million families. Only a small percentage of recipients have been removed from the rolls because of the new time limits on benefits.

There is a darker side, however. Most of the people who left the welfare rolls in the late 1990s could find only low-wage jobs in industries such as food service, home health care, and retail sales. In the early 2000s, an economic downturn caused many of these types of jobs to disappear. At the same time, wages grew more slowly or actually fell. As a result, many of those who left the welfare rolls continue to live in or near poverty. Research suggests that the current emphasis of welfare reform on reducing the welfare rolls rather than on reducing poverty hampers the long-term economic well-being of former recipients (McNamara 2004).

Consequences of Stratification

The unequal distribution of income, wealth, power, and prestige has additional consequences for people. Two broad categories of class-related social consequences are *life chances* and *lifestyle* (Charles Tilly 1999).

Life Chances

Life chances refers to the likelihood of possessing the good things in life: health, happiness, education, wealth, legal protection, and even life itself. The probability of acquiring and maintaining the material and nonmaterial rewards of life is significantly affected by social class level. Power, prestige, and economic rewards increase with social class level. The same is true for education, the single most important gateway to these rewards. But additional more subtle life chances exist, and these, too, vary with social class (Putnam 1996; Pennar 1997). These less obvious life chances are in good part the product of inequality in the distribution of education, power, prestige, and economic rewards.

How does social class affect more subtle life chances? The probability of possessing life itself—the most fundamental, precious life chance—declines with social class level. Whether measured by the death rate or by life expectancy, the likelihood of a longer life is enhanced as people move up a stratification structure (Pear 2008). This disparity is generated by differences in such things as neighborhood safety, possession of health insurance, quality of medical care, value placed on medical attention, and concern with proper nutrition.

In light of differences in life expectancy, it is not surprising that physical health is affected by social class level. Those lower in a stratification structure are more likely to be sick or disabled and to receive poorer medical treatment once they are ill (Califano 1985; R. Wilkinson 1986; Conrad and Kern 1990). It is no different for mental health. Persons at lower class levels have a greater probability of becoming mentally disturbed and are less likely to receive therapeutic help, adequate or otherwise (Goodman et al. 1983; Miech and Hauser 1998).

Innumerable other life chance inequalities exist. For example, the poor often pay more for the same goods and services. They are more likely to get caught for committing a crime, and when they are apprehended, they stand a greater chance of being convicted and serving prison time. Critics offer a dramatic example of class-related inferior public service when they contend the federal government neglected the poor in New Orleans (mostly African Americans) in the evacuation process prior to Hurricane Katrina and in subsequent efforts to deal with the devastating consequences.

F E E D B A C K

1. _____ poverty is defined as the absence of enough money to secure life's necessities.
2. _____ poverty is measured by comparing the economic condition of those at the bottom of a society with that of other segments of the population.
3. Which of the following is *not* one of the major categories of the poor in the United States?
   a. children under eighteen
   b able-bodied men who refuse to work
   c. the elderly
   d. the disabled
   e. people who live alone or with nonrelatives
4. According to the ideology of _____, those at the bottom of the stratification structure are there because of their own inadequacies.
5. The philosophy behind the Economic Opportunity Act of 1964 was
   a. to provide temporary relief for the poor.
   b. to make the poor dependent on the federal government.
   c. to make it easy for the poor to receive help.
   d. to help poor people help themselves.
6. The problem with the poor is that they have lower economic and educational aspirations than the nonpoor. T or F?

Answers: 1. Absolute; 2. Relative; 3. b; 4. Individualism; 5. d; 6. F

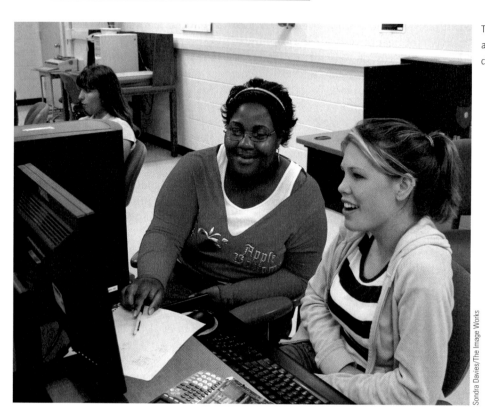

These community college students are attempting to improve their life chances and lifestyle.

Sondra Davies/The Image Works

Lifestyle

How does social class affect lifestyle? The rich and the poor are separated by more than money. Social class differences in lifestyle can be observed in many areas of American life, including education, marital and family relations, child rearing, political attitudes and behavior, and religious affiliation.

People in higher social classes tend to marry later, experience lower divorce rates, and have better marital adjustment (Kitson and Raschke 1981; Fergusson, Horwood, and Shannon 1984). While working-class and middle-class parents are more alike in child-rearing practices than in the past, some differences remain. Compared with the middle class, lower-class

parents tend to be less attentive to their children's social and emotional needs; in disciplining, they are more inclined to use physical punishment rather than reasoning. Middle-class parents are more interested in helping their children develop such traits as concern for others, self-control, and curiosity. Finally, working-class parents emphasize conformity, orderliness, and neatness, whereas middle-class parents stress self-direction, freedom, initiative, and creativity (Kohn 1963, 1977, 1990; Wright and Wright 1976; Williamson 1984).

The incidence of voting and political involvement increases with social class level. The middle and upper classes are more likely to be Republicans, whereas the working and lower classes tend to register as Democrats. Those in the lower class tend to be more liberal than

FEEDBACK

Identify the following as examples of life chances (LC) or lifestyle (LS).
_____ a. divorce rate
_____ b. quality of neighborhood street repair
_____ c. life expectancy
_____ d. child-rearing practices
_____ e. voting behavior
_____ f. church attendance

Answers: a. (LS), b. (LC), c. (LC), d. (LS), e. (LS), f. (LS)

middle- and upper-class Americans are on economic issues, but more conservative on social issues. Thus, lower-class people are more in favor of labor unions, government control of business, and social welfare programs but tend to exhibit less tolerance and sympathy than do higher social classes for social issues involving civil rights and international affairs (Wolfinger and Rosenstone 1980; Szymanski 1983).

The rate of church membership and attendance is lowest at the extremes of the stratification structure. Episcopalian, Congregational, and Presbyterian churches are significantly less populated by members of the lower class. Lower-class Americans lean more toward Baptist and fundamentalist churches. Methodist and Lutheran denominations fall in between (see Chapter 15).

Social Mobility

Social mobility refers to the movement of individuals or groups within a stratification structure. In the United States, the term *mobility* implies an elevation in social class level. Beyond this, it is possible either to move down in social class or to move with little or no change in social class.

Types of Social Mobility

Social mobility can be horizontal or vertical. Both types of mobility can be measured either within the career of an individual—**intragenerational mobility**—or from one generation to the next—**intergenerational mobility**.

How does horizontal mobility differ from vertical mobility? A change from one occupation to another at the same general status level is called **horizontal mobility**. Examples of intragenerational horizontal mobility are familiar: an army captain becomes a public school teacher, a minister becomes a psychologist, a restaurant server becomes a taxi driver. The daughter of an attorney who becomes an engineer illustrates intergenerational horizontal mobility. Because horizontal mobility involves no real change in occupational status or social class, sociologists are not generally interested in investigating it.

Vertical mobility, however, is investigated extensively. In **vertical mobility**, occupational status or social class moves upward or downward. Vertical mobility can also be intragenerational or intergenerational. The simplest way to measure intragenerational mobility is to compare an individual's present occupation with his or her first one. Someone who began as a dockworker and later became an insurance salesperson has experienced upward intragenerational mobility. Intergenerational mobility involves the comparison of a parent's (or

grandparent's) occupation with the child's occupation. If a plumber's daughter becomes a physician, upward intergenerational mobility has occurred. If a lawyer's son becomes a carpenter, downward intergenerational mobility has occurred.

Caste and Open Class Stratification Systems

The extent of vertical mobility varies from society to society. Some societies have considerable mobility; others have little or none. This is the major difference between caste systems and open class systems.

How much mobility can we expect in a caste system? In a **caste system**, there is no social mobility because social status is inherited and cannot be changed. In a caste system, statuses (including occupations) are ascribed or assigned at birth; individuals cannot change their statuses through any efforts of their own. By reason of religious, biological, magical, or legal justification, those in one caste are allowed to marry only within their own caste and must limit relationships of all types with those below and above them in the stratification structure. Apartheid, as practiced in South Africa before the election of Nelson Mandela, was a caste system based on race.

The caste system in India is based on occupation and religion and is as complex as it is rigid. Hindus believe that there are four primary caste categories, ranked according to their degree of religious purity. The Brahmin, the top caste, is composed of priests and scholars. Next come the Kshatriyas—noble and warrior caste. Merchants form the third caste, called the Vaishyas. Finally, there is the Shudra caste, containing farmers, menial workers, and craftspeople. Actually, there is a fifth category called the "untouchables"—Indians thought to be so impure that any physical contact contaminates the religious purity of higher caste members.

How is the caste system kept intact? Traditional rules in India prevented movement into a higher caste. Not only were members of different castes not permitted to eat together, but higher-caste people would hardly accept anything to eat or drink from lower-caste persons. Untouchables, who had to live apart from everyone else, couldn't even drink water from the wells used by higher castes. Although the long-standing legal prohibition against dating or marrying someone in a higher caste no longer exists, such crossings are still infrequent. Another sign of the breaking up of the caste system is the rise of a woman named Mayawati, who went from the former "untouchable" community to candidate for prime minister of India (Chu 2008). The caste system has not completely disappeared due to the

Jeff J. Mitchell/Getty Images

Using Microsoft founder Bill Gates as an example promotes the myth of unlimited upward social mobility in American society.

power of those in the higher castes, who use their political power, wealth, and prestige to retard change.

How does an open class system differ from a caste system? In an **open class system,** an individual's social status is based on merit and individual effort. Individuals move up and down the stratification structure as their abilities, education, resources, and commitment to work permit. Although inequality exists in an open class system, it is supposed to be based on differences in monetary worth and personal accomplishment.

Just as some social mobility occurs in an actual caste system, the opportunity for upward mobility is sometimes denied individuals or groups within an open class stratification structure. Prominent illustrations in American society today—which is relatively, but not purely, an open class system—are African Americans, Native Americans, and Latinos (see Chapter 9).

Extent of Upward Mobility

How much upward mobility exists in America? The classic Horatio Alger stories—in which a down-on-his-luck boy makes good through honesty, pluck, and diligence—embody the long-standing belief that America is the land of opportunity. The only thing standing between any American citizen and success, it is argued, is talent, a strong work ethic, and perseverance (Kluegel and

Smith 1990). The careers of people like Oprah Winfrey, Tiger Woods, and Bill Gates appear to support this idea of unlimited mobility in American society. In reality, though, these individuals are exceptions to the rule. Upward mobility has never been as great as the rags-to-riches myth would have us believe (Schwarz 1997).

This does not mean that upward mobility does not occur in American society. It does mean, however, that great leaps in social class level are rare. Earlier studies, comparing fathers' and sons' occupational level, consistently showed that upward mobility, when it occurred, was usually slight (Sorokin 1927; Jackson and Crockett 1964; Blau and Duncan 1967; Featherman 1971; F. J. Davis 1982; D. A. Gilbert 2008). For example, the son of a janitor becomes a bus driver, or the daughter of a retail sales worker becomes a commercial artist.

More recent research measuring social mobility using income levels (rather than occupations) shows the same pattern of intergeneration inheritance and it holds true for both daughters and sons (Isaacs 2007a). Children from homes on the bottom of the income structure are likely (42 percent) to remain at that level as adults. Those born to parents on the top income rung are also quite likely to stay there (39 percent). This is consistent with other research showing that one of the most dependable predictors of an American child's economic future is the characteristics of his or her parents, a factor children cannot control (Sawhill and Morton 2007).

But, even when mobility does occur, the move is generally stunted because of the sons of lower manual workers (operatives, service workers, and laborers) moved to upper-white collar (professional or managerial) positions, while 36 percent followed their fathers into unskilled manual jobs. On the other hand, more than 40 percent of the sons of upper-white collar fathers remained at the same high occupational level as their fathers. Only 15 percent fell to lower manual work. Despite some mobility, Gilbert (2008) found the same pattern of inheritance among fathers and daughters. As an illustration, 54 percent of the daughters of upper-white collar fathers remained in managerial and professional jobs. Though one-third of daughters of lower-manual workers had managerial or professional jobs, most tended to be in clerical, sales, technical, service, and factory jobs.

Why is upward intergenerational mobility so limited? Mobility is heavily influenced by the extent of "intergeneration assistance" (Holland 2007). Children from more affluent families enjoy more aid in the form of college tuition payments, down payments for homes, capital to generate a career, or inheritance of an existing business. Because these offspring emerge from college with little or no debt, they often hit the ground running. Beyond financial assistance, the social class level of one's parents affects things like attitudes toward education, contacts for networking, knowledge of economic opportunities, and optimism about careers. Consequently, children from poorer families are less likely to move up and children from wealthier families are less likely to move down.

Why, then, is there any social mobility at all? According to Peter Blau and Otis Dudley Duncan (1967), there are three basic determinants of upward mobility. First, as less qualified workers take lower-level jobs, the more qualified people fill higher-level positions. Second, because the higher social classes have fewer children, members of the lower social classes are needed to fill higher-level positions. Third, as an economy advances, technology creates higher-level jobs faster than lower-level jobs. Shrinking job opportunities at the lower end preclude the children of lower-level manual and farm workers from following in their parents' occupational footsteps. They have to prepare themselves to move up somewhat. This type of social mobility, which occurs because of changes in the distribution of occupational opportunities, is called **structural mobility** (Hope 1982; DiPrete 1996; Sobel, Becker, and Minick 1998; Isaacs 2007a).

One example of structural mobility in the United States is the change in labor markets due to the new technology and the globalization of business. Emboldened by computer-driven production, improved means of communication, and better transportation, many U.S. companies are moving their manufacturing operations overseas as they seek to lower costs and increase profits. And they are doing so often. As a result, many blue-collar jobs are being transferred to lower-paid international workers, forcing Americans upward, even if only slightly.

Of course, on the downside, with fewer higher-paying manufacturing jobs, and without the education needed to perform the new, more technologically sophisticated and financially rewarding jobs, many U.S. workers are being forced to take lower-paying jobs (structural downward mobility). Compared to their parents, more U.S. workers are now experiencing downward mobility (Newman 1999; Bernhardt et al. 2001; Frank 2007; Isaacs 2007a, b).

Still, don't Americans experience more upward mobility than people in other countries? In fact, the reverse seems to be true. Using parents' and children's incomes as an indicator of mobility, data show that many countries, including Denmark, Norway, Finland, Canada, Sweden, Germany, and France permit greater mobility than the United States. For example, Canadians have 2.5 times more mobility than Americans. (Isaacs 2007a; Sawhill and Morton 2007; see Figure 8.6).

Why is upward mobility more difficult for Americans today? Gilbert (2008) compared the occupational mobility of younger workers in the 1970s with workers around 2000.

Over this period of time, upward mobility declined and downward mobility increased. He attributes this change to a reduction in upward structural mobility as the earlier expansion of higher-level jobs and shrinkage of some lower-level jobs waned.

Holland (2007) identifies three other factors making it harder to get ahead today than was true for earlier generations of working Americans. He discusses educational differences, the accelerating loss of high-paying manufacturing jobs, and the weaker social safety net in the United States.

The United States now has a relatively inferior K–12 educational system as measured by international assessments, increased need for remedial work in college, and employers' complaints about workers' basic skills. A weak educational system disadvantages students from poorer homes. Because, for example, the U.S. education system is largely funded by state and local property taxes, quality of education is determined by the wealth of the community in which children live.

On the college scene, no one has to tell students that tuition costs are going through the roof. With average tuition costs for a year of college exceeding $10,000, a public university education is tougher to finance than it was a generation ago. This is due, in large part, to the fact that the United States is the only advanced country without direct federal government involvement in its educational system. This makes higher quality education even more dependent on family resources.

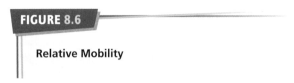

Relative Mobility

This figure graphically displays the comparative mobility rates in the economically developed world. Explain the results.

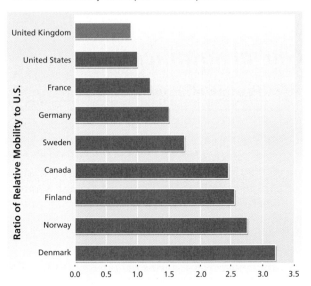

Source: Isabel Sawhill and Joe E. Morton. "Economic Mobility: Is the American Dream Alive and Well?" Wahington DC: the Pew

FEEDBACK

1. _____ refers to the movement of individuals or groups within a stratification structure.
2. Match the major types of social mobility with the illustrations.
   - ____ a. intergenerational mobility
   - ____ b. vertical mobility
   - ____ c. horizontal mobility

   (1) A restaurant waiter becomes a taxi driver
   (2) An auto worker becomes a manager
   (3) The daughter of a hairdresser becomes a college professor
3. In a _____ system, social status is inherited at birth from one's parents and cannot be changed.
4. In an _____ system, rank based on the stratification structure is achieved.
5. Upward mobility in America usually involves only a small improvement in status. T or F?
6. The son of a steelworker finds that the steel industry has shrunk so much that a job in the mill is not available. Instead of following in his father's footsteps, he attends a community college and becomes a computer operator. This is an example of
   a. horizontal mobility.
   b. intragenerational mobility.
   c. caste mobility.
   d. structural mobility.

Answers: 1. Social mobility; 2. a. (3), b. (2), c. (1); 3. caste; 4. open class; 5. T; 6. d

The loss of higher-paying manufacturing jobs also disadvantages the lower classes for upward mobility. At one time, these higher-paying manufacturing jobs were a stepping-stone into the middle class because working parents could afford to support a son or daughter in college. The college-educated sons and daughters of high school-educated parents could more easily become teachers, doctors, and lawyers.

Finally, there is an inverse relationship between the quality of a country's safety net and the extent to which children of working families can experience upward mobility. In the event of job loss and/or serious illness, American parents with a weak or nonexistent safety net (e.g., unemployment and health care benefits) find it even more challenging to educate their children. Children in advanced countries with a stronger safety net, then, enjoy an advantage over U.S. children.

Global Stratification

Identification of Economies

Thus far, the focus has been on stratification within societies. But scarce desirables—income, wealth, prestige, and power—are also differentially distributed among nations. The gross domestic product (GDP) of a country—the total value of the goods and services it produces in one year—is a reasonably good indicator for classifying countries into economic categories. GDP is useful both because it is a single, measurable economic indicator and because a nation's economic rank is highly correlated with the extent of its prestige and power (*World Bank Atlas* 1999).

Predictably, levels of income inequality vary around the world, from low income inequality in Sweden and Ukraine to moderate income inequality in the United States to high income inequality in Bolivia (see Figure 8.7). The United States' score of 41 reflects greater income inequality than exists in most developed countries.

Where are the high-income economies, and how wealthy are they? The richest countries, nearly all capitalistic, include countries such as England, Germany, France, Norway, Finland, Denmark, Sweden, and Switzerland in western Europe; the United States and Canada in North America; Australia and New Zealand in Oceania; and Japan in Asia. A few oil-rich nations, such as Kuwait and the United Arab Emirates, also have high-income economies. High-income economies, spanning approximately one-quarter of the Earth's land surface, have only 15 percent of the world's population (approximately 800 million people). Yet, these countries are in control of most of the wealth. Due to this concentration of wealth, the standard of living of nearly all the people living in these countries, even most of the poor people, is higher than that for the average person in low-income economies (Schnitzer 2000).

What is the nature of middle-income economies? Middle-income economies include, but are not limited to, nations historically founded on or influenced by socialist or communist economies. This encompasses members of the former Soviet Union, many of them renamed and now part of the new Commonwealth of Independent States that was created in 1991. Other countries that fell into the Soviet orbit but are not part of the Commonwealth of Independent States include Poland, Bulgaria, the Czech Republic, the Slovak Republic, Hungary, Romania, Estonia, Latvia, Lithuania, and Cuba.

FIGURE 8.7

Levels of Income Inequality Vary Around the World

The numbers in this figure represent points on the Gini index. Income inequality increases as the numbers increase. A score of zero represents perfect income equality (everyone has the same income), while a score of 100 corresponds to perfect income inequality (all the income is held by only one person). The median score across the world's countries is 39. Are you surprised by any of the countries' scores? Explain.

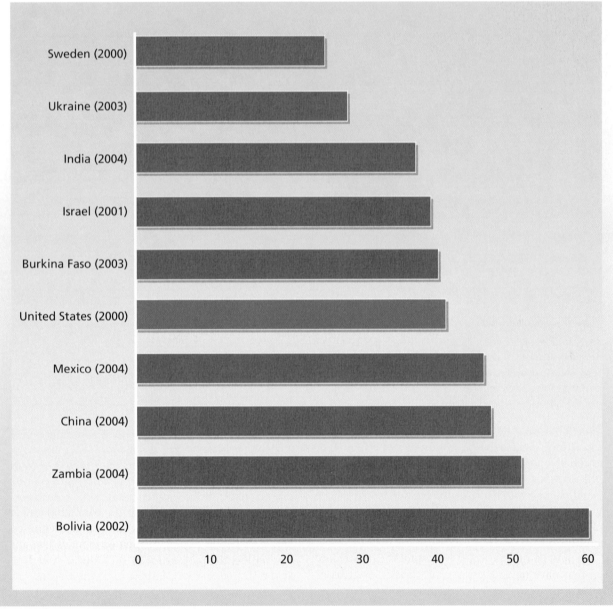

Source: World Bank, World Development Indicators, online database (http://publications.worldbank.org/WDI/, accessed May 25, 2007). It also appears on the *2007 World Population Data Sheet.* (Washington, DC: Population Reference Bureau), 2007.

From the end of World War II until 1989, all these countries, under the political, economic, and military control of the former Soviet Union, constituted the Eastern bloc in the Cold War against the West. These countries are no more densely populated than high-income economies—they occupy 15 percent of the Earth's land with only 10 percent of the world's population (about 500 million people). Although more prosperous than low-income countries, citizens of middle-income economies do not enjoy the standard of living characteristic of high-income economies. The extent of industrialization in these countries is variable. Most inhabitants of these countries live in urban areas; in comparison to countries with high-income economies, though, a greater percentage of their people live in rural areas and participate in agricultural economic activities.

WORLDVIEW 8.1

Social Classes in World Perspective

The World Bank, whose business it is to dispense economic advice and loans to low- and middle-income countries, continuously monitors income levels around the world. The map below displays the World Bank's classification of countries in terms of per capita income. For example, the United States is one of the few nations with a "high-income" economy—one that has a gross national income (GNI) per capita of $10,066 or more.

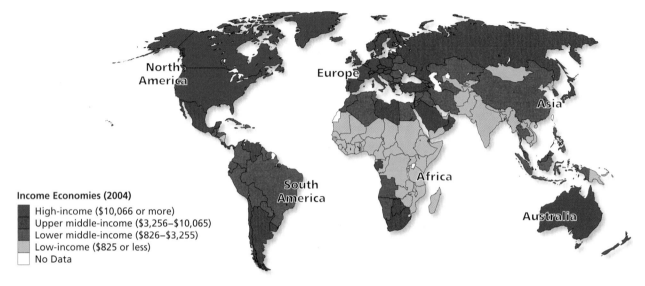

Income Economies (2004)

- ■ High-income ($10,066 or more)
- ■ Upper middle-income ($3,256–$10,065)
- ■ Lower middle-income ($826–$3,255)
- ■ Low-income ($825 or less)
- ☐ No Data

Source: World Bank, 2005.

Interpret the Map

1. Explain the definitions of the four income groups.
2. Identify two countries within each of the four income categories.
3. Why do you think the U.S. is one of the few countries to fall within the high-income category?
4. Compare poverty in the United States to poverty in low-income economies.

In 1989, eastern European countries (freed of Soviet domination, its socialist economy, and government-controlled bureaucracy) turned to capitalism and its market mechanism. A major source of current political and economic turmoil in the former Soviet Union revolves around the struggle between the socialist and capitalist roads to economic development (see Chapter 14).

The middle-income category contains other countries that have never been part of the old Soviet bloc. These countries are spread from Mexico, Latin America, and South America to parts of Africa to Thailand, Malaysia, and New Guinea in the Far East.

What is distinctive about low-income economies? The economic base of low-income economies is primarily agricultural. These countries span the globe from south of the United States border to Africa to China to Indonesia. Population density within these countries as a whole is very high. In fact, just over two-thirds of the world's population (more than 5 billion people) live on 60 percent of the Earth's land (Livernash and Rodenburg 1998).

No single economic system is typical of these economies. Capitalism, socialism, and various combinations of each are practiced. If these countries share no economic ideology, they do share one important economic characteristic: they are unimaginably poor. Problems of the poor in high-income countries, which have already been explored in this chapter, are real and significant. But most people in low-income economies live annually on less than 7 percent of the official poverty line in the United States (U.S. Bureau of the Census 2001). The plight of the poor in low-income countries is particularly bleak because they are trapped in a double bind. Their populations are exploding (due to declining infant mortality rates while they maintain their traditionally high birth rate).

Global Poverty

Poverty persists despite a $400-trillion-plus global economy. Three international reports detail the extent and distribution of global poverty (*Human Development Report* 1999; *World Development Report* 2001; *Least Developed Countries Report* 2002).

What is the state of global poverty at the beginning of the twenty-first century? Despite progress, one-quarter of the Earth's population remains in extreme poverty. More than one-half of the world's population (3.3 billion) lives below the internationally established poverty line of less than $2 a day (*World Population Data Sheet* 2007). Poverty by this standard is most extensive in South Asia, sub-Saharan Africa, Latin America, and the Caribbean (*World Development Report* 2001; Sheehan 2002).

In the more economically developed world, the greatest increase in poverty in the last ten years is found in eastern European countries and among member nations of the former Soviet Union. Whereas poverty in these areas was formerly not widespread, about a third of these people (120 million) now have incomes below $4 daily. Even in industrial nations, more than 100 million people subsist below the established poverty line.

According to the Food and Agriculture Organization of the United Nations, advancement in reducing world hunger has come to a virtual halt. Since 1990–1992, the number of undernourished people in the world has decreased by only 2.5 million per year. Even this slight reduction is due to significant progress in a few large nations. For example, if the progress in China is left out of the calculation, the number of undernourished people in the rest of the developing world has risen by more than 50 million since 1990–1992. And almost one-fourth of the world's children under age 5 are underweight. The rapid rise in grain prices in 2008 threatens to result in the world's most extreme food crisis in a generation (*State of Food Insecurity in the World* 2002; *World Population Data Sheet* 2007; Faiola 2008).

This slum housing in Indonesia illustrates the subsistence level of living in low-income economies.

Has there been no progress in combating global poverty? Poverty has declined in many areas of the world (Bhalla 2002; *World Development Report* 2003). In fact, in all regions of the world, the past fifty years has seen poverty decrease more than in the preceding five centuries. In less than twenty years, China and fourteen other countries, whose populations exceed 1.6 billion, have cut in half the proportion of their people existing in poverty. An additional ten nations, who number almost 1 billion, have cut their poverty rate by one-fourth. Still, poverty is pervasive throughout the world and may increase as a result of slowed economic growth and persisting conflict within and between countries (*Least Developed Countries Report* 2002; Weisbrot 2002; Wermuth 2002; Renner 2003).

FEEDBACK

1. Classify the following examples as high-income (HI), middle-income (MI), or low-income (LI) economies.
   ____ a. Commonwealth of Independent States
   ____ b. agricultural economic base
   ____ c. movement toward capitalism
   ____ d. 800 million people
   ____ e. double population bind
   ____ f. one-fourth of the Earth's land surface
2. Do you think that classifying all the nations of the world into three broad categories leads to some distortion of reality? Explain.
3. What proportion of the Earth's population is considered poor today?
   a. one-tenth
   b. one-fifth
   c. one-fourth
   d. one-third
4. Over the past fifty years, global poverty has actually decreased. T or F?

Answers: 1. a. (MI), b. (LI), c. (MI), d. (HI), e. (LI), f. (HI); 3. c; 4. T

REVIEW GUIDE

SUMMARY

1. There is a universal tendency for humans to create inequality. When individuals are ranked by wealth, prestige, and power, social stratification exists. Any stratification structure is composed of either social classes or segments of a population whose members hold a similar share of community resources. What's more, these class members share attitudes, values, norms, and an identifiable lifestyle.

2. A stratification structure is based on economic, power, and prestige dimensions. Income inequality has always existed in American society, but it has increased dramatically since 1980. Prestige, especially in developed societies, is based primarily on occupation. Occupations with the greatest prestige generally pay the most; require the greatest amount of training, skill, and ability; provide the most power; and are considered the most important.

3. Those individuals who are ranked high on one dimension of stratification are usually high on the other dimensions. Wealth, power, and prestige tend to go together.

4. Functionalists contend that stratification is necessary. It motivates people to prepare themselves for difficult and important jobs, and it motivates them to perform well once they are in those jobs. According to the conflict theory of stratification, inequality exists because some people have more power than others and are willing to use it to promote their self-interest.

5. The functionalist and conflict theories shed light on the reasons for stratification. Symbolic interactionism aids in understanding the manner in which stratification structures are perpetuated. This perspective emphasizes the legitimization of a stratification structure through symbols. These symbols, of course, are learned through socialization.

6. Although most sociologists believe that class lines in America are difficult to draw, they agree that the following social classes have been identified: the upper class, the middle class, the working class, and the lower class. Sociologists agree that the poor exist as an underclass.

7. Poverty can be measured in two ways—absolute or relative. Absolute poverty is determined by annual income.

Anything below a determined (absolute) amount is categorized as poverty. Relative poverty is a comparative measure; it contrasts income groups at the bottom of the stratification structure with those above them. Because of the continuing influence of social Darwinism, Americans tend to blame the poor for their own situation, despite evidence to the contrary.

8. The federal government became serious about combating poverty during the mid-1960s. There have been objections, from some quarters, ever since. The debate on the role of government continues.

9. Americans do not wish to abolish all welfare benefits, but they do want a system that forces the able-bodied off the rolls. In 1996, the U.S. Congress passed a sweeping welfare reform act that dramatically changed the relationship between the poor and their government.

10. Social classes can be thought of as subcultures. For this reason, distinctive patterns of thinking, feeling, and behaving are associated with the various social classes. Both life chances and lifestyle characteristics are affected by one's social class level.

11. Social mobility, the movement of individuals or groups within the stratification structure, is usually measured by changes in occupational status. Sometimes movement within a stratification structure is horizontal. Sociologists, however, are much more interested in upward or downward (vertical) mobility, whether within an individual's lifetime or from one generation to the next.

12. Caste societies, such as India's, permit little vertical mobility; open class societies, such as those in industrialized countries, allow considerable upward mobility. Even in open class societies, however, an upward move tends to be a small one.

13. Nations are also stratified. Richer nations, primarily in the West, are classified as high-income economies. Other less developed industrialized nations, including member nations of the former Soviet Union, can be classified as middle-income economies. The rest of the world, poor and resting primarily on an agricultural base, is classified as low-income economies. In these societies, poverty is widespread.

LEARNING OBJECTIVES REVIEW

- Explain the relationship between social stratification and social class.
- Compare and contrast the three dimensions of stratification.
- State the major differences between the functionalist, conflict, and symbolic interactionist approaches to social stratification.
- Identify the distinguishing characteristics of the major social classes in America.
- Discuss the extent of poverty and perceptions of poverty in the United States.
- Evaluate the U.S. commitment to poverty programs.
- Outline some of the consequences of social stratification.
- Describe upward and downward social mobility in the United States.
- Discuss the major features of global stratification.

CONCEPT REVIEW

Match the following concepts with the definitions listed below them.

_____ a. life chances
_____ b. horizontal mobility
_____ c. caste system
_____ d. intergenerational mobility

_____ e. relative poverty
_____ f. power
_____ g. social mobility
_____ h. wealth

_____ i. conspicuous consumption
_____ j. structural mobility

1. all the economic resources possessed by an individual or group
2. the consumption of goods and services to display one's wealth to others
3. social mobility that takes place from one generation to the next
4. the ability to control the behavior of others, even against their will
5. movement of individuals or groups within a stratification structure
6. the type of stratification structure in which there is no social mobility because social status is inherited and cannot subsequently be changed

7. the likelihood of securing the "good things in life," such as housing, education, good health, and food
8. the measurement of poverty by a comparison of the economic condition of those at the bottom of a society with that of other segments of the population
9. a change from one occupation to another at the same general status level
10. mobility that occurs because of changes in the distribution of occupational opportunities

CRITICAL-THINKING QUESTIONS

1. Stratification can occur in any social setting. Think about a group with which you are familiar that is stratified. Discuss the "unequal shares of scarce desirables" used to rank members.

_____

_____

2. Consider the existence of a class at the bottom of the American social stratification structure. Compare the explanations for the enduring presence of this class that would be given by functionalism and conflict theory.

_____

_____

3. More and more of the poor in the United States are women and children. Relate the feminization of pov-

erty to the prevailing view of the poor held by most Americans. Is there an inconsistency? Why or why not?

_____

_____

4. You are a member of a particular social class. Discuss your life chances and lifestyle in sociological terms.

_____

_____

5. Analyze the social mobility that has occurred in your family across as many generations as you can. Use specific sociological concepts in your analysis.

_____

_____

MULTIPLE-CHOICE QUESTIONS

1. **The creation of layers of people possessing unequal shares of scarce desirables is known as**
   a. social layering.
   b. social grading.
   c. social tracking.
   d. social stratification.
   e. resource accumulation.

2. **A _____ is a segment of a population whose members hold a relatively similar share of scarce desirables and who share attitudes, values, norms, and identifiable lifestyles.**
   a. social network
   b. social nexus
   c. social group
   d. social category
   e. social class

3. Which of the following statements regarding Marx's analysis of class differences is false?

   a. All aspects of capitalistic societies are economically based.

   b. The capitalists control all social institutions, which they use to their advantage.

   c. Capitalists will both rule and exploit the working class because they own the means of production.

   d. Prestige is the prime determinant of social stratification.

   e. Several social classes existed in nineteenth-century industrial society.

4. In contrast to Karl Marx, Max Weber placed emphasis on

   a. the consequences of people's relationship to the economic institution.

   b. the economic dimension as an independent variable.

   c. the development of a single social class.

   d. social stratification from an evolutionary perspective.

   e. the "culture of poverty."

5. Which is the true statement regarding economic inequality in the United States?

   a. There is surprisingly little economic inequality in the United States.

   b. The majority of wealth in the United States is held by the richest fifth of the population.

   c. The middle fifth of the American population holds approximately half of the wealth.

   d. Economic inequality is declining in the United States today.

   e. The majority of wealth in the United States is controlled by religious organizations.

6. Max Weber argued that economic success and power do not always overlap completely. Which of the following statements does *not* reflect his line of argument?

   a. Money and ownership of the means of production are not the only types of resources that can be used as a basis of power.

   b. The fact that money can be used to exert power does not mean that it will be used in this way.

   c. An individual's power is a reflection of his or her relationship to the means of production.

   d. Power is also attached to the social positions we hold.

   e. Individuals can overcome a scarcity of resources if they have large numbers of people on their side.

7. The most stable source of prestige in modern society is associated with

   a. occupation.

   b. education.

   c. social class.

   d. physical beauty.

   e. athletic performance.

8. According to the text, which of the following statements best describes the relationship among the economic, power, and prestige dimensions of stratification?

   a. Individuals who are low on the prestige dimension tend to be high on the other two dimensions.

   b. Individuals who are high on the economic dimension tend to be low on the other two dimensions.

   c. Individuals who are high on one dimension tend to be high on the other two dimensions.

   d. Individuals who are low on the power dimension tend to be high on the other two dimensions.

   e. Sociologists now know that it is impossible to generalize about the relationships among the economic, power, and prestige dimensions of stratification.

9. According to the functionalist theory of stratification, _____ is necessary to ensure that the most important positions are filled by the most qualified people.

   a. social equality

   b. comparable worth

   c. conflict

   d. cooperation

   e. social inequality

10. _____ refers to a sense of identification with the goals and interests of the members of one's social class.

   a. Class consciousness

   b. Class diversion

   c. Goal placement

   d. Class preference

   e. Class knowledge

11. The absence of enough money to secure life's necessities is known as

   a. relative poverty.

   b. chronic poverty.

   c. undeserved poverty.

   d. lifestyle deprivation.

   e. absolute poverty.

12. _____ mobility is measured by an upward or downward change in occupational status over an individual's lifetime.

   a. Intergenerational

   b. Horizontal

   c. Lateral

   d. Intragenerational

   e. Caste

13. The world poverty rate currently stands at _____ of the world's population.

   a. one-eighth

   b. one-quarter

   c. one-third

   d. one-half

   e. three-fourths

FEEDBACK REVIEW

True-False

1. The top 1 percent of American households hold nearly 40 percent of the total wealth. T or F?
2. In America, working-class individuals who fail to experience upward mobility claim that it is the system rather than themselves. T or F?

Fill in the Blank

3. The most stable source of prestige in modern societies is associated with _____.
4. The conflict theory of stratification is based on Marx's ideas about class _____.
5. A major difference between the upper upper class and the lower upper class is that the latter has relied on _____ rather than birth.

Multiple Choice

6. **Which of the following is not a criticism of the functionalist theory of stratification?**
   a. There are people at the top whose jobs do not seem very important.
   b. It has a bias toward cultural relativism.
   c. It overlooks the barriers to competition.
   d. It glosses over the inheritance of social class level.

7. **The son of a steelworker finds that the steel industry has shrunk so much that a job in the mill is not available. Instead of following in his father's footsteps, he attends a community college and becomes a computer operator. This is an example of**
   a. horizontal mobility.
   b. intragenerational mobility.
   c. caste mobility.
   d. structural mobility.

Matching

8. Match the concepts with the examples.
   ____ a. life chances
   ____ b. power
   ____ c. prestige
   ____ d. conspicuous consumption
   (1) the respect accorded doctors
   (2) a politician conforming to the interests of a lobby
   (3) living in a neighborhood with good police protection
   (4) a lavish wedding

9. Match the theories of stratification with the examples.
   ____ a. functionalist theory
   ____ b. conflict theory
   ____ c. symbolic interactionism
   (1) Corporate executives make more money because they determine who gets what in their organizations.
   (2) Engineers make more money than butlers because of the education they possess.
   (3) Ghetto children tend to have low self-esteem.

10. Match the major types of social mobility with the illustrations.
    ____ a. intergenerational mobility
    ____ b. vertical mobility
    ____ c. horizontal mobility
    (1) A restaurant waiter becomes a taxi driver
    (2) An auto worker becomes a manager
    (3) The daughter of a hairdresser becomes a college professor

GRAPHIC REVIEW

Figure 8.3 graphically portrays the American class structure.

1. Write a paragraph using the information in Figure 8.3 to describe the American class structure.

   _____

   _____

   _____

2. Discuss the most surprising findings in Figure 8.3.

   _____

   _____

   _____

ANSWER KEY

Concept Review	Multiple Choice	Feedback Review	
a. 7	1. d	1. T	10. a. 3
b. 9	2. e	2. F	b. 2
c. 6	3. d	3. occupations	c. 1
d. 3	4. a	4. conflict	
e. 8	5. b	5. achievement	
f. 4	6. b	6. b	
g. 5	7. a	7. d	
h. 1	8. c	8. a. 3	
i. 2	9. e	b. 2	
j. 10	10. a	c. 1	
	11. e	d. 4	
	12. d	9. a. 2	
	13. b	b. 1	
		c. 3	

Inequalities of Race and Ethnicity

9

Lynne Sladky/AP Photo

LEARNING OBJECTIVES

- Distinguish among the concepts of minority, race, and ethnicity.
- Describe patterns of racial and ethnic relations.
- Differentiate prejudice from discrimination.
- Illustrate the different views of prejudice and discrimination taken by functionalists, conflict theorists, and symbolic interactionists.
- Describe, relative to the white majority, the condition of minorities in the United States.
- Describe the increasing racial and ethnic diversity in America.

Have African Americans reached income parity with white Americans? Many people believe that as a result of the 1960s civil rights legislation, African American and white family incomes are now approaching comparable levels. In reality, evidence reveals a very different picture. The average income of African American households is considerably less than that of white households. Moreover, at each level of education (the gateway to jobs), African American males gain less income than do white males. On average, for example, white high school graduates can expect to earn annually nearly as much as African American college graduates.

Like all minorities, African Americans suffer the consequences of prejudice and discrimination. Latinos, Native Americans, Jewish Americans, white ethnics, and other minorities have felt the same sting of prejudice and unequal treatment. More recently, attention has been called to other categories of people who are subject to the fallout from prejudice and discrimination, particularly women and the elderly. Inequalities of age and gender are covered in subsequent chapters.

Racial and Ethnic Minorities

All Americans are either immigrants or the descendants of immigrants. Even Native Americans are thought to have migrated to this continent many centuries ago. In part because of this large, diverse migration to the United States, prejudice and discrimination have developed toward minority groups. The emergence of minorities is a recurring theme in the American experience, as immigrant and native groups alike have encountered barriers to full integration into American society. Due to the nature of prejudice and discrimination, much of the research on minorities is approached from the conflict perspective.

The Definition of Minority

Customarily, a minority refers to a relatively small number of people. A sociological definition of a minority, however, can apply to a people numerically larger than others in a society. Before the end of apartheid, for exam-

ple, blacks in South Africa were a minority—even though they outnumbered whites. Women in the United States are referred to as a minority, and they outnumber men. Obviously, for sociologists, something more than size distinguishes a minority (M. E. Williams 1997).

If small numbers do not necessarily make a minority, what does? Louis Wirth identifies the most important characteristics of a **minority**:

We may define a minority as a group of people who, because of their physical or cultural characteristics, are singled out from the others in the society in which they live for differential and unequal treatment, and who therefore regard themselves as objects of collective discrimination. The existence of a minority in a society implies the existence of a corresponding dominant group with higher social status and greater privileges. Minority carries with it the exclusion from full participation in the life of the society. (Wirth 1945:347)

This definition expresses several key ideas. First, a minority has distinctive physical or cultural characteristics. Such characteristics can be used as a reminder that the minority person is different. This is more evident for physical characteristics such as skin color, facial features, or disabilities. Minority cultural characteristics might include accent, religion, language, or parentage. Culturally, a minority member can pass as a member of the majority by a change in name, the loss of an accent, or the adoption of the majority's culture. Experience suggests, however, that where differences are not sufficiently visible to allow easy identification, other means of identification may be imposed. During the Nazi regime, for example, Jewish Germans were forced to wear yellow stars to separate them from non-Jewish Germans.

Second, minority status is reflected in a society's stratification structure. Almost any society has desired goods, services, and privileges. Largely because the majority is the dominating group, it holds an unequal share of the desired goods, services, and privileges. Further, minority members have less access to the desirable resources. Minorities, for example, often have difficulty getting good jobs because of inferior schooling or discriminatory hiring.

Third, the distinctive cultural or physical traits of a minority can be judged by the majority to be inferior to their own and can be used to justify unequal treatment. In other words, the alleged inferiority of a minority can become part of the majority's ideology—a set of ideas used to justify and defend the interests and actions of those in power (Eagleton 1994). A majority practices job discrimination more easily if, for example, its ideology depicts a minority as shiftless or lazy.

Fourth, because members of a minority regard themselves as objects of discrimination, they have a sense of

How many races are represented in this photo? On what basis did you make that determination?

common identity. Within the minority there is a "consciousness of kind." It is from this sense of common identity that a "we" and "they" vocabulary is accepted within the minority. This vocabulary reflects a strong sense of solidarity and loyalty.

Finally, membership in a minority is ascribed. People do not make an effort to join a minority; they become members by birth. Because membership in a minority is an ascribed status, it is not easily shed. One does nothing to achieve this status, and there is little one can do to avoid it. Although some members of ethnic minorities may, through great effort and a name change, leave this ascribed status behind, it is nearly impossible for members of racial minorities to do so.

Throughout history, various peoples have been treated as minorities. For this reason, it is important to understand the nature of race and ethnicity (Smaje 2000; Parrillo 2005a).

The Significance of Race

What is race? A **race** is a category of people who are alleged to share certain biologically inherited physical characteristics that are considered socially important within a society. Biologists have used such physical characteristics as skin color, hair color and texture, facial features, head form, eye color, and height to create broad racial classifications. The most commonly used system of racial classification has three major racial categories—Negroid, Mongoloid, and Caucasoid—along with some unclassified racial categories.

Are racial classifications valid? Sociologists now consider racial classifications, first developed by nineteenth-century biologists, arbitrary and misleading. Ashley Montagu (2000) refers to race as "man's most dangerous myth." Although certain physical features have been associated with particular races, scientists have known for a long time that there is no such thing as a "pure" race. This long-standing belief has been reinforced recently through DNA studies. Geneticist Kenneth Kidd of Yale University concludes from the DNA samples he has investigated that there is

a virtual continuum of genetic variation around the world. There's no place where you can draw a line and say there's a major difference on one side of the line from what's on the other side. If one is talking about a distinct, discrete, identifiable population, . . . There's no such thing as race in [modern] *Homo sapiens.* (Quoted in Marshall 1998:654)

Consequently and frequently, features—or markers—typical of one race show up in other races. For example, some people born into African American families are assumed to be white because of their facial features and light skin color (Appiah and Gutmann 1998).

Even defenders of racial classification acknowledge that their number of identifiable categories exceeds thirty. If racial classification categories are this numerous, scientists conclude, they lose any meaningful capacity to differentiate physical characteristics (Dobzhansky 1970; Hirschfeld 1998; S. Olson 2002, 2003; Wells and Read 2003).

If arbitrary racial classifications are accepted, social definitions can then be imposed on rather superficial physical differences. For example, in the United States prior to the Civil War, a person born of African American and white parents was traditionally considered "black." If "white" blood were merely a matter of biology, there would have been no legal penalty attached to biracial offspring who claimed to be Caucasian. However, attempting to "pass for white" carried heavy legal sanctions. Some southern and border states defined a person as legally African if one of the great-grandparents was African (one-eighth) or if one of the great-great-grandparents was African (one-sixteenth). In practice, though, it took far less than one-sixteenth of the blood to be "black": anyone with any known African ancestry was usually considered black. This was known as the "one-drop" rule (Spickard 1992; Korgen 1998). Absent the socially imposed restrictions attached to "black" blood, these individuals could have classified themselves as "white" if they had preferred. Fortunately, the legal system in the United States now recognizes the arbitrariness of these racial labels, and parents are able to decide for themselves into which census category to classify their children. For the first time, the 2000 Census allowed residents to select more than one racial category.

What is the relationship between alleged racial physical characteristics and mental capacity? Racist thinkers, from Count de Gobineau in the mid-nineteenth century to Adolf Hitler in the twentieth century, have attempted to link physical differences among people to innate mental and physical superiority or inferiority. According to past research, physical characteristics are superior only in the sense that they provide advantages for living in particular environments. For example, a narrow opening between eyelids protects against bright light or extreme cold such as found in Siberia or Alaska. A darker skin is better able to withstand a hot sun. But these physical differences are controlled by a very few genes. In fact, geneticists claim that there is more genetic difference between a tall person and a short person than between two people of different races who are the same height. Only about six genes in the human cell control skin color, while a person's height is affected by dozens of genes. Thus, a six-foot white male is closer genetically to a black male of similar height than to a five-foot white male.

The bulk of scientific evidence does not support **racism**: ideas that attempt to connect biological characteristics with innate racial superiority or inferiority (van den Berghe 1978; Wieviorka 1995; Hurley 1998). Research has found, for example, no biologically inherited differences in intelligence among the various races. Most social scientists endorse the following statement from a UNESCO Statement on Race, prepared by a distinguished group of social scientists:

> *According to present knowledge, there is no proof that the groups of mankind differ in their innate mental characteristics, whether in respect of intelligence or temperament. The scientific evidence indicates that the range of mental capacities in all ethnic [or racial] groups is much the same.* (Klineberg 1950:466)

A subsequent review of research supports UNESCO's earlier conclusion ("Statement on Race and Intelligence" 1969). Both of these reports attribute any differences in measured IQ among racial groups to differences in social environment, training, and education.

While the idea of innate racial differences has generally been pronounced dead by social scientists, it has not been buried. Race as a biological reality is being resuscitated by recent DNA research (Harmon 2007; Kashef 2007; Saletan 2007). When scientists first decoded the human genome in 2000, they offered it as evidence of the essential similarity among all humans. A comparison of any two people revealed a 99 percent DNA overlap. Lately, DNA research has begun exploring the possible importance of the remaining one percent difference. Noting some continental variations in DNA, these studies are again raising the question of biologically based racial differences. Relevance of these small DNA differences to the concept of race remains highly doubtful.

What do social scientists mean by "social" races? The consequences of racial discrimination based on alleged differences are real. Those people professing innate racial differences can make judgments regarding superiority/inferiority and then attempt to justify prejudice and discrimination. For this reason, many social scientists have become interested in "social" races (J. E. Farley 1995; Banton 1998; Feagin and McKinney 2005). African Americans, Native Americans, Latinos, and Asian Americans are examples of socially defined races that, in America, often continue to face barriers. It seems reasonable to conclude that race remains an influential part of the American experience (Omi and Winant 1994; Feagin 2001; Loury 2003).

The Significance of Ethnicity

A people can be considered a racial category at one time and not at another time. The Irish and Italians, for example, were once considered racial groups by Anglo-Saxon Americans. Later, as it became clear that they had no distinctive physical characteristics, the Irish and Italians became ethnic minorities.

What is an ethnic minority? The term *ethnicity* comes from the Greek word *ethnos*, originally meaning "people" or "nation." Thus, the original Greek word referred to cultural and national identity. Today an **ethnic minority** is socially identified by its unique characteristics related to culture or nationality. Just as physical characteristics define racial minorities, cultural differences define ethnic minorities.

Because of their differences from the host culture, ethnic minorities are subcultures. They have a way of life that is based on their own language, religion, values, beliefs, norms, and customs. Like any subculture, they are part of the larger culture—their members work in the majority (or host) economy, send their children through the dominant educational system, and are subject to the laws of the land—but they are also separate from the larger culture. This separation persists either because an ethnic minority wishes to maintain its cultural and national origins or because the majority erects barriers that prevent the ethnic group from blending in with the larger culture. Michael Novak (1996), himself a Slovak, has contended that members of white ethnic minorities from southern and eastern Europe—Poles, Slavs, Italians, Greeks—have not been able to blend completely into American society because they are more culturally different from white Anglo-Saxon Protestants (WASPs) than are immigrants from western and northern Europe. Western Europeans, for example, have an alphabet similar to English and have religions similar to Americans.

How are ethnic minorities received? Negative attitudes toward ethnic minorities exist in part because of ethnocentrism (the judgment of others based on one's own cultural standards). Many of the majority—out of loyalty to and preference for their own values, beliefs, and norms—may respond to other cultural ethnic views as inferior. Because members of ethnic minorities do not meet the majority's definition for appropriate ways of behaving, it may be assumed that something is wrong with the ethnic members. Ethnocentric judgments, of course, are often expressed via prejudice and discrimination. Jews, for example, have had to contend with prejudice and discrimination based on their religious beliefs. The heated immigration debate is fueled by negative stereotypes of Latino immigrants such as the false claim that they rely heavily on welfare.

A society's attitudes toward ethnic (and racial) minorities vary from minority to minority and from time to time. Figure 9.1 illustrates the opinions of many Caucasian Americans toward a variety of immigrant minorities. Clearly, older immigrant minorities (Irish and Polish) are viewed more favorably than later arrivals (Cubans, Haitians, Iranians). Whether you are Caucasian American or a member of an ethnic or racial minority, you may want to compare your attitudes with those depicted in Figure 9.1.

Recently, in some quarters, Latino immigrants and Muslims have become a target of intense negative social reactions. Latinos are said, by a vocal segment of Americans, to be threatening jobs, overwhelming social services, and making streets unsafe. Because of often intense prejudice, more than one-half of Latino adults worry about deportation for themselves or someone close to them (Martin and Midgley 2006; Carroll

2007; Reich 2008). Substantial pockets of Americans have negative feelings or prejudices for Muslims and support extra security measures with Muslims to combat terrorism (Saad 2008). A majority of Muslim Americans thinks it is more difficult to be a Muslim in the United States since the September 11, 2001, terrorist attacks ("Muslim Americans: Middle Class and Mostly Mainstream," 2007).

Patterns of Racial and Ethnic Relations

When people of various racial and ethnic backgrounds interact, a wide range of outcomes is possible (Wallace 1997). It is helpful, however, to divide these many outcomes into two major types: patterns of assimilation (minority groups are accepted) and patterns of conflict (minority groups are rejected).

Assimilation involves "those processes whereby groups with distinctive identities become culturally and socially fused together" (Vander Zanden 1990:274). It is the integration of a racial or ethnic minority into a society, where minority members are given full participation in all aspects of the society. The classic works in the study of assimilation are those of Milton Gordon, who has defined three basic assimilation patterns in American society: Anglo-conformity, melting pot, and cultural pluralism (M. M. Gordon 1964, 1978).

What is the most common pattern of assimilation? *Anglo* is a prefix indicating an American of English descent. So in *Anglo-conformity*, traditional American institutions are maintained. Basic to this pattern of assimilation

---

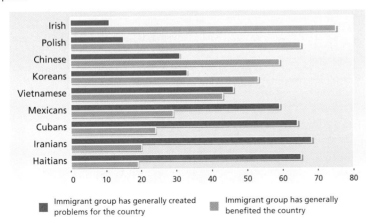

FIGURE 9.1

Attitudes of Americans Toward Immigrant Minorities

This figure displays the attitudes toward immigrant groups in the United States. Do these attitudes coincide with your own attitudes? With those of your friends? Explain.

Source: George H. Gallup Jr., *The Gallup Poll: Public Opinion 1993* (Wilmington, DE: Scholarly Resources, Inc., 1994), pp. 250-251. Copyright © 1994 by Scholarly Resources, Inc. Reprinted by permission of Scholarly Resources, Inc.

is the acceptance of immigrants as long as they conform to the "accepted standards" of the host society. Anglo-conformity has been the most prevalent pattern of assimilation in America. It is the least egalitarian assimilation pattern because the immigrant minority is required to conform and, by implication, either give up or suppress its own values.

Is America more like a melting pot or a tossed salad? A second pattern of assimilation is the *melting pot*, in which all ethnic and racial minorities blend together. Israel Zangwill's drama *The Melting Pot* expresses the goal of

> the completion of a vast American symphony which will express [the immigrant's] deeply felt conception of [the] adopted country as a divinely appointed crucible in which all the ethnic divisions of mankind will divest themselves of their ancient animosities and differences and become fused into one group, signifying the brotherhood of man. (M. M. Gordon 1978:193)

Although the ethnic and racial pot did simmer, especially in American cities, there is a question as to how much melting of differences and resulting fusion have really taken place. Instead of a melting pot, many sociologists are now using the idea of a "tossed salad," where traditions and cultures exist side by side.

The third assimilation pattern is *cultural pluralism*. Unlike the ideal of a unified culture underlying the Anglo-conformity and melting-pot patterns, cultural pluralism recognizes the immigrants' desire to maintain at least a remnant of their "old" ways. Ethnic enclaves, or settlements, characterized American cities of the past two centuries, and many survive today. Such enclaves allowed immigrants to maintain many of their traditional ways while accommodating American values and norms.

Accommodation, an extreme form of cultural pluralism, occurs when a minority, despite its familiarity with the beliefs, norms, and values of the dominant culture, maintains its own culturally unique way of life. An accommodated minority learns to deal with the dominant culture when necessary, but remains independent in language and culture. Cubans in Miami are an example of a distinct community within a larger community (K. L. Wilson and Martin 1982). Cultural pluralism emphasizes the ethnic and racial diversity that still characterizes American society and is a more democratic and egalitarian pattern of assimilation than either Anglo-conformity or the melting pot. Like the other two patterns, however, cultural pluralism also overlooks the continuing inequities that characterize American racial and ethnic relations today. Each of these assimilation patterns stresses a kind of egalitarian outcome and thus omits patterns of conflict.

What are the patterns of conflict? Conflict patterns are associated with the dominance of a majority over racial and ethnic minorities. Three basic patterns of conflict are genocide, population transfer, and subjugation (Mason 1970).

Genocide, the most tragic expression of dominance, refers to politically motivated mass murder of most or all of a targeted population (Fein 1993; Chirot and Seligman 2001). One of the least-known examples occurred between 1492 and 1650 when Christopher Columbus totally eliminated 250,000 Arawak Indians of the Bahamas (Zinn 2005). One of the best-known examples is the Holocaust, Adolph Hitler's attempt to destroy all European Jews during the 1930s and 1940s. (See Figure 9.2.) The Nazis succeeded in killing some 6 million Jews before World War II ended in 1945. Less known is the Rape of Nanking, begun in 1937, during which the Japanese massacred an estimated 260,000 to 350,000 Chinese men, women, and children (Chang 1998). Tragically, genocide campaigns are more common in world history than might be supposed. The Serbian campaigns of "ethnic cleansing" against the Muslims in Bosnia and Kosovo and the ethnic slaugh-

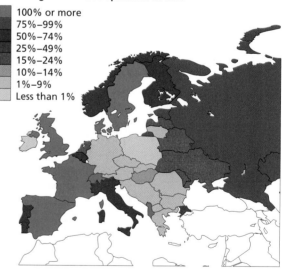

FIGURE 9.2

Impact of the Holocaust

Jewish Population in 1992 as a
Percentage of Jewish Population in 1930

- 100% or more
- 75%–99%
- 50%–74%
- 25%–49%
- 15%–24%
- 10%–14%
- 1%–9%
- Less than 1%

One of the worst examples of genocide was the Nazis' attempt, in the 1930s and 1940s, to exterminate the European Jewish population. This map shows the decline in Jewish population in European countries as a result of the Holocaust.

FEEDBACK

1. Which of the following is *not* always a characteristic of a minority?
   a. distinctive physical or cultural characteristics
   b. smaller in number than the majority
   c. dominated by the majority
   d. denied equal treatment
   e. a sense of collective identity
2. _____ is an ideology that links a people's physical characteristics with their alleged intellectual superiority or inferiority.
3. A(n) _____ is a people within a larger culture that is treated as a minority because of its distinctive cultural characteristics.
4. _____ refers to those processes whereby groups with distinctive identities become culturally and socially fused.
5. The pattern of conflict in which one people is forced to move to a remote location is called _____.

Answers: 1. b; 2. Racism; 3. ethnic group; 4. Assimilation; 5. population transfer

ter of anywhere from 200,000 to 500,000 people by the Tutsi tribe of Rwanda are instances of genocide (Prunier 1997; Marks 1999; Naimark 2002; S. Power 2001, 2003). An even more recent instance of genocide is the killing of more than 400,000 blacks by Arab militias in the Darfur region of Sudan, Africa (Caplan 2006). An incipient episode of genocide may be brewing in Kenya, Africa. In early January 2008, a national election triggered a wave of house burnings and killings of members of the ethic group the Kikuyus (McCrummen 2008).

Population transfer is another way in which a society dominates a minority. In this pattern of conflict, a minority is forced either to move to a remote location or to leave entirely the territory controlled by the majority. Population transfer was the most common policy used against Native Americans, especially during the late 1800s. The Cherokees, for example, were once a single nation in the southeastern United States until 20,000 Cherokees were forcibly removed to Oklahoma reservations, where they became dependent on the U.S. government. Four thousand of these Cherokees died because of harsh conditions along the Trail of Tears. Although Cherokee reservations can be found today in North Carolina and Oklahoma, many Native Americans have migrated to cities to find work, a move that in itself threatens the survival of some tribes. In another example, some believe that the 1999 Serbian aggression in Kosovo was actually closer to an attempt at population transfer than it was to genocide, because the Serbian intent was not to annihilate the Albanians but to move them (Wiesel 1999).

Subjugation is the most common pattern of conflict and most clearly reflects the characteristics of a minority relationship: the majority enjoys greater access to the culture and lifestyle of the larger society. Inequities appear in such areas as power, economics, and education as well as in other important indicators of the qual-ity of life, such as health and longevity. In subjugation, the majority and the minorities may live in the same general area and may participate together in at least some aspects of social life, but members of the minority remain clearly subordinate. Subjugation may be based on the law—**de jure subjugation**—as in the segregation of African Americans following Reconstruction in the United States or as in the former apartheid system of law in South Africa. It may also be based in actual everyday practice—**de facto subjugation**—regardless of what the law is. In this case, subjugation is visible primarily in indicators like political underrepresentation and low occupational achievement. The continuing practice of housing discrimination in the United States is an illustration of subjugation. This practice has existed over many decades, even in the face of direct efforts, including open housing laws, designed to overturn it. African Americans, Latinos, and other minorities still experience considerable discrimination when they attempt to rent or purchase a home.

At any given time, the patterns of race and ethnic relations are many, diverse, and often coexisting. Also, in the history of relations between two peoples, we can find wide variation through time in the ways the people have interacted. Many factors can influence the pattern of racial and ethnic relations adopted: the nature of the first contact, the reasons for contact and interaction, visibility of minority groups, views held by respective members, and general social conditions.

Theories of Prejudice and Discrimination

A variety of theoretical perspectives has been used to explain prejudice and discrimination. Before introducing these theories, however, the nature of prejudice and discrimination needs to be discussed.

The Nature of Prejudice and Discrimination

How is a sociological definition of prejudice different from the popular view of the term? Prejudice is often viewed as a general characteristic of human relationships. Students can be prejudiced against professors, professors against university administrators, Catholics against Protestants, and employees against employers. In sociology, however, the concept of **prejudice** has a more precise meaning. It refers to widely held preconceptions of a group (minority or majority) and its individual members. Because such attitudes are based on strong emotions and rooted in unchallenged ideas, they are difficult to change, even in the face of overwhelming invalidating evidence. New information that contradicts one's prejudices tends to be denied because of selective perception. It is easier to explain an individual who doesn't fit the stereotype as an exception rather than reexamine a whole set of established beliefs. Consequently, prejudiced attitudes generally are not altered either by new personal experiences or by favorable accounts from others. An Anglo student may prefer to believe that an Asian classmate who does poorly in math is the "exception to the rule" that Asians have a "gift" for mathematics.

Prejudice involves an either/or type of logic: a minority (or majority) is either good or bad, and it is assumed that every member possesses the characteristics attributed to the group. Prejudice, then, involves overgeneralization based on biased or insufficient information. Such overgeneralizations are usually defended by one of two responses: citing limited personal experiences with minority members or reciting stories told by others about their experiences.

How is discrimination different from prejudice? Whereas prejudice refers to biased attitudes, **discrimination** refers to unequal treatment. Discrimination can take various forms: avoiding social contact with members of minority groups, denying them positions that carry authority, or blocking their access to exclusive neighborhoods. It can also involve such extremes as physically attacking or killing targeted people (Allport 1979).

What is the relationship between prejudice and discrimination? Prejudice is usually considered the cause of discrimination. Although it usually does, prejudice need not lead to discrimination. The Roman High Empire (98 to 180 A.D.) considered conquered people to be barbarians and themselves as "Heaven's representatives among mankind." Yet Romans of this period were very tolerant of the same people, making them Roman citizens and permitting them extraordinary upward mobility (Chua 2007). Discrimination may also cause prejudice. For example, unskilled workers may believe that their jobs are in jeopardy because of the massive immigration of a new ethnic group. This fear of economic threat may lead to unequal treatment of the ethnic members. To justify this discrimination, threatened workers may attempt to show why a minority deserves unequal treatment. This occurs in part through the application of stereotypes. This is exactly what happened to Chinese Americans during the nineteenth and early twentieth centuries (Lyman 1974).

What is a stereotype? Like prejudice, stereotyping can appear throughout society. In the United States, for example, athletes are thought to be "all brawn and no brain," and politicians are assumed to be corrupt. As used in sociology, a **stereotype** is a set of ideas based on distortion, exaggeration, and oversimplification that is applied to all members of a social category (Stangor 2000; Loury 2003: Pickering 2001). Prominent radio shock jock Don Imus lost his job after referring to some members of the Rutgers University women's basketball team as "nappy headed hoes." His thirty-year run was ended temporarily (another company subsequently brought him back) after considerable public protest to CBS (Carter and Steinberg 2007).

Prejudice can lead to discrimination such as the act committed by the person(s) outside this Jewish cemetery.

Spinning a Web of Hate

White supremacists, neo-Nazis, and other hate groups have discovered the Internet as an effective means of spreading hatred of Jews, African Americans, homosexuals, and fundamentalist Christians (Sandberg 1999). The Web boosts the recruiting efforts of these groups from leaflets and small meetings to a mass audience—more than 100 million worldwide. Reaching this audience is inexpensive and allows hate groups to present their message unedited and uncensored.

If you had browsed the World Wide Web in 1995, you would have found only one hate site, Stormfront. According to the Anti-Defamation League, there are now 2,000 websites advocating racism, anti-Semitism, and violence. Aryan Nation identifies Jews as the natural enemy of whites; White Pride Network offers a racist joke center; Posse Comitatus defends alleged abortion clinic bomber Eric Robert Rudolph; World Church of the Creator is violently anti-Christian.

Because they use high technology to deliver their message to a mass audience, organized racists are no longer easily identifiable. While members of hate groups used to be recognized by their white hoods or neo-Nazi swastikas, they can now just as easily be wearing business suits instead of brown shirts. The Southern Poverty Law Center is especially concerned about the repackaging of hate-based ideologies to make them appear more respectable to mainstream America. To reach the young, hate websites offer such child-friendly attractions as crossword puzzles, jokes, cartoons, coloring books, contests, games, and interactive comic strips.

One of the most prominent racist groups is the Council of Conservative Citizens (CCC). This organization, founded in 1985, now has more than 15,000 members in at least twenty-two states. Its members include scores of state legislators in Mississippi, and it has sought the political support of several state governors and U.S. senators and representatives. The CCC is an outgrowth of the White Citizens Councils from the 1950s and 1960s, whose primary purpose was to block the civil rights movement and school desegregation. While the CCC disavows racism, billing itself as "a conserva-

tive organization that tries to defend the Constitution," its website regularly publishes racist material. One of its featured columnists, H. Millard, made this prediction about the outcome of the interracial melting pot: "What will emerge will be just a slimy brown mass of glop. The genocide being carried out against white people hasn't come with marching armies. . . . Genocide via the bedroom chamber is just as long-lasting as genocide via the gas chamber."

Not all hate group activity comes from white supremacists who target African Americans. The Southern Poverty Law Center also tracks the activities of black separatists and documents several recent hate crimes committed by blacks against whites. In addition, the continued immigration of Asians and Central and South Americans is drawing the angry attention of hate groups of all types. More information on hate group activities can be found at the Southern Poverty Law Center website: www.splcenter.org.

Think About It
When the economy is not performing well, membership in hate groups rises; membership declines when the economy is doing well. Relate this fluctuating membership pattern to scapegoating you may have observed or heard about.

Stereotypes may be used as a justification for prejudiced attitudes and discrimination. For example, very early relationships between the colonists and Native Americans were relatively peaceful and cooperative, but as the population of the colonies grew, conflicts over land and resources became more frequent and intense. To justify expansion into Indian territory, the colonists nurtured a stereotype of Native Americans as lying, thieving, murdering, heathen savages. This picture was frequently not held by the trappers and traders who moved individually among the Indians and, because of friendly contact, tended to judge the Indian differently. But the prejudice of the farmer–settler prevailed, leading to the continued seizure of Native American lands with a minimum of compensation and with the dramatic reduction of the Native

American population from as many as 5 to 7 million to as few as a quarter million in 1890 (Snipp 1996; Ellis 2007).

Discrimination based on stereotypes is not merely past history. Take, for example, a field experiment designed to measure racial discrimination in the labor market. Marianne Bertrand and Sendhil Mullainathan (2002) sent résumés in response to help-wanted ads appearing in Boston and Chicago newspapers. To manipulate the perception of race, each résumé was randomly assigned either a very white-sounding name or a very African American–sounding name. Résumés with white names drew about 50 percent more callbacks than those with African American names. In addition, the researchers introduced variations in the quality of credentials. Among whites, the higher-quality résumés elicited

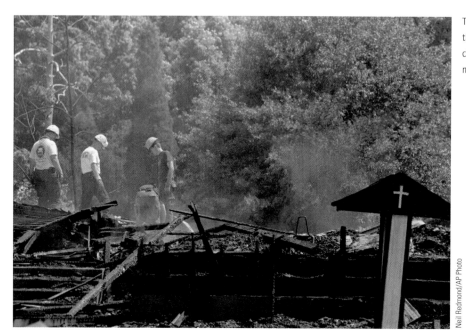

These arson investigators are searching through the ashes of an African American church in North Carolina. What would make this case of arson a hate crime?

Neil Redmond/AP Photo

30 percent more callbacks than the lower quality ones. Among African Americans, in contrast, the higher-quality résumés did not elicit significantly more call-backs. In short, whites benefited from improving their credentials; African Americans did not.

What is a hate crime? In 1999, James Byrd Jr., an African American from Texas, was chained to a pickup truck and then dragged to death. That same year saw Matthew Shepard, a gay college student at the University of Wyoming, tied to a fence and beaten to death. Both murders were hate crimes. A **hate crime** is a criminal act motivated by prejudice (F. M. Lawrence 2002; Perry 2000; Jenness and Grattet 2001). Hate or "bias" crimes involve discrimination related to race, religion, sexual orientation, national origin, or ancestry (Levin and McDevitt 2002; Swain 2002). Victims include, but are not limited to, African Americans, Native Americans, Latinos, Asian Americans, Jews, Muslims, gay men and women, and people with disabilities (Green, McFalls, and Smith 2001; Levin and McDevitt 2002; F. M. Lawrence 2002; Perry 2000, 2003).

The term *hate crime* is relatively new; the behavior is not. The federal government has kept statistics since 1900. In 2006, there were 7,722 criminal incidents as a result of bias against a particular race, religion, sexual orientation, ethnicity/nation origin, or physical or mental disability (U.S. Department of Justice 2007). As of 2008, forty states had enacted hate crime laws.

Obviously, prejudice and discrimination cannot be accounted for by any single factor. The causes are varied. Psychology, functionalism, conflict theory, and symbolic interactionism each contribute to the under-standing of prejudice and discrimination.

The Psychological Perspective

Psychological explanations of prejudice and discrimination focus on the prejudiced person's personality, how it developed, and how it functions in the present. Questions like these have been asked about prejudiced persons: What was their relationship with their parents or with their significant others? What are their values, attitudes, and beliefs? How high is their self-esteem? Two prominent psychological explanations of prejudice and discrimination are the frustration-aggression explanation and the authoritarian personality.

How can frustration lead to prejudice and discrimination? According to the *frustration-aggression* explanation, prejudice and discrimination are the products of deep-seated hostility and aggression that stem from frustration. Aggression is most likely when built-up hostility cannot be directed at the actual source of frustration. Pent-up hostility and frustration may subsequently be redirected toward some substitute object that is less threatening than the one causing the frustration. These substitute objects, known as **scapegoats**, serve as convenient and less feared targets on which to place the blame for one's own troubles, frustrations, failures, or sense of guilt.

John Dollard (Dollard et al. 1939), originator of the frustration-aggression explanation, contended that the frustrations experienced by the Germans after World War I help account for their acceptance of anti-Semitism. The loss of the war, the disappearance of international prestige, the forced acceptance of the Treaty of Versailles, and a ruined economy all contributed to tremendous frustration among the German people. Because direct aggression against the Allies was not an alternative, the Germans channeled their aggression toward other targets, one of which was the Jews. The best scapegoats are those who have already been singled out by the majority for unequal treatment and who therefore have the least chance to defend themselves.

What personality type may tend toward prejudiced attitudes? Another psychological explanation contends that there is a personality type—the **authoritarian personality**—that tends to be more prejudiced than other types. The authoritarian personality is characterized by excessive conformity; submissiveness to authority figures; inflexibility; repression of impulses, desires, and ideas; fearfulness; and arrogance toward persons or groups thought to be inferior. T. W. Adorno and his colleagues, the creators of this theory, summarized it this way:

> *The most crucial result of the [study of the authoritarian personality] is the demonstration of the close correspondence in the type of approach and outlook a subject is likely to have in a great variety of areas, ranging from the most intimate features of family and sex adjustment through relationships to other people in general, to religion, and to social and political philosophy. Thus a basically hierarchical, authoritarian, exploitative parent–child relationship is apt to carry over into a power-oriented, exploitatively dependent attitude toward one's sex partner and one's God and may well culminate in a political philosophy and social outlook which has no room for anything but a desperate clinging to what appears to be strong and a disdainful rejection of whatever is relegated to the bottom.*
> (Adorno et al. 1950:971)

The Functionalist Perspective

Functionalists have focused on the dysfunctions of prejudice and discrimination, and as the last half of this chapter will demonstrate, the negative consequences of prejudice and discrimination are wide and deep. The social, political, educational, and economic costs of the exploitation and oppression of minorities are extremely high. Furthermore, the safety and stability of the larger society are at risk because violence periodically erupts between the groups.

Do functionalists see any positive contributions of prejudice and discrimination? Because functionalists look for positive contributions that each aspect of a culture makes to that society's stability and continuity, they have identified some potential benefits. As noted in Chapter 7, Emile Durkheim identified the ways in which deviance contributes to the cohesiveness of society. The functionalist theory of stratification, as shown in Chapter 8, contends that social inequality helps society to channel the most qualified people into the most important posi-

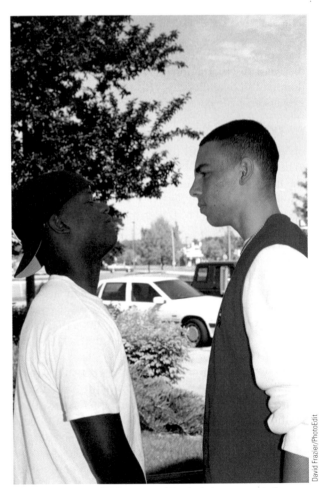

Functionalists see conflict between racial and ethnic groups as dysfunctional. Why?

David Frazier/PhotoEdit

tions and to ensure that people in these positions are motivated to perform their tasks competently. Another potential benefit is the social solidarity of the majority. This happens because the attempted exclusion of outside groups rests in part on ethnocentrism—the tendency to judge others based on one's own cultural standards. Once a majority is convinced of its superiority, it can unite around its own way of life. This strengthens the boundaries of the majority.

But what is the downside of ethnocentrism? As you know, what is functional for one part of society may be dysfunctional for other parts. The price of national unity via ethnocentrism is extremely high for its targets. Passage of the U.S. Constitution in 1789 permitted the retention of slavery. This consigned black slaves to nearly another 100 years of bondage. Documentation of the persistent damage of this extreme oppression to African Americans is covered in this chapter. This early sense of national unity has also been expensive to the entire society. Consequences of the Civil War, which saw the country violently divided in the nineteenth century, still reverberate through regional factions.

The Conflict Perspective

According to conflict theory, a majority uses prejudice and discrimination as weapons of power in the domination of a minority. This theory, then, traces the existence of prejudice and discrimination to majority interests rather than to personality needs. Domination by the majority is often motivated by its desire to gain or increase its control over scarce goods and services. When those with power in a society are able to persuade the majority that a minority *should* be subjugated, the minority have been effectively eliminated or neutralized as competitors (W. J. Wilson 1973).

According to the conflict perspective, how do minority groups view each other, and what effect does this have for the majority? Members of a majority tend to think of minorities as people unlike them but similar to one another. Despite being a common target of the majority's prejudice and discrimination, minorities tend to view one another as competitors, rather than as allies, in their struggle for scarce resources (Olzak and Nagel 1986). Conflict among minorities, particularly African Americans and Latinos, is increasing in the United States as whites leave cities and African Americans assume political power. On the one hand, many urban blacks believe that Latinos are benefiting from the civil rights movement waged by African Americans. On the other hand, many Latinos believe that African Americans are using their political clout to push an agenda that favors blacks. A Latino member of Chicago's board of education has charged the five African American members on the board with imposing "apartheid" on Latino children. The board members had voted to send more African American children to integrated schools while disregarding the overcrowding suffered by Latino students. It remains to be seen if urban African Americans and urban Latinos will become allies for their mutual welfare or if they will engage in fierce conflict over the scarce resources available to them.

From a Marxian viewpoint, the ruling class (capitalists) benefits from a continuation of conflict among minority groups. If various minorities fight with each other and attribute their job losses to each other, then the ruling elite can, with little or no resistance, downsize, move jobs offshore, and replace workers with technology. In other words, so long as the conflict is not too extreme, capitalists are the beneficiary of a divided working class.

Prejudice and discrimination cannot be accounted for solely by functionalist and conflict explanations. These theories do not explore the factors involved in learning prejudice and discrimination.

The Symbolic Interactionist Perspective

According to the symbolic interactionist perspective, prejudice and discrimination, just like other aspects of culture, are acquired through socialization. Members of a society learn to be prejudiced in much the same way that they learn to be patriotic (Van Ausdale and Feagin 2001). Gordon Allport (1979) described two stages in the learning of prejudice. In the first stage—*the pregeneralized learning period*—children may have been exposed to prejudice, but they have not yet learned to categorize people. The idea that Asian Americans form a distinct category is beyond their understanding. Later, when children enter the second stage of learning prejudice—*total rejection*—they are able to use physical clues to sort people into groups. At this point, if children hear their parents systematically denigrate a minority, they will reject all members of the group on all counts and in all situations. By this stage, the child has learned the name of the minority he or she is supposed to dislike and can identify individuals who belong to it. A nine-year-old Kosovar Albanian child who wants to "kill Serbs" when he grows up is in the stage of total rejection (Nordland 1999).

TABLE 9.1

FOCUS ON THEORETICAL PERSPECTIVES: Prejudice and Discrimination

This table illustrates how a particular theoretical perspective views a central sociological concept. Switch the concepts around and illustrate how each theoretical perspective would view a different concept. For example, discuss some functions and dysfunctions of differential power or the self-fulfilling prophecy.

Theoretical Perspective	Concept	Example
Functionalism	Ethnocentrism	• White colonists used negative stereotypes as a justification for taking Native American land.
Conflict theory	Differential power	• African Americans accuse Latinos of using their political clout to win advantages for themselves.
Symbolic interactionism	Self-fulfilling prophecy	• Members of a minority fail because of low expectations they have for their own success.

Language itself can provide a context supportive of prejudice and discrimination (James M. Jones 1972). For example, in Anglo culture, although it is good to be "in the black" financially, many references using the term *black* are negative. Such terms as *blackball*, *blacklist*, and *black mark* illustrate the negative connotations associated with the term *black*. During the 1960s, militant African American groups coined the slogan "Black is beautiful" to combat this negative connotation. By comparison, it is difficult to think of similar instances in which *white* is used negatively. Such cultural referents, learned as a part of everyday culture, create a context in which some people may erroneously assume that the well-dressed African American male standing outside a fine restaurant is the doorman rather than a customer. Such responses support the continuation of prejudice and discrimination.

Symbolic interactionists also underscore the labeling process. In 1996, Texaco Inc. agreed to pay more than $115 million as reparation for the economic effects of racism within the company. This settlement was fueled in part by a tape recording made by a high corporate official. On this tape, one executive appeared to impose racial labels (Solomon 1996).

What's more, symbolic interactionism underlies the concept of the **self-fulfilling prophecy**—when an expectation leads to behavior that then causes the expectation to become a reality (Merton 1968). For example, if two nations are convinced they are going to war, they may engage in hostile interaction that actually leads to war. Similarly, if members of a minority are constantly treated as if they are less intelligent than the majority, the minority members may eventually accept this limitation. This, in turn, may lead them to place less emphasis on education. Then, indeed, the minority members appear to themselves and others as less intelligent. Or members of a minority may be socialized to believe that they are not capable of holding important positions in a society and may begin to accept this definition. Given this negative image, and the lack of opportunity to develop their abilities, members of minorities may become locked in lower-level jobs.

Institutionalized Discrimination

Although it is popular to think of the United States as a society in which everyone has an equal chance to achieve their chosen goals, such freedom has always been limited. This is the "American dilemma" of which Gunnar Myrdal wrote with regard to African Americans (Myrdal 1944; Obie 1996). At various times in American history, such practices as slavery and the internment of Japanese Americans during World War II have reflected the open and legal practice of discrimination against members of various minorities (Muller 2007). Virtually all minorities have encountered such practices to a greater or lesser degree (Feagin and Feagin 1986; Luhman 2002; Schaefer 2007).

With the passage during the 1960s of a series of civil rights laws, many Americans felt that racism was going to be eradicated. Although these statutes did stop many discriminatory practices, many people in this country still suffer from **institutionalized discrimination**. This type of discrimination is the result of unfair practices that are part of the structure of society and have grown out of traditionally accepted behaviors. The legacy of 300 years of discrimination is not easily erased from American life and continues to be publicly exposed. For example, blacks in New Orleans (mostly poor) were provided less protection and suffered disproportionately more deaths in the wake of Hurricane Katrina.

How does direct institutionalized discrimination differ from indirect institutionalized discrimination? To help clarify discriminatory practices, Joe and Clairece Feagin (1986) distinguish between direct and indirect institutionalized discrimination. **Direct institutionalized discrimination** refers to organizational or community actions intended to deprive a racial or ethnic minority of its rights. **Indirect institutionalized discrimination** refers to unintentional behavior that negatively affects a minority. Examples of direct

FEEDBACK

1. _____ refers to the unequal treatment of individuals based on their membership in a minority.
2. A _____ is a set of ideas based on distortion, exaggeration, and oversimplification that is applied to all members of a social category.
3. The _____ is characterized by excessive conformity; submissiveness to authority figures; inflexibility; repression of impulses, desires, and ideas; fearfulness; and arrogance toward those thought to be inferior.
4. According to Gordon Allport, children learning prejudice move from a pregeneralized learning period to a stage of total _____.
5. A _____ occurs when an expectation leads to behavior that causes the expectation to become a reality.

Answers: 1. Discrimination; 2. stereotype; 3. authoritarian personality; 4. rejection; 5. self-fulfilling prophecy institutionalized discrimination are laws that segregated African Americans or that denied Mexican American children the right to speak their native language in public schools. High school exit exams provide an example of indirect institutionalized discrimination. When required for graduation, these exit exams are disadvantageous to minority students, who are more likely to drop out because they are less likely than white and Asian American students to pass on their first attempt. Seniority systems (promotion and pay increase with years of service) discriminate against the promotion of newly hired people—many of whom are minority workers. Because of past institutionalized discrimination, members of minorities are just now beginning to enter jobs with seniority systems. Because they have to wait their turn for advancement and because they have fewer years of service than members of the majority, who have been in the system for years, members of minorities often find that their chances for quick promotion are slight, even though the seniority systems may not have been intentionally designed to obstruct minority progress.

Another example of institutionalized discrimination can be found in public education. Schools with large numbers of minority students are more common in large urban areas than in the wealthier suburbs. This is partly because of "white flight" to the suburbs and housing discrimination against minorities. As a result, minority children are more concentrated in school districts with a tax base too low to provide resources equal to those in the suburbs. This lack of funding has many repercussions: teachers in minority schools receive fewer opportunities for training; textbooks, when students have them, are outdated; there is little, if any, money for new technology; and buildings are badly in need of repair. Moreover, parental and community support is generally not as strong in these school districts as in the suburbs.

Institutionalized discrimination in the United States is reflected in the experiences of minorities—African Americans, Latinos, Native Americans, Asian Americans, white ethnics, and Jewish Americans. For each minority, the social and economic costs of discrimination are enormous (Andrews 1996; Aguirre and Turner 1997; Spencer 1997; Pollard 1999; Smelser, Wilson, and Mitchell 2001).

African Americans

African Americans make up one of the two largest minority groups in the United States, numbering just over 40 million, or 13.3 percent of the total population (see Table 9.2). They are also one of the oldest minorities, first brought to America as indentured servants in 1619.

What are the barriers to African American assimilation? Because African Americans are physically identifiable, and because of negative stereotypes based on physical characteristics, it has been very difficult for them to assimilate. A second reason for the continuing minority status of African Americans lies in the historical nature of slavery, which can be traced to the first shipment of slaves to Jamestown, Virginia, in 1619 (Zinn 2005). According to Frank Tannenbaum (1947), the attitudes and practices associated with *manumission*—the transition from slave to free person—are the most important aspects of any slave system. Despite an increase in the practice of manumission following the Revolutionary War, slavery in the United States, particularly in the South, was generally a permanent condition. Unlike

Institutionalized discrimination has contributed to the deterioration of many inner-city public schools, such as the Thomas Alva Edison high school in Philadelphia.

Peter Byron/PhotoEdit

Katherine S. Newman—No Shame in My Game

Katherine Newman has created a rich portrait of minimum-wage workers employed in four fast-food restaurants in central Harlem. These people we label the "working poor"—they hold jobs and pay taxes, but they do not earn enough money to buy the basic necessities of life.

Her research involved 300 New Yorkers. Two hundred worked in four large fast-food restaurants in Harlem. The others could not find jobs but had applied at these four restaurants.

Newman's research team used a variety of research techniques. All respondents answered a survey, and half gave complete life histories in three- to four-hour interviews. After collecting this information, Newman's research team worked in the restaurants for four months. Twelve workers opened their lives to up-close observation. They were observed at home, holiday celebrations, movies, schools, and churches. The twelve also kept personal diaries for a year and agreed to let their friends be interviewed.

In the following passage, Newman argues that the working poor share the same basic values as white middle-class society. Therefore, there is shame in a cultural view that defines employment in fast-food jobs as degrading. The working poor do not prefer such work; it is one of the few employment opportunities open to them.

Swallowing ridicule would be a hardship for almost anyone in this culture, but it is particularly hard on minority youth in the inner city. They have already logged four or five years' worth of interracial and cross-class friction by the time they get behind a Burger Barn [a fast-food restaurant in Harlem] cash register. More likely than not, they have also learned from peers that self-respecting people don't allow themselves to be "dissed" without striking back. Yet this is precisely what they must do if they are going to survive in the workplace.

This is one of the main reasons why these [fast-food] jobs carry such a powerful stigma in American popular culture: they fly in the face of a national attraction to autonomy, independence, and the individual's "right" to respond in kind when dignity is threatened. In ghetto communities, this stigma is even more powerful because—ironically—it is in these enclaves that this mainstream value of independence is most vigorously elaborated and embellished. Film characters, rap stars, and local idols base their claim to notoriety on standing above the crowd, going their own way, being free of the ties that bind ordinary mortals. There are white parallels, to be sure, but this is a powerful genre of icons in the black community, not because it is a disconnected subculture but because it is an intensified version of a perfectly recognizable American middle-class and working-class fixation.

It is therefore noteworthy that thousands upon thousands of minority teens, young adults, and even middle-aged adults line up for jobs that will subject them, at least potentially, to a kind of character assassination. They do so not because they start the job-seeking process with a different set of values, one that can withstand society's contempt for fast food workers. They take these jobs because in so many inner-city communities, there is nothing better in the offing. In general, they have already tried to get better jobs and have failed, landing at the door of Burger Barn as a last resort. . . .

The stigma also stems from the low social status of the people who hold these jobs: minorities, teenagers, immigrants who often speak halting English, those with little education, and (increasingly in affluent communities afflicted with labor shortages) the elderly. To the extent that the prestige of a job refracts the social characteristic of its average incumbents, fast-food jobs are hobbled by the perception that people with better choices would never purposely opt for a "McJob." . . . There is no quicker way to indicate that a person is barely deserving of notice than to point out he or she holds a "chump change" job at Kentucky Fried Chicken or Burger King.

Ghetto youth are particularly sensitive to the status degradation entailed in stigmatized employment. As Elijah Anderson . . . and others have pointed out, a high premium is placed on independence, autonomy, and respect among minority youth in inner-city communities—particularly by young men. No small amount of mayhem is committed every year in the name of injured pride. Hence jobs that routinely demand displays of deference force those who hold them to violate "macho" behavior codes that are central to the definition of teen culture. There are, therefore, considerable social risks involved in seeking a fast food job in the first place, one that the employees and job-seekers are keenly aware of from the very beginning of their search for employment.

It is hard to know the extent to which this stigma discourages young people in places like central Harlem from knocking on the door of a fast food restaurant. It is clear that the other choices aren't much better and that necessity drives thousands, if not millions, of teens and older job-seekers to ignore the stigma or learn to live with it. But no one enters the central Harlem job market without having to face this gauntlet.

Source: Katherine S. Newman, *No Shame in My Game: Working Poor in the Inner City* (New York: Alfred A. Knopf, 1999), pp. 93, 95. Reprinted with permission of Alfred A. Knopf, a division of Random House, Inc.

Evaluate the Research

1. Do you think that Newman is using the functionalist perspective or the conflict perspective? Explain.

2. Analyze this study from the viewpoint of symbolic interactionism.

3. Suppose you were a politician about to speak to a group strongly opposed to government support of poor people. How would you use Newman's research to counter their position?

TABLE 9.2

U.S. Resident Minority Populations

This table shows the percentage increase in the larger minority populations in the United States from 2000 to 2003. Are you surprised by the growth of any group?

Minority group	Number (in millions) 2003	Percent change 2000–2003	Percent of total population 2006
African American (non-Latino)	40,200	3.7	13.3
Latino	44,300	13.0	14.8
Asian (non-Latino)	14,900	12.7	5.0
Native American/Alaska Native (non-Latino)	5,500	4.0	1.5
Native Hawaiian/ Other Pacific Islander (non-Latino)	391	6.4	Less than 1 percent

Source: U.S. Bureau of the Census, *Statistical Abstract of the United States, 2004-2005.* U.S. Bureau of the Census, "Minority Population Tops 100 Million." http://www.census.gov/Press-Release/www/releases/archives/population/_00048.html

African slavery, support for manumission was weak, and there were numerous barriers to it. Freed slaves constantly ran the risk of being returned to slavery. For instance, in Maryland in 1717, any freed slave who married a white person was returned to slavery for life. In Virginia, an emancipated slave could be sold into slavery again if he or she remained in the state after one year. Those slaves who did manage to gain and keep their freedom were barred from voting, holding public office, and pursuing most jobs (C. R. Johnson and Smith 1998; O. Patterson 1998).

Due to the unique nature of the American form of slavery, slaves freed before the Civil War were denied opportunities for upward social mobility. According to Richard Hofstadter:

The Anglo-Americans of the North American mainland quickly became committed to sharp race separation, took a forbidding view of manumission, defined mulattoes simply as Negroes, and made outcasts of free Negroes. Hence there was as little upward mobility from slavery as possible, especially in the Southern colonies and states, and even where masters chose to manumit slaves. (Hofstadter 1973:114)

The denial of opportunity for upward mobility did not end even though the Civil War was over and slavery was legally abolished by the Thirteenth Amendment (1865). By the late 1800s, de jure segregation (the separation of blacks and whites based on law) was insti-

tutionalized throughout the country, especially in the South. Such practices continued until the late 1960s, when they were made illegal by the passage of civil rights legislation and by court decisions. In a very real sense, then, African Americans have been legally free in much of the United States for little more than thirty years. The gap between African Americans and whites

More African Americans are moving into the middle class. Still, their average income lags far behind that of whites.

in education, income, and employment represents the legacy of centuries of prejudice and discrimination (Klinker and Smith 1999; Anderson and Massey 2001; Feagin 2001; Feagin, Vera, and Batur 2001).

What is the economic situation for African Americans today? The median income of African American households ($30,134) is substantially below that of white households ($48,977; see Figure 9.3). The poverty rate for African Americans (24.7 percent) is about triple that of whites (see Table 9.3). Recent experience carries both good and bad news for African Americans. While African American workers have made significant economic gains since 1960, this advancement has not closed the gulf between black and white Americans (B. Anderson 2002). African American income remains at just more than two-thirds that of whites. For every $100 an average white family earns, an average African American family earns $61 (U.S. Bureau of the Census 2007a). Thus, as noted in Using the Sociological Imagination at the beginning of this chapter, the average African American income in the United States is far from equal to the average income for whites.

A recent study of the economic mobility of black and white children provides some additional corroborative details (Isaacs 2007c). A higher percentage of white children exceed their parents' income than black children at a similar point in the income distribution (68 percent versus 31 percent). White children are more likely to move up the economic ladder while black children are more likely to move down. Thirty-seven percent of white children from middle-income homes move up, compared with only 17 percent of black children. Almost half (45 percent) of black children from middle class homes fall to the bottom of the income distribution, while the same is true for only 16 percent of white children. Finally, black children from poor families are more likely to remain poor in adulthood than poor white children (54 percent versus 31 percent).

Not surprisingly, there is also a vast gulf in wealth (home and car equity, net business assets, net liquid assets). The average African American family holds less than one-quarter of the wealth of the average white family.

How do African Americans fare in the job market? According to recent "social audit" studies, African Americans experience employment discrimination despite stronger legal prohibitions. Audit studies involve sending white and minority researchers with comparable résumés to the same firms, applying for the same jobs. Research consistently shows that employers are more likely to interview or hire white applicants. This discrimination now takes more sophisticated and covert forms. For example, discriminatory employers interview or offer jobs to all white applicants before tapping their list of black applicants. Discriminatory firms also offer white applicants higher salaries and higher-status positions (Herring 2002). Such employment practices perpetuate the long-standing minority overrepresentation in low-prestige, low-paying jobs. About one-third of African American men are employed in the highest occupational categories: professional, managerial, technical, and administrative, whereas about one-half of white men have jobs in these categories. Similarly, about 60 percent of African American women are employed in these occupational categories, compared to three-fourths of white women. African Americans are almost

FIGURE 9.3

Majority and Minority Median Household Incomes

Explain why sociologists consider Asian Americans a minority group despite their relatively high annual income.

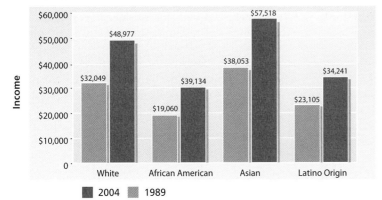

Source: U.S. Bureau of the Census, 2005.

TABLE 9.3

Conforming and Deviant Responses in Merton's Strain Theory

This figure presents some important social and economic characteristics of the majority and larger minorities in the U.S. Can you make sociological generalizations about income level and education based on these data?

	Whites	African Americans	Latinos	Native Americans	Asian Americans
Percent of Families in Poverty	8.6%	24.7%	21.9%	24.0%	9.8%
Median Income	$48,977	$30,134	$33,241	$33,132	$57,518
Percent with High School Diploma	85.1%	80.0%	57.0%	76.0%	87.6%
Percent with College Degree	27.6%	17.3%	4.4%	14.0%	49.8%

Source: U.S. Bureau of the Census, 2005.

twice as likely as whites are to work in low-level service jobs.

New long-term economic trends threaten to make matters worse. These trends include a shift from higher-paying manufacturing jobs to lower-paying service jobs, technological replacement of workers, and transfer of high-wage jobs to low-wage countries. Reduced job opportunities for African Americans and other minorities, especially in segregated neighborhoods, are an important negative consequence.

Patterns of unemployment also affect the economic status of African Americans. Jobless rates among African Americans are more than double those of whites (U.S. Department of Labor 2008). And these rates do not take an account of all unemployed persons. Traditional unemployment rates are based on the number of unemployed people who are looking for jobs. They do not include so-called **hidden unemployment**—discouraged workers who have stopped looking or part-time workers who would prefer full-time jobs. When hidden unemployment is considered, the jobless rate for African Americans approaches one in four workers, the rate during the Great Depression of the 1930s.

It is among African American teenagers that the greatest unemployment problem exists. According to official statistics, about one out of every three African American teenagers is unsuccessfully looking for full-time work (U.S. Department of Labor 2001a). With hidden unemployment taken into account, it is estimated that over 40 percent of all African American teenagers are unemployed. Consequently, thousands of African American youths are becoming adults without the job experience vital to securing good employment (Jacob

1994; Pearlstein and Brown 1994; "World Without Work" 1999).

Are there gender differences among African Americans? Knowledge of the nationwide gap between men and women would lead you to answer yes. And you would be right. The long-standing advantage of black males over black females, however, appears to be eroding somewhat. At every educational level, black men still earn more than black women do. But African American women with college degrees earn substantially more than the median for all African American men. That is not true for white men and women. And African American college-educated women earn as much as white women with college degrees. Given that a college education opens more doors for African American women, the additional good news for them is their increasing educational level. While more than a third of African American women enter college, only a fourth of black men do. Contrast this to 1970, when African American men and women had a roughly equal level of college enrollment (U.S. Bureau of the Census 2001; Cose 2003).

What are the educational opportunities for African Americans? Education is the traditional American path to economic gain and occupational prestige. The educational story for African Americans is mixed. As of 2004, 85 percent of whites had finished high school compared with 80 percent of African Americans. Similarly, whereas almost 28 percent of whites had completed college, only 17 percent of African Americans had done so. (See Table 9.3.) Moreover, higher educational attainment doesn't pay off for African Americans as it does for whites. Although income tends to rise with educational level for both African Americans and whites, it increases much less for African American men (and for women of both races) than for white men. At each level of schooling, African American men tend to gain less than their white peers do. (See Table 9.3.)

What is the African American presence in politics? African Americans have enjoyed some political success since 1970. This progress is due in part to legal changes that occurred in the 1950s and 1960s, including the increase in African American voter registration. Also, African Americans constitute a growing political majority in many large cities. Consequently, more than 5,400 African Americans are serving as city and county officials, up from 715 in 1970. There are more than 9,000 African American elected officials in the United States, a sixfold increase since 1970 (U.S. Bureau of the Census 2007c).

Some see the slow emergence of a "biracial politics"—election of African Americans in predominantly white areas—as a hopeful sign (Kilson 2002). According to Richard Zweigenhaft and William Domhoff, some African Americans, though still vastly underrepresented, have entered the "power elite" of America:

> *Although the power elite is still composed primarily of Christian white men, there are now . . . blacks . . . on the boards of the country's largest corporations; presidential cabinets are far more diverse than was the case forty years ago; and the highest ranks of the military are no longer filled solely by white men.* (Zweigenhaft and Domhoff 2006:176)

There are several prominent examples of blacks recently entering the political elite. Colin Powell and Condoleezza Rice both served as Secretary of State in George W. Bush's administration. Many took Barack Obama's performance in the 2008 presidential election as a sign of at least some decline in the significance of race in American society. Of course, even if Obama were to be elected president this would not signal a postracist America.

It is in the statehouses (only two African American governors have been elected) and at the national level that African Americans have experienced the least political gain. Less than 1 percent of national political offices are held by African Americans. This is true even though there have been more African Americans in the U.S. House of Representatives in the last four Congresses (42 in 2008) than at any other time in American history. This is more than double the number of African American House members in 1981. Still, blacks comprise 13.3 percent of the U.S. population and only 10 percent of the House. In 2008, there was only one African American in the U.S. Senate. Although African Americans are an emerging political force, they must accomplish much before they can claim proportionate political representation.

In fact, a 1995 U.S. Supreme Court decision appears to represent a roadblock to political power for African Americans. The Court ruled that race could no longer be the "predominant factor" in determining congressional districts (Fineman 1995). The Court created, at the same time, a legal test that makes it difficult, if not impossible, to keep in place government programs that give an advantage to minorities and women. In fact, "antipreference" forces are making significant progress toward repealing affirmative action laws. The Universities of California and Texas have abandoned all racially based admissions preferences, and the state of California has repealed its affirmative action laws.

Voting patterns reflect another dimension of political inequality among African Americans. Because politicians respond most favorably and consistently to the desires of those who vote, it is in the interest of African Americans to make their presence known on election day. Yet, the African American voting registration rate is typically several percentage points lower than the white registration rate (61 and 71 percent, respectively; Breed 2008). Adding significance is the fact that African Americans are swamped, numerically, by whites.

African Americans vote at a lower rate than whites for at least two reasons. First, African Americans are disproportionately represented in the working class,

Barack Obama made helping working families a top priority prior to his run for the presidency of the United States.

among the working poor, and in the underclass—the socioeconomic categories least likely to vote. Second, because African Americans have much less confidence in the political system than whites, they are more likely to believe that their votes will count less. This distrust of politics is nurtured by voting experiences like the 2000 presidential election. Intense nationwide scrutiny was given to the undercounting of low-income minority voters in Florida (Parker and Eisler 2001). Unfortunately, Florida is not the exception. Undercounting of low-income and minority voters is a nationwide pattern. Voters in low-income, high-minority congressional districts had greater than triple the likelihood of having their votes discarded than were voters in affluent, low-minority districts (Dooley 2001). In 2008, the U.S. Supreme Court heard a case on a 2005 Indiana law requiring voters to show photo identification before being permitted to vote, a law that critics claim will discourage minority voters (Urbina 2008). Given this situation, it is easy to understand why many African Americans believe that they live in a "still-white" society (Roediger 2003).

What is meant by "two black Americas"? During the 1960s and 1970s, some occupational progress was made. The number of African Americans in professional and technical occupations—doctors, engineers, lawyers, teachers, writers—increased by 128 percent. The number of African American managers or officials is more than twice as high as it was in 1960. Due to the recent upward mobility of educated African Americans, some scholars predict the emergence of two black Americas—a black **underclass** composed of the permanently poor trapped in inner-city ghettos and a growing black middle class (W. J. Wilson 1984; Landry 1988; Kilson 1998; Pattillo-McCoy 2000). In Richard Freeman's (1977) terms, a black elite is said to be emerging in America. David Swinton offers this view of the mixed state of African Americans:

> *The empirical evidence does show that some individual blacks have made impressive economic gains. In fact, there has been an increase in the proportion of blacks who can be classified as upper middle class. This limited upward mobility for the few cannot offset the stagnation and decline experienced by the larger numbers of black Americans whose economic status has deteriorated. The central tendency for the group as a whole is revealed by the trends in the averages. And these trends tell a consistent story of stagnation or decline.* (Swinton 1989:130)

U.S. Census data confirm the existence of two black Americas—the rich and the poor. The number of African American households earning a minimum of $50,000 annually has more than tripled since 1970. About 30 percent of African American households earn between $25,000 and $50,000 each year. As shown earlier, the percentage of poor African Americans, however, continues to dwarf that of whites.

Is the significance of race declining for African Americans? The argument for the declining significance of race is applied to both successful African Americans and the black underclass. Let's examine each in turn.

Although African Americans remain behind whites on all dimensions of stratification, gains made since the 1960s have led some to conclude that race is of declining significance in America (Cose 1995; Thernstrom and Thernstrom 1999). According to some analysts, race is now less important than resources in determining the life chances of young, well-educated African Americans. Well-educated African Americans, it is argued, can now compete on equal ground with whites. African Americans with finances and education are said to be no longer shut out because of their color (W. J. Wilson 1990, 1993a, 1997, 2002).

William Julius Wilson concurs that race is now less important than economic class for successful African Americans. Wilson goes further by saying that race is declining in significance even for the African American urban poor who are part of the underclass—those in poverty who are either continuously unemployed or underemployed due to the absence of job opportunities and/or required job skills. In some of his early work on this topic, Wilson stated it this way:

> *The recent mobility patterns of blacks lend strong support to the view that economic class is clearly more important than race in predetermining job placement and occupational mobility. In the economic realm, then, the black experience has moved historically from economic racial oppression experienced by virtually all blacks to economic subordination for the black underclass.* (W. J. Wilson 1980:152)

In his book *The Truly Disadvantaged*, Wilson (1990) elaborates on the thesis of the declining significance of race in the United States. It is true, Wilson acknowledges, that many African Americans (and other minority group members) are in the underclass as a result of historical patterns of discrimination. This, however, is in the past. Moreover, the argument continues, African Americans remaining in the underclass are there not because of prejudice and discrimination, but because of features of the American economy. Inner-city African Americans are adversely affected by the deindustrialization of the American economy. In the past, African Americans were drawn to Detroit, Chicago, and other large cities by the availability of high wages and stable employment in factories. These manufacturing jobs, which traditionally required little education, have gone to low-income countries where labor is cheap and labor unions are nonexistent. Consequently, today's inner-city poor are denied the type of jobs used for upward mobility. A second contributing economic factor is the internal movement of business and industry from central cities to the suburbs. A third economic factor contributing to the perpetuation of the underclass is the movement of upwardly mobile African Americans to the suburbs. Lacking the positive role models available in the past, the inner-city poor now see drug dealing, prostitution, and other illegal economic activities as the primary avenues to success (Vobejda 1991; O. Clayton 1996; Massey and Denton 1996; Small and Newman 2001).

If economic factors are perpetuating the underclass, Wilson argues, solutions are to be found in the economic sphere. Wilson calls for a federal economic policy designed to create a higher employment rate and better jobs for all Americans. Wilson looks to such things as higher-paying jobs, improved education, job training, relocation support to encourage members of the underclass to secure and keep decent jobs, publicly supported day care for the low-income employed, a markedly higher minimum wage, and medical insurance for the employed in low-paying jobs (W. J. Wilson 1993b, 1997).

Use of the term *underclass* has been questioned (M. B. Katz 1993). Wilson, himself, suggests substituting the term *ghetto poor* because the concept of underclass seems to be a code word for inner-city African Americans and is often used by journalists to highlight unflattering behavior in the ghetto (W. J. Wilson 1991). Herbert Gans (1990) sees the term *underclass* as a pejorative, value-laden term now being used to describe the "undeserving" poor. Joining a growing number of social scientists, Gans believes that the term *underclass* is hopelessly mired in ideological connotations of undeservingness and blameworthiness. These social scientists, including Wilson, do not wish to obscure the harsh reality faced by the urban poor by continuing to use the term *underclass* if such use contributes to victim blaming. It remains to be seen whether the term *underclass* will lose its currency in sociological terminology. At any rate, some term, perhaps Wilson's *ghetto poor*, will be utilized to capture the problems associated with the inner-city poor.

There is some evidence that young African American college graduates are reaching an economic par with their white peers and that many African Americans in general have made significant gains in the past twenty years (R. Farley 1984; Payne 1998). It is not certain, however, that this situation is a permanent one. The strides since the 1960s may simply be the result of tremendous changes in the past two decades that will not be maintained in the future.

Many critics disagree with the idea that race is a declining force for African Americans, whether or not progress has been made (Thomas and Hughes 1986; Carnoy 1994; Hacker 1995; Cancio, Evans, and Maume 1996; Hughes and Thomas 1998; Pulera 2002). Doris Wilkinson (1995) believes that discrimination still

determines the life chances of African Americans. The current racial dominance, she argues, is simply more sophisticated and subtle. Others point to inferior education in America's inner cities, a situation tied in part to racial discrimination (see Chapter 13). Still other critics point to housing discrimination and segregation as an enduring barrier to large-scale African American migration to the suburbs, where many jobs are located (Massey and Denton 1987; Massey 1990; Massey and Eggers 1990; Pattillo-McCoy 2000). According to the 2000 Census, U.S. inner cities are more segregated today than they were before segregation was outlawed in the 1960s (Ware and Allen 2002). Others point to the high rate of incarceration among black males, who are eight times more likely to enter prison than white males (Western and Pettit 2002). Finally, some argue that the concept of a black underclass applies to rural areas. According to this viewpoint, the black underclass is more densely concentrated in rural parts of the South than in the inner cities of the North (O'Hare 1992a, 1992b; O'Hare and Curry-White 1992).

Latinos

How diverse is the Latino population? Latinos in the United States number more than 44 million (14.8 percent of the population). They are composed of many diverse ethnic minorities, including Mexican Americans, Puerto Ricans, Cubans, and increasing numbers of people from Central and South American countries (see Figure 9.4). High birth rates and immigration rates combine to make Latinos one of the most rapidly growing minorities in the United States. In fact, early in this century, Latinos will become America's largest ethnic minority. By 2050, the Latino population is projected to reach nearly 100 million, constituting 24 percent of the U.S. population (del Pinal and Singer 1997; Pollard 1999; Suarez-Orozco and Paez 2002; U.S. Bureau of the Census 2006).

Just less than two-thirds of Latinos today are of Mexican descent; Puerto Ricans make up a little less than one-tenth the total Latino population. Most Puerto Ricans are concentrated in or near New York City, although greater geographical dispersion is occurring. Cubans make up the third most populous category of Latinos, with about 1 million people, most of whom are located in the Miami area (*2002 National Survey of Latinos* 2002).

Latino peoples are diverse. Each group came to the United States under different circumstances and retains a sense of its own identity and separateness. In addition, there are significant internal differences within individual Latino minorities (Delgado and Stefancic 1998; Suarez-Orozco and Paez 2002). For example, the first large group of Cuban immigrants to enter the United States included successful middle- and upper-class people who fled from

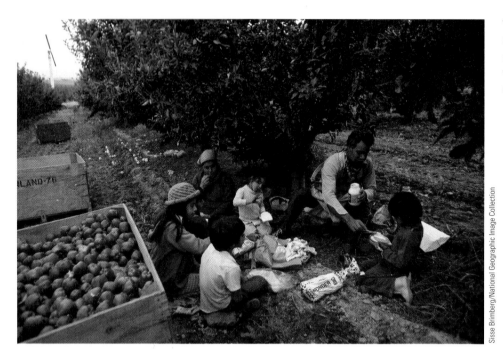

The meager wages earned by migrant farm laborers keep many children out of school where they could receive an education.

Sisse Brimberg/National Geographic Image Collection

Cuba when Fidel Castro instituted a communist government there in the late 1950s. These Cuban Americans were substantially more educated than later Cuban immigrants were (Darder and Torres 1997; Stavans 2001).

What is the socioeconomic condition of Latinos? Latinos fall behind white Americans in formal education. The median educational level for the total American population is about 12.7 years, or high school graduation, but for all Latinos, it is 11.7 years. Although in 2005, 57 percent of Latinos age twenty-five or over had completed high school, 85 percent of non-Latino whites had done so (see Table 9.3). Mexican Americans have the lowest levels of educational attainment, Cubans the highest. This is due to the migration pattern and the history of these peoples. Mexican immigrants came to the United States with little education; they were unable to speak English and were willing to accept the lowest-level jobs. Cubans, on the other hand, tended to be middle- and upper-middle-class immigrants who had been educated and economically successful in an economy similar to that of the United States (Valdivieso and Davis 1988; Delgado and Stefancic 1998).

As shown in Figure 9.3, the average income for Latinos ($34,241) is now higher than that of African Americans, but significantly lower than that of non-Latino whites ($48,977). Cubans are the most affluent Latinos, but even their median income is only about 75 percent that of whites. The poorest among the large Latino groups are the Puerto Ricans, whose income is about 51 percent that of whites. Approximately 22 percent of Latino families were living below the poverty level in 2004, compared with 8.6 percent of non-Latino whites (see Table 9.3). Puerto Rican families have the highest poverty rate, Cubans the lowest.

As one might expect from these income figures, Latinos tend to be concentrated in lower-paying and lower-status jobs as semiskilled workers and unskilled laborers. Farmwork is still common, especially among Mexican Americans. Cuban men belong to the only Latino minority that approaches the occupational profile of Anglos (Moore and Pachon 1985; Trueba 1999).

The increase in Latino-owned businesses must be noted. In 1969, there were only 100,000 Latino-owned

FIGURE 9.4

Composition of the Latino Population in the United States, 2006

The Latino population will clearly become the largest minority in the United States during this century. This figure displays the diversity within that population itself. Do you see any reason why the composition of the Latino population will change over the next fifty years?

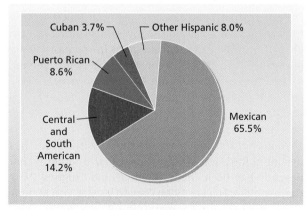

Cuban 3.7% — Other Hispanic 8.0%

Puerto Rican 8.6%

Central and South American 14.2%

Mexican 65.5%

Source: U.S. Census Bureau, *Current Population Survey*; Annual Social and Economic Supplement 2006. http://www.census.gov/population/socdemo/hispanic/cps2006/CPS_Powerpoint_2006.pdf

businesses, with less than $5 billion in revenue. By 2000, there were about 800,000 Latino-owned firms with revenues totaling nearly $50 billion.

No Latino minority is very close to the national average of 67 percent home ownership. Even the relatively well-off Cubans show a figure of only 52 percent. Moreover, Latino economic progress was curtailed by the recession beginning in 2001 (Lowell 2002).

How do Latinos stand politically? Politically, Latinos are becoming a more nationally visible force shaping American politics (Rodriguez 1999; de La Isla 2003). As of 2008, Latinos held twenty-two seats in the U.S. House of Representatives—a tripling from 1981. Of these members of Congress, the majority were Mexican Americans. In 2002, Cuban American Robert Menendez became the Democrat caucus chair, the highest post ever held in Congress by a Latino. On the state and local level, there were almost 6,000 elected Latino public officials. Latinos, of course, still have a long way to go to achieve political parity. In 2008, Latinos represented only 5 percent of the U.S. House of Representatives (twenty-two members, as just discussed) and had only three members in the U.S. Senate, despite constituting almost 15 percent of the population. Issues of education and immigration, as well as income and the quality of life, promise to keep Latinos politically active (Vazque and Torres 2002).

Native Americans

What is the composition of the Native American population? If a single word can be used to describe Native Americans, it is *diversity*: they are divided into 500 (or so) tribes and bands. This diversity is generally unrecognized because of stereotypes perpetuated by old Hollywood films and cheap paperback adventures of the old West. Most Americans don't know that the Navajo and Sioux are entirely different nations with different cultures, social structures, and problems. Tribal groups are as different from one another as they are from the dominant culture, or as different as Anglo Americans are from Italians or Brazilians. Today, Native Americans number around 4.5 million, three-fourths of whom do not live on reservations (Lobo and Peters 2001; U.S. Bureau of the Census 2007a).

What has been the U.S. policy toward Native Americans? Ambivalence in government policy toward Native Americans was present from the founding. George Washington attempted to establish government policies promoting a just accommodation with the Native Americans. Only a few years later, Thomas Jefferson encouraged state governors and military commanders to sign supply contracts designed to impoverish the eastern tribes, thereby accelerating their elimination or removal to the western land newly acquired by the

Louisiana Purchase (Ellis 2007). Over the years, the federal government has vacillated between a policy of paternalism (i.e., domination and care) and a policy of neglect. In the past, impoverishment, oppression, and deceit have typically, although not always, characterized relations between the government of the United States and various tribal governments (Marger 2008). Jonathan Turner sums up both past and present government policy toward Native Americans:

> The legacy of economic exploitation, especially the great "land grabs" by whites in this century, has forced Indians into urban areas because they can no longer support themselves on their depleted reservations. Yet the legacy of isolation on the reservation prevents many Indians from making the cultural and psychological transition to urban, industrial life. And the burden of change has been placed on the individual Indian, for white institutions—from factories to labor unions and welfare agencies—display little flexibility in adjusting to Indian patterns. The contemporary Indian is therefore faced with impoverishment no matter what course of action he chooses: to stay on the reservations results in poverty, but to leave the reservation and encounter the hostile white economic system in urban areas also results in poverty. (Turner 1976:185)

Both the paternalistic and neglectful approaches have left Native Americans outside the mainstream of social and economic opportunity (K. M. Dudley 1997; T. R. Johnson 2000).

What is the current situation of Native Americans? This neglectful approach is evident in the fact that the U.S. Census Bureau does not regularly report data for Native Americans. Other than a special study performed in the early 1990s, little data are available to describe their current status. Thus, although many of the descriptions that follow are dated, they are the most current available.

Abject poverty remains a major fact of life among Native Americans, especially for many who remain on reservations. Approximately 24 percent of the Native American population live below the poverty line (see Table 9.3). The median income is $33,132 per year, and about 15 percent have an income greater than $50,000 per year, as compared with some 30 percent of the white population.

A gap in education also exists. Of Native Americans twenty-five years of age or older, 76 percent were high school graduates, compared with 85 percent of the non-Latino white U.S. population. Only 14 percent had completed four years or more of college, compared to almost 27 percent of the white population.

Native Americans have made only scant penetration into the upper levels of the occupational structure. Although some gains were made during the 1960s and 1970s, only 20 percent of all employed Native American men and women currently hold professional, managerial, or administrative positions. One-third are in blue-collar jobs (craftsworkers, supervisors, operatives, and nonfarm laborers). Currently, there is no Native American member of the U.S. Senate and only two in the House of Representatives.

Are conditions better or worse on the reservations? For the approximately one-fourth of Native Americans living on reservations, the situation is considerably worse than for those living off the reservations. One-half of those on reservations live below the poverty line (compared with 24 percent for the total Native American population). The rate of college education for Native Americans living on reservations is only about half that of those living off reservations.

A relatively recent development on reservations is the introduction of casino-type gaming establishments. Given the traditionally poor social and economic conditions on reservations, it is not surprising that many Native Americans embrace the gaming industry as a source of income. Because only a few of the tribes received the bulk of this money, its long-term effects remain to be seen.

Is there any momentum for change? Recently, the relationship between the U.S. government and various tribal governments has been changing. According to Vine Deloria and Clifford Lytle, government policy now grants tribes special political status. As a result, many tribes have proclaimed their nationhood, demanding some form of representation in the United Nations (Deloria and Lytle 1984; Wilkins 2001).

Deloria and Lytle cite several other possible directions for change. There is, first of all, the need for reform in tribal governments themselves to allow continuity with the past as well as movement into the future. There is also a need to establish a place for Native American cultures in the modern world. Another need is the creation of relationships with federal, state, and local governments based on mutual respect and parity. Change will not come easily, given the tradition of inequality and discrimination that has characterized Native American life.

Asian Americans

Is the Asian American population increasing, decreasing, or stabilizing? Nearly 15 million Asians live in the United States, comprising 3.5 percent of the total population. Asian Americans, the fastest-growing minority in the United States, increased in population by 49 percent between 1990 and 2000. In 1984 alone, more Asian immigrants entered the United States than during the entire thirty years between 1930 and 1960. In 1996, approximately 220,000 Asians were permitted to immigrate to the United States. The Asian American population is predicted to double by 2025 and more than triple by 2050.

Like Latinos, Asians come from many different national and ethnic backgrounds. The largest groups are from China, the Philippines, Japan, India, Korea, and Vietnam. If a success story can be told for any minority groups in America, those groups are the Chinese and Japanese Americans, particularly the latter (S. M. Lee 1998). Even for Chinese and Japanese Americans, however, the road has not been smooth (Kitano and Daniels 2005: Espiritu 1996; Gatewood and Zhou 2000; Wu and Song 2000; Kibria 2003).

How have Chinese Americans fared over the years? Heavy Chinese immigration to California in 1848 caused a furor. Americans feared that the presence of another race would encourage the institution of slavery, cause internal discord, and introduce incurable diseases. Although Chinese labor was accepted, even exploited, in the mining and railroad industries in the West, there was fear that the Chinese immigrants would migrate across the country. Added to this fear was the belief that the Chinese would deprive white men of work in the declining mining and railroad industries. These fears, combined with long-standing prejudice—that existed even prior to their immigration—resulted in an anti-Chinese movement that lasted into the first part of the twentieth century. During this period of more than fifty years, there was much violence against the Chinese, including riots, lynchings, and massacres. In the late 1880s, when a group of Chinese in Tonapah, Nevada, asked for time to collect their belongings before being expelled from the town, they were robbed and beaten; one of them died as a result (Lyman 1974).

Finally, the "Chinese question" became a political matter of high priority. Many state laws were created to restrict Chinese from holding jobs that could deprive whites of employment. The Chinese Exclusion Act passed by the U.S. Congress in 1882 prohibited, for a ten-year period, the entry of any skilled or unskilled Chinese laborers or miners. Strict federal legislation against Chinese immigration continued to be passed until after 1940. During this period, Chinese Americans were driven into large urban ghettos known as Chinatowns, where they are still concentrated today.

Although Chinese Americans in many ways remain isolated from American life, their situation began to improve after 1940 (Loo 1998). American-born Chinese college graduates began to enter professional occupations, and Chinese American scholars and scientists began to make publicly recognized contributions to science and the arts. Their dedication to hard work and education and their contributions to American society have been widely recognized.

What has been the history of Japanese Americans? The earliest relations between Americans and Japanese were positive (unlike the Chinese experience). Early diplomatic relations were warm and cordial. Beginning in 1885, large numbers of Japanese men immigrated to the West Coast of the United States, but the timing for this massive immigration was wrong. The entry of the Japanese came on the heels of America's attempt to exclude Chinese immigrants. Although the Japanese suffered prejudice and discrimination during these early years, they moved from being laborers in certain industries (railroad, canning, logging, mining, meat packing) to being successful farmers (Kitano 2005).

When the Japanese began to compete with white farmers, however, anti-Japanese legislation was passed. The California Alien Land Bill of 1913, for example, permitted Japanese to lease farmland for a maximum of three years; it did not allow land they owned to be inherited by their families. In 1924, the U.S. Congress halted all Japanese immigration, and the 126,000 Japanese already in the United States became targets for prejudice, discrimination, stereotyping, and scapegoating. In 1942, Japan's attack on Pearl Harbor brought the United States into World War II. Wartime hysteria generated fear of a possible Japanese invasion. This led President Roosevelt to issue Executive Order 9066, which sent more than 110,000 of the 126,000 Japanese in America—two-thirds of whom were American citizens—into internment camps (euphemistically referred to as "relocation centers") away from the West Coast (Nagata 1993; Robinson 2001). It was argued that Japanese Americans posed a security threat during World War II. Even though the same argument could have been made about German and Italian immigrants—their countries were also at war with the United States—they were not relocated (U.S. Commission on Wartime Relocation and Internment of Civilians 1983). Eventually, in 1987, the Supreme Court ruled that the internment of Japanese Americans was "based upon racism rather than military necessity."

Despite their internment, Japanese Americans have not had to overcome the centuries of prejudice and discrimination endured by African Americans and Native Americans. Nevertheless, they have overcome great hardship and have become one of the most successful racial minorities in the United States. Through an emphasis on education and hard work, Japanese Americans are an amazing story of economic and occupational success (Montero 1981; Zweigenhaft and Domhoff 2006).

Why have Asian Americans been so successful? Asian Americans have been particularly successful at using the educational system for upward mobility (S. M. Lee 1998; Marger 2008). This fact is reflected in the academic achievement of school-age Asian Americans, whose average Scholastic Aptitude Test (SAT) scores are 45 points higher than that of the general high school population. Furthermore, about 50 percent of Asian Americans have completed four years of college, compared with 27 percent of whites (see Table 9.3).

Some claim that Asian Americans are academically successful because they are innately more intelligent than other Americans. Most scholars believe that Asian academic excellence is due to culture, socialization, and influence of the family (Gibson and Ogbu 1991; Caplan, Choy, and Whitmore 1992). Most Asians see education as the key to success. Moreover, in the Confucian ethic (which is so much a part of the Chinese, Japanese, Vietnamese, and Korean cultures), academic excellence is the only way of repaying the debt owed to parents. According to school principal Norman Silber, "Our Asian kids have terrific motivation. They feel it is a disgrace to themselves and their families if they don't succeed" (McGrath 1983:52). In addition, Asian immigrants are accustomed to a tougher academic regimen. For example, the average number of school days in America is 180, but it is 225 days in Japan. Whatever the reasons, Asian Americans have been successful in part because of their willingness to work hard inside and outside of school (Barrett 1990).

The struggle, of course, is not over. Although Asian Americans have been more successful than other American racial minorities, they still feel the effects of prejudice and discrimination. In fact, the relative success of Asian Americans has hurt them in some ways. The stereotype of success has led the public to conclude erroneously that Asian Americans are all well educated, are overrepresented in higher-level occupations, and are making as much money as white Americans—maybe even more. The facts speak otherwise. Vast socioeconomic differences exist among groups within Asian American communities. Popular emphasis on success has led to a disregard of those Asian Americans who have not done very well (Buckley 1991). Moreover, the accent on success camouflages the fact that even "successful" Asian Americans are worse off than are similarly educated and employed white Americans. Due to the stereotype of success, many experts believe that Asian Americans have been neglected and ignored by government agencies, educational institutions, private corporations, and other sectors of society (U.S. Commission on Civil Rights 1980).

In 1996, Gary Locke became the first Asian American governor on the U.S. mainland (Puente 1996). In 2002, there was a combined total of five Asian Americans holding seats in the House of Representatives and two in the Senate, including the first ever Korean American to serve in Congress.

White Ethnics

White ethnics are descendants from what parts of the world? White ethnics are the descendants of immigrants from eastern and southern European nations, particularly Italy and Poland, but they also include Greek, Irish, and Slavic peoples (Rubin 1994). Most of them are blue-collar workers living in small communities surrounding large cities in the eastern half of the United States (Palen 2007). During the 1960s, white ethnics gained the undeserved reputation of being conservative, racist, pro-war, and "hardhats."

What are white ethnics really like? This portrait of white ethnics is not an accurate one. In fact, the evidence is just the contrary. Surveys conducted during the 1960s showed white ethnics to be more against the Vietnam War than white Anglo-Saxon Protestants were. Catholic blue-collar workers were found to be more liberal than either Protestant blue-collar workers or the country as a whole: they were more likely to favor a guaranteed annual wage, more likely to vote for an African American presidential candidate, and more concerned about the environment. Finally, white ethnics, when compared with WASPs, were more likely to be sympathetic to government help for the poor and were more in favor of integration (Greeley 1974).

According to David Featherman (1971), white ethnics have not been long-term victims of occupational and income discrimination, although it existed from the start for certain minorities such as Italian Americans and Irish Americans. In fact, Andrew Greeley (1976) contended that white ethnics have ultimately been so successful that it is inaccurate to label them working class, as some have done. White ethnic success has not generally been recognized, according to Greeley, because America's elite are not willing to abandon the myth of the blue-collar ethnic.

Despite their relative success, many white ethnics have in recent years become notably conscious of their cultural and national origins (R. T. Schaeffer 2007) and have formed a white ethnic "roots" movement. The new trend toward white ethnic identity was influenced by the black power movement of the 1960s, when many African Americans expressed a desire to preserve their cultural and racial identities. Many white ethnics believe that "white ethnicity is beautiful" and that the price of abandoning one's cultural and national roots is simply too high.

Lillian Rubin (1994) links the continuing accent on white ethnicity to the economic decline of white ethnics over the past twenty years and the rising demands of minorities. White ethnics, she contends, are attempting to establish a public identity that enables them to take a seat at the multicultural table.

On the other hand, some sociologists contend that white ethnicity is fading. According to Richard Alba (1985, 1990), the remaining relatively small number of white ethnics may not survive long. Ethnic identity, he argues, cannot be maintained in the face of disappearing ethnic families, neighborhoods, and communities. Mary Waters (1990) agrees with Alba's description and sees this as the "twilight" of white ethnics. In researching white ethnics, she found ethnicity not very important to those she studied. Although her respondents felt that being Italian, Polish, or Irish might make one distinctive in some way, ethnicity had little effect on where they lived, whom they married, or what they did for a living.

Jewish Americans

The United States and Israel are virtually tied for the largest Jewish population at 5.3 million each. Between them, they hold more than 80 percent of the world's Jews (*American Jewish Year Book* 2006). The majority of Jewish Americans are concentrated in several northeastern states, including New York, New Jersey, Massachusetts, and Connecticut, although a sizable number live in California, Florida, and Pennsylvania. The most recent immigrants—primarily well educated and highly skilled—come from Israel and the former Soviet Union. It is estimated that about 10 percent of all Jewish Americans have been in the United States less than ten years.

The first Jewish immigrants landed in New Amsterdam in 1654. All of the colonies discouraged them from immigrating. The colonies, in fact, legally prohibited them from holding political office or voting. Still, by the 1840s, Jews began to arrive in large numbers. Anti-Semitism reached its peak in the 1920s and 1930s, particularly among America's upper and upper middle classes. New York was often called "Jew York," and Jewish Americans were subjected to occupational and social segregation (F. J. Davis 1978; Parrillo 2005b).

How do Jewish Americans demonstrate success, and how do they respond socially? Throughout the first half of the twentieth century, Jewish Americans were almost totally excluded from top positions in most major industries, denied membership in social and recreational organizations, and subjected to quotas in colleges and universities. Although limited in their opportunities for movement into the top circles of the most economically powerful American corporate entities, Jewish Americans are one of the most successful ethnic groups ever to migrate to the United States. This is partly because Jewish men, particularly later immigrants, entered the country as skilled workers in a proportion far above the norm for immigrant groups. In addition to bringing skills necessary to fit into industrial society, they valued hard work, sacrifice, perseverance, family responsibility, and education. Although working at skilled blue-collar jobs, Jewish parents urged their children to pursue higher education, to become teachers, lawyers, and doctors. Consequently, Jewish Americans represent an above-average proportion of college graduates; they comprise 2 percent of the population, but account for 5 percent of college graduates. In 2008, there were eleven Jews in the U.S. Senate, twenty-six Jews in the U.S. House of Representatives, and two Jews on the U.S. Supreme Court.

Because of the exclusion they experienced, Jewish Americans in the past have tended to remain socially isolated from the rest of American society. This is undergoing some change. Whereas 10 percent of Jewish Americans who married did so outside the faith in 1965, 52 percent now do so. And in 2000 Joe Lieberman became the first Jewish American to run, as a major political party candidate, for the vice presidency of the United States. Although the social isolation of Jewish Americans has diminished somewhat, some separation will probably continue for two basic reasons. First, anti-Semitic sentiment appears to be increasing again. And second, many Jewish Americans do not want to abandon their cultural roots.

Beyond Direct Institutionalized Discrimination

The socioeconomic gap between many minorities in the United States and the general white population is largely due to past and present prejudice and discrimination. But the situation is now even more complex. Minorities are concentrated at the bottom of the stratification structure partly due to the nature of the economy, a factor that operates beyond the influence of direct, or consciously intended, institutional discrimination. This factor has been explored through the context of the **dual labor market**—the existence of a split between core and peripheral segments of the economy and the division of the labor force into preferred and marginalized workers (Hodson and Sullivan 2008). This perspective, which attributes persisting racial inequality to economic factors, originated in an effort to understand the continuing survival of large-scale racial (and gender) inequalities in the face of programs designed to overcome such disparities (Hodson and Kaufman 1982).

What is the dual labor market theory? Traditionally, it has been assumed that only one labor market existed in which all workers competed for jobs. It was also assumed that occupational advancement was based on hard work, education, and training. These assumptions have been successfully challenged by advocates of conflict theory. The existence of a dual labor market means that the rewards for hard work, education, and training vary in different segments of the labor market. Workers involved in the core sector of the labor market—durable manufacturing and petroleum industries, for example—enjoy high wages, good opportunities for advancement, and job security. Those involved in the peripheral sector, including such industries as textile manufacturing and retail trades, are employed in low-paying jobs with little hope for advancement. African Americans and other minority peoples tend to be trapped in these secondary labor markets and lack the resources to alter their situation. Thus, it is concluded, historical practices of racial and ethnic

---

FEEDBACK

1. _____ discrimination refers to unintentional organizational or community actions that negatively affect a racial or ethnic minority.
2. The attitudes and practices associated with _____ in the American slave system help account for the long-term prejudice and discrimination experienced by African Americans.
3. The evidence clearly shows that race is declining in importance in America. T or F?
4. Which of the following Latino ethnic minorities is in the best socioeconomic condition?
   a. Puerto Ricans
   b. Cubans
   c. Mexican Americans
5. The federal government's policy toward Native Americans has vacillated between _____ and almost total neglect.
6. Contrary to popular opinion, white ethnics in the United States are politically more liberal than Protestant blue-collar workers. T or F?
7. Anti-Semitism in America during the 1920s and 1930s was most prevalent among the _____ and upper middle classes.
8. Within the dual labor market, minorities tend to be disproportionately trapped in the _____ sector.

Answers: 1. Indirect institutionalized; 2. manumission; 3. F; 4. b; 5. paternalism; 6. T; 7. upper; 8. secondary or peripheral discrimination now interact with contemporary economic processes to lock a significant and growing number of minority peoples out of the core economy, thereby reducing the likelihood of any improvement in their life chances and lifestyles (Bonacich 1976; Szymanski 1976; Cummings 1980; Lord and Falk 1982).

How does indirect institutionalized discrimination operate in the dual labor market? Minority groups, who are overrepresented in the peripheral sector of the American economy and consequently tend to hold low-skill, low-paying jobs, are subjected to indirect institutionalized discrimination. A long-standing division of labor automatically reduces minority access to jobs in the core sector despite their education and training. In addition, minority members are disproportionately shut out of the more desirable jobs and out of the education and training needed to move into core sector jobs on the grounds that they actually prefer, and are better suited to, the marginal jobs they hold (Hodson and Sullivan 2008).

Increasing Racial and Ethnic Diversity in the United States

Foreign-born populations are rising in both industrialized and developing countries. Many countries in western Europe have seen an increase in flux of migrants from both former colonies in Asia and African (to the Netherlands) and eastern Europe (to Ireland). Several developing countries such as Costa Rica are a magnet for both refugees fleeing political upheaval in neighboring countries and those looking for work. Botswana is drawing political refugees and economic migrants from southern African neighbors (*World Population Data Sheet* 2007).

The United States and Canada have long attracted immigrants. It probably is not surprising that the Latino surnames Garcia and Rodriquez are now among the top ten surnames in the United States (Roberts 2007). In fact, the United States is moving from a predominately white population of Western origin and culture to a society comprising larger numbers of racial and ethnic minorities. This growth and increasing diversity of the minority population is one of the most important developments in the United States.

Minorities' Share of the Total Population

In 2006, the combined number of African Americans, Latinos, Asian Americans, and Native Americans exceeded 100 million, up from just under 10 million in 1900 and 21 million in 1960. If these minority groups

resided in a separate country, they would be the twelfth most populous nation in the world. Great Britain, France, Italy, and Spain would all be smaller. There are now more minority members than there were people in the United States in 1910 (Martin and Midgley 2006; U.S. Bureau of the Census 2007b).

Figure 9.5 shows the share of the American population that minorities represented in 2000 and that they are expected to represent in 2050. For most of the twentieth century, America's minority population constituted a relatively constant percentage of the total population—the share of the population represented by minorities increased from 13.1 to only 14.9 percent between 1900 and 1960. Since then, however, the minority proportion of the population has increased to about 30 percent and, as noted earlier, is expected to constitute one-half of all Americans by 2050 (see Figure 9.5).

Relative Growth of Minority Groups

Whereas the total population of the United States is expected to increase by 23 percent between 2000 and 2025, the minority population is projected to grow 64 percent. The predominant minority in the United States for most of the twentieth century was African American. Except for Native Americans, a rather dramatic change in the relative growth rate of minorities is expected between 2000 and 2050. Although the African American popula-

---

FIGURE 9.5

The U.S. Population by Race and Ethnicity, 2000 and 2050

The racial and ethnic composition is expected to look very different by 2050. Discuss some social consequences of this changing composition.

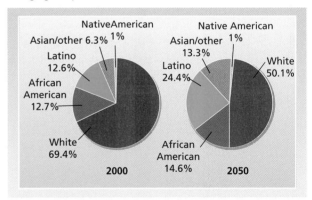

Note: Percentages do not total 100% due to rounding. The White, African American, and Asian/other categories exclude Latinos, who may be of any race.
Source: U.S. Bureau of the Census 2004; Philip Martin and Elizabeth Midgley, "Immigration: Shaping and Reshaping America," *Population Bulletin* 61 (December 2006): 16–17.

WORLDVIEW 9.1

Ethnic Diversity The degree of worldwide ethnic diversity is displayed in this map. The color code indicates the proportion of a national population represented by members of an ethnic, racial, or national minority.

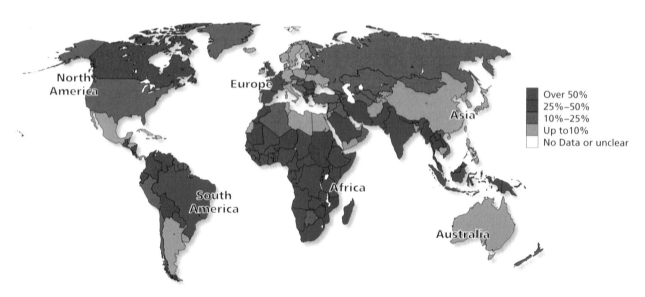

Over 50%
25%–50%
10%–25%
Up to10%
No Data or unclear

Source: Barry Turner (ed.), *The Statesman Yearbook* (New York: St. Martin's Press, 1996–1997).

Interpret the Map

1. Were you surprised by the comparative degree of ethnic diversity between any particular countries? Describe some differences between your expectations regarding national ethnic diversity and the actual degree of diversity.

2. Write a paragraph detailing your conclusions regarding the extent of ethnic diversity across the globe.

3. Do you believe that the extent of ethnic diversity will increase or decrease during the twenty-first century? Explain your position.

tion is expected to expand by about 60 percent, Latinos are projected to increase by almost 200 percent, Native Americans are expected to grow by almost 58 percent, and Asian Americans will have mushroomed by over 240 percent (Riche 2001; see Figure 9.5).

These changes in the relative size and diversity of the minority population promise to alter the nature of American society. Social, political, and economic reconsideration of racial and ethnic minorities seems inevitable.

FEEDBACK

1. About _____ percent of all Americans will be members of minority groups in 2050.
2. Rank order (1–4) the following minority groups to indicate their relative proportion of the total American minority population projected for 2050.
   ____ a. Asian Americans
   ____ b. Native Americans
   ____ c. Latinos
   ____ d. African Americans

Answers: 1. 50; 2. a. (3), b. (4), c. (1), d. (2)

SUMMARY

1. A minority consists of more than small numbers. A minority possesses some distinctive physical or cultural characteristics, is dominated by the majority, and is denied equal treatment. Minorities also have a sense of collective identity, tend to marry within their own kind, and inherit their minority status at birth.

2. A race is a category of people who share certain biologically inherited physical features. Racists use these physical characteristics as an index of a race's superiority or inferiority. Despite the lack of scientific support for this viewpoint, prejudice and discrimination are often justified by alleged differences in intelligence and ability.

3. Ethnic minorities have distinct subcultures. Unique cultural characteristics of ethnic minorities are used by the majority as a justification for prejudice and discrimination.

4. Patterns of racial and ethnic relations assume two general forms: assimilation and conflict. Patterns of assimilation include Anglo-conformity, melting pot, and cultural pluralism. Genocide, population transfer, and subjugation are the major patterns of conflict.

5. Prejudice—negative attitudes toward some minority—is difficult to change because prejudiced individuals reject information that contradicts their existing attitudes. Whereas *prejudice* refers to attitudes, *discrimination* refers to behavior. When members of a minority are denied equal treatment, they are being discriminated against. Although prejudice usually leads to discrimination, there are exceptions. In fact, in some instances, discrimination against a minority may lead to the creation of prejudiced attitudes, usually embodied in racial or ethnic stereotypes.

6. Many attempts have been made to explain the existence of prejudice and discrimination. Some psychologists attribute prejudice and discrimination to deep-seated frustration. Hostility and frustration are said to be released on scapegoats who had nothing to do with causing their development in the first place. Other psychologists point to the existence of a personality type—the authoritarian personality—that tends to be very prejudiced against those thought to be inferior. Although functionalism emphasizes the negative consequences of prejudice and discrimination, it also points to some benefits. According to conflict theory, prejudice and discrimination are used by the majority as weapons of power to dominate subordinate minorities. Symbolic interactionism concentrates on the learning of prejudice and discrimination through socialization.

7. Virtually all minorities have felt the consequences of institutionalized discrimination. Institutionalized discrimination may be direct, as in segregation laws; or it may be indirect, as in the form of seniority systems that prevent upward mobility of newly hired minorities.

8. Institutionalized discrimination has been harmful to America's minorities. As a result, African Americans, Native Americans, Latinos, some Asian Americans, and white ethnics lag behind the white majority occupationally, economically, and educationally. Successful cases do exist, but the gains of all minorities remain fragile.

9. Some scholars contend that the disadvantages that minorities are currently experiencing are due to economic as well as social factors. Because of the dual labor market, minorities are trapped in the peripheral labor market, which offers low-paying, dead-end jobs.

10. Racial and ethnic diversity is increasing dramatically in the United States. By the middle of the twenty-first century, one-half of the entire American population will be minority group members. There are differential growth rates among minorities. African Americans, traditionally the predominant minority in the United States, will be surpassed in numbers by Latinos in the early part of the twenty-first century.

LEARNING OBJECTIVES REVIEW

- Distinguish among the concepts of minority, race, and ethnicity.
- Describe patterns of racial and ethnic relations.
- Differentiate prejudice from discrimination.
- Illustrate the different views of prejudice and discrimination taken by functionalists, conflict theorists, and symbolic interactionists.
- Describe, relative to the white majority, the condition of minorities in the United States.
- Describe the increasing racial and ethnic diversity in America.

CONCEPT REVIEW

Match the following concepts with the definitions listed below them.

____ a. ethnic minority
____ b. underclass
____ c. race

____ d. minority
____ e. de facto subjugation
____ f. assimilation

____ g. self-fulfilling prophecy
____ h. discrimination
____ i. ideology

1. the integration of a racial or ethnic minority into a society
2. subjugation based on common, everyday social practices
3. unequal treatment of individuals based on their minority membership
4. a people who are socially identified and set apart by others and themselves on the basis of their unique cultural or nationality characteristics
5. a set of ideas used to justify and defend the interests and actions of those in power in a society

6. a people who possess some distinctive physical or cultural characteristics, are dominated by the majority, and are denied equal treatment
7. a distinct category of people who are alleged to share certain biologically inherited physical characteristics
8. when an expectation leads to behavior that causes the expectation to become a reality
9. those in poverty who are continuously unemployed or underemployed because of the absence of job opportunities and/or required job skills

CRITICAL-THINKING QUESTIONS

1. Evaluate the definition of minority used by sociologists.

2. Do you think that both the pattern of assimilation and the pattern of conflict accurately describe the state of racial relations between blacks and whites in America today? State your position using terms and evidence from the text.

3. ____ Compare and contrast the functionalist and conflict perspectives on prejudice and discrimination. Is one a better explanation than the other? Are they complementary? Defend your conclusion.

4. Taking into account the past and present situation of African Americans in the United States, construct a line of argument for or against affirmative action. Be specific.

MULTIPLE-CHOICE QUESTIONS

1. ____ **A set of ideas used to justify and defend the majority's interests and actions is known as**
   a. an ideology.
   b. a doctrine.
   c. a dogma.
   d. a code of ethics.
   e. a philosophy.
2. **People who possess some distinctive physical or cultural characteristics, are dominated by the majority, and are denied equal treatment are known as**
   a. a minority.
   b. a race.
   c. an ethnic group.
   d. a subculture.
   e. a caste.

3. **Which of the following statements *best* represents contemporary social scientists' beliefs regarding differences in intelligence among various races?**
   a. The weight of current scientific evidence supports a connection between genetically determined physical characteristics and innate superiority or inferiority of certain races.
   b. There is no scientific proof that races differ in their innate mental characteristics.
   c. Recent reports attribute existing differences in measured IQ to racial differences rather than to the social environment.
   d. Social scientists are now becoming more interested in biological races than in social races.
   e. Social scientists have stopped doing research on what they consider a closed issue.

4. _____ is a pattern of assimilation in which immigrants are allowed to maintain their traditional cultural ways while at the same time learning the values and norms of the host culture.

a. The melting pot
b. Anglo-conformity
c. Cultural pluralism
d. Accommodation
e. Population transfer

5. **The unequal treatment of individuals based on their membership in some minority is known as**

a. stereotyping.
b. prejudice.
c. racism.
d. the ideal-type method.
e. discrimination.

6. **Believing that "all politicians are corrupt" is an example of**

a. racism.
b. stereotyping.
c. discrimination.
d. prejudice.

7. **The _____ contends that barring others from social, political, and economic opportunities strengthens the boundaries of the majority.**

a. conflict perspective
b. psychological perspective
c. interactionist perspective
d. functionalist perspective

8. **Two nations that are convinced they are going to war engage in hostile interaction that causes a war to occur. This is an example of**

a. stereotyping.
b. discrimination.
c. a self-fulfilling prophecy.
d. agitation.

9. **Forcing Native Americans onto reservations is an example of**

a. Anglo-conformity.
b. subjugation.
c. demographic transition.
d. partition.
e. population transfer.

10. **According to the text, which of the following statements regarding the current condition of African Americans is *false*?**

a. Their median income is substantially below that of white households.
b. Their poverty rate is four times that of whites.
c. African American median family income decreased in the 1980s.
d. Less than one-third of African American men are employed in the highest occupational categories.
e. Young African American families hold less than one-fifth of the wealth of young white families.

11. **The United States' policy toward the Native Americans over the years has**

a. been consistently paternalistic.
b. been consistently neglectful.
c. vacillated between paternalism and neglect.
d. attempted to unite the various tribes so they can help themselves.
e. been based on the concept of workfare.

12. **According to the text, Asian Americans have been especially successful because they**

a. possess superior intelligence.
b. profit from illegal activities such as gambling and prostitution.
c. refuse to work at jobs that do not pay well.
d. embrace education as the key to success.
e. place little value on human life.

13. **The success of Jewish Americans can be attributed to all of the following *except***

a. their willingness to work hard.
b. their emphasis on family responsibility.
c. the high value they place on education.
d. the skills many Jewish immigrants brought with them.
e. their easy assimilation into mainstream American society.

14. **According to dual labor market theory, minority workers tend to be trapped in the _____ labor market, with low-paying jobs and little hope for advancement.**

a. secondary or peripheral
b. core
c. primary
d. inferior
e. malfunctioning

FEEDBACK REVIEW

True-False

1. The evidence clearly shows that race is declining in importance in America. T or F?
2. Contrary to popular opinion, white ethnics in the United States are politically more liberal than Protestant blue-collar workers. T or F?

Fill in the Blank

3. According to Gordon Allport, children learning prejudice move from a pregeneralized learning period to a stage of total _____.

4. The attitudes and practices associated with _____ in the American slave system help account for the long-term prejudice and discrimination experienced by African Americans.

5. Almost _____ percent of all Americans will be members of minority groups in 2050.

Multiple Choice

6. Which of the following is not always a characteristic of a minority?
   a. distinctive physical or cultural characteristics
   b. smaller in number than the majority
   c. dominated by the majority
   d. denied equal treatment
   e. a sense of collective identity

Matching

7. Rank order (1–4) the following minority groups to indicate their relative proportion of the total American minority population projected for 2050.
   ____ a.   Asian Americans
   ____ b.   Native Americans
   ____ c.   Latinos
   ____ d.   African Americans

GRAPHIC REVIEW

Some attitudes of Americans toward various immigrant minorities are summarized in Figure 9.1.

1. If you had to write a one-sentence summary of Figure 9.1 for the CNN evening news, what would you say?

   _____
   _____
   _____

2. What do the data in Figure 9.1 suggest regarding the future of ethnic relations in the United States?

   _____
   _____
   _____
   _____

ANSWER KEY

Concept Review	Multiple Choice	Feedback Review
a. 4	1. a	1. F
b. 9	2. a	2. T
c. 7	3. b	3. rejection
d. 6	4. c	4. manumission
e. 2	5. e	5. 50
f. 1	6. b	6. b
g. 8	7. d	7. a. 3
h. 10	8. c	b. 4
i. 3	9. e	c. 1
j. 5	10. b	d. 2
	11. c	
	12. d	
	13. e	
	14. a	

10

Inequalities of Gender

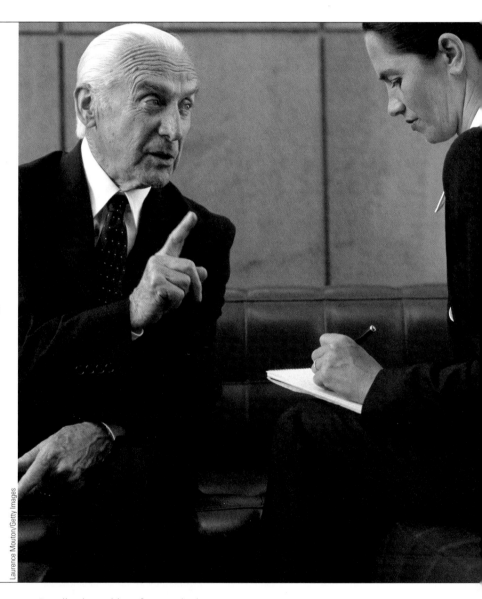

Laurence Mouton/Getty Images

How well does the United States compare with other industrialized countries on gender income equality? Even if Americans recognize that women are not paid on a par with men, most believe that at the very least the United States is making more progress than other countries. On the contrary, among modern countries, America is surprisingly near the bottom in male–female income parity. Only Luxembourg and Japan have wider gender gaps than the United States. Whereas French women in manufacturing jobs, for example, earn 93 percent of the wages paid men, females in the United States earn only 80 percent of the wages paid men—for the same work.

Throughout history, men have predominantly performed the social, political, and economic functions outside the home, while women have assumed responsibility for child-care and household tasks. Although this division of labor does not necessarily imply any status difference between the sexes, in practice it certainly has this effect. The female domestic tasks are undervalued in industrial societies, where women—thought to be dependent, passive, and deferring—have usually been considered subordinate to independent, aggressive, and strong men. Historically, the division of labor based on sex has almost always led to gender inequality. And this chapter will demonstrate other disadvantages attached to the female gender role as well. But first let's delve into the cultural and social underpinnings of gender roles.

Antecedents of Gender

In the past, people believed that anatomy was destiny. Behavioral differences between men and women were popularly attributed to **sex**—the biological distinction between male and female. Males were assumed to be naturally more aggressive than women and to be built for providing and protecting. Thought of as being naturally more passive, females were believed to be designed for domestic work (Valian 1999). This way of thinking is called **biological determinism**—the attribution of behavioral differences to inherited physical characteristics. If this popular conception were true, all men and all women in all societies would behave uniformly in their unique biologically determined ways because of inborn biological forces beyond their control.

As will be shown, biological determinism lacks scientific validation. Although biology may create some behavioral tendencies in the sexes, such tendencies are so weak that they are easily overridden by cultural and social influences (Ridley 1996; Sapolsky 1997; Bearman and Bruchner 2002; Powell and Graves 2002). Therefore, **gender**—the expectations and behaviors associated with a sex category within a society—is acquired through socialization (Jackson and Scott 2001).

From the moment of birth, males and females are treated differently. Parents stress the characteristics and behavior that fit the society's image of the ideal male or female, including modes of dress, ways of walking and talking, play activities, and life aspirations. In response, girls and boys gradually conform to these definitions and learn to behave as expected. From this process comes **gender identity**—an awareness of being masculine or feminine, based on culture. Margaret Andersen succinctly captures the difference between sex and gender:

> The terms sex and gender have particular definitions in sociological work. Sex refers to the biological identity of the person and is meant to signify the fact that one is either male or female. . . . Gender refers to the socially learned behaviors and expectations that are associated with the two sexes. Thus, whereas "maleness" and "femaleness" are biological facts, becoming a woman or becoming a man is a cultural process. Like race and class, gender is a social category that establishes, in large measure, our life chances and directs our social relations with others. Sociologists distinguish sex and gender to emphasize that gender is a cultural, not a biological, phenomenon. (Andersen 2005:31)

Controversy over explanations for gender differences persists; the nature (biology) versus nurture (socialization) distinction remains at the heart of the debate (J. Q. Wilson 1993; Canary et al. 1998). All the same, a notable difference exists today: research is beginning to replace bias and opinion. Explanations for masculinity and femininity are now based on research rather than on tradition and common knowledge. This research investigates the biological and cultural antecedents of gender (Moir 1991; Hrdy 1999; Andersen 2005).

Biological Evidence

In what ways are men and women biologically different? The obvious biological differences between the sexes include distinctive muscle and bone structure and fatty tissue composition. The differences in reproductive organs, however, are much more important because they result in certain facts of life: only men can impregnate; only women are able to produce an ovum, carry and nurture the developing fetus, and give birth; and only women can secrete milk for nursing infants. In addition, the genetic composition of the body cells of men and women is different. All human beings have twenty-three pairs of chromosomes (the components of cells that determine heredity), but one of those pairs is sex distinctive. Males have an X and a Y chromosome in the pair that determines sex, and females have two X chromosomes. At about the eighth week of prenatal development, the information coded into the chromosomes determines whether the embryo—which is still sexually neutral—will begin to develop testicles or ovaries. These reproductive glands discharge the hormones that are characteristic of either females or males: estrogen and progesterone in females, testosterone and androgens in males. These hormone combinations influence development in both males and females throughout life (Geary 1998).

Some subsequent research indicates that the brains of men and women are slightly different in structure (Gur et al. 1995; Fisher 1999). For example, men show more activity in a region of the brain thought to be more tied to adaptive, evolutionary responses such as fighting. Women have more activity in a newer region of the brain; this region is more highly developed and thought to be linked to emotional expression. The female brain is less specialized than the male brain. Women tend to use both sides of the brain simultaneously when performing a task. For example, whereas men tend to process verbal tasks on the left side of the brain, women are more likely to use both sides. Women tend to use both ears when listening, and men tend to use the right ear. And while males are more muscular and stronger as a group than females, females have greater physical endurance and generally outlive males.

Do biological differences necessarily lead to differences in social behavior? Biological determinists point to research indicating that men and women in dozens of different cultures (at varying stages of economic development) are associated with some distinctly different ways of behaving. For example, men and women differ in what they look for in romantic and sexual partners. Men value physical appearance more than women do. Women place more emphasis on social class and income. Men tend to prefer slightly younger mates, while women favor slightly older ones. In addition, males in general tend more toward physical aggressiveness in conflict situations (Buss 1994a, 1994b, 1995; Maccoby 1997; Wodak 1997; Buss, Malamuth, and Winstead 1998).

On the other hand, researchers investigating behavioral differences between the sexes have been unable to consistently establish biological differences as an independent variable (Fausto-Sterling 1987; Stockard and Johnson 1992; Kimmel 2000; Andersen 2005). One researcher's findings tend to contradict another's. Furthermore, many studies seek to find differences between males and females but ignore the similarities. To compound the problem, researchers often fail to document the variation in characteristics that exists *within* sex categories. Some men, for example, tend to be submissive, weak, and noncompetitive, and some women are aggressive, strong, and competitive. Either may be assertive, strong, and sensitive.

Cultural Evidence

Is there cross-cultural evidence that gender-related behavior is not solely the product of biology? In her classic study of three primitive New Guinean peoples, Margaret Mead (1950) demonstrated the influence of culture and socialization on gender role behavior. Among the Arapesh, Mead found that both males and females were conditioned to be cooperative, unaggressive, and empathetic. Both men and women in this tribe behaved in a way that is consistent with the more traditional concept of the female gender role. Among the Mundugumor, both men and women were trained to be "masculine"—they were aggressive, ruthless, and unresponsive to the needs of others. In the Tchambuli tribe, the gender roles were the opposite of those found in Western society. Women were dominant, impersonal, and aggressive, and men were dependent and submissive. On the basis of this evidence, Mead concluded that human nature is sufficiently malleable to rule out biological determination of gender roles. Cross-cultural research since Mead's landmark work has clearly substantiated her findings that gender roles are not fixed at birth (Janssen-Jurreit 1982; Montagu 2000).

For example, look at the gender behavior of the five-nation League of the Iroquois, which was in territory now part of Pennsylvania and upper New York. In a matrilineal society (family line traced through females) the Iroquois women enjoyed power and respect. Senior women chose the male members of the ruling council, and if the men failed to follow the women's wishes, the males were removed from office. Male dominancy and female subordination were absent in a culture where women took charge of village affairs, including military activities (Zinn 2005).

Studies have also been conducted on infants whose parents intentionally treated their children as if they belonged to the opposite gender. Apparently, individuals can fairly easily be socialized into the gender of the opposite sex. What's more, after a few years, these children resist switching back. These studies indicate that biological tendencies can be greatly influenced by culture and society (Schwartz 1987; L. Shapiro 1990; Ridley 1996; Sapolsky 1997).

What are the most common male and female roles? Societies such as the Arapesh, Mundugumor, and Tchambuli are exceptions to the rule. The general pattern in cross-cultural studies of gender roles in preliterate societies is one of male dominance and female nurturance (Zelditch 1955; Barry, Bacon, and Child 1957; Collins 1971; Reiss 1976; G. R. Lee 1977; Sinclair 1979). Although both men and women perform economic tasks that provide for family welfare, women's tasks usually involve domestic chores, child rearing, and emotional harmony. Men, on the other hand, are more likely to provide financially for the family and to represent the family in activities outside the home (Tilly and Scott 1978). This is probably because, beginning with early hunting societies, men have had the advantage of greater size and physical strength, and women have had the capability of childbearing and breast-feeding. Larger physical size has led to social, economic, and political advantages for men. Rarely, if ever, have these advantages been shared equally with women.

Do gender definitions change? Cultural definitions of gender roles can and do change. In his book *American Manhood*, E. Anthony Rotundo (1993) identifies three distinct culturally created conceptions of manhood. These definitions and redefinitions of gender roles developed in the North American middle class prior to the twentieth century. The first phase, *communal manhood*, was developed in the socially integrated society of colonial New England. During this period, the definition of manhood was embedded in one's obligations to community. A man's social class at birth gave him his place in the community. He fulfilled himself through public usefulness more than through economic success. As head of the household, a man expressed his value to his community and bestowed a social identity on his wife and children. During this era, before 1800, men were considered superior to women. In particular, men were held to be more virtuous. Thought to possess greater powers of reason, men were considered better able than women were to control passions such as ambition and envy.

The second phase, *self-made manhood*, became the dominant male gender definition in the first decades of the nineteenth century. Social status at birth was replaced with a social identity based on personal achievement. A man's work role became his source of male identity. "Male" passions were allowed more fullness of expression. Because a man was supposed to prove his superiority, the drive for dominance was viewed as virtuous. As males turned from the tradition of public usefulness to the pursuit of self-interest, the nature of women was redefined. Women were attributed a stronger moral sense than men and were assigned the role of protecting the common good in an age of rampant male individualism. Despite her moral superiority, a woman's social identity still depended on her husband's place in society, and she was denied participation in the pursuit

Although gender stereotypes keep some people from pursuing certain activities, there are actually very few behaviors that cannot be performed by either sex.

of individualism. Her primary purposes were to control men's passions, to make others happy, and to take care of her husband and children.

Late in the nineteenth century there came another shift. The third phase, *passionate manhood*, emphasized not just achievement, but ambition, combativeness, and aggression. Male toughness was admired, tenderness scorned. The gender definition of men and women changed over these periods, but there was a constant: the exclusion of women from economic, political, and

Gender-Based Hierarchy

In the not-too-distant past, most doctors were men, who worked closely on a daily basis with female nurses and receptionists who were clearly subordinate to them. This pattern has not disappeared despite the influx of women into the ranks of physicians. In many occupational settings today, most of the executives, supervisors, or higher-level professionals are men, assisted by female secretaries, clerks, aides, or lower-level managers . . .

When women enter a workplace, they frequently find a male hierarchy already established. Whether a woman's entry creates tensions for herself or others in the workplace presumably depends on the level of the job she takes and the source of any authority inherent in the job. She may come in at a subordinate level as a clerk or receptionist. But if she comes in at the same level as male coworkers, she will be faced with the unaccustomed process of jockeying for position among them, and competing with them for the attention and approval of the people higher up in the hierarchy. If she comes in as a manager or supervisor who has male subordinates, she must learn how to deal with people who may want her job, or who may find it difficult to adjust to being supervised by a woman. Some men believe so strongly in male superiority that they resist women's advancement up the workplace hierarchy. Such men sometimes say quite explicitly that it would be "an insult to their intelligence" to be supervised by a woman.

. . . And there are fairly widespread male beliefs concerning the "natural" (male-dominant) relation between the sexes. In adulthood there is a strong stereotype associating power with masculinity . . . just as there was in childhood, so that women in supervisory positions generate ambivalent reactions in men: are they to treat this woman as a powerful person or a feminine person? If she is seen as feminine, then a man with traditional attitudes might feel it is appropriate to be protective and chivalrous, or at least courteous, while at the same time failing to take her seriously where work-related matters are concerned; clearly, he would expect to be the person who "takes charge" when they interact. Can he forget that a female supervisor is female and adapt himself to a situation where she is the one who takes charge? It is much easier for men—perhaps for many women too—to slip into a traditional male boss/female secretary or male doctor/female nurse kind of work relationship in which the "appropriate" power relationships between the sexes are maintained in the workplace hierarchy. Such traditional attitudes may be weakening, but they are still prevalent enough to impede the promotion of women in many situations.

Think About It
Talk to several men and women with work experience. Ask them a few open-ended questions that you make up to test Maccoby's contention. In your verbal or written report, be specific about similarities or differences in male and female answers.

Source: Excerpted from Eleanor Maccoby, *The Two Sexes: Growing Up Apart, Coming Together.* Cambridge, MA: Howard University Press, 1997, pp. 247–248.

---

social power. In these arenas, women were consistently relegated to a subservient position.

Where does the nature/nurture debate stand? The nature/nurture controversy has unfortunately been posed as an either/or situation by both the mass media and some researchers. In general, researchers investigating behavioral differences between the sexes have not been able to prove that any particular behavior has a biological cause (Fausto-Sterling 1987; Stockard and Johnson 1992; Andersen 2005). In this light, it is reasonable to credit *both* biology and culture for gender differences. Biological characteristics exist, but they can be modified through social influences (Parker and Easton 1998). In other words, men and women can learn to be submissive or aggressive by mirroring the behaviors of influen-

---

1. _____ is the biological distinction between male and female.
2. _____ refers to the expectations and behaviors associated with a sex category within a society.
3. Gender is acquired through the process of _____.
4. _____ is an awareness of being masculine or feminine.
5. Societies that do not follow traditional gender definitions are exceptions to the rule. T or F?

Answers: 1. Sex; 2. Gender; 3. socialization; 4. Gender identity; 5. T tial role models, such as parents or siblings. Also, this is a good time to remind ourselves that human behavior is the result of multiple causes.

Theoretical Perspectives on Gender

The functionalist perspective focuses on the origins of gender differences, whereas conflict theory concentrates on the reasons gender differences continue to exist. Symbolic interactionism attempts to explain the ways in which gender differences are acquired.

Functionalism and Gender

If some pattern of behavior is not beneficial to society, functionalists argue, it will ultimately cease to be important. According to functionalism, the division of responsibilities between males and females survived because it was beneficial for human living. Early humans found sex-based division of labor efficient. In part because of their size and muscular strength, men performed hunting and defense tasks. In addition, men were assigned these tasks because men were more expendable than women. Because these duties were more dangerous and because one male was enough to ensure sufficient procreation to sustain the survival of a group (the same is not true of a single woman with several men), it hurt the group's chances of survival less to lose a man. Women were assigned domestic and child-care tasks. Because this sexual division of labor promoted the survival of the species, assert the functionalists, it was retained.

Although giving credit to the functionalist perspective for its insights into the original emergence of gender roles, some critics have contended that it is too conservative and inadequate to explain recent changes in gender (Peplau 1983). Gerald Maxwell (1975), like most sociologists, concedes that the traditional division of labor may have been appropriate at one time. Today, however, argues Maxwell, rapid social change has undermined the value of functionalist theory in understanding gender. Maxwell believes that it is to the advantage of modern society to treat people based on their ability rather than on an outdated conception of instrumental and expressive roles. According to Maxwell and other critics, then, the contribution of the functionalist view of gender is restricted to an explanation of the development of gender.

In response to Maxwell's criticism, functionalists now recognize that the traditional division of labor has dysfunctions, especially for modern society. Consequently, they point to harmful gender definitions resulting from this division of labor. These dysfunctions are examined later in the discussion on gender inequality.

Conflict Theory and Gender

Conflict advocates are concerned not only with the origins of the traditional conception of gender but also with gender's persistence (Andersen and Collins 2007). The focus here is on the latter. According to conflict theory, men and women have differential access to the necessary resources for outside-the-home success. It is to the advantage of men (higher-status members of the society) and the disadvantage of women (lower-status members of the society) that women not gain access to political, economic, and social resources. By keeping the traditional division of labor intact, men can maintain the status quo and preserve the privileges they enjoy.

Perhaps the most recent example of maintaining the gender status quo was found in Afghanistan, when the then ruling Taliban militia practiced "gender apartheid." This gender war trapped women in a way of life unknown elsewhere in the modern world (O'Dwyer 1999; Rashid 2001). The Taliban prohibited girls from attending school, banned women from all work outside the home, punished women who left home without a male relative, enforced the wearing of black sacks over the body with only a wire mesh for vision, required windows of houses to be painted black, forced women

This family is structured along traditional gender roles. How would the functionalist interpret this arrangement?

to wear silent shoes, and required women to remain mute in public.

Conflict theorists see traditional gender roles as outdated. Although these conventional gender roles may have been appropriate in hunting and gathering, horticultural, and agricultural societies, they are inappropriate for the industrial and postindustrial era. Male physical strength may have been important when hunting was the major means of subsistence, but work in modern society does not position men with that advantage. In addition, demographic changes make women more available for outside work. Women are marrying later, are having fewer children, are younger when their last child leaves home, are remaining single in greater numbers, and are increasingly choosing to be single parents. According to conflict theorists, women who prefer careers in fields formerly reserved for men have every right to make that choice, whether or not it is "functional" for society. For example, any potential harm to the family must now be weighed against the damage done to women who are not permitted to develop their occupational potential. According to conflict advocates, research suggests that there are benefits to the family associated with working wives and mothers (Hoffman and Nye 1975; Moen 1978; Kate 1998; Zuo and Tang 2000; Hochschild 2001). Also, they say, the barriers blocking women from entry into higher-level jobs waste talents and skills that modern societies need.

Marxist and socialist feminists see the position of women in capitalist society as the result of two interrelated influences: patriarchal (male-dominated) institutions and the historical development of industrial capitalism. Historically, women have been relegated to subordinate roles because their unpaid labor in the home is needed to produce the future labor force. Home work provides for the care and feeding of workers, the creation of new workers and consumers (through childbearing and child rearing), and the creation of people who can devote their time to the consumption activities necessary to keep factories running and workers working. In addition, women provide a reserve army of cheap labor for times of crisis (e.g., war) and for expansion of the labor force when more workers are needed. Thus, the subordination of women is seen as a key component in the maintenance of the political and economic institutions in capitalist society (Sokoloff 1980; Chafetz 1984, 1990; Jaggar and Rothenberg 1993; Nierenberg 2002).

How is sexual harassment related to conflict theory? The conflict perspective is built into the concept of **sexual harassment**—the use of one's superior power in making unwelcome sexual advances. Sexual harassment may be experienced as raw superior physical power. This was clearly the case in the 1992 Tailhook scandal in which many women Navy personnel, including officers, were sexually harassed by naval aviators at a Las Vegas convention. The subsequent investigation, however, revealed a pattern of sexual harassment in the Navy that went beyond the physical power of an individual or the collective physical strength of a group.

Sexual harassment in all branches of the military has subsequently been revealed to be quite high (Thomas and Vistica 1997). According to a 1995 U.S. Army survey, 4 percent of all female soldiers reported that they had been the victim of an actual or attempted rape or sexual assault within the previous year. That is almost ten times greater than the rate of rape and sexual assault in the general population. The highly publicized court cases during the 1990s involving the sexual abuse of military recruits by their superiors were apparently only the tip of an ugly and frightening iceberg. In fact, evidence of the prevalence of sexual harassment in the military surfaced again in early 2003 at the Air Force Academy, when the country learned of more than fifty reported cases of male cadet rape or sexual assault of female cadets. As the number of cases increased, the Air Force Secretary James G. Roche admitted that the actual number of rapes is probably three times the reported number ("Air Force Reports 54 Rapes, Assault" 2003; Lorch 2003). Many observers consider this assessment as far too conservative.

In addition to physical superiority, sexual harassment can be a matter of social and economic power. A prominent case illustrates the difficulties involved in legally establishing sexual harassment. In 1991 Anita Hill, a law professor, brought the issue of sexual harassment to public attention when she testified in televised Senate Judiciary Committee confirmation hearings of now Supreme Court Justice Clarence Thomas. Professor Hill accused Justice Thomas of sexually harassing her during the time she worked on his staff. The Senate committee, then totally composed of men, initially discounted the sexual harassment charge and subsequently confirmed Justice Thomas. The truth was not determined.

Symbolic Interactionism and Gender

Symbolic interactionists focus on **gender socialization**—the social process in which boys learn to act the way society assumes boys will behave and girls learn to act in ways society expects of them. This process occurs in large part through interaction with parents, teachers, and peers as well as through the mass media. This socialization process is very powerful, with gender being incorporated into the self-concept through role taking and the looking-glass self.

How do parents contribute to gender socialization? Parents are vital in gender socialization because they transfer values and attitudes regarding the ways boys and girls should behave. The learning of gender begins at birth and is well established by the time the child is two and a half years old (Mussen 1969; Frieze et al. 1978; Davies 1990). Actually, parents begin laying the groundwork for gender socialization before the child is born. Parents often painstakingly select a name to represent the gender-related characteristics they expect their child to possess. Immediately after the baby's birth, friends and relatives give gifts "appropriate" to the child's sex—blue or pink blankets, the baseball playsuits or frilly dresses, balls and trucks or dolls. Numerous studies report that gender definitions lead to differential treatment of boys and girls. Studies of infant care have found that girls are cuddled more, talked to more, and handled more gently. Parents expect boys to be more assertive, and they discourage them from clinging.

Parental differential treatment of the young does not stop at infancy. For example, the bedrooms of preschoolers reflect masculine and feminine themes. Girls' rooms contain dolls and household items, whereas the rooms of boys are more likely to contain guns, footballs, and trucks. Gender is even taught and reinforced in the assignment of family chores. In an investigation of almost 700 children between the ages of two and seventeen, Lynn White and David Brinkerhoff (1981) found that boys were often given "masculine" jobs, such as cutting grass and shoveling snow, whereas girls were more often assigned "feminine" chores, such as washing dishes and cleaning up the house.

As parents respond to their children's behavior, they usually intentionally as well as unconsciously continue to transfer their gender-related values. Parents often even evaluate children according to their level of conformity to gender definition.

In what ways do schools reinforce gender socialization? Although the most critical period of gender socialization occurs within the family during early childhood, gender socialization occurs through interaction in schools as well (K. A. Martin 1998). Teachers may not recognize that they encourage different behaviors from boys and girls, but observation of preschool teachers reveals a clear sex-based difference in many teacher–student relationships. This pattern continues into the elementary school years. Myra and David Sadker, in an extensive study of fourth-, sixth-, and eighth-grade students, found boys to be more assertive in class: boys were eight times more likely than girls to call out answers, whereas girls sat patiently with their hands raised. They linked this classroom behavior to the differential treatment given boys and girls by teachers. Teachers, report these researchers, are more likely to accept the answers given by boys who call out answers. Girls who call out in class are given such messages as, "In this class we don't shout out answers, we raise our hands." According to Sadker and Sadker, the message is subtle and powerful: "Boys

should be academically assertive and grab teacher attention; girls should act like ladies and keep quiet" (Sadker and Sadker 1985:56).

In their later book *Failing at Fairness*, the Sadkers (1995) examine sexism from elementary school through college. The conclusion is consistent with their earlier findings: through differential treatment, America's schools often shortchange females; gender bias results in an inferior education for girls. Academically, girls typically outperform boys in the early years of school. Through the transmission of gender role values, well-intentioned teachers often dampen female competitiveness. Girls, the study concludes, are systematically taught passivity, a dislike of math and science, and a deference to the alleged superior abilities of boys. Females carry this impairment into adult life—into the working world, where its disadvantageous effects take full toll.

How do peers contribute to gender socialization? Even children in day-care centers and nursery schools are aware of and encourage gender definitions. When three-year-olds behave according to their gender definition, their peers encourage, praise, and imitate them. Children who imitate the opposite sex meet considerable opposition within the group. This opposition includes criticism as well as efforts to modify the behavior (Lamb, Easterbrooks, and Holden 1980).

Interaction among adolescent peers also affects gender conceptions because adolescence is the time when an individual's identity is being firmly established, and identity is closely linked with definitions of masculinity and femininity (Erikson 1973, 1982; Adler and Adler 1998). Because adolescents want approval from their peers, acceptance or rejection by one's friends can significantly influence a teenager's self-concept. Teens who most closely mirror traditional gender definitions—such as football players and cheerleaders—are generally given the greatest respect, whereas "feminine" boys and "masculine" girls are often accorded low status. This peer group pressure encourages teenage conformity to society's idealized role models. To do otherwise is to risk rejection and a significant loss of self-esteem.

What is the role of the media in gender socialization? All members of society are subject to the continual bombardment of images from television, books, magazines, radio, movies, and advertising. For an extreme example, in the video game *Grand Theft Auto: Vice City*, players can have sex with a female prostitute and earn additional points by beating and murdering her. Many studies document the distorted view of women's roles prevalent in the media and the ways in which this distorted view affects people. In general, the media present the most stereotypical version of gender definitions, thus reflecting and reinforcing the limits on the

Jeff Olson/HO, Sports Illustrated/AP Photo

Most Americans accept this conception of beauty. Because slimness is fundamental to this conception, women tend to compare themselves to fashion models.

options available to both sexes (Craig 1992; Kilbourne 2001; Shields and Heinecken 2001; Dines and Humez 2002; Gauntlett 2002).

In the 1970s, illustrated children's books presented a conservative, traditional image of the sexes, with women underrepresented in titles and as major characters. One study examining more than 2,700 children's stories found 5 stories about boys for every 2 stories about girls, 119 biographies of men compared with 27 biographies of women, and 65 stories degrading girls compared with 2 degrading boys (American Association of School Administrators 1974). In a study of a library display of children's books, Deane Gersoni-Stavn (1974) found less than half included pictures of women. Of these books, 80 percent included a picture of a woman wearing an apron. In the few books that pictured women working outside the home, the jobs were traditionally feminine occupations, such as teaching and nursing, or unusual female roles, such as queen or trapeze artist. In a content analysis of several prizewinning children's books—the Caldecott Medal winners for the years 1967–1971—Leonore Weitzman and Deborah Eifler (1972) found a distinct gender bias. Human male characters outnumbered human female characters by 11 to 1; the ratio of male to female non-human animals was 95 to 1. Males were the central characters who dominated the action, whereas females were cast in passive roles.

But haven't things changed dramatically? A later study of Caldecott winners published between 1972 and 1979 showed some improvement: human males appeared more often than females by a ratio of 1.8 to 1; non-human male animals outweighed females by 2.66 to 1 (Kolbe and LaVoie 1981). A gender bias reminiscent of the Weitzman and Eifler study was still apparent in these books, however. The characters in seventeen of the nineteen books portrayed the traditional gender definitions, and not one female character was shown to be employed outside the home.

When researchers later replicated Weitzman and Eifler's study, they also found only slight improvement. Boys were still depicted as competitive, active, and creative; girls were cast as submissive, dependent, and passive (J. A. Williams et al. 1987; McDonald 1989;

TABLE 10.1

FOCUS ON THEORETICAL PERSPECTIVES: Gender Inequality

Each of the major theoretical perspectives can focus on gender inequality in its own unique way. Explain why the examples given fit each theoretical perspective. How would each of the other theories approach the same social arrangement differently?

Theoretical Perspective	Social Arrangement	Example
Functionalism	• Gender-based division of labor	• Women are expected to perform household tasks for the benefit of society.
Conflict theory	• Patriarchy (male domination)	• Women are denied high-status occupations for the benefit of men.
Symbolic interactionism	• Favoring males over females in the classroom	• Few females believe they can become scientists.

1. According to the _____ perspective, the division of labor based on sex has survived because it is beneficial and efficient for human living.
2. According to the _____ perspective, traditional gender definitions exist because they provide greater rewards and privileges to men than to women.
3. _____ is the theoretical perspective that attempts to explain how gender is acquired.
4. Which of the following is *not* a true statement?
   a. Boys are less assertive in class than girls.
   b. Illustrated children's books still tend to present very traditional gender definitions.
   c. Parents' perceptions of the physical characteristics of their children are heavily influenced by their children's gender.
   d. Nursery school children who fail to conform to their "appropriate" gender are met with resistance from their peers.

Answers: 1. functionalist; 2. conflict; 3. Symbolic interactionism; 4. a

Bigler and Liber 1990). Although male and female characters were about the same numerically and one-third of the principal characters were female, only one female, a waitress, was shown working for pay. Despite the increased appearance of females, they still were unequally represented in active roles. The power of children's literature to influence young minds is now being recognized. Although this heightened sensitivity is currently reflected in the publication of more books with tempered depictions of the sexes, not all traces of gender bias have disappeared (Purcell and Stewart 1990). When Myra and David Sadker (1995) examined the content of grade-school textbooks, they found that the ratio of boys/men to girls/women in the language arts textbooks ranged from a ratio of 2 to 1 to a ratio of 3 to 1 in favor of males. And out of a 631-page textbook on world history, only 7 pages related to women.

Harlequin romance books are another example of the staying power of gender role stereotypes in books. In excess of 160 million of these books are published in more than 20 languages. In fact, Harlequin romance book sales account for almost one-third of trade paperbacks sold in the United States and Canada annually. In Harlequin books, women are consistently depicted as more sexually passive; men are the sexual aggressors (Grescoe 1997).

Television continues to reflect traditional gender roles. Despite a reduction in gender bias, most of those women on television who are portrayed as working outside the home are in female-dominated jobs. Male characters are either heroes or villains, whereas females tend to be adulterers or victims. In general, television consistently depicts men as aggressive, independent, and in charge of events. Women are portrayed as dependent and passive. There are signs of change in current television programming. Female stars of shows such as *Sex and the City, Alias, Desperate Housewives,* and *CSI* are certainly not playing traditional female roles.

Nowhere in the mass media are gender role stereotypes more prevalent than in advertising. Gender is used to promote everything from automobiles to underwear. Despite an increasing number of gender role reversal ads, men usually appear in advertising as symbols of dominance and aggression and women are portrayed as submissive (Thomas and Treiber 2000). This depiction actually starts targeting audiences quite early in life. Advertising designed for children uses the same stereotypes without sexual overtones—girls are more often cast as frivolous and helpless, and boys are more likely to play dominant and aggressive roles (Browne 1998).

Terry Fruch and Paul McGhee (1975) found that children who viewed television in excess of twenty-five hours per week had more traditional gender conceptions than children viewing television ten or fewer hours a week. Similar findings have been reported among preschoolers and young adults (Gross and Jeffries-Fox 1978; Eisenstock 1984). Thus, the gender images seen on television still appear to restrict the gender definitions of viewers (Garst and Bodenhausen 1997; MacKay and Covell 1997; Eagley and Wood 1999).

Sex Stereotypes and Gender Roles

Sex Stereotypes

In Chapter 9, **stereotype** was defined as a set of ideas based on distortion, exaggeration, and oversimplification that is applied to all members of a social category. A stereotype does not take into account individual differences, so every member of the group is assumed to have the same traits, strengths, and weaknesses. The stereotype is a standard used to judge individuals. Men in

American society are expected to be virile, brave, sexually aggressive, unemotional, logical, rational, mechanical, practical, dominating, independent, aggressive, confident, and competitive. Women are expected to be the opposite: weak, fearful, sexually passive, emotional, insecure, sentimental, "arty," dependent, submissive, modest, shy, and noncompetitive (Williams and Best 1990). In American society, these stereotypes reflect a belief in the innate superiority of men and the inferiority of women.

Stereotypes do not allow for a range of alternatives. One label is applied to all members of the group. Moreover, stereotypes tend to be exaggerations or extremes infrequently found in real life. For example, consider the stereotype of men offered by Marc Fasteau:

> *The male machine is a special kind of being, different from women, children, and men who don't measure up. He is functional, designed mainly for work. He is programmed to tackle jobs, override obstacles, attack problems, overcome difficulties, and always seize the offensive. He will take on any task that can be presented to him in a competitive framework. His most positive reinforcement is victory.*

> *He has armor plating that is virtually impregnable. His circuits are never scrambled or overrun by irrelevant personal signals. He dominates and outperforms his fellows, although without excessive flashing of lights or clashing of gears. His relationship with other male machines is one of respect but not intimacy; it is difficult for him to connect his internal circuits to those of others. In fact, his internal circuitry is something of a mystery to him and is maintained primarily by humans of the opposite sex.* (Fasteau 1975:60)

Gender Roles

Gender roles involve culturally based expectations associated with each sex. They represent an ideal, because in reality people do not always behave as expected. Although mothers are expected to be patient and loving with their children, many are not. Consequently, individuals can decide to emphasize different aspects of their gender according to their preferences and abilities. Thus, as noted earlier, there is a wide range of actual feminine and masculine behavior. Not all men try to be strong, fearless, and aggressive; not all women attempt to be sweet, submissive, and deferring.

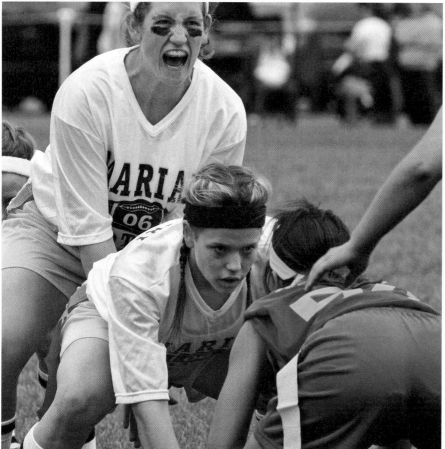

These active girls do not fit the stereotypical image of male football player/female cheerleader.

Jim West/Image Works

Role Conflict and Role Strain

What female role conflicts and strains are created by gender roles? Women suffer from conflicts and strains built right into the female role. And, a changing society is creating even greater conflicts. Because married women are increasingly working outside the home (see Figure 10.1), they must balance the requirements of work with the demands of housework and child care (Burros 1993; Hochschild 2003). Married women must juggle, first, the expectation that they will move from one place to another as they or their husbands climb occupationally and, second, their concern for the effects of rootlessness on their children.

While most married women now share responsibility for the breadwinner role, men, by and large, have been slow to share domestic responsibilities. Furthermore, no other institution in society—whether public or private—has stepped in to help working women. As a result, women have shouldered the responsibilities of work and family largely on their own (Reskin and Padavic 2002).

According to past research, housewives report higher rates of illness than women in the labor force (Nathanson 1980). This apparent positive health effect of women working outside the home, however, may now be offset by the increasing stress that accompanies the combined demands of work and family (Rosenfield 1989). In addition, greater stress and depression occur among women whose husbands do not support their working than among women with supportive husbands (Ulbrich 1988). Poorer health is also more likely among minority working women with husbands opposed to their employment.

What male role conflicts and strains are created by gender roles and sex stereotypes? Like women, men experience conflicts and strains within the traditional masculine role. Many working-class men, whose wives must work just to help support their families, are bothered by a feeling of having failed at their task of provider:

> *A guy should be able to support his wife and kids. But that's not the way it is these days, is it? I don't know anybody who can support a family anymore, do you? . . . Well, I guess those rich guys can, but not some ordinary Joe like me.* (Rubin 1994:78)

Men are expected to be occupationally and economically successful while simultaneously devoting time to their wives and children. Men are encouraged to admire other women sexually, but they are expected to remain faithful to their wives. The "macho man" ideal conflicts with women's preferences for warm, gentle, romantic lovers. Men also have to handle the inconsistency of being cool and unemotional at work but loving at home (Doyle 1995).

FIGURE 10.1

Marital Status of Women in the U.S. Civilian Labor Force

These data show the relationship between marital status and employment. Write a paragraph describing the important trends you can see.

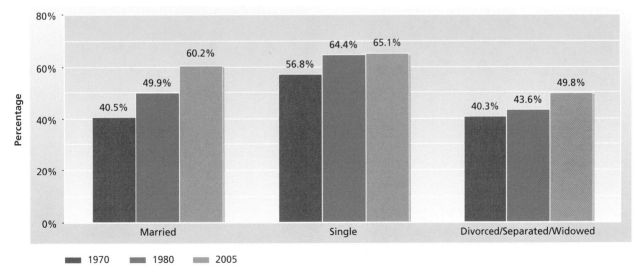

Source: U.S. Bureau of the Census, *Statistical Abstract of the United States, 2008* (Washington, DC: U.S. Government Printing Office, 2007), table 580.

1. Following are several statements about the elderly. Identify each statement with one major theoretical perspective: functionalism (F), conflict theory (C), or symbolic interactionism (S).

_____ a. Ageism results in part from an oversupply of labor.

_____ b. The young may wish to distance themselves from the elderly to avoid facing their own mortality.

_____ c. The stigma attached to aging promotes a low self-concept among the elderly.

_____ d. Ageism is associated with industrialization.

_____ e. Older people are stereotyped.

_____ f. Ageism exists in part because older workers are more expensive than younger ones.

Answers: 1. a. (C), b. (F), c. (S), d. (F), e. (S), f. (C)

Role conflict and strain for men are intensifying in today's environment of changing gender roles. Although many women want to be liberated from male domination, they sometimes still expect men to be protective of them. This is partly because appropriate gender role behavior is currently conditioned by the situation. For example, some women want to be treated as equal to men while at work but treated in more traditional ways off the job. Other women resent any attempt by men to treat them differently from the way they would treat other men. Consequently, men are often confused about what women expect. Moreover, although many men wish to shed the John Wayne image, they are less certain than women about the role model that should replace it (Wetcher, Barker, and McCaughtry 1991; Faludi 2000).

Why do strains associated with changing gender roles tend to be greater within working-class families? In *Worlds of Pain*, Lillian Rubin (1992) reports that working-class Americans advocate more traditional gender roles; yet, for economic reasons, the women are more likely to work outside the home. She also reports that working-class husbands with wives in the labor force do not participate very much in domestic tasks. This assertion has since been bolstered by other research reporting that less educated, lower-income men are less likely to share household labor when their wives work outside the home (Miller and Garrison 1982; Sweet, Bumpass, and Call 1988). In addition, Rubin concludes, working-class husbands of working women see their wives' employment as a challenge to their self-esteem and masculinity. The contradiction between traditional gender role attitudes and the sharing of the provider, household, and child-care roles among working-class couples, Rubin asserts, leads to personal strain. Rubin wrote that, among other things, this contradiction led working-class men to demand more submission and obedience from their wives and children in order to bolster their own egos and sense of masculinity.

In *Families on the Faultline*, Rubin (1994) reports that although more contemporary working-class men now believe in a woman's right to full independence, most still want women with sufficient dependency and need to make them feel manly. Nearly all the working-class men Rubin studied did some work inside the family—washing dishes, vacuuming, grocery shopping—but only 16 percent shared family work equally with their wives. Almost all of these men, however, either worked different shifts than their wives or were unemployed. The high price paid for the retention of traditional sex stereotypes and gender roles remains much higher for working-class women.

Gender Inequality

Women as a Minority Group

Most scientists consider biological determinism a moral threat because historically it has been used to rationalize the treatment of some categories of people as infe-

It is still unusual to see a woman, such as Inda Nooyi, serve as chairman and CEO of a corporation like PepsiCo.

Harriet Bradley—Men's Work, Women's Work

In a 1989 study, Harriet Bradley focused on occupational sex segregation and job sex-typing in Britain after the Industrial Revolution, around 1750. She presented case studies of various industries and occupations—food and raw material production, agriculture, fishing, mining, manufacturing, pottery, hosiery, shoemaking, shopwork, medicine, and teaching. In these studies, Bradley documented the development of the division of labor along gender lines. Her database consisted of a combination of secondary sources in the form of previously published research, along with new material she collected from primary sources. This study is an excellent example of the integration of sociology and history.

According to Bradley, pervasive sex-typing of jobs—the allocation of certain work-related activities to men and others to women—results in men and women rarely being allocated the same types of jobs. Even when men and women are working side by side, Bradley points out, they are doing different things. This allocation of certain work to men and certain work to women, she argues, is based on a

simple principle: men occupy jobs that place them in control, and women do work that requires obedience. In the fields, men cut and women gather; in the factory, men stamp out parts and women sew them together; in the office, men handle accounts and women do the typing and filing.

The social definition of what is "women's work" and what is "men's work" may vary from time to time and place to place. Jobs thought of as the province of women may have historically been assigned to men, or vice versa. Prior to the industrialization of the cotton industry, for example, men were weavers, while women did the spinning. These roles were flip-flopped when power-driven machinery replaced hand labor. What does not vary, Bradley asserts, is the assignment of jobs based on the principle of men as superordinate and women as subordinate. Weavers had higher status prior to industrialization and lower status afterward.

Although occupational sex segregation and job sex-typing existed in preindustrial society, Bradley contends that capitalism dramatically increased both. From the case studies, she concludes that the tremendous expansion of capitalist production in the 1880s and 1890s created the foundation for the current patterns of gender job segregation and job sex-typing.

During the capitalist expansion, according to Bradley, hiring women in jobs labeled as "women's work" served the interests of capitalists as

well as male workers. Capitalists could increase their profits by hiring women for lower wages because of the low value placed on female labor. In addition, capitalists used women to challenge the control male workers held as a result of their expertise in preindustrial craft production techniques. As men fought back by demanding that certain jobs remain "men's work," occupational sex segregation deepened; and women were forced into low-paying "female" jobs. Males and females, Bradley concludes, continued to be steered into socially appropriate lines of work. This played into their desire to assert unique sexual identities. Gender-based ideologies emphasize sex differences. She quotes from C. Cockburn:

> What a man "is," what a woman "is," what is right and proper, what is possible and impossible, what should be hoped and what should be feared. The hegemonic ideology of masculinism involves a definition of men and women as different, contrasted, complementary and unequal. It is powerful and it deforms both men and women. (C. Cockburn 1986:85)

Evaluate the Research

1. Which theoretical framework do you think Bradley used? Explain your choice.
2. Do you think that the current changes in gender relations in the United States will alter the state of occupational sex segregation described by Bradley? Support your conclusion.

rior. This view, in short, has led to racism and sexism. Minorities suffer the effects of racism; women are hurt by **sexism**: a set of beliefs, norms, and values used to justify sexual inequality. Sexist ideology—the belief that men are naturally superior to women—has been used and is still being used to justify men's leadership and power positions in the economic, social, and political spheres of society. It explains women's dependence

on men as well (Clayton 1996; Swann, Langlois, and Gilbert 1998; Benokraitis 2004).

Although there is no factual basis to support the maintenance of sexism, its influence remains. Women are easily identified, biologically and culturally, and are subject to many forms of discrimination (Blau and Ehrenberg 1997; McCall 2001; The World's Women 2000).

But isn't sex discrimination disappearing? The answer is yes and no. Some segments of American society have more positive attitudes about women. And a few women now hold positions traditionally reserved for men. For example, there are now ten female CEOs of Fortune 500 companies, and 16 percent of corporate officers of these companies are female. Women now head a number of top universities, including Princeton, Brown, Duke, Chicago, and Michigan. The percentage of women in the Army went from 9.4 percent in 1981 to 15.5 percent in 2001. Women also comprise 18.3 percent of the Air Force and 13.3 percent of the Navy. And approximately 130 women are Army combat pilots, some of whom fought or are fighting in the Iraq war (Gibbs 2003). Still, a careful examination reveals many gaps in social rights, privileges, and rewards for women in the United States (Valian 1999). These gaps, although they have closed somewhat in recent years, are reflected in the continuing occupational, economic, legal, and political inequality experienced by American women (Bianchi and Spain 1996; Riley 1997; Lorber 2005).

Occupational and Economic Inequality

By far the most important U.S. labor market development over the past thirty years has been the dramatic increase in the number and proportion of women in the workforce. From 1972 through 1989, more than a million additional women entered the labor force each year. In 1960, 23 million women worked outside the home for pay; by 2000, the figure had reached nearly 66 million. In 2006, 55 percent of women worked outside the home, compared with 82 percent of men (see Figure 10.2). Women are expected to represent approximately 47 percent of the U.S. labor force in 2012 (see Figure 10.3).

The increase in labor force participation among women twenty-five to fifty-four years of age has been particularly steep. For example, the labor force participation rate of married women with children under six years of age has skyrocketed from 19 percent in 1960 to 37 percent in 1975 and reached about 60 percent in 2005 (U.S. Bureau of the Census 2007c).

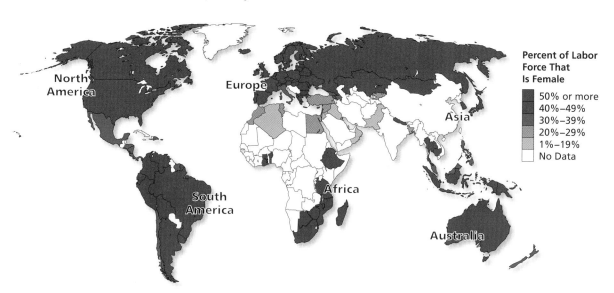

WORLDVIEW 10.1

Women in the Workplace

In most countries of the world, fewer women than men are employed in the labor force. In addition, the higher-paying jobs and better opportunities still go to men. This map shows the percentage of women in various national labor forces.

Percent of Labor Force That Is Female

- 50% or more
- 40%–49%
- 30%–39%
- 20%–29%
- 1%–19%
- No Data

Source: United Nations Statistics Division, *Statistics and Indicators on Women and Men*, 2005.

Interpret the Map

1. The countries with the highest percentage of women in the workplace include both highly developed countries such as the U.S., Germany, and France and relatively undeveloped countries such as Mongolia, Tanzania, and Ghana. Can you think of some reasons for this?

2. Create a graph representing any other patterns you may find in this map.

FIGURE 10.2

U.S. Labor Force Participation Rates by Sex: 1890–2006

This figure tracks changes in the percentage of women and men in the U.S. labor force since 1890. The participation gap between the sexes has changed radically. Describe this shift using specific percentages.

Source: U.S. Department of Labor; Bureau of Labor Statistics, *Employment in Perspective: Women in the Labor Force*, No. 865 (Washington DC: U.S. Government Printing Office, 1993), p. 1; U.S. Department of Labor, Bureau of Labor Statistics, *Employment and Earnings*, (Washington, DC: U.S. Government Printing Office, 1997), p. 6; and U.S. Bureau of the Census, *Statistical Abstract of the United States, 2001* (Washington, DC: U.S. Government Printing Office, 2001), p. 367, U.S. Bureau of the Census, *Statistical Abstract of the United States; 2008*. (Washington, DC: U.S. Government Printing Office, 2007, table 572.)

FIGURE 10.3

Composition of the U.S. Labor Force, by Sex: 1870–2012

As this figure demonstrates, the male-female composition of the U.S. labor force has steadily moved toward parity. The female percentage of the labor force has moved from less than 15 percent in 1870 to more than 46 percent today. What do you think is the most important social consequence of this change?

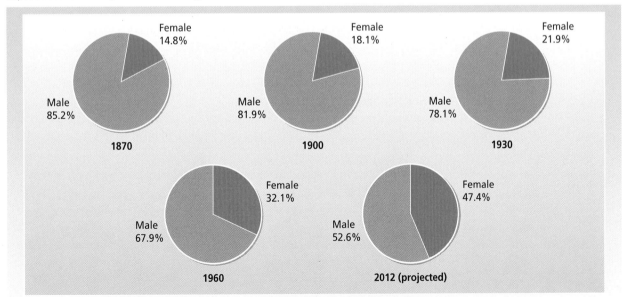

Source: U.S. Department of Labor, Bureau of Labor Statistics, 2005.

Has increased labor force participation eliminated occupational inequality? Despite their increased participation in the labor force, women are still concentrated in different occupations than men. This is known as **occupational sex segregation** (Reskin 1993; J. Farley 1995; Yamagata et al. 1997; Perlmann and Margo 2001; Tomaskovic-Devey and Skaggs 2002). Women are underrepresented in high-status occupations and overrepresented in low-status occupations. To begin with, females fill only about 14 percent of architecture and engineering positions and about one-third of attorney jobs. By contrast, women occupy nearly all of the pink-collar jobs—secretaries, clerks, stenographers—whose purpose is to support those higher up the occupational ladder (U.S. Bureau of the Census 2007c). Moreover, when women are in high-status occupational groups, they are concentrated in lower-prestige, lower-paid jobs. Even within female-dominated occupations, men fill a disproportionate share of higher positions. Occupational sex segregation, then, produces a **dual labor market**—a split of the economy into a core segment of preferred workers and a peripheral segment of the marginalized workers.

Have women and men reached financial equality? Earnings inequality has followed a pattern since the 1960s. After increasing in the late 1960s, earnings inequality stabilized for most of the 1970s, decreased during the 1980s, and has increased somewhat since (Grubb and Wilson 1992; Wasserman 2007). Thus, there remains a wide discrepancy between the earnings of American women and men. By 2005, women who worked full-time earned only 80 cents for every dollar earned by men. To put it another way, women now work about seven days to earn as much as men earn in five days (U.S. Department of Labor 2006). On the one hand, there is good news: this salary gap has decreased since 1980, when women were earning 64 percent as much as men. On the other hand, there is bad news: the overall salary gap in 2006 was still too close to what it was in 1955 (see Figure 10.4).

In virtually every occupational category, men outstrip the earning power of women. The earning gap persists, regardless of educational attainment. Women in the same professional occupations as men earn less than their male counterparts (see Figure 10.5). This is true even for women who have pursued careers on a full-time basis for all of their adult lives. Furthermore, males in female-dominated occupations typically earn more than women.

How do race and ethnicity affect women's earnings compared with men's? As mentioned earlier, women in the United States earn about 80 percent of what American men earn. Comparing groups of women by race and ethnicity, however, presents another picture. White women, for example, earn about 82 percent of what men earn, while the earnings for African American women are only about 71 percent of the earnings for men. Asian women have the highest female-to-male

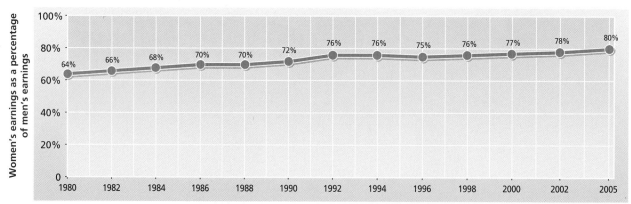

FIGURE 10.4

What Women Earn Compared to Men

This figure traces the ratio of women's to men's earnings since 1980. Discuss two important conclusions you can make from these data. Use material in the text to help.

Source: U.S. Department of Labor, Bureau of Labor Statistics, 2005, 2006.

INTERNET LINK   For additional information on employment and earnings of women and men in the United States, you can visit the Web site of the Department of Labor, Bureau of Labor Statistics, at http://stats.bls.gov/

FIGURE 10.5

Female-to-Male Earnings: 2004

On average, women in the U.S. earn about 80 cents for every dollar a man earns. In what way do the data in this figure support the contention that gender inequality is real?

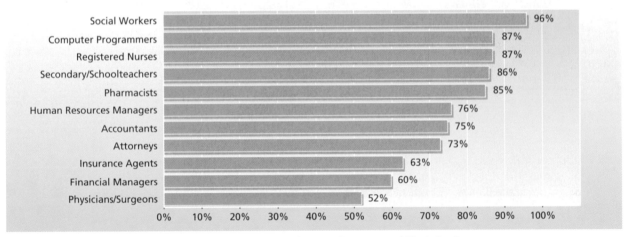

Women's Earnings as a Percentage of Men's Earnings

Source: U.S. Department of Labor, Bueau of Labor Statistics, 2005.

earnings ratio—nearly 86 percent. Latino women, on the other hand, have the largest earnings gap; they early only about 59 cents for every dollar earned by men (see Figure 10.6).

Is there hope of closing the salary gap? Although the salary gap will likely remain indefinitely, there are a few signs of hope. Probably the most encouraging word is

that two-thirds of the women currently entering the labor force are taking jobs historically dominated by men. For instance, as more women became computer programmers over the past few years, their earnings rose from 70 percent to almost 85 percent of what their male counterparts earned.

More generally, as the percentage of females in professional, managerial, and technical occupations

FIGURE 10.6

Women's Wages Compared with Men's Wages, by Race and Ethnicity

This figure compares what women of various races and ethnicities earn compared to men. Which group of women has the highest female-to-male earnings ratio?

Source: U.S. Department of Labor, Bureau of Labor Statistics, 2005.

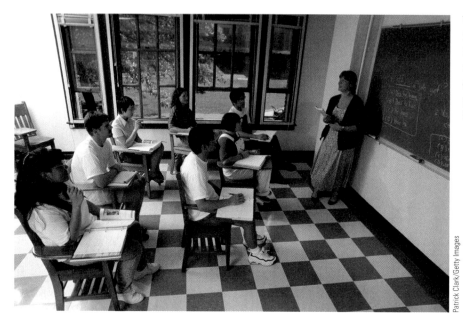

The income of American working women is three-quarters that of men. The gap in earnings between men and women is not obliterated even when education and occupation are comparable. This female professor in all probability is paid less than her male counterparts.

Patrick Clark/Getty Images

has increased, the female-to-male income ratio has improved in a number of these occupations. Also, increasing numbers of women are leaving the wage system behind as they become entrepreneurs. Excluding large corporations, women now own about one-fourth of all businesses. However, females are still more likely than males to hold jobs that pay relatively low wages; and as you have seen, the differential in female-to-male earnings persists within occupational categories.

Why does the salary gap persist? Research on wage differences has shown that even taking into account all possible factors that might reasonably account for wage differences between women and men—such as age, experience, amount of absenteeism, and educational background—a large difference remains. With the elimination of all other explanations, the conclusion reached from these studies is that sex discrimination is the major source of economic differences.

One of the most serious indicators of economic inequality is the disproportionate number of women living in poverty. More than half of poor families have a female head of household. Individuals living in female-headed households are about three times as likely to live below the poverty level; female-headed households among minority groups are about four times as likely to live in poverty. This, of course, reflects the feminization of poverty discusses in Chapter 8. This trend is fueled by increases in teenage pregnancy, singlehood, and divorce.

How do American women fare globally? As noted in Using the Sociological Imagination at the beginning of this chapter, women in the United States do not fare very well economically relative to women in other developed countries. Here, of course, we are talking about relative earning power, not absolute dollar amounts. American women in nonagricultural employment earn only 80 percent of that earned by men, compared with 87 percent in Costa Rica and 93 percent in France (see Figure 10.7).

What about the comparative economic success of men and women? Men and women in the United States tend to have levels of income similar to their own parents with one exception. Daughters of low-income parents experience less income mobility than sons of low-income parents.

It must be noted that women in their thirties today make considerably more in inflation-adjusted dollars than women of the same age in their mothers' generation. Despite making more than their female contemporaries, men in their thirties today make less than men in their fathers' generation. Also men's employment rates, hours worked, and wages have been flat or declining between 1974 and 2004, while women have seen a substantial increase in all three components (Isaccs 2007b).

Legal and Political Inequality

Sexism is similarly built into America's legal and political institutions. National, state, and local legal codes reflect sexual bias; and important differences exist between women and men in positions of political power.

Is sexism built into the law? According to the U.S. Civil Rights Commission, more than 800 sections of the U.S. legal code are sexually biased. Legal inequities and complexities are compounded by overlapping state

FIGURE 10.7

Women's Wages Compared with Men's Wages in Nonagricultural Employment, 2003

This figure reveals where the United States stands worldwide in wage parity. Explain this fact based on information in this chapter.

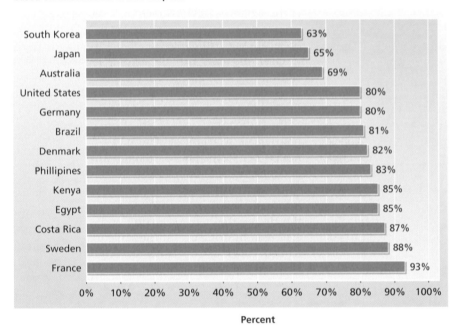

Percent

Source: International Labour Office, Bureau of Statistics, http://laborsta.ilo.org/ (accessed 18 October 2003).

and local jurisdictions. Because state and local statutes apply as long as they do not conflict with federal law, the legal situation for women varies greatly from one place to another. For example, some states have enacted state versions of the Equal Rights Amendment, which, if enacted, would enter this language into the United States Constitution: "Equality of rights under the law shall not be denied or abridged by the United States or by any state on the basis of sex" (Andersen 2005:292). Other states, however, have retained archaic nineteenth-century laws based on the idea that women are the property of either a husband or a father. Georgia, for example, has an 1863 law that "defines a woman's legal existence as 'merged in the husband.'"

The dependent status of women permeates the law. A U.S. Supreme Court decision refused to grant women the legal guarantee of health insurance benefits for pregnancy-related medical costs, despite the fact that medical coverage for medical conditions unique to men—such as prostate problems and vasectomies—was routinely provided. At the state level, some states have traditionally refused women the right to keep their own surnames after marriage. Other states have had "protective legislation" restricting women's rights. Supporters of these laws viewed them as safeguards against the abuse and exploitation of women. However, the end result was that women were denied certain jobs, many

of which were better paid than more traditional occupations for women (Andersen 2005). Such protective legislation limited the number of hours women could work, the conditions under which they could work (such as barring women from toxic areas because of potential birth defects in their children), and the kinds of work they could do (such as setting a thirty-pound limit on the amount of weight a woman could lift).

Due to the passage of Title VII of the Civil Rights Act of 1964, such laws have been preempted; but despite the prohibition of such laws, the practices still linger. Moreover, the Family and Medical Leave Act of 1993, which requires that employees be given up to twelve weeks without pay for childbirth, adoption, serious illness, and so forth, inadvertently affects women negatively. Because women are more likely than men to take parental leave, this legislation gives employers another reason to hire men rather than women.

There are gender differences in criminal law as well. Certain crimes are typically associated with one gender or the other. For example, laws against prostitution are generally enforced only against women, while male customers go free. On the other hand, sexual relations with a minor female, even though she consents, may result in prosecution of the man for statutory rape.

Rape has been one of the most controversial legal areas. Nowhere is sexism more explicitly built into the language and application of the law. For example, no other crime besides rape requires independent corroboration that a crime has been committed. There are two assumptions of many rape laws: women will lie about the occurrence of rape, and unless evidence of rape is produced, no prosecution can occur. For women raped without witnesses (presumably the majority of cases), or for women who submit to rape due to the threat of violence, there is often no legal protection. Furthermore, when cases actually do come to trial, the victim is often treated more harshly than the defendant. Many observers have described rape trials as trials of the victims, with evidence of the victims' past sexual activity used to undermine their testimony and acquit the accused. Even convicted rapists have been allowed to go free on the grounds of provocation. A notorious case involved a Wisconsin judge who refused to send a convicted rapist to prison on the grounds that the way women dress today causes men to lose control.

How does sexism relate to political power? One possible explanation for legal bias is the underrepresentation of women in the lawmaking institutions. In cities of more than 30,000 people, only 16 percent had female mayors in 2008. In the same year, less than a fourth of state legislators were women (see Figure 10.8 and Table 10.2). The number of female governors, lieutenant governors, and attorneys general stood at just above 25 percent.

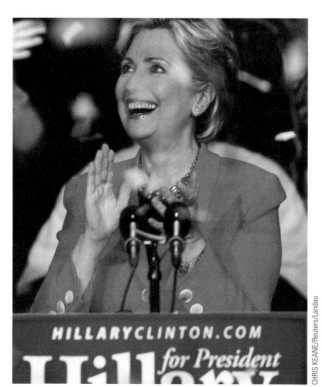

American women are still not well represented in political office. Hillary Clinton, who unsuccessfully ran for the presidential nomination in the Democratic party, was right to say that the number of votes she received in the primaries represent 18 million cracks in the glass ceiling.

CHRIS KEANE/Reuters/Landau

A record number of women serve in the U.S. Congress. Still, while women comprise more than one-half of the U.S. population, they occupied only about 16 percent of the seats in the U. S. House of Representatives (seventy members) in 2008. And although the number of female U.S. Senators has increased from two to sixteen over the last ten years, women still represented just 16 percent of the Senate in 2008. Women in Congress have seldom risen to positions of power. Only 14 women hold House or Senate leadership roles (see Table 10.2).

Over the past two decades, a few women have reached the national political stature previously held exclusively by men. In 1988, Geraldine Ferraro became the first female vice presidential candidate in the history of the United States; Madeleine Albright was named the first female secretary of state in 1996; Elizabeth Dole ran for her party's presidential nomination in 2000. In 2002, Democrat Nancy Pelosi became the first female House minority whip and was subsequently elected House minority leader. In 2006, she became Speaker of the House, the highest-ranking position a female has ever held in Congress. At the same time, as the newly named Chair of the House Republican Conference, Deborah Pryce became the highest-ranking Republican woman ever to serve in Congress. After her election as senator from New York, Hillary Clinton became the first woman to enter the presidential primaries as a leading candidate. In 2008, Sarah Palin became the first woman to run for Vice President of the U.S. on the Republican ticket. The outcome of the 2008 presidential election will indicate something about any declining significance (not to say disappearance) of gender in American politics (Falk 2008).

Female exclusion from political power is not restricted to elective offices. The record for women in appointed offices is likewise poor. Although there have been recent increases in the number of appointments, the total is extremely small. When President Jimmy Carter appointed four women to his cabinet, he quadrupled the number ever to have held such positions at one time. President Bill Clinton, almost twenty years later, appointed thirteen women to cabinet posts, and George W. Bush has appointed five women to cabinet-level positions. Still, only twenty-nine women have ever served as cabinet officers. President Ronald Reagan appointed the first woman Supreme Court Justice, Sandra Day O'Connor, in 1981; and President Clinton elevated Ruth Bader Ginsburg to the high court in 1993. Upon her retirement, Sandra Day O'Connor was replaced by a man. Only a small percentage of federal judges are women. Even though women comprise more than 40 percent of the federal workforce, they have only recently begun to gain access to higher pay grades (S. Barr 1999). About 40 percent of women are employed in midlevel grades. Women hold less than one-third of the supervisor and manager positions and account for only one-fifth of senior executives.

FIGURE 10.8

Percentage of Women in Elective Offices

This graph shows the degree of female success in the elective politics from 1979 to today. Explain why the figures are still so low.

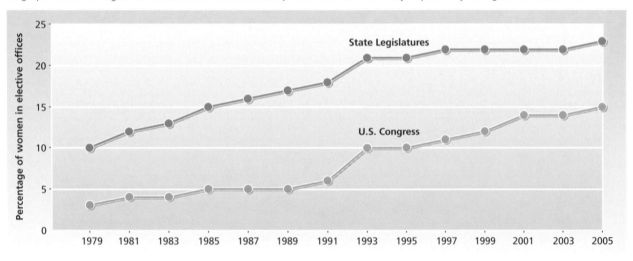

Source: Center for American Women and Politics, National Information Bank on Women in Public Office, Eagleton Institute of Politics, Rutgers University.

The political position of women in the United States is among the worst in the Western world. With some notable exceptions, western European nations have much greater female political participation. Margaret Thatcher became England's first female prime minister in 1979, and Angela Merkel took power in 2005 as Germany's first female chancellor. In 2005 female leaders were also elected for the first time in Liberia and Chile. In 2007, there were eight female presidents and five female prime ministers.

SNAPSHOT OF AMERICA 10.1

Women In Public Office, 2004

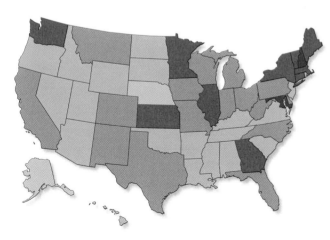

Number of Women Holding State Public Offices

- 60 or above
- 45–59
- 30–44
- 15–29
- 14 or less

Source: U.S. Bureau of the Census, Statistical Abstract of the United States, 2006.

Hillary Clinton's bid for the presidency as a U.S. Senator is one reflection of the relatively recent accessibility of political office to women. The map at left shows the number of women holding state public offices.

Interpret the Map

1. Identify any patterns you see in the distribution of female office holders across the United States.
2. Are you surprised or disappointed at the place of women in state politics? Explain your answer.

1. _____ is a set of beliefs, norms, and values used to justify sexual inequality.
2. On the average, women earn about _____ of what men earn.
   a. three-fourths
   b. one-half
   c. two-thirds
   d. 95 percent
3. Despite Title IX, females are still not represented more equally in the collegiate sports power structure. T or F?
4. Although women are discriminated against in civil law, they are not at a disadvantage in criminal law. T or F?
5. The exclusion of women from political power applies mainly to elective offices. T or F?

Answers: 1. Sexism; 2. a; 3. T; 4. F; 5. F

On the positive side, 89 percent of men and 88 percent of women now say they would vote for a woman as their political party's nominee (Newport and Carroll 2007). When Gallup first asked this question in 1937, only 33 percent of Americans said they would vote for a female presidential candidate. On the downside, about half of the American public believes that a man would be a better president than would a woman (W. W. Simmons 2001).

Richard Zweigenhaft and William Domhoff (2006) do point out that women are now part of the power elite. The power elite is no longer the exclusively male enclave it used to be. Still, women are seriously underrepresented. And most of those women who do make it come predominantly from upper-class backgrounds.

Changing Gender Roles

Gender Roles and Social Change

What are the possibilities for change in gender roles? What are the consequences of change? One way to begin answering these questions is to look at the factors discouraging and encouraging gender role changes.

Why is there resistance to change? First, although men may not be aware that their suppression of women is to their own advantage, like all dominant groups, they can see little reason to change the way things have always been. Second, even if men were willing to relinquish their dominant position, it would be costly to them to do so. Third, some women resist giving up some advantages they see for themselves in the traditional gender role arrangement. Some fear losing the option to be full-time homemakers and do not want to relinquish their husbands' legal obligation to provide for them and their children. Others do not wish to lose conventional male courtesies and protection. Finally, an ideology still exists to support members of both sexes who wish to maintain the traditional gender roles.

Resistance to gender role change can hardly be surprising to anyone who understands the durability of deeply entrenched culture (Risman 1998). Traditional gender roles are embedded in the long-standing cultural myth of male superiority. Again, the idea of male superiority, revealed as a myth today, actually began as a myth in the first writings of the ancient Greeks sometime in the seventh or eighth century B.C. In his account of the birth of the Greek gods and the social structure of daily life, the poet Hesiod, who was writing before the Greeks used prose, chronicled the creation of the myth of male superiority (Hesiod 700 B.C./1983). Zeus, the king of gods, decided to act harshly after another god, his son Prometheus, went too far in his constant deviousness by giving fire to man. In addition to requiring hard work for subsistence living, the myth goes, Zeus created woman as the carrier of all the evils that she eventually unleashed on men. Even in rural Greece today, peasants only reluctantly acknowledge that they have a female child, and it is not uncommon for a man to say that he has three children (sons) and one girl (Hesiod 1983:97–98). Considered one of the more moderate Islamic nations, Jordan remains a patriarchy. Daughters are killed or abused by their fathers if they "dishonor" the family. Jordanian wives can renew their passports only with their husbands' permission (Khouri 2005).

It was, of course, from this myth, as well as similar ideas from many early cultures around the world (O'Brien and Palmer 1993), that male supremacy was founded. This supremacy underlies the establishment of **patriarchy**—a hierarchical social structure in which women are dominated by men (see Chapter 12).

What are some of the factors promoting change? There are a number of agents promoting change in gender roles. Demographic trends—longer life expectancy, smaller families, higher education, increased numbers of jobs requiring skilled labor—have had a profound impact on everyone's lives. For women, these trends have led to their increased participation in activities outside the home—in school, work, politics, and voluntary organizations. In 2005, 60 percent of all married women held jobs outside the home. Another source of change in gender roles is the feminist movement. Although the goals and objectives of the feminist movement appear radical to many people, its central ideas are not new.

The Feminist Movement

What is feminism? **Feminism** is a social movement aimed at the achievement of sexual equality—socially, legally, politically, and economically. Recent feminism is best viewed as the reemergence of a long historical struggle that began as far back as 1792 with the publication in England of *A Vindication of the Rights of Women*, by Mary Wollstonecraft. The movement picked up tempo in the United States in 1848, when a group of women abolitionists drew up the *Declaration of Sentiments*, which demanded equal rights for women. By the end of the nineteenth century, feminism was a mass movement (Kraditor 1971; S. M. Evans 1989; West and Blumberg 1990; Tobias 1997; Andersen 2005).

The women's movement almost perished in 1920 when, with the passage of the Nineteenth Amendment, women gained the right to vote. However, many of the problems that had prompted feminism in the first place—the economic dependence of women and the restriction of their full participation in work and politics—remained. These remaining problems, coupled with other developments, such as effective birth control technology and the growth of certain sectors of the labor market (clerical, sales, technical), kept the women's movement alive (V. Taylor 1989).

During the 1960s, what appeared to be a new social movement, the women's movement, gained national attention through its efforts for gender role and societal changes. Beginning with a few isolated voices, such as the publication of Betty Friedan's (1963) enormously influential book *The Feminine Mystique*, the women's movement gained momentum. By the 1970s, it was a mass movement with widespread support (S. M. Evans 1980).

Advocates of feminism offer several basic rationales for the movement.

1. Male-oriented social, legal, political, and economic structures have inhibited women in their development as complete human beings.
2. The results of early socialization, as well as some blatant forms of discrimination, have restricted women's options and personal growth. (More recently, it has been recognized that men are also the victims of limited options based on gender roles, although with different outcomes.)
3. Problems created by sexism are not created by individuals and cannot be solved by individuals. Like racism, the problem of sexism is embedded in the entire structure of society, and therefore its eradication requires changes in social structures as well as in individuals. To obtain such

change, group action is required (Ferree and Hess 1994; Tobias 1997).

Gender Roles in the Future

Can we predict the future of gender roles? Making predictions is risky. For one thing, an economic recession or depression could seriously curtail the occupational progress of women, who could easily fall victim once again to the "last hired, first fired" policy. For another, people's attitudes toward gender roles are inconsistent and unstable. For example, although polls show that most Americans favor equal pay for equal work, there is far less agreement on whether or not sex bias actually exists in American society. Consequently, there is disagreement as to whether special steps should be taken—such as affirmative action programs and quotas. Because some Americans support traditional gender roles and others favor a "unisexual" society, it is difficult to know in which direction the society will go.

The ups and downs of the Equal Rights Amendment (ERA) are a good indicator of the uncertainties involved. First introduced in the U.S. Congress in 1923, it was almost fifty years before the ERA passed both the Senate and the House of Representatives and was sent to the individual states for ratification. As of 1979, thirty-five of the necessary thirty-eight states had ratified it. Opponents from a wide variety of right-wing political and religious groups mobilized opposition to the ERA on grounds ranging from fear of women being drafted to the specter of unisex toilets. The strong opposition of President Reagan and many of the conservative politicians elected in 1980 signaled the defeat of the amendment in 1982. Efforts were immediately under way, however, to reintroduce it.

The 1980s and early 1990s were a time of both right-wing backlash against the aims of the feminist movement and increased mobilization of supporters (Faludi 2000). Efforts to ban abortion, to restrict access to contraception, and to weaken Title IX regulations (which ensure equality in education) are all attempts at reasserting traditional gender roles. At the same time, feminists have mobilized, with increasing militancy, around such issues as spouse abuse, rape, sexual harassment, and pension discrimination (Richardson and Taylor 1983; A. L. Taylor 1983; Andersen 2005). Such inconsistencies illustrate the problem of forecasting the future.

Notwithstanding these problems, some speculation can be made. The changes that have taken place provide a foundation for discussing the future of gender roles and the consequences of gender role changes.

What could gender roles be like in the future? Women are living longer, bearing fewer children, becoming better educated, and entering the labor force in greater numbers than at any other time in history (Reskin and

Padavic 2002). If these trends continue—and there is every reason to believe that they will—certain gender role changes should occur. For example, more and more women can be expected to plan for and pursue careers. No longer will female gender roles define work and career as unfeminine pursuits, nor will certain occupations be considered off-limits to women. The increasing enrollment of women in medical, law, and business schools indicates that this development is already under way.

Women may be placing less emphasis on children and marriage as standards for femininity. Men may become less preoccupied with work and success. Many observers have speculated on the nature of gender roles in future generations. Robert Hefner and Rebecca Meda (1979) have summarized the various alternatives in this way:

- *Emergent pluralism.* There would be no restrictions on the options available to either sex. Individuals following the traditional gender roles would exist alongside those who are pursuing a different course.
- *Conservative pluralism.* Men's and women's roles would be different but valued equally. In other words, men might still have major economic responsibility for the family, and women would perform housework and child care. Or men might be doctors and lawyers, and women would work as nurses and secretaries. However, all roles, whether male or female, would be considered equally important and rewarded accordingly.
- *Melting pot.* There would be no important differences in gender roles for men and women. Male

and female roles would be combined to create androgynous individuals who combined the strengths of the two traditional roles, but without their weaknesses. Both men and women would be strong and self-sufficient and yet capable of expressing emotion and sensitivity.

- *Assimilation to the male model of success.* Women would be encouraged to follow the traditional masculine gender role if they wanted to participate in traditionally male activities. Women could become doctors, lawyers, and businesspersons as long as they behaved just like men. This alternative would assume that feminine traits are only a hindrance and can make no contribution to the performance of societal roles.
- *Female exclusion.* There would be a continuation of traditional gender roles. On the assumption that sex differences are biological, men and women would be excluded from exchanging tasks. In particular, women would be discouraged from assuming male roles, and their own tasks would still be under-rewarded.

It is difficult to predict which of these gender role patterns will prevail in the future because unforeseen events could result in the reversal of present trends. However, the changes that have already occurred suggest a future in which more alternatives will be available to both men and women. If so, the most restrictive pattern (female exclusion) should continue to break down, and a variety of factors and efforts will probably be directed toward creating an atmosphere that permits individuals to choose from a number of options (emergent pluralism).

FEEDBACK

1. The women's movement can be traced as far back as the _____ century.
2. _____ is a social movement aimed at the achievement of sexual equality.
3. A central assumption of the feminist movement is that the problems created by sexism can be solved by individuals. T or F?
4. Match the gender role alternatives for the future with the descriptions beside them.

____ a. emergent pluralism	(1) Male and female gender roles are basically identical.
____ b. conservative pluralism	(2) Traditional gender roles exist along with alternatives.
____ c. melting pot	(3) Traditional gender roles continue.
____ d. assimilation to the male model of success	(4) Women are socialized into the masculine role.
____ e. female exclusion	(5) Men's and women's roles are different but equally valued.

Answers: 1. eighteenth; 2. Feminism; 3. F; 4. a. (2), b. (5), c. (1), d. (4), e. (3)

SUMMARY

1. All societies have definitions of gender—expectations and behaviors associated with a sex category. Through socialization, members of a society acquire an awareness of themselves as masculine or feminine.

2. Aside from obvious biological differences, such as genital characteristics and muscular structure, there are no established, genetically based behavior differences between the sexes. Behavioral differences between men and women are culturally and socially created.

3. Anthropological research reveals cultural diversity in gender characteristics. All in all, though, the traditional gender roles remain the predominant pattern.

4. Functionalists and conflict theorists have different explanations for traditional gender definitions. According to functionalists, a division of labor based on sex was a convenient and obvious solution for preindustrial societies. Conflict theorists, who are more concerned with the continued existence of traditional roles than with their origins, contend that men want to maintain the status quo; they want to protect their dominance of women and society. According to conflict advocates, traditional gender definitions are not appropriate in the modern world.

5. According to symbolic interactionism, gender definitions are imparted through the socialization process. They are learned and reinforced through interaction with parents,

teachers, peers, and the media. This interaction is especially important because gender becomes intertwined with the self-concepts of those affected by it. Gender socialization, which begins before birth and continues throughout life, occurs through elements of the mass media such as books, television, and advertising.

6. Gender roles permit some deviation from the ideal, but sex stereotypes are labels applied to all members of each sex. Sex stereotypes are employed to encourage all men to be masculine and all women to be feminine and allow for no variation from the generalized picture.

7. Sex stereotypes and gender roles are damaging to both men and women. Among other things, they intensify role conflict and strain for men and women.

8. Although both men and women are damaged by sexism, it is women who feel its full effect. Women, like other minorities, are subject to prejudice and discrimination occupationally, economically, athletically, politically, legally, and socially.

9. There are several trends promoting gender role changes, but there are also barriers. Consequently, it is risky to make concrete predictions. Most likely, the exclusion of women from traditionally male domains will continue to diminish, and both sexes will increasingly be able to choose from a variety of gender role alternatives.

LEARNING OBJECTIVES REVIEW

- Distinguish among the concepts of sex, gender, and gender identity.
- Demonstrate the relative contributions of biology and culture to gender formation.
- Outline the perspectives of gender expressed by functionalists, conflict theorists, and symbolic interactionists.
- Describe the position of women in the United States with respect to work, law, politics, and sports.
- Discuss factors promoting change in gender roles as well as factors promoting resistance to change in traditional gender roles.
- Describe the future of gender roles.

CONCEPT REVIEW

Match the following concepts with the definitions listed below them.

____ a. sex
____ b. gender
____ c. patriarchy

____ d. feminism
____ e. occupational sex segregation
____ f. sexual harassment

____ g. dual labor market

1. a split between core and peripheral segments of the economy and the division of the labor force into preferred and marginalized workers

2. the type of society that is controlled by men who use their power, prestige, and economic advantage to dominate women

3. a social movement aimed at the achievement of sexual equality—socially, legally, politically, and economically

4. the expectations and behaviors associated with a sex category within a society

5. the concentration of women in different occupations than men

6. the biological distinction between male and female

7. the use of one's superior power in making unwanted sexual advances

CRITICAL-THINKING QUESTIONS

1. Suppose that one of your high school teachers invites you back to speak to her class on the nature versus nurture gender debate. Write a brief essay you could use to summarize the scientific knowledge and to initiate class discussion.

   _____

   _____

2. Describe the major influences on the development of your gender identity. Use the categories of influence mentioned in the chapter in laying out the specifics of your experiences.

   _____

   _____

3. Some sociologists draw a parallel between racism and sexism. Discuss the consequences of sex stereotypes to substantiate or refute this assertion.

   _____

   _____

4. Draw on the information in this chapter to discuss the need for affirmative action programs. Substantiate a case either for or against continuation of these programs.

   _____

5. Use information throughout this chapter to predict the future of gender roles. Choose one of the five alternatives for the future of gender roles as the framework for your analysis.

   _____

   _____

MULTIPLE-CHOICE QUESTIONS

1. Whereas _____ is a biological term, _____ is a psychological and cultural term.
   a. gender; sex
   b. identity; role
   c. gender; gender identity
   d. sex; gender
   e. phenotype; genotype

2. **The debate over explanations for sex differences persists, with the _____ distinction still at the heart of the controversy.**
   a. nature versus nurture
   b. culture versus socialization
   c. innate forces versus genetic forces
   d. gender roles versus gender identity
   e. testosterone versus estrogen

3. **Although given credit for its insights into the original emergence of gender roles, functionalism is often criticized on the grounds that**
   a. it tends to support a male bias.
   b. it is too conservative and inadequate to explain recent changes in gender definitions.
   c. its proposition that gender roles are functional is detrimental to women.
   d. it does not address the issue of gender-based inequality.
   e. it fails to take the biological basis of gender roles into account.

4. **According to conflict theory, gender roles persist because**
   a. men and women have differential access to the resources needed to succeed outside the home.
   b. they are beneficial and efficient for human living.
   c. gender roles are incorporated into the self-concept through role taking and the looking-glass self.
   d. they are essential to human survival.
   e. women are naturally better suited to perform certain tasks.

5. **Symbolic interactionists believe that traditional gender roles persist because**
   a. men and women have differential access to the resources needed to succeed outside the home.
   b. they are beneficial and efficient for human living.
   c. gender is incorporated into the self-concept through role taking and the looking-glass self.
   d. they are essential for human survival.
   e. of innate biological differences.

6. **Which is the true statement regarding the contribution of peers to gender socialization?**
   a. Nursery schools now attempt to ignore gender roles and do not encourage conformity to them.
   b. Teenagers who most closely fulfill gender role expectations are given the greatest respect.
   c. Acceptance or rejection by peers is having less and less influence on the adolescent's self-concept.

d. Boys are more concerned than girls with gender role socialization.

e. In schools today, children are often encouraged to act out the roles appropriate to the opposite sex.

7. **Women are hurt by the set of beliefs, norms, and values used to justify sexual inequality, known as**
   a. sexism.
   b. prejudice.
   c. gender identity.
   d. discrimination.
   e. gender roles.

8. **On the average, women in the United States are now earning _____ cents for every dollar earned by men.**
   a. 45
   b. 52
   c. 60
   d. 76
   e. 83

9. **According to the text, the political position of women in the United States is**
   a. among the worst in the Western world.
   b. more favorable than that of women in Scandinavian countries.
   c. harmed by "women's lib."

   d. actually superior to that of men because women vote in greater numbers.
   e. measured by the dependency ratio.

10. **The women's movement was an identifiable mass movement by the end of the _____ century.**
    a. sixteenth
    b. seventeenth
    c. eighteenth
    d. nineteenth
    e. twentieth

11. **The social movement aimed at the achievement of sexual equality is generally known as**
    a. sexism.
    b. feminism.
    c. genderism.
    d. communism.
    e. comparable worth.

12. **According to the text, the most likely form of gender roles in the future is**
    a. conservative pluralism.
    b. the melting pot.
    c. assimilation to the male model of success.
    d. female inclusion.
    e. emergent pluralism.

FEEDBACK REVIEW

True-False

1. Societies that do not follow traditional gender roles are exceptions to the rule. T or F?
2. Sex stereotypes and gender roles are essentially the same thing. T or F?

Fill in the Blank

3. Gender is acquired through the process of _____.
4. According to the _____ perspective, the division of labor based on sex has survived because it is beneficial and efficient for human living.
5. According to the _____ perspective, traditional gender definitions exist because they provide greater rewards and privileges to men than to women.
6. _____ is the theoretical perspective that attempts to explain how gender is acquired.

Multiple Choice

7. Which of the following is *not* a true statement?
   a. Boys are less assertive in class than girls.
   b. Illustrated children's books present very traditional gender definitions.
   c. Parents' perceptions of the physical characteristics of their children are influenced very little by their children's gender.

   d. Nursery school children who fail to conform to their "appropriate" gender are met with resistance from their peers.

8. The idea that victory is the most positive reinforcement for all men is an example of
   a. sex.
   b. a sex stereotype.
   c. a gender role.
   d. a value uniquely held by men.

Matching

Match the gender role alternatives for the future with the descriptions beside them.

_____ a. emergent pluralism
_____ b. conservative pluralism
_____ c. melting pot
_____ d. assimilation to the male model of success
_____ e. female exclusion

(1) Male and female gender roles are basically identical.
(2) Traditional gender roles exist along with alternatives.
(3) Traditional gender roles continue.
(4) Women are socialized into the masculine role.
(5) Men's and women's roles are different but equally valued.

GRAPHIC REVIEW

Figure 10.4 displays American women's median annual earnings as a percentage of men's from 1955 to 2000.

1. How would you explain this trend to your sixteen-year-old daughter (if you had one)?

2. Assuming that the upward trend since 1980 continues, which of the five gender role patterns outlined in the text would be most descriptive? Why?

ANSWER KEY

Concept Review	Multiple Choice	Feedback Review
a. 6	1. d	1. T
b. 4	2. a	2. F
c. 2	3. b	3. socialization
d. 3	4. a	4. functionalist
e. 5	5. c	5. conflict
f. 7	6. b	6. Symbolic interactionism
g. 1	7. a	7. a
	8. d	8. b
	9. a	9. a. 2
	10. d	b. 5
	11. b	c. 1
	12. e	d. 4
		e. 3

Family

11

Jim Craigmyle/Getty Images

LEARNING OBJECTIVES

- Describe the types of family structure, dimensions of family structure, societal norms for mate selection, and types of marital arrangements.
- Compare and contrast views of the family proposed by functionalists, conflict theorists, and symbolic interactionists.
- Outline the extent and cause of divorce in America.
- Give an overview of family violence in the United States.
- Describe contemporary alternatives to the traditional nuclear family structure.
- Discuss the future of the family in the United States.

U.S. divorce rates are soaring! Family life is on the ropes. Yes? No? Films, television, music, and print media suggest that America's family life is in jeopardy. Actually, data on divorce provide grounds for some optimism. Although a dramatic rise in the divorce rate did begin in the early 1960s, the divorce rate began a decline in 1985.

The facts regarding divorce are not as clear-cut as commonly believed, and the family is not as simple an institution as it appears. It is a complex social unit with many dimensions. In addition, there is great diversity in the family structure—across various societies and even within one society (United Nations 1996; Sussman, Steinmetz, and Peterson 1999).

Marriage and Family Defined

In the simplest terms, a **marriage** is a legal union based on mutual rights and obligations, and a **family** is a group of people related by marriage, blood, or adoption. These brief definitions mask some modern complexities (Eshleman and Bulcroft 2005; Knox and Schacht 2008). By tradition, a marriage in the United States involves a legal relationship between a man and a woman. The recent push for legal recognition of same-sex couples attempts to challenge this tradition. So far, only Massachusetts and California recognize as legitimate a marriage between two men or two women. Other states have tried and failed to legalize gay marriage. A few states recognize "civil unions," some of which grant legal rights associated with heterosexual married couples. The traditional definition of family also fails to include cohabiting couples, which, along with same-sex domestic partners, we cover later in this chapter.

There are several types of families and types of marriages. Some of these appear in all cultures, others appear only in certain cultures at certain times.

What types of families are common to all cultures? The family of orientation is the family a person is born into, or the family of birth. It provides children with a name, an identity, and a heritage. In other words, it gives the child an ascribed status in the community. The family of birth "orients" children to their neigh-borhood, community, and society. The forms you fill out to register for college typically ask: What is your father's name? What is your mother's name? Where do you live? Where do your parents work? The answers to these questions indicate who you are and to whom you belong. The family of orientation locates you in the world. The **family of procreation** is established upon marriage. The marriage ceremony signifies that it is legal (officially sanctioned) for a couple to have offspring and to give the children a family name. The family of procreation, then, becomes the family of orientation for the children created from that marriage.

All known societies have families and marriage; the permutations, however, are staggering. Exposure to these variations requires something of an anthropological excursion.

Cross-Cultural Analysis of Family and Marriage

The structure of family and marriage involves many variations within many cultures. This cross-cultural analysis examines the predominant family and marriage forms in modern society as well as an appreciation for alternative lifestyles.

Variations in Types of Family Structure

The **nuclear family**, the smallest group of individuals that can legitimately be called a family, is generally composed of a mother, a father, and any children. Because the nuclear family is usually based on marriage, it is sometimes called the *conjugal family*. However, the two are not always synonymous. A nuclear family can be composed of a single parent and children or of a brother and sister. The **extended family** consists of two or more adult generations of the same family whose members share economic resources and live in a common household. Extended families may contain grandparents, children, grandchildren, aunts, uncles, and so forth. Because it is based on blood ties, the extended family is often identified as the *consanguine family*.

In general, extended families are most characteristic of preindustrial societies and rural parts of industrial societies. This is not to say, however, that nuclear families were not common in preliterate societies or in preindustrial Europe (Nydeggar 1985). Moreover, it would be inaccurate to conclude that the nuclear family—most characteristic of modern society—exists in isolation from a larger kinship network. Both of these issues—the association of the nuclear family with industrialization and the question of the isolation of the modern nuclear family—require elaboration.

In terms of structure, how would sociologists distinguish these two families?

How are industrialization and the nuclear family related? As Figure 11.1 indicates, there is a curvilinear relationship between types of family structure and industrialization. Before humans began to domesticate animals and cultivate crops, most economies were based on hunting and gathering. Small bands of nuclear families followed herds of animals and changing seasons, moving around constantly, never staying long in any one place. When humans learned to domesticate animals to help with tilling the soil and cultivating crops (about 9,000 years ago), they no longer needed to be mobile to maintain a food supply. Families began to farm, settle down, and establish roots. Large families were then needed to plow, harvest, and fulfill the many other tasks needed to live. Thus, the emergence of family farms and permanent residences encouraged the development of the extended family. When agriculture became the dominant economic base, the extended family became the major type of family. In general, when societies move from agricultural economies to industrialized ones, the extended family is replaced by the nuclear family because large families are no longer needed to supply the necessities of life (W. J. Goode 1970).

What accounts for the relationship between industrialization and nuclear family structure? One factor is increasing geographical mobility. People in rural areas must leave their relatives to secure industrial employment in cities. Furthermore, for occupational reasons, those already living in urban areas come to accept frequent changes in residence. Either way, geographical mobility makes it difficult to maintain close ties with large numbers of relatives; usually, only the most immediate

kin can be taken along. Also, before industrialization, family members were expected to take care of the sick and the elderly. (Actually, they scarcely had a choice, because there were few outside resources to rely on.) With the growth of special-purpose organizations and government programs, however, the sick can be cared for in hospitals—which insurance companies often pay for—and the elderly can live in assisted living homes—paid for in part by their Social Security payments. In short, because outsiders have taken over many of the

FIGURE 11.1

Economy and Family Structure

The nature of family structure varies with the type of economy. Agricultural society promotes the extended family because of the need for labor from a large number of people. The nuclear family is more compatible with hunting and gathering societies and modern societies.

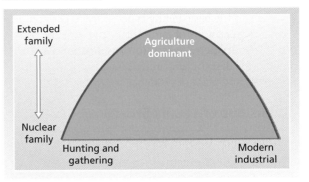

Source: Richard R. Clayton, *The Family, Marriage, and Social Change*, 2d ed. (Lexington, MA: D.C. Heath, 1979), p. 95. Reprinted by permission of the publisher.

services formerly performed by the family, relatives do not have to depend on one another as much as they once did. Thus, certain conditions associated with industrialization are more compatible with the nuclear family than with the extended family.

How isolated is the modern nuclear family? Although families in industrialized societies tend to have a nuclear structure and maintain separate residences, there is evidence that they are not as isolated from their relatives as was once believed. Early research depicting the American nuclear family as isolated was conducted prior to some recent changes—changes that are now occurring in other industrialized societies as well. For one thing, modern transportation and communication make it easier for relatives to maintain family ties even though separated geographically. Second, because of metropolitan growth, children no longer need to leave their parents and other relatives to go to college or to pursue their occupations.

Early on, Marvin Sussman (1953) observed considerable cooperative effort and mutual aid between adults in urban nuclear families and their parents. According to Sussman, families of orientation and procreation are linked together through a kin network. He found, for example, parents offering a variety of services to their married children, including financial aid, home repair and maintenance, baby-sitting, and various kinds of help during illnesses. Subsequent researchers have also found this network of kin relationships (Litwak 1960; Adams 1968, 1970; Troll 1971). A study of Muncie, Indiana, by Theodore Caplow and his colleagues (1982) showed that a high proportion of the close kin of the residents surveyed lived within 50 miles of the city. Most of the residents with parents still living saw their parents at least once a month, and nearly half saw them either weekly or more often. Apparently, the overwhelming majority of Americans maintain frequent contact with a substantial number of relatives. At the very least, it must be concluded that the modern family has either a modified nuclear or extended family structure that falls somewhere between the isolated nuclear family and the classic extended family (De Luca 2001).

Whatever the basic family type—nuclear or extended—several dimensions of family structure must be recognized. These patterns take into consideration three circumstances: descent and inheritance, authority within the family, and place of residence.

Dimensions of Family Structure

Who owns what? Different arrangements exist for determining descent (who becomes head of the family) and inheritance (who owns family property). If the arrangement is **patrilineal**, descent and inheritance are passed from the father to his male descendants. Iran, Iraq, and Tikopia in the Western Pacific are examples of patrilineal societies. Should descent and inheritance be transmitted from the mother to her female descendants, the arrangement is a **matrilineal** one. Some Native American tribes, such as the five belonging to the League of the Iroquois in Pennsylvania and upper New York, were matrilineal (Zinn 2005). In other societies, descent and inheritance are bilateral, passed equally through both parents. Thus, both the father's and mother's relatives are accepted equally as part of the kinship structure. For most American families, this is the situation today.

Who has authority? With **patriarchal control**, the oldest man living in a household has authority over the rest of the family members—as in Jordan, Iraq, and China (Khouri 2005). In the purest form of patriarchy, the father is the absolute ruler. Usually, patriarchal families are extended families structured along bloodlines. So rare is **matriarchal control,** in which the oldest woman living in a household holds the authority, that controversy exists over whether any society has ever had a genuinely matriarchal family structure. With **democratic control**, authority is split evenly between husband and wife. Many families in the Scandinavian countries and in the United States follow the democratic model.

Who lives where? In a nuclear family, a married couple establishes a new residence of its own. Such a residential arrangement is a **neolocal residence**. Extended families, of course, have different residence norms. A **patrilocal residence,** as in premodern China, calls for living with or near the husband's parents. Residing with or near the wife's parents is expected under a **matrilocal residence**, such as the Nayar caste of Kerala in southern India.

Mate Selection

In the United States, most people assume they will have the freedom to choose their mate. It is true that American parents generally no longer arrange marriages for their children. American women no longer fill hope chests with quilts, china, silverware, and other household goods in anticipation of marriage. An American man is not expected to negotiate, with a future father-in-law, a price for the woman he wishes to marry. Therefore, we assume that freedom of choice prevails in the selection of a marriage partner. Actually, this is a false assumption. All societies, including the United States, have norms regarding who may marry whom.

What does society tell us about whom to marry? Exogamy refers to mate selection norms requiring individuals to marry someone outside their kind. Exogamous norms are usually referred to as incest taboos. In the classic Chinese culture, for example, two people with identical family names could not marry unless their family lines were known to have diverged at least five generations previously (Queen, Habenstein, and Quadagno 1985). In the United States, you are not legally permitted to marry a son or daughter, a brother or sister, a mother or father, a niece or nephew, or an aunt or uncle. In twenty-nine states, marriage to a first cousin is prohibited, and it is illegal to marry a former mother-in-law or father-in-law. Incest is almost universally prohibited. In fact, incest has seldom existed as an established pattern of mate selection. Royalty in ancient Europe, Hawaii, Egypt, and Peru were prominent exceptions. Even in these instances, most members of the royal families chose partners to whom they were not blood related.

Are these two just-married individuals in a homogenous or heterogamous relationship? Explain.

Endogamy involves mate selection norms that require (or at least encourage) individuals to marry within their own kind. In the United States, an endogamous norm requires that marriage partners be of the same race, ethnicity, religion, and social class. The degree to which these endogamous norms are enforced is reflected in the small percentage of interracial and interfaith marriages (2 percent).

What is the difference between endogamy and homogamy? While endogamy refers to cultural pressure to marry someone similar, **homogamy** refers to the tendency to marry someone similar to oneself based on personal preference. Such choices are made because of the greater likelihood of being attracted to and becoming intimate with someone like ourselves. Some factors leading to homogamous choice are age, educational background, degree of physical attractiveness, shared interests, and extent of religious commitment. Of course, endogamy and homogamy can converge when norms encouraging mate selection with one's race, religion, or social class are so completely internalized that they determine personal preference. But people do tend to marry someone based on one or more shared characteristics. Most marriages in the United States occur between individuals who are about the same age. Most people who have never before married become married to someone who also has always been single. Divorced people tend to marry those who have been previously married. Finally, people tend to choose marriage partners from their own communities or neighborhoods (Knox and Schacht 2008).

Although still the exception in the United States, **heterogamy** is rising. In heterogamous marriages, partners are dissimilar with respect to some important social characteristics. Presently, in America, more marriages are crossing traditional barriers of age, race, social class, and ethnicity. This trend is due to several factors. America has become a bit more racially and ethnically integrated. Although they represent only 2 percent of all marriages in the United States, mixed marriages have doubled since 1980 (Suro 1999). (Figure 11.2 shows the racial and ethnic breakdown of intergroup marriages in 2001.) Also, class lines are crossed with greater frequency because more Americans of all social classes are attending college together. Finally, norms separating age groups have weakened.

As you can see, who may marry whom in a society is not completely a matter of choice. The field of eligible marital partners is limited by norms of mate selection (Kephart and Jedlicka 1997; Kennedy 2003; Romano 2006).

Types of Marriages

Mention a wedding, and most Americans think of a bride dressed in a long, white gown walking down an aisle; the husband-to-be waiting for his bride at the end

Nruboc/Dreamstime LLC

FIGURE 11.2

Mixed Marriages and Intergroup Married Couples in the United States

Although only 5 percent of marriages in the U.S. are mixed, the number has quadrupled since 1980.

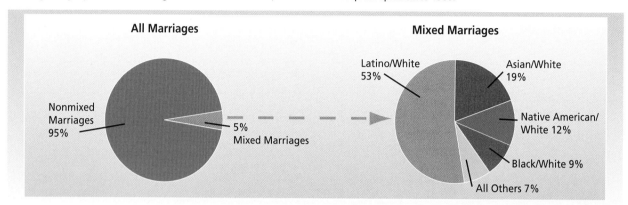

Source: American Demographics, Population Reference Bureau, 1998; Miliken Institute, 2001.

of the aisle; the repetition of ancient vows to love and honor; an exchange of rings; and a kiss. In addition, though rarely, some modern couples get married while standing on mountaintops, while diving out of planes, and while swimming underwater. In other cultures, the wedding ceremony looks even more different. Here is a description of part of the ceremony among the Reindeer Tungus of Siberia:

After the groom's gifts have been presented, the bride's dowry is loaded onto reindeer and carried to the groom's lodge. There the climax of the ceremony takes place. The bride takes the wife's place—that is, at the right side of the entrance of the lodge—and members of both families sit around in a circle. The groom enters and follows the bride around the circle, greeting each guest, while the guests, in their turn, kiss the bride on the mouth and hands. Finally, the go-betweens spit three times on the bride's hands, and the couple are formally husband and wife. More feasting and revelry bring the day to a close. (Ember and Ember 1999:310–311)

Whatever form it takes, the marriage ceremony is an important ritual announcing to the world that a man and woman have become husband and wife, that a new family has been formed, and that any children born to the couple can legitimately inherit the family name and property. The traditional marital arrangement in the United States is merely one of several possibilities.

What are the possible types of marriage? Monogamy—the marriage of one man to only one woman at a time—is the most widely practiced form of marriage in the world today. In fact, it is the only form of marriage that is legally acceptable in the United States and most other societies. Polygamy, on the other hand, which is the marriage of a male or female to multiple partners of the other sex, takes two forms: polygyny and polyandry. **Polygyny** is the marriage of one man to two or more women at the same time. The most obvious example of polygyny can be found in the Old Testament of the Bible, in which King Solomon is reported to have had 700 wives and 300 concubines. George Murdock (1957) found that polygyny was practiced in 75 percent of the 575 preliterate societies he studied. However, polygyny is not legal in any Western society (Paddock 1999). As stated in its 1948 statement on human rights, the United Nations explicitly frowns on the practice of polygyny. **Polyandry**—the marriage of one woman to two or more men at the same time—is an even rarer form of marriage, found exclusively in the South Asian countries of Tibet, Nepal, and India (Stockard 2002). It is known to have been common in only two societies: in Tibet and among the Todas of India (Queen, Habenstein, and Quadagno 1985). Where polyandry has existed, it usually has consisted of several brothers sharing a wife (fraternal polyandry). Instances of a father sharing a wife with his sons have been reported, but not with any degree of regularity. A group marriage consists of two or more men married to two or more women at the same time. This form of marriage is also rare. In fact, researchers cannot identify even one society in which group marriage has been the normal form. There have, however, been communes that have practiced group marriage (Berger, Hackett, and Millar 1972; Kanter 1972; Ferrar 1977).

This analysis of the family and marriage is complicated because of the large number of concepts. Table 11.1 summarizes these concepts within the context of the traditional (extended) and modern (nuclear) family structures.

F E E D B A C K

1. A _____ is a group of people related by marriage, blood, or adoption.
2. A family of _____ is the family into which a person is born.
3. A family of _____ is the family established upon marriage.
4. The _____ family consists of two or more generations of the same family sharing a common household and economic resources.
5. There is a _____ relationship between types of family structure and industrialization.
6. Industrialization promotes the shift from the extended family to the _____ family.
7. In a _____ family, the oldest man living in the household controls the rest of the family members.
8. Indicate which type of mate selection norm is reflected in each of the following situations.

    ____ a. Jews are supposed to marry Jews.   (1) exogamy
    ____ b. A father is not permitted to marry his daughter. (2) endogamy
    ____ c. Members of the same social class marry.   (3) homogamy

9. _____ is the marriage of a woman to two or more men at the same time.

Answers: 1. family; 2. orientation; 3. procreation; 4. extended; 5. curvilinear; 6. nuclear; 7. patriarchal; 8. a. (2), b. (1), c. (3); 9. Polyandry

Theoretical Perspectives and the Family

The family is as central to society as it is complex. It is not surprising that each of the major theoretical perspectives sheds light on this institution.

Functionalism

You were born into a family with whom you have lived for almost twenty years. You are likely to marry, and despite the high rate of divorce, the overwhelming majority of your adult life will be spent within a family.

The family performs many essential functions. First, it provides the initial learning experiences that make people human. Second, it generally provides a supportive and loving atmosphere that fulfills basic human social and emotional needs. Third, it is the only legitimate source of reproduction for a society. Fourth, it regulates sexual activity. Fifth, it places people in a social class at birth. Finally, the family serves an important economic function. Because of these vital functions, the family is often referred to as the basic institution of society. Each function deserves elaboration.

TABLE 11.1

Major Characteristics of Extended and Nuclear Families

Examine the major characteristics of extended and nuclear family structures. Do you live in a nuclear or extended family? Explain by describing your family based on each characteristic.

Characteristics	Extended Family	Nuclear Family
Family structure	Extended (also some nuclear)	Nuclear (also some extended)
Basis of family bond	Blood (consanguine)	Marriage (conjugal)
Line of descent and inheritance	Patrilineal (male lineage)	Bilateral (dual parental lineage)
	Matrilineal (female lineage)	
Locus of control	Patriarchal (male dominance)	Democratic (sexes share power)
	Matriarchal (female dominance)	
Place of residence	Patrilocal (husband's parents)	Neolocal (independent)
	Matrilocal (wife's parents)	
Marriage structure	Monogamy (one spouse)	Monogamy (one spouse)
	Polygyny (several wives)	
	Polyandry (several husbands)	
	Group (several husbands and wives)	

How does the family contribute to the socialization process? A major function of the family begins immediately after a baby has been born. In addition to caring for an infant's physical needs, parents begin the vital process of teaching the child necessities for participating in society. During the first year, the infant begins to mimic words and, later, sentences. During the second and third years, parents begin to teach the child customary ways of behaving. By providing role models, training, and education, the process of family socialization continues as the child enters new stages of development.

What is the socioemotional function of the family? Another major function of the family is "socioemotional maintenance" (Nimkoff 1965). This begins in the first year with intimate holding, touching, stroking, and talking to the baby. Generally, the family provides the one supportive environment in society in which an individual is unconditionally accepted and loved. Family members accept one another as they are; every member is special and unique. Without this care and affection provided by parents (or other adults), children will not develop normally. They may have low self-esteem, fear rejection, feel insecure, and eventually find it difficult to adjust to marriage or to express affection to their own children. In fact, some young children who do not receive love and attention become retarded or die (see Chapter 4). Even individuals who are well integrated into society require support when adjusting to changing norms and when developing and continuing healthy relationships. Here again, the family can provide socioeconomic maintenance.

What is the reproductive function of the family? Society cannot survive without new members. The family's reproductive function provides an orderly mechanism for producing generation after generation. So important is this function that for many cultures and religions, it is the primary purpose for sexual relations. In many societies in developing nations, the failure of a wife to bear children can lead to divorce. Residents of places like the Punjab region of northern India, for example, view children as an economic necessity. The significance of having children is also seen in the hundreds of rituals, customs, and traditions associated with pregnancy and childbirth in virtually all cultures around the world. It is customary in the United States, for instance, to honor the mother-to-be with a baby shower. After birth, babies are displayed in hospital nurseries so that friends and relatives can marvel at them. Among the Sirono Indians of eastern Bolivia, there are three days of rituals celebrating the birth of a child. These rituals are designed to protect the life of the infant and to ensure the child's good health (Holmberg 1969).

In what ways does the family regulate sexual activity? In no known society are people given total sexual freedom. Even in sexually permissive societies, like that of the Hopi Indians, there are rules about mating and marrying (Queen, Habenstein, and Quadagno 1985). The function of sexual regulation is usually assigned to the family.

Norms regarding sexual activities vary from place to place, but whatever the norms, it is almost always up to the family to enforce them. To illustrate, in the Trobriand Islands off the coast of Papua New Guinea, all children are encouraged to engage in premarital sex (Malinowski 1929). Thus, much of childhood is spent going from partner to partner, as a bee goes from flower to flower. By contrast, other societies, like those in Iran or Afganistan, go to great lengths to prevent any contact between unrelated single males and females. In Muslim countries, for instance, the bride's family must guarantee that she is a virgin; the honor of the family rests on that guarantee (E. L. Peters 1965). In the United States, the norm has traditionally fallen between these two extremes.

Sexual norms also change within a society. In colonial America, sex outside of marriage was considered a sin. But after Alfred Kinsey's studies of sexual behavior hit the

Stephen Wilkes/Getty Images

More than 25 percent of working wives earn more than their husbands. How do you think the fact that the wives out earn their husbands changes the balance of power and responsibility in these marriages?

TABLE 11.2

American Youths Grade Their Parents

In a national survey, Americans in the seventh through the twelfth grades were asked to "grade" their mothers and fathers. The results are shown below. The left-hand column lists various aspects of child rearing, and the remaining columns indicate the percentage of students who assigned each grade. For example, on the dimension "Raising me with good values," 69 percent gave their fathers an A, 17 percent a B, and so forth.

Grading Dad — **Assigned Grade**

Aspect of Child Rearing	A	B	C	D	F
Raising me with good values	69%	17%	8%	4%	2%
Appreciating me for who I am	58%	21%	11%	8%	2%
Encouraging me to enjoy learning	58%	24%	12%	4%	2%
Making me feel important and loved	57%	22%	13%	6%	2%
Being able to go to important events	55%	22%	13%	5%	5%
Being there for me when I am sick	52%	20%	16%	8%	4%
Establishing traditions with me	41%	26%	15%	11%	7%
Being involved in school life	38%	24%	19%	12%	7%
Being someone to go to when upset	38%	22%	15%	12%	13%
Controlling his temper	31%	27%	20%	10%	12%
Knowing what goes on with me	31%	30%	17%	12%	10%

Grading Mom

Aspect of Child Rearing	A	B	C	D	F
Beign there for me when I am sick	81%	11%	5%	2%	1%
Raising me with good values	74%	15%	6%	3%	2%
Making me feel important and loved	64%	20%	10%	5%	1%
Being able to go to important events	64%	20%	10%	3%	3%
Appreciating me for who I am	63%	18%	8%	6%	5%
Encouraging me to enjoy learning	59%	23%	12%	3%	3%
Being involved in school life	46%	25%	13%	10%	6%
Being someone to go to when upset	46%	22%	13%	8%	9%
Spending time talking with me	43%	33%	14%	6%	4%
Establishing traditions with me	38%	29%	17%	10%	6%
Knowing what goes on with me	35%	31%	15%	10%	9%
Controlling her temper	29%	28%	19%	12%	11%

1. Based on this data, what conclusions would you draw about the closeness of families in America?
2. Select the three aspects of child rearing you think are the most important, and compare the grade you would give your parent or parents on these aspects with the grades in this national sample.

Source: Ellen Galinsky, *Ask the Children* (New York: Harper Collins, 2000).

American public like a bombshell (Kinsey, Pomeroy, and Martin 1948; Kinsey and Gebhard 1953), most experts agreed that a sexual revolution had occurred.

How does the family transmit social status? As noted in Chapter 8, an individual's place in the stratification structure is largely assigned at birth. If people move up or down the stratification structure, it is usually only a very slight move. Transmission of social class occurs primarily through the family. This is partly because families provide economic resources that open and close occupational doors. The sons and daughters of high-income professionals, for example, are more likely to attend college and graduate school than are the children of blue-collar workers. Consequently, the children of professionals are more likely as adults to enter professional occupations. The family also transmits values that later affect social status. The children of professionals, to cite only one illustration, tend to feel a greater need to pursue a college degree than do their counterparts from blue-collar families. In these and many other ways, the family affects the placement of children in the stratification structure.

What is the economic function of the family? At one time, families were self-sufficient economic units whose members all contributed to the production of needed goods—growing food, making cloth, and taking care of livestock. Malls, corporations, and supermarkets were not part of the scene. In contrast, the modern American family is a unit of consumption rather than production. Adult members, increasingly including working mothers, work outside the home and pool their resources. But the end result is the same: the family provides the economic necessities of life.

Conflict Theory

Functionalism emphasizes the ways in which the family attempts to contribute to the order and stability of society. The conflict perspective draws attention to social concerns such as sexual inequality.

How was the conflict perspective first applied to the family? The interest of conflict theorists in family gender inequality can be traced to Friedrich Engels, a collaborator of Karl Marx. Engels saw women's oppression as the result, first, of women's loss of a productive role, and second, of men's interest—with the development of private property and surplus wealth—in utilizing monogamy to confirm paternity (Engels 2001; originally published in 1884). To Engels, then, the family has historically been used to maintain male domination of females.

How do the ideas of early feminist writers fit with Engels's approach? Some feminists share with conflict theorists the belief that families reflect social inequality and experience conflict over resources (Benokraitis 2007). Because these feminists focus on patriarchy (control by men over women), they point out that the domination by men over women predates private ownership of property and capitalism (P. Mann 1994). Conflict in the family, according to these feminists, is primarily the result of women's attempt to gain more power within the family structure.

Do conflict theories still endorse this viewpoint? This theme, which is still being advanced by conflict theorists, bears on at least three issues: the domination of women by men, family rules of power and inheritance, and the male-dominated economic division of labor. Because most family structure throughout history has been patriarchal, women have traditionally been considered the property of men—first of fathers, then of husbands. Only since 1920 have American women been permitted to vote in federal elections (after a tremendous struggle), and only since 1974 have women gained the right to make contracts and obtain credit independently of husbands or fathers. Even so, the connotation of women as the property of men often continues, despite the trend toward gender equality in advanced industrial societies (Collins 1971). Moreover, because most arrangements have been patrilineal as well, male control of family members and male ownership of property typically have been transmitted through male bloodlines. This has created built-in gender inequality in most family systems because male dominance is considered "natural" and "legitimate."

The predominant patrilineal and patriarchal family patterns are also related to the traditional economic division of labor, which again perpetuates gender inequality. In the traditional division of labor, males work outside the home for finances to support the family, while

---

TABLE 11.3

FOCUS ON THEORETICAL PERSPECTIVES: Perspectives on the Family

Both functionalism and conflict theory are more concerned with the ways social norms affect the nature of the family. Symbolic interactionism tends to examine the relationship of the self to the family. If functionalism and conflict theory were used to focus on the self, what examples would you give?

Theory	Topic	Example
Functionalism	Sex norms	Children are taught that sexual activity should be reserved for married couples.
Conflict theory	Male dominance	Husbands use their economic power to control the ways money is spent.
Symbolic interactionism	Developing self-esteem	A child abused by her parents learns to dislike herself.

FEEDBACK

1. Which of the following functions of the family is not shared with any other institution?
   a. socialization
   b. reproduction
   c. socioemotional maintenance
   d. sexual regulation

2. Match the following examples with the major theoretical perspectives:

   ____ a. a father "giving away" the bride   (1) functionalism
   ____ b. sexual regulation   (2) conflict theory
   ____ c. development of self-concept   (3) symbolic interactionism
   ____ d. a newly married couple adjusting to each other
   ____ e. child abuse
   ____ f. social class passed from one generation to another

Answers: 1. b; 2. a. (2), b. (1), c. (3), d. (3), e. (2), f. (1)

women remain at home to prepare meals, keep house, and care for the children. Women are unpaid laborers who make it possible for men to earn wages. This economic power enables males to maintain dominance over the economically dependent wife and mother.

These traditional patterns have been altered by industrialization. But despite some changes, the persistent social, economic, occupational, political, and legal inequality of women continues, as we saw in Chapter 10.

Symbolic Interactionism

What is the relevance of symbolic interactionism to the study of the family? Symbolic interactionism is frequently used in the study of marriage and the family (Waller and Hill 1951; Leslie and Korman 1989). According to this perspective, a key to understanding behavior within the family lies both in the interactions among family members and in the meanings that members assign to these interactions. Within the family, all the major concepts of symbolic interactionism can be observed—socialization, looking-glass self, role taking, primary group, reference group, significant others, symbolic interpretation of events, and symbolic communication (Waller and Hill 1951; Leslie and Korman 1989).

Socialization begins within the family. As family members share meanings and feelings, children develop self-concepts and learn to put themselves mentally in the place of others. Interaction with adults permits children to acquire the personality and social characteristics associated with human beings. With the repertoire of personality and social capabilities learned within the family, children are able to develop further as they interact with people outside the home.

According to symbolic interactionists, relationships within the family are constantly being redefined. A newly married couple will spend many months (perhaps years) testing their new relationship. As time

passes, the initial relationship changes, along with some aspects of the partners' personalities—including self-concepts. These changes occur as the partners grapple with such problem issues as distribution of responsibilities, personality clashes, and in-law distractions. With the arrival of children comes a new set of adjustments. Parental views may differ on number of children desired, child-rearing practices, and education for the children. The situation is made even more complex by the new member(s) of the family, who must also become part of the interaction patterns.

Family and Marriage in the United States

The Nature of the American Family

The **marriage rate**—the number of marriages per year for every 1,000 members of a population—has fluctuated in the United States since 1940. As shown in Figure 11.3, the marriage rate peaked at over 16 per 1,000 immediately following World War II. Since then, the marriage rate, with some ups and downs, has been cut in half.

Some claim that there is no "typical American family." After all, the United States is a large, diverse society in which various groups—reflecting vastly different cultural heritages—have blended together. The early white settlers came primarily from Holland and England. Not long after they arrived, many Africans were brought in as slaves. Subsequently came waves of immigrants from northern Europe, Ireland, Italy, the Slavic and Baltic countries, China, Japan, Southeast Asia, and, more recently, countries in Latin America.

Still, there are more similarities than differences among American families. As the various ethnic groups

FIGURE 11.3

Divorce and Marriage Rates: 1940–2005

Can you apply what you learned in history to interpret this chart? (a) What happened in the mid–1940s that caused the dramatic rise in the marriage rate during this period? (b) Why do you think the marriage rate dropped so low in the 1950s? (c) What are some possible reasons the divorce rate peaked in the 1980s?

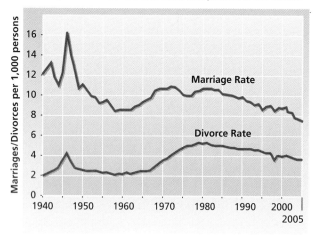

Sources: National Vital Statistics Reports; *Statistical Abstract of the United States.*

assimilate into American life, most families tend to follow the American pattern: containing only the parents and children in the same household (nuclear), tracing lineage and passing inheritance equally through both parents (bilateral), dividing family decision making equally (democratic), establishing a separate residence (neolocal), and following the norm of one woman married to one man at any particular time (monogamous).

A large family was once the norm in American society. In 1971, almost three-fourths of American adults thought that three or more children would be the most desirable number (Roper Organization 1985). In 2007, the average number of children born to American women was 2.1, enough to replace the country's population for the first time in 35 years (*World Population Data Sheet* 2007).

Why have large families declined in popularity? Many factors contribute to the smaller American family. Both men and women are marrying later and staying in school longer. Increasing numbers of women are in the labor force, for financial rewards as well as personal fulfillment (McLaughlin and Associates 1988). Legalized contraception and abortion have also reduced family size. Two additional factors promoting a lower fertility rate are the high cost of children and the removal of the stigma on childless marriages. (More is said about childless marriages later in the chapter.)

Americans are increasingly choosing alternatives to the family pattern of mother (as homemaker), father (as breadwinner), and children: slightly less than one-fourth of American households consist of a married couple with children under age eighteen. These new family and marriage arrangements include single-parent families, childless marriages, dual-employed marriages, single life, cohabitation, and gay and lesbian couples, all to be discussed later in the chapter.

Romantic Love as the Basis for Marriage

In a recent poll of the American public, over 80 percent of both men and women rated "being in love" as the most vital reason to marry. The relationship between love and marriage, though, has not always been viewed in this way. Among the British feudal aristocracy, romantic love was a game of pursuit played outside of marriage. Marriage was not thought to be compatible with deeply romantic feelings. In ancient Japan, love was considered a barrier to the arrangement of marriages by parents. Among Hindus in India today, parents or other relatives are expected to find suitable mates for the young. Criteria for mate selection include caste, wealth, family reputation, and appearance. Love is not absent in Hindu marriages, but love follows marriage rather than the other way around (F. D. Cox 2005).

Are modern marriages based on romantic love alone? In modern societies, romantic love is almost always stated as a condition for marriage, but it is seldom the only one. People marry for many reasons, and love may or may not be one of them. They may marry to obtain regular sex partners, to legitimate living together, to enter a rich and powerful family, or to advance a career. One of the strongest motivations for marriage is conformity. Throughout our formative years, we are told what to do: "Feed yourself," "Join the Girl Scouts," "Learn to drive," "Go to college," "Get a job." Many parents also expect their children to marry after a certain age and worry about them—perhaps even pressure them—if they remain single very long. Peers are another source of pressure. Because well over 90 percent of all adults in the United States do marry, conformity must certainly be a motivating factor (Saad 2006a). In fact, most Americans are married by their mid-twenties.

Although Americans typically believe that a marriage that is not based on romantic love cannot last, it may be more accurate to say that a marriage based only on romantic love is almost sure to fail. According to research, love may be a good start, but it is only the beginning. For a marriage to last, a couple must build a relationship that goes beyond romantic love (J. F. Crosby 1985).

WORLDVIEW 11.1

Percentage of Women Aged 15–19 Ever Married

The age at which Americans marry has been rising for decades. Consequently, only a very small percentage of American teenagers are married. This is not the case in all countries around the world, however. This map shows the percentage of teenage women in each country who have ever been married.

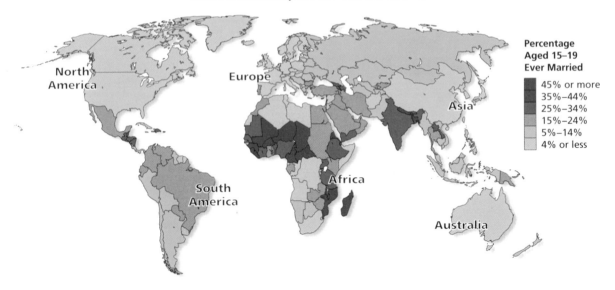

Percentage Aged 15–19 Ever Married

- 45% or more
- 35%–44%
- 25%–34%
- 15%–24%
- 5%–14%
- 4% or less

Source: United Nations Statistics Division, 2005.

Interpret the Map

1. What two general statements can you make about the information shown on the map?

2. How might a high percentage of teenage marriages affect a country's population size?

Divorce

What is the current divorce situation? The **divorce rate** is the number of divorces annually for every 1,000 members of the population. Except for a peak and decline after World War II, the divorce rate in the United States increased slowly between 1860 and the early 1960s. A dramatic increase in the divorce rate occurred over the next twenty years. It more than doubled from 2.2 in 1960 to 5.3 in 1981. Since then, as indicated in Using the Sociological Imagination at the beginning of this chapter, the rate has leveled off, declining slightly since 1985. The divorce rate, which peaked at 5.3 in 1981, stood at just less than 4 in 2005 (see Figure 11.3).

In 1940, the marriage rate was 12.1 per 1,000 people, whereas the divorce rate was only 2 per 1,000 persons. Now the rate of divorce is half the rate of marriage. Based on these data, it is correct to say only that the divorce rate in the United States is now half its mar-

riage rate. It is wrong to conclude that half of all marriages end in divorce, because the divorce and marriage rates are snapshots of a single year and do not take into account the total number of divorces and intact marriages from all years past.

This is why sociologists also employ the **divorce ratio**—the number of divorced persons in the population divided by the number of persons who are married and living with their spouses. In the United States in 2006, there were 23 million divorced persons and 128 million persons married and living with their spouses. Dividing the number of divorced persons by the number of married persons yields a divorce ratio of 180. This divorce ratio is five times the 1960 ratio of 35. Women generally have a higher divorce ratio than men (204 versus 153 in 2006), reflecting the greater tendency of men to get remarried after divorce and to do so more quickly than women (U.S. Bureau of the Census 2007c).

These figures provide some basis for optimism for the future of the American family, just as the divorce rates of recent years do. Between 1960 and 1970,

Looking for Mr. or Ms. Right

This activity will give you some ideas for evaluating whether a current boyfriend or girlfriend is a good candidate for a successful long-term relationship.

From the list on the right (and on a separate sheet of paper), list the ten most important qualities to you. (Number 1 as the most important, number 2 the next more important, and so forth.) Then fold your paper in half. In the right-hand column, either have your partner fill out the questionnaire or rank the characteristics yourself according to how you think your partner would.

Which of the items listed on the right do you think are the most important in predicting marital success? According to research, the last seven items (17–23) are the most important. High compatibility between you and your partner on these seven characteristics would probably increase your chances of marital success. A low degree of matching does not, of course, ensure an unhappy marriage or a divorce, but it does suggest areas that may cause problems in the future.

Adapted from the Department of Human Development and Family Studies, Colorado State University.

Think About It

Do you think that the qualities listed in the questionnaire are relevant to your choosing a wife or a husband? Why or why not? Are there characteristics more important to you and your friends? Explain.

I am looking for a partner who . . .

	Partner	Self	
1.	_____	_____	is honest and truthful.
2.	_____	_____	is fun to be with.
3.	_____	_____	is of the same educational background.
4.	_____	_____	will take care of me.
5.	_____	_____	wants to have children.
6.	_____	_____	communicates well with me.
7.	_____	_____	will share household jobs and tasks.
8.	_____	_____	is a good friend with whom I can talk.
9.	_____	_____	is of the same religious background.
10.	_____	_____	makes decisions.
11.	_____	_____	earns good money.
12.	_____	_____	is physically attractive.
13.	_____	_____	is in love with me and I with him/her.
14.	_____	_____	encourages me to be my own person.
15.	_____	_____	has interests like mine in making money and having fun.
16.	_____	_____	makes me feel important.
17.	_____	_____	is faithful.
18.	_____	_____	shares mutual interests in home, children, romantic love, and religion.
19.	_____	_____	has had a happy childhood with happily married parents.
20.	_____	_____	is emotionally mature.
21.	_____	_____	is prepared to support a family.
22.	_____	_____	is interested in waiting to marry until age twenty-two or older.
23.	_____	_____	wants a six-month to two-year engagement period.

the divorce ratio in the United States increased by an annual average of 1.2 (from 35 in 1960 to 47 in 1970). During the 1970s, the average annual increase rose sharply to 5.2 (from 47 in 1970 to 100 in 1980). Since 1980, the average annual increase has actually declined. In other words, although the divorce ratio is still rising each year, its rate of increase has declined since 1980 (Cherlin 2002). Still, the U.S. divorce rate is high among industrialized countries (Ahlburg and DeVita 1992; F. D. Cox 2005).

Why the high U.S. divorce rate? There are no easy answers to this question; like family and marriage, divorce is a complex matter. Those who study marriage and divorce have sought answers at both individual and societal levels.

At the individual level, the following factors seem to be associated with divorce (Knox and Schacht 2008). First, the earlier one marries, the greater the likelihood of divorce. Second, the longer a couple has been married, the lower the probability that their marriage will end in divorce. The average length of first marriages ending in divorce in the United States is about six years. However, the largest number of first marriages ending in divorce occurs between the second and third anniversaries. This suggests that poor decisions in mate selection may be as important as marital conflict. Third, as might be expected, divorce is related to the nature and quality of the marital relationship. The more respect and flexibility taking place between partners, the lower the chance of divorce.

At the societal level, several factors seem to affect the divorce rate. First, divorce rates increase during economic prosperity and decrease during economic recession or depression. This is probably because economic prosperity permits people to concentrate on issues beyond survival and to consider options other than marriage for their personal happiness. Second, the rise in the divorce rate in the United States in the 1960s reflects the passage of the baby boom generation into the marriageable ages. Not only are there greater numbers of them, but baby boomers as a group are much more forgiving of divorce and remarriage than are earlier generations. Third, because women are more economically independent and because they have a wider choice of child care options, they are less hesitant to dissolve a bad marriage. Finally, American values and attitudes about marriage and divorce have changed. The stigma once associated with divorce is much weaker today (R. Farley 1998).

What about future divorce rates? The recent decline in the U.S. divorce rate and divorce ratio may continue for several reasons. First, we know that the later people marry, the less likely they are to divorce. (This happens in part because more mature individuals have more realistic expectations about their relationships and because they have fewer economic and career problems that create emotional strain.) The average age at first marriage is, in fact, increasing in the United States (see Figure 11.4). In 1970, the average age at first marriage was 23.2 for men and 20.8 for women. The average age has increased to 26.8 for men and 25.1 for women. This trend is likely to continue well into the twenty-first century. Second, the average age of the population of the United States is increasing as baby boomers grow older. It was this exceptionally large generation that set records for divorce in the late 1960s and 1970s. Baby boomers now range in age from late thirties to mid-fifties, removing them from the age bracket that produces the highest divorce rates. Third, American couples are having fewer children and the children are spaced farther apart. This reduces pressure.

The recent decline in the U.S. divorce rate is encouraging. At the very least, it is the first break in a long-term increase. Still, many experts are cautious about predicting the future, pointing to the fact that the extent of the divorce rate is cyclical and that the recent decline may be only temporary.

Is the rising divorce rate unique to the United States? The mass media often leave the impression that the United States is experiencing a rising divorce rate due to social factors peculiar to this country. In fact, since 1960 the divorce rate has risen more than fivefold in Canada and the United Kingdom, almost tripled in France, and doubled in Germany and Sweden. The divorce rate has also gone up in Japan but to a lesser degree (see Figure 11.5). On the other hand, countries in Asia, Central America, and South America have significantly lower divorce rates.

To understand this worldwide variation in divorce rates, it is necessary to identify the major socioeconomic factors influencing these rates. Increasing prosperity, to cite one important factor, is associated with higher divorce rates. Consider China, whose divorce rate has risen with its recent economic development. China's divorce rate is now twice as high as it was in 1990. Because of social and economic development, more women are initiating divorce as their economic opportunities and rights increase.

The correlation between economic development and divorce rate is far from perfect. For example, the divorce rate is higher in Belarus despite its lower economic development, while the divorce rate is lower in a highly prosperous country such as Japan. This is because the divorce rate is affected by additional social factors.

FIGURE 11.4

Median Age at First Marriage

This figure displays changes in the median age at first marriage in the United States since 1900. The marrying age for both men and women has been on the increase since the 1970s. Discuss the implications of this trend for the future divorce rate.

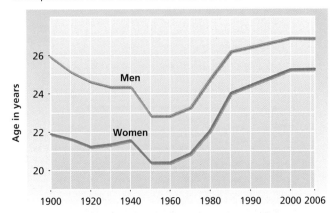

Source: U.S. Bureau of the Census, *Current Population Reports*, Series P20-537, "America's Families and Living Arrangements: 2000," and earlier reports; at www.census.gov/prod/2001pubs/p20-537.pdf. *The World Almanac and Book of Facts 2008* (New York: World Almanac Books, 2008, p.158).

Religion is a second factor affecting divorce rates, particularly in less developed countries. The predominance of Roman Catholicism in Central and South America, for example, diminishes the frequency of divorce. Religion also affects divorce, independently of level of socioeconomic development, as seen in Roman Catholic dominated Italy and Ireland.

A third factor retarding the divorce rate is a patriarchal power structure. The more patriarchal a society, the lower its divorce rate. In many Middle Eastern countries, for example, a husband can dissolve his marriage by saying "I divorce thee" three times in the presence of witnesses. A woman wishing a divorce, however, must convince a court of her husband's economic neglect and his morally corrupting influence on the family. And only recently have Egyptian women even been allowed to file for divorce in the absence of extraordinary maltreatment. Unless they give considerable wealth to a willing husband, Taiwanese women are discouraged from seeking divorces because their children will automatically be in the sole custody of their husbands (Chang 1998).

Family Violence

Americans, like the rest of the world, have traditionally denied the existence of widespread family violence (*The World's Women* 2005). Its occurrence was mistakenly associated most often with lower-class families,

---

FIGURE 11.5

Divorce Rates by Country

This figure compares divorce rates by country. Using the two extremes (Belgium and Turkey), describe to a friend the relative condition of marriage in these two countries. Also describe a few instances where the divorce rates surprised you.

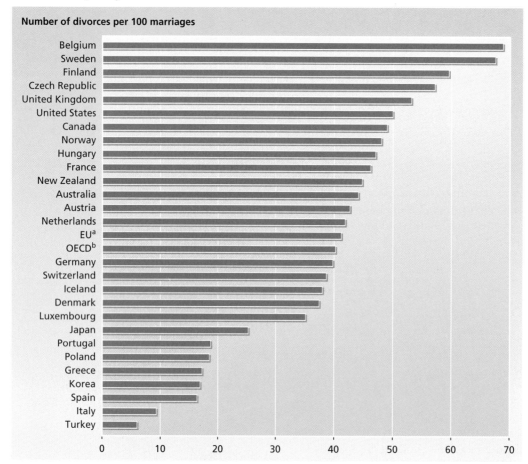

Number of divorces per 100 marriages

Sources: Eurostat and national sources (see the annex on the Internet). *Society at a Glance: OECD Social Indicators* (Paris, France: Organisation for Economic Co-operation and Development, 2001), p. 33.
[a]EU = European Union
[b]OECD = Organisation for Economic Co-operation and Development; members include Australia, Austria, Belgium, Canada, the Czech Republic, Denmark, Finland, France, Germany, Greece, Hungary, Iceland, Ireland, Italy, Japan, Korea, Luxembourg, Mexico, the Netherlands, New Zealand, Norway, Poland, Portugal, the Slovak Republic, Spain, Sweden, Switzerland, Turkey, the United Kingdom, the United States.

In patriarchal societies, the divorce rate is lower. Why?

Arabian Eye/Getty Images

with maybe a few exceptional cases in the middle and upper classes.

There is a good reason for this erroneous picture. Early research was done using law enforcement and public medical records. Because law enforcement officials and public hospitals were more likely to be aware of lower-class family violence, the statistics were skewed toward these families. Middle- and upper-class families disproportionately conceal their domestic violence (S. Weitzman 2000).

However, the 1990s brought heightened public attention to domestic violence (Ingrassia and Beck 1994; Smolowe 1994; Jasinski 2000). The mass media made every effort to give Americans the details of several dramatic legal cases. Erik and Lyle Menendez admitted planning and carrying out the murder of parents they accused of abusing them. In a televised trial, O. J. Simpson was tried for the wrongful death of his wife, Nicole Brown Simpson, and her friend, Ron Goldman. Evidence indicated that Simpson had abused his wife when they were married. And claiming past abuse, humiliation, and forced sex, Lorena Bobbitt severed her husband's penis while he slept (Kaplan 1994). In 2008, four children, ages five to seventeen, were found dead in their home. Their mother was charged with murder.

How widespread is family violence? Although the family does, as the functionalists contend, usually provide a safe and warm emotional haven, it can also be a hostile environment. Assault and murder are more likely to be the acts of someone living in the home. More than one-fifth of all reported cases of aggravated assault involve domestic violence. The situation is actually much worse because most episodes of domestic violence go unreported. Perhaps as little as 10 percent of information about such cases reaches the authorities (Hampton et

al. 1993; Gelles 1997; Gelles et al. 2004). Domestic violence is not reserved for any particular members of the family. It involves children, spouses, siblings, and older people ("Violence Against Women" 2002; Barnett, Perrin, and Miller-Perrin 2004).

Parental abuse of children may begin in the womb. A large percentage of assaults on wives by their husbands occur during pregnancy; birth defects and miscarriages are common results. In fact, such abuse may lead to more birth defects than all other childhood diseases combined. Nor do all children grow up in a loving home, nurtured by caring parents (Miller-Perrin and Perrin 2006). About one-quarter of adults in the United States report being physically abused as children. In most cases, physical violence involves a slap, a shove, or a severe spanking. However, kicking, biting, punching, beating, and threatening with a weapon are part of abusive violence on the part of many parents. Furthermore, according to estimates, one of every four girls and one in ten boys are victims of sexual aggression either within the home or outside. Child sexual abuse goes beyond physical contact. Some children are forced into pornography or are made to view pornography in the presence of the abuser. What's worse, the abuser is usually someone the child trusts—a parent, friend of the family, child caregiver, brother. About 900,000 children in the United States were determined to be victims of child abuse in 2005. More than 40 percent of the victims were five years of age or younger (U.S. Department of Health and Human Services 2007).

As many as one-half of all married women in the United States are victims of spousal violence (Gibbs 1993b). At least 4 million women are battered by their husbands annually, probably many more. More than 4,000 women each year are beaten to death. Rape and sexual assault are common. As many as 14 percent of married women are sexually attacked by their husbands every year. The extent of physical abuse is underestimated, in part because three-fourths of spousal violence occurs during separation or after divorce and most research is conducted among married couples. (See Table 11.4 for data on levels of domestic violence against women in selected countries.)

What does research reveal about husband abuse? Husband abuse is frequently overlooked (Steinmetz 1988; Wiehe 1990). Although marital relationships in the United States are generally male dominated, there seems to be spousal equality in the use of physical violence (Straus and Steinmetz 2006). Patricia Pearson and Michael Finley (1997) found that almost one-third of the husbands in their survey had acted violently against their wives and that wives were almost as likely to have used physical violence against their husbands. Other studies also show that husbands and wives assault each other at about the same rate (Hotaling et al. 1988; Flynn 1990). Much of the violence on the part of women, however, involves self-protection or retaliation. Whoever initiates the violence, women are more likely to suffer greater injury because the average man is bigger, stronger, and more physically aggressive (Gelles 1997).

Is abuse always physical? Domestic violence, of course, is not limited to physical abuse. Verbal and psychological abuse are a part of many families (M. D. Schwartz 1997; Jacobson and Gottman 1998; Knox and Schacht 2008). This is seen in parent–child relationships as well as spousal relationships. Furthermore, psychologists report that the feelings of self-hate and worthlessness that are often the effects of such abuse can be as damaging as physical wounds. In addition, neglect, a condition of being ignored rather than abused, occurs at four times the rate of physical abuse, involving more than half a million children in 2005 (Helfer, Kempe, and Krugman 1999; U.S. Bureau of the Census 2007c).

What other forms of family violence are of concern? Probably the most frequent and most tolerated violence in the family occurs between children. Much violence occurs between children who are too young to channel their frustrations in nonphysical channels. Abuse among siblings may be based on rivalry, jealousy, disagreements over personal possessions, or incest. Although violence among siblings declines somewhat with age, it does not disappear. Sibling violence appears to be prevalent and on the rise. According to one study, violence in the family is most likely to come from children (Hotaling et al. 1988).

Little is known about the abuse of elderly people because less research has been done in this area. Abuse of older people usually takes the form of physical violence, psychological mistreatment, economic manipulation, or neglect. Normally, the abuser is a family member (Pillemer and Finkelhor 1988). Estimates of elder abuse range from 500,000 to 2.5 million cases annually (Gelles 1997). There is fear of an increase in abuse of older people as the population bulge, represented by the baby boomers, enters old age. For one thing, there will be fewer working adults to support a growing, aging segment of the population (Kart 1990). The most likely to be abused are white Protestant women over seventy-five years of age who are mentally, physically, or financially disadvantaged (Garbarino 1989). Older women represent a higher proportion of the abused, in part because they live longer than men. There are simply more elderly women with greater dependency needs. Abuse of the elderly is partly due to the psychological, physical, and economic strain of caring for elderly persons. These strains of caretaking may lead to hostility and depression, followed by violence. Elder abuse may involve a dependent adult child who has moved back into his or her parents' home as a result of divorce or financial problems (Gelles 1997), or it may occur in nursing homes.

The high incidence of divorce, violence, and abuse has motivated sociologists to examine the strength and durability of the American family. One area of research focuses on what is being called *family resiliency*.

Family Resiliency

What is family resiliency? Synonyms for the word resilient include *flexible, supple,* and *buoyant.* Thus, resiliency implies an ability to cope successfully with shock. **Family resiliency** refers to the family's capacity to emerge from crises as stronger and more resourceful. A resilient family is able to flourish despite distress.

What factors promote family resiliency? Sociologists have identified four sets of factors promoting resiliency. First, family resiliency is enhanced by *individual* characteristics such as self-esteem, autonomy, sense of humor, and problem-solving skills. Second, resiliency is engendered by *family* characteristics such as emotional support, commitment, warmth, affection, and cohesion. Third, resiliency is nurtured by *community* characteristics such as ample opportunities for participation in community life, emphasis on helping others, avenues for communication with friends and adults, and availability of youth activities. Fourth, resiliency is increased via family-friendly *public policy.* Because it is the most abstract factor, public policy requires some elaboration.

What role does public policy play in promoting family resiliency? **Public policy** is a broad course of governmental action expressed in specific laws, programs,

TABLE 11.4

Events of Domestic Violence Against Women in Selected Countries

Levels of domestic violence against women clearly vary from country to country.

Country	Reported Abuse
Industrialized Countries	
Canada	27% of ever-married/common law–partnered women report being physically assaulted by a current or former partner since the age of sixteen.
Norway	18% of women report being hit or physically abused by a male partner.
Switzerland	21% of women report being physically assaulted.
United Kingdom	30% of women had been punched or slapped by a partner or ex-partner in their lifetimes.
United States	22% of women report at least one episode of physical violence from their partner.
Asia and the Pacific	
Korea	38% of wives report being physically abused by their spouses in the last year.
Thailand	20% of women report physical assault by a partner in their current relationship.
Middle East	
Egypt	34% of women report being beaten by their husbands at some point in their marriage.
Israel	32% of women report at least one episode of physical abuse by their partners during the last twelve months; 30% report sexual coercion by their husbands in the last year.
Africa	
Kenya	42% of women report ever being beaten by a partner; of those, 58% report that they were beaten often or sometimes.
Ethiopia	45% of women report ever being beaten by a partner; 10% report abuse within the last year.
Latin America and the Caribbean	
Chile	26% report at least one episode of violence by a partner, 11% report at least one episode of severe violence, and 15% of women report at least one episode of less severe violence.
Colombia	19% of women have been physically assaulted by their partners in their current relationships.
Mexico	27% report at least one episode of physical violence by a partner; 13% report physical violence within the last year.

Source: World Health Organization, 2002.

and initiatives; it is created and changed in democratic societies through the interplay of government, lobbyists, and interest groups (Patterson 2006; Bardes, Shelly, and Schmidt 2007; Parenti 2007). The making of public policy involves trade-offs, compromises, winners, and losers. The 1993 Family and Medical Leave Act granted American workers the right to twelve weeks of unpaid leave for family emergencies without the threat of job loss. But the United States remains the only major industrialized country without paid maternity leave. Germany, for example, provides new parents with four-teen weeks of fully paid leave (Knox and Schacht 2008). In the United States, this lack of a universal public policy places the American family at a disadvantage in coping with the unique pressures of modern life.

For example, the lack of a national health insurance program has resulted in some 40 million uninsured Americans with millions more underinsured. This adds to the stress of low-income, poor, and unemployed families already disproportionately characterized by high rates of divorce, violence, and abuse. The United States also reduces family resiliency through the absence of paid parental leaves surrounding childbirth. Longer paid family leaves promote such vital conditions as improved maternal health, lower infant mortality, enhanced infant and child development, and greater likelihood of women resuming employment after childbirth. These conditions, in turn, encourage family resiliency.

Lifestyle Variations

American marital and familial arrangements are undergoing dramatic change. Although the future of the American family is difficult to predict, an examination

of current alternatives to the traditional nuclear family structure will provide some hints. Social approval of most of these alternative lifestyles is a continuing long-term trend, although certain divergences continue to be spurned by some Americans (Thornton and Young-DeMarco 2001; Skolnick and Skolnick 2006).

Blended Families

The relatively high divorce rate in the United States has led to the creation of the **blended family**—a family formed when at least one of the partners in the marriage has been married before and has one or more children from a previous marriage. This type of family can

Murray A. Straus, David B. Sugarman, and Jean Giles-Sims —Spanking and Antisocial Behavior

Many children in the United States experience spanking and other legal forms of physical punishment from their parents. In the 1980s, more than 90 percent of parents used corporal punishment on young children, and more than half persisted in its use during the early teen years. Although high, this extent of corporal punishment was less than in the 1950s (99 percent) and in the mid-1970s (97 percent). Despite a further decline since 1985, nearly all American children still experience some form of corporal punishment.

The use of corporal punishment to correct or control the behavior of children is a part of American culture. "Spare the rod and spoil the child" is a warning deep in our national consciousness. Murray Straus, David

Sugarman, and Jean Giles-Sims (1997), however, present evidence contradicting the supposed salutary effects of corporal punishment on children's behavior.

These researchers interviewed more than 800 mothers of children, aged six to nine years, in a national longitudinal (involving before and after measurement) study. This research compared parents' use of corporal punishment with antisocial behavior in children. The study defined corporal punishment as "the use of physical force with the intention of causing a child to experience pain, but not injury, for the purpose of correction or control of the child's behavior" (Straus et al. 1997:761). Slapping a child's hand or buttocks and squeezing a child's arm are examples. A measure of antisocial behavior was based on the mothers' reports of their children's behavior: "cheats or tells lies," "bullies or is cruel or mean to others," "does not feel sorry after misbehaving," "breaks things deliberately," "is disobedient at school," and "has trouble getting along with teachers."

Because this was a longitudinal study, data on the frequency of parents' use of corporal punishment was collected *prior* to reports on subsequent antisocial behavior. Contrary to common expectation, Straus and colleagues found that the higher the use of corporal punishment, the higher

the level of antisocial behavior two years later.

Straus and his colleagues suggest that the reduction or elimination of corporal punishment could have significant benefits for lowering antisocial behavior in children. In addition, given the research indicating a relationship between childhood antisocial behavior and violence on the one hand and other crime in adulthood on the other, society at large could benefit from abandoning the use of corporal punishment in child rearing.

Evaluate the Research

1. Explain why a longitudinal study inspires greater confidence that a relationship is causal.
2. Suppose that you are on a panel reporting on child rearing to the President of the United States. Describe the study you would conduct on a possible relationship between childhood corporal punishment and adult crime.
3. How do you anticipate corporally punished children will discipline their own children later in life?
4. Describe what you think would be more effective means of discipline.
5. How were you disciplined? How will you discipline your children? Why?

FEEDBACK

FEEDBACK

1. Which half of each of the following pairs best describes the typical American family?
   a. nuclear/extended
   b. patrilineal/bilateral
   c. authoritarian/democratic
   d. neolocal/matrilocal
   e. polygynous/monogamous
2. The idea of love as a crucial factor in the mate selection process did not become common practice until which century?
   a. seventeenth
   b. eighteenth
   c. nineteenth
   d. twentieth
3. The _____ is the number of divorces in a particular year per 1,000 members of the total population.
4. Which of the following is not stated in the text as one of the factors associated with divorce?
   a. decline of religious influence
   b. age at first marriage
   c. length of marriage
   d. economic conditions
5. Contrary to what one would expect from the traditional male dominance in American family life, women are spousal abusers almost as frequently as men are. T or F?
6. _____ refers to the family's capacity to emerge from crises as stronger and more resourceful.

Answers: 1. a. nuclear, b. bilateral, c. democratic, d. neolocal, e. monogamous; 2. d; 3. divorce rate; 4. a; 5. T; 6. Family resiliency

become extremely complicated (Ganong and Coleman 1994; Herbert 1999). Consider the number and complexity of relationships in the following blended family:

> Former husband (with two children in the custody of their natural mother) marries new wife with two children in her custody. They have two children. Former wife also remarries man with two children, one in his custody and one in the custody of his former wife, who has also remarried and had a child with her second husband who also has custody of one child from his previous marriage. The former husband's parents are also divorced and both have remarried. Thus, when he remarries, his children have two complete sets of grandparents on his side, plus one set on the mother's side, plus perhaps two sets on the stepfather's side. (F. D. Cox 2005:530)

Part of this complexity arises because more marriages are ending through divorce rather than death. For children, a parent is being added, not replaced. Along with the third parent come his or her relatives. Blended families, then, create a new type of extended family, a family that is not based strictly on blood relationships. As the preceding excerpt shows, it is possible for a child in a blended family to have eight grandparents, if each of the child's biological parents remarries. Although not all blended families are this complicated,

it is significant that about 39 percent of households in the United States contain biologically unrelated individuals (Barnes et al. 1998).

What are the reasons for greater instability in blended families? Although many blended families are harmonious, especially if they make adjustments during the first few years, children from previous marriages are one factor in the higher divorce rate among second marriages (Baca Zinn and Eitzen 2006). Sociologists point to three major problems facing blended families: financial difficulties, stepchildren's antagonism, and unclear roles (Lamanna, Riedmann, and Riedmann 2005).

- *Financial difficulties.* Incomes are lower in stepfamilies. This is compounded by the financial demands from both the former and present families. Remarried husbands are often legally obligated to support children from their previous marriage. Second wives may resent the income denied their own children to support children from that previous marriage.
- *Stepchildren's antagonism.* Hoping for a reunion of their original parents, stepchildren may attempt to undermine the new marriage. Even five years after divorce, about a third of stepchildren continue to strongly disapprove of their original parents'

divorce. This is especially true for teenagers, who can be very critical of their stepparent's values and personality.

- *Unclear roles*. The roles of stepparents are often vague and ambiguous. Stepchildren often don't consider their parent's new spouse as a "real" father or mother. It is also uncertain to stepparents or stepchildren how much power the new spouse really has. As a result, questions of control and discipline can become very contentious, especially when teenagers are involved.

Single-Parent Families

How widespread are single-parent families? **Single-parent families**, on the rise since 1970, now account for slightly more than one out of four American families. The percentage of families with children headed by a single parent has more than doubled in the last thirty years, going from 12 percent in 1970 to 28 percent in 2006. The greatest proportion of children in single-parent families live with their mother (23 percent). Only 5 percent of children living with one parent are in a male-headed household (*Child Trends DataBank* 2002; U.S. Bureau of the Census 2007c).

In 2003, 62 percent of all African American families were headed by a single parent, compared with 35 percent among Latinos and 27 percent among whites. African American and Latino children are more likely than white children to live alone with their mothers because of high divorce and out-of-wedlock birth rates, combined with lower rates of marriage and remarriage. While the proportion of one-parent families has increased for all races since 1970, the gap between them has remained relatively constant (see Figure 11.6).

To what degree is social class an ingredient in single-parent households headed by women? Although, in cases of separation and divorce, courts today are more sensitive to fathers' claims, women in all social classes are still more likely to win custody of their children. Frequently, though, there are factors unique to poor single-parent women. The poor single-parent households are generally created by unwed mothers or by women abandoned by their husbands and/or the fathers of their children. And poor women are able to marry (or remarry) at a very low rate.

Although significantly fewer, there is an increasing number of well-educated, professional women who head single-parent households. With the stigma of unwed motherhood declining, more affluent unmarried women are choosing to have children and to care for them alone. In addition, these women have the economic resources to support an independent family. Finally, well-educated women are adopting higher standards for selecting husbands (Seligmann 1999; Hertz 2006).

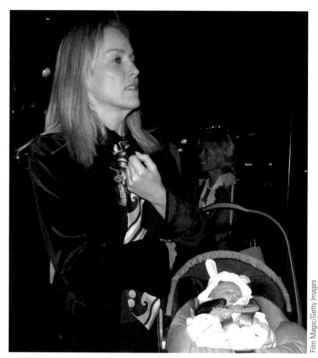

A debate exists over the appropriateness of celebrities, like Sharon Stone, choosing to be single mothers.

What are the effects of single-parent families? Females heading single-parent households face considerable time and economic constraints (Kissman and Allen 1993). Single working parents can seldom provide the time, attention, and guidance to their children that two parents can give, although research has been inconsistent on the effects of this situation (F. D. Cox 2005). One study concluded that children in female-headed families are as socially, intellectually, and psychologi-

FIGURE 11.6

Percentage of Single-Parent Families: 1970–2003

This graph compares the percentage of African American, Latino, and white families that have never married or have one parent. What generalization can you make from this data?

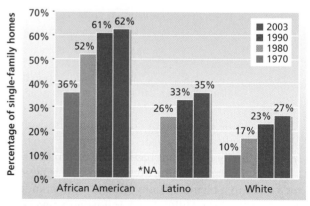

*Latino data not available for 1970
Source: U.S. Bureau of the Census, *Statistical Abstract of the United States*, 2004–2005.

cally well adjusted as children from two-parent families (Cashion 1982). In contrast, most of the research indicates an increase in negative effects on children. Sheppard Kellam, Margaret Ensminger, and R. Jay Turner (1977) found that children from one-parent families consistently score lower on tests of psychological well-being and are less able to adapt in social settings, such as school, than their counterparts from two-parent or two-adult families. They concluded that the problem lies not so much in the absence of the father, but in the aloneness of the mother. Children living in two-adult families fared almost as well as children from two-parent families. Researchers now generally agree that growing up in a fatherless house places children at risk (McLanahan 2002).

Adolescents who live with one parent or with a step-parent have much higher rates of deviant behavior than adolescents living with both natural parents (Dornbush et al. 1985; Popenoe 1999). David Popenoe's study (1999) of a national sample of twelve- to seventeen-year-olds indicates that arrests, school discipline, truancy, run-aways, and smoking occur more often in single-parent and stepparent families, regardless of income, race, or ethnic background. Daniel Mueller and Philip Cooper (1986), in a study of nineteen- to thirty-four-year-olds, found that compared to individuals from two-parent families, persons raised by single parents (especially single mothers) tend to have lower educational, occupational, and economic success. Adults who grew up in single-parent homes are also more likely to be divorced, have an illegitimate child, be receiving psychological counseling, be school truants, and use alcohol and drugs (Bianchi 1990; P. Taylor 1991; McLanahan and Sandefur 1996).

Are the negative effects for these children actually due to divorce and living in single-parent homes? Some researchers contend that the achievement and behavioral problems noted here may not be attributable to single-parent families alone. Such factors as poor parent–child relationships, failure to communicate, lack of economic resources, and family disharmony may be more important than the structure of the family (Demo and Acock 1991). In a series of longitudinal studies of children (one of the studies compared children in the United States and Great Britain), some researchers conclude that the adjustment problems we associate with divorce and single-parent families may actually be reflections of conditions existing prior to separation or divorce (Block, Block, and Gjerde 1986; Cherlin et al. 1991; Furstenberg and Cherlin 1991).

On one point all researchers agree. Female single parents and their children suffer from insufficient economic resources. Although single mothers comprise

SNAPSHOT OF AMERICA **11.1**

Grandparents As Child Caregivers

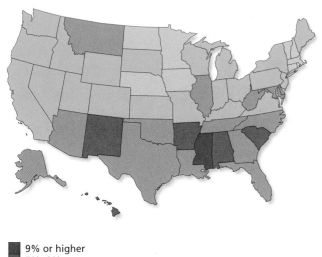

9% or higher
7%–8%
5%–6%
3%–4%
1%–2%

Source: 2005 Kids Count Data Sheet. Baltimore, MD: The Annie E. Casey Foundation, 2005.

This map shows, by state, the percentage of grandparents acting as caregivers of their grandchildren.

Interpret the Map
1. Describe any patterns you detect in this map. Explain.
2. Do you expect the role of grandparents as caregivers to increase, decrease, or stay at about the same level? Defend your position using other information in this chapter.
3. From what you have learned about childhood socialization in this course, discuss the likely consequences of grandparents serving as caregivers of their grandchildren.

only about one-fifth of all family households, they account for about half of households with minor children with income below the poverty line. Because only about 16 percent of the single-parent families headed by men are below the poverty line, experts are now writing about the feminization of poverty (see Chapters 8 and 10). Earning a living is a problem for these single-parent mothers, not only due to the difficulty of arranging for child care (which all working mothers must face) but also because most of these women do not have the educational credentials to do well in the job market. Many were housewives or part-time workers before a divorce; others are young, uneducated women who never married. Thus, if they do not get an education or marry a solidly employed male, these mothers are unlikely to work their way out of poverty. Their children, of course, suffer the consequences along with them.

Childless Marriages

In the past, there was a stigma attached to marriages without children. Due in part to the assumption that all women have a maternal instinct, married women without children were seen as failing to fulfill their biological destiny. In addition, as we learned in Chapter 5, children of that time were needed for farm labor. As the traditional negative societal response to childless marriages diminishes, more American couples are choosing not to have children.

To what extent are married couples choosing to be childless? Around one-fourth of ever-married American women are childless. This is up from about 15 percent in 1970. Experts predict that the number of married couples choosing never to have children is likely to increase (Benokraitis 2007).

Why are some married women choosing not to have children? The reasons are diverse. An important factor, of course, is the disappearing social stigma attached to childless marriages. Also, some women are so involved in their careers that they see children either as an impediment or as too important to shortchange if brought into the world. Sometimes women want their independence in order to pursue personal goals. According to others, today's world is a questionable place into which to bring a child. Some women simply do not enjoy the presence of children. For some couples, having children has been delayed so long that it becomes difficult to make the necessary adjustments for raising a family. Finally, some childless marriages are not a result of choice but are due to physical or mental limitations.

Aren't childless couples missing an important part of life? Because of the strains placed on a marriage, the presence of children tends to reduce marital happiness and satis-faction. Marital satisfaction is higher at the start of marriage, lowest during children's teenage years, and higher again when the children leave home. It is true, however, that among childless couples who want children, marital happiness is generally lower than for married couples with children (Singh and Williams 1981). This is not the case among women who prefer not to have children. Most voluntarily childless couples appear to be happier and more satisfied with their marriages and lives than couples with children (Polonko, Scanzoni, and Teachman 1982; Callan 1983; F. D. Cox 2005).

Dual-Employed Marriages

In a **dual-employed marriage,** both husband and wife are in the labor market. This is quite different from the marriage in which the wife works at a job merely to supplement the family income, a pattern long observed in the lower and working classes (Bielby and Bielby 1989). Although a relatively new trend, the dual-employed marriage is now the norm, as suggested in Chapter 10 (Hertz and Marshall 2001; Moen 2003).

Do dual-employed marriages create special strains? Women in dual-employed marriages are apparently expected to handle the bulk of household tasks and child-care responsibilities in addition to the pursuit of full-time jobs (Goldscheider and Waite 1991; Thompson and Walker 1991; Spain and Bianchi 1996). Combining employment with child-care and domestic tasks, married working women work about fifteen hours more a week than men do. This additional month of twenty-four-hour days a year Arlie Hochschild (1997; Hochschild and Machung 2003) calls the "second shift." Although men now spend an average of sixteen hours per week (up from twelve hours in 1965) in household and child-care duties, women clearly bear the larger burden (Williams, Swisher, and Stalcup 1997).

In addition to this greater workload, women in dual-employed marriages often must cope with role conflict. They are torn between the time requirements of their jobs and their desire to spend more time with their children and husbands (Rubenstein 1991).

Despite their general unwillingness to assume household responsibilities equal to their wives, men in dual-employed marriages also may feel the negative effects of conflicting roles and excessive demands on their time (Moen 1992). In addition, having an employed wife, particularly if she earns more, may not fit with a man's image of himself as provider (Menaghan and Parcel 1991; G. Spitze 1991).

The challenge in the future will be to find ways of reducing the strains placed on members of dual-employed families (L. A. Gilbert 1993). Some pressure on working women could be relieved if men became more participative in domestic duties. Many experts believe that the combined pressures of job and family will lead to accommodative changes in the workplace. Alterations in company policies and practices will likely involve more flexible working schedules, family leaves for both males and females, work-at-home situations, and on-site child-care centers (Deutschman 1991; Labich 1991).

Some evidence suggests that men are assuming a more active role in child care. By 1991, 20 percent of all children under age five were being cared for by their fathers, up from 15 percent only three years earlier (O'Connell 1993). This mirrors the long-term trend toward approval of gender equality in families (Thornton and Young-DeMarco 2001). It could signal even greater movement toward willingness on the part of working fathers to offer more help at home.

How do women, children, and men feel about their dual employment? Thus far, emphasis has been placed on the strains experienced by dual-employed couples. There can be a positive side. On balance, most recent research reveals beneficial effects for the psychological well-being of women (Moen 1992; Crosby 1993; F. D. Cox 2005). Working outside the home can provide a wider set of social relationships and greater feelings of control, independence, and heightened self-esteem. Employment also appears to provide a socioemotional cushion for women when their children leave home.

Dual employed couples face special strains. It hurts these parents to leave their baby with a nanny. The baby doesn't look too happy either.

Michael Newman/PhotoEdit

Compared with women who do not work outside the home, employed women tend to have more alternative channels for self-expression (Adelmann et al. 1989; Wolfe 1997).

As economic pressures are eased, more discretionary money is available for purchases that improve the quality of life of all family members. In addition, sons and daughters of working mothers can benefit in non-economic ways. Daughters of working mothers may see themselves as working adults, as capable of being economically independent, and as benefiting from further education. Sons may choose wives with similar attitudes toward education and employment.

For men, benefits of a dual-employed marriage may include freedom from sole economic responsibility, increased opportunity for job changes, and opportunity for continuing an education. Men with employed wives can often share the triumphs and defeats of the day with someone who is in a similar situation. If their wives are happier working outside the home, husbands also enjoy a better marital relationship. Those husbands who take advantage of the opportunity can form a closer relationship with their children by being more active parents (Booth and Crouter 1998).

Whatever happens between partners, there seems to be a role here for government and business. As Table 11.5 shows, when compared with the United States, other industrialized countries have been more diligent in requiring employers to participate in relieving the strains associated with dual employment. Sweden has been the most advanced. Following the birth of an employee's child, Swedish employers are required to provide thirty-eight weeks of work furlough at 90 percent of pay for either the father or mother and up to one year at a further-reduced compensation rate.

In 1993, under the Clinton administration, the Family and Medical Leave Act was passed. Prior to this federal law, no employers had been required to offer family leave for childbirth or medical emergency. This act requires that employers with fifty or more employees provide up to twelve weeks of unpaid leave for either the father or mother of a newborn child, the adoption of a child, the placement of a foster child, or for some other family member's medical emergency. A serious health condition is broadly defined to include illness, injury, and mental problems. The employer must permit the employee to return to the same or an equivalent job and must continue group health insurance for the employee during the leave period (Snarr 1993; Stranger et al. 1993).

There are several shortcomings to this step, which is relatively modest in comparison to government policies in other industrialized societies. First, because most American workers are employed in organizations with fewer than fifty employees, this benefit is unavailable to them. Second, less than 40 percent of dual-employed

TABLE 11.5

Paternal and Maternity Leave Policies, United States vs. Selected Countries

This table compares the paternal and maternity leave policies of selected countries. Does the rank of the United States among these industrialized countries surprise you? Why?

Country	Leave Duration (Weeks)	Percentage of Pay/Weeks	Recipient
Canada	17–41	60%/15 weeks	Mother
Italy	22–48	80%/22 weeks	Mother
West Germany	52	100%/14–18 weeks	Mother or father
Sweden	12–52	90%/38 weeks	Mother or father
Austria	16–52	100%/20 weeks	Mother
Chile	18	100%/18 weeks	Mother
United States	12	0	Mother or father

Source: John J. Sweeney and Karen Nussbaum, *Solutions for the New Work Force* (Cabin John, MD: Seven Locks Press, 1989), p. 108. Reprinted by permission.

couples can afford to take an unpaid six-week leave. Third, employees must have been employed by their company for at least twelve months to be eligible (Shaller and Qualiana 1993; Wisensale 2001; Reskin and Padavic 2002).

Full-time work outside the home has declined in popularity among mothers over the past ten years ("Fewer Mothers Prefer Full-Time Work" 2007). In 1997, 32 percent of working mothers with children under seventeen years of age said that full-time work is ideal for them. Ten years later only 21 percent stated the same preference. Sixty percent of today's working mothers prefer part-time work, up from 48 percent in 1997. Reversing a twenty-five year trend, there is a recent decline in employment among women with infants under one year of age, from 60 percent in 1998 to 55 percent in 2006. It would be premature to predict a continuing decrease in working mothers with infants. Moreover, there has been no decrease in employment among mothers of older children (Bachu and O'Connell 2001; Knox and Schacht 2008).

Single Life

To what extent are Americans remaining single? The increased age at first marriage for both sexes and the high divorce rate have combined to create an increase in the percentage of single adults. In 1990, about 18 million women and 22 million men over the age of eighteen had never been married. By 2006, these numbers had increased to 25 million for women and almost 30 million for men. More than 30 million Americans over the age of fifteen now live alone, an increase of almost 50 percent since 1980. Although many of these people will eventually marry, an increasing percentage will remain single all their lives (U.S. Bureau of the Census 2007c).

Why are more Americans choosing to live alone? Remaining single has always been an alternative, but it has carried a stigma. During colonial days, bachelors were taxed more heavily than married men; spinsters were viewed as millstones around the necks of their families. Adults were expected to marry and to begin having children as soon as morally and physically possible. Failure to marry was seen as a form of inadequacy and deviance.

The stigma attached to remaining single has faded since colonial times, especially in the past two decades, and more Americans are either choosing this alternative or at least marrying later than previous generations. In addition to the lifting of the stigma, there are other factors contributing to the popularity of singlehood. More single Americans are choosing to remain childless, to obtain sexual gratification outside of marriage, to pursue careers, to rear children from a former marriage, to adopt children, to have strictly homosexual relationships, or to rear out-of-wedlock children (DePaulo 2007).

Will the current trend continue? It is too early to predict confidently whether the increase in singlehood among the young will eventually lead to a decline in marriage at all ages. It is safe to say that singlehood is an increasingly popular alternative to traditional marriage. This does not necessarily mean a rejection of marriage, but it does imply a desire to expand the period of "freedom" after leaving home and an unwillingness to rush into the responsibilities of early marriage and parenthood.

Cohabitation

What is cohabitation, and how widespread is it? Cohabit-ation—living with someone in a marriagelike arrangement without the legal obligations and responsibilities of formal marriage—has emerged as an alternative to traditional monogamy. It is still not known if cohabitation has increased because more Americans are delaying marriage or because cohabitation is being substituted for marriage.

It is currently impossible to predict accurately whether cohabitation will ever become a commonly accepted practice in the mate selection process, something experienced by virtually everyone. Certainly, as the average age at first marriage goes up, increased toleration for this lifestyle is a possible outcome; given the high divorce rate, it may be that cohabitation will become a more popular practice in the selection of a second marital partner as well. Cohabitation has risen markedly since 1970, a trend that continues. The number of American adults cohabiting increased from 439,000 to nearly 6 million between 1960 and 2005, accounting for about one-half of unmarried householders (U.S. Bureau of the Census 2007c). Growing acceptability of cohabitation among younger Americans is reflected in a study of more than 1,000 undergraduates. Eighteen percent had already

While marriage is still a thriving institution, more American couples, like these young people, are choosing to delay or to reject marriage through cohabitation.

cohabited and more than 70 percent indicated that they would live with a sexual partner unmarried (Knox and Schracht 2008). This tracks national data indicating that one-third of American women age fifteen to forty-four report that they have cohabited at some time in their lives. For women age twenty-five to twenty-nine, the figure reaches nearly 50 percent.

Cohabitation has risen among people of all ages and all marital statuses, but particularly among the young and the divorced. About 75 percent of all unmarried-couple households are maintained by someone under forty-five years of age, and about one-third involve at least one child under fifteen (Peterson 2000; U.S. Bureau of the Census 2001).

Is cohabitation a workable alternative to marriage? Research reports on cohabitation are not encouraging. Only about 25 percent of cohabiting couples stay together more than four years, reflecting a lower level of certainty about commitment than is found in marriage. This lack of commitment is probably an important reason for the lower satisfaction among cohabiting couples than among married couples (Nock 1995). Another factor is the higher rate of abuse among cohabiting women than among married, divorced, or separated women. There is no evidence that the marriages of those cohabiting before marriage are any stronger than those who did not and some evidence that those cohabiting before are more likely to divorce after marriage (F. D. Cox 2005; Poponoe 2007).

Same-Sex Domestic Partners

How prevalent is homosexual cohabitation? Because of the stigma surrounding homosexuality, it is impossible to know what proportion of the American population is gay. Consequently, it is also difficult to obtain an accurate count of the frequency of same-sex cohabitation. According to the 2000 Census, there are 600,000 same-sex couples cohabiting, surely an underestimation. We do know that the incidence of same-sex partners living together is increasing, although their number remains small compared with heterosexual marriages.

How do Americans feel about same-sex marriage? Fifty-three percent of Americans oppose legalizing gay marriage. Forty-eight percent of Americans also oppose civil unions. As of 2008, almost all states had "Defense of Marriage Acts" or constitutional amendments prohibiting same-sex marriage.

Several recent legal, social, and political occurrences reflect some change in the social landscape for same-sex partners. In 1996, the U.S. Supreme Court ruled that gays have equal rights under the Constitution (E. Thomas 2003). The Supreme Court also struck down the nation's antisodomy laws in 2003.

The percentage of Americans approving of same-sex marriage with all the rights attached to traditional marriage has risen from 27 percent in 1996 to 46 percent in 2007. This change is especially evident in the more liberal attitudes of young people (Nagourney and Thee 2007).

Several states have enacted more liberal legislation bearing on this issue. So far, California and Massachusetts are the only states that have legalized gay marriage. In 2000, Vermont became the first state to approve civil unions. Those who enter this type of arrangement in Vermont have the same rights as traditional married couples in such matters as taxes, health benefits, and inheritance (Rosenberg 2000). Some other states have legalized civil unions or "domestic relationships" for same-sex couples: Connecticut, New Hampshire, New Jersey, Maine, California, Washington, and Hawaii. Iowa briefly permitted same-sex marriage when a county judge overturned a state law banning gay marriage. This window was closed after four hours when the judge ordered the delay of granting licenses until the Iowa Supreme Court decides whether to consider an appeal. Finally, same-sex couples from New Mexico can marry in Massachusetts because their home state has not banned same-sex marriage.

Homosexual families—same-sex partners living together with children—are also increasing in number, although their number is small compared with heterosexual marriages. Homosexual families are created in several ways. A divorced person may take his or her children into a new homosexual relationship. In fact, about 15 percent of lesbian mothers are currently given custody of their children, and homosexual fathers are beginning to seek custody as well. Homosexual couples are also adopting children, a movement encouraged by a 1979 New York case in which the court gave permanent custody of a thirteen-year-old boy to an openly homosexual minister who had been caring for the boy (Maddox 1982). Even before this case, welfare agencies in large cities were placing orphaned children with homosexual couples. Although it is as yet uncommon, lesbian couples can purchase frozen semen from sperm banks and have their families by artificial insemination, and homosexual male couples can hire surrogate mothers who are inseminated with the couple's mixed semen.

Debate may become even more complicated if recent laboratory research on animals proves applicable to humans. In 2003, scientists reported turning ordinary mouse embryo cells into egg cells in laboratory dishes. In 2007, researchers unveiled a technique for turning ordinary skin cells of mice into stem cells (Stolberg 2007; Weiss 2007a) If the same can be done with human cells, then gay male couples could produce children. This would make it possible for male gay partners whose ordinary cells are transformed into eggs to become biological mothers even though gestation would take place in females. There are ever more extreme complications.

If it ever becomes possible to make sperm from stem cells, then a lesbian couple could reproduce a child in complete independence from males (R. Weiss 2003). And researchers are now attempting to create life forms from completely artificial DNA (Weiss 2007b).

Other countries are beginning to legalize same-sex marriages. In 2000, the Netherlands became the first country to legalize same-sex marriage, with the same rights enjoyed by heterosexual married couples. Belgium followed the Netherlands in 2003. Canada and Spain did the same in 2005. Finally, Uruguay became the first Latin American country to legalize gay civil unions in 2007.

Adult Children Returning Home

Young American adults (eighteen- to thirty-four-year-olds) have a much higher probability of living in their parents' home than they did thirty years ago. For example, the percentage of adults aged eighteen to twenty-four living at home increased from 43 percent in 1960 to 57 percent in 2000 (Bianchi and Casper 2000). Moreover, 26 percent of adults eighteen to thirty-four years old now live with their parents (U.S. Bureau of the Census 2001). These offspring have been labeled *boomerang kids* because they are returning to their point of origin (Quinn 1993; Goldscheider and Goldscheider 1994).

Why are more adult children returning home? Several factors combine to produce this higher proportion of young adults living with their parents (Ward, Logan, and Spitze 1992; F. D. Cox 2005). Because young adults are marrying later, more stay at home longer. In addition, more of them are continuing their education and find living at home the best solution to the challenges of self-support and school expenses. Many young adults return home (or remain home) even after completing their education because the high cost of living outstrips their earning capacity. The high divorce rate also increases the proportion of young adults living at home because parents tend to accept their children in the home after a failed marriage.

What are some possible consequences of this boomerang effect? An added financial burden can create significant strain for older parents whose adult children live at home—costs associated with education, day-to-day living, and perhaps even a grandchild or two. Many parents complain that their adult children do not share in expenses, fail to help around the house, rob them of their privacy, and prevent them from developing relationships with spouses and friends. It is not surprising that higher marital dissatisfaction among middle-aged parents is associated with adult children living at home (Glick and Lin 1986).

Adult children who find themselves in this situation could suffer as well. Adult children who have returned home normally do so from necessity rather than choice. They are likely to be having difficulties associated with balancing school and work, making their way economically, forming a family, or surviving the aftermath of a divorce. They know the burden they represent. In addition, returning home usually means forfeiting some freedom and being subject to unwanted parental control.

Despite these strains, most families appear to adjust well with the return of older children (Mitchell and Gee 1996). This is especially true when the returning older child contributes financially and helps with household duties.

The Sandwich Generation

The current middle-aged generation faces problems associated with new family forms. Due to prolonged life expectancy, refusal to place parents in nursing homes, and fewer siblings to share the burden, more middle-aged adults are finding mothers or fathers (their own or their spouse's) living with them (Knox and Schacht 2008). *Sandwich generation* is the term applied to those adults caught between caring for their parents and caring for the family they formed after leaving home. Sandwiching, of course, can occur whether or not the elderly parent(s) lives with the younger couple.

What are the repercussions? On the positive side, elderly parents unable to take care of themselves usually receive better care from those who love them and feel responsible for their well-being. Older children may enjoy taking care of those who reared them. The caregiving is usually not one-way. Aging parents can offer emotional support and financial resources.

There are also negative repercussions. Taking care of an elderly parent is often not easy. A parent with severe arthritis or Alzheimer's disease demands close and constant attention. Younger adults in this situation may resent the social and emotional intrusion in their own family life. Guilt feelings, then, may arise because of this resentment, and the younger individuals may suffer a loss of self-esteem because they perceive themselves to be selfish. The older parents, too, may feel guilt and anger about the burden they are placing on those they love. They also experience stress and depression from the problems they see their adult children, their spouses, and their grandchildren undergoing (Pillemer and Suiter 1991).

One additional negative consequence deserves special mention. The burden of caring for an aging parent falls much more heavily on women. According to one study, about two-thirds of unpaid caregivers are female (Knox and Schacht 2008). It is typically daughters who take charge of the care for an aging parent. The average woman in the United States is now likely to devote more years caring for her aging parents than she did caring for her own children.

Looking Forward

In early 2000, Darva Conger and Rick Rockwell, a couple who had never met, married as part of a television contest called *Who Wants to Marry a Millionaire?* The marriage was annulled less than two months later. Most Americans shook their heads, wondering if these events marked the final stages of deterioration for the American family. While this was truly a bizarre event, it is not representative of the state of the American family (C. Wilson 2000).

Belief that the family will continue is found even in the most futuristic views.

If the frequency of marriage and remarriage is any indication, the nuclear family is not disappearing. More than 90 percent of men and women in the United States marry sometime during their lives, and about three-fourths of men and women between the ages of twenty-five and sixty-five are married at any given time (Goldstein and Kenney 2001). It is estimated that two-thirds of all divorced persons remarry. In fact, 45 percent of all weddings each year are remarriages. Sixty percent of second marriages end in divorce, but three-fourths of people twice divorced marry a third or even a fourth time. This is known as *serial monogamy*, an alternative that is on the rise.

Despite all the current experimentation with alternatives, the nuclear family remains the most popular choice among Americans. In 2005, 51 percent of all households were composed of married couples. Moreover, just over 67 percent of children under 18 live with two parents (McFalls 2007; U.S. Bureau of the Census 2007c).

The nuclear family, then, is not being abandoned. Contrary to a long-standing fear, most Americans are not avoiding marriage permanently; they are simply postponing it or sampling it often (Popenoe 2007).

This is not to say that change in the American family is not occurring. The proportion of all households

with the traditional husband as wage earner, wife as homemaker, and two children is expected to continue to account for less than one-fourth of all American households. Continued increases are expected for other family lifestyles such as the dual-employed family and the single-parent family (see Figure 11.7). As sociologist Judith Stacey notes, there is no longer "a single culturally dominant family pattern to which the majority of Americans conform or aspire" (1990:17).

But, change is not the equivalent of obliteration. Contrary to Arlie Hochschild's (1997) highly influential thesis, a shift toward women's view of work as more satisfying than family life does not appear to be the case. In the first longitudinal study on this question, K. Jill Kiecolt (2003) reports an increase in satisfaction with home over work among women since 1973. And a relatively stable proportion of younger Americans continue to value marriage and family life, complete with children (Thornton and Young-DeMarco 2001).

Although variations will become more prevalent and socially acceptable, the family system is not about to disappear from American society. In the United States, as well as in other modern societies, the nuclear family will remain the bedrock of the family system, despite the growing diversity in family form (Wolfe 1999; Bianchi and Casper 2000; Knox and Schacht 2008).

FIGURE 11.7

Families in the Labor Force: 1940–2000

The nature of families in the United States has changed since 1940. What do you believe will be the most significant trend by 2025? Explain.

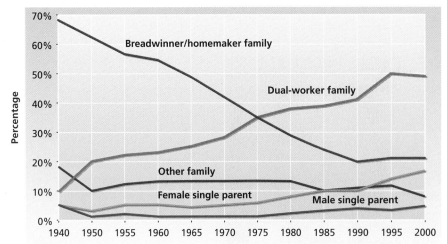

Sources: Howard V. Hayghe, "Family Members in the Work Force," *Monthly Labor Review* 113 (March 1990):16; and U.S. Bureau of the Census, *Statistical Abstract of the United States: 2001* (Washington, DC: U.S. Government Printing Office, 2001), p. 374.

INTERNET LINK   Considerable information on the marital status and living arrangements in the United States can be accessed from the website of the U.S. Bureau of the Census at www.census.gov/population/www/socdemo/ms-la.html

1. A family composed of at least one remarried man or woman with at least one child from a previous marriage is a _____ family.
2. Although the proportion of single-parent families has increased for both blacks and whites since 1970, the gap between them has
   a. decreased somewhat.
   b. increased significantly.
   c. remained the same.
   d. decreased significantly.
3. Voluntarily child-free couples are generally
   a. less educated than couples with children.
   b. less emotional than couples with children.
   c. less satisfied with their marriages and lives than couples with children.
   d. more satisfied with their marriages and lives than couples with children.
4. In a _____ marriage, both husband and wife are in the labor market.
5. In 1997, approximately _____ percent of women and _____ percent of men in America between the ages of twenty-five and twenty-nine had never married.
6. _____ involves living with someone of the opposite sex in a marriagelike arrangement without the legal obligations and responsibilities of formal marriage.
7. The percentage of adults aged eighteen to twenty-four who live with their parents now stands at more than _____ percent.
8. Although the nuclear family is not disappearing, Americans are spending less and less of their lives married. T or F?

Answers: 1. blended; 2. c; 3. d; 4. dual-employed; 5. 35, 51; 6. Cohabitation; 7. 53; 8. T

SUMMARY

1. A family is a group of people related by marriage, blood, or adoption. Each of us may belong to two families—the family into which we were born and the family we create upon marrying.
2. The extended family and the nuclear family are two basic types of family structure. A curvilinear relationship exists between family structure and industrialization. In hunting and gathering societies, the nuclear family was the most prevalent type. With the rise of agriculture as the means of subsistence, the extended family came to prevail. The nuclear family regained popularity in modern industrial society.
3. Whether the family is nuclear or extended, there are several important dimensions of family structure. These dimensions pertain to descent and inheritance, family authority, and residence pattern.
4. Mate selection is never completely a matter of individual choice. Exogamous, endogamous, and homogamous mate selection norms exist in all societies.
5. There are four basic types of marriages: one man and one woman may marry, one man and several women may marry, one woman and several men may marry, or two or more men may marry two or more women.
6. In all societies, the family has been the most important institution. It is the institution that produces new generations, socializes the young, provides care and affection,

regulates sexual behavior, transmits social status, and provides economic support.
7. Although functionalism emphasizes the benefits of the family for society, the conflict perspective depicts the traditional family structure as the instrument of male domination over women. Evidence of this domination, they assert, is reflected in the traditional ownership of women by men, family rules of power and inheritance, and the male-dominated economic division of labor.
8. Symbolic interactionism is used frequently in the study of the family. It is within the family that socialization of children begins and children develop a self-concept. Most of the interactions within families can be analyzed within this theoretical perspective.
9. Modern marriages are primarily based on love. This is a relatively new development in the formation of marriages.
10. The divorce rate in the United States rose dramatically in recent times. Factors promoting divorce include the quality of the marital relationship, the increasing economic independence of women, and the social acceptability of divorce. The dissolution of marriage, however, is not expected to continue to escalate. In fact, the divorce rate has declined slightly.
11. Although the American family provides social and emotional support for its members, violence in this intimate setting is not uncommon. Spouses use violence against

each other, parents abuse children, siblings are physically violent with each other, and the elderly are abused.

12. Problems such as the high divorce rate and frequent family violence and abuse have raised the question of the strength and durability of the American family. Some researchers approach this area of research via the concept of family resiliency—the family's capacity to emerge from crises as stronger and more resourceful.

13. The nuclear family is not going to be replaced on any broad scale. Still, there is considerable experimentation with new patterns of marriage and family: blended families, single-parent families, child-free marriages, people remaining single, cohabitation, gay and lesbian couples, adult children returning home, and the sandwich generation. In addition, the American family is being affected by the rise of the dual-employed marital arrangement.

LEARNING OBJECTIVES REVIEW

- Describe the types of family structure, dimensions of family structure, societal norms for mate selection, and types of marital arrangements.
- Compare and contrast views of the family proposed by functionalists, conflict theorists, and symbolic interactionists.
- Outline the extent and cause of divorce in America.
- Give an overview of family violence in the United States.
- Describe contemporary alternatives to the traditional nuclear family structure.
- Discuss the future of the family in the United States.

CONCEPT REVIEW

Match the following concepts with the definitions listed below them.

____ a. matrilinear descent
____ b. exogamy
____ c. nuclear family
____ d. family
____ e. polyandry

____ f. blended family
____ g. marriage
____ h. democratic control
____ i. patrilineal descent
____ j. family of procreation

____ k. divorce ratio
____ l. monogamy
____ m. family resiliency

1. composed of a remarried man or woman and at least one child from a previous marriage
2. the form of control in which authority is split evenly between husband and wife
3. the number of divorced persons per 1,000 persons who are married and living with their spouses
4. mate selection norms requiring individuals to marry someone outside their kind
5. a group of people related by marriage, blood, or adoption
6. the family group established upon marriage
7. a heterosexual union in which public approval is given to sexual activity, having children, and assuming mutual rights and obligations

8. the familial arrangement in which descent and inheritance are passed from the mother to her female descendants
9. the form of marriage in which one man is married to only one woman at a time
10. the smallest group of individuals (mother, father, and children) that can be called a family
11. the familial arrangement in which descent and inheritance are passed from the father to his male descendants
12. the form of marriage in which one woman is married to two or more men at the same time
13. the family's capacity to survive crises

CRITICAL-THINKING QUESTIONS

1. The text outlines six functions of the family. Describe how you have experienced these functions within your family of orientation. Be specific.

2. Does conflict theory explain any of the social relationships you have seen in your family or the families of your friends? Explain.

3. Focusing on the looking-glass self, show how symbolic interactionism can be used to explain your experiences in your family of orientation. Provide examples that apply to you or other members of your family.

4. What do you think the divorce rate trend line in the United States will be in the twenty-first century? Develop your answer within the context of information in this chapter.

5. Given all the current marital and familial lifestyle variations in the United States today, many observers argue that the nuclear family is going to become a choice of the minority of Americans. Do you agree or disagree? Use data and social trends to support your position.

MULTIPLE-CHOICE QUESTIONS

1. The family that provides one with a name, an identity, and a heritage is known as the
   a. matrilineal family.
   b. family of orientation.
   c. patriarchal family.
   d. nuclear family.
   e. neolocality family.

2. The _____ family consists of two or more adult generations of the same family sharing a common household and economic resources.
   a. nuclear
   b. extended
   c. adult–child
   d. patriarchal
   e. matriarchal

3. As societies move from an agricultural to an industrial base, the _____ family tends to be replaced by the _____ family.
   a. nuclear; extended
   b. matriarchal; patriarchal
   c. matrilocal; patrilocal
   d. extended; nuclear
   e. democratic; patriarchal

4. The family arrangement in which descent and inheritance are transmitted from the mother to her female descendants is called a
   a. female-centric family.
   b. matrilineal family.
   c. neolineal family.
   d. matrilocal family.
   e. bilateral family.

5. A residential arrangement in which a married couple establishes a new residence of its own is known as a(n) _____ arrangement.
   a. bilateral
   b. democratic
   c. matrilocal
   d. neolocal
   e. independent

6. A mate selection arrangement in which norms encourage individuals to marry within their own kind is known as
   a. endogamy.
   b. exogamy.
   c. monogamy.
   d. polyandry.
   e. incest.

7. Which of the following is the most widely practiced form of marriage around the world today?
   a. polygyny
   b. polygamy
   c. polyandry
   d. monogamy
   e. group marriage

8. Which of the following is not considered a vital function of the family?
   a. the provision of initial learning experiences that make people human
   b. the provision of a warm and loving atmosphere that fulfills basic human social and emotional needs
   c. the legitimization of reproduction in a society
   d. the regulation of sexual activity
   e. the promotion of social mobility

9. According to conflict theorists,
   a. the family has historically been an important means of maintaining male dominance over females.
   b. the family in modern society is helping to shatter male dominance.
   c. relationships within the family are failing to undergo the proper kinds of changes.
   d. male dominance over females is due to the dynamics of intimate and personal relationships within the family.
   e. the family remains the one place in society in which an individual is unconditionally accepted.

10. The number of divorced persons, per 1,000 persons in the population, who are married and living with their spouses is known as the
    a. divorce rate.

b. divorce ratio.

c. marital failure rate.

d. marriage–divorce ratio.

e. marital success ratio.

11. According to research cited in the text, voluntarily child-free couples appear to be

    a. pathologically self-centered.

    b. more committed to religion than couples with children.

    c. declining in number compared to couples with children.

    d. lacking in self-esteem.

    e. more satisfied with their marriages and their lives than couples with children.

12. _____ involves living with someone in a marriagelike arrangement without the legal obligations and responsibilities of formal marriage.

    a. Dual-consent marriage

    b. Blended marriage

    c. Duolocal residence

    d. Homogamy

    e. Cohabitation

13. It is estimated that about _____ percent of the U.S. population is gay or lesbian.

    a. 2

    b. 5

c. 10

d. 12

e. 20

14. Approximately _____ percent of American adults between the ages of eighteen and thirty-four live at home with their parents.

    a. 12

    b. 19

    c. 26

    d. 31

    e. 35

15. Which of the following is the most accurate statement regarding the future of the American family?

    a. The nuclear family is no longer the most popular choice.

    b. Childless couples are now so prevalent that the growth of the population is being seriously affected.

    c. Most Americans are not avoiding marriage permanently.

    d. Homosexual marriage is undermining the institution of the family.

    e. Americans are spending more and more of their lives married.

FEEDBACK REVIEW

True-False

1. Contrary to what one would expect from the traditional male dominance in American family life, women are spousal abusers almost as frequently as men are. T or F?

2. Although the nuclear family is not disappearing, Americans are spending less and less of their lives married. T or F?

Fill in the Blank

3. Industrialization promotes the shift from the extended family to the _____ family.

4. In a _____ family, the oldest man living in the household controls the rest of the family members.

5. In 1997, approximately _____ percent of women and _____ percent of men in America between the ages of twenty-five and twenty-nine had never married.

Multiple Choice

6. Which of the following functions of the family is not shared with any other institution?

    a. socialization

    b. reproduction

    c. socioemotional maintenance

    d. sexual regulation

7. Which of the following is not stated in the text as one of the factors associated with divorce?

    a. decline of religious influence

    b. age at first marriage

    c. length of marriage

    d. economic conditions

8. Although the proportion of single-parent families has increased for both blacks and whites since 1970, the gap between them has

    a. decreased somewhat.

    b. increased significantly.

    c. remained the same.

    d. decreased significantly.

Matching

9. Indicate which type of mate selection norm is reflected in each of the following situations.

    ____ a. Jews are supposed to marry Jews.

    ____ b. A father is not permitted to marry his daughter.

    ____ c. Members of the same social class marry.

    (1) exogamy

    (2) endogamy

    (3) homogamy

10. Match the following examples with the major theoretical perspectives.

    ____ a. a father "giving away" the bride

    ____ b. sexual regulation

    ____ c. development of self-concept

    ____ d. a newly married couple adjusting to each other

    ____ e. child abuse

    ____ f. social class passed from one generation to another

    (1) functionalism

    (2) conflict theory

    (3) symbolic interactionism

GRAPHIC REVIEW

U.S. trends in age at first marriage for males and females are graphically depicted in Figure 11.4.

1. If this upward trend in age at first marriage for both sexes continues, what do you predict will happen to the divorce rate?

_____

_____

2. Explain the reasons for the relationship between age at first marriage and the divorce rate you assumed in answering the first question.

_____

_____

_____

REVIEW GUIDE

ANSWER KEY

Concept Review	Multiple Choice	Feedback Review
a. 8	1. b	1. T
b. 4	2. b	2. T
c. 10	3. d	3. nuclear
d. 5	4. b	4. patriarchal
e. 12	5. d	5. 35, 51
f. 1	6. a	6. b
g. 7	7. d	7. a
h. 2	8. e	8. c
i. 11	9. a	9. a. 2
j. 6	10. b	b. 1
k. 3	11. e	c. 3
l. 9	12. e	10. a. 2
m. 13	13. c	b. 1
	14. c	c. 3
	15. c	d. 3
		e. 2
		f. 1

12 Education

Digital Vision/Getty Images

LEARNING OBJECTIVES

- Describe the relationship between industrialization and education.
- Discuss schools as bureaucracies and the attempts to debureaucratize education in the United States.
- Outline the basic functions (manifest and latent) of the institution of education.
- Evaluate the meritocratic model of public education.
- Discuss educational inequality.
- Describe the ways in which schools socialize students.
- Identify and describe the dominant issues in higher education.

Is federally funded Head Start another example of government waste? Americans tend to believe that the federal "bureaucracy," including its efforts in Head Start, is doomed to failure. This is government "interference" in our lives. Research, to the contrary, has consistently shown the Head Start program to produce positive benefits for underprivileged children: improvements in intelligence tests, general ability tests, and learning readiness.

Education serves many functions in modern society, including the promotion of a common identity among all members of society. Due to educational inequality among minority groups, the achievement of this integrative function has been retarded in the United States. Although Head Start has helped in this regard, barriers in the educational system remain. Before turning to the issue of educational inequality, however, some other topics require coverage, beginning with the development and structure of the American educational institution.

The Development and Structure of Education

The basic purpose of education is the transmission of knowledge. Education was originally a family responsibility, but industrialization changed that, dramatically.

Industrialization and Education

Why did schools emerge as the primary educational agent? Before industrialization, the family, the rural community, and the church were the major socializing groups in a child's life. The family taught children the values, norms, and farming skills needed to survive in an agricultural society; as societies industrialized, however, this knowledge increasingly lost its relevance. And as knowledge increased in volume and complexity, family members gradually lost the ability to adequately educate their children.

As a consequence, the educational institution developed. Many leaders of the American Revolution created blueprints for publicly financed school systems. These plans for state-sponsored education for all citizens finally bore fruit in the common school movement of the early nineteenth century (G. S. Wood 2003).

What was the objective of early schools? The founding fathers considered education an investment in democracy because it would create a literate, active, and informed public. Schools, they believed, would create some minimal "democracy of knowledge," thus preventing the privileged few from having a monopoly on knowledge. The teaching of democratic values and marketable skills became especially popular in urban areas inundated by the many ethnic groups who immigrated to the United States between 1850 and 1910. Labor leaders, fearing the loss of adult jobs to child laborers, urged city governments to keep children in school. Many new questions about education were raised. Should children attend school with other children like themselves? Should neighborhoods operate schools? Should "professional" teachers be paid for their work? Should education be viewed as charity for poor children? On the whole, there was no consensus.

Eventually, some agreement about the mission of education was reached. Most late-nineteenth-century proponents of public education believed the objective of the early school to be

not so much . . . intellectual culture, as the regulation of the feelings and dispositions, the extirpation of vicious propensities, the preoccupation of the wilderness of the young heart with seeds and germs of moral beauty, and the formation of a lovely and virtuous character by the habitual practice of cleanliness, delicacy, refinement, good temper, gentleness, kindness, justice, and truth. (M. B. Katz 1975:31)

Thus, public schooling in nineteenth-century America was not designed primarily to teach sophisticated intellectual skills; it was a mechanism of social control. Otherwise, it was feared, children would grow up "vicious" or the "wilderness of the young heart" would lack "seeds of moral beauty."

How did education develop after the turn of the twentieth century? Just as farming slowly gave way to manual factory labor in the latter part of the nineteenth century, relatively simple manual skills began to be less and less important after the beginning of the twentieth century. New middle-management and clerical jobs were seen as highly desirable by many Americans seeking to better their own lives and, more important, their children's future. Business leaders and parents began to demand more advanced training in schools. School administrators responded by developing secondary schools that emphasized advanced learning. Figure 12.1 indicates the percentage of Americans (including major racial

High School Graduates, 1970 and 2003

Displayed in this figure are the percentages, by racial and ethnic category, of persons 25 years old and older who have completed high school. Note that the proportion of high school graduates in each group has increased sharply between 1970 and 2003. As a result, each of these groups is placing more pressure on public schools to accommodate the needs of their members.

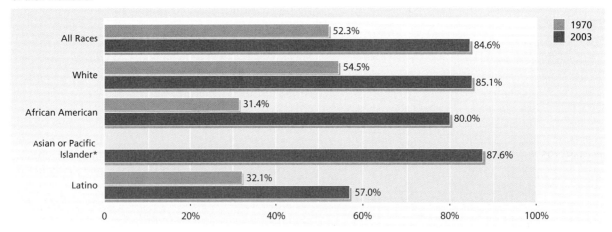

*Note: No data available for Asian or Pacific Islander for 1970.
Source: U.S. Census Bureau, *Statistical Abstract of the United States*, 2004–2005.

and ethnic groups) over twenty-five years of age who have at least completed high school.

Bureaucracy in Education

Organizationally, early school administration was based on the factory model of education. Children were to be educated in much the same way as cars were mass-produced:

> Schooling came to be seen as work or the preparation for work; schools were pictured as factories, educators as industrial managers, and students as the raw materials to be inducted into the production process. The ideology of school management was recast in the mold of the business corporation, and the character of education was shaped after the image of industrial production. (Cohen and Lazerson 1972:47)

Although this philosophy softened considerably, the concept of bureaucratized public education remains prevalent, with its specialization, rules and procedures, and impersonality.

In what ways are schools bureaucratic? In keeping with the bureaucratic model, the process of educating children is divided into specializations: administrators who run schools; teachers who either concentrate on one or two topics or teach a specific grade level; specialists who purchase materials; specialists who test and coun-

sel students; and specialists who manage libraries. There are even specialists who drive buses, serve lunches, and type memos.

According to the bureaucratic model, education for large numbers of students can be accomplished most efficiently when the students are homogeneous in development and ability. In this way, a teacher can focus attention on one set of learning materials for large numbers of students. Age-graded classrooms, in which all students receive the same instruction, reflect the impersonal, bureaucratic nature of schools.

It is also thought more efficient (the ultimate goal of a bureaucracy) to have all teachers of a given subject teaching the same, or at least similar, material. Materials can be approved and purchased in bulk, and testing can be standardized. In addition, if parents need to know what subjects are being taught at any given time, school administrators can readily advise them. This practice also allows students to transfer from one school to another without losing continuity of curriculum. Rules and procedures exist to ensure that this happens.

Schools are also part of a much larger bureaucratic system. This system begins with the federal government and progresses layer by layer through state and local governments.

Critics of the bureaucratic model claim that children are not inorganic materials to be processed on an assembly line. Children are human beings who come into school with previous knowledge and who interact socially and emotionally with other students. Moreover, it is hard to determine whether children have been "processed" correctly. How can we know what takes place inside a child's head? According to critics of formal schooling, the school's bureaucratic nature is unable to respond to the expressive, creative, and emotional needs of individual children.

Reforms in the Classroom

Are there alternatives to the bureaucratic model? American educational reform occurs in cycles. The U.S. progressive education movement of the 1920s and 1930s was a reaction to the Victorian authoritarianism of early-nineteenth-century schools. Educational philosopher John Dewey (1859–1952) led the progressive education movement, with child-centered focus and an emphasis on work-related knowledge (Dewey 2001). This movement almost disappeared in the 1950s but had a resurgence in the 1960s and 1970s, in the humanistic education movement. The humanistic movement advocated such steps as the elimination of restrictive rules and codes and the involvement of students in the educational process. The aim of the humanistic movement was to create a more democratic, student-focused learning environment (Ballantine and Spade 2007). This movement has spawned a number of educational philosophies aimed at classroom reform. Three formulations of the humanistic educational impulse are the open classroom, cooperative learning, and the integrative curriculum.

What is the open classroom? The **open classroom** is a nonbureaucratic approach to education based on democratic relationships, flexibility, and noncompetitiveness:

> In the 1960s, advocates of the open classroom proposed alternatives to the bureaucratically organized school. They agreed on an underlying philosophy: Children are naturally curious and motivated to learn by their own interests and desires. The most important condition for nurturing this natural interest is freedom, supported by adults who enrich the environment and offer help. In contrast, coercion and regimentation only inhibit emotional and intellectual development. (Graubard 1972:352)

The open classroom philosophy advocates elimination of the following customs: the sharp authoritarian line between teachers and students, the predetermined curriculum for all children of a given age, the constant comparison of students' performance, the use of competition as a motivator, and the grouping of children according to performance and ability. In short, schools adhering to the open classroom approach wish to avoid the bureaucratic features of traditional public schools (Silberman 1973; Kozol 1985; Kohl 1988; Holt 1995).

The open classroom approach was not fulfilled in a clear-cut way. Success or failure was affected by the availability of money, student–teacher ratio, and teacher endorsement. However, approaches to school structure continue to be debated (Hallinan 1995). Some argue that schools must operate bureaucratically because limited money requires the less expensive mass instruction. Others claim that concern for children should supersede administrative needs and that individualized programs developing the child's whole personality should be encouraged. According to another viewpoint, the bureaucratic nature of schools prepares children for the bureaucratic society to which they must ultimately adapt.

Although further experimentation continues and appears to be warranted, it remains to be seen whether Americans are in the mood for it. In fact, the "back-to-basics" movement, which emerged in the mid-1980s and 1990s, conflicts directly with the open classroom philosophy. The back-to-basics movement represents an attempt to reinstate a traditional curriculum ("reading, writing, and arithmetic") based on more bureaucratic methods.

What is the back-to-basics movement? In 1983, America received an "educational wakeup call" when the National Commission on Excellence in Education issued a report dramatically entitled *A Nation at Risk* (*A Nation at Risk* 1984). Catching the attention of politicians and the general public, the report warned of a "rising tide of mediocrity" in America's schools. Because of deficiencies in its educational system, the report claimed, America is at risk of being overtaken by some of its economic world competitors (D. P. Gardner 1983). As evidence, the report pointed to several indicators:

- Twenty-three million American adults, 40 percent of minority young people, and 13 percent of all seventeen-year-olds are functionally illiterate.
- Scholastic Aptitude Test (SAT) scores dropped dramatically between the early 1960s and 1980.
- Only one-third of all seventeen-year-olds can solve a mathematics problem requiring several steps.
- American children are being outperformed by children of other nations on achievement tests.

Unlike the recommendations of the progressive and humanistic reform movements, most of the solutions offered by the commission tended toward the bureaucratic. The report urged a return to basic skills such as reading and mathematics. High school graduation requirements should be strengthened to include four years of English, three years of mathematics, three years of science, three years of social studies, and a half-year of computer science. School days, the school year, or both should be lengthened. Standardized achievement tests should be administered as students move from one level of schooling to another. High school students should be given significantly more homework than they are assigned at present. Discipline should be tightened through the development and enforcement of codes for student conduct.

The commission's report stirred intellectual debate (Adelson 1984; Bunzel 1985; Gross and Gross 1985) and aroused the public. Americans generally favor the commission's recommendations, especially its call for more homework and greater discipline (Barrett 1990).

In a series of books, English professor E. D. Hirsch contends that American schoolchildren are not being taught the knowledge they need in today's world (Hirsch 1987, 1999, 2004). Hirsch has labeled American schoolchildren "culturally illiterate." Surprising numbers of schoolchildren don't know when the American Revolution occurred, who George Washington Carver was, or where Washington, D.C., is located. Although concerned for all Americans, Hirsch is particularly concerned that the failure of schools, and the accompanying cultural illiteracy, is dooming children from poor and illiterate homes to perpetual poverty. Keeping poor and illiterate children in the same condition as their parents is, for Hirsch, an unacceptable failure of American schools.

Late in 1987, U.S. Secretary of Education William Bennett, under President Ronald Reagan, proposed a back-to-basics curriculum for American high schools. Emphasis was placed on a minimum core curriculum that all students must take (Bennett 1988). In 1991, the first President Bush's secretary of education, Lamar Alexander, formulated a national educational reform plan with six goals (*America 2000* 1991):

1. Every child must begin school ready to learn.
2. The national high school graduation rate must be 90 percent.
3. Competence in core subjects must be shown after grades 4, 8, and 12.
4. American students should be the best educated in the world in math and science.
5. All adults must be literate and possess the skills necessary for citizenship and competition in a global economy.
6. Schools should be free of drugs and violence.

Implementation of the plan included a longer school year, national standards and state exams, and merit pay for teachers.

The latest manifestation of the back-to-basics philosophy is the "high-stakes" testing movement. As part of this movement, President George W. Bush signed in January 2002 a bill he considers the cornerstone of his administration's domestic policy. Known as No Child Left Behind, this federal law requires that each student racial and demographic subgroup improve each year (grades 3–8) on mandatory standardized tests. Schools failing to accomplish this improvement for two years running must help students transfer to better schools and use public money for tutoring students. A school that continues to fall short of the federal standard must either replace its principal and teachers or reopen as a charter school (Fletcher 2003; *From the Capital to the Classroom* 2003).

To date, there is no evidence that the back-to-basics movement, on its own, can live up to its advance billing (Kantrowitz 1993a; Hancock and Wingert 1995; Wingert 1996). (Some successes have been seen when parents and community members become part of the overall approach.) Evidence indicates that implementation of some of the solutions actually inflicted greater harm. The dropout rate increased as students' time increased in school, on testing, and on homework assignments. The low self-esteem of many low-performing students was exacerbated. Stricter grading with increased academic requirements discouraged low-income and minority students even more. Bureaucratically oriented solutions appear to fail as the cure-all they seemed to be. Despite the efforts, it appears that the nation remains at risk (Wagner 2003).

Public opinion indicates perceived shortcomings of the most recent back-to-basics reform. Americans showed very mixed feelings in 2008 as Congress prepared to debate renewal of the No Child Left Behind Act. Fifty-eight percent of Americans polled believed that the law either made schools worse or had no impact; 34 percent said that the law improved schools. Of Americans with children in public school, only 30 percent claimed that No Child Left Behind upgraded their children's schools ("No Child Left Behind Gets Mixed Grades," 2008).

How does cooperative learning contribute to a less bureaucratic classroom? Evidence indicates that a less bureaucratic approach to education should not be dismissed (R. A. Horwitz 1979; Hurn 1993). In fact, the open classroom philosophy that became prominent in the 1960s and 1970s is currently being implemented again (emerging alongside the back-to-basics movement). The philosophy this time is in the form of cooperative learning. **Cooperative learning** takes place in a nonbureaucratic classroom structure in which students study in groups, with teachers acting as guides rather than as the controlling agents:

> *Cooperative learning approaches are based in part on the assumption that students learn best when actively working with others in small heterogeneous groups. . . . It occurs when teachers have students work together in small groups on a task toward a group goal—a single product (a set of answers, a project, a report, a piece or collection of creative writing, etc.) or achieving as high a group average as pos-*

Are Americans Losing Their Minds?

That America is dumbing down is not a new charge. In 1837, Ralph Waldo Emerson observed that "The mind of this country, taught to aim at low objects, eats upon itself." So begins Susan Jacoby's current examination of anti-intellectualism and anti-rationalism in America (Jacoby 2008a, 2008b).

Jacoby identifies Richard Hofstadter's (1966) *Anti-Intellectualism in American Life* as the classic book in this area. Hofstadter saw anti-intellectualism as cyclical, waxing and waning as broad social forces change. Jacoby believes that the prevalence and depth of "dumbness" in our culture would shock even Hofstadter if he were still alive.

Jacoby offers some facts to bolster her claim of a current wave of anti-intellectualism. For example:

Reading has declined not only among the poorly educated, according to a report last year by the National Endowment for the Arts.

In 1982, 82 percent of college graduates read novels or poems for pleasure; two decades later, only 67 percent did. And more than 40 percent of Americans under 44 did not read a single book—fiction or nonfiction—over the course of a year. The proportion of 17-year-olds who read nothing (unless required to do so for school) more than doubled between 1984 and 2004. (Jacoby 2008a)

Of the several broad social forces promoting anti-intellectualism today, Jacoby sees the triumph of video culture over print culture as paramount. With rising use of personal computers, instant messages, video games, YouTube, Facebook, and MySpace, we have seen a decline in book, newspaper, and magazine reading. This trend, Jacoby contends, is most obvious among the young.

Some supporters of the new technology see the ability of children to spend long periods of time in front of video games and computers as a sign of an ability to focus (S. Johnson 2006). Jacoby attempts to debunk this contention through research on children. She cites a study of babies between eight and sixteen months who watch videos such as *Baby Einstein*. This University of Washington research team reported that children this age recognize an average of six to eight fewer words for every hour spent watching videos each day. A study of toddlers ages one to three at Children's Hospital in Seattle found a positive

correlation between the amount of television watched and attention problems at age seven. For each hour of television watched daily, the chances of developing attention-deficit hyperactivity disorder rose by almost 10 percent. This disorder is associated with difficulties in concentrating, organizing, and controlling impulsive behavior (Brody 2004).

It is not just that 20 percent of Americans believe the sun revolves around the Earth, or that almost half of Americans between ages eighteen and twenty-four do not care to know the locations of countries where significant news is being made. Really worrying Jacoby is what she sees as "the alarming number of Americans who have smugly concluded that they do not need to know such things in the first place" (Jacoby 2008a:37). It is arrogance about the lack of knowledge that she considers anti-intellectual or anti-rationalist.

Think About It

1. As a member of the first generation to experience video culture as Jacoby describes it, do you agree or disagree with her thesis? Defend your answer.
2. As a future (or present) parent, do you think it is necessary to protect your children from video culture? Why or why not?

---

sible on a test—then reward the entire group on the basis of the quality or quantity of its product according to a set standard of success. (Oakes 1985:208, 210)

According to the cooperative learning method, students learn more if they are actively involved with others in the classroom (Sizer 1996). The traditional teacher-centered approach rewards students for being passive recipients of information and requires them to compete with others for grades and teacher recognition. Cooperative learning, with its accent on teamwork rather than individual performance, is designed to concentrate on results rather than on performance

comparisons among students. Cooperation replaces competition. Students typically work in small groups on a specific task. Credit for the task is given only if all group members do their part.

Some benefits of the cooperative learning approach have been documented. Uncooperativeness and stress among students is reduced; academic performance is elevated; students have more positive attitudes toward school; racial and ethnic antagonism decreases; and self-esteem increases (Oakes 1985; Aronson and Gonzalez 1988; Children's Defense Fund 1991).

Although Japanese schools have been using group learning in the early grades for years, resistance remains

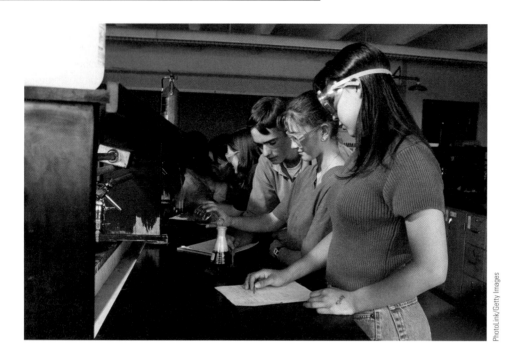

Working cooperatively in groups was one of the most democratic school reforms of the twentieth century.

PhotoLink/Getty Images

strong among many American teachers (Kantrowitz 1993b). They fear loss of control with younger children and disciplinary problems among older students, especially teenagers. Many American teachers and principals believe that mapping a student's progress is more difficult in a cooperative learning setting. This approach also requires more preparatory work for teachers, who must plan lessons allowing for less frequent teacher intrusion. Grading is also more complex when blending an individual and group evaluation.

What is the integrative curriculum? The **integrative curriculum**, another extension of the open classroom, is an approach to education based on student–teacher collaboration. In the traditional classroom, the curriculum is predetermined for students; in the integrative curriculum, the curriculum is created by students and teachers together. Because students are asked to participate in curriculum design and content, the integrative curriculum is democratic in nature. Giving students such power obviously deviates from the traditional subject-centered curriculum (Barr 1995). Rather, subject matter is selected and organized around certain real-world themes or concepts, as exemplified in a Washington State sixth-grade unit of study on water quality:

> The unit became a part of an actual water quality project that originated in the Great Lakes region of the United States but now spans the globe. Lessons were organized around the actual work of determining water quality in Puget Sound. These lessons culminated in students' reporting to community groups about the quality of the water. In this way learning was relevant to a real-world problem that the students contributed to solving. (Simmons and El-Hindi 1998:33)

Instruction in this unit was organized around hands-on experience and a "multiple intelligence" approach, recognizing that students in a classroom are not all identical. Students bring to any unit of study a variety of learning styles, interests, and abilities, and different units of study will engage students in varying ways.

Competitors to the Traditional Public School

While the debate over the most effective classroom methods continues, educators and politicians are looking for answers beyond the classroom—to the way schools are organized, funded, and administered. Meanwhile a school choice movement promotes the idea that the best way to improve schools is by using the *free enterprise model*—creating competition for the public school system. According to school choice advocates, parents and students should be able to select the school that best fits their needs and provides the greatest educational benefits. There are several approaches to the goal of school choice: the voucher system, charter schools, for-profit schools, and magnet schools. Results are mixed (Fuller and Elmore 1996).

What is a voucher system? Proponents of **vouchers** contend that the government should make available to families with school-age children a sum of money that they can use in private or parochial schools. Families who chose a public school would pay nothing, just as in the current system. Parents who chose a religious or other private school would have the tuition paid up to the amount of the government voucher and would have to make up the difference (if any) between the amount of the voucher and the tuition amount. A voucher plan in Cleveland, for example, provided publicly funded scholarships of just over $2,000 annually to almost 4,000 city children. Most parents chose to spend the money at private schools rather than to keep their children in public schools. In Wisconsin, adoption of a voucher plan ($5,300 per voucher for students whose parents earn less than $30,000 for a family of four) increased from fewer than 10 schools to 103 (M. A. Fletcher 2001).

Proponents point to the freedom vouchers give parents, the increased quality that would presumably come from schools competing for students, and the ability to place students in educational environments best suited to their particular needs. Critics fear that free choice of school would make urban schools even more of a dumping ground for the poor and minorities, because poor and minority parents in the inner cities cannot make up the difference between the amount of the voucher and the cost of sending their children to higher-quality schools. Critics contend that the use of vouchers will erode national commitment to public education, leaving the public school system in even worse shape (Kozol 1992; Henig 1994; Martinez et al. 1995; Schrag 1999; Worchol 2007).

Claiming that vouchers violate separation of church and state, courts traditionally ruled voucher systems unconstitutional. However, a 1998 U.S. Supreme Court ruling validated a Wisconsin law allowing state money for low-income students to attend either private or parochial schools (Perry and McGraw 1999). And in a 2001 decision, the Supreme Court declared the Cleveland vouchers to be neutral with respect to religion because parents still retained choice of where and how to school their children (Morse 2002).

Evidence on the effectiveness of the voucher system is inconsistent. Although some voucher programs have improved student test scores (compared with public schools), other programs produced no improvement (Toch and Cohen 1998; Moe 2001; Witte 2000; *School Vouchers* 2001; Blum and Mathews 2003). Rand Corporation researchers conclude that many of the key questions about vouchers still await answers. With regard to academic achievement, these researchers found that privately funded small-scale experimental voucher programs aimed at low-income students may have a modest academic benefit for African American students after one to two years in these programs. While even this effect must be tentatively reported, no conclusion at all can be drawn regarding children of other racial or ethnic groups (Gill et al. 2001).

The harshest critics draw several negative conclusions: public schools lose money to private voucher schools, making public education less able to be successful; voucher programs do not necessarily lead to higher academic performance; and voucher schools are not accountable to taxpayers (Miner 2000). American public opinion appears to be swinging toward the critics. With increased approval of public schools, support of vouchers declined from 44 percent in 1997 to 22 percent in 2004 (Ray 2004). More generally, the momentum of the school choice movement is declining.

What are charter schools? Charter schools are publicly funded schools operated like private schools by public schoolteachers and administrators. Freed of answering to local school boards, charter schools have the latitude to use nontraditional or traditional teaching methods and to shape their own curricula.

The Mosaica Academy in Bensalem, Pennsylvania, which opened in 1998, is deliberately not organized along public school lines. The school day is nearly two hours longer than at public school, and the school year is twenty days longer. This school created its own curriculum with the goal of immersing students in a study of civilizations that developed over a period of 4,000 years (Symonds 2000).

The number of charter schools exceeded 2,100 in 2002, with more than 500,000 students in thirty-seven states and Washington, D.C. After ten years of charter school operations, the American Federation of Teachers (AFT) has concluded that such schools have not fulfilled their promise. Among other problems, the AFT reported that charter schools spend more on administration and less on instruction than public schools do; student performance is generally no better, and often worse; and charter schools are more homogeneous in race and class than their public counterparts (Wingert 2002; Dillon 2007).

Like vouchers, for charter schools the jury is still out. Some charter schools have improved education, but others have not (Schorr 2000). The Rand researchers conclude that it is too soon to know if the fears of opponents or the hopes of supporters will be confirmed (Gill et al. 2001).

What is the nature of for-profit schools? Some reformers do not believe that local or federal government is capable of improving the educational system. Government, they say, is too wasteful and ineffective. Why not look to business and market forces to solve the problems facing schools today? **For-profit schools** would be supported by government funds but run by private companies. By borrowing from modern business practices, the argument goes, these schools could be efficient, productive, and cost effective. Marketplace forces are alleged to ensure that the "best" schools would survive.

The most comprehensive for-profit organization is Edison, which launched a $40 million, three-year campaign in 1992 to develop its program. Edison schools feature challenging curricula, along with a schedule that has children in school almost a third

1. The _____ institution developed to fulfill society's need to transmit knowledge to the young.
2. Contemporary public education is important primarily because of
   a. the quality of teachers in modern schools.
   b. efficient operation of school bureaucracies.
   c. historical changes in the economic and social structures of industrial societies.
   d. lack of qualified individual tutors.
3. Early public schooling in America was viewed as a mechanism of _____.
4. Schools in America today are based on the _____ model of organization.
5. _____ attempt to avoid most of the bureaucratic features of traditional public schools.
6. _____ is a nonbureaucratic classroom structure in which students study in groups with teachers acting as guides rather than as the controlling agents determining all activities.
7. A school that concentrates on science in its curriculum is known as a(n) _____ school.
   a. magnet
   b. for-profit
   c. integrative
   d. charter school

Answers: 1. educational; 2. c; 3. social control; 4. bureaucratic; 5. Open classrooms; 6. Cooperative learning; 7. a

longer than in the average public school. Beginning in the third grade, students are equipped with a computer and modem that they can take home to access the Edison's intranet system (Symonds 2000). In 2002, Philadelphia privatized forty-two schools, twenty of which were turned over to Edison (Steinberg 2002; Winters 2002).

Critics of this approach are bothered by the idea of mixing profit and public service. What happens to the students when their needs are weighed against the profit margin? Will for-profit schools skimp on equipment, services, and training? Another problem involves oversight: with a for-profit system, voters lose the power to influence officials and educational policy (Saltman 2001, 2007). In one of the few studies of for-profit schools, researchers found no increase, over a three-year period, in student achievement outcomes in the first ten Edison schools opened in 1995 and 1996 (Miron and Applegate 2000).

What is unique about magnet schools? **Magnet schools** are public schools that attempt to achieve excellence by specializing in a particular area. One school may emphasize the arts; another may stress science. If their public school district includes magnet schools, families can decide which particular school best suits their children's needs. Magnet schools are designed to enhance school quality and promote school desegregation. They have emerged as a significant factor in urban education (Blank, LeVine, and Steel 1996).

This urban magnet school emphasizing the arts is one alternative to the public school system.

Amy Conn-Gutierrez/AP Photo

The Functionalist Perspective

Functions of Education

According to the functionalists, social institutions develop because they meet one or more of society's basic needs. The educational institution performs several vital functions in modern society: cultural transmission; social integration; selection and screening of talent; promotion of personal growth and development; and the dissemination, preservation, and creation of knowledge.

How does the educational institution transmit culture? Cultural transmission must occur if a society is to endure. Indeed, the educational institution perpetuates a society's cultural heritage from one generation to the next. When we see evidence of the teaching of Marxist-Leninist doctrine in communist societies, we call it indoctrination. We are less aware of the process of indoctrination when, through the study of Washington, Jefferson, and Lincoln, we Americans learn about the merits of democracy. Thus, although schools teach basic academic skills such as reading, writing, and mathematics, they also transmit culture by instilling in students the basic values, norms, beliefs, and attitudes of the society. The value of competition, for example, is taught through an emphasis on grades, sports, and school spirit. (More is said about this function later.)

How does education contribute to social integration? Although television is now a strong competitor, the educational system remains the major agent of social integration. Formal education transforms a diverse population into a community with a common identity. Learning an official language, sharing in national history and patriotic themes, and being exposed to similar informational sequences facilitate a shared identity. The result is a society with relatively homogeneous values, norms, beliefs, and attitudes. Even newly arrived immigrant children without the ability to speak and write English attend local schools and soon learn to participate in the American way of life.

The current debate on bilingual education touches on this function. People who emphasize the need for recognizing and honoring cultural diversity (*multiculturalists*) usually support instruction in the student's own language, at least for some period of time. Opponents of bilingual education argue that it hinders the development of a common American cultural identity and that it has not been proven to help students succeed academically. A conservative political backlash to the bilingual approach has led twenty-three states to adopt English as their official language.

How do schools select and screen students? Scores on intelligence and achievement tests are used for grouping students. The stated purpose of testing (administered for more than fifty years now) is to identify an individual's talents and aptitudes. Test scores are also used for **tracking**—placing students in curricula consistent with the school's expectations for the students' eventual occupations (Oakes 1985; Oakes and Lipton 1996; Taylor, Peplau, and Sears 1997). Counselors use test scores and early performance records to predict careers for which individuals may be best suited. Counselors usually feel a moral obligation to ensure that students pursue careers consistent with their abilities. (The concept of tracking plays an important role in the conflict perspective of education, discussed shortly.)

How do schools promote personal growth and development? Schools expose students to a wide variety of perspectives and experiences that encourage them to develop creativity, verbal skills, artistic expression, intellectual accomplishment, and cultural tolerance. In this way, education provides an environment in which individuals can improve the quality of their lives. In addition, schools promote personal growth and development by preparing students for their occupational pursuits.

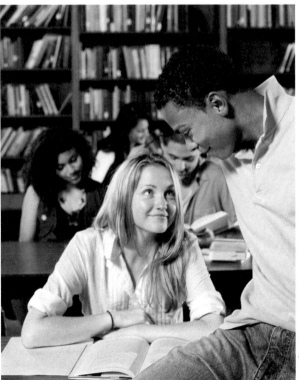

Do you consider the student opportunity to flirt with the opposite sex a positive or negative function of schools?

How does education encourage the dissemination, preservation, and creation of knowledge? Educational organizations disseminate knowledge not only via the classroom but through contributions such as written or audiovisual materials that can reach hundreds of thousands of people. Preservation of knowledge is perpetuated by the transmission of knowledge from generation to generation. The task of preserving knowledge is also achieved by such activities as deciphering ancient manuscripts; protecting artifacts; and recording knowledge in written, video, or audio form. *Innovation*—the creation or discovery of new knowledge through research or creative thinking—can take place at any level within the educational system, but it traditionally receives more emphasis at the university level.

Latent Functions of Education

The functions of schools discussed to this point are recognized and intended—**manifest functions**. The educational institution has other functions as well that are unrecognized and unintended—**latent functions**.

What are the latent functions of education? We do not usually count among the functions of schools the provision of day-care facilities for dual-employed couples or single parents. Nor do we think that our tax money goes to schools so that people can find marriage partners. Also, schools are not consciously designed to prevent delinquency by holding juveniles indoors during the daytime. Nor are schools intended as training grounds for athletes. Finally, schools are not generally recognized for inoculating the discipline needed to follow orders in a bureaucratized society (DeYoung 1989; also see Doing Research in Chapter 4).

Each of the latent functions just mentioned, although unrecognized and unintended, is considered a positive contribution to society. But some consequences, of course, are negative, or dysfunctional. Determining what is dysfunctional depends on the reference point. Tracking, for example, has just been presented as a positive function of the educational institution. Most would agree with this assessment. Most see this function as a manifest, positive function of schools, because it seems to fulfill a societal need. According to the conflict perspective, however, tracking serves a latent dysfunction:

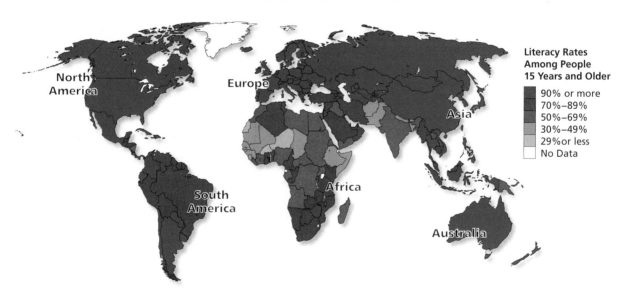

WORLDVIEW 12.1

Literacy Rates

One of the functions of education is to promote literacy—the key to continued learning, problem solving, and information analysis. This map shows rates of literacy among persons fifteen years of age and older in various countries of the world. Because of cultural norms and discrimination, more men than women are able to read and write.

Literacy Rates Among People 15 Years and Older

- 90% or more
- 70%–89%
- 50%–69%
- 30%–49%
- 29% or less
- No Data

Source: U.S. Central Intelligence Agency, *The World Factbook, 2005*.

Interpret the Map

1. Do you see a pattern in the rates of literacy? Explain.

2. How does the United States measure up?

it is a means for those benefiting from the stratification structure to continue to benefit. In other words, tracking is one mechanism for perpetuating the social class structure. Moreover, evidence suggests that tracking is harmful to those placed on "slower" tracks (Hurn 1993; Oakes and Lipton 1996). The conflict perspective of education is presented in the following discussions on meritocracy, educational inequality, and cognitive testing. These topics are central to the relationship between education and social inequality, a topic of research in the sociology of education since World War II (Hallinan 1988).

The Conflict Perspective

Meritocracy

In a **meritocracy**, social status is based on ability and achievement rather than social class or parental status (M. Young 1967; Bell 1976). According to the meritocratic model, all individuals have an equal chance to succeed and to develop their abilities for the benefit of themselves and their society. A meritocratic society is free of barriers that might prevent individuals from developing their talents. Meritocracy, then, by definition, involves competition. For this reason, sports are seen as the ultimate meritocracy. Although some sports have glaring shortcomings in this regard, they do fit very closely the definition of **competition** as "a social process that occurs when rewards are given to people on the basis of how their performances compare with the performances of others doing the same task or participating in the same event" (Coakley 2006:78).

Is America a meritocracy? Many Americans believe that the United States is a meritocratic society. However, a number of recurring problems suggest that the model does not always work the way we think it does (Lieberman 1993; Lemann 2000).

Although America purports to be a meritocratic society, researchers recognize the existence of barriers to true merit-based achievement. For example, critics argue that public education primarily serves the economic elite. Because schools are not equal for children of all social classes, students do not all have an equal starting point (Bowles and Gintis 1976; Giroux 1983; Gittell 1998).

Randall Collins (2007) contends that America is not a meritocratic society, but is composed of various status groups competing with one another for wealth, power, and prestige. **Status groups** are made up of individuals who share a sense of status equality based on their participation in a common culture, with similar manners, hobbies, opinions, and values. In their struggle for advantage in society, individuals frequently turn to others like themselves as a resource in their struggle with other status groups. In America, for example, white Anglo-Saxon Protestant males dominate most of the positions of wealth, power, and prestige. Collins argues that this elite has structured public schools to reinforce the shared culture of their group. Entry into higher-status positions requires acceptance of their values. People from higher-status homes who have been educated in "better" schools, for example, have an easier time getting into the higher-level jobs. Those who are from lower social classes; who are of a different color or sex; or who do not possess the appropriate values, manners, language, and dress find it more difficult to enter better colleges and universities and to obtain higher-level jobs.

In an earlier book, Collins (1979) takes the conflict interpretation even further with the concept of **credentialism**—the idea that credentials (educational degrees) are unnecessarily required for many jobs. Why, he asks, must American physicians be trained almost into middle age when communist countries can produce qualified doctors in a relatively short time? Credentials are required, according to Collins, so that those who can afford them can maintain a strong hold on the elite occupations. Otherwise, why would people spend so much time in school learning things that are useless to them on the job (Labaree 1997)?

Racial and ethnic minorities face related barriers to achievement. An important example is the lower performance on college entrance examinations.

How do minorities perform on college entrance exams, and why? African Americans, Latinos, and Native Americans have lower average scores on the Scholastic Assessment Test (SAT) than whites (see Table 12.1). Sociologists attribute this fact, in part, to the differences in school quality noted earlier. And both school quality and SAT performance are related to social class levels of parents. Children from upper-class and upper-middle-class families attend more affluent schools. These children also have higher SAT scores. Social class clearly affects SAT performance.

How does SAT performance influence occupational and economic achievement? The SAT, originally created in 1926, was subsequently used to enable talented youth, regardless of social class background, to attend premier colleges and universities (Lemann 2000). Ironically, as we have just seen, social class is a major factor in SAT performance. Consequently, social class (through SAT performance) still influences who will attend the elite institutions, which are the gateway to America's higher social classes.

But don't the rewards accompanying high SAT performance indicate that America is a meritocracy? On the surface, it does seem that merit is being rewarded. After

TABLE 12.1

SAT Scores by Race and Ethnicity

An examination of this table reveals the gap in average SAT scores for whites and Asian Americans versus African Americans, Latinos, and Native Americans. Interpret these data as a conflict theorist would in the context of the U.S. as a meritocracy.

Racial/Ethnic Category	SAT Verbal Mean Scores	SAT Math Mean Scores	Totals
Native American or Alaskan Native	498	488	986
Asian, Asian American, or Pacific Islander	511	577	1088
African American	433	427	860
Latino Background	453	458	911
Mexican American	460	452	912
Puerto Rican	463	465	928
Latin American			
South American, Central American, or Other Latino			
White (excluding Latino origin)	532	531	1063

Source: The College Board, 2005.

all, it is those who do better academically who enjoy higher levels of success.

But there are two problems. First, the advantage some people have because of their parents' social class can create an uneven playing field; talent in the lower social classes often does not get recognized and developed. Second, the assumption that SAT performance measures academic ability and the likelihood of success in both college and life requires examination. For example, African American students who attend the most prestigious schools—including students with SAT scores below 1000—complete college at a higher rate than black students attending less rigorous institutions. They are also more likely to go on to graduate or professional schools (Bowen and Bok 2000).

At the least, these findings raise doubts. Recognizing this, an official at the Educational Testing Service (ETS)—developer and marketer of the SAT—announced in 1999 that ETS was creating a "Strivers" score. The idea was to adjust a student's SAT score to factor in social class as well as racial and ethnic characteristics thought to place him or her at a competitive disadvantage. Any student whose original score exceeded by 200 points the score predicted for his or her social class or racial or ethnic category would be considered a striver. The striver's score would be made available for colleges and universities to use, if they desired, in their admissions decisions (Glazer 1999; Wildavsky 1999). The proposal was quickly withdrawn after a firestorm of criticism. But, concern did not die. By 2005 the SAT is supposed to be transformed from an IQ-like test to a measure of academic preparedness (J. E. Barnes 2002).

Of course, the basic problem is not an SAT score deficiency (Darragh Johnson 2002). The larger issue is educational inequality.

Educational Inequality

How has the definition of educational inequality changed? From the beginning, American public education stressed educational equality (Demaine 2001). That originally meant that all children must be provided a free public education; that all children, regardless of social and economic background, must be exposed to the same curriculum; that all children within a given locality must attend the same school; and that all schools within a locality should be equally financed by local tax dollars (Coleman 1969; E. W. Gordon 1972; Brint and Brint 1998).

Despite this ideal, argue conflict theorists, not all children have been considered deserving of a formal education. Native Americans and African Americans were specifically excluded. During the 1850s, it was believed that slaves could not and should not be educated. It was not until slavery was abolished and former slaves began to assume industrial jobs that African Americans eventually were included among those Americans thought to be educable. Even then, African Americans were not given equal educational treatment. In the *Plessy v. Ferguson* decision of 1896, the Supreme Court made "separate but

Social class is a strong predictor of success on the SATs. How is race related to social class?

equal" part of America's legal framework. This doctrine held that equal treatment was being given when African Americans and whites had substantially equal facilities, even when these facilities were separate.

It was not until 1954 that the separate-but-equal doctrine was seriously challenged by the judicial system. In the *Brown v. Board of Education of Topeka* decision of 1954, Chief Justice Earl Warren wrote, as part of the Court's opinion:

> *Does segregation of children in public schools solely on the basis of race, even though the physical facilities and other "tangible" factors may be equal, deprive the children of the minority group of equal educational opportunities? We believe that it does.*

Warren went on to say that the separate-but-equal doctrine had no place in America's public educational structure and that separate educational facilities are inherently unequal. Still, it took the civil rights movement of the 1950s and 1960s to make the long-existing educational inequality for African Americans and other minorities a socially recognized problem for America. Thus the Civil Rights Act of 1964 prohibits discrimination on the grounds of race, color, religion, or national origin in any educational program or activity receiving financial support from the federal government.

What is meant today by educational equality? The most recent notion of **educational equality** addresses equality in relation to "effects" of schooling. Educational equality exists when schooling produces the same results (achievement and attitudes) for lower-class and minority children as it does for other children. Results, not resources, are the test of educational equality (Coleman et al. 1966; Hallinan 2006).

In 1964, James Coleman was commissioned by Congress to conduct a survey on discrimination and educational opportunities. From information on approximately 600,000 students and 60,000 teachers from some 4,000 schools, Coleman concluded that some consistent differences in black and white schools did exist. On the average, African American students had fewer school facilities: laboratories for physics, chemistry, and languages were less plentiful; textbooks were in shorter supply; and fewer books per pupil were available in school libraries. Coleman reported that minorities also suffered in relation to the majority when it came to curricular and extracurricular matters: African Americans in secondary schools were less likely to attend accredited schools; college-preparatory and accelerated curricula were less accessible to minorities; and whites generally enjoyed better access to academically related extracurricular activities, such as debate teams and student newspapers. An important source of these differences was the inequality of school financing. Using the survey data collected by Coleman, Christopher Jencks (1981) reported that about 15 to 20 percent more funds were spent for the average white schoolchild per year than for the average African American student.

It was not surprising that Coleman found some differences in resources allocated to majority and minority schools. It was surprising, however, that the differences in resources accounted for only a small part of the differences in majority and minority student performance on standardized achievement tests. According to Coleman, if educational equality is to exist, schools must overcome the educational inadequacies children bring with them from home. It is not enough, therefore, to provide minority children with equal school resources (Coleman 1968).

SNAPSHOT OF AMERICA 12.1

School Expenditures

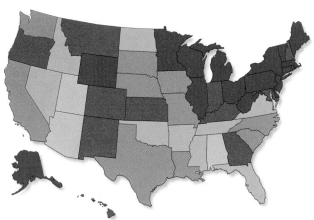

**Public School Expenditures
Per Pupil, 2003**

- $9,000 or more
- $8,000 to $8,999
- $7,000 to $7,999
- $6,000 to $6,999
- $5,000 to $5,999

Source: U.S. Bureau of the Census, *Statistical Abstract of the United States*, 2004–2005.

Everyone has heard "You get what you pay for." Because of this idea, many people use the amount of money spent on public schools as a measure of the quality of education. The accompanying map shows that some states spend more than twice as much per student as other states.

Interpret the Map
1. How does your state compare with other states in school expenditures?
2. Do you see a regional pattern in public school funding? Explain.
3. What other factors might you want to know to determine if the amount of money spent affected the quality of education delivered?

SNAPSHOT OF AMERICA 12.2

Median Household Income

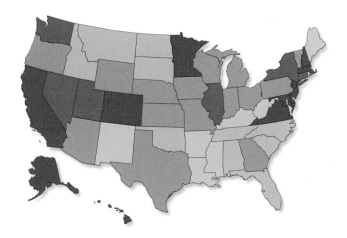

- $50,000 or more
- $45,000 to $49,999
- $40,000 to $44,999
- $35,000 to $39,999
- $30,000 to $34,999

Source: U.S. Bureau of the Census, *Statistical Abstract of the United States*, 2006.

This map displays the median household income by state. You would expect to see a positive relationship between state median household income and state school expenditures.

Interpret the Map
1. Looking at Snapshots of America 12.1 and 12.2, how would you describe the correlation between state median household income and state school expenditures? Explain your answer.
2. How does your state compare with other states with respect to median household income and school expenditures?
3. What is the relationship between income and school expenditures in your community? Be as specific as you can in your answer.

Such logic supported the mandatory busing programs instituted in urban areas from the late 1960s to the present. As interpreted by the courts, public schools must provide not only equal resources such as texts, teachers, and buildings, but also equal "social climates," which are the result of students' personalities and motivational factors. Because students influence one another's learning capabilities, school districts must provide similar types of students within each school based on racial, social class, and ethnic characteristics (Coleman 1990).

How do teacher evaluation and tracking affect educational equality? According to Ray Rist (1977), even the best-intentioned teachers often evaluate students' potential based on their social class, their racial characteristics, and their ethnic heritage rather than on their objective abilities. There is evidence to support this allegation. For example, in his classic study of "Elmstown's" youth, August Hollingshead (1949) found that when middle- and upper-class children had difficulty in class, their parents were usually called in to discuss learning problems. Similar learning difficulties, however, were interpreted as behavioral problems when they occurred among lower-class children. Aaron Cicourel and John Kitsuse (1963) later found the same phenomenon in their study of high school counseling procedures. When test scores were consistent with the counselor's expectations, there was no problem. The counselor, however, was much more willing to find excuses for poor performance among high-status students than among low-status ones. Rist (1977) reported a similar pattern among classroom teachers. When asked questions in class, the students perceived as more capable by teachers were given more time to answer than "less able" students were given. Also, lower-status students were more likely to be considered by teachers as less capable.

Most high schools have a series of curricula (tracks): college preparatory, commercial, vocational, and general for those who are undecided about their occupational futures (Kilgore 1991; Gamoran 1992). Researchers report that social class and race heavily influence the placement of students in these curricula—regardless of their intelligence or past academic achievement (Oakes and Lipton 1996; Taylor, Peplau, and Sears 1997).

Subsequent levels of academic performance have been influenced more by students' tracks than by their intelligence or current scholastic performance. Regardless of earlier school performance or intelligence, the overall academic performance of those bound for college increases, whereas the performance of those on a noncollege track decreases. Through tracking, then, schools may deny educational equality to many students, especially minority children, who often come from disadvantaged families (Oakes 1985; Gibson and Ogbu 1991). In part because tracking has been linked to educational inequality, American high schools have

been retreating from the practice. Some research, however, casts doubt on the idea that detracking promotes greater educational equality (Lucas 1999). Rather than assigned courses being based on students' course-specific abilities, mathematical achievement has become a gateway to college-prep courses.

Testing Cognitive Ability

The technical term for intelligence, today, is **cognitive ability**—the capacity for thinking abstractly. Dating back to the turn of the twentieth century, there has been a tradition in schools to attempt to measure cognitive ability. Because cognitive ability testing is an important element in sorting and tracking students, it contributes to educational inequality. Whenever cognitive ability tests are discussed, the question of inherited intelligence always arises.

Is intelligence inherited? In the past, some people assumed that individual and group differences in measured intellectual ability were due to genetic differences. This assumption, of course, underlies social Darwinism.

A few researchers still take this viewpoint. More than thirty years ago Arthur Jensen (1969), an educational psychologist, suggested that the lower average intelligence score among African American children may be due to heredity. A more recent book by Richard Herrnstein and Charles Murray (1996), entitled *The Bell Curve*, is also in the tradition of linking intelligence to heredity. According to these authors, humans inherit 60 to 70 percent of their intelligence level. Herrnstein and Murray further contend that the fact of inherited intelligence makes largely futile the efforts to help the disadvantaged through programs such as Head Start and affirmative action.

Social scientists have generally criticized these advocates of genetic differences in intelligence for failing to consider adequately the effects of the social, psychological, and economic climate experienced by children of various racial and ethnic minorities. Even those social scientists who grant genetics an important role in intelligence criticize both the interpretations of the evidence and the public policy solutions contained in *The Bell Curve*. They point to the body of research that runs counter to Herrnstein and Murray's thesis. More specifically, they see intelligence not as an issue of nature versus nurture, but as a matter of genetics and environment (Morganthau 1994; L. Wright 1995; Fischer et al. 1996). We know, for example, that city dwellers usually score higher on intelligence tests than do people in rural areas, that African Americans with higher socioeconomic status score higher than lower-status African Americans, and that middle-class African American children score about as high as middle-class white children. We also have discovered that as people get older, they usually score higher on intelligence tests. These findings, and others like them, have led researchers to conclude that environmental factors affect test performance at least as much as genetic factors (Samuda 1975; Schiff and Lewontin 1987; Jencks and Phillips 1998; Arrow, Durlauf, and Bowles 2000). One of these environmental factors is a *cultural bias* in the measurement of cognitive ability.

How can intelligence tests be culturally biased? Robert Williams (1974) is one of many early social scientists who argued that intelligence tests have a **cultural bias**—an unfair measure of cognitive abilities for people in certain social categories. Specifically, intelligence tests are said to be culturally biased because they are designed for middle-class children and because the results measure learning and environment as much as intellectual ability. Consider this intelligence test item cited by Daniel Levine and Rayna Levine (1996:67):

A symphony is to a composer as a book is to what?
☐ *paper* ☐ *sculptor* ☐ *author* ☐ *musician* ☐ *man*

According to critics, higher-income children find this question easier to answer correctly than lower-income children because higher social class members are more likely to have been exposed to information about classical music. (The same charge was made by critics of a recent SAT question that used a Bentley automobile as its illustration.) Several studies indicate that because most intelligence tests assume fluency in English, minorities cannot respond as well. Some researchers suggest that many urban African American students are superior to their white classmates on several dimensions of verbal capacity, but this ability is not recognized, because intelligence tests do not measure these areas (S. J. Gould 1981; Goleman 1988; Hurn 1993).

The testing situation itself can affect performance. Low-income and minority students, for example, score higher on intelligence tests when tested by adult members of their own race or income group. Apparently, they feel threatened when tested in a strange environment by someone dissimilar to them. Middle-class children are frequently eager to take the tests because they have been taught the importance of tests and academic competitiveness. Because low-income children do not recognize the importance of tests and have not been taught to be academically competitive, they ignore some of the questions or look for something more interesting to do. Nutrition also seems to play a role in test performance. Low-income children with poor diets may do less than their best when they are hungry or when they lack particular types of food over long periods of time.

Promoting Equality in Education

Although it is difficult to overcome the barriers of economics and social class, policy makers and educators are exploring avenues for educational equality. Four of the avenues—school desegregation, compensatory education, community control, and private schooling—are discussed here.

Does school desegregation improve academic achievement of minority children? The Coleman Report supports the idea that it does. Specifically, James Coleman et al. (1966) reported that African American children in classrooms that are more than one-half white scored higher on achievement tests. The finding that minority children perform better academically when placed in desegregated, middle-class schools has been verified by several other studies. In a broad review of research on school desegregation, Meyer Weinberg (1970) concluded that African American children consistently experience higher academic achievement in desegregated schools. It does not seem coincidental that the period of higher school desegregation witnessed major academic progress for African American students, and that the resegregation of the 1990s is associated with a growing gap in racial academic achievement (Kaufman 2000, 2001).

On the other hand, several studies have shown that racially balanced classrooms can have either positive or negative effects on the academic achievement of minority children. Mere physical desegregation without adequate support may have detrimental effects on children. However, desegregated classrooms with an atmosphere of respect and acceptance promote improved academic performance (U.S. Commission on Civil Rights 1967; I. Katz 1968; Longshore 1982a, 1982b; Patchen 1982; Orfield et al. 1992). For example, a study of seniors at a racially and ethnically diverse high school in Cambridge, Massachusetts, revealed a positive effect of diversity on higher-education aspirations ("The Impact of Racial and Ethnic Diversity on Educational Outcomes" 2002).

Some postschooling benefits of desegregation include higher occupational positions and higher incomes for minority students who attended desegregated public schools than for minority students who attended segregated schools. These differences cannot be attributed solely to differences in educational attainment. Rather, increased social contact with middle-class students also contributes to college attendance and better employment opportunities, because middle-class students provide norms of behavior, dress, and language often required by employers in the middle-class hiring world (Crain 1970; Crain and Weisman 1972; W. D. Hawley 1985). Today, more than ever, the social and economic benefits of school equality are an important consideration. Good jobs require postsecondary education. Those well-paying factory jobs that require minimal education are a shrinking part of the economy (Orfield 2001).

Some evidence suggests that school desegregation may also lead to better racial and ethnic relations because students are exposed to people of different backgrounds (W. D. Hawley 1985; Hawley and Smylie 1988; Kurlaender and Yun 2000; Eaton 2001). Support for this finding also appeared in the aforementioned study of seniors in a racially and ethnically diverse high school. These students felt well prepared for life as adults in a diverse community, indicated increased understanding of different points of view, and expressed enhanced appreciation of the background of other groups ("The Impact of Racial and Ethnic Diversity on Educational Outcomes" 2002).

On such evidence rests the promise of **multiculturalism**: an approach to education in which the curriculum accents the viewpoints, experiences, and contributions of minorities—including women as well as ethnic and racial minorities (Katsiaficas and Kiros 1998; Watson 2000). Multiculturalism attempts to dispel stereotypes and to make the traditions of minorities valuable assets for the broader culture. According to an earlier study, school attendance and academic performance have increased in some schools with multicultural curricula (Marriott 1991).

As noted earlier, public schools in the United States are drifting back toward segregation. In a reversal of the trend toward integration from the mid-1960s until 1988, African American and Latino youngsters are becoming racially and ethnically segregated. Since 1986, school districts have tended to show lower levels of interracial exposure (Frankenberg and Lee 2002). This intensification of school resegregation throughout the 1990s is hardly surprising given the three major U.S. Supreme Court decisions permitting a shift back toward segregated neighborhood schools. Around three-fourths of African American and Latino students are now in classrooms with a predominantly minority enrollment (Orfield and Eaton 1997; Orfield 2001; Orfield and Lee 2006; Frankenberg and Orfield 2007).

Does compensatory education increase school achievement for disadvantaged children? It appears so (Entwisle and Alexander 1989, 1993; Ramey and Campbell 1991; Zigler and Styfco 1993; Campbell and Ramey 1994). The best-known attempt at remedial education is the Head Start program, a federally supported program for preschoolers from disadvantaged homes. Head Start originated as part of the Office of Economic Opportunity in 1965. It was conceived, in part, as a means of preparing disadvantaged preschoolers for entering the public school system. Its goal is to provide disadvantaged children an equal opportunity to develop their potential (Waldman 1990–1991). Initial studies on the progress of Head Start report positive long-term results. Low-income youngsters between the ages of nine and nineteen who had been in preschool compensatory programs performed better in school, had higher achievement test scores, and were more motivated academically than low-income youth who had not been in compensatory education programs (Bruner 1982; Etzioni 1982).

Follow-up research also supports the benefits of Head Start (Schweinhart 1992; Zigler and Muenchow

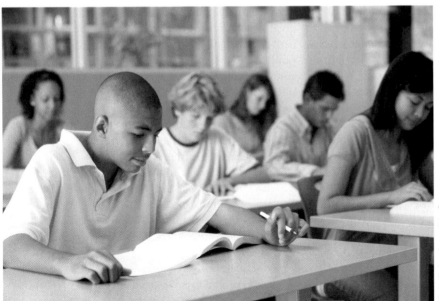

Some evidence suggests that ethnic and racial integration in schools improves attendance and academic achievement. Despite these potential advantages, over two-thirds of African American and Latino students are in segregated classrooms.

Monkey Business Images/Index Open

1992; K. Mills 1998). For example, in one study children entering a Head Start program scored lower on intelligence than a group of their peers, but follow-up research revealed benefits for those Head Start students: better school attendance, completion of high school at a higher rate, and entrance into the workforce in greater proportion (Passell 1994).

As noted in Using the Sociological Imagination at the beginning of this chapter, children's scores on IQ tests, general ability tests, and learning readiness assessments significantly improve after exposure to Head Start programs. Studies report that the scores of Head Start participants eventually reach the national average on tests of general ability and learning readiness. Attitudinal, motivational, and social progress have been observed as well.

Politicians, both Republicans and Democrats, currently recognize the benefits of Head Start. After overwhelming support in both the House and Senate, new Head Start authorization legislation was signed by President Clinton on May 19, 1994. Under this legislation, funding was significantly increased, and coverage was extended to approximately 800,000 children (Dewar and Vobejda 1994).

What is the relationship between community control and educational equality? The authority to finance and administer elementary and high schools lies within the states rather than with the federal government. The states in turn delegate most of this authority to local school systems. Ultimately, then, the power rests with school administrators who carry out policies established by the local school boards.

The local school board is thought to represent the interests of the community residents while, it is assumed, administrators represent the professional needs of teachers and the educational needs of students. These assumptions have been challenged. It is argued that the current school governance arrangement has become so centralized and bureaucratic that accountability to parents and neighborhood is rare. It is also argued that political decentralization in urban areas, which would place direct control of each school in the hands of neighborhood residents, is desirable because the current remote, bureaucratic structure fails to meet the needs of children—needs that vary from one neighborhood to another. This is held to be a particular disadvantage by minorities, who charge that schools are oriented toward white students' needs and are not responsive to the problems of minority students.

Although many school administrators have agreed to share major educational decisions with parents, it remains to be seen whether they are willing to share their power as well. There is also the question of whether neighborhood schools can raise enough money to ensure quality education. Finally, there is the possibility that community control will increase school segregation.

Do private schools offer a better teaching environment for the disadvantaged? Amid the controversy surrounding educational opportunity in public schools has come some interesting research by James Coleman and his colleagues (Coleman, Hoffer, and Kilgore 1981; Coleman, Kilgore, and Hoffer 1982; Coleman et al. 1982a, 1982b). Coleman and his colleagues attempted to investigate whether actual achievement among students attending private schools was better than that among public school students. Much to the dismay of many public school educators, the private schools studied produced higher student achievement levels in mathematics and vocabulary, even when controlled for background characteristics of students. Of particular interest to the researchers was the observation that minority children in these private schools, particularly Catholic schools, performed better than their counterparts in public schools. The authors argued that higher academic standards, stricter discipline, more homework, and the accompanying student perceptions of academic emphasis accounted for the differences in achievement levels. These findings were used by Coleman and his associates to argue for changes in public school policies that could enhance achievement levels for all students and to suggest that the growing private school movement might have beneficial consequences for many of America's children.

The results of the Coleman study have been criticized. Arthur Goldberger and Glen Cain (1981, 1982) accuse the researchers of being biased toward private schools. Others contend that Coleman has overstated his conclusions. The government agency sponsoring the Coleman research, for example, contends that the performance comparisons are inaccurate because a greater percentage of private school students (70 percent) pursue college-bound academic programs than do public school pupils (about one-third). After a reanalysis of his data, Coleman agreed that precollegiate public school seniors perform as well on standardized tests as Catholic school seniors.

These criticisms should not be allowed to overshadow the results of the Coleman study. Diane Ravitch places the debate into proper focus:

> What it [the Coleman study] does show is that good schools have an orderly climate, disciplinary policies that students and administrators believe to be fair and effective, high enrollment in academic courses, regular assignment of homework, and lower incidence of student absenteeism, class-cutting, and other misbehavior. These conditions are found more often in private than in public schools (though they are not present in all private schools, and they are not absent in all public schools). (Ravitch 1982:10)

1. _____ refers to a social system in which social status is a function of ability and achievement rather than background or parental status.
2. The idea that some educational requirements are necessary for many jobs today is called
   a. status grouping.
   b. credentialism.
   c. meritocracy.
   d. conflict.
3. A significantly larger proportion of higher social class children attend college than lower-class children. This is largely due to differences in ability among social classes. T or F?
4. Educational equality exists when schooling produces the same _____ for lower-class children and minorities as it does for other children.
5. Most sociologists agree that IQ test scores do not measure inherited learning ability. T or F?
6. _____ is an approach to education in which the curriculum accents the viewpoints, experiences, and contributions of women and diverse ethnic and racial minorities.
7. Compensatory education appears to benefit deprived children. T or F?

Answers: 1. Meritocracy; 2. b; 3. F; 4. results; 5. T; 6. Multiculturalism; 7.T.

Symbolic Interactionism

Schools are in the business of transmitting culture, and it is in this socialization that symbolic interactionists are interested. Socialization takes place primarily through three sources: the hidden curriculum, textbooks, and teachers.

The Hidden Curriculum

How does the hidden curriculum contribute to socialization? Modern society places considerable emphasis on the verbal, mathematical, and writing skills an adult needs to obtain a job, read a newspaper, balance a checkbook, and compute income taxes. However, schools teach much more than these basic academic skills. They also, through a **hidden curriculum**, transmit to children a variety of nonacademic values, norms, beliefs, and attitudes (Sadker and Sadker 1995; Ballantine and Spade 2007). The hidden curriculum teaches children skills such as discipline, order, cooperativeness, and conformity. These skills are thought to be necessary for success in modern bureaucratic society, whether one becomes a doctor, a college president, a computer programmer, or an assembly-line worker. Over the years, schools, for example, socialize children for the transition from their closely knit, cooperative families to the loosely knit, competitive adult occupational world. The school provides systematic practice for children to operate independently in the pursuit of personal and academic achievement. The values of conformity and achievement are emphasized through individual testing and grading. Because teachers evaluate young people as students—not as relatives, friends, or equals—students participate in a model for future secondary relationships (e.g., employer–employee, salesperson–customer, or lawyer–client).

Textbooks and Teachers

How do textbooks socialize children? A by-product of the hidden curriculum is the development of patriotism and a sense of civic duty. Partly for this reason, courses such as history and government are generally slanted in favor of a particular view of history. Accounts of the American Revolution, for example, are not the same in British and American textbooks. Few societies are willing to admit imperfections, and this reluctance is often reinforced in school materials. Most U.S. history textbooks, for example, have not accurately portrayed the government's treatment of Native American peoples. And the resistance to critical accounts of history is especially apparent when a teacher attempts to introduce a controversial book, such as Alice Walker's *The Color Purple*.

Textbooks implicitly convey values and beliefs. Sociologists interested in gender stereotypes point out that older primary textbooks tended to present men in challenging and aggressive activities while portraying women as homemakers, mothers, nurses, and secretaries. Women were not only placed in traditional roles, they appeared far less frequently than men. Underprivileged children who compared their homes with middle-class homes felt out of place. Parents of low-income or inner-city children complained that such pictures of middle-class life harmed their children (Trimble 1988; Gibson and Ogbu 1991). Today, active parent groups, minority

George Ritzer—The McDonaldization of Higher Education

Max Weber, you remember, was concerned that rationalization would go too far. He feared that rationalization would spread to all aspects of society, resulting in an "iron cage of rationality." This dehumanizing cage of rationality, Weber asserted, would eventually deprive everyone of freedom, autonomy, and individuality.

George Ritzer uses the metaphor of the "McDonaldization of society" to illustrate Weber's image of the iron cage. McDonaldization, Ritzer states, is the process of increasing efficiency, calculability, predictability, and control by replacing human technology with nonhuman technology. *Efficiency* refers to the relationship between effort and result. An organization is most efficient when the maximum results are achieved with minimum effort. For example, fast-food restaurants are efficient in part because they transfer work usually done by employees to customers. *Calculability* involves estimation based on probabilities. High calculability exists when the output, cost, and effort associated with products can be predicted. A McDonald's manager trains employees to make each Big Mac within a rigid time limit. *Predictability* pertains to consistency of results. Predictability exists when products turn out as planned. Big Macs are the same everywhere. *Control* is increased by replacing human activity with technology. McDonald's drink machines stop after a cup has been filled to its prescribed limit. McDonald's restaurants, Ritzer argues, ideally reflect the dehumanizing effects of extreme rationality.

The fast-food restaurant is often a dehumanizing setting in which to eat or work. People lining up for a burger, or waiting in the drive-through line, often feel as if they are dining on an assembly line, and those who prepare the burgers often appear to be working on a burger assembly line. Assembly lines are hardly human settings in which to eat, and they have shown to be inhuman settings in which to work. (Ritzer 2004:12)

Ritzer (1998) uses a wide variety of materials to document the McDonaldization of many important aspects of American society. His sources of information include newspapers, books, magazines, and industry publications. As a college student, you will find Ritzer's conclusions on the "McUniversity" of immediate interest. Higher education, Ritzer concludes, is well on the way to becoming another consumer item. Increasingly, students and parents view a college degree as a necessary credential to compete successfully in the job market.

This consumer orientation, Ritzer asserts, can be observed on most college campuses in the United States. For example, students want education to be conveniently located and available as long as possible each day. They seek inexpensive parking, efficient service, and short waiting lines. Students want a high-quality service at the lowest cost. A "best buy" label in national academic rankings catches the attention of parents and students.

Public colleges and universities, Ritzer contends, are responding to this consumer orientation. They are doing so in part because government funding for higher education is becoming scarcer. To meet reduced funding, colleges and universities are cutting costs and paying more attention to "customers." For example, Ritzer points to student unions, many of which are being transformed into mini malls with fast-food restaurants, video games, and ATMs.

Ritzer predicts that a far-reaching customer-oriented tactic will be to

McDonaldize through new technology. The McUniversity will still have a central campus, but it will also have convenient satellite locations in community colleges, high schools, businesses, and malls:

Students will "drop by," for a course or two. Parking lots will be adjacent to McUniversity's satellites (as they are to fast-food restaurants) to make access easy. (Ritzer 1998:156)

McDonaldization, Ritzer contends, will dehumanize the process of education. Most instructors at satellites will be part-timers hired to teach one or more courses. They will come and go quickly, depriving students of the opportunity to form relationships with permanent, tenured faculty. To make the courses alike from satellite to satellite, course content, requirements, and materials will be highly standardized, losing the flavor individual professors bring to their classes. Students will not be able to choose a particular instructor for a course because there will be only one per satellite. In fact, there may be no teacher physically present at all. More courses will be delivered by professors televised from distant places.

It is not, Ritzer states, that colleges and universities will become a shopping mall or a chain of fast-food restaurants. Institutions of higher education will retain many traditional aspects. But there will undoubtedly be a significant degree of McDonaldization.

Evaluate the Research

1. Ritzer's study is based on one of the three theoretical perspectives. Explain which one you think it is and why.
2. Could you use another perspective instead? Why or why not?
3. Analyze the functions and dysfunctions of the McDonaldization of higher education.

TABLE 12.2

Investigating Education

This table illustrates differences in the ways the major theoretical perspectives investigate education as a social institution. It is, of course, possible for a theoretical perspective to study education using one of the concepts associated in this table with another perspective. Explain, for example, how conflict theory would interpret the hidden curriculum and tracking.

Theoretical Perspective	Concept	Example
Functionalism	Tracking	• Schools shape the occupational future of children by placing them in educational programs based on test scores and early school performance.
Conflict theory	Meritocracy	• Students attending better schools have an occupational advantage over students from poorer schools.
Symbolic interactionism	Hidden curriculum	• Schools teach children the values of conformity and achievement.

special interest groups, and state boards of education are working with textbook authors and publishers to ensure that a more balanced picture of society is presented to students.

How do teachers socialize children? Teachers, usually a child's first authority figures outside the family, intentionally socialize students by having them perform academic tasks in predetermined ways. But teachers unintentionally affect children as well, as can be seen in a study by Robert Rosenthal and Lenore Jacobson (1989). In their book *Pygmalion in the Classroom*, Rosenthal and Jacobson demonstrate the **self-fulfilling prophecy**—when an expectation leads to behavior that causes the expectation to become a reality (Merton 1968). (Self-fulfilling prophecy was discussed in Chapter 9.) In their study, elementary schoolteachers were given a list of children in their classrooms who, according to the researchers, were soon to blossom intellectually. Actually, these children were picked at random from the school roster and were no different from the student body in general. At the end of the year, the authors found that this randomly selected group of children had scored significantly higher on the same intelligence test they had taken at the beginning of the year. Their classmates as a group did not. According to Rosenthal and Jacobson, the teachers expected these "late bloomers" to spurt academically. Consequently, the teachers treated these students as if they were special. This behavior by the teachers encouraged the students to become higher academic achievers (see Doing Research). Eleanor Leacock (1969) also found the self-fulfilling prophecy at work in a study of second and fifth graders in black and white low- and middle-income schools. Both studies demonstrated the transmission of negative self-impressions as well

as positive ones. Subsequent research generally supports dramatic effects of teacher (and parent) expectations (Gamoran 1986). To quote the Children's Defense Fund: "Educational research shows that children adjust their performance and self-image to meet adult expectations, and children who are placed in lower ability classes soon come to believe that they cannot excel in school" (Children's Defense Fund 1991:80).

How do teachers foster sexism? Following a long line of researchers, Myra Sadker and David Sadker (1995) contend that well-meaning teachers unconsciously transmit sexist assumptions, basing these assumptions on stereotypes for appropriate gender behavior. America's teachers, therefore, are often unfair to girls, crippling them through differential classroom treatment. Girls, for example, learn to talk softly, avoid certain subjects (especially math and science), defer to the alleged intellectual superiority of boys, and emphasize appearance over intelligence.

Moreover, in a coeducational setting, girls, compared with boys, are

• Five times less likely to receive the most attention from teachers.
• Three times less likely to be praised.
• Eight times less likely to call out in class.
• Three times less talkative in class.
• Half as likely to demand help or attention.
• Half as likely to be called on in class.

Throughout their book, Sadker and Sadker emphasize that boys are more talkative in class, raise their hands more often, move around more, argue with teachers more, and get more of the teachers' attention than do girls. All of this tends to make young women mature

Catherine Ledner/Getty Images

What evidence has been presented about the advantage of single-gender schools?

as second-class citizens, these researchers conclude, and deprives the world of important contributions women can make.

But what about the upbeat media reports? Writers who paint an optimistic picture fail to produce convincing evidence (Deak 1998). Inequalities remain. There is objective evidence that girls are guided in school toward traditional female jobs and away from high-paying, powerful, and prestigious jobs in science, technology, and engineering (Millicent 1992). True, significantly more high school girls want to go into engineering today than in the past. But five times more men than women receive bachelor's degrees in engineering.

Significantly, these gender-based discrepancies cannot be explained by ability differences. Girls perform almost as well as boys on math and science tests (O'Sullivan, Reese, and Mazzeo 1997). Girls score higher than boys at reading and writing at all grade levels and are more likely to attend college (Greenwald et al. 1999). Moreover, females fare better in single-gender schools and single-gender classes in coeducational schools:

> Girls in these situations, in general, get better grades, report that they learn more and are more positive about the learning situation, have higher self-esteem, and more often move on to advanced courses than do girls in regular coeducational situations. (Deak 1998:19–20)

FEEDBACK

1. According to symbolic interactionists, schools are very much in the business of transmitting _____.
2. The _____ teaches children such things as discipline, order, cooperativeness, and conformity—skills thought to be necessary for success in a modern bureaucratic society.
3. _____ are the first authority figures children encounter on a daily basis outside the family.
4. The _____ occurs when an expectation leads to behavior that causes the expectation to become a reality.
5. Match the following research topics with the three theoretical perspectives.

    ____ a. functionalism   (1) requirement of unnecessary credentials by schools

    ____ b. conflict theory   (2) benefit of schools for society

    ____ c. symbolic interactionism   (3) effects of teacher reactions on the self-esteem of students

Answers: 1. culture; 2. hidden curriculum; 3. Teachers; 4. self-fulfilling prophecy; 5. a. (2), b. (1), c. (3)

Higher Education

College Attendance

More than 20 million students are enrolled in more than 3,600 American colleges and universities. Immediately following World War II, as a result of the G.I. Bill and the influx of veterans, enrollment in institutions of higher education began to grow; it has not stopped (see Figure 12.2). Since 1980, the percentage of high school graduates attending college has increased from 49 percent to 69 percent (U.S. Bureau of the Census 2007c). This increase has occurred despite predictions of a downturn in college enrollment—a prediction based on a decline in the number of eighteen- to twenty-four-year-olds, the traditional college-age category.

Why didn't college enrollment drop as anticipated? Several trends account for the increase. First, despite a decline in the eighteen- to twenty-four-year-old population, a substantially larger proportion of this age category is attending college. In addition, the number of students over age twenty-four attending college, particularly women, has continued to grow, and the number of students remaining in college after their first year has increased.

At least three factors account for these trends. Fearful of declining enrollments, colleges and universities mounted strong student recruitment campaigns. This stimulated student demand and enlarged university applicant pools. Also, low-paying service jobs increasingly replace high-paying manufacturing jobs, resulting in fewer attractive noneducational options for those with only high school diplomas. With the loss of high-wage unionized jobs in blue-collar industries such as automobiles and steel, the incentive to attend college increases. Finally, the increase in the number of community and branch colleges offers an opportunity for many who could not otherwise continue their education (Arenson 2008).

What can we expect in the future? A new baby boom is about to hit colleges and universities. The leading edge of the children and grandchildren of baby boomers—the so-called "baby boomlet"—is beginning to work its way into colleges. This is only the beginning. The year 2009 is predicted to have the largest high school graduating class in the history of the United States—some 3.2 million students. The college enrollment of 13 million

---

FIGURE 12.2

College Enrollment in the United States: 1965–2016

This figure traces the increase in college enrollment in the United States since 1965. Explain why college enrollment continues to increase.

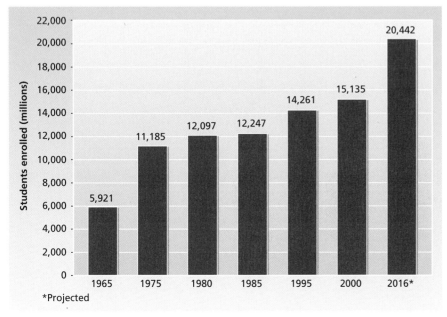

*Projected

Source: U.S. Bureau of the Census, *Statistical Abstract of the United States: 2001* (Washington, DC: U.S. Government Printing Office, 2001), table 205, *U.S. Bureau of the Census: 2008* (Washington, DC: Government Printing Office, 2007), table 207.

in 1988 reached more than 17 million in 2005, a 28 percent increase (U.S. Bureau of the Census 2007c).

At the same time that college enrollments are accelerating—across the country—state budget shortfalls are spawning deep cuts in higher education. The impending train wreck between student demand and limited college financial resources will last at least until 2012. Colleges will scramble to revamp their enrollment policies, to expand services, to create adequate housing and class space, and to generate new sources of funding. College students and their parents will meet head-on with rising tuition and fees as well as more intense competition at nationally reputed schools (DeBarros 2003).

Increased tuition can burden community college students even further. A tuition increase accounts for a larger share of available personal financial resources among students who are employed and older than twenty-one. And most community college students are beyond the typical college age, supporting themselves. Independent from their families and denied financial aid because they have incomes, many community college students find this additional obstacle the last straw; they drop out. Others will reduce the number of hours they take each term and/or work more hours. Either of these alternatives further retards their often already slow progress caused by work and/or family responsibilities.

Education, Income, and Occupation

Because of advanced industrialization, education is a ticket to better jobs and higher incomes. A college degree has a significant payoff, financially and occupationally (Guinzburg 1983; Katchadourian and Boli 1994).

What is the economic payoff? As first noted in Chapter 2, college education results in higher annual income, regardless of race or gender. Among African American and white males who have a college education or more, annual average income is more than double that of high school dropouts. White male high school graduates earn annually about 65 percent of that earned by white college graduates. African American males with high school diplomas earn about 58 percent of that earned by African American male college graduates. White female high school graduates earn each year only about 59 percent of that earned by white college graduates. African American females with high school diplomas earn only 50 percent of the earnings of African American female college graduates (see Figure 12.3).

The downside of the data in Figure 12.3 is the comparison between African Americans and whites and between males and females. Although African Americans and women benefit from a college educa-

---

FIGURE 12.3

Median Annual Income by Gender, Race, and Education

Clearly, this figure documents the income advantage that white males in the United States have over white females and African Americans of both sexes. Explain how this situation can be used to challenge the existence of a true meritocracy.

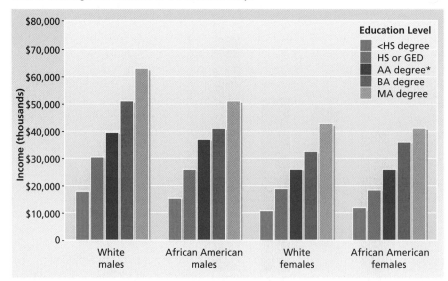

*An Associate Arts or AA degree is a two-year college program.
Source: U.S. Bureau of the Census, 2004.

INTERNET LINK To learn more about income in the United States, visit the U.S. Bureau of the Census Web site at www.census.gov. On its home page, select "Subjects A to Z," then select "I" ("Income") and then your choice of topics.

tion, they clearly suffer in comparison to white males with identical educational credentials. (See Chapters 9 and 10 for more detailed analyses of the socioeconomic situation of African Americans and women in the United States.)

How strong is the relationship between occupational status and education? If anything, the benefits of college are even more apparent occupationally. For African Americans, whites, males, and females, occupational status (measured by blue-collar versus white-collar status) increases dramatically with higher education. The percentage for all race and gender combinations employed in white-collar occupations rises dramatically with educational level, whereas the percentages in blue-collar jobs decline sharply. Although African Americans and females are more concentrated in lower-level jobs within the higher-status occupations, the significant relationship between occupational status and higher education cannot be dismissed (D. A. Gilbert 2008).

Changes in Higher Education

How would you describe university environments: static or fluid? College studies and college life have changed dramatically during the past forty years. College faculties and programs reached their peak in America during the fifteen-year period before the war in Vietnam. During this period, institutions of higher learning were very selective, academic programs were extremely rigorous, and college faculties had extensive control over the direction of student learning and development (Riesman 1980).

A shift from academic merit to student consumerism began in the mid-1960s, when most American undergraduate and graduate programs began to package their curricula more attractively. As a result, colleges and universities are losing some of their ability to shape academic and intellectual development. Students increasingly demand courses and programs that are more beneficial for employment. This is evident in the rise of vocational programs in community colleges, the shift of students from liberal arts into business and economics, and grade inflation.

Stanley Aronowitz and Henry Giroux (1994) are also concerned about the influence of labor market considerations. They point to the rise of business enrollments and the concomitant decline in the liberal arts majors. They see social science and humanity departments becoming service departments for nonmajors. They witness the diminution of graduate programs and faculty in liberal arts. As a result, they contend, a generation of liberal arts scholars has been lost.

An explosion was set off in 1987 when Allan Bloom, a University of Chicago professor, published *The Closing of the American Mind*. This best-selling book became the touchstone for those who believe that higher education

is on its way to intellectual chaos. According to these critics of higher education, academia is now under the control of administrators and faculty who were part of the radical left in the 1960s. Universities, they believe, are using a radical egalitarian philosophy to elevate works of popular culture to the level of Platonic studies. The most frequently cited illustration is a Stanford University faculty decision, in 1988, to drop from a required Western culture course some traditionally classic works by white male authors in favor of some women and minority authors. A basic fear of these academic critics is that the gender, racial, and ethnic background of authors is supplementing merit as a basis for designing programs, courses, and reading lists. This is a debate that promises to continue for some time.

Another problem in higher education—declining public financial support—threatens the traditional high quality and low cost of public higher education. The Association of Governing Boards of Universities and Colleges describes the situation as "the new depression in higher education." Since World War II, the public, through its tax dollars, has supported higher education. That commitment is currently diminishing. Even the budgets of the most prestigious public universities have been cut, causing them to become semipublic institutions at best. The University of California's system, consistently one of the world's finest, has seen the proportion of its state-funded financial base shrink from a high of 70 percent thirty years ago to less than 25 percent today. The University of Michigan now receives less than 20 percent of its funding from the state; the University of Colorado receives only 10 percent of its budget from the state.

State colleges and universities have responded to reduced state funding by increasing both private funding and tuition charges. As a result, the cost of a college education is skyrocketing (Morganthau and Nayyar 1996; Symonds 2003). The low-cost, high-quality system of higher education that was provided the veterans after World War II, as well as their "baby boomer" children who were born between 1946 and 1964, is being denied the postboomer generation. In response to recent public outcry, tuition increases have leveled off, going from an increase of 9 percent in 2003 to an increase of 6 percent in 2007 (College Board 2007).

Parents and the general public are in shock. Between 1981 and 2001, tuition increased 112 percent at private colleges and universities and 106 percent in public institutions, whereas family income rose only 27 percent (College Board 2001). In 2007, a year at a public four-year university cost an average of $13,589, while a year at a private university cost $32,307 (College Board 2007). With college tuition rising at two times the rate of inflation, and with the current no-new-taxes mood of the country, the call is for universities to restructure. If companies such as IBM and Xerox can downsize and do more with less, the argument goes, so can public

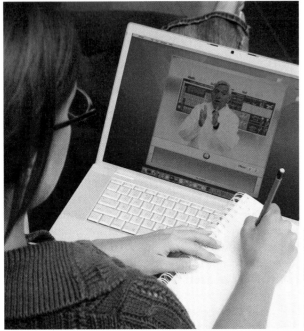

Colleges and universities are changing from being a completely geographically bound institution to a more virtual one.

colleges and universities. "Time to Prune the Ivy" was the clever title of an article in *BusinessWeek* (Farrell 1993). Higher education and the public are obviously in conflict. The taxpayers are generally opposed to tax increases, and university administrators and professors fear a decline in quality due to restructuring and downsizing. Is the future at risk for a country that has built its prosperity on higher education?

An important casualty of reduced public funding may be the democratization of higher education achieved during the twentieth century. Prior to World War II, colleges were attended by only a small elite group. Through extensive federal and state investment after 1945, the United States became the first country to support mass higher education. As government finan-cial backing declines, the costs of higher education are increasingly shifted onto students and their parents. While most students feel the pinch, the most damaged are those from low-income, immigrant, and minority families. Reduced access jeopardizes the promise of higher education as a means to upward social mobility (Symonds 2003).

A new relationship between leading research univer-sities and business is emerging. In the past, large corpo-rations—like RCA, Xerox, and the old AT&T—housed internal research facilities. In essence, these corporate labs were research universities within private compa-nies. Replacing these vanishing labs is a leading group of universities opening their research facilities to corpo-rations. Business views this relationship as cost efficient and more productive. Universities use the collaboration to bring in significant amounts of money. Stanford, for example, will receive $100 million from ExxonMobil over a ten-year period (Zachary 2007). Many professors, at universities where such deals have been made, do not welcome this move. They worry about contamination of the research process by the profit motive. They fear that their disinterested pursuit of knowledge will be corrupted by the corporate desire for self-interested answers.

Finally, the technologically driven information age is transforming colleges and universities from a geo-graphically bound campus to a virtual one. Although the traditional residence-based system will continue, higher education will increasingly be characterized by geographically dispersed faculty and students. Distance learning will be necessary both because technology makes it possible and because higher education is being redefined from a set number of semesters to a lifelong process of learning. For example, anyone can view one of Yale University's most popular philosophy professors sitting on his desk, talking about death. MIT is offering 1,800 classes on line, practically the entire curriculum open to all at no charge. Of the 35 million people who have tried MIT's online courses, almost half are simply people just curious to learn (Kinzie 2007). This trend, reinforced by a rapidly changing global economy, will surely accelerate (Brody 1997; Ritzer 1998).

FEEDBACK

1. As expected, college enrollment in the United States declined in the 1980s because of a decrease in the number of eighteen- to twenty-four-year-olds. T or F?
2. Although a college education pays off financially for white males, it does not benefit African American men and women. T or F?
3. Occupational status gains produced by a college education are less apparent than the relationship between income and higher education. T or F?
4. According to David Riesman, most colleges and universities have shifted from academic merit to _____.
5. Public commitment to higher education is as strong for the postboomer generation as it was for their parents. T or F?

Answers: 1. F; 2. F; 3. F; 4. student consumerism; 5. F

SUMMARY

1. With industrialization and urbanization, the ability of the family to prepare its young for the future declined. As a result, formal schooling became necessary. The early emphasis in American schools was on "civilizing" the young; but after the turn of the twentieth century, the emphasis shifted to education for jobs.

2. Early schools were modeled after businesses and have increasingly become more bureaucratic. Advocates of open classrooms and cooperative learning contend that bureaucratically run schools fail to take into account the emotional and creative needs of individual children. The evidence suggests that some aspects of these alternatives provide benefits that should not be overlooked. The back-to-basics movement, however, seems to clash with the open classroom philosophy.

3. The debate over the best classroom methods has led to the research for alternatives. Major competitors to the traditional public school are vouchers, charter schools, for-profit schools, and magnet schools.

4. Functionalists see the emergence of the educational institution as a response to society's needs. The manifest functions of education include the transmission of culture; social integration; selection and screening of talent; promotion of personal growth and development; and the dissemination, preservation, and creation of knowledge. Schools also serve latent functions.

5. America is supposed to be a meritocracy in which social status is achieved rather than determined by family background. Advocates of the conflict perspective point to flaws in the meritocratic model, contending that the elite use the educational institution to preserve their privileged position.

6. Many of the flaws in the meritocratic model appear within the context of educational inequality, which exists when schooling produces different results for lower-class and minority children than it does for other children. Schools themselves promote educational inequality by reacting to students' social class backgrounds and racial or ethnic characteristics rather than to their abilities and potential. Schools use intelligence testing—which ignores the environmental deficiencies of lower-class students—to channel students into noncollege tracks.

7. Although there is disagreement, some evidence suggests that school desegregation reduces educational inequality. Three other approaches to promoting educational equality are compensatory education, increased community control of schools, and private schooling.

8. Symbolic interactionists emphasize the nonacademic socialization that occurs within schools. Through the hidden curriculum, children are taught a variety of values, norms, beliefs, and attitudes thought to be desirable by a society. This nonacademic socialization is accomplished in large part through textbooks and teachers. Much of this socialization is designed to help young people make the transition from home to the larger society.

9. Contrary to expectation, college enrollment has increased. This occurred partly because of active college recruiting and partly because of the decline of high-paying unionized manufacturing jobs. College education still pays off handsomely in the United States. Both income and occupational level increase with higher education.

10. Critics are disturbed about changes in higher education that began in the mid-1960s. Some see a shift from academic merit to student consumerism. The financial crisis in higher education looms as a genuine threat to the long-standing reliance of American society on an affordable, high-quality system of public higher education. Some observers contend that colleges and universities will have to redesign themselves to adapt to the information age.

LEARNING OBJECTIVES REVIEW

- Describe the relationship between industrialization and education.
- Discuss schools as bureaucracies and the attempts to debureaucratize education in the United States.
- Outline the basic functions (manifest and latent) of the institution of education.
- Evaluate the meritocratic model of public education.
- Discuss educational inequality.
- Describe the ways in which schools socialize students.
- Identify and describe the dominant issues in higher education.

CONCEPT REVIEW

Match the following concepts with the definitions listed below them.

____ a. multiculturalism
____ b. hidden curriculum
____ c. vouchers

____ d. cooperative learning
____ e. self-fulfilling prophecy
____ f. dysfunction

____ g. manifest function

1. a negative consequence of some element of a society
2. the nonacademic agenda in schools that teaches children skills thought to be necessary for success in modern bureaucratic society
3. an intended and recognized consequence of some element of a society
4. an approach to education in which the curriculum accents the viewpoints, experiences, and contributions of women and diverse ethnic and racial minorities

5. when an expectation leads to behavior that causes the expectation to become a reality
6. a nonbureaucratic classroom structure in which students study in groups with teachers acting as guides rather than as the controlling agents determining all activities
7. tax allowances granted to cover part of the costs for an academic year at the school selected

CRITICAL-THINKING QUESTIONS

1. The open classroom philosophy represents an alternative to the bureaucratic model of education. Use the symbolic interactionist perspective to convince a local school board of the desirability of adopting the open classroom approach.

_____

_____

2. Assume for the moment that schools cannot provide educational equality. Analyze the consequences of the persistence of educational inequality within the context of the functions of education discussed in the text.

_____

_____

3. With the hidden curriculum in mind, think back to your early school years. Provide specific examples from your own experience to support the existence of the hidden curriculum.

_____

_____

4. Identify what you believe are the two biggest issues facing higher education in the twenty-first century. Develop recommendations for solving these problems.

_____

_____

MULTIPLE-CHOICE QUESTIONS

1. **Schools emerged as the primary educational agent in industrializing societies because**
   a. the skills needed for survival in an agricultural society became increasingly important.
   b. knowledge increased in volume and complexity and family members gradually lost the ability to educate their children adequately.
   c. too many parents were qualified to teach their children, and not enough parents were willing to work in agriculture.
   d. parents no longer received the income they needed to educate their children at home.
   e. the birth rate increased dramatically in the early 1800s.

2. **According to the text, critics of the bureaucratic model of education claim that**
   a. bureaucracies are notoriously inefficient.
   b. the U.S. school system should have more "vouchers."
   c. teachers should treat all students the same.
   d. administrators are useless and should therefore be eliminated.
   e. it is hard to determine whether children have been "processed" correctly.

3. **Education contributes to social integration by**
   a. allowing individuals to learn several languages.
   b. exposing individuals to similar educational traditions.
   c. exposing individuals to different ethnic and cultural traditions.
   d. presenting various religious doctrines.

e. promoting personal growth and individual development.

4. **Schools promote personal growth and development by**
   a. preparing students to assume a place in the occupational world.
   b. discouraging emotional experiences.
   c. tracking lower-class students into vocational programs.
   d. discouraging group activities.
   e. teaching students the importance of following orders in a bureaucratic society.

5. **The school's efforts at the dissemination of knowledge most commonly take the form of**
   a. field trips outside the classroom.
   b. using television as a teaching tool.
   c. classroom teaching.
   d. personal experience.
   e. homework assignments.

6. **According to Randall Collins, credentials are required for many jobs because**
   a. they allow those who can afford to get them a strong hold on elite occupations.
   b. they perpetuate a meritocratic society.
   c. they fulfill a vital service for society by verifying educational attainment.
   d. they prevent the best talent from being lost.
   e. they provide worthwhile goals for those who lack advanced degrees.

7. **A significantly larger proportion of upper-class children attend college than do lower-class children. According to the text, this is due in part to differences in**
   a. academic ability.
   b. economic resources.
   c. desire for intellectual growth.
   d. motivation to learn.
   e. study habits.

8. **The most recent concept of educational equality involves**
   a. equality of resources.
   b. equality of opportunity.
   c. equality in busing.
   d. equality in effects of schooling.
   e. equal education for equal ability.

9. **Which of the following is a *false* statement?**
   a. City dwellers usually score higher on intelligence tests than people in rural areas.
   b. Performance on IQ tests increases with age.
   c. Middle-class African American children score about as high on intelligence tests as middle-class white children.
   d. Intelligence tests are culturally biased.
   e. Recent evidence has convinced most sociologists that about half of human intelligence is genetically based.

10. **Central to _____ is the idea that special preschool programs can help deprived children overcome an inferior intellectual environment.**
    a. forced busing
    b. tracking based on standardized testing
    c. expanding the scope of the Special Olympics
    d. postsecondary school remedial reading
    e. compensatory education

11. **In his study on the value of a private school education, James Coleman found that**
    a. public schools produce higher student achievement levels.
    b. private schools produce higher student achievement levels.
    c. students in public and private schools produce similar levels of achievement.
    d. minority children in public schools perform better than minority children in private schools.
    e. minority students perform equally well in private and public schools.

12. **When girls are successfully taught in school to see their self-worth in relation to their physical appearance rather than in relation to their abilities to contribute to society, the _____ has been at work.**
    a. functionalist perspective
    b. self-fulfilling prophecy
    c. "blooming" effect
    d. multicultural perspective
    e. policy of tracking

13. **Which of the following is a *false* statement?**
    a. A college education results in higher annual income regardless of race and gender.
    b. College-educated African American males and white males make more than double the annual income of high school dropouts.
    c. African American females earn annually three times as much as white high school dropouts.
    d. African American male high school graduates earn annually over one-half of that earned by African American college graduates.
    e. A college education pays off economically for white Americans, but not for African Americans.

14. **In his analysis of higher education, David Riesman contends that**
    a. university administrators are not paying enough attention to keeping educational organizations in business.
    b. universities should devote more attention to intercollegiate athletics in order to retain alumni interest and financial contributions.
    c. universities have increasingly lost their ability to shape development along academic and intellectual avenues.
    d. universities are now under the control of administrators who were part of the radical left of the 1960s.
    e. universities should do more to get students involved in activist causes and social movements.

FEEDBACK REVIEW

True-False

1. Most sociologists agree that IQ test scores do not measure inherited learning ability. T or F?
2. Public commitment to higher education is as strong for the postboomer generation as it was for their parents. T or F?

Fill in the Blank

3. Early public schooling in America was viewed as a mechanism of _____.
4. Educational equality exists when schooling produces the same _____ for lower-class children and minorities as it does for other children.
5. According to symbolic interactionists, schools are very much in the business of transmitting _____.
6. According to David Riesman, most colleges and universities have shifted from academic merit to _____.

Multiple Choice

7. Contemporary public education is important primarily because of
   a. the quality of teachers in modern schools.
   b. efficient operation of school bureaucracies.
   c. historical changes in the economic and social structures of industrial societies.
   d. lack of qualified individual tutors.
8. The idea that some educational requirements are unnecessary for many jobs today is called
   a. status grouping.
   b. credentialism.
   c. meritocracy.
   d. conflict.

Matching

9. Match the following research topics with the three theoretical perspectives:
   ____ a. functionalism
   ____ b. conflict theory
   ____ c. symbolic interactionism
   (1) requirement of unnecessary credentials by schools
   (2) benefit of schools for society
   (3) effects of teacher reactions on the self-esteem of students

GRAPHIC REVIEW

The trend in educational level among whites and major minority groups since 1970 is displayed in Figure 12.1. Answer these questions, referring to the data in this figure.

1. Relate the educational levels in the various categories to the process of industrialization.

   _____
   _____
   _____

2. Would functionalists and conflict theorists interpret the gap between the educational level of whites and the educational attainment of African Americans and Latinos differently, or in the same way? Explain.

   _____
   _____
   _____

ANSWER KEY

Concept Review	Multiple Choice		Feedback Review
a. 4	1. b	12. b	1. T
b. 2	2. e	13. e	2. F
c. 7	3. b	14. c	3. social control
d. 6	4. a		4. results
e. 5	5. c		5. culture
f. 1	6. a		6. student consumerism
g. 3	7. b		7. c
	8. d		8. b
	9. e		9. a. 2
	10. e		b. 1
	11. b		c. 3

Political and Economic Institutions

13

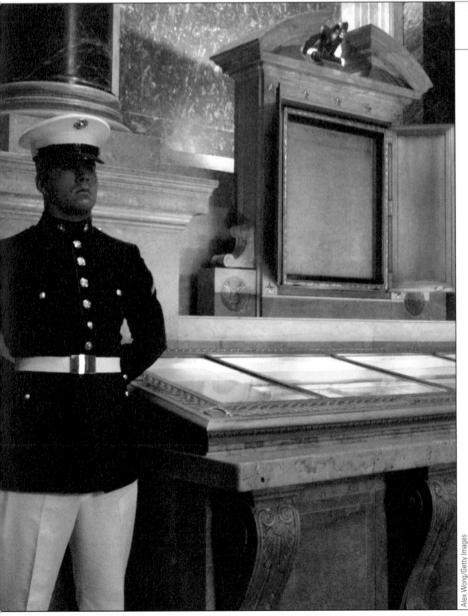

Alex Wong/Getty Images

LEARNING OBJECTIVES

- Distinguish among power, coercion, and authority, and identify three basic forms of authority.
- Discuss the nation-state, comparing and contrasting the functionalist and conflict approaches to this form of political authority.
- Identify the major differences among democracy, totalitarianism, and authoritarianism.

- Differentiate the views of functionalism and conflict theory on the distribution of power in America.
- Describe the major characteristics of capitalism, socialism, and mixed economic systems.
- Discuss the effects of the modern corporation on American society.

- Describe America's changing workforce composition and occupational structure.
- Discuss corporate downsizing and its consequences.

Do Japanese or Americans work more hours per week? Less than a decade ago, Americans looked at workers in Japan with awe. Stories of men and women slaving away for ten or more hours per day, six days per week, made the rounds of corporate America and surprised many American workers. The reality today is different. The average American works longer than workers in any other industrialized country, including Japan. Between 1977 and 2000, the average workweek among salaried American workers lengthened from forty-three to forty-eight hours. In that same period, the number of workers putting in more than fifty hours per week went from 24 percent to 37 percent. In fact, Americans work an equivalent of eight weeks longer every year than western Europeans. In this chapter on politics and the economy, we examine where, why, and how Americans vote and work.

Power and Authority

All societies must deal with the challenges of economics and politics. Economic issues exist because of the need to produce goods and services. The institution that carries out the production and distribution of goods and services is the **economy**. Because economic decisions affect organizations and the general public, severe conflicts often arise. Every society must develop some means of handling these conflicts, and such decisions are usually made by those responsible for the general welfare of the society. This responsibility lies in the institution through which power is obtained and exercised—the **political institution**.

The 1997 strike by the Teamsters Union against United Parcel Service (UPS) illustrates the interaction between the economy and the political institution. This labor confrontation was based on a strong objection to the UPS corporate policy of downsizing—changing permanent jobs into temporary ones. UPS, by asking President Clinton to intervene early in the dispute, demonstrated the intimate interdependence between business and government in modern society. The interplay of these two institutions will become apparent in this chapter.

Defining Power and Authority

Max Weber (1946), a founder of political sociology, defined **power** as the ability to control the behavior of others, even against their will. Power can exist in one's personal appeal or magnetism. For example, in his classic study of power and decision making, Floyd Hunter (1953) wrote about James Treat, a man whose extensive power was not evident from his simple, nonthreatening appearance. Treat's power was based not on his visible personal characteristics but on his ability to influence the actions of others—including Georgia's governor. His power existed in his relationships with others.

Weber recognized another form of power, power through force, or **coercion**. A blackmailer might extort money from a politician. A government might confiscate, without compensation, the property of one of its citizens. In neither case do the victims recognize this use of power as valid. Moreover, because victims of coercive power see it as illegitimate, they normally are resentful and wish to fight back. Weber recognized that a political system based on coercive power is inherently unstable. Because force produces compliance only in the short run and only when those in power can exercise surveillance, Weber believed that a political institution must rest on a more stable form of power. This more stable form of power is **authority**—power accepted as legitimate by those subjected to it. For example, students take exams and accept the evaluations they receive because they believe that their teachers have the right to determine grades. Citizens pay taxes because they believe that their government has the right to collect the money. Because authority is based on others' acceptance, it is the most stable form of power. Its existence reduces the likelihood that coercion will be needed to maintain order.

Forms of Authority

Weber identified three forms of authority—*charismatic*, *traditional*, and *rational-legal*. Those invested with these forms of authority can expect compliance from others, who recognize authority figures as holders of legitimate power.

What is charismatic authority? **Charismatic authority** arises from a leader's personal characteristics. Charismatic leaders can lead because of their magnetic personalities or the feelings of trust they inspire. Martin Luther King Jr., Nelson Mandela, and César Chávez are examples of charismatic leaders. The power of a charismatic figure may be unlimited while it lasts; but for modern nation-states, charismatic authority alone is too unstable to provide a permanent basis of power. It is linked to an individual and is therefore difficult to transfer to another. When charismatic leaders die, they relinquish their authority, leaving no traditional norms for the continuation of power. Adolf Hitler, himself a charismatic leader, made a futile attempt at the end of

World War II to name his successor. But as historian John Toland has noted:

> Hitler's death brought an abrupt, absolute end to National Socialism. Without its only true leader, it burst like a bubble. . . . What had appeared to be the most powerful and fearsome political force of the twentieth century vanished overnight. No other leader's death since Napoleon had so completely obliterated a regime. (Toland 1976:892)

So, even governments controlled by charismatic leaders must eventually come to rely on other types of authority. The two alternatives identified by Weber are traditional authority and rational-legal authority.

What is traditional authority? In the past, most states relied on **traditional authority**, authority in which the legitimacy of a leader is rooted in custom. Early kings, for example, often claimed rule by divine right. Orderly perpetuation was a key benefit. This peaceful transfer of power was possible because only a few individuals were eligible to become the next ruler. Until recently, custom was sufficiently strong to legitimate one's inherited right to rule. Because, for example, the kings in eighteenth-century Europe could depend on the custom of loyalty from their subjects, their traditional authority provided a stable political foundation—more stability than charismatic authority could have provided.

What is rational-legal authority? Most modern governments are based on a system of **rational-legal authority**; the power of government officials is based on their offices. Those who hold government office are expected to operate on the basis of specific rules and procedures that define and limit their rights and responsibilities.

Because rational-legal authority is invested in positions rather than in individuals, persons lose their authority when they leave their formal positions of power. Citizens are expected to follow the directives of the political leaders, and leaders are expected to stay within the boundaries of their legal authority. Even presidents (e.g., Richard Nixon) can lose their power if excessive abuse is made public. Legal authority thus limits the power of government officials.

How are these forms of authority ideal-typical? In the real world, people may hold more than one form of authority. Weber knew this, but he intentionally distinguished these three forms of authority through use of the ideal-type method defined in Chapter 6. As you may recall, this method involves isolating the most basic characteristics of some social entity. The purpose here is to crystallize the meaning of some social entity by removing that entity from the context in which it is embedded with other entities. Stripped of the surrounding complexity, the essential characteristics of a social entity can be identified.

Although the power of any American president is fundamentally grounded in rational-legal authority, authority forms are somewhat mixed. The concept of the "imperial presidency," for example, suggests that some element of traditional authority is attached to the office. Particular presidents (Theodore Roosevelt, John Kennedy, Ronald Reagan) may also hold charismatic authority, whereas other presidents (Gerald Ford, Jimmy Carter) may not.

No matter the form, because authority is based on others' acceptance, it is the most stable expression of power. This is particularly true of modern nation-states.

FEEDBACK

1. The institution within which power is obtained and exercised is the _____ institution.
2. _____ is the ability to control the behavior of others, even against their will.
3. Power that is considered legitimate by those subjected to it is called _____.
4. On which of the following types of authority are kingships based?
   a. charismatic
   b. traditional
   c. rational-legal
5. Which of the following types of authority places the strongest limits on government officials' freedom of action?
   a. charismatic
   b. traditional
   c. rational-legal

Answers: 1. political; 2. Power; 3. authority; 4. b; 5. c

The Nation-State

Distinguishing between Government and Nation-State

By combining several independent political units, a nation-state consolidates power over a large geographical area. The United States, for instance, became a nation through the Constitutional Convention of 1787, whereby particular powers of individual states were transferred to the federal government.

Historically, the nation-state is a recent development. For 99 percent of human history, people lived in small hunting and gathering bands, and there was little need for large legal systems of control. Even the Roman Empire, with a few exceptions, consisted of self-governing cities (A. H. M. Jones 1966; Service 1975; Bottomore 1993; P. James 1996). Nation-states began to emerge in Europe during the fifteenth century, but as recently as 100 years ago, there were still politically unclaimed territories in the world. Many people lived without formal political rule. No more. Presently, the Earth is divided into more than 100 nation-states, and nearly everyone is a member of one of them. One is born, lives, and dies only by the official recognition of a nation-state (Alford and Friedland 1985; Skocpol 1985; Skocpol and Amenta 1986; P. James 1996).

What is a nation-state? Although the nature of nation-states may vary, the **nation-state** is always the political authority over a specified territory (Charles Tilly 1990). One key characteristic of a nation-state is the absolute sovereignty it has over its citizens. Citizens can appeal no higher than the laws of the state, and the state has supreme authority over its citizens. Only state representatives have a legal right to act on behalf of the state. Only they, for example, can impose taxes, imprison criminals, or declare war. A second key characteristic of a nation-state is its devotion to **nationalism**—a people's commitment to a common destiny based on a recognition of a common past and a vision of a shared future (Stoessinger 1981).

It is important to distinguish between a state, which has ultimate authority over a territory, and a **government**—the political structure that rules a nation. All societies have government; arrangements for settling disputes are always necessary. A state exists as an entity, however, only if governmental functions are separated from other institutions, such as the family and religion. As societies modernize, states become increasingly common; governmental functions are handled by groups of officials whose roles in society are distinctly governmental.

The emergence of the state is associated with advances in agriculture and industry. In hunting and gathering societies, most decisions could be reached by consensus. Agricultural and industrial developments, however, led to surpluses that could be converted into wealth and power for certain segments of society. When surpluses lead to conflict over who will control them, a clearly defined state is needed to maintain order (Nolan and Lenski 2005). Both functionalists and conflict theorists have explanations for the emergence of nation-states.

The behavior of these Chinese citizens at the opening of the 2008 Olympics is a display of nationalism.

Nick Laham/Getty Images

A Functionalist View of the Nation-State

According to functionalism, nation-states are formed because they are needed by society. Unless a strong state exists, argued English philosopher Thomas Hobbes (1958; originally published in 1651), people will act only according to their own selfish interests. Without a state to control these selfish impulses, there will be a "war of every one against every one" and life will be "solitary, poor, nasty, brutish, and short." To escape such chaos, people agree to create nation-states on the basis of a "social contract" of cooperation (Lilla 2007).

A similar argument was developed by Emile Durkheim (1964; originally published in 1893). He feared the rapid and extensive social change experienced within modern societies. Unless sufficient controls are developed, admonished Durkheim, modern societies will experience disruptive internal conflicts. Although attempting to act in their self-interest, people will actually create a situation in which disruption prevents the achievement of personal goals. For Durkheim, the role of the state is to provide centralized regulation of economic life. Like Hobbes, Durkheim believed that the absence of a state invites chaos.

Talcott Parsons (1959) agrees with Hobbes; humans must have external constraints. Through social constraints, a government can perform necessary functions. For example, a system of laws can be developed and enforced. Government officials can mobilize resources to plan, coordinate, and achieve commonly shared goals. They can formulate policy decisions reflecting society's highest values. State representatives can oversee relationships with other nation-states—whether political, economic, or military.

A Conflict Perspective of the Nation-State

According to the conflict perspective, nation-states exist primarily to serve the interests of a society's elite. One of Hobbes's earliest critics was the French philosopher Jean-Jacques Rousseau. In the original state of nature, Rousseau argued, people were noble savages living in harmony with their environment. They were probably socially indifferent rather than at war with one another. There was no reason to start fighting until the creation of private property led to the rise of social inequality. At this point, Rousseau agreed, a social contract may indeed restore order by creating a state. This social contract, however, carries advantages primarily for the privileged. He described the contract in this fashion:

You need me, for I am rich and you are poor. Let us therefore make a contract with one another. I will do you the honor to permit you to serve me under the condition that you give me what little you have left for the trouble I shall take in commanding you. (quoted in Cassirer 1951:260)

Rousseau argued for a new social contract in which obedience to the privileged class is replaced by obedience to the "common will," a will aimed at the welfare of all, rich and poor alike.

Karl Marx offers the most influential expression of the conflict perspective. The nature of a society, Marx contended, is determined by the relations between those who own property and those who do not. Industrial societies, in Marx's view, are dominated by the relatively small number of people who own factories, machines, and other productive resources. As noted in Chapter 1, those who own productive property are the ruling class, and those who do not own property are the workers. The **bourgeoisie** (the few in the ruling class) and the **proletariat** (the mass of workers) form the major social classes in capitalist societies.

According to Marx, political life is a reflection of relations between these two social classes. Because the ruling class controls the resources, the government serves its needs: "The executive of the modern State is but a committee for managing the affairs of the whole bourgeoisie. . . ." (Tucker 1972:337; originally published by Marx in 1867). The state, through the power of force, provides a means for controlling workers. In most cases, however, the use of force is unnecessary, because the state draws on common values and experiences to promote an ideology that encourages general acceptance of the system. Marx used the term **false consciousness** to describe acceptance of a system that works against one's own interests.

According to Marx, then, those in charge of the state in capitalist societies—the government, military, police, bureaucracy—make decisions to benefit the ruling class. It is only through the revolution of the proletariat that the state will serve the interests of everyone. A short-term "dictatorship of the proletariat" had to exist during the transition from the capitalist state to communism, the final stage of historical development. Because Marx spent his time analyzing the problems of capitalist society, he did not refine a conception of the communist society. It does seem, however, that the state under communism had to be nonauthoritarian and nonbureaucratic and to exist for purely administrative reasons. Ownership of the means of production, including land, had to be in public hands, not in the hands of a few. To ensure that state functionaries did not form a new elite class, their pay had to be kept commensurate with those outside government.

Marx failed to foresee the difficulties in reducing the power of state bureaucracy and in persuading government officials to relinquish their positions and power. Modern conflict theorists broaden Marx's concept of

FEEDBACK

1. A _____ is the political entity that holds authority over a specified territory.
2. A _____ is the political structure that rules a nation.
3. According to Thomas Hobbes and Emile Durkheim, nation-states exist
   a. to serve the interests of powerful, privileged social classes.
   b. because of custom.
   c. to control the selfish impulses that would otherwise make social order impossible.
4. According to Jean-Jacques Rousseau, it is primarily the privileged classes who benefit from the existence of the state. T or F?
5. According to Karl Marx, those in charge of the state in capitalist societies make decisions that serve the interests of the owners of the means of production. He used the term _____ to describe acceptance of a system that works against one's own interests.

Answers: 1. nation-state; 2. government; 3. c; 4. T; 5. false consciousness

class conflict to encompass social conflict of all types, whether between classes or between political interest groups.

Political Systems

Nation-states embrace two polar types of political systems—*democracy* and *totalitarianism*. Because few societies are pure enough to maintain all characteristics of each polar type, the real world contains variations between the two (Orum 1983).

Democracy

One conception, inspired by the ancient Greeks, assumes a form of self-government that involves the political participation of *all* citizens. This classical conception of democracy applies both to Athens in the fifth century B.C. and to old New England town meetings. A second, more familiar conception, based on realities of modern times, conceives **democracy** as a system of elected officials responsible for fulfilling majority wishes. The United States is the oldest representative democracy (republic) in the world (Ellis 2002).

What assumptions are made in a representative democracy? Representative democracy is based on two assumptions. The first assumption recognizes that not everyone in modern society can be actively involved in all political decision making. Citizens are not expected to be involved deeply in politics; they merely need to vote. Given the minute division of labor in contemporary society, it would be hopelessly inefficient for all qualified citizens to be preoccupied with political activities. Who would manage companies, teach school, design buildings, prepare taxes, rear children? The classical practice of democracy in ancient Greece could exist only because the masses (women and

slaves), excluded from the political process, were available to perform nonpolitical functions. In the second assumption of representative democracy, politicians who fail to satisfy the wishes of the majority will not be elected (or reelected).

In the two-party system of the United States, we have a "winner take all" form of representative democracy. Here, the party with the most votes wins the election. In many European countries, where third-party systems are common, parties participate in the government to the extent that they win representation in general elections. For example, one party might win 40 percent of the vote and control 40 percent of the legislature. Three other parties might take 20 percent each and still control a combined 60 percent of the legislature. This proportional representation system seems to be more democratic; it tends to encourage compromises and cooperation. Governments formed under this system can be fragile, however, and shifting political alliances may force new elections, even after relatively short periods of time.

Is democracy increasing throughout the world? The collapse of Soviet communism and the end of the Cold War created opportunities for more democratic societies. Over the past twenty years, eighty-one countries advanced toward democracy with civilian governments replacing thirty-three military dictatorships. And 140 of the world's nearly 200 countries have multiparty elections, the highest number in history (*Human Development Report* 2002). Still, there is little evidence of sustained worldwide progress (see Worldview 13.1). Only about half of the eighty-one new democracies are fully democratic: some did not continue their transition to democracy, others slipped back into authoritarianism or war. For the most part, "free" political systems are associated with advanced economic development and are found primarily in a few nations: Japan; western Europe; and former

WORLDVIEW 13.1

Political Freedom

Democracy is unevenly distributed worldwide. The accompanying map classifies countries according to three degrees of political freedom: free, partly free, and not free.

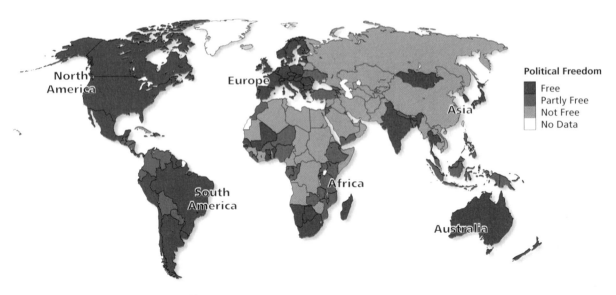

Political Freedom

- Free
- Partly Free
- Not Free
- No Data

Source: Freedom House, Washington, D.C., 2005.

Interpret the Map

1. Do you see any pattern in the degree of political freedom around the world?

2. Predict how political freedom around the world will change in the next fifty years. Explain you predictions.

European colonies, notably Canada, Australia, some Latin American countries, several African nations, and the United States (Karatnycky 1995; Vanhanen 1997; *Human Development Report* 2002).

Totalitarianism

What is the nature of totalitarianism? **Totalitarianism**, which lies at the other end of the political spectrum from democracy, is the type of political system in which a ruler with absolute power attempts to control all aspects of social life (see Figure 13.1). Totalitarian states are replete with "not-free" characteristics: a detailed ideology designed to encompass all phases of individuals' lives, a single political party typically controlled by one person, a well-coordinated system of terror, total control of all means of communication, a monopoly over military resources, and a planned economy directed by a state bureaucracy (Friedrich and Brzezinski 1965). Classic illustrations of totalitarian states are Nazi Germany and the former Soviet Union. A brief description of Nazi Germany in relation to its totalitarian characteristics is enlightening.

How extreme was Hitler's Nazi government? Hitler's National Socialist (Nazi) government, which came to power in Germany in the early 1930s, offers a good example of the way a totalitarian system works. Hitler, in his autobiography, *Mein Kampf*, laid the ideological foundation for the National Socialist Party. At the center of this ideology is the myth of Aryan racial superiority and its role in saving the world, in part by destroying

FIGURE 13.1

The Political Continuum

Political institutions offer varying degrees of freedom for their members.

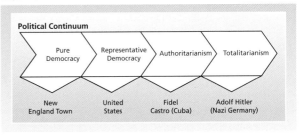

Political Continuum

| Pure Democracy | Representative Democracy | Authoritarianism | Totalitarianism |

New England Town — United States — Fidel Castro (Cuba) — Adolf Hitler (Nazi Germany)

AP Phtos

At the opposite end of the political spectrum from democracy is totalitarianism, the type of political system in which an absolute ruler controls all aspects of society. This photo chillingly illustrates a totalitarian state.

whose terrorist practices squelched all dissenters and eliminated state enemies. With few exceptions, all news media were either seized by the party or eliminated. Hitler dominated the armed forces, often devising his own military plans over the opposition of his generals. Nazi Germany's four-year economic plans included strategies for budgets, production, organization of factories, and forced labor.

Authoritarianism

Because authoritarianism is a residual (leftover) category between democracy and totalitarianism, it is more difficult to define than either polar type. As a type of political system, it is, of course, closer to totalitarianism than to democracy. **Authoritarianism** refers to a political system controlled by nonelected rulers who generally permit some degree of individual freedom (Linz 1964). Countless regimes have leaned toward totalitarianism but have fallen short of all its defining characteristics. These authoritarian governments include monarchies (King Fahd of Saudi Arabia), military seizures of power (Pakistan's General Musharraf Parvez), and dictatorships (Saddam Hussein of Iraq).

Political Power in American Society

Jews. Despite presenting a facade of democracy, Hitler and the National Socialist Party held the power. In fact, Hitler helped solidify his own power by setting up identical offices in the state bureaucracy and the party. The ensuing competition allowed Hitler to divide and conquer any opposition. Hitler's absolute control was also bolstered by the Gestapo (secret police) and SS troops,

Political elitism is inevitable, even in democratic societies, because the public, which cannot govern directly, must delegate political decisions and governing responsibilities to representatives. By this act of delegation, an elite group is created (Dye 2001). Still, citizens in democratic societies are supposed to be able to influence the actions of their representatives, through voting (usually via political parties) and through joining

F E E D B A C K

1. _____ is the type of political system in which elected officials are held responsible, through the mechanism of voting, for fulfilling the wishes of the majority of the electorate.

2. The United States is characterized as a
   a. classical democracy.
   b. lineage democracy.
   c. representative democracy.
   d. democracy of hegemony.

3. The type of political system in which a ruler with absolute power attempts to control all phases of social life is called _____.

4. _____ refers to the type of political system controlled by nonelected rulers who generally permit some degree of individual freedom.

Answers: 1. Democracy; 2. c; 3. totalitarianism; 4. Authoritarianism interest groups. Voting is covered next, followed by a discussion of interest groups within the context of the pluralist model.

The Vote

Like all contemporary democracies, the United States emphasizes political participation through voting. Voting is supposed to be an important source of power for citizens, and it does in fact enable people to remove incompetent, corrupt, or insensitive officials from office and to influence issues at the local, state, and national levels. Unfortunately, American political parties are more concerned with winning elections than with maintaining a clear-cut ideological position, contributing to a politics of "blandness."

How much actual power do Americans exercise through the ballot box? In current U.S. practice, voting has severe limitations as a means of exercising power (see Snapshot 13.1). In the first place, the range of candidates from which to choose is restricted, because of the power of political parties—organizations designed to gain control of the government through the election of candidates to public office (Bardes, Shelley, and Schmidt 2007). Because the United States has only two major political parties, and because only candidates endorsed by those parties have any chance of winning a major political office, the political views of candidates

for any particular office often resemble each other more than they differ. Consequently, many groups are not effectively represented (S. Hill 2003). Second, the limited choice of political candidates is reinforced by the high cost of political campaigns. Only people who are wealthy themselves or who are able to attract large contributions from others can mount effective campaigns. For example, by the summer of 2004, both George W. Bush and John F. Kerry raised more than $300 million for each of their primary and general election presidential election campaigns (Federal Election Commission 2006). Not only are such people often not representative of the public, but they must face powerful contributors who expect favors once their candidates are in office (VandeHei and Hamburger 2002; Dye 2006).

A third limitation lies in the way power is exercised by those at the top of the American political structure. According to Howard Zinn (2003), the ballot box is a poor tool for coping with such matters as racial discrimination, economic justice, and the formation of foreign policy. Singling out foreign policy as the most clear-cut example, he documents the powerlessness of voters with respect to waging war. The Constitution gives Congress the power to declare war, but Zinn cites examples throughout U.S. history of presidents waging war on the advice of a few *appointed* officials, even after campaigning as peacemakers (Dreyfuss and Vest 2004). And once war is declared, presidents are rarely seriously challenged by elected representatives or the people.

SNAPSHOT OF AMERICA 13.1

Distribution of Electoral College Votes

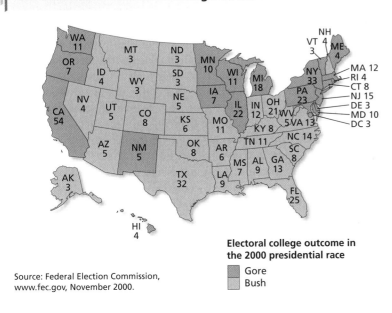

Electoral college outcome in the 2000 presidential race

Gore
Bush

Source: Federal Election Commission, www.fec.gov, November 2000.

As the American population learned during the 2000 presidential election, it is the electoral college rather than the popular vote that determines the winner. This map shows the distribution of electoral votes by state within the context of whether Bush or Gore won the state.

Interpret the Map

1. What does the distribution of electoral votes across the country (without respect to the candidates) say about political power in the United States? Explain.

2. Look now at the states Bush and Gore won. What does the pattern of states won by each candidate indicate about differences in the characteristics of Republican and Democrat supporters? Explain.

FIGURE 13.2

Voter Participation in Presidential Elections: 1932–2004

Is there any correlation between the confidence level and the voter participation rate? Explain.

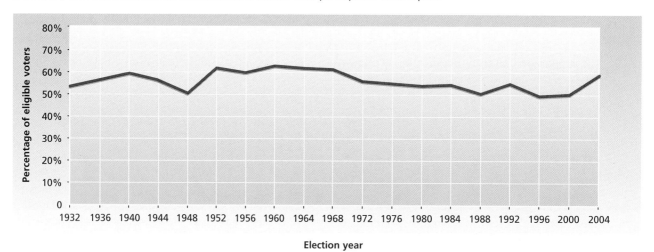

Source: U.S. Bureau of the Census, *Statistical Abstract of the United States, 2004–2005* and *Current Population Survey* (November 2004).

How fully do Americans take advantage of the voting privilege? American interest in voting is low, partly because of the public's relatively low confidence in political leaders (Jones 2006, 2007a). In the 2004 presidential election, 58 percent of eligible U.S. voters exercised their right. This means that only about a fourth of the American people voted in the 2004 presidential election, a proportion somewhat higher than the 20 percent who elected President Reagan in 1980 (C. Lewis 2004; see Figure 13.2). U.S. voter participation is even lower in off-year (nonpresidential) elections. Only 44 percent of eligible voters went to the polls in 2006 (U.S. Bureau of the Census 2007c). In fact, the United States has one of the lowest voter turnout rates in the industrialized world (Center for Voting and Democracy 2002; T. E. Patterson 2003).

In general, the lower social classes and the working class in the United States tend to vote in smaller proportions than the middle and upper classes. Members of minorities, people with less education, and people with less income are less likely to vote in both congressional and presidential elections (Walsh 2000; U.S. Bureau of the Census 2007a).

On what do Americans base their vote? Most attitudes and beliefs that are expressed as political opinions are gained through a learning process called *political socialization.* This can be formal, as in government class, or informal, as through the family, the media, economic status, and education. Studies have shown that most political socialization is informal:

- *The family.* Children learn political attitudes the same way they learn values and norms, by listening to everyday conversations and by watching the actions of other family members. In one study, more high school students could identify their parents' political party affiliation than any other of their parents' attitudes or beliefs.

- *Education.* Level of education influences political knowledge and participation. For example, more highly educated men and women tend to show more knowledge about politics and policy. They also tend to vote and participate more often in politics.

- *Mass media.* Television is the leading source of political and public affairs information for most of us. Television and other mass media determine which issues, events, and personalities are in the public eye. By publicizing some issues and ignoring others and by giving some stories high priority and others low priority, the media decide the relative importance of issues. While it is clear that the media play an important role in shaping public opinion, the extent of that role is unclear. Studies indicate that the mass media have the greatest effect on those people who have not yet formed an opinion about the issue being discussed.

- *Economic status and occupation.* Economic status clearly influences political views. Disadvantaged people are more likely to favor government assistance programs than wealthier people, for example. Similarly, where you work affects how you vote. Corporate managers are more likely than hourly factory workers to favor tax shelters and aid to businesses.

- *Age and gender.* On most issues, young adults are a bit more liberal than older Americans. Young adults, for example, tend to be more progressive than older persons on racial and gender equality. Women tend to be more liberal on abortion rights, women's rights, health care, and government-supported child care.

Pluralism: The Functionalist Approach

In a democratic society, two major models of political power are evident—*pluralism* and *elitism*. According to **pluralism**, decision making is the result of competition, bargaining, and compromise among diverse special interest groups. In this view, no one group holds the majority of power; rather, power is widely distributed throughout a society or community. The second major model of political power, elitism, is in direct contrast to pluralism. According to **elitism**, a community or society is controlled from the top by a few individuals or organizations. Whereas the elitist model is based on the conflict perspective, the pluralist model is consistent with the functionalist perspective.

How do pluralists view the distribution of power? According to David Riesman (2001), major political decisions in the United States are made via competition among special interest groups, each of which has its own stake in the issue. In addition to attaining their own ends, interest groups try to protect themselves from opposing interest groups (Ethridge and Handelman 2008).

What is an interest group? An **interest group** is a group organized to influence political decision making. Group members share one or more goals—either of their own members (as in the case of the National Rifle Association) or of a larger segment of society (as in the case of ecology-oriented groups such as the Sierra Club).

Interest groups are not new to American politics. In the nineteenth century, they were active in extending women's rights and promoting the abolition of slavery. Furthermore, a history of twentieth-century America could not be written without describing a wide range of interest groups, such as the Women's Christian Temperance Union and the early labor union movement. The 1960s—with controversies surrounding civil rights, the Vietnam War, the environment, the women's movement, and corporate power—strengthened many interest groups and led to the creation of new ones (Clemens 1997).

New interest groups are born all the time. The environmental lobby is one example. There were relatively few environmental interest groups prior to the passage of major environment legislation (such as the Clean Water Act) in the 1960s. The success of the then existing environmental groups that lobbied for this legislation spawned additional groups, now numbering three times the total number of groups before the legislation. This added clout produced additional environmental legislation (e.g., the 1990 Clean Air Act Amendments), which subsequently led to the creation of still other interest groups (Bardes, Shelley, and Schmidt 2007). The popularity of Al Gore's global warming film, *An Inconvenient Truth*, and his subsequent receipt of the Nobel Prize, has provided even greater impetus to the environment lobby. The health-care industry is another

In the movie "An Inconvenient Truth" Al Gore makes the case for global warming. Does he represent an interest group?

Supplied by FilmStills.net/Digital Railroad

TABLE 13.1

Types of Interest Groups

The U.S. government is influenced by a wide variety of interest groups. This table illustrates the most important types of interest groups. Do you believe that the influence of all these interest groups promotes or hinders democracy? Explain your answer using conflict theory or functionalism.

Organization	Membership	Objectives
ECONOMIC GROUPS		
Business Groups		
National Association of Manufacturers (NAM)	• 14,000 manufacturing firms	• Restrain labor unions and reduce federal taxes on corporations
U.S. Chamber of Commerce	• 200,000 medium-sized and small businesses	• Lobby for policies favorable for all businesses
American Petroleum Institute	• 9,400 members	• Seek legislation favorable to the oil industry
National Association of Home Builders	• 200,000 members	• Work for a friendly regulatory environment for the housing industry
Agricultural Groups		
National Farmers Union	• 300,000 farm and ranch families	• Represent family farms
American Farm Bureau Federation	• 3 million members	• Lobby for agribusiness and owners of large farms
Professional Groups		
American Medical Association (AMA)	• 271,000 members	• Oppose government involvement in medical practices
American Bar Association	• 360,000 members	• Protection of the civil and political rights of all U.S. citizens
American Association of University Professors	• 44,000 members	• Dedicated to the defense of academic freedom and tenure and to the promotion of sound academic standards and due process

illustration. In response to the managed health-care industry (representing its stockholders), the American Medical Association in 1999 endorsed unionization to represent physician interests. Predictably, consumers will organize to represent their interests (Lore 1999).

Gary Wasserman illustrated interest group bargaining and compromise:

When Ralph Nader, the consumer advocate, proposes that automobile makers be required to install more safety devices such as air bags in their cars, numerous interests get involved in turning that proposal into law. The car manufacturers worry about the increased costs resulting in fewer sales and reduced profits, and they may try to limit the safety proposals. Labor unions may want to make sure that the higher costs do not result in lower wages. Insurance companies may be interested in how greater safety will affect the claims they have to pay out. Oil companies may worry about the effect on gasoline consumption if fewer cars are sold because of their increased cost.

Citizens' groups like Common Cause may try to influence the legislation so that it provides the greatest protection for the consumer. The appropriate committees of the House and Senate and the relevant parts of the bureaucracy will weigh the competing arguments and pressures as they consider bills covering automobile safety. The resulting legislation will reflect the relative power of the competing groups as well as the compromises they have reached among themselves. (Wasserman 2007:254)

In all of this, responsibility falls to government leaders for balancing the public welfare with the desires of special interest groups. (See Table 13.1 for a sample of interest groups and their primary issues.)

How prevalent are interest groups? Alexis de Tocqueville depicted Americans as a nation of joiners:

As soon as several Americans have conceived a sentiment or an idea that they want to produce before the world, they seek

TABLE 13.1

Types of Interest Groups (continued)

Organization	Membership	Objectives
Labor Groups		
AFL-CIO	• Over 100 affiliated-unions (14 million members)	• Lobby to protect member unions from business influence in government
United Mine Workers	• 130,000 members	• Advance the goals of mine workers
National Air Traffic Controllers Association (NATCA)	• 14,000 members	• Shape policy decisions affecting air traffic controllers
CITIZENS' GROUPS		
Public Interest Groups		
League of Women's Voters	• 1,100 local leagues, 130,000 members and supporters	• Work on the drive to simplify voter registration and related issues
Common Cause	• 265,000 members	• Advocate political reform such as more restrictive campaign financial laws
Public Citizen	• 100,000 members	• Focus on consumer issues such as auto safety and pollution control
Single-Issue Groups		
Sierra Club	• 550,000 members	• Advocate the protection of scenic areas
National Audubon Society	• 550,000 members	• Dedicated to the conservation and restoration of natural ecosystems
Greenpeace USA	• 2,100,000 members	• Pursue environmental protection
Ideological Groups		
Americans for Democratic Action (ADA)	• 65,000 members	• Support liberal positions on social, economic, and foreign policy issues
Christian Coalition	• 2,000,000 members	• Promote the restoration of "Christian values"
National Organization for Women (NOW)	• 250,000 members	• Protect the rights of women

Sources: Adapted from Thomas E. Patterson, *The American Democracy*, 8th ed. (Boston: McGraw-Hill, 2006).

each other out, and when found, they unite. Thenceforth they are no longer isolated individuals, but a power conspicuous from the distance whose actions serve as an example; when it speaks, men listen. (Tocqueville 1955:488; originally published in 1835)

The presence of more than 100,000 associations in the United States today, however, might have surprised even Tocqueville. About two-thirds of all Americans belong to some voluntary formal association. Most of these, of course, are not interest groups attempting to affect government policy. The Washington, D.C., telephone directory contains listings for some 3,000 organizations whose names begin with "National Association of." These organizations, along with many other interest groups, energetically push the economic interests of their members. In Washington, the number of lobbyists attempting to influence Congress is estimated at about 27,000 and rising (E. Brown 2006).

From a study of New Haven, Connecticut, Robert Dahl (2005) made an interesting observation: power in New Haven was not concentrated in the hands of one elite group. The groups, for example, trying to influence political decisions on public education were not the same groups competing, bargaining, and compromising on the issue of highway construction. Power was sufficiently dispersed that few segments of the community were without power. This system of checks and balances worked on the local level just the way James Madison, in *The Federalist Papers*, predicted it would in a large republic (Ellis 2002).

Pluralists point to the beneficiaries of the 1997 tax-cut bill. Tax breaks came not only to the wealthy, such as Microsoft shareholders, but also to groups with more modest resources, such as churches and people with mental disabilities (P. Hill 1997). We can further examine the pluralist model by focusing on political action committees.

Political Action Committees

American political parties were influential in the past because they controlled the nominations for major offices, because they oversaw campaign financing, and because many citizens tended to vote the straight party ticket. These bases of political party power appear, however, to be eroding. Fewer than one-fourth of Americans indicate that party affiliation makes a great deal of difference in their decision to vote for a candidate for Congress, whereas more than one-third say that party affiliation would have little or no influence on their vote. More than one-third of Americans consider themselves "independents." With the decline of citizen participation in political parties, political interest groups are currently most often represented by political action committees that in turn lobby the political parties (Greider 1993; Birnbaum 2000; M. Green 2002).

What are political action committees? **Political action committees (PACs)** are organizations established by interest groups to raise and distribute funds for the purpose of electing or defeating candidates. PACs can channel up to $5,000 per election (primary, general, special) into any candidate's election campaign and up to $15,000 annually to a national political party. Moreover, there is no legal restriction on the number of PACs that may support a candidate.

Since PACs were authorized in the mid-1970s, their number has mushroomed. Whereas only about 600 political action committees existed in 1974, there are now more than 4,000 (see Figure 13.3). Moreover, contributions from PACs are beginning to dwarf the role of political parties in financing political campaigns. PACs contribute more than ten times as much support to congressional candidates as do political parties. Dollar amounts of PAC contributions to federal candidates (so-called hard money, limited by legal restrictions) increased from just under $12 million in 1974 to almost $270 million in 2003–2004 (U.S. Bureau of the Census 2007c).

For a complete report of campaign contributions, one has to add the "soft money" (unregulated contributions) given to political parties during a presidential election cycle. Records were set in the spring of 2000 when the Republican party raised more than $21 million in two weeks from donors at fund-raising dinners—only to be topped by the Democratic Party, which held a $26.5 million fund-raising affair. During the 2004 presidential election cycle, the Republican and Democratic parties, combined, raised more than $1 billion (Federal Election Committee 2006).

A relatively recent type of advocacy group called 527s (because they are tax-exempt organizations under section 527 of the Internal Revenue Code) raise money for such political activities as voter mobilization efforts and issue advocacy. Although 527s may be PACs, many raise only unlimited soft money, which cannot be used to expressly favor the election or defeat of a federal candidate.

During the 2004 election cycle, 527s raised and spent more than a half-billion dollars, double the amount spent during the 2002 cycle. The 2004 election marked the first time they made a considerable impact on the results of a national election. For example, the 527 pop-

---

FIGURE 13.3

Growth in the Number of PACs: 1974–2006

This figure graphically depicts the sharp increase in the number of PACs in the United States. Evaluate the presence of PACs within the context of pluralism and elitism.

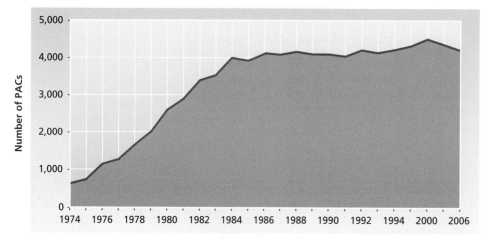

Source: Federal Election Commission, 2007; U.S. Bureau of the Census, *Statistical Abstract of the United States: 2008* (Washington, DC: U.S. Government Printing Office, 2007c), Table 408.

ularly known as the Swift Boat Veterans played a major role in the 2004 presidential election through its persistent questioning of John Kerry's military honors.

PACs (and interest groups in general) also lobby for political decisions in their favor. Lobbyists attempt to gain the ear of politicians through a smorgasbord of gifts (Williamson 2007a). The extent to which these activities occur is reflected in the cases of Randy "Duke" Cunningham and Jack Abramoff. In 2005, Cunningham, an eight-term U.S. representative, was sentenced to eight years in prison after pleading guilty to charges of conspiracy to commit bribery, fraud, and tax evasion (Hettena 2007). In 2006, Abramoff pled guilty to illegally soliciting Republican lawmakers' votes with a wide variety of gifts, including expense-free trips to foreign golf resorts, jobs for their spouses, and skybox sports seats and was sentenced to ten years in prison.

Do PACs contribute to pluralism? PACs were originally designed to provide a mechanism for small contributors to become involved in politics, to allow candidates without personal wealth to run for office, and to encourage independence from the viewpoints of political parties. Critics are concerned, however, that PACs may be undermining pluralism, because most PACs are connected with business-related interests (Eismeier and Pollock 1988; Watzman, Youngclaus, and Shecter 1997; Bardes, Shelley, and Schmidt 2007).

Charles Lewis (1998) makes a strong case for the influence of business-related PACs on members of Congress. His book, *The Buying of Congress*, is replete with purported links between business-related PAC money and congressional votes. As one illustration, consider the possible political influence by business-related PACs associated with the Tax Reform Act of 1986. As the congressional tax-writing panels were considering tax reform, PACs contributed $6.7 million—$119,643 per member—to fifty-six members of the House Ways and Means and Senate Finance Committees. This was two and a half times more money than the members of those same committees received from PACs two years earlier.

There are more recent examples. In 2007, the sugar industry successfully campaigned to preserve its $1 billion, ten-year government subsidy. During the housing boom, the subprime mortgage industry successfully avoided the enactment of laws regulating what proved to be ruinous lending practices to home buyers (Morgan 2007; Simpson 2007).

Finances heavily influence who can run for office and who wins elections in the United States (M. Green 2002; K. Phillips 2002; Palast 2004). According to the Federal Election Commission (FEC), about 90 percent of the primary races in the 2006 congressional primary elections were won by candidates who raised the most money. In fact, victorious candidates raised four times as much as their opponents. And, as in past elections,

most of this money was given by relatively few large donors, many of whom were out-of-state residents. In the 2006 midterm elections, slightly more than 95 percent of U.S. House races and 75 percent of Senate races were won by the candidate who spent the most money. At that same time, more than 170 members of Congress were millionaires (Keller 2002).

Worse yet were the so-called stealth PACs—shell groups that raised millions from individuals and corporations without revealing their funding sources. These organizations emerged as a way to take advantage of loopholes in campaign and tax legislation. Unlike conventional PACs, these new groups did not have to report any of their activities to the public or to government oversight organizations. A business group favoring whale hunting, for example, could call itself "Save the Whales" while spending lots of money attempting to block, repeal, or soften legislation actually designed to protect whales. The most widely known, issue-advocacy PAC was Republicans for Clean Air (formed by two supporters of George W. Bush), which spent $2.5 million against Bush's rival John McCain during the 2000 presidential primaries (Isikoff 2000). In the first alteration of the campaign finance laws in twenty years, Congress in 2000 passed legislation forcing stealth PACs to disclose their political donors and spending. After Pesident Clinton signed the bill, these PACs could no longer work in the shadows for partisan purposes (Dewar 2000; Trister 2000). Further reforms were passed in March 2002, when the Bipartisan Campaign Reform Act (more popularly known as the McCain-Feingold bill) was approved. By ruling the McCain-Feingold ban on unlimited soft money contributions to political parties unconstitutional in early 2003, a federal district court set this latest campaign reform act on a path to the Supreme Court, which upheld its constitutionality (Lane 2003; Lewis and Oppel 2003). This bill bans soft money contributions to national parties (Miller and Penniman 2002). Before the ink dried on McCain-Feingold, both Republicans and Democrats were establishing new, unaffiliated organizations to collect and spend the unlimited contributions (soft money) now legally denied them (Edsall 2002a).

Does this mean that the pluralist model is invalid? Despite the mixed results of PACs, pluralism has not disappeared from the American political landscape. Individuals in American society are represented to the extent that they belong to groups—whether political parties or interest groups—that are capable of influencing political decision making. Therefore, interest groups are one of the most important avenues for political representation. Groups representing civil rights, the environment, and consumer concerns have won some concessions that otherwise would not have been obtained. They have accomplished this through lobbying, making

TABLE 13.2

FOCUS ON THEORETICAL PERSPECTIVES: Characteristics of Two Models of Political Power

This table illustrates the way the functionalist and conflict perspectives view political power. Several key features of the political system are compared. Which theory do you think best describes power in the United States? Explain.

Characteristics	FUNCTIONALIST PERSPECTIVE Pluralist Model	CONFLICT PERSPECTIVE Power Elite Model
Who exercises power?	Bargaining and compromising interest groups	National political, economic, and military leaders
What is the source of power?	Resources of interest groups	Leadership positions in major institutions
Where is power located?	Spread widely among interest groups	Concentrated in hands of elites
How much influence do nonelites have?	Considerable influence on public policy	Very little influence on public policy
What is the basis for public policy decisions?	Goals and values are shared by the general public	Preferences of the elites

contributions to political campaigns, gathering information, generating publicity to express their viewpoint, and filing lawsuits. The National Rifle Association, for example, launches major letter-writing campaigns whenever gun control legislation is introduced. Many environmental organizations have taken legal action against companies that violate pollution laws and have forced some government regulatory agencies to investigate illegal environmental practices.

Doesn't the public have any influence over its government? Political pressure is routinely channeled through voting, despite the limitations of the ballot box. Two recent cases, moreover, demonstrate that an aroused public can have a political voice that is heard above the din of special interest groups. The popular will was credited with the passage of the Brady handgun bill in 1993 and the ban on the sale of assault weapons in 1994, despite strong opposition by the National Rifle Association (Lacayo 1993; J. Alter 1994). The tobacco lobby is also on the wrong end of public opinion. For example, nearly all of the American public believes that smoking should either be banned from workplaces or restricted to special areas. In part as a response to public opinion, steps are being taken to control cigarette smoking. Consider the national level alone. In 1994, President Clinton signed a bill prohibiting smoking in all public and some private schools. The U.S. Department of Defense, in that same year, banned smoking in all military workspaces, whether they are on military bases or inside battlefield tanks. The Smoke-Free Environment Act bans smoking in all buildings entered by ten or more people each day. All structures—bars, restaurants, workplaces—except homes would be smoke-free zones. Violators would face fines of up to $5,000 a day. In 1994,

a panel of the Food and Drug Administration concluded that nicotine is an addictive drug, setting the stage for possible regulation of nicotine products (Infante 1994). Regulation of nicotine has been publicly supported by the American Medical Association. State and local governments are also becoming more aggressive, as are many private companies such as McDonald's and many owners of professional baseball franchises (C. J. Farley 1994; Solomon 1994). During 1997, a group of state attorneys general negotiated a settlement with the tobacco companies that called both for the payment of $368 billion and for measures designed to reduce smoking by children. For their part, the tobacco companies would be granted immunity from future civil lawsuits claiming damages (J. Hall 1997). The U.S. Congress failed to approve the settlement, but in 1999 state attorneys general negotiated a civil settlement for all states; the settlement was worth $206 billion.

These two illustrations, of course, are not meant to pronounce dead the gun and tobacco lobbies. According to functionalism and the pluralist model, interest groups will continue their influence. They are part of the fabric of the American political system. At the same time, claim the functionalists, don't assume that the American public is excluded from all important political decisions, because the public can express its will through a strong collective voice. There is, however, a competing viewpoint (J. Q. Wilson 1995).

Elitism: The Conflict Approach

According to elitism, power is concentrated in the hands of a relatively few individuals and organizations. Fear of elitism is as old as the American Revolution. The intellectual leaders of the Revolutionary generation, being familiar with old Europe, held concentrated power as the paramount threat to the new republic. John Adams, the second president of the United States, expressed this fear: "When economic power became concentrated in a few hands, then political power flowed to those possessors and away from the citizens, ultimately resulting in an oligarchy or tyranny." Abraham Lincoln shared this concern. The Civil War had enthroned corporations, he said, ensuring an era of corruption at the top. The concern about elitism informs the conflict perspective (Conason 2008).

What is the power elite? The pluralist model has been dominant in American social science since the mid-twentieth century. But C. Wright Mills (1956) and others (F. Hunter 1953, 1980) developed an influential alternative view. According to Mills, the United States no longer has separate economic, political, and military leaders; rather, the prominent people in each sphere overlap to form a unified elite. Whereas most pluralists saw two levels of power—special interest groups and the public—Mills outlined three levels. Overshadowing special interest groups and the public is the **power elite**—a unified coalition of top military, corporate, and government leaders (the executive branch in particular). Mills placed U.S. senators and representatives

From your knowledge of John Adams and the other founders, do you think they were acting as functionalists or conflict theorists?

National Portrait Gallery, Smithsonian Institution/Art Resource, Inc.

at the middle levels of power rather than at the higher levels. It is at the higher levels where decisions are made regarding national policy, war, and domestic affairs.

According to Mills, members of the power elite share common interests and similar social and economic backgrounds. Members of the elite tend to be educated in select boarding schools, military academies, and Ivy League schools; belong to the Episcopalian and Presbyterian churches; and come from upper-class families. Members of the power elite have known each other for a long time, have mutual acquaintances of long standing, share many values and attitudes, and intermarry. All this makes it easier for them to coordinate their actions and obtain what they want.

Although advocates of elitism do not always use Marxian analysis, their explanations have often been influenced by Marx's thinking. Like Marx, elitist theorists tend to see political leaders as captives of business and corporate interests (Wolfe 1999). Pluralists, who are more likely to have been influenced by classical theorists of democracy than by Marx, view political leaders as referees among competing interest groups.

In a democratic society, in which power is supposed to rest with the people, the powerful prefer that their activities not be widely advertised. Power, according to conflict theory, is exercised behind the scenes through informal channels. Although the invisibility of power is an obstacle, it does not stop social scientists from investigating it (Zweigenhaft and Domhoff 2006; Parenti 2007).

Is there a national power elite in America? G. William Domhoff (1990, 2005) has spent much of his career investigating a ruling class, as described by Mills. Using such measures as membership in private schools, prestigious urban clubs, and the Social Register, Domhoff concluded that the very top of the upper class actually consists of less than 1 percent of the population (perhaps as little as 0.5 percent). Thomas Dye corroborates this conclusion. He reviewed studies of eminent leaders in political, military, and corporate positions. These institutional leaders are not typical Americans: "They are recruited from the well-educated, prestigiously employed, older, affluent, urban, white, Anglo-Saxon, upper- and upper-middle-class male populations of the nation" (Dye 2001:189). For example, about half of the elite group graduates from only twelve prestigious private universities, only 5 percent are female, and only one-fourth graduates from public universities.

The extensive socioeconomic homogeneity among military, political, and corporate institutional leaders does not *prove* the existence of the power elite. To make a convincing case for the existence of a power elite, it is necessary to document the presence of an elite group that is unified and that cooperates for the promotion of its shared interests.

F
E
E
D
B
A
C
K

1. The model of political power known as _____ depicts political decision making as the outcome of competition, bargaining, and compromise among diverse special interest groups.
2. The model of political power known as _____ contends that political control is held by a unified and enduring few.
3. The majority of political action committees (PACs) are now associated with
   a. labor unions.
   b. citizen groups.
   c. political parties.
   d. corporations and other businesses.
4. Interest groups are an important means by which the nonelite can influence political decision making. T or F?
5. The operation of PACs in America today invalidates the pluralist model. T or F?
6. According to C. Wright Mills, which of the following is *not* part of the power elite?
   a. military organizations
   b. U.S. senators and representatives
   c. large corporations
   d. executive branch of the government

Answers: 1. pluralism; 2. elitism; 3. d; 4. T; 5. F; 6. b

Although it is impossible to definitively identify a national power elite, conflict supporters offer evidence for its existence. Thomas Dye (2001) identified 7,314 first-rate national positions in fourteen different institutions: education, industrial corporations, banking, investments, insurance, mass media, law, government, utilities, foundations, the military, civic/cultural organizations, transportation, and communications. The "elite" are those individuals who occupy the top-drawer positions in these institutional sectors. They control one-half of America's industrial and financial assets, almost one-half of the assets of private foundations, and more than two-thirds of the assets of private universities. In addition, they control the mass media, the most prestigious civic and cultural organizations, and all three branches of the federal government.

At the time of Dye's study, these 7,314 top positions were occupied by 5,778 individuals. The existence of fewer occupants than positions means that some individuals held more than one top position. Approximately 15 percent of the elite held more than one top position at a time. What's more, almost one-third of all top positions were interrelated because some individuals held three or more positions. Some even held six or more top positions at the same time. The members of this inner group of multiple position holders are situated to allow communication with one another and to coordinate the activities of a wide variety of powerful institutions. Still, it must be remembered that 85 percent of the top position holders were not interrelated. While this leaves considerable room for disunity and conflict among the top institutional leaders at any given time, this latitude was offset by the fact that some of these individuals held top leadership positions in a number of institutional areas over their careers. A study by Kevin Hallock found 9,804 director seats in 602 large American corporations occupied by 7,519 persons (Hallock 1997). In addition, we know that prominent people move from business to government, from government to universities, from universities to foundations, and so on (Useem 1984; B. Mintz and Schwartz 1985; Zweigenhaft and Domhoff 2006).

Economic Systems

The earlier discussion of interest groups established the fact that many groups are tied to corporate and other business interests. Discussions of elitist and pluralist models revealed that business interests play an important role in the distribution of power. These two points illustrate the close link between the polity and the economy. Societies differ, however, regarding the kinds of political and economic systems they develop. Variations in economies play particularly important roles in relationships between nations because fundamental differences in worldviews are likely to underlie differing solutions to problems associated with the production and distribution of goods and services. Because preindustrial societies were covered in Chapter 5, industrial societies are featured in this chapter.

What is the basic distinction between capitalism and socialism? **Capitalism**, as an economic system, is founded on two assumptions: the sanctity of private property and the right of individuals to profit from their labor. Successful individuals, according to capitalists, deserve to own and control land, factories, raw materials, and the tools of production. In addition, economic success is most likely promoted by free competition with no government interference. Other economic systems assume that property belongs to the society. These economies are carefully planned by a centralized government. This latter type of economy, in which production is owned by the people and the economy is controlled by the government, is called **socialism** (Gregory and Stuart 1989; Schnitzer 2000).

When we speak of the political/economic systems of capitalism and socialism, we are referring to ideals rather than actual operating systems. As we saw in Chapter 6, much can be learned by considering ideal types that involve an isolation of the most basic characteristics of some social entity. No society is likely to embody all the principles described in the ideal type, but a contrast of ideal types can help us understand how societies become very different by placing different degrees of emphasis on particular sets of ideals (Schnitzer 2000).

Capitalism

Capitalist societies are founded on the belief that private individuals and organizations have the right to pursue their own private gain and that society will benefit from their activities. Although economic activity under capitalism is motivated by the pursuit of profit, the individuals and organizations that are pursuing profit also run the risk of losing money:

> *In a typical year [in the United States] about four out of ten corporations report net losses. Of ten business firms started in an average year, five close down within two years, and eight within ten years—lack of success being the main reason.* (Ebenstein, Ebenstein, and Fogelman 2000:132)

This combination of profit seeking and risk taking is possible only under certain economic ownership arrangements.

Who owns the property in a capitalist system? Under capitalism, most property belongs to private individuals and organizations rather than to the state or the community. Private ownership is based on a conviction that people have an inalienable right to hold and to control their own property. While advocates of capitalism recognize that private ownership leads to concentrations of power, they believe that distributing power among a large number of owners trumps concentrating power in the state. Moreover, because of private ownership, the pursuit of profit is possible. And, defenders of capitalism argue, private ownership benefits society.

How is capitalism thought to benefit society? According to Adam Smith, eighteenth-century Scottish social philosopher discussed in Chapter 1, a combination of both private ownership of property and the pursuit of profit brings advantages to society. Because of competition from other capitalists, Smith stated, individual capitalists will always be motivated to provide the goods and services desired by the public at prices the public is willing and able to pay. Capitalists who produce inferior goods or who charge too much will soon be out of business because the public will turn to their competitors. The public, Smith reasoned, will benefit through the "invisible hand" of market forces. Not only will the public receive high-quality goods and services at reasonable prices, but capitalists will always be searching for new products and new technologies to reduce their costs. As a result, capitalist societies will use resources efficiently. Efforts by government to control the economy will distort the economy's ability to regulate itself for the benefit of everyone.

Do contemporary capitalist economies work this perfectly? We cannot say how perfectly or imperfectly a purely capitalist economy operates, because none exists. At the same time, the United States is generally considered the purest existing example of a capitalist system. Most Americans view their own affluence as evidence that capitalism works.

What are some deficiencies of the capitalist model? In practice, there are important deviations from the ideal capitalist model, one of which involves the tendency to form a **monopoly**, a single company controlling a particular market, or an **oligopoly,** a combination of companies controlling a market. This muzzling of competition is reflected in several ways. When capitalist organizations experience success, they tend to grow until they become giants within their particular industries. When this happens, new organizations and entrepreneurs find it difficult to enter the industry. There is little hope of competing on an equal basis. Worse yet, the creation of monopolies and oligopolies permits price fixing, leaving consumers with the choice of either buying or not buying a good or service (Ebenstein, Ebenstein, and Fogelman 2000).

A recent example of monopolistic practices in the U.S. economy involves the Microsoft Corporation. Microsoft manufactures, among other products, the Windows operating system—by far the most popular operating system for personal computers. Computer manufacturers typically include Windows on the machines they sell. In the 1990s, Microsoft began to insist that these manufacturers include its Internet browser, Explorer, on their computers as well. Not only that, but the manufacturers were not to install another browser in addition to Explorer. If they refused, Microsoft would withhold their license to sell Windows on their computers. Because Microsoft had so much power over computer manufacturers, other makers of Internet browsers, such as Netscape, were essentially excluded from the market

TABLE 13.3

Examples of Government Economic and Regulatory Assistance

The government is extensively involved in the U.S. economy. What would Adam Smith say about this?

Public Utilities are often owned and operated by state or local governments.

The agricultural industry feels the influence of government through price controls and embargoes on exports to other countries.

Antitrust legislation exists to control the growth of corporations.

The federal government is heavily involved in the defense industry.

Business could not survive without publicly financed roadways, airports, and waterways.

Publicly funded public schools, colleges, and universities supply business with a skilled workforce and provide basic research for product development.

The U.S. military protects American international business interests.

Government supports business through tax breaks.

Legislation requires labor and business to obey labor laws.

(Chandbasekaran 1999). Eventually, the federal government took Microsoft to court, where it was decided that Microsoft was engaging in monopolistic practices. Microsoft reached a settlement with the federal government and nine state governments in November 2001 (*The Economist* 2002).

Also, the rise of widespread stock ownership can affect the profit motive. When a firm is owned and operated by one or a few individuals, the motivation to make a profit is usually very high. When a company has thousands of shareholders and the managers are not the owners, the profit motive may have to compete with other motives, such as company growth or self-aggrandizement by the managers.

How does the government contribute to the U.S. economy? Adam Smith is often misinterpreted as saying that government should take a hands-off approach where the economy is concerned. Although Smith did strongly oppose overregulation by government, he reserved a place for it. In fact, the U.S. government has always been involved in the workings of the economy.

The Constitution expressly provided a role for the national government in the promotion of a sound economy. Government functions included the regulation of commerce, development of a strong currency, creation of uniform standards for commerce, and the provision of a stable system of credit. In 1789, Congress supported our shipping industry through a tariff on goods imported by foreign ships, and since this initial move into the economy, the federal government has continued to augment business, labor, and agriculture. For example, The Federal Reserve took several steps in 2008 to bail out Wall Street investment bankers. In an effort to prevent a widespread meltdown in financial markets, the Fed approved a $30 billion credit line to underwrite the purchase of a failing Bear Stearns, one of the largest firms on Wall Street (Andrews 2008). U.S. labor is strengthened by the government through minimum wages, maximum working hours, health and safety conditions, and unemployment support. The agricultural industry feels the influence of government through price controls, price supports, and grain embargoes against other countries. Small farmers and agribusinesses receive financial assistance that runs into the billions of dollars each year (Katznelson, Kesselman, and Draper 2005; T. E. Patterson 2006). See Table 13.3 for additional examples of government economic and regulatory assistance.

Socialism

The quest for private property and the pursuit of profit, according to many critics of capitalism, leads to serious inequities because workers are paid less than the value of their products. In keeping with Karl Marx, contradictions within the capitalist system would lead to its overthrow by the proletariat, who would replace it first with socialism and later with communism. Marx, however, was not the first to propose a socialist system. Plato's *Republic* and some portions of the Bible also contain elements of socialist thought.

Who owns property in a socialist system? According to socialist thinkers, the means of production must be removed from the hands of wealthy capitalists. The state, as the people's representative, should own and control property. The state is expected to ensure all members of society a share in the monetary benefits.

How is socialism thought to benefit society? Capitalist thinkers, of course, see a role for the state in a capitalist economy—military protection, a system of criminal justice, and a limited number of public works and institutions (e.g., highways and public schools). Socialist thinkers envision a far more ambitious role for the state. They see state control of the economy as a means for the people to maintain control over the production and distribution of goods.

Socialist theory also points to important benefits for workers. Whereas workers under capitalism receive wages below the value that their labor produces and experience little control over their work, workers under socialism should profit because both the state and the workplace exist for their benefit (J. E. Elliott 1985).

Do contemporary socialist economies work this perfectly? Cases of pure socialism have been as rare as cases of pure capitalism. In the socialist economy of the former Soviet Union, for example, some agricultural and professional work was performed privately by individuals who worked for a profit, and significant portions of housing were privately owned as well. Managers received salaries that were considerably higher than those received by workers, and managers were eligible for bonuses such as automobiles and housing.

Private enterprise existed in Poland under Russian communist rule. Service businesses, such as restaurants and hotels, had a significant degree of private ownership. Hotels, in fact, were typically built and managed by multinational chains. Because Poles could travel abroad, they formed business relationships, learned about capitalist methods, imported goods to fill demand, and brought back hard currency. They then used the hard currency earned abroad to create private businesses (Schnitzer 2000).

Some socialist societies even introduced economic incentives for higher productivity—a step that would seem to encourage the workers' desire for profit. Socialist systems have not been successful in eliminating income inequalities, nor have they been able to develop overall economic plans that guarantee sustained economic growth.

Mixed Economic Systems

Because capitalist and socialist economic systems are based on different assumptions, their relationships are often strained. The historic tension between the United States and Russia illustrates the point. This conflict is often predicated on seeing the other as a pure case. In reality, however, every nation violates some of the assumptions of its economic system, and most nations fall between the extremes of capitalism and socialism. Countries in Western Europe, for example, have developed capitalist economic systems in which both public and private ownership play important roles. In these nations, highly strategic industries (banks, transportation, communications, and some others) are owned and operated by the state, while other industries are privately owned but are more closely regulated than in the United States (Esping-Andersen 1996; C. D. Harris 1997; Ollman 1998; Amsden 2003).

Is there movement toward mixed systems in socialist societies? As the former Soviet Union lost control over its republics and eastern Europe, many of the formally socialist countries began to move toward a market system (Elster, Offe, and Preuss 1998; Stark and Bruszt 1998). In Hungary, state-owned enterprises have been allowed to become privately owned companies. More than 1 million Hungarians have been given the right to buy land, businesses, buildings, or other property taken over by the Russians in 1949. Nearly all of the state-owned small businesses are now in the hands of private owners. Agricultural cooperatives have also been privatized (Schnitzer 2000). In 1991, Cuba's Communist Party adopted some degree of capitalism by permitting handymen, plumbers, carpenters, and other tradespeople to work for profit.

Economic change has not been limited to Soviet-bloc countries. In 2002, the Communist nation of North Korea announced the creation of a free capitalist zone near its Chinese border. Freed of central government control for fifty years, this zone will solicit private capital from its neighbors and the West (French 2002). After the death of Mao Zedong, the longtime communist leader of the People's Republic of China, the Chinese began market-related economic reforms. The changes initiated in 1979—restoration of financial incentives, free-market prices, education, and economic relations with the West—accelerated in the early 1980s. Private enterprise was legitimated in the Chinese Constitution in 1999. Market forces increasingly replace central economic planning as farmers are permitted to work the land freely (rather than pay the government a predetermined amount of produce), markets and local entrepreneurship are allowed to develop, foreign businesses are encouraged to open facilities on the mainland, and capitalists are permitted to join the Community Party (Nee and Stark 1989; Ollman 1998; Muldavin 1999; Pomfret 2001).

High inflation and public demand for greater political freedoms (to match the newly granted economic liberation in the late 1980s) led to political turmoil. Communist hardliner Li Peng gained control after a popular revolt in Tiananmen Square in June 1989 was suppressed. Mainland China, nevertheless, did not revert to central economic planning. It continues its rapid move toward what its now-deceased leader, Deng Xiaoping, called the creation of a "socialist market economy" (Nelan 1993a, 1993b; Schell 1997). When ownership of the port city of Hong Kong reverted to China in 1997, the Chinese government pledged not to interfere with the city's capitalistic economy and so far has lived up to that promise (Kraar 1997). Most recently, China's Community Party began a path to "state capitalism." In this system, the government promotes the success of selected companies via supports such as tax breaks, contracts, and land deals. The payback from these companies comes in the form of party and government loyalty, taxes, and kickbacks (Pomfret 2002). In 2002, China's party revised its constitution to permit membership of private entrepreneurs, formerly considered capitalist "exploiters" (Liu 2002).

According to some scholars, the demise of communism ensures the triumph of capitalism. The twenty-first century, it is argued, will see a global economy dominated by capitalism (Thurow 2003a; Fukuyama 2006; Reich 1993, 2001, 2007). Whether or not socialism passes from the economic and political scene, capitalism is clearly spreading.

The Global Cultures of Capitalism

What are the cultures of capitalism? Although several variants of capitalism can be identified (Hampden-Turner and Trompenaars 1995), they can be collapsed into two basic types. **Individualistic capitalism** is founded on the principles of self-interest, the free market, profit maximization, and the highest return possible on stockholder investment. This variant of capitalism dominated the world in the nineteenth century (the United Kingdom) and twentieth century (the United States). European and Japanese capitalism of the late twentieth century is known as **communitarian capitalism**—the type of capitalism that emphasizes the interests of employees, customers, and society. As will be seen, these two types of capitalism deserve the label "cultures of capitalism" because each rests on a distinctive value base.

What is culturally distinctive about individualistic capitalism? Individualistic capitalism is based on the ideas of John Locke and Adam Smith. Locke, a seventeenth-century philosopher who strongly influenced the leaders of the American war for independence and the nature of the U.S. Constitution, placed individual rights, including the right to own property, at the heart of his *Second Treatise of Government* (1980; originally published in 1690). Government is needed to protect the property of individuals from others, Locke argued, but it must have very limited powers. Although Smith has been widely misinterpreted (Shepard, Shepard, and Wokutch 1991; Werhane 1991; Muller 1993), some of his ideas were used to shape individualistic capitalism and continue to be used to maintain it. The concept of the "invisible hand," mentioned only once in his *Wealth of Nations* (2000; originally published in 1776), has been particularly influential in individualistic capitalism. According to the popular interpretation, if individuals are released to pursue their own self-interest, their separate activities will, through the free-market mechanism, automatically benefit everyone else by increasing the supply of consumable goods and services.

These cultural underpinnings are easy to observe in individualistic capitalism (Thurow 2003a). Both individuals and firms have their own economic strategies for success. The individual wants to prosper, and firms wish to satisfy their stockholders. Firms seek to maximize profits because shareholders invest to maximize profits. Customers and employees are the firm's means to reach the goal of higher shareholder return. Because higher wages cut into profits, firms try to keep them as low as possible and lay off employees whenever necessary. Short-term thinking in the name of profit maximization is the norm. Employees are expected to leave one job for another if they can get higher wages. Employees and employers have no mutual obligations outside the legal employment contract.

Within individualistic capitalism, government is not to interfere with the workings of the free market. Government is supposed to make minimal rules and act as an umpire to settle disputes (M. Friedman 2002). There is no place for government in investment funding or economic planning. As long as government protects private property rights, the pursuit of profit maximization will ensure the greatest prosperity.

What is the value base of communitarian capitalism? Communitarianism, not to be confused with communism, is a form of "soft capitalism" also rooted in the ideas of a philosopher. Jean-Jacques Rousseau, writing in the eighteenth century, was, like Locke, a social contract theorist. Both believed that individuals forfeit some of their freedom to obtain protection from a central government. Locke saw the social contract as an individualistic device, but Rousseau thought it a communitarian one. Locke's notion of individualism appeared in the seventeenth century and influenced the American Revolution. Rousseau's idea of the "General Will" emerged in the eighteenth century and helped to fuel the French Revolution. The General Will always puts the community first: "The French state thus became an instrument of intervention in the interest of the General Will, to plan and guide the nation's institutions" (Lodge 1975:13).

As this Yangshuo KFC restaurant reflects, elements of capitalism are being introduced into China.

Lee Snider/The Image Works

FEEDBACK

1. Which of the following is *not* characteristic of capitalist thought?
    a. Private individuals have a right to pursue private gain.
    b. Society will benefit from attempts to make a profit.
    c. The state must make an active attempt to control the economy.
    d. Individuals pursuing a profit must be willing to risk losing money.
2. Under _____, most property belongs to private individuals and organizations rather than to the state or the community as a whole.
3. As capitalist systems mature,
    a. the economy is increasingly dominated by small businesses.
    b. the willingness to take risks increases.
    c. consumer goods become very scarce.
    d. many industries are dominated by a few large firms.
4. According to economic theorists, which system should have the greatest degree of control by the state?
    a. capitalism
    b. socialism
    c. democratic socialism
    d. welfare capitalism
5. Proponents of socialism contend that it will
    a. prevent workers from exerting significant control over social policy.
    b. lead to increased profits for owners.
    c. allow workers to exert significant control over work organizations.
    d. result in very inactive governments.
6. Currently, there seems to be a movement toward mixed economic systems in socialist societies. T or F?
7. Several characteristics of capitalist economic systems are listed here. Indicate in the space beside each characteristic whether it best describes individualistic capitalism (IC) or communitarian capitalism (CC).
    ____ a. government and business cooperation
    ____ b. high job mobility
    ____ c. profit maximization
    ____ d. more value placed on the interests of employees and consumers
    ____ e. self-interest paramount
    ____ f. long-range economic planning

Answers: 1. c; 2. capitalism; 3. d; 4. b; 5. c; 6. T; 7. a. (CC), b. (IC), c. (IC), d. (CC), e. (IC), f. (CC)

Japan shares a communitarian viewpoint with France and many other western European countries. They each emphasize cooperation, interpersonal harmony, and the subordination of the individual to the community. Roots of Japanese communitarianism, however, lie in its own philosophical and religious ideas and in its unique demographic and geographical situation.

In communitarian capitalism, individuals and firms also have strategies (Thurow 2003a). But given the cultural foundations just outlined, those strategies are very different from strategies in individualistic capitalism. Individuals in communitarian capitalism attempt to select the best firm and then work hard to be part of the company team; personal success or failure is identified with the fate of the company. Job switching is not widespread. In fact, in many Japanese firms, the person who voluntarily leaves a company is a "traitor" ("Graduates Take Rites of Passage into Japanese Corporate Life" 1991). At the same time, both Japanese and German workers are less often laid off.

Communitarian firms place the interests of shareholders behind those of their employees and customers. Because employees are of first importance, high wages are a top priority. Maintenance of high wages and job security supersede profits. Long-range planning is emphasized over profit maximization, and shareholders, in fact, earn relatively low returns.

In both Europe and Japan, the government is expected to play a significant role in economic funding, planning, and growth. There is cooperation between business and government as well as between business and labor. Labor officials sit on the boards of directors of German firms, and German banks are major stockholders in German companies. Job training is the responsibility of both business and government. In short, there is no line of demarcation between private and public interests.

The contrast between the two cultures is conceptually clear, although in actual practice, it is more complicated. Individualistic capitalists in Britain, Holland, and the United States believe that by concentrating on individual self-interest, they will automatically better serve customers and society. Communitarian or so-called soft capitalists, on the other hand, assume that by concentrating on serving customers and society, they will automatically serve their own interests (Kuttner 2007; Hartmann, Palast, and Miller 2008; Phillips 2008).

Which culture of capitalism is superior? Some scholars believe that in the long run, communitarian capitalism outperforms individualistic capitalism (Lodge and Vogel 1987; Albert 1993; Thurow 2003a). In fact, according to George Lodge, the American economy has already moved, albeit kicking and screaming, toward communitarian capitalism (Lodge 1986). However, only future unfolding events will answer our question.

The Corporation

Within capitalist and mixed economic systems, corporations are extremely important actors. In their provocative book, *America Inc.*, Morton Mintz and Jerry Cohen contend that large corporations now own and operate the United States. Government power is a derivative of corporate power:

> *The danger that this superconcentration poses to our economic, political and social structure cannot be overestimated. Concentration of this magnitude is likely to eliminate existing and future competition. It increases the possibility for reciprocity and other forms of unfair buyer-seller leverage. It creates nation-wide marketing, managerial and financing structures whose enormous physical and psychological resources pose substantial barriers to smaller firms wishing to participate in a competitive market. . . . [This] superconcentration creates a "community of interest" which discourages competition among large firms and establishes a tone in the marketplace for more and more mergers. This leaves us with the unacceptable probability that the nation's manufacturing and financial assets will continue to be concentrated in the hands of fewer and fewer people.* (Mintz and Cohen 1972:39)

The political and economic effects of modern corporations are seen at the local, national, and international levels, a point discussed later. For now, let's examine the nature of modern corporations.

The Nature of Modern Corporations

The economy of the United States has experienced substantial and continual growth. The gross domestic product (GDP)—the domestic output of goods and services—has steadily increased since 1933. This increase in the GDP was accompanied by an increasing domination of the economy by large business corporations. While the emergence of giant business organizations contributed to economic growth, it also created problems. These problems will be better understood if we first examine the nature and extent of modern corporations.

What is a corporation? A **corporation** is an organization owned by shareholders who have limited liability and limited control over organizational affairs. A key feature is the separation of ownership from control (Berle and Means 1968; originally published in 1932). Although shareholders receive reports of gains and losses from the corporation's transactions, they are not legally liable for them. And although shareholders are formally entitled to elect a board of directors, in reality they routinely approve the candidates recommended by the existing board. The actual control of a corporation rests with the board of directors and the corporate managers. Some critics contend, however, that this widely assumed separation between ownership and management is a myth.

How important are corporations today? At the beginning of the twentieth century, only a few American industries, such as railroads, shipping, steel, oil, and mining, were organized as corporations (Roy 1999; Perrow 2005). Today, U.S. corporations dominate the economy, providing everything from diapers to retirement communities.

Total corporate assets are concentrated in the hands of a relatively few giant corporations. The top 100 corporations, which account for less than 0.1 percent of all corporations in the United States, control more than 9 percent of total corporate assets; the top 200 corporations control nearly 11 percent. This has changed little since 1960. There is also a very high concentration of assets *within* the top corporations. The top 100 corporations have three-fourths of the assets held by the top 500 corporations, and the top 200 corporations possess 88 percent of the assets held by the top 500 corporations. Although only 0.2 percent of American corporations have assets of $250 million or more, these corporations control more than 81 percent of the corporate assets in the United States. In sharp contrast, the 98.5 percent of corporations with assets of less than $10 million control only about 7 percent of the nation's corporate assets.

The Effects of Modern Corporations

Corporations represent massive concentrations of economic resources. Because of their economic muscle, corporations such as Microsoft, AT&T, and General Motors make their voices heard by government decision makers. Many government policies, such as those regarding consumer safety, tax laws, and relationships with other nations, reflect corporate influence.

How do corporations affect political decision making? The tremendous political influence of corporate officials exists for several reasons. First, corporate leaders develop influential political and social ties. James Treat, the businessman whom Floyd Hunter found to be an influential adviser to Georgia's politicians, described his status in these terms:

> I am asked occasionally to appear before state legislative committees, but I am not registered as a lobbyist. I just happen to know a lot of people, and I get called in on a lot of things. Sometimes, I think, on too many! (F. Hunter 1953:164)

Treat, of course, knew a multitude of people—and exerted influence—primarily because of his business position and the business activities of his relatives. On the federal level, many influential, wealthy individuals alternate between high corporate and political positions. For example, Robert McNamara was president of the Ford Motor Company before becoming secretary of defense in the Kennedy administration; Michael Blumenthal was chairman of Bendix Corporation before becoming Jimmy Carter's secretary of the treasury; and Donald Regan was chairman of the board and chief executive officer (CEO) of Merrill Lynch before entering Ronald Reagan's cabinet as secretary of the treasury. In addition to being a CEO himself, George W. Bush picked a former CEO as his running mate and appointed three former CEOs to his cabinet and one as his White House Chief of Staff (M. Moore 2001; Page 2002). The movement between government and corporate positions occurs on lower levels as well. For example, after authoring the 2003 Medicare drug program sought by the drug companies, U.S. Representative Billy Tauson left Congress to become the chief lobbyist for the pharmaceutical industry.

Second, corporate officials—because of their personal wealth and organizational connections—are able to reward or punish elected government officials through investment decisions. It is unlikely, for example, that Detroit's $500 million Renaissance Center would have been constructed without the personal and financial support of Henry Ford II. On a more ominous note, financially troubled International Harvester extracted considerable concessions from both Springfield, Ohio, and Fort Wayne, Indiana, by announcing that it would close its plant in one of the two cities. The decision to allow the Springfield plant to remain open was reached only after the city found $30 million in local and state funds to buy the plant and lease it back to the company.

Third, in addition to affecting political processes through investment decisions, wealthy members of society, including corporate officials, affect a political candidate's chance of being elected in the first place. State and national politicians hesitate to jeopardize their chances for election (or reelection) by offending corporate officials or wealthy potential contributors, who are frequently the same people (Keen 2002).

Fourth, the political clout of large corporations is multiplied because of **interlocking directorates**— members of corporations sitting on one another's boards of directors. Although interlocking directorates are illegal if the corporations are competitors, this does not prevent noncompeting corporations with common interests from forming interlocking directorates. For example, members of the General Motors board of directors are also on the boards of many other corporations, including Eastman Kodak, Bristol-Myers Squibb, and Merck and Company. Moreover, these interlocks are only the beginning. The corporations having interlocking directors with General Motors have interlocks with hundreds of other corporations. In addition, there are indirect interlocks that permit two board members of competing firms to be members of a third corporation. It is not difficult to imagine the political power emanating from this spider's web of interlocks among America's most powerful corporations (Mintz and Schwartz 1981a, 1981b, 1985; Useem 1984; Dye 2001).

Fifth, the political power of corporations is enhanced through conglomerates. A **conglomerate** is a network of unrelated businesses operating under a single corporate umbrella. The parent corporation does not actually produce a product or provide a service; it simply has controlling interest in a set of diverse enterprises. RJR Nabisco Inc., for example, owns companies in such different areas as tobacco, pet foods, candy, cigarettes, food products, bubble gum, research, and technology. A list of the company's North American subsidiaries covers almost an entire page in *Who Owns Whom* (*LexisNexis Corporate Affiliations* 2008).

The informal contacts and global power, then, give corporations important advantages in the political decision-making process. And although corporate officials may disagree with one another and may not share all of the same goals, they have the ability, when they see their interests threatened, to effectively use their organizational resources. This influence is reflected in what critics call **corporate welfare**—economic benefits government regularly gives to corporations. Corporate welfare comes in many forms, including tax breaks, subsidies, and lax regulation and law enforcement for white-collar crimes. Mark Zepezauer and Arthur Naiman (1996) estimate that the government hands out about half a trillion dollars annually to big business and wealthy individuals.

Finally, the political influence of corporations is not confined to their countries of origin. The world is increasingly influenced by **multinational corporations**, firms in highly industrialized societies with operating facilities throughout the world. Multinationals are concentrated in a few industries, each controlled by a small number of giant companies. The oil and automobile industries alone account for about half of all multinational activity. Although multinational corporations existed for centuries, they became more powerful after World War II. Improvements in communication and transportation technology allowed them to exert wide control over their worldwide operations.

How economically powerful are multinational corporations? Of the world's hundred largest political and economic entities, fifty-one are multinational corporations rather than nation-states. Seven corporations based in the United States—ExxonMobil, IBM, General Motors, Ford Motors, AT&T, Wal-Mart Stores, and General Electric—have sales volumes on a par with the gross national products of the world's forty most productive nations. Nations ranking below both General Motors and Ford Motors include Argentina, South Africa, Venezuela, Greece, Portugal, Israel, and Ireland (Steedley 1994). If the 300 largest multinational corporations were to band together, their combined economic resources would be exceeded only by the combined wealth of the United States and Japan. Figure 13.4 compares some multinational corporations with selected nations.

What are the negative effects of multinational corporations? Defenders of multinationals argue that the corporations provide developing countries access to technology, capital, foreign markets, and products that would otherwise be unavailable. Critics claim that multinationals actually harm economies of host nations—exploiting their natural resources, disrupting local economies, introducing inappropriate technologies and products, and increasing income inequality. Multinationals, these critics note, often rely on inexpensive labor or abundant raw materials in the developing nations while returning their profits to corporate headquarters and shareholders in rich nations. Multinationals' domination of their industries make it difficult for the economically developing nations to establish new companies that can effectively compete with the multinationals. As a result, multinationals may retard rather than promote the economic development of some regions of the world.

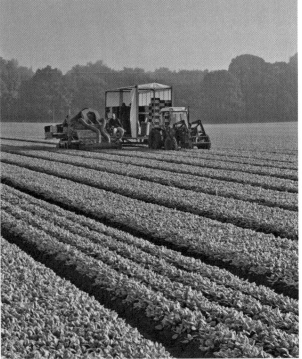

Corporations dominate the American economic system and influence the economies of nations worldwide. Even farming, historically a family business, is not immune to the corporate form of organization.

Ludwig Haskins/Digital Railroad

FEEDBACK

1. A(n) _____ is an organization owned by shareholders who have limited liability and limited control over organizational affairs.
2. Which of the following is *not* one of the reasons behind corporate political influence?
   a. Corporate managers and directors have informal and organizational ties that lead them to be consulted about matters of political policy.
   b. Many corporate officials have reward and coercive power over elected government officials.
   c. Many elected officials are also top corporate officials.
   d. Interlocking directorates exist.
   e. Conglomerates are present.
3. Indirect interlocks permit two board members of different corporations to sit on each other's boards of directors. T or F?
4. Firms with operating facilities in several different countries are called _____.
5. Multinational corporations are concentrated in a few industries, each dominated by a small number of giant companies. T or F?
6. Multinational corporations
   a. are easily manipulated by the governments of low-economy nations.
   b. sometimes create problems involving unsafe working conditions and inappropriate products.
   c. are known to have overthrown large numbers of low-economy governments.
   d. carefully avoid environmental damage in their areas of operation.

Answers: 1. corporation; 2. c; 3. F; 4. multinational corporations; 5. T; 6. b

Although the evidence is not always clear, these firms do possess huge concentrations of power whose abuses cannot always be checked, even by national governments. According to critics, evidence suggests that executives of International Telephone and Telegraph Corporation (ITT) worked with elements of the U.S. government in an effort to overthrow the government of Chile, headed by Salvador Allende (J. Anderson 1974; Wise 1976). Although such actions may be atypical of multinationals, they illustrate the difficulties faced by a government wishing to control the multinationals within its borders.

Environmental impacts are similarly problematic. For example, in Jamaica such companies as Alcoa and Revere are polluting the lakes with "red mud," a by-product of aluminum-producing activities. Union Carbide and Sun Oil refineries operating in the Caribbean are linked to increased incidences of respiratory disease. Products marketed by multinationals can also create problems when they are incompatible with local conditions. For example, an infant formula marketed by Nestlé led to significant illness and death in several low-economy nations when the dry formula was mixed with contaminated local water supply.

Finally, multinationals often move operations to less developed countries that offer cheap labor and weak regulatory controls. This could severely affect the U.S. economy, as well as economies of other developed nations.

FIGURE 13.4

Total Revenue of Multinational Corporations versus National Gross Domestic Products

This table compares the revenue of selected multinational corporations to the gross domestic product (value of all goods produced and consumed domestically) of some countries in 2003. Were you surprised by any of the information?

Wal-mart Stores, Inc. $258.7 Billion	vs.	Bangladesh $258.8 Billion
Home Depot $64.8 Billion	vs.	Syria $58.0 Billion
IBM $89.1 Billion	vs.	New Zealand $85.3 Billion
Exxon Mobil $213.2 Billion	vs.	Greece $212.2 Billion
Bank of America $48.1 Billion	vs.	Ethiopia $48.5 Billion
General Electric Co. $134.2 Billion	vs.	Israel $120.6 Billion

Source: *The World Almanac and Book of Facts, 2005.*

Work in the Contemporary Economy

The United States economy is dynamic. This is reflected in the radical alterations projected for the workforce composition. Also, current change in the occupational structure is expected to accelerate, with some serious consequences (Rifkin 1995; Milkman 1997; Mazaar 1999; Hodson and Sullivan 2008).

The Changing Workforce

To understand work in modern society, it is necessary to be familiar with the three basic economic sectors: primary, secondary, and tertiary.

How do the economic sectors differ? The **primary sector** of an economy depends on the natural environment to produce economic goods. The types of jobs in this sector vary widely—farmer, miner, fishery worker, timber worker, cattle rancher. In the **secondary sector**, manufactured products are made from raw materials. Occupations in this sector include blue-collar workers of all types, from those who produce computers to those who turn out *Grand Theft Auto* video games as well as associated white-collar employees in management and office positions. Employees in the **tertiary sector** provide services. If you went to class, filled your car with gas, stopped by the bank, and visited your doctor today, you spent most of your time and money in the tertiary (service) sector. Other service industries include insurance, real estate, retail sales, and entertainment. We generally refer to people in these industries as white-collar workers because they are the predominant type of employee in the tertiary sector.

Historically, how have the three sectors evolved? Obviously, the primary sector dominated the preindustrial economy. At that stage of economic development, most citizens were involved in the battle for economic survival. Very few people could be spared for religious, medical, or educational purposes. And physical goods were made by hand.

This balance began to change with the mechanization of farming in the agricultural economy. Mechanical innovations (cotton gin, plow, tractor), along with the application of science (seed production, fertilization, and crop rotation) reduced the proportion of the labor force needed in the primary sector. During the 1800s, the average farmer could feed five workers or so. Today, the figure is eighty. Not surprisingly, the proportion of workers in the farming sector has declined from almost 40 percent in 1900 to about 2 percent today.

Another change began occurring along with the mechanization of farming. With other technological advancements (power looms, motors of all types, electrical power) came the shift of agricultural workers from farms to factories, ushering in the secondary sector. As Figure 13.5 indicates, the percentage of the U.S. labor force engaged in blue-collar jobs reached almost 40 percent in 1900.

Just as in agriculture, technological developments in the secondary sector permitted greater production with fewer workers. After World War II, the fastest-growing occupations were the white-collar jobs—managers, professionals, sales workers, clerical workers. In 1956, white-collar workers for the first time accounted for a larger proportion of the U.S. labor force than did blue-collar workers. In manufacturing industries, white-collar workers tripled the number of blue-collar workers.

Technological progress did not stop with the secondary sector. As relative growth of workers in goods-producing jobs was decreasing, the demand for labor in the tertiary section was increasing. Fueled by computer technology, the U.S. economy moved from a manufacturing base to a "knowledge" base (see Chapter 5). Demand for people who could manage information and deliver services became the new workforce hallmark. This development, along with the growth of white-collar employment in the secondary sector, increased the proportion of white-collar workers in the United States from just below 30 percent in 1930 to about 77 percent today (see Figure 13.5). The ever-increasing number of women employed outside the home has also influenced the American workplace.

FIGURE 13.5

Distribution of Workers by Occupational Category

This table shows the percentage of American workers who are employed in each occupational category. It also shows how each occupational category is broken down by gender.

Occupational category	Percent of total	Percent male	Percent female
Management, professional, and related occupations	34.9	49.7	50.3
Service occupations	16.3	43.2	56.8
Sales and office occupations	25.4	36.1	63.9
Natural resources, construction, and maintenance occupations	10.5	95.6	4.4
Production, transportation, and material moving occupations	12.9	77.0	23.0

Source: U.S. Department of Labor, Bureau of Labor Statistics, 2004.

The American economy is shifting from assembly-line production with blue-collar workers to a service economy with white-collar employees.

Occupational Structure

We read in Chapter 9 about America's dual labor market. The emphasis there was on institutionalized discrimination suffered by minorities. Actually, the implications of the dual labor market are much broader.

What is the nature of the two-tiered occupational structure? One tier—the **core tier**—contains dominant positions in large firms. Computer technology, pharmaceutical, and aerospace firms are prime examples. Between 30 and 40 percent of U.S. workers are in the core. The **peripheral tier** is composed of jobs in smaller firms that either are competing for business left over from core firms or are engaged in less profitable industries such as agriculture, textiles, and small-scale retail trade. Most U.S. workers—60 to 70 percent—are employed in the peripheral tier.

Historically, jobs in the core paid more, offered better benefits, and provided longer-term employment. This is not surprising, since the firms involved are large and highly profitable. Peripheral jobs are characterized by low pay, little or no benefits, and short-term employment (Weakliem 1990; Hudson 2000). These features follow from the weaker competitive position and the smaller size of the employing firms.

How are the core and peripheral tiers changing? During the last twenty years, the core industries scaled back, laying off experienced workers. As early as 1983, for example, a steel mill in Hibbing, Minnesota, that once employed 4,400 people had a payroll of only 650 ("Left Out" 1983). Since 1983, the Weirton Steel Company

cut its production capacity by 30 percent and laid off more than half of its workforce. In fact, since 1979, more than 43 million jobs have been eliminated in the United States. More than 570,000 job cuts were announced in the United States in 1998, more than half of which occurred in manufacturing plants (McNamee and Muller 1998; Riederer 1999). American corporations created more than 21,000 mass layoffs in 2001, affecting nearly 2.5 million workers (Marchese 2002). Of course, as these top-tier jobs disappear, peripheral jobs become a larger share of the total jobs.

Newer industries, such as those based on microchip technology, are beginning to replace manufacturing jobs in some areas. These industries, however, provide few jobs suited to the skills and backgrounds of laid-off manufacturing workers. Moreover, most jobs in high-tech industries pay minimal wages and offer few chances for promotion. Responsible positions with high pay are held by a very small proportion of high-tech employees.

What does this mean for U.S. workers? The U.S. economy is losing higher-paying jobs and gaining lower-paying jobs. This helps explain why, since the 1970s, most workers are losing economic ground. Whereas one American worker could support a family thirty years ago, the dual-employed married couple is the norm today.

This process, known as "downwaging," is expected to continue, propelled, in part, by the movement off-shore of higher wage blue-collar jobs and higher income white-collar jobs as a result of globalization policies such as NAFTA (North American Free Trade Agreement; Thurow 2003a). Of the ten occupations projected to

Employee Rights

The Supreme Court of the United States has historically granted employers a great deal of power over their employees. In 1878, a New York company posted a list of rules that told employees, among other things, "On the Sabbath, everyone is expected to be in the Lord's House" and "All employees are expected to be in bed by 10:00 p.m." At the turn of the nineteenth century, Henry Ford's automobile workers were carefully watched by management for signs of bad character. Many Ford Motor Company employees lost their jobs for smoking, drinking, or criticizing the firm.

Even today, some employee rights are curtailed at work. The Constitution, for example, protects free speech for all citizens. Employees, however, can be prevented from printing and distributing a critical newsletter to customers of their companies. Of recent concern is the right of employers to track workers' movements on the Internet and to read personal e-mails.

Today, a growing employee rights movement is pushing for greater political and legal protection on the job. Here is a partial list of the rights that many workers feel should be theirs today.

- The right to a job.
- The right to protection from arbitrary or sudden termination.
- The right to privacy of possessions and person in the workplace, including freedom from arbitrary searches, use of polygraphs, surreptitious surveillance, and intrusive psychological or medical testing.
- The right to a clean, healthy, and safe environment on the job, including freedom from undue stress, sexual harassment, cigarette smoke, and exposure to toxic substances.
- The right to be informed of records and information kept and to have access to personnel files.
- The right to freedom of action, association, and lifestyle when off duty.
- The right to freedom of conscience and to inform government or media about illegal or socially harmful corporate actions.
- The right to due process for grievances against the employer.

Many of these rights already exist; others need to be discussed with employers. There is one thing most employees and employers agree on, however. If employees take a balanced approach to pursuing their rights on the job, both individuals and organizations will benefit.

Think About It

1. Some observers believe that violations of employee rights contradict the rational-legal basis of organizational authority. Do you agree?
2. Discuss the above list of rights with some workers. Ask them if they know whether or not these rights exist in their workplaces. Are there any rights not on the list that they believe should be added?

---

expand between 1994 and 2005, five pay below the poverty level. Only two of the top ten shrinking jobs fail to meet the poverty threshold. Many sociologists believe that these trends of job loss and downwaging will threaten the American Dream (Newman 1993; Barlett and Steele 1996; see Doing Research).

How useful is the dual labor market perspective? The dual labor market perspective has been criticized for at least two reasons: because some of its assumptions have not been clarified and because it does not always provide accurate predictors of labor market experiences. Studies within each sector reveal considerable variation in experiences of individuals. Despite a need for refinement, however, dual labor market theory alerts us to shifts in industries and firms that will be offering jobs in the future (Hodson and Kaufman 1982; M. R. Kelley 1990; Sakamoto and Chen 1991; Hudson 2000).

Downsizing and Contingent Employment

Clearly, the American labor landscape is changing, due in part to corporate management decisions. Two interrelated steps by management—downsizing and contingent employment—deserve special attention (Mazaar 1999).

Why do corporations downsize? **Downsizing** refers to the reduction in a corporation's workforce designed to cut costs, increase profits, and enhance stock values. To justify downsizing, business leaders point to several factors: rising health-care costs, the need to become leaner and less hierarchically structured in an increasingly competitive global environment, and the necessity to eliminate levels of middle management that can be replaced by computers and information technology (Gleckman et al. 1993; Huber and Korn 1997).

To what extent is downsizing occurring? Downsizing became popular during the 1990s (Boroughs 1996; Sloan 1996; Belton 1999; Burke and Cooper 2000; Cohen and Thomas 2001). Since 1985, it is estimated that more than 8 million employees were downsized, half of whom held white-collar jobs. In 2001, corporate layoffs increased nearly 40 percent over the previous year (Marchese 2002). For a multitude of reasons, the momentum for downsizing and layoffs continues. In 2005 General Motors announced plans to close twelve factories and eliminate 30,000 factory jobs. Ford Motor Company followed with the intention of eliminating 25,000 jobs and closing fourteen manufacturing plants in North America. In fact, between 2000 and 2006, the auto industry cut 200,000 jobs.

How do employers approach downsizing? Part of the motivation for downsizing is top management's belief that their companies employ a surplus of people and that work can be done by fewer employees without reduction in efficiency and effectiveness. The desire to shrink the size of the workforce, however, is not the only incentive. Part of the impetus for downsizing is the belief that work done by permanent, full-time employees can be done in other, less expensive ways.

Profitable companies are increasingly using the services of **contingent employees**—individuals hired on a part-time, short-term, or contract basis. These contingent employees are hired to replace full-time employees ("The New World of Work" 1994; T. S. Moore 1996; Cappelli et al. 1997; Sullivan 2004). The contingent label is appropriate because their employment is contingent on daily, weekly, or seasonal company need. They may also be referred to as "marginal" workers because they are out of the traditional system of job and economic security. Some even refer to them as "disposable" workers (Kirkpatrick 1988; Castro 1993).

Contingent employees now make up more than one-third of the U.S. labor force (more than 40 million people). Temporary employment has increased more than 250 percent since 1982. Temporary employees earn about 60 percent of the amount earned by the average full-time worker and are not entitled to health insurance or retirement benefits. The contingent part of the labor force has traditionally included groups from the bottom of the occupational structure—minorities, women, youth, the elderly. Contingent employees are often recent immigrants who will work for less pay and lower benefits. Most contingent workers have been in low-level, low-paying clerical, retail, and service jobs. Increasingly, however, downsizing is affecting higher-level white-collar people—engineers, accountants, nurses, managers, and physicians (Heckscher 1995; Rifkin 1995; Altman 2002).

In addition to domestic employment, contingent employment also takes the form of job *offshoring*, or moving operations outside the United States. Initially, offshoring involved only factory workers. But white-collar jobs are no longer immune. Increasingly information technology, legal, accounting, customer service, and insurance jobs are leaving the U.S. for places like India and Mexico. According to a recent survey by the Pew Research Center, 77 percent of Americans believe that outsourcing jobs to other countries is hurting American workers. There are no signs that this concern is deterring corporate offshoring.

What will be the effects of downsizing and contingent employment? These trends may portend a fundamental change in the U.S. labor market (Harrison 1994; Heckscher 1995; Osterman et al. 2001; Baumol, Blinder, and Wolff 2005). For some it marks the dawn of a brave new workplace of individual entrepreneurialism (Kiechel 1993; Sherman 1993b). Others insist that it constitutes a fundamental change in the implied social contract between businesses and employees (K. M. Dudley 1997; Wallulis 1998; Fraser 2002). As evidence of a firm's need to be lean, adaptable, and flexible, Daniel Bell (1993), representative of the entrepreneurial view, cites the decline of the American steel industry. On the other hand, Robert Kuttner (1993a, 1993b) points out, these corporate actions are fracturing job security, with its underlying social contract. Similarly, Robert Reich (1993, 1997), former secretary of the U.S. Department of Labor under President Clinton, expresses concern that these actions will create greater polarization between those who control capital and those who do not. As such critics believe, the disposable workforce is the most important trend in business today; it is fundamentally changing the relationship between Americans and their employers. Downsizing through domestic contingent workers and offshoring, they contend, will threaten worker faith in a valued future relationship with a firm (Wolman and Colamosca 1997; Armour 2002; Greider 2002, 2004).

For corporations as well, the disadvantages of extensive downsizing may outweigh the advantages. Many companies that downsized did not increase their earnings and are now less competitive (Baumohl 1993). Stress among managers who make and carry out downsizing decisions is an increasing problem. A *Fortune* magazine article entitled "Burned-Out Bosses" concludes with this warning:

> *Constant restructuring has become a fact of business life in this era of change. Well and good, but companies that don't acknowledge the stress that survivors undergo and support those who are in danger of burning out may find that their glistening, reengineered enterprises end up being run by charred wrecks.* (L. Smith 1994:52)

Kathryn Marie Dudley—The End of the Line

Because she grew up near Chrysler's auto plant in Kenosha, Wisconsin, researcher Kathryn Marie Dudley had an interest in studying the cultural fallout from the plant's closing in 1988. She offers Kenosha as a representative example of the cultural consequences of the current restructuring of the U.S. economy. As Dudley indicates, the plant relocations, downsizings, and job eliminations undertaken by large corporations over the past few decades are seen as part of the shift from industrial to postindustrial society:

What was once a fundamental segment of the American economic structure—heavy industry and durable goods manufacturing—has now become a marginal part of the national portfolio. As this sector of the economy gives way to the new "knowledge industries," workers in this sector are being superseded as well. In America's new image of itself as a postindustrial society, individuals still employed in basic manufacturing industries look like global benchwarmers in the competitive markets of the modern world. (Dudley 1997:161)

Dudley's research is a case study of a large plant in a one-industry community experiencing the ongoing shrinking of core jobs in postindustrial society.

During the year following the shutdown, Dudley conducted in-depth interviews with autoworkers and with a wide variety of professionals in the Kenosha area. Interview questions were open-ended to give informants freedom to roam where their thoughts and feelings took them. Dudley's only restriction was that the interviews be geared to the cultural meaning of repercussions suffered by the community because of its declining employment base.

For Dudley, the demolition of the auto plant was a metaphor for the dismantled lifestyle of U.S. blue-collar workers in core manufacturing industries. These increasingly displaced blue-collar workers, contends Dudley, find themselves caught between two interpretations of success. On the one hand, middle-class professionals justify their place in society by reference to their educational credentials and "thinking" jobs. Blue-collar workers, on the other hand, legitimize their place in society by the high market value placed on their hard labor. One ex-autoworker, whom Dudley calls Al Tirpak, captures the idea beautifully:

We're worth fifteen dollars an hour because we're producing a product that can be sold on the market that'll produce that fifteen dollars an hour. . . . I don't know if you want to [base a person's value] strictly on education. You can send someone to school for twelve years and they can still be doing something that's socially undesirable and not very worthwhile for society. I don't know if they should get paid just because they had an education. In my mind, yuppie means young unproductive parasite. We're gonna have an awful lot of yuppies here in Kenosha that say they are doing something worthwhile when, really, they aren't. (Dudley 1997:169)

Due to the massive loss of high-paying factory jobs, Dudley contends that the blue-collar vision of success is coming to "the end of the line." They have lost their cultural niche in a postindustrial world where work is based on education and the application of knowledge.

Dudley documents the blue-collar workers' view of this new reality. From her extensive interviews, she constructs a portrait of their struggle to preserve their cultural traditions in a social world that is removing the type of employment on which these traditions were built. The penalty for not creating new cultural supports for a sense of social worth, Dudley concludes, will be life in a state of confusion with a sense of failure.

Evaluate the Research

1. What does Dudley mean when she says that the blue-collar workers caught in the wave of corporation restructuring are experiencing the shift from industrial to postindustrial society?
2. Do you think that Dudley's research methods are strong enough to support her conclusion?
3. Do you believe that Dudley can be objective in this study of her hometown? Explain.

According to a survey of more than 25,000 employees across the United States, there is a growing split in employees' attitudes toward their work and their employers (*Global Workforce Study* 2007). Although employees express higher job satisfaction, there are indicators of eroding trust in management. Workers seem to be losing faith in management's commitment to them. For instance, only about 40 percent of those surveyed believe that their company considers employees' interests. A sense of injustice is heightened by the growing compensation gap between top corporate executives and employees (Kuttner 2007; Hartmann 2008).

While median CEO annual compensation tripled to $6 million between 1993 and 2002, pay for all workers increased only a third (R. J. Samuelson 2003). Ronald J. Carty, then CEO of American Airlines, was forced in early 2003 to resign following disclosure of company plans to pay retention bonuses to executives and shield their pensions in the event of bankruptcy while asking employees to absorb up to 23 percent pay cuts (Goo 2003). Trust and loyalty are difficult to maintain when employees do not believe that their companies' policies treat them fairly (S. Greenhouse 2002).

We do not know the full consequences that downsizing and contingent employment will have on either companies or employees. What will happen to employee loyalty, performance, and product quality? Will trust between managers and employees decline even further (see Figure 13.6)? Will the survivors of downsizing suffer more illnesses due to greater workloads and higher stress? Will there be a deluge of lawsuits based on charges of wrongful discharge?

One conclusion seems clear: if expected economic benefits are to be realized, management will at the very least have to acknowledge the problems of survivors and initiate appropriate remedial responses (Sloan 1996; Cappelli et al. 1997). The needs are basic: maintaining honest and constant communication between top management and survivors of layoffs, setting goals with survivors, further empowering survivors on the job, exercising accessible and sensitive leadership, and creating a positive organizational culture for survivors (O'Neill and Lenn 1995; Snizek and Kestel 1999).

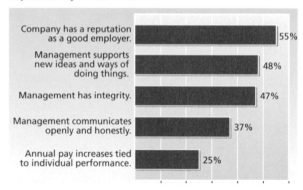

FIGURE 13.6

Evidence of Declining Trust in Management

According to this graph, employee trust in management is relatively low. Are any of these factors affecting trust more important to you than others?

Factor	%
Company has a reputation as a good employer.	55%
Management supports new ideas and ways of doing things.	48%
Management has integrity.	47%
Management communicates openly and honestly.	37%
Annual pay increases tied to individual performance.	25%

FEEDBACK

1. Which of the following is *not* one of the major labor-force trends discussed in the text?
   a. decrease in primary sector employment
   b. decline in agricultural employment
   c. increased employment of service workers
   d. increase in blue-collar jobs
   e. increase in white-collar proportion of the labor force
2. Most of the jobs that were added to the American economy over the past twenty years were in which of the following sectors?
   a. service and white-collar
   b. manufacturing
   c. mining
   d. skilled blue-collar
3. The _____ industrial sector contains large, profitable firms with dominant positions within major industries.
4. The _____ industrial sector contains small, less profitable firms operating in more competitive industries.
5. _____ is the decision on the part of top management to reduce its workforce to cut costs, increase profits, and enhance stock values.

Answers: 1. d; 2. a; 3. core; 4. peripheral; 5. Downsizing

SUMMARY

1. The economic and political institutions are closely related. The economy is the institution designed for the production and distribution of goods and services, and it is in the political institution where power is obtained and exercised.

2. Power is the ability to impose one's will on others whether or not they wish to comply. Authority is power accepted as legitimate by those subjected to it. Max Weber identified three types of authority: charismatic authority, which is too unstable; traditional authority, which is too outdated to serve as the primary basis of power for modern nation-states; and rational-legal authority, which is the authority of most nation-states today. The power of most modern government officials, then, is limited by legal rules and regulations.

3. A nation-state is the political entity that holds authority over a specified territory; a government is the political structure that rules a nation. The nation-state is relatively new in human history, emerging with the rise of agriculture and industrial production. According to functionalism, the nation-state developed to maintain social order. From the conflict perspective, nation-states exist primarily to promote the interests of the societal elite.

4. Democracy and totalitarianism are the two polar types of political systems. In modern societies, democracies are representative with minimal citizen involvement in political affairs. Totalitarian political systems have absolute rulers who control all aspects of social life. Between these two polar types lies authoritarianism. This third type of political system involves nonelected rulers who possess absolute control but frequently permit some individual freedom.

5. Voting seems to have limitations as an effective means for the nonelite to influence political decision making. This is partly because America does not have a high voter participation rate and partly because the most disadvantaged tend to be nonvoters.

6. The two major models of power distribution are pluralism and elitism. Pluralists, whose view is associated with functionalism, depict power in the United States as widely distributed among diverse special interest groups. Interest groups appear to be on the rise as the public becomes more politically sophisticated and as the influence of political parties declines. The growing importance of political action committees (PACs) reflects the same social conditions. Some powerful interest groups, notably those PACs associated with business interests, thwart public opinion by using their power for the attainment of narrow, selfish goals. Still, interest groups provide one

of the few mechanisms for the representation of non-elite segments of a society. Some of these interest groups have successfully challenged the political and economic establishment.

7. Advocates of the conflict perspective contend that American society is controlled by a unified and enduring elite. This is a "power elite" comprised of a unified coalition of top military, corporate, and government leaders. Although some evidence documents the existence of a national power elite, its presence cannot be definitively asserted.

8. Economies differ in their organization and underlying assumptions. Capitalist economies are based on private property and the pursuit of profit without government interference. In socialist economies, the means of production are owned by the people, and government has an active role in planning and controlling the economy. There is movement toward capitalism in socialist countries. Capitalist and socialist ideologies appear to be becoming less of a barrier in the relationships among nations.

9. Individualistic capitalism and communitarian capitalism are the two basic variants of capitalism. Self-interest, the free market, profit maximization, and highest return possible on stockholder investment are foundational principles for individualistic capitalism. In communitarian capitalism, the emphasis is on the interests of employees, customers, and society.

10. The rise of the corporate economy has intensified the link between political and economic institutions. Some critics contend that the political institution in the United States is now controlled by vast corporations whose economic assets are owned by a relatively small segment of the population.

11. Corporations are rapidly increasing in size, number, and power. Their influence is felt domestically and internationally. Corporations affect domestic political decision making and also influence the political and economic institutions of countries around the world.

12. Workers in the contemporary economy face a changing labor force composition. The occupational structure is changing as traditional industrial jobs are replaced with white-collar and service jobs. More American corporations are downsizing and replacing full-time employees with contingent, or temporary, workers. Granted, the full implications of these changes for workers and for the future of American society are not yet known, but evidence is beginning to establish some negative consequences.

LEARNING OBJECTIVES REVIEW

- Distinguish among power, coercion, and authority, and identify three basic forms of authority.
- Discuss the nation-state, comparing and contrasting the functionalist and conflict approaches to this form of political authority.

- Identify the major differences among democracy, totalitarianism, and authoritarianism.
- Differentiate the views of functionalism and conflict theory on the distribution of power in America.
- Describe the major characteristics of capitalism, socialism, and mixed economic systems.
- Discuss the effects of the modern corporation on American society.
- Describe America's changing workforce composition and occupational structure.
- Discuss corporate downsizing and its consequences.

CONCEPT REVIEW

Match the following concepts with the definitions listed below them.

____ a. individualistic capitalism	____ g. monopoly	____ m. rational-legal authority
____ b. political action committees	____ h. downsizing	____ n. bourgeoisie
____ c. oligopoly	____ i. communitarian capitalism	____ o. elitism
____ d. government	____ j. authoritarianism	____ p. interest group
____ e. nation-state	____ k. traditional authority	____ q. corporation
____ f. democracy	____ l. power elite	

1. the type of political system controlled by nonelected rulers who generally permit some degree of individual freedom
2. members of a society who own productive property
3. the type of political system in which elected officials are held responsible for fulfilling the goals of the majority of the electorate
4. the theory of power distribution that sees society in the control of a few individuals and organizations
5. a group organized to achieve some specific, shared goals by influencing political decision making
6. a situation in which a single company controls a market
7. the political entity that holds authority over a specified territory
8. organizations established by interest groups for the purpose of raising and distributing funds to selected political candidates

9. a unified coalition of top military, corporate, and government leaders
10. authority based on rules and procedures associated with political offices
11. the political structure that rules a nation
12. legitimate power rooted in custom
13. the type of capitalism that emphasizes the interests of employees, customers, and society
14. the decision of top management to reduce its workforce to cut costs, increase profits, and enhance stock values
15. the type of capitalism founded on the principles of self-interest, the free market, profit maximization, and the highest return possible on stockholder investment
16. a situation in which a combination of companies controls a market
17. an organization owned by shareholders who have limited liability and limited control over organizational affairs

CRITICAL-THINKING QUESTIONS

1. Like all organizations, universities are based on some form of authority. Discuss each of the three types of authority, and show which type is most basic to the university structure. Be careful to show why each type of authority is applicable or inapplicable to the university's organizational structure.

   _____

   _____

2. Functionalism and conflict theory each have a unique perspective on the nature and purpose of the nation-state. Do you think that one of these perspectives more accurately describes American society than the other? Why or why not?

   _____

   _____

3. Define the concept of the power elite. Is American society best characterized as a pluralist society or a society controlled by a power elite? Defend your position.

   _____

   _____

4. Distinguish between the two global cultures of capitalism. Do you envision a convergence of these two approaches to capitalism within the rapidly developing global marketplace? Support your viewpoint.

   _____

   _____

5. American corporations have in recent years turned to the practices of downsizing and contingent employment. State the reasons for this managerial strategy, and discuss whether or not it will promote the organizational goals it is designed to reach.

   _____

   _____

MULTIPLE-CHOICE QUESTIONS

1. Max Weber defined _____ as the ability to control the behavior of others, even against their will.
   a. authority
   b. charisma
   c. power
   d. elitism
   e. influence

2. _____ is power accepted as legitimate by those subjected to it.
   a. Pluralism
   b. Authority
   c. Elitism
   d. Influence
   e. Coercion

3. The political structure that rules people in a specific territory is known as a
   a. Leviathan.
   b. nation-state.
   c. democracy.
   d. political party.
   e. government.

4. According to the conflict perspective, nation-states exist primarily to
   a. serve the interests of a society's elite.
   b. eliminate tensions between social classes.
   c. solve the Hobbesian problem of order.
   d. perpetuate competition for new economic markets.
   e. establish the legitimacy of charismatic authority.

5. What two types of political systems exist in nation-states?
   a. democracy and authoritarianism
   b. totalitarianism and authoritarianism
   c. Republican and Democratic
   d. communism and Republican
   e. democracy and totalitarianism

6. According to _____, a community or society is controlled from the top by a few individuals or organizations.
   a. symbolic interactionism
   b. elitism
   c. communism
   d. functionalism
   e. pluralism

7. Decision making resulting from competition, bargaining, and compromise among diverse special interest groups is based on the _____ model.
   a. elitist
   b. laissez faire
   c. conflict
   d. pluralist
   e. functional

8. According to C. Wright Mills, power in the United States is controlled by
   a. the people through a democratic process.
   b. labor unions that are able to exert covert influence on management and the general public.
   c. top military, corporate, and government leaders.
   d. *Fortune* 100 companies.
   e. competing interest groups.

9. A system founded on the belief that the means of production should be owned by the people as a whole and that government should have an active role in planning and controlling the economy is known as
   a. socialism.
   b. capitalism.
   c. communism.
   d. industrialism.
   e. pluralism.

10. As capitalist systems mature,
   a. consumer goods come under the influence of the law of diminishing returns.
   b. the willingness to take economic risks increases.
   c. the desire to turn to a mixed economic system increases.
   d. many industries are dominated by a few large firms.
   e. many small businesses are able to compete with larger firms.

11. An organizational form in which shareholders have limited liability and limited control over organizational affairs is known as a(n)
   a. corporation.
   b. cooperative.
   c. unilateral organization.
   d. voluntary organization.
   e. employees organization.

12. Which of the following statements is *false*?
   a. Communist countries have not been slow to move to some sort of market system after the fall of the former Soviet Union.
   b. Economic change in recent years has not been limited to Soviet-bloc economies.
   c. Cuba has now adopted some degree of capitalism.
   d. Capitalism now threatens to triumph over communism.
   e. China, with the repossession of Hong Kong, is reverting to a planned economy.

13. According to the text, the workforce hallmark today is
   a. a decline in knowledge-based workers.
   b. a reduction in white-collar workers.
   c. an increase in blue-collar workers.
   d. the management and delivery of services.
   e. the resurgence of agricultural employment due to corporate farming.

14. Which of the following generalizations regarding corporate downsizing in the United States now appears to be *true*?
    a. Extensive downsizing is a fading corporate strategy.
    b. Contingent employment has passed its peak.
    c. The disadvantages of extensive downsizing may outweigh the advantages.
    d. Management no longer has to be concerned about problems associated with downsizing and contingent employment because employees have become accustomed to these practices.
    e. Since 1990, job loss has been receding from the pinnacle of employee fears.

FEEDBACK REVIEW

True-False

1. According to Jean-Jacques Rousseau, it is primarily the privileged classes who benefit from the existence of the state. T or F?

2. The operation of PACs in America today invalidates the pluralist model. T or F?

Fill in the Blank

3. According to Karl Marx, those in charge of the state in capitalist societies make decisions that serve the interests of the owners of the means of production. He used the term _____ to describe acceptance of a system that works against one's own interests.

4. _____ refers to the type of political system controlled by nonelected rulers who generally permit some degree of individual freedom.

5. Firms with operating facilities in several different countries are called _____.

6. The _____ industrial sector contains large, profitable firms with dominant positions within major industries.

Multiple Choice

7. **Most political action committees (PACs) are now associated with**
    a. labor unions.
    b. citizen groups.
    c. political parties.
    d. corporations and other businesses.

8. **Which of the following is not one of the reasons behind corporate political influence?**
    a. Corporate managers and directors have informal and organizational ties that lead them to be consulted about matters of political policy.
    b. Many corporate officials have reward and coercive power over elected government officials.
    c. Many elected officials are also top corporate officials.
    d. Interlocking directorates exist.
    e. Conglomerates are present.

9. **Which of the following is *not* one of the major labor force trends discussed in the text?**
    a. decrease in primary sector employment
    b. decline in agricultural employment
    c. increased employment of service workers
    d. increase in blue-collar jobs
    e. increase in white-collar proportion of the labor force

Matching

10. Several characteristics of capitalist economic systems are listed here. Indicate in the space beside each characteristic whether it best describes individualistic capitalism (IC) or communitarian capitalism (CC).
    ____ a. government and business cooperation
    ____ b. high job mobility
    ____ c. profit maximization
    ____ d. more value placed on the interests of employees and consumers
    ____ e. self-interest paramount
    ____ f. long-range economic planning

GRAPHIC REVIEW

Figure 13.5 displays changes in the composition of the U.S. labor force since 1900.

1. Which occupational category had the steepest decline? Describe this change between 1900 and 2000.

   _____
   _____
   _____

2. In which occupational category has growth been the most rapid? Describe this change.

   _____
   _____
   _____

3. Discuss the most important implications regarding employment in the United States that can be suggested from these data.

   _____
   _____
   _____

ANSWER KEY

Concept Review	Multiple Choice	Feedback Review
a. 16	1. c	1. T
b. 8	2. b	2. F
c. 17	3. e	3. false consciousness
d. 12	4. a	4. Authoritarianism
e. 7	5. e	5. multinational corporations
f. 3	6. b	6. core
g. 6	7. d	7. d
h. 15	8. c	8. c
i. 14	9. a	9. d
j. 1	10. d	10. a. CC
k. 13	11. a	b. IC
l. 10	12. e	c. IC
m. 11	13. d	d. CC
n. 9	14. c	e. IC
o. 2		f. CC
p. 4		
q. 5		
r. 18		

Religion

Rolmat/Dreamstime, LLC/ARTEKI/Used under license from Shutterstock;Mikhail Nekrasov/Used under license from Shutterstock

LEARNING OBJECTIVES

- Explain the sociological meaning of religion.
- Demonstrate the different views of religion taken by functionalists and conflict theorists.
- Distinguish among the basic types of religious organization.
- Discuss the meaning and nature of religiosity.

- Define secularization, and describe its relationship to religiosity in the United States.
- Differentiate between civil and invisible religion in America.
- Describe the current resurgence of religious fundamentalism in the United States.

- Identify a wide variety of religious movements in America.
- Outline the relationship of baby boomers to religion.

USING THE SOCIOLOGICAL IMAGINATION

Is secularization destroying religion in the United States? Many self-proclaimed prognosticators fear the decline of religion in the United States. On the contrary, evidence reveals that America, compared with other industrialized nations, remains a fairly religious nation.

Religion in America, then, remains a viable topic to explore. Before doing so, we will view religion within the context of sociology, define religion as an institution, examine religion from the functionalist and conflict perspectives, explore religious organization, and identify the ways people express their religious beliefs.

Religion, Science, and Sociology

Religion and Science

Can religion and science coexist? Because religion involves matters beyond human observation and because science is all about observation, these two institutions can potentially conflict. It is not that scientists cannot also be religious. Rather, scientists separate their professional work (based on reason and observation) from their religious life (based on faith and the unobservable).

Science and religion do conflict, however, when people try to mix the two. Some American religious fundamentalists do just that when they argue that *creationism* (explanation of the natural world through the Bible) is as scientifically based as evolutionary biology. The evolution versus creation debate has intensified in several states (Nelson 2002).

Will science replace religion? Science need not, and doubtless will not, eradicate religion. Because all societies have some form of religion, we can assume that humans seek answers to some fundamental questions about existence: Is there a higher purpose to life? Why does life unfold for us in particular ways? Is there an afterlife? Answers to such questions are simply beyond science. We can therefore expect religion to endure as long as human life exists. Furthermore, it is likely that science and religion will continue on their historical path of separation and coexistence.

Religion and Sociology

The study of religion attracted the attention of sociologists in the nineteenth century. These early sociologists were interested in the place religion holds in social life. Auguste Comte, the man who coined the term *sociology*, saw sociology as a new religion (a religion of science) and sociologists as high priests. Other pioneering sociologists, such as Emile Durkheim, Max Weber, and Karl Marx, viewed the practice of religion as a key to some of the mysteries of social life. Religion, they believed, could help explain the presence of social order and the cultural variation of societies. Despite this early scholarly interest, the sociological study of religion disappeared during the first half of the twentieth century. Only after World War II did social scientists again become interested in the scientific study of religion (Sharot 2001).

Sociologists of religion neither explain nor endorse particular religions but rather are interested in all sociological aspects of religion. This interest must begin with a sociological definition of religion.

What is the sociological meaning of religion? **Religion** is a unified system of beliefs and practices relating to sacred things. This definition comes from Emile Durkheim, whose work was based on studies of the Australian aborigines in the late nineteenth century. According to Durkheim, every society distinguishes between the **sacred**—entities that are set apart and given a special meaning that transcends immediate human existence—and the **profane**—nonsacred aspects of life. (Profane in this context does not mean unholy or defiled, but simply commonplace and not involving the supernatural.)

Sacred things take on a public character that makes them appear important in themselves; the profane do not. Temple Mount in Jerusalem is a particularly illuminating example because its sacred meaning is not the same for the three religions that revere it. Jews hold sacred the Temple Mount "Wailing Wall" because it is the remnant of the First and Second Temples destroyed by the Romans in 70 A.D. Christians consider it holy because it was a site of Jesus' preaching. For Muslims, it is sacred as a place from which Mohammed ascended to heaven. Because Hindus do not share any of these religious beliefs, Temple Mount is part of their nonsacred, or profane, world.

Interestingly, some nonreligious aspects of culture can assume something of a sacred character. Babe Ruth's bat illustrates the difference between the sacred and the profane:

When Babe Ruth was a living idol to baseball fans, the bat he used to slug his home runs was definitely a profane object. It was Ruth's personal instrument and had little social value in itself. Today, however, one of Ruth's bats is

Buddhas, like this one at a Mongolian monastery, are sacred objects in the Far East and Southeast Asia, and wherever Buddhists live. What makes an object sacred?

Christine/Dreamstime LLC

enshrined in the Baseball Hall of Fame. It is no longer used by anyone. It stands, rather, as an object which in itself represents the values, sentiments, power, and beliefs of all members of the baseball community. What was formerly a profane object is now in the process of gaining some of the qualities of a sacred object. (Cuzzort and King 1976:27)

Babe Ruth's bat illustrates two points about the sociological study of religion. First, a profane object can become sacred, and vice versa. Second, sociologists can deal with religion without becoming involved in theological issues. By focusing on the cultural and social aspects of religion, sociologists avoid questions about the ultimate validity of any particular religion. This point needs elaboration.

How do sociologists avoid theological issues? Religion involves a *transcendent reality*—a set of meanings attached to a world beyond human observation (Lilla 2007). Because sociologists must deal with this dimension of unobservability, their task is particularly difficult. For example, in this realm, sociologists face unique problems: Can an empirical science actually study the

unobservable? Can scientific objectivity be maintained while investigating something as value laden as religion? Can sociologists retain religious faith after scientifically probing religion? Each of these questions deserves a brief answer.

Obviously, sociologists cannot study the unobservable. As part of an empirical science, sociologists who investigate religion must formulate theories and hypotheses on aspects of religion that can be measured and observed. Consequently, sociologists avoid the strictly spiritual side of religion and focus on social aspects of religion that can be measured and observed.

Objectivity, as we saw in Chapter 2, is a goal for which scientists strive. Because there are so many emotional overtones in the study of religion, researchers must be especially vigilant of their own biases. Thus, sociologists investigating religion do not attempt to determine the truth or falseness of any religion. Because definitions of good or bad, truth or falsity are avoided, theological issues lie beyond the sociology of religion. Sociologists can point out that conceptions of God, religious beliefs, and religious practices vary from culture to culture without attempting to make value judgments of any of

1. A _____ is a unified system of beliefs and practices relative to sacred things.
2. _____ things are set apart and given special meaning that transcends immediate existence.
3. Sociologists who study religion are normally forced to abandon their religious beliefs and convictions. T or F?

Answers: 1. religion; 2. Sacred; 3. F

the beliefs and practices involved. Because the validity of any religion is founded in faith, sociologists restrict themselves to the social manifestations of religion.

For this reason, it is possible for sociologists to retain a personal religious faith while investigating the social dimensions of religion. In fact, sociologists, like other people, vary in their religious orientations. They have varying degrees of religious commitment, exhibit myriad religious behaviors, and belong to quite different religious organizations (see Figure 14.1 and Table 14.1).

Theoretical Perspectives

The functionalist and conflict perspectives have different views of the sociological dimensions of religion. We begin with functionalism, which examines the role of religion in relation to the functions it serves for society.

Functionalism and Religion

We know from archaeological discoveries and anthropological observations that religion has existed in some form in virtually all societies (Nielsen et al. 1993). The earliest evidence of religion has been traced as far back as 50,000 B.C., when humans had already begun to bury their dead—a practice that suggests belief in some existence after death. In early Rome, there were specific gods for objects and events—a god of trees, a god of money, a goddess of fever. According to the ancient Hebrew religious texts (both the Book of Genesis and the Book of Leviticus), pigs were unclean animals whose pollution would spread to all who touched or tasted them. To the tribes of New Guinea and the Pacific Melanesian Islands, on the other hand, pigs are holy creatures worthy of ancestral sacrifice (M. Harris 1974).

Emile Durkheim, the first sociologist to examine religion scientifically, wondered why it is that all soci-

FIGURE 14.1

Division of World Population by Religions*

This graph compares the number of all religious believers belonging to a particular religion to the total estimated world population.

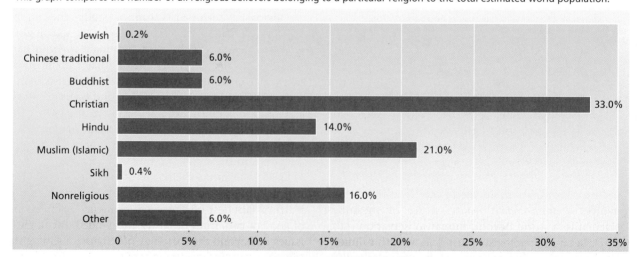

Religion	Percent
Jewish	0.2%
Chinese traditional	6.0%
Buddhist	6.0%
Christian	33.0%
Hindu	14.0%
Muslim (Islamic)	21.0%
Sikh	0.4%
Nonreligious	16.0%
Other	6.0%

Percent of world population

*Figures shown are estimates and do not total to 100 due to rounding errors.

TABLE 14.1

Major World Religions

This table summarizes characteristics and beliefs of the major world religions being widely practiced today.

Religion	Origination	Key Figure	Beliefs	Main Geographic Areas	Number of Followers
Hinduism	Before 2000 B.C.	Unknown	• Of many gods, Brahma is the creator of the universe. Life is determined by the law of karma (the spiritual force generated by one's own actions, which determines one's next reincarnation).	India	837,262,000
Judaism	Before 1200 B.C.	Abraham	• The one true God has established a covenant with the people of Israel, who are called to lives of justice, mercy, and obedience to God.	Israel, eastern Europe, USA	14,551,000
Buddhism	About 500 B.C.	Siddhartha Gautama	• The existence of God is not assumed. Through adherence to the Eightfold Path (correct thought and behavior), one can escape from desire and suffering and achieve nirvana (a state of bliss reached through extreme denial of the self).	Far East, Southeast Asia	372,974,000
Confucianism	About 500 B.C.	Confucius	• The Analects (sayings of Confucius) stress moral conduct and virtuous human relationships.	China	6,425,000
Christianity	About A.D. 1	Jesus Christ	• Jesus is the Son of the one true God. Through God's grace and profession of faith, people have eternal life with God.	Europe, North America, South America	2,069,883,000
Islam	About A.D. 600	Muhammad	• Muhammad received the Koran (holy scriptures) from the one true God. Believers go to an eternal Garden of Eden.	Africa, Middle East, Southeast Asia	1,254,222,000

eties have some form of religion. In one of his books, *The Elementary Forms of Religious Life* (1995; originally published in 1915), Durkheim offered a sociological explanation for the ubiquity of religion, an explanation rooted in the function religion performs for society. The essential function of religion, he believed, was to provide through sacred symbols a mirror for members of society to see themselves. Through religious rituals, people worship their society and thereby remind themselves of their shared past and future existence. Following Durkheim's lead, sociologists identify the following social functions of religion: (1) to legitimate social arrangements, (2) to encourage a sense of social unity, (3) to provide a sense of meaning, and (4) to promote a sense of belonging.

How does religion legitimate social arrangements? *Legitimation* justifies and explains the status quo. It explains why a society is—and should be—the way it is. Legitimations tell us why some people have power and others do not, and why some are rich and others poor. Many such legitimations are based on religion. According to Durkheim, legitimation is the central function of religion. For him, ideas of the sacred within a society represent the society itself. When people think of their souls, Durkheim believed, they are really thinking of the social element within themselves—an element that is superior to them and will live on after them.

If, as Durkheim believed, religion is a representation of society, its power to provide legitimation for society is considerable. Through its system of beliefs, a religion explains the nature of social life, the existence of evil, and the imperfect nature of the world. Religion has been used to justify slavery in America (many ministers in the nineteenth-century South preached that African Americans were descendants of Ham, a son of Noah whose ancestors were condemned to slavery) and male superiority (Eve was created from one of Adam's ribs, and it was her encouragement that led Adam to eat the forbidden apple). One of the most important and widely agreed-on social functions of religion, then, is the legitimation or justification of social arrangements.

How does religion promote social unity? Religion, according to Durkheim, is a glue that holds society together. Without religion, society would be chaotic. As Cuzzort and King have stated, Durkheim "provided the greatest justification for religious doctrine ever granted by a social scientist when he claimed that all societies must have religious commitments. Without religious dedication there is no social order" (Cuzzort and King 1976:43). In some cases, however—Northern Ireland being one example—religion provokes societies to fragment, even to the point of civil war. Thus, whereas it is accurate to say that religion is usually a source of social unity, it can also divide a society (Lilla 2007).

How does religion provide a sense of meaning? Religion not only legitimates existing social arrangements and encourages social unity but also provides individuals meaning beyond day-to-day life. People mark important events in life—birth, sexual maturity, marriage, death—with religious ceremonies and explain such events in religious terms. Religion gives believers cosmic significance and gives eternal significance to a short and uncertain earthly existence.

How does religion promote a sense of belonging? Religious organizations provide opportunities for people to share significant commonalties—ideas, a way of life, an ethnic background. Religion supplies a kind of group identity. People usually join religious organizations freely and feel a degree of influence within these organizations. For many people in modern society, membership in a religious organization provides a sense of community. This counteracts depersonalization, powerlessness, and rootlessness.

Of course, religious organizations are not unique in providing a sense of belonging. Fraternal groups, clubs, and other voluntary associations do the same thing. Because of religion's other functions, however, the sense of belonging to a religious organization can be particularly appealing.

Conflict Theory and Religion

As noted, Durkheim believed that religion supports societies by providing legitimation for social arrangements. Durkheim's ideas are still being debated, however, and other sociological theorists have identified quite different relationships between religion and society.

Georg Friedrich Hegel, an eighteenth-century German philosopher, believed that ideas and beliefs, including religious ones, determine the nature of social life. Ideas and beliefs, Hegel thought, cause social life to be structured in certain ways. Hegel's position was very popular within the religious and intellectual establishment of his day, but it was not without critics. By the early nineteenth century, a group of critics known as the young Hegelians had begun to develop a radical makeover of Hegel's position. They claimed to set Hegel "on his head." Social structure determines ideas and beliefs, not the other way around. More specifically, the economic structure of a society determines the nature of the society's religion. Karl Marx, deciding the group had gone too far the other way, developed a view encompassing both ideas and structure.

How did Marx view religion? Marx believed that once people create a unified system of sacred beliefs and practices, they act as if it were something beyond their control. They become "alienated" from the religious system they set up. Although humans have the power to change (or, better yet, in Marx's mind, to abandon) the religion they have created, they don't do so because they see it as a binding force to which they must conform.

Religion, Marx wrote, is used by the ruling class to justify its economic, political, and social advantage over the oppressed. The elite attribute poverty, degradation, and misery to God's will. So to eliminate inequalities and injustices is to tamper with God's plan. Change is sacrilegious to the elite. Many illustrations of ties between the elite and religion exist; in fact, a close relationship between the dominant classes and the church has been the rule in human history. The Catholicism of medieval Europe, for example, was so closely identified with the upper classes that the lower and lower middle classes welcomed the Protestant Reformation. In his classic study *Millhands and Preachers*, Liston Pope (1942) documented a case of religious control by the managers of a mill in a North Carolina community (Gastonia):

The churches were inextricably bound to mill management by their finances if not by their ideology. Organized religion was a key instrument in creating a manipulable labor force and in resisting the union's effort to break with the exploitative status quo. Ministers not only justified management practices; they were mute in reaction to the violent recriminations exerted by the community against the union organizers. (Pope 1942:xx)

We know that religion is an important part of almost all societies because of the religious artifacts that have been left behind. Consider this painting of the Pope being carried on his papal throne around 1600.

Hutton Archive/Getty Images

In a restudy of Gastonia, churches were still providing latent support for long-standing economic relations in the community.

In addition, Marx contended, the oppressed employ religion to explain and justify their deplorable existence and to provide hope for a better life after death. In this way, they can find self-fulfillment in their oppression. Marx thought of religion as a narcotic for the oppressed:

> *Religious suffering is at the same time an expression of real suffering and a protest against real suffering. Religion is the sign of the oppressed creature, the sentiment of a heartless world, and the soul of soulless condition. It is the opium of the people.* (quoted in Bottomore 1963:43–44; originally published by Karl Marx in 1844)

If people abandoned the religion of their oppressors, Marx alleged, they would turn from the comforts offered by an afterlife and instead concentrate on the present. This would lead them to reject their oppressed economic condition and decide to do something about it. Nonetheless, Marx recognized that it would be difficult for the oppressed to view their religion as a manipulation of the elite. A contemporary scholar of ideology highlights the latter point:

> *The most efficient oppressor is the one who persuades his underlings to love, desire and identify with his power, and any practice of political emancipation thus involves that most difficult of all forms of liberation, freeing ourselves from ourselves.* (Eagleton 1994:xiii–xiv)

Capitalism and the Protestant Ethic

How did Weber view the place of religion in social life? Whereas Marx contended that religion retards social change because the elite wish to maintain the status quo, Max Weber (1958; originally published in 1904) supposed that religion sometimes encourages social change. Weber looked for a case that would demonstrate his point and found it in the relationship between the ideas of early Protestantism and the rise of capitalism.

Weber began with the fact that capitalism was widespread only in northwestern Europe and America. Why had this happened? Why had capitalism not emerged as the basic economic system in other parts of the world? Through an extensive study of various religions of the world, Weber (1951, 1952, 1958, 1964) found a possible answer in the basic compatibility between what he termed the spirit of capitalism and the Protestant ethic.

What is the relationship between the spirit of capitalism and the Protestant ethic? Capitalism, Weber noted, involved a radical redefinition of work. With capitalism, work became a moral obligation rather than a mere necessity. Capitalism also required the reinvestment of profits, if businesses were to grow. Because capitalism involved the reinvestment of capital, investment for the future was more important than immediate consumption. All of this Weber called the **spirit of capitalism**.

Although most major religions did not define hard work as an obligation or demand the reinvestment of capital for further profits (rather than for immediate enjoyment), some Protestant sects did. Here, then, embodied in Protestantism, was a cluster of values, norms, beliefs, and attitudes that favored the emergence of modern capitalism. Weber referred to this cluster of values and attitudes stressing hard work, thrift, and discipline as the **Protestant ethic**.

What is the nature of the Protestant ethic? Weber found the roots of the Protestant ethic in the seventeenth-century Puritan theology of Calvinism. Of particular importance was the Calvinist belief that every person's eternal fate was unalterably predetermined by God, at birth—predestination. Society, the Calvinists believed, was divided into the few who were elected to salvation and the many who were predestined to spend eternity in hell. Individuals could do nothing to alter their fate.

Calvinism taught its followers to look for signs of God's blessing because, according to Calvin, God identifies his elect by rewarding them in this world. This was translated into the belief that the more successful people were in this life, the more sure they were of being a member of God's select few. In addition, consumption beyond necessity was considered sinful; those who engaged in self-pleasure were agents of the devil. Finally, Calvinists believed that there was an underlying purpose of life: glorification of God on earth through one's occupational calling. Because everyone's material rewards were actually God's and the purpose of life was to glorify God, profits should be multiplied (through reinvestment) rather than used in the pursuit of personal pleasures.

The theology of sixteenth-century theologian John Calvin formed the basis for the Protestant ethic.

Other Protestant religious beliefs may have contributed to the development of capitalism. For example, Protestants endorsed the idea that salvation came from unselfish good works here on earth. Protestants were encouraged to do more and more work while receiving less and less of the material rewards.

Does the Protestant ethic still exist? The debate over Weber's Protestant ethic continues (MacKinnon 1988a, 1988b; Zaret 1992; Lehrmann and Roth 1993; R. F. Hamilton 1996; J. Cohen 2002). However, any link between Protestantism and economic behavior that may have existed 200 years ago appears to be rather weak today. Surveys show that Protestants do not value success and hard work any more than Catholics do. Moreover, Jews have higher incomes and are more successful occupationally than Protestants. If the Protestant ethic still exists, it does not appear to be the exclusive property of Protestants. As a matter of fact, it may not be the property of any religious group.

Did Protestantism precipitate the development of capitalism? Weber did not contend that capitalism developed *because* of Protestantism. (Capitalism had actually flourished among Catholics in some parts of the world before the Protestant Reformation.) He did, however, suggest that early Protestantism seems to have provided a social environment *conducive* to the emergence of this economic system. Weber's emphasis on the influence of ideas on social structure remains important as a counterpoint to Marx's accent on the effect of social structure (economy) on ideas. He recognized that capitalism could exist apart from Protestant dogma. Weber himself pointed out that capitalism helped destroy many of the foundations of Protestantism. He was well aware that from the seventeenth century on, religion gradually lost its hold over capitalist society. Jere Cohen (2002) concludes that Weber's exaggeration of the effects of Protestantism on the development of modern capitalism should not cause us to overlook the real economic impact of Protestantism.

Gender and Religion

Conflict theorists view religion as a primary source of division and strife within societies and between societies. They point out that most wars are grounded in religious conflict. The Arab–Israeli–United States conflict is only one of the most current illustrations. War, however, is only one manifestation of religious-based conflict. Less visible forms of conflict are generated by the efforts of religious elites to maintain control over an entire society or subgroups within a society. An important instance is male domination of women that exists in most of the world's major religions (McGuire 2001).

For starters, feminists see the assumed maleness of the Supreme Being, in all major religions, as blatant sexism.

WORLDVIEW 14.1

Gender Inequality in Religion

Women have been fighting for equal rights in all aspects of society—religion as well as government and business. In some religions women have equal status within their orders. Other religions see feminism as a "Western" issue and irrelevant to their faiths. This map shows how major denominations in each country view the status of women.

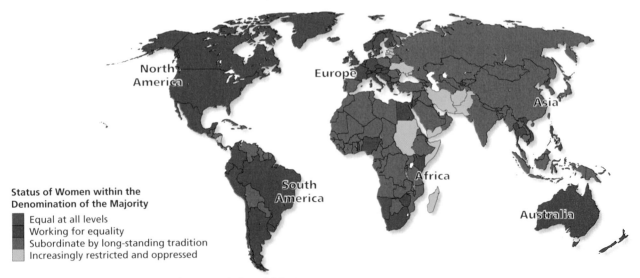

Status of Women within the Denomination of the Majority

- Equal at all levels
- Working for equality
- Subordinate by long-standing tradition
- Increasingly restricted and oppressed

Adapted from *The State of Religion Atlas,* New York: Simon & Schuster.

Interpret the Map

1. Do you see any patterns of inequality in women's rights in religion? Explain.
2. Where in the world would a woman be most likely to head an entire religious organization? Support your answer. Do some research to see if this has occurred. If it has not, explain why.
3. How does the United States compare with the Scandinavian countries in terms of gender equality? Why do you think this difference exists?
4. How would you explain the mixed status of women in India?

Female inferiority and subordination, they argue, is prominent in all sacred texts of the major religions. Orthodox Jewish males thank God daily that they were not born female. In the Muslim Koran, the superiority of men over women comes from God-given gender differences. The account of creation in the Christian Bible blames Eve for the eating of forbidden fruit and announces an eternal punishment for women: "In sorrow thou shalt bring forth children; and thy desire shall be to thy husband, and he shall rule over you" (Genesis 3:16).

Despite the plethora of positive statements about women attributed to Jesus in the New Testament, the negative view in Genesis has historically dominated the Christian view of gender. And even after the many significant contributions women made to American religious denominations over the centuries, gender inequalities in modern religious practices and organization continue to reflect the maintenance of male patriarchy (Benowitz and Vowels 1998).

Since the early 1980s, feminists have worked against male patriarchy in Christian religious organizations. Their results are meager. This is partly because many major religious organizations, including Roman Catholics, Eastern Orthodox churches, conservative and Orthodox Jews, and Missouri Synod Lutherans, continue to bar female ordination. And women are at an extreme disadvantage even within religious organizations that accept their ordination. Although the increasing number of female seminarians and ordained ministers is one important consequence of feminist efforts to eradicate sexism in churches, their number remains small. Even after ordination, women face a glass ceiling, salaries below those of male ministers, and assignment to small congregations. These conditions frustrate female ministers as well as discourage women from preparing for ordination.

An Evaluation

Interpretations of the relationship between religion and social change, as the preceding paragraphs show, are varied. Marx, writing from the conflict perspective, believed that religion retards social change in part because of its ties to an economic elite. Unlike Marx, Weber argued that religion could encourage social change by supporting new social arrangements.

Although the two views appear contradictory, each may be viable under particular conditions. Meredith McGuire (2001) suggested that the relationship between religion and social change depends on several situations: quality of religious beliefs and practices existing in a society, dominant ways of thinking in a culture, social location of religion within a larger society, and internal structure of religious organizations and movements. For example, McGuire has argued that religion is likely to promote change if it provides a critical standard against which the established social system can be measured. When religion provides such a standard, it allows the social critic to say, "This is what we say is the right way to act, but look how far our group's actions are from these standards!" For example, during the American civil rights movement religious leaders pointed to American values of equality and opportunity and their incompatibility with racial discrimination.

The degree of a society's network for change is another factor. In many Latin American countries, for example, the Roman Catholic Church plays an important role in social change and protest because it is the sole dissenting grassroots organization with an established network and sympathetic leanings. When other channels for action exist—such as interest groups—religion may play a minor role in social change.

The relationship of religion to social change, then, is quite variable. Although the interpretations developed by Marx and Weber call attention to specific aspects of that relationship, no single interpretation appears applicable to all situations.

Symbolic Interactionism and Religion

Sociologist Peter Berger (1990) captured the relationship between religion and **symbolic interactionism** in his book *The Sacred Canopy*. In this book, Berger explores the idea that humans create from their religious traditions a canopy, or cover, of symbolic meanings to lay over the secular world. These meanings are used to illuminate the difference between the sacred and the profane and to guide everyday social interaction. A canopy of religious beliefs, rituals, and ideas provides stability and security in a changing and uncertain existence.

Symbolic interactionism, for example, helps us understand the expression "There are no atheists in foxholes." Insecurity and uncertainty, of course, are at a peak in the life-and-death situation of war, and the desire to regain security and certainty is a natural human response. Religious meanings, especially those related to a possible afterlife, can offer some relief. Like the Japanese kamikaze pilots in World War II, Middle Eastern terrorists can infuse into suicidal behavior the meaning of a reward beyond life. Less dramatically, people enduring troubled marriages can be strengthened by their commitment to uphold their holy vows of matrimony spoken in a place of worship.

So far, the interplay between religion and society has been portrayed as one-way. Religion does bestow meanings on social life, but the reverse is also true; social change within a society also influences religious beliefs, rituals, and ideas. As scientific knowledge began to increase with the Enlightenment, religious beliefs in the West also began to change. Gradually, the belief that devils inhabited witches weakened. An understanding of the universe slowly emptied the prisons of scientists who disputed the theological truism of an earth-centered solar system. In our own time, most Protestants now commit the former heresy of practicing birth control, without negative church sanction. Given the widespread practice of birth control among the Catholic laity, one can almost imagine the distant acceptance of birth control methods by the Church itself. And the general rejection of female ministers within Protestant faiths seems certain to be eroded by the tide of gender equality (Eck 2002).

Each of the three major theoretical perspectives aids in the scientific study of religion. Table 14.2 shows the unique light each perspective sheds.

TABLE 14.2

Selected Socioeconomic Characteristics of Elderly Americans

From this table, some comparative statements can be made about the economic situation of elderly Americans. What are two important statements that can be supported by these data?

	Whites	African Americans	Latinos	Women	Men
Medium Income	$16,200	$11,500	$10,400	$12,100	$12,200
Percent in Poverty	8%	24%	19%	12%	7%

Source: CRS Report for Congress, 2005.

1. According to Karl Marx, religion is a force working for
   a. the good of all people.
   b. the good of the proletariat.
   c. maintenance of the status quo.
   d. the creation of conflict between the classes.
2. According to Marx, religion is created by the _____ to justify its economic, political, and social advantages over the oppressed.
3. The _____ is the belief that work is an obligation and that capital should be reinvested for further profits rather than used for immediate enjoyment.
4. According to Max Weber,
   a. early Protestantism provided a social environment conducive to capitalism.
   b. capitalism developed because of Protestantism.
   c. capitalism strengthened Protestantism.
   d. religion will always be linked with economy.
5. Religion
   a. has been found to promote social change under almost all conditions.
   b. is unrelated to social change.
   c. promotes social change only under certain conditions.
   d. always supports the interests of underprivileged groups in society.

Answers: 1. c; 2. elite; 3. Protestant ethic; 4. a; 5. c

Religion: Structure and Practice

Religious Organization

Major religious faiths in the world include Christianity, Judaism, Islam, Hinduism, and Buddhism. Because most people practice religion through some organizational structure, religious organization is an important component of the sociological study of religion. Early scholars identified four basic types of religious organization: church, denomination, sect, and cult (Troeltsch 1931; Niebuhr 1968; originally published in 1929). A religious faith may be practiced through any one of these types of religious organizations.

How do sociologists distinguish among the basic types of religious organization? In this context, **church** is a life-encompassing religious organization to which all members of a society belong. This type of religious organization exists when religion and the state are closely intertwined. King Henry VIII's divorce troubles stemmed from the Roman Catholic Church's hold on sixteenth-century England. In Elizabethan England, Anglican Archbishop Richard Hooker of the Church of England wrote that "there is not any man of the Church of England but the same man is also a member of the commonwealth; nor any man a member of

the commonwealth which is not also of the Church of England."

A **denomination** is one of several religious organizations that most members of a society accept as legitimate. Because denominations are not tied to the state, membership in them is voluntary, and competition among them for members is socially acceptable. Being one religious organization among many, a denomination generally accepts the values and norms of the secular society and the state, although it may at times be in opposition to them. Most American "churches"—Methodist, Episcopalian, Presbyterian, Baptist, Roman Catholic, Reformed Jewish—are actually denominations.

A **sect** is a religious organization formed when members of an existing religious organization break away in an attempt to reform the "parent" group. Generally, sect members believe that some valuable beliefs or traditions were lost by the parent organization, and they form their own group to save these features. Thus, they see themselves not as establishing a new religious faith but as redeeming an existing one. The withdrawal of a sect from the parent group is usually psychological, but some sects go further and form communal groups apart from the larger society. The Pilgrims, who landed in Jamestown in 1609, wished to reform the Church of England from which they had separated. Another example is the Amish, a sect formed in 1693 when a Swiss bishop named Jacob Amman broke from the Mennonite Church in Europe (Kraybill and Olshan 1994). Less extreme sects in the United States today include the Seventh-Day Adventists, the Quakers, and the Assemblies of God.

The Amish are a religious sect. How does a sect differ from a church, denomination, or cult?

A **cult**, by contrast, is a religious organization whose characteristics are not drawn from existing religious traditions within a society (Karson 2000; Kaplan and Loow 2002). Whether imported from outside the society or created within the society, cults bring something new to the larger religious environment. We may think of cults as engaging in extreme behavior. The world was shocked in 1997, for example, when reports came of the ritualistic suicides of thirty-nine members of the Heaven's Gate cult in California (E. Thomas 1997). Dwarfing this incident was the mass killing of approximately 1,000 members of the Ugandan cult called the Movement for the Restoration of the Ten Commandments of God ("Cult Killings Exceed Jonestown Toll" 2000). Cults do not usually appear in such an extreme and bizarre form, however. More conventional examples of cults are the Unification Church, the Divine Light Mission, and the Church of Scientology (C. S. Clark 1993).

Although the term *cult* covers a wide range of groups and organizations, the religious cults in the United States today share several characteristics. At their center are an authoritarian structure, rejection of the secular world's laws and ways, strict discipline of adherents, rigidity in thinking, conviction of sole possession of truth and wisdom, belief in the group's moral superiority, and discouragement of individualism (Appel 1983; J. J. Collins 1991). More is said about cults later.

Religiosity

For many years, Charles Glock and Rodney Stark (1965, 1968) have explored religious experiences in relation to **religiosity**—the ways in which people express their religious interests and convictions. These authors' work focuses on the types of religious attitudes and behavior people display in their everyday lives.

How do people display religiosity? Glock and Stark distinguish five dimensions of religiosity: belief, ritual, intellectual, experience, and consequences. *Belief* refers to some idea about what a person considers to be true. People may, for example, believe that Christ died on the cross for everyone, or that there is no god but Allah. A *ritual* is a religious practice that members of a religion are expected to perform. A ritual may be private (personal prayer) or public (attending mass). The *intellectual* dimension of religiosity may involve knowledge of the Bible or an interest in such religious aspects of human existence as evil, suffering, and death. Religious persons are expected to be informed about their faith. *Experience* encompasses certain feelings attached to religious expression. This dimension is the hardest to measure, but James Davidson (1975) contended that it can be ascertained by asking people whether or not they think specific religious experiences are desirable and whether or not they have had any specific religious experiences. *Consequences* are the decisions and commitments people make as a result of religious beliefs, rituals, knowledge, or experiences. Consequences may be social (opposing or supporting capital punishment or abortion) or personal (restricting one's sex life to marriage or telling the truth regardless of the cost).

Why use a dimensional approach to religiosity? Although the dimensions of religiosity may be related to one another (Clayton and Gladden 1974), they may also operate independently at times. People who attend a place of worship (ritual dimension), for example, may know little about religious doctrines (intellectual dimension). And some people who worship regularly may believe that people who do not frequent religious services will spend eternity in hell (belief dimension).

Understanding the Danger of Cults

In late November 1978, news began to arrive in the United States that a semireligious, socialistic colony in Guyana, South America, headed by the Reverend Jim Jones—founder of the California-based People's Temple—had been the scene of a shocking suicide-murder rite in which some 900 people died from cyanide poisoning. Many Americans wondered how people could have become involved in something like that.

Some dismissed the participants as ignorant or mentally unbalanced. But as more news came out, it became known that many of the members were fairly well-educated young people and that Jones was trusted and respected by some members of the California political establishment. We also learned that such events, although rare, have occurred before.

Why are people willing to join extremist religious groups? Sociology can help us understand the motivations.

- Most converts to extremist religious groups seek friendship, companionship, acceptance, warmth, and recognition. These groups can provide a supportive community that helps overcome past loneliness and isolation. They can provide emotional ties that converts have not found at home, school, church, or work. Many

groups even adopt kinship terms to give recruits new identities to separate them from their former lives.

- Most extremist religious groups emphasize immediate experience and emotional gratification. Converts "feel" religion rather than merely think about it. Whether by meditation, speaking in tongues, or singing hymns, followers have frequent and intense emotional experiences they have not found elsewhere.

- Extremist religious groups emphasize security through strict authority. Under a firm authority structure and a clear, simple set of beliefs and rules, converts have something in which they can believe. Converts think they can exchange uncertainty, doubt, and confusion for trust and assurance through absolute obedience.

- Extremist religious groups claim to offer authenticity and naturalness in an "artificial" world. By emphasizing such things as natural foods, communal living apart from civilization, and a uniform dress code, these groups attempt to show they are not part of the flawed outside world.

Religious movements may not actually be able to meet their followers' needs any better than the outside world. Many of these religious groups lead to disillusionment, frustration, and bitterness when members realize that they cannot completely escape the outside world, which is full of uncertainty, confusion, fuzzy choices, and shades of gray. Moreover, many of these religious groups have joined the consumer society they profess to deplore, attractively packaging and selling themselves to the public. Not only may the new religious groups not solve the problems

people in modern society must face, many are as inauthentic as they accuse society of being.

Some key questions exist to evaluate the authenticity of any religious group's claims. For purposes of self-protection, these questions should be answered carefully before committing to an extremist religious group.

- Does it require that you cut yourself off from family and friends?
- Does it consider drugs to be a major vehicle for true religious experiences?
- Is corporal punishment or intensive, hours-long psychological conditioning a part of its program?
- Does it claim to have special knowledge that can be revealed only to insiders?

If the answer to any one of these questions is yes, you stand a chance of getting "hooked." If the answers to several of these questions are positive, the chances of getting hooked increase dramatically.

Think About It

1. Do you agree or disagree with the reasons given for why people join extremist religious groups? Discuss each reason and explain why you agree or disagree.
2. Can you think of other reasons people may be attracted to such groups? Show that any reason you identify does not fit into one of the four reasons stated.
3. If you had a friend considering membership in an extremist religious group, how would you use the information in this Sociology and You section to discourage him or her?

SNAPSHOT OF AMERICA 14.1

Public Charities

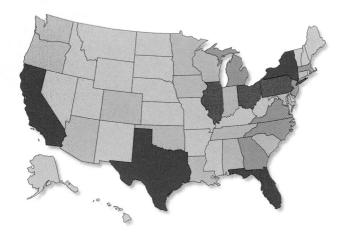

Number of Public Charities

- 50,000 or more
- 40,000–49,999
- 30,000–39,999
- 20,000–29,999
- 10,000–19,999
- 1,000–9,999

Source: National Center for Charitable Statistics, 2004.

No matter their religion, Americans tend to believe it is important to help people less fortunate than themselves. In 2004 alone, individuals and businesses in the United States contributed nearly $1.4 trillion to more than 820,000 public charities. This map shows the number of public charities based in each state.

Interpret the Map

1. Which states have the most public charities? Why do you think these states have the most charities?
2. About how many public charities are based in your state? In neighboring states?

FEEDBACK

1. Match the following descriptions with the types of religious organizations.
   - ____ a. church
   - ____ b. sect
   - ____ c. denomination
   - ____ d. cult
   (1) a religious organization whose characteristics are not drawn from existing religious traditions within a society
   (2) an exclusive and cohesive religious organization based on the desire to reform the beliefs and practices of another religious organization
   (3) a life-encompassing religious organization to which all members of a society belong
   (4) one of several religious organizations within a society that are considered legitimate
2. _____ refers to the ways people express their religious interests and convictions.
3. Match the dimensions of religiosity with the examples beside them.
   - ____ a. belief   (1) prayer in church
   - ____ b. ritual   (2) knowing the content of the Koran
   - ____ c. intellectual   (3) having seen an angel
   - ____ d. experience   (4) conviction that the Bible is divinely inspired
   - ____ e. consequences   (5) marching in support of prayer in schools

Answers: 1. a. (3), b. (2), c. (4), d. (1); 2. Religiosity; 3. a. (4), b. (1), c. (2), d. (3), e. (5)

Religion in the United States

The nature rites of South Sea Islanders and the well-ordered church of the Middle Ages were relatively simple and stable forms of religion. By comparison, advanced industrial society presents a challenge to the sociological study of religion.

The Development of Religion in America

Although the search for religious freedom was only one of the reasons the Puritans came to America, there was a genuine religious element in the American colonization and revolution (Brauer 1976; Marty 1985). Robert Bellah has described the American religious connection this way:

> *In the beginning, and to some extent ever since, Americans have interpreted their history as having religious meaning. They saw themselves as being a "people" in the classical and biblical sense of the word. They hoped they were a people of God.* (Bellah et al. 1991:2)

Religion in the colonies was varied, but beginning in the second half of the eighteenth century, the approach to religion was increasingly influenced by the leaders of the Enlightenment—Bentham, Voltaire, Diderot, Hume, and many others—in Europe and the British Isles. The same ideas that inspired the questioning of arbitrary colonial government and the writing of the Declaration of Independence also led to a critical examination of religion and its relationship to the state. Although the framers of the U.S. Constitution seldom raised arguments against religious faith, they were sharply critical of any entanglement between religion and the state. Indeed, the ideas of separation of church and state and freedom of religious expression are cornerstones of American life. Despite this tradition, people in the United States have experienced incidents of religious persecution, including some directed at immigrant groups.

Secularization: Real or Apparent?

Religion has been an important influence during most of human history, including the history of the United States. However, the industrialization (and accompanying secularization) of many nations has diluted the influence of religion. As Figure 14.2 shows, religiosity tends to decline as wealth (per capita GDP) increases, except for the United States. More will be said about this fact, but first secularization needs to be defined.

What is secularization? **Secularization** is a profane process through which the sacred loses influence over society. Through this process, other social institutions are emptied of religious content and freed from religious control (B. Wilson 1982; Swatos and Olson 2000; Lilla 2007). Religion itself becomes a specialized, isolated institution. This is in sharp contrast to preindustrial society, where religion and social life were inseparable. Hester Prynne, for example, the central character of Nathaniel Hawthorne's novel *The Scarlet Letter*, was forced by religious leaders and peers in her Puritan community to live forever with shame and ostracism for her act of adultery. Religion and social life were so intertwined in seventeenth-century America that the discovery of Hester's adultery led to her loss of a normal social existence.

How is secularizaton influencing religiosity in the United States? Evidence is mixed concerning the relative importance of religion in the United States today. There are some indications of a diminution of the importance of religion in the United States. The percentage of Americans claiming that religion is very important in their lives declined from 75 percent in 1952 to 56 percent in 2007, increased a bit since hitting a low of 52 percent in 1978, and is now declining a bit (see Figure 14.3). The Princeton Religion Index, comprised of six to eight leading indicators, also shows a decline since the 1940s. Fourteen percent of the public in 1957 indicated that religion was losing influence on American life; in 2007, 61 percent of the public saw a lessening of religion's influence on American life (Lyons 2005; Newport 2007c).

Finally, while only 16 percent of Americans claim no preference for a religious faith, it is more than double the percentage who claim not to have been affiliated with a particular faith as children. The religiously unaffiliated are now the country's fourth largest "religious group." And there appears to be evidence of a further generational shift today. Young adults between eighteen and twenty-nine are much more likely than these seventy or over to indicate a lack of affiliation with any particular religion (25 percent versus 8 percent; Banerjee 2008; J. Green 2008).

On the other hand, researchers characterize America as being unusually religious among industrialized nations (Ingelhart and Baker 2000; Kohut et al. 2000; Gallup 2002). There are more than 300 recognized denominations and sects and thousands of independent congregations in the United States (Linder 2008; see Figure 14.4). Only 16 percent of the American population is without affiliation with some religious faith, and about 78 percent identify themselves with the Christian religion. About six out of ten Americans have some specific religious association, and about half of these claim to be active in their congregations. Only 17 percent of adult Americans say they never attend religious services, which means that eight out of ten Americans attend worship services at least "seldom." About a third say they attend once a week and 12 percent say they attend almost every week.

FIGURE 14.2

Wealth and Religiosity

This figure shows at least two things. First, the religiosity of countries decreases as wealth increases. Second, the relatively high degree of religiosity in the United States makes it an outlier among Western nations. Why do you think this is so? Do you think the situation will remain the same or change? Why?

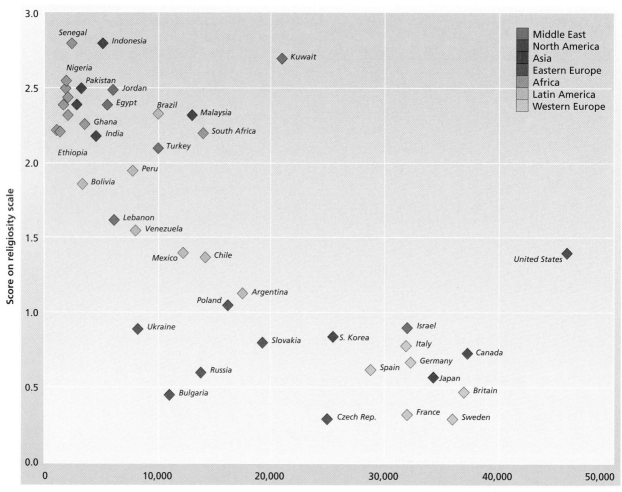

Per capita GDP (in U.S. dollars, adjusted for price differences between countries)

Source: "Among Wealthy Nations...U.S. Stands Alone in its Embrace of Religion." Washington, DC: The Pew Center Research Center for the People and the Press, Pew Global Attitudes Project, 2002. This graphic is also adapted from Alan Wolfe, "And the Winner Is..." *The Atlantic* (March 2008):59.

So, about 44 percent purport to be frequent worshippers (Newport 2007).

As to the future, American teenagers tend to support traditional religious beliefs on a par with adults. Ninety-eight percent of American teenagers believe in God or a universal spirit, 67 percent believe in life after death, 91 percent believe in heaven, and 76 percent believe in hell. Seventy-six percent believe in the existence of angels (Gallup 2002).

Why is there a continuing interest in religion? Theodore Caplow and his colleagues (Caplow, Bahr, and Chadwick 1983) suggest that religion is sought as a buffer against the insatiable demands of the state. According to Gallup, Americans are searching for spiritual moorings for three

basic reasons: the threat of nuclear war, loneliness, and disenchantment with what is perceived as a society without rules. Several sociologists contend that religion remains important in helping many Americans deal with problems of grief, misery, fear of the unknown, hopelessness, hunger, death, and the search for meaning beyond life in this world (Wind and Lewis 1994; Greeley 1996; Stark and Finke 2000).

Does this mean that secularization is not an important force in American society today? No, it does not. As Robert Wuthnow (1985) emphasizes, when Weber, Durkheim, and Marx wrote about secularization, they were not thinking of short-term, individual religious attitudes and behavior such as belief in God or synagogue atten-

FIGURE 14.3

Percentage of Americans Saying Religion Is Very Important in Their Lives: 1952–2007

This figure tracks changes in the percentage of Americans who say that religion is very important in their lives. Why do you think the percentage was so high in the early 1950s? What prediction do you make for the next ten years?

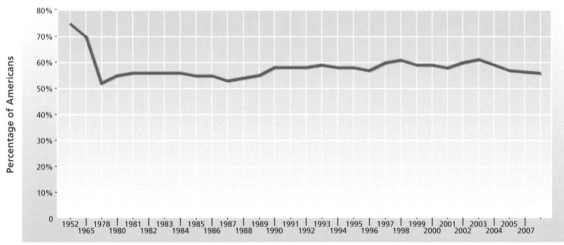

Source: The Gallup Organization, Gallup polls on religion.

INTERNET LINK The Gallup Organization compiles a tremendous amount of data on the American public's views. You can go to its website at www.gallup.com/

FIGURE 14.4

Membership in Selected Religious Organizations in the United States

On the basis of these data, how would you describe the religious composition of the U.S.?

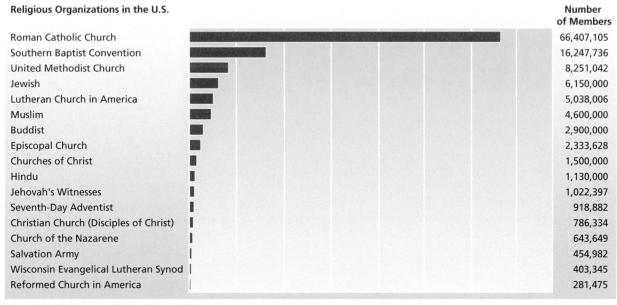

Religious Organizations in the U.S.	Number of Members
Roman Catholic Church	66,407,105
Southern Baptist Convention	16,247,736
United Methodist Church	8,251,042
Jewish	6,150,000
Lutheran Church in America	5,038,006
Muslim	4,600,000
Buddist	2,900,000
Episcopal Church	2,333,628
Churches of Christ	1,500,000
Hindu	1,130,000
Jehovah's Witnesses	1,022,397
Seventh-Day Adventist	918,882
Christian Church (Disciples of Christ)	786,334
Church of the Nazarene	643,649
Salvation Army	454,982
Wisconsin Evangelical Lutheran Synod	403,345
Reformed Church in America	281,475

Source: *World Almanac and Book of Facts, 2005.*

dance. They were observing major historical trends in European history since the Middle Ages, such as the separation of church and state. From this long-term, large-scale perspective, secularization accompanied modernization throughout the industrial world, including the United States (Bruce 2002; Chaves 2002).

In addition to reducing the social influence of religion, secularization promotes the mixing of the sacred and the profane (Fenn 1978). American places of worship are becoming more secular, and the American clergy are increasingly involved in community affairs. Religious leaders were heavily involved in the civil rights movement and the Vietnam War protests of the 1960s (Hadden and Rymph 1973) and are still concerned with such political issues as world peace and poverty. Although the clergy's participation in public affairs has receded in favor of parishioners' spiritual needs, the clergy remain involved in solving such social problems as drug abuse, juvenile delinquency, hunger, malnutrition, homelessness, and school violence. Secular society is entering religion in other ways as well. Worship services, for example, use "worldly" musical instruments such as saxophones and electric guitars, as well as jazz and rock liturgies. The televangelists of the "electronic church" use many marketing and show-business techniques to attract viewers and bring in vast amounts of money (Hadden and Swann 1981; Hoover 1988).

According to C. Kirk Hadaway, Penny Long Marler, and Mark Chaves (1993), characterizations of religious commitment in America typically rely on poll data. Self-reports of church attendance, they contend, inflate estimates of church attendance. Comparing church attendance rates based on actual counts of those who attend church with self-reported rates of church attendance, these researchers conclude that religious service attendance rates for Protestants and Catholics in the United States are actually about one-half the generally cited rates.

Robert Wuthnow (1976, 1978, 1990) argues that secularization, like increased religious fervor, should be seen as a process that fluctuates in strength. He claims that the religious revival of the 1950s waned because of opposition to it within the countercultural generation of the 1960s. Wuthnow thinks that a return to religious commitment may occur as successive age groups, unaffected by the counterculture, mature.

Civil and Invisible Religion

There is yet other evidence of the enduring religious influence on American society. This can be seen in the existence of civil religion and invisible religion.

What is civil religion? Every American president's inaugural address (except Washington's second one) has made reference to God. Similar religious references exist throughout American life: Thanksgiving Day is a day of public thanksgiving and prayer; our currency proclaims "In God We Trust"; the Pledge of Allegiance to our flag contains the phrase "under God"; many formal public occasions open with prayer. Even voluntary organizations such as the Rotary Club and the Boy Scouts carry out this religious theme.

None of these public religious allusions, however, are supposed to involve any specific religion. References are not made to Jesus Christ, Moses, or the Virgin Mary but are made to the general concept of God, a concept compatible with nearly all religions. **Civil religion,** then, is a public religion that expresses a strong tie between a deity and a culture; it is broad enough to encompass almost the entire nation. Robert Bellah (1967, 1992; Bellah and Hammond 1980; Lilla 2007), the sociologist most closely identified with the concept of civil religion, has summarized it this way:

The separation of church and state [in America] has not denied the political realm a religious dimension. Although matters of personal religious beliefs, worship, and association are considered to be strictly private affairs, there are,

Megachurches are among the fastest growing faith groups in the United States.

at the same time, certain common elements of religious orientation that the great majority of Americans share. These have played a crucial role in the development of American institutions and still provide a religious dimension for the whole fabric of American life, including the political sphere. This public religious dimension is expressed in a set of beliefs, symbols, and rituals that I am calling the American civil religion. (Bellah and Hammond 1980:171)

The concept of civil religion continues to generate debate (Gehrig 1981; Demerath and Williams 1985; Audi and Wolterstorff 1996; Horsley 2003), but many studies support Bellah's claim for the existence of a civil religion in the United States (Wimberley et al. 1976; Christenson and Wimberley 1978; Wimberley and Christenson 1980; Kearl and Rinaldi 1983; Warner 1994; Heclo and McClay 2002). Martin Marty (1985) points out that civil religion is alive and well today, thanks in part to its adoption by the conservative evangelical churches, which in the 1980s linked God and nationalism. The public outcry over a federal court's ruling that bans the phrase "under God" from the Pledge of Allegiance provides strong evidence of civil religion in the United States.

What is invisible religion? Those focusing only on traditional religious beliefs and practices underestimate the religiosity of Americans (Luckmann 1967; Yinger 1969, 1970). Many Americans who do not belong to churches or sects nevertheless practice a type of religion. These individuals, according to Luckmann, practice an **invisible religion**—a private religion that is substituted for formal religious organizations, practices, and beliefs.

In a study of 208 households in Baton Rouge, Louisiana, Richard Machalek and Michael Martin (1976) found support for the existence of an invisible religion, even in a region of the United States that is defined as deeply religious in the conventional sense. The authors asked participants in their survey to list the most important concerns of life, beyond immediate daily problems, and to indicate the means they used to cope with them. They found that 18 percent of the respondents cited such traditional religious concerns as life after death, salvation, and the existence of heaven. More significantly, 8 percent listed a variety of less religiously conventional concerns, such as consideration and love for others, the need for less selfishness in the world, national corruption, lack of political leadership, poverty, the need for social equality, and the meaning of life. One-third of the sample relied on conventional religious practices—prayer, Bible study, formalized church activities, and informal religious study groups—to help them handle their daily concerns. For two-thirds of the sample, coping strategies were not related to institutionalized religion. They relied on such humanistic strategies as reflection and meditation, participation in civic organizations, and informal discussion groups composed of family and friends.

Even people who ordinarily remain outside formal religion tend to observe religious ceremony during such rites of passage as birth, marriage, and death. For example, Gorer (1965) found that although 25 percent of his sample did not believe in life after death, 98 percent arranged religious rites when they buried their relatives.

Whether or not the United States is as religious as in the past, it is clear that religious effects abound throughout the American way of life. Americans join congregations, enter synagogues, and attend services in comparatively large numbers, and civil religion and private religion influence their lives daily. Moreover, there has recently been a revival of religious fundamentalism (Riesebrodt 1998).

The Resurgence of Fundamentalism

Since the late 1960s, most mainline American Protestant denominations—Methodists, Lutherans, Presbyterians, Episcopalians—have either declined in membership or fought to hold their own. In contrast, contemporary fundamentalist denominations—the conservatives of Protestantism—have been growing. Fundamentalists (also referred to as born-again or evangelical Christians) exist in all Protestant organizations, but they are predominantly found in such religious bodies as the Assemblies of God, the Seventh-Day Adventists, the Southern Baptists, and the Jehovah's Witnesses (Finke and Stark 1992; Aldridge 2000; Antoun 2001). About half of Protestants identify themselves as born-again or evangelical Christians (Gallup 2002; Green 2008).

It is, of course, inaccurate to limit fundamentalism to Protestants alone. Fundamentalism is found in all religions, including the Roman Catholic (21 percent), Jewish (7 percent), and Mormon (24 percent) faiths. In this section, however, we will concentrate exclusively on Protestant fundamentalism.

What is the nature of fundamentalism? Fundamentalism is based on the rejection of secularization and the corresponding close adherence to traditional beliefs, rituals, and doctrines. It is not surprising that most religious fundamentalists are politically conservative, given that the roots of contemporary religious fundamentalism can be found in the latter part of the nineteenth century. Two issues disturbed the early fundamentalists. First, fundamentalists were concerned about the spread of secularism. Science was challenging the Bible as a source of truth, Marxism was portraying religion as an opiate for the masses, Darwinism was challenging the biblical interpretation of creation, and religion was losing its traditional influence on social institutions. Second, fundamentalists rejected the movement away from the traditional message of Christianity toward an accent on social service (Johnstone 2006).

The theological agenda of today's fundamentalists is very close to that of their nineteenth-century forebears. Today's Protestant fundamentalists believe in the literal

Eduardo Sverdin/Digital Railroad

Reverend John Hagee epitomizes fundamentalism with his strong resistance of secularization and rigid adherence to traditional beliefs, rituals, and doctrines.

truth of the Scriptures, being "born again" through acceptance of Jesus Christ as the Son of God, the responsibility of all believers to give witness for God, the presence of Satan as an active force for evil, and the destruction of the world prior to the Messiah's return to establish his kingdom on earth (H. Cox 1996; Hunt, Hamilton, and Walter 1998; C. Smith and Emerson 1998).

Are all fundamentalists alike? Variations exist, of course, among fundamentalists. The term *evangelical*, for example, is often used as a synonym for fundamentalism. However, when compared with evangelicals, most fundamentalists are more theologically rigid and less accommodating in their opposition to the rest of the world. Most evangelicals are on the political right or in the center, but some are on the ideological left. Evangelicals share much of the traditional theology of fundamentalism, but often reject its political and social conservatism (Quebedeaux 1978; J. D. Hunter 1987; Roof 2001). Other religious organizations that share in much of the fundamentalist theology have some unique beliefs and practices of their own (C. Smith 2000). Neo-Pentecostalism—or the charismatic movement, as it is sometimes called—has occurred for the most part within traditional religious organizations, particularly the Roman Catholic and Episcopal churches. Those involved in this movement often speak of being "born again" or of receiving "the baptism of the Holy Spirit." But central to most neo-Pentecostal groups is the experience of "speaking in tongues" *(glossolalia)*, which believers claim is a direct gift of the Holy Spirit (H. Cox 1992).

Why has fundamentalism reappeared? There are several reasons for the revival of fundamentalism. First, many Americans feel their world is out of control. The social order of the 1950s has been shattered by a string of traumatic events beginning with the civil rights movement and progressing through campus violence, political assassinations, the Vietnam War, Watergate, and destruction of the World Trade Center towers. Fundamental religion, with its absolute answers and promise of eternal life, provides a strong anchor in a confusing, bewildering world. Second, by placing emphasis on warmth, love, and caring, fundamentalist churches provide solace to people who are witnessing and experiencing the weakening of family and community ties. Mainline churches are more formal and impersonal. Third, fundamentalist churches offer what they consider a more purely sacred environment, in contrast to mainline denominations that fundamentalists see as accommodating to secular society. Finally, the electronic church has articulated the appropriate messages to a receptive nationwide audience. As part of the mass media, the electronic church has been an important contributing factor in the growth of religious fundamentalism (Hadden and Swann 1981). The place of the electronic church deserves some elaboration.

What has been the place of the electronic church in the resurgence of fundamentalism? In the past, religious evangelists typically traveled from community to community and preached to the faithful in tents, open fields, or rented meeting halls. Modern technology has changed all that. Television, radio, and direct-mail campaigns via computers have made it possible for religious broadcasters to reach millions of people easily, rapidly, and repetitively. Ministers, particularly from the fundamentalist flank of Protestantism, have created vast and profitable television ministries. It is estimated that religious organizations own more than 1,400 radio stations and 60 television stations. According to the National Religious Broadcasters Association, there are about 350 television ministries. From the thousands of weekly hours of religious programming come more than $500 million tax-free each year (Johnstone 2006).

Sexual and financial scandals surrounding Jim Bakker, Jimmy Swaggart, and the Catholic Church have damaged the religious institution. The proportion of Americans indicating that they have "a great deal" or "quite a lot" of confidence in religion as an institution decreased from 66 percent in 1985 to 53 percent in 2004, a drop of 13 percentage points (Lyons 2005). Although Americans are generally satisfied with the extent of religion's influence, they also would choose less rather than more influence (Newport 2007a).

Fenggang Yang and Helen Rose Ebaugh— The Secularization of Religion?

Yang and Ebaugh (2001) challenge the long-standing view that religious pluralism accompanying modernization secularizes religions. They present evidence for a new view: instead of promoting the decline of religions, religious pluralism leads to their revitalization. Organizational and theological transformations provoked by exposure to a wide variety of religions are said to energize religions.

Their research is based on the study of immigrants who came to the United States after 1965. These "new immigrants" arrive in much larger numbers from Asia and Latin America than from Europe, the predominant home of "old immigrants." Whereas earlier European immigrants left mainly Judeo-Christian countries, many of the new immigrants from Asia are Muslim, Hindu, Buddhist, or adherents to other religions (Eck 2002). Even immigrants from South and Central America grew up with forms of Catholicism and Protestantism distinct from U.S. versions.

Data came from thirteen immigrant churches, ranging from Buddhist temples to Protestant churches. In 1997–1998, intensive interviews were conducted with clerical and lay leaders, new immigrants, and more established residents.

According to Yang and Ebaugh, these churches have prospered by adopting new organizational and theological forms. New immigrant religions are becoming congregational, theologically pure, and open to outsiders.

Adopting Congregation Forms

New immigrants often abandon the religious institutional forms brought from their home countries in favor of U.S. Protestant-style congregations, which are communities that gather voluntarily. Rather than remaining for life in their religions of birth, new immigrants feel freer to join or leave a religious group. Also, these new immigrants had been taught reliance on periodic visits to formal religious leaders at their often-isolated temples and monasteries. These immigrants now emphasize lay leadership within their local congregations, and they are providing social services and recreational centers.

Returning to Theological Roots

New immigrants come from societies dominated by one or two religions. In such societies, religion is intertwined with the culture. Consequently, being a Muslim in one country may be different in certain ways from practicing Islam in another. For example, in the United States, Pakistani Muslim men, accustomed to praying with their caps on, find themselves worshipping with Arab Muslims, who pray bareheaded. Which custom does one's pluralistic congregation adopt? Rather than attempting to make a host of such decisions, American Muslim congregations return to the original teachings of Prophet Muhammad, which they share worldwide. Various culturally based adaptations of Islam are dropped as congregations seek commonalities. Reference to original theological foundations provides culturally diverse Muslim congregations with a unified rationale for determining what aspects of their faith to keep and to discard.

Openness to Outsiders

Diverse congregations forge a distilled theological base. By reducing religious idiosyncrasies, new immigrant congregations can incorporate members from their own faith as well as from other religious traditions. Not only can Pakistani and Arab Muslims worship together, they can attract native-born American converts. This is appealing because it speeds up the process of Americanization.

Evaluate the Research

1. Why would these organizational and theological processes of change work to prevent a weakening of the new immigrants' religions?
2. Explain the interrelationships among these three processes of change.
3. Discuss the link between these processes of change and secularization.

Religious Movements in the United States

Other religious movements have emerged alongside the resurgence of fundamentalism (Heelas 1996; Bainbridge 1997; Stone 1997; Bromley 1998; Zellner and Petrowsky 1998). Speaking broadly, these religious movements embrace such varied organizations as the Jesus People (on the fringe of the Christian establishment), the Guru Maharaj Ji, the Hare Krishnas, the Messianic Jews, and politically oriented movements such as the Christian Coalition.

Religious cults are an interesting part of these religious movements. Three examples, varying in their closeness to Christianity, are the Raelians, the Unification Church, and Scientology.

Who are the Raelians? In January 2003, a religious cult known as the Raelians proclaimed to the world their successful attempt to clone the first human being (named "Eve"). The group's promise to verify this claim with DNA evidence from the baby and the egg donor, however, was never kept. Claiming 55,000 adherents in more than 80 countries, this cult was founded by Claude Vorilhon (a.k.a. Raël), a French-born Canadian auto racer whose messianic vision was sparked by being aboard an alien spaceship for a trip to a distant planet whose inhabitants created life on earth through cloning themselves in a laboratory 25,000 years ago. Raël's mission is to replace the "myth of God" with a scientific vision of creation. As part of this mission, the Raelians see human cloning as a path to achieving external life. The Raelians are preparing a nucleus of attractive females ("Order of the Angels") to welcome the return of their alien creators sometime prior to 2035 (Adler 2003).

What is the Unification Church? One of the most controversial religious movements is the Reverend Sun Myung Moon's Unification Church, a religion whose adherents are generally (and negatively) referred to as Moonies. Moon is a Korean industrialist turned prophet–minister–messiah–newspaper publisher who proclaims a religion that is a combination of Protestantism and anticommunism (Barker 1984).

Several interesting sociological observations can be made about the Unification Church. First—as in some of the other new mystical movements, such as the Hare Krishnas—the Unification Church expects total conformity of its members. Second, despite parental fears, its morality is generally conservative, sometimes rigidly so. More often than not, it endorses such basic American values as a powerful God, a strong sense of family, and a belief in individualism. Third, the more the Unification Church grows, the more it resembles a denomination. For example, it now has its own seminary with faculty members who belong to respected national professional societies.

Although the Unification Church is small, it has attracted considerable attention. It forced society to deal with the legal rights of converts whose family members believe their kin were brainwashed (Richardson 1982; W. Shepherd 1982; Bromley and Shupe 1984). The Unification Church has been in the public spotlight because of its economic activities. The Reverend Moon's movement now controls more than $10 billion in business assets around the world. These assets include an automobile plant in Asia, the University of Bridgeport (Connecticut), the *Washington Times* newspaper, pharmaceutical companies, banks, publishing ventures, and other enterprises.

What is Scientology? Scientology, a spiritual therapy inspired by the late science fiction writer L. Ron Hubbard, has gained notoriety because of the behavior of one of its famous adherents, movie star Tom Cruise.

Hubbard claimed to have discovered the therapy and technique of Dianetics, a new approach to mental health that was received with faddish seriousness in the 1950s. Later, Hubbard developed an elaborate system of Scientology that included far more metaphysical speculation, particularly the doctrine of reincarnation. Hubbard and his followers claim the ability to perform mental and physical healing. Its practitioners charge for their services; they have very little formal religious ritual; and, like the Unification Church and some other religious movements, they demand absolute conformity to stated doctrines. In recent years, however, the requirement that followers sever all non-Scientological connections was relaxed (Hubbard 1998).

Are these religious movements unique? These religious movements resemble older religions in many ways. The New Testament, for example, demands that Christians place the gospel ahead of family and friends. And throughout the history of Western Christendom, movements have periodically arisen that demand absolute conformity, self-denial, and separation from family and friends who could not be converted (Beckford 1985). Similarly, the political activities of religious fundamentalists such as Jerry Falwell and Pat Robertson have many parallels in American history, and political involvement by fundamentalist religious groups also occurs in other nations (Marty 1980; Yinger and Cutler 1982). Nor are exotic cults new to our nation. Consider this example from the early 1880s:

> Some frontier religions were mildly eccentric, while others were extremely bizarre. Possibly the most flamboyant messiah of all was Isaac Bullard. Clad in nothing but a bearskin loincloth, he led his troop of adherents from Vermont through New York and Ohio and finally south to Missouri, proclaiming a religion compounded of free love, communism and dirt—Bullard often boasting that he hadn't washed in seven years. (Wise 1976:20)

In late November 1978, news began to arrive in the United States, first slowly and then with a tragic rapidity, that a semireligious, socialistic colony in Guyana, South America, headed by the Reverend Jim Jones—founder of the California-based People's Temple—had been the scene of one of the most shocking suicide-murder rites in recent times. The bizarre accounts appearing in the media prompted many people to ask how anyone could have become involved in the cult.

Some tried to answer the question by dismissing the participants as ignorant or mentally unbalanced individuals. But as more facts were uncovered, people learned that many of the members were fairly well-educated young people and that Jones was trusted and respected by many in the California political establishment. They also learned that such events, although rare, had occurred before.

Moreover, it would happen again. Like Jim Jones, David Koresh was a charismatic cult leader. In 1993, under FBI assault, he and many of his Branch Davidians (some sixty people, including seventeen children) went up in flames in their Waco, Texas, compound. Koresh saw himself as Jesus Christ in sinful form, and his devoted followers probably believed they were on their way to join Jesus in heaven after the Apocalypse (Edmonds and Potok 1993; Gibbs 1993a).

Why are people willing to go to extremes? Sociology cannot answer this question, but it can help us understand why people may be disposed to join religious cults.

Why have these religious movements emerged? Reasons for movements closely related to fundamentalism were covered in the discussion of the revival of fundamentalism. However, such movements should not be identified with the likes of the People's Temple of Jim Jones or David Koresh's Branch Davidians. In this section we look exclusively at some explanations for the current rise of cults and why people join them.

Thomas Robbins, Dick Anthony, and James Richardson (1978) contend that cults are the product of a value crisis or normative breakdown in modern industrial society. In this view, some people attempt to escape moral ambigu-

ity—a sense of personal helplessness, doubt, and uncertainty—by grasping onto a new "truth."

According to Charles Glock and Robert Bellah, interest in cults is a reaction to the excessive emphasis on self-interest in American Protestantism. The exaggeration of the parts of the Bible emphasizing self-interest has resulted in a failure to provide either social benefits or a sense of personal worth in complex urban society. The religious yearning seen in cults is geared toward socialism and mysticism, which are incompatible with American denominations that emerged from an agricultural or frontier environment:

> *The demand for immediate, powerful, and deep religious experience, which was part of the turn away from future-oriented instrumentalism toward present meaning and fulfillment, could on the whole not be met by the [traditional] religious bodies. . . . In many ways Asian spirituality provided a more thorough contrast to the rejected utilitarian individualism than did biblical religion. To external achievement it posed inner experience; to the exploitation of nature, harmony with nature; to the impersonal organization, intense relation to a guru.* (Glock and Bellah 1976:340–341)

Harvey Cox (1977) identifies four seductive features of religious cults. First, most of the cult converts are

No one has brought Scientology to public attention more than actor Tom Cruise, a vocal advocate for the movement.

Mark Mainz/Getty Images

looking for friendship, companionship, acceptance, warmth, and recognition. The cult provides a supportive community that helps overcome past loneliness and isolation. The Eastern religious cults provide emotional ties that converts cannot find at home, school, church, or work. Many of them even use kinship terms to give recruits new identities—sister, brother, Hare Krishna. Often a convert is renamed. Entertainer Steve Allen's son Brian joined the Church of Jesus Christ at Armageddon in Seattle and was "reborn" under the name Logic Israel.

Second, most of the Eastern religious cults emphasize immediate experience and emotional gratification rather than deliberation and rational argument. Converts report "experiencing" religion rather than merely thinking about it. Whether by meditation, speaking in tongues, or singing hymns, adherents have the frequent and intense emotional experiences they could not find elsewhere.

Third, Eastern religious cults emphasize authority. By having a firm authority structure and a clear, simple set of beliefs and rules, they offer converts something in which to believe. Converts profess to exchange uncertainty, doubt, and confusion for trust and assurance.

Finally, these cults purport to offer authenticity and naturalness in an otherwise artificial world. By emphasizing natural foods, communal living apart from "civilization," a uniform dress code, and sometimes nudity, these groups attempt to show that they are not part of the "plastic society."

Social Correlates of Religion

Social class and politics are two important social correlates of religious preference. Before discussing them, however, it will be helpful to glance at a picture of religious affiliation in the United States.

What are the religious preferences of Americans? About 84 percent of Americans profess some religious affiliation (Linder 2008). Although there are more than 300 denominations and sects in the United States, Americans are largely Protestant (51 percent) and belong to a few major denominations—Baptist (20 percent), Methodist (10 percent), Lutheran (6 percent), Presbyterian (4 percent), and Episcopalian (4 percent). Fourteen percent prefer other Protestant denominations. Catholics constitute a relatively large proportion of the American population (24 percent) and Jews a small proportion (2 percent). Figure 14.4 lists many of the religious organizations in the United States with memberships above 300,000.

What is the relationship between social class and religious characteristics? There are marked differences in social class (as measured by education and income) among the adherents of various religions in the United States. Generally speaking, Presbyterians, Episcopalians, and Jews are on the top of the stratification structure. Below them are Lutherans, Catholics, and Methodists. When measured by education and income, Baptists, on the average, come out the lowest. Because these are average figures, there are, of course, many individual exceptions to these rankings.

The social class differences are due partly to self-selection: people tend to prefer churches with members who have socioeconomic characteristics similar to their own. The socioeconomic differences occur partly because religious organizations place varying emphasis on worldly success versus otherworldly rewards (Niebuhr 1968). Finally, a historical dimension exists. With the exception of the Jewish groups, the higher-status denominations have experienced this social benefit for a long time. Thus, members of these denominations have had tremendous socioeconomic advantages. Lutherans and Roman Catholics, in contrast, tend to be later immigrants who have not had as many opportunities to achieve higher status (Johnstone 2006).

Differences in religiosity exist between the upper and lower classes. Reflecting Marx's view of religion as an opiate of the masses, many early sociologists saw religion primarily as a source of solace for the deprived poor. This entrenched view was challenged by research in the 1940s showing higher denominational affiliation among the wealthy. Further research in the 1950s and 1960s revealed religion as important at both extremes of the stratification structure. It is just that the upper and lower classes express their religious beliefs in different ways. The upper classes display their religiosity through church membership, church attendance, and observance of ritual, whereas lower-class people more often pray privately and have emotional religious experiences (Stark and Bainbridge 1985).

What is the relationship between religion and politics? Because American religious groups vary widely in socioeconomic characteristics, we must be careful in interpreting empirical correlations between religion and politics (Audi and Wolterstorff 1996). Some of the apparent influence of religion on political attitudes and behavior is in part due to social class.

Followers of the Jewish faith are particularly aligned with the Democratic Party; they are followed in strength of support by Catholics and Protestants. This is predictable, because Protestants generally are more politically conservative than Catholics or Jews, and the Democratic Party is generally not associated with political conservatism. Of the major Protestant denominations, the greatest support for the Republican Party is among Episcopalians and Presbyterians, which is hardly surprising, because the upper classes are more likely to be identified with the Republican Party.

There are some contradictions in this general pattern. Despite their affiliation with the more conservative Republican Party, Episcopalians and Presbyterians are less socially conservative than Baptists, who have been the strongest supporters of the Democratic Party of all Protestant denominations, especially in the South.

Religion and the Baby Boomers

What is unique about baby boomers? Two points need to be established about America's baby boomers, the generation born between 1946 and 1964 (whose oldest members are now in their sixties). First, there are a lot of them. Their number—72 million—represents about 36 percent of the adult population and about 26 percent of the entire population of the United States. When baby boomers say "Jump," American society generally replies "How high?" America's schools, political system, mass media, and housing industry have adapted to accommodate their needs and take advantage of their consumption potential at every stage of the life cycle. Second, having grown up during a period of startling change, baby boomers are not the homogeneous generation that their collective label implies. Their ranks include all social classes, ethnic groups, and races. They are also separated by gender. Despite this diversity, baby boomers have a shared identity sufficient to view them as a socially meaningful generation.

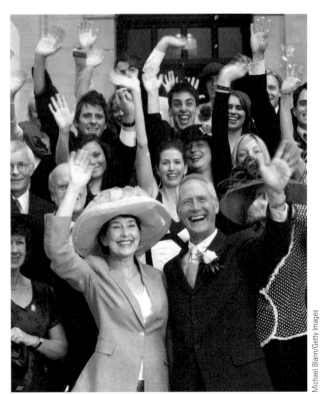

Some evidence suggests that many baby boomers, who left their churches in large numbers earlier, are returning to their religious roots. This trend, however, has not produced the shock wave baby boomers typically cause when they touch an institution.

FIGURE 14.5

Age Cohorts and Attitudes Toward Organized Religion

As this figure reveals, religious attitudes in the United States are changing. How do you think your generation would compare to pre–baby boomers and baby boomers?

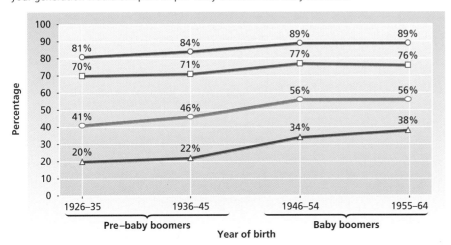

Source: "Age Cohorts and Attitudes Toward Organized Religion," *A Generation of Seekers,* by Wade Clark Roof. Copyright © 1993 by Wade Clark Roof. Reprinted by permission of HarperCollins Publishers, Inc.

What are the baby boomer effects on religion? The religious institution first felt the effects of the baby boomers when their parents swelled membership rolls by faithfully bringing them to houses of worship. The second shock came when the baby boomers, after leaving home, dropped out of their religious organizations, in record numbers (Bouvier and De Vita 1991). As revealed in Figure 14.5, the two **age cohorts**—persons born during the same time period in a particular population—born prior to and during World War II (1926–1935 and 1936–1945) have a much stronger orientation toward organized religion than the baby boomers' birth cohorts (1946–1954 and 1955–1964). These differences in religious attitudes help explain why the bulk of baby boomers left their religious organizations after leaving home and why a substantial proportion still do not return.

According to some researchers, large numbers of baby boomers are reverting to the spiritual roots they abandoned. If that is true, American religion is once again about to feel the shift made by the baby boom generation. Organized religion may be in for another landscaping.

About one-third of the baby boomers never left their synagogue or denomination, and more than 40 percent remain dropouts from organized religion. It is the returning one-fourth that is altering the nature of formal religion in America and the religious practices of millions of its inhabitants. These are some of the observations made by Wade Clark Roof, who conducted an extensive survey on religion among the baby boomers (Roof 1994, 2001; Roozen, Carroll, and Roof 1998).

Why are many baby boomers returning to church? Baby boomers are at the age when preceding generations of Americans have typically become more active in organized religion. According to a Gallup poll, many baby boomers are following this pattern (Gallup 2002).

Although some baby boomers are returning to religious organizations for needed social support following divorces or severe career disappointments, most are doing what parents normally do—showing concern about the well-being of their children. Many baby boomer parents believe it their duty to expose their children to religious tenets. Later, the children can, as a matter of their own choice, embrace or reject the religious influence (Aldridge 2000). Others see their children without exposure to moral values except within a religious context. For some, the church or synagogue is an instrument for reinforcing values they are attempting to instill in their children. Still others are looking to organized religion for the sense of community it can provide in a highly individualistic world.

FEEDBACK

1. _____ is based on the rejection of the effects of modern society on traditional religious beliefs, rituals, and doctrines.
2. Which of the following is *not* one of the reasons presented for the resurgence of fundamentalism in the United States?
   a. fear that the world is out of control
   b. provision of solace by fundamentalist organizations
   c. the electronic church
   d. the rise of invisible religion
3. The origins of religious movements in the U.S. appear to lie in
   a. social conditions in modern industrial society.
   b. desecularization.
   c. the rising influence of the Middle East.
   d. the decline of civil religion.
4. Which of these groupings of religious organizations is at the top of America's stratification structure?
   a. Jewish, Presbyterian, Episcopalian
   b. Lutheran, Jewish, Catholic
   c. Catholic, Episcopalian, Methodist
   d. Methodist, Baptist, Presbyterian
5. Which of the following is the major reason many baby boomers are turning to churches?
   a. concern for the ethical and religious training of their children
   b. need for comfort following divorces
   c. guilt over letting their own religiously oriented parents down
   d. fear of dying without practicing a religious faith

Answers: 1. Fundamentalism; 2. d; 3. a; 4. a; 5. a

SUMMARY

1. Religion is a unified system of beliefs and practices about sacred things. From his studies of Australian aborigines, Emile Durkheim concluded that every religion distinguishes between the sacred and the profane.

2. Because religion involves a set of meanings attached to a world beyond human observation, sociologists who study it scientifically face some unique problems. They do not evaluate the validity of various religions, leaving theological issues to theologians and philosophers.

3. Religion is found in nearly all societies. Its importance is due in part to the functions it serves. Religion legitimates the structure of society, promotes social unity, and provides a sense of meaning and belonging.

4. Writing from the conflict perspective, Karl Marx contended that religion is used by the elite to justify and maintain their power, and it is used by the oppressed to soothe their misery and maintain hope for improvement. Religion, in his view, is the opiate of the masses.

5. According to the conflict interpretation, religion retards social change because those in power wish to maintain the status quo. Max Weber, on the other hand, believed that religion could promote social change. He attempted to support this idea by connecting the Protestant ethic with the rise of capitalism. The Protestant ethic emphasizes hard work and reinvestment for profit, characteristics that were highly compatible with the needs of emerging capitalism.

6. Most of the world's major religions suppress women. Starting in the early 1980s, feminists have challenged the male patriarchy characteristic of Christian religious organizations.

7. Two of the most important components of religion are religious organization and religiosity. The major forms of religious organization are churches, denominations, sects, and cults. The ways people express their religious interests and convictions (religiosity) can be analyzed in relation to five dimensions: belief, ritual, intellectual, experience, and consequences.

8. Through secularization, the sacred and the profane are increasingly intermixed in modern society. The existence of secularization does not mean, however, that a society is not religious. Although many observers have noted a decline in the role of religion in the United States, others contend that religion may be stronger than supposed. For one thing, Americans practice both a civil (public) religion and an invisible (private) religion.

9. Moreover, there has been a revival of religious fundamentalism in the United States. This fundamentalism, which is occurring largely outside the mainline religious organizations, shares a theological conservatism with nineteenth-century American fundamentalism. In addition, contemporary fundamentalism has its own conservative social and political agenda.

10. The resurgence of religious fundamentalism is due to several factors, not the least of which is the electronic church. Television, radio, and computers (for direct-mail campaigns) have enabled fundamentalist organizations to spread their messages to vast numbers of people on a continuous basis. It remains to be seen whether recent scandals among some televangelists will damage the electronic church and/or fundamentalism itself.

11. The rise of other religious movements since the 1960s may also be an indication of a religious revival. The Raelians, the Unification Church, Scientology, and other religious movements may be a response to contemporary feelings of confusion, doubt, helplessness, and uncertainty.

12. Two important social correlates of religion are social class and politics. The major religious faiths can be ranked based on educational level, income, and occupational prestige. Also, the upper and lower classes express their religiosity in quite different ways.

13. The baby boomers, a generation that has influenced all aspects of American social life since they began being born in 1946, are having a new effect on organized religion. Having dropped out of churches early in their lives, many are returning, primarily out of concern for the ethical and religious training of their children.

LEARNING OBJECTIVES REVIEW

- Explain the sociological meaning of religion.
- Demonstrate the different views of religion taken by functionalists and conflict theorists.
- Distinguish among the basic types of religious organization.
- Discuss the meaning and nature of religiosity.
- Define secularization, and describe its relationship to religiosity in the United States.
- Differentiate between civil and invisible religion in America.
- Describe the current resurgence of religious fundamentalism in the United States.
- Identify a wide variety of religious movements in America.
- Outline the relationship of baby boomers to religion.

CONCEPT REVIEW

Match the following concepts with the definitions listed below them.

____ a.   secularization   ____ d.   age cohort   ____ g.   denomination
____ b.   invisible religion   ____ e.   profane
____ c.   sacred   ____ f.   civil religion

1. persons born during the same time period in a particular population
2. a public religion that expresses a strong tie between a deity and a society's way of life
3. the process whereby religion loses its influence over society as a whole
4. one of several religious organizations that members of a society accept as legitimate
5. a private religion that is substituted for formal religious organizations, practices, and beliefs in a secular society
6. nonsacred aspects of our lives
7. things that are set apart and given a special meaning that transcends immediate human existence

CRITICAL-THINKING QUESTIONS

1. Does the distinction between the sacred and the profane still have validity in America today? Explain why or why not, using examples.

_____

_____

2. Consider the four functions of religion identified by sociologists. Which of these functions are being performed by religion in the contemporary United States? Illustrate each.

_____

_____

3. Describe Weber's Protestant ethic. Explain why Marx could not have endorsed the relationship between the Protestant ethic and the spirit of capitalism proposed by Weber.

_____

_____

_____

4. What do you think is the relationship, at the present time, between secularization and religiosity in the United States? Develop your answer with material from the text.

_____

_____

5. Evaluate the place of civil religion and invisible religion in America today.

_____

_____

6. Identify what you think are the three most important effects of fundamentalism on religion in contemporary America. Support your case with specifics.

_____

_____

MULTIPLE-CHOICE QUESTIONS

1. Which of the following statements is *false*?
   a. A profane object can become a sacred object.
   b. Profane objects do not take on a public or social character that makes them appear important in themselves.
   c. Religion is a unified system of beliefs and practices relative to sacred things.
   d. In their research, sociologists must avoid the strictly spiritual side of religion.
   e. Sociologists cannot successfully avoid questions about the ultimate validity of any particular religion.

2. Aspects of life that are everyday, ordinary, and taken for granted are labeled
   a. sacred.
   b. transcendental.
   c. profane.
   d. legitimate.
   e. material.

3. Archaeological discoveries and anthropological observations have demonstrated that religion
   a. began around the eleventh century.
   b. tends to dissolve into a limited number of basic denominations.

c. has been more successful at governing societies than political parties.

d. always contains an element of salvation and afterlife.

e. has existed in some form in nearly all known societies.

4. **Emile Durkheim contended that _____ was the central function of religion.**

   a. sacredness

   b. supernaturalism

   c. legitimation

   d. transcendentalism

   e. salvation

5. **Religious ceremonies marking important events in life, such as birth, sexual maturity, marriage, and death, illustrate the contribution of religion to**

   a. a sense of unity.

   b. a sense of anomie.

   c. a sense of time.

   d. a sense of meaning.

   e. the legitimation of social arrangements.

6. **Karl Marx contended that religion _____ social change.**

   a. promotes

   b. retards

   c. refracts

   d. serves no function in

   e. precludes

7. **The Calvinist belief that every person's eternal fate is unalterably predetermined by God at birth is known as**

   a. sacred legitimation.

   b. transcendentalism.

   c. predestination.

   d. deism.

   e. immaculate salvation.

8. **According to Max Weber,**

   a. early Protestantism provided a social environment conducive to capitalism.

   b. religion tends to promote the economy in most societies.

   c. capitalism developed as a direct result of Protestantism.

   d. Protestantism developed as a direct result of capitalism.

   e. religion serves as an economic opiate for the masses.

9. **The ways in which people express their religious interests and convictions is called _____ by sociologists.**

   a. piety

   b. religiosity

   c. worship

   d. secularization

   e. ritualism

10. **_____ is the process through which religion loses influence over society as a whole.**

    a. Dereligiosity

    b. Transcendentalism

    c. Segmentation

    d. Secularization

    e. Pragmatism

11. **According to studies of religiosity in the United States,**

    a. secularization is destroying religion in the United States.

    b. only half of all Americans now believe in God or a universal spirit.

    c. only 20 percent of all Americans now believe in life after death.

    d. the proportion of regular churchgoers has declined sharply.

    e. compared with other industrialized countries, America still appears to be a fairly religious nation.

12. **Researchers cited in the text attribute Americans' continuing interest in religion to three basic reasons: the threat of nuclear war, loneliness, and**

    a. the need to impress others with their piety.

    b. disenchantment with what they perceive as an anomic society.

    c. basic spiritual and psychological needs.

    d. the absence of competing institutions to fulfill their needs.

    e. guilt and fear of divine punishment after death.

13. **The involvement of contemporary clergy in community affairs demonstrates**

    a. the mixing of the sacred and the profane.

    b. their resistance to secularization.

    c. the strong tie between a deity and a society's way of life.

    d. the demise of civil religion.

    e. the increasing significance of religion.

14. **Many of the religious movements in the United States have emerged alongside the resurgence of**

    a. fundamentalism.

    b. escapism.

    c. desecularization.

    d. invisible religion.

    e. civil religion.

15. **Which of the following statements about baby boomers is correct?**

    a. About one-third of baby boomers never left their churches.

    b. Baby boomers are religiously heterogeneous.

    c. Older baby boomers with children are more involved in church than younger members of their generation.

    d. Baby boomers do not appear to be returning to the spiritual roots they had abandoned.

    e. The bulk of baby boomers abandoned their churches after leaving home.

FEEDBACK REVIEW

True-False

1. Sociologists who study religion are normally forced to abandon their religious beliefs and convictions. T or F?

Fill in the Blank

2. According to Marx, religion is created by the _____ to justify its economic, political, and social advantages over the oppressed.

3. _____ is based on the rejection of the effects of modern society on traditional religious beliefs, rituals, and doctrines.

Multiple Choice

4. **According to Karl Marx, religion is a force working for**
   a. the good of all people.
   b. the good of the proletariat.
   c. maintenance of the status quo.
   d. the creation of conflict between the classes.

5. **According to Max Weber,**
   a. early Protestantism provided a social environment conducive to capitalism.
   b. capitalism developed because of Protestantism.
   c. capitalism strengthened Protestantism.
   d. religion will always be linked with economy.

6. **Which of the following is not one of the reasons presented for the resurgence of fundamentalism in the United States?**
   a. fear that the world is out of control
   b. provision of solace by fundamentalist organizations
   c. the electronic church
   d. the rise of invisible religion

7. **The origins of religious movements appear to lie in**
   a. social conditions in modern industrial society.
   b. desecularization.
   c. the rising influence of the Middle East.
   d. the decline of civil religion.

8. **Which of these groupings of religious organizations is at the top of America's stratification structure?**
   a. Jewish, Presbyterian, Episcopalian
   b. Lutheran, Jewish, Catholic
   c. Catholic, Episcopalian, Methodist
   d. Methodist, Baptist, Presbyterian

9. **Which of the following is the major reason many baby boomers are turning to churches?**
   a. concern for the ethical and religious training of their children
   b. need for comfort following divorces
   c. guilt over letting their own religiously oriented parents down
   d. fear of dying without practicing a religious faith

Matching

10. Match the dimensions of religiosity with the examples beside them.
   ____ a. belief   (1) prayer in church
   ____ b. ritual   (2) knowing the content of the Koran
   ____ c. intellectual   (3) having seen an angel
   ____ d. experience   (4) conviction that the Bible is divinely inspired
   ____ e. consequences   (5) marching in support of prayer in school

GRAPHIC REVIEW

Data in Figure 14.5 compares attitudes of pre–baby boomers and baby boomers toward organized religion.

1. Did the change in attitudes toward organized religion begin with the baby boomers? Explain your answer by using the statistics presented.

   _____
   _____
   _____

2. The typical analysis contrasts the attitudes and behavior of pre–baby boomers with baby boomers. Use the data in Figure 14.4 to discuss differences in attitudes toward organized religion *within* each generation.

   _____
   _____
   _____

ANSWER KEY

Concept Review

a. 3
b. 5
c. 7
d. 1
e. 6
f. 2
g. 4

Multiple Choice

1. e
2. c
3. e
4. c
5. d
6. b
7. c
8. a
9. b
10. d
11. e
12. b
13. a
14. a
15. d

Feedback Review

1. F
2. elite
3. Fundamentalism
4. c
5. a
6. d
7. a
8. a
9. a
10. a. 4
    b. 1
    c. 2
    d. 3
    e. 5

15

Health Care and Aging

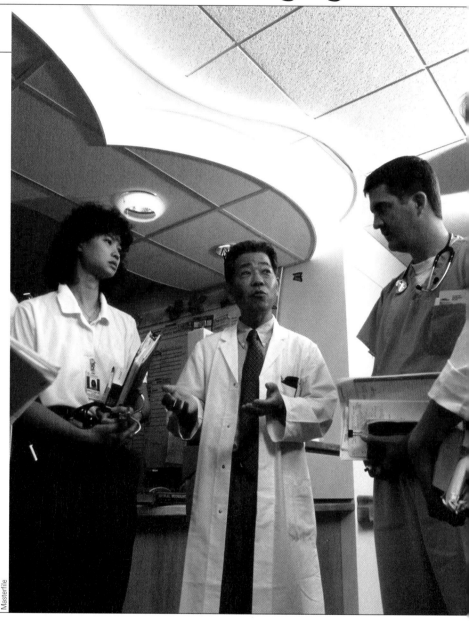

Masterfile

436

Is Canada our only model for a national health-care system? Canada is not the only industrialized country with a health-care system organized and operated by the federal government. In fact, the United States is the only highly developed nation that does *not* provide a federally centralized, free (or inexpensive) health insurance program, a service for all citizens on an equal access basis.

To this point, we have explored the most thoroughly developed institutions in modern society—family, education, polity, economy, and religion. Neither modern society nor sociology is static, however. As society changes, sociology expands to encompass the study of new social phenomena. Several additional institutions, such as health care and sports, are beginning to assume increasing relevance in contemporary society. In this chapter we explore one of these emerging institutions—the health-care system. Sport is covered in the next chapter.

noted the concern Americans tend to express for their well-being. But even he might be overwhelmed by the current mania over physical health among Americans. Burgeoning industries supply the needs of joggers, weight lifters, health food devotees, tennis players, and aerobic dancers.

This American propensity for personal health is embedded in culture. Americans tend to respond aggressively to their health-care needs. This aggressiveness, contends Lynn Payer (1996), is reflected in U.S. medical practice as compared with other industrialized countries. Compared with their European colleagues, for example, American physicians practice extremely aggressive medicine, and their patients expect it. Thus, American physicians are much more likely than their European counterparts are to perform surgery, run tests, and prolong life, even for individuals facing imminent death.

The Nature of the Health-Care System

The **health-care system** embraces the professional services, organizations, training academies, and technological resources committed to the treatment, management, and prevention of disease. The American health-care system constitutes a very large sector of the total national economy. Its growing claim on national resources reveals it to be an institution of increasing import. As shown in Figure 15.1, Americans spent almost $2 trillion on health care in 2005, up from approximately $75 billion in 1970. Since 1980, total health expenditures increased more than 400 percent. When viewed as a proportion of the gross domestic product (GDP)—the monetary value of all the goods and services produced by the economy during a given year—these numbers reveal even more clearly the size of the health-care system. Health-care expenditures now account for about 16 percent of the GDP; the comparable figure in 1970 was about 7 percent (see Figure 15.2).

Health Care as a Social Institution

Health Care and Society

Health is an issue in every society. Whether they rely on shamans, barber-surgeons, magic, rituals, herbs, or hospitals, people develop cultural patterns to cope with sickness and death. In the 1830s, Alexis de Tocqueville

Americans have always been health conscious, a point made by Alexis de Tocqueville in the 1830s. Yoga has recently become a popular form of exercise in the United States.

FIGURE 15.1

Rising Health-Care Expenditures in the United States: 1970-2016

This figure shows the steep increase in health-care expenditures in the United States since 1970. What do you think accounts for this trend?

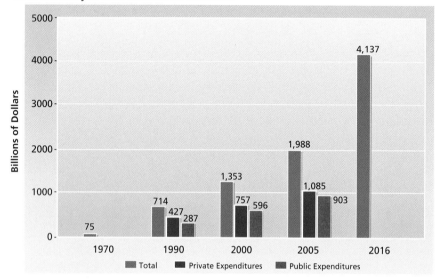

Source: U.S. Department of the Census, *Statistical Abstract of the United States: 2008* (Washington, D.C.: U.S. Government Printing Office, 2007) table 123; *The World Almanac and Book of Facts 2008* (New York: World Almanac Books, 2007), p.132.

FIGURE 15.2

United States Health-Care Expenditures as Percentage of GDP: 1970–2005

This figure displays the increase in U.S. health-care expenditures as a percentage of GDP. Do you think anything should be done about this development? Why or why not?

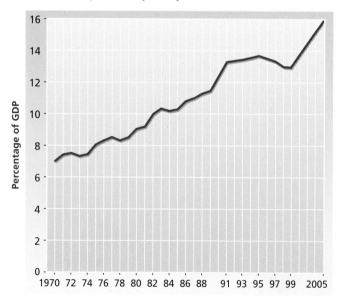

Sources: U.S. Department of Health and Human Services, *Health United States: 1992*, Public Health Service, Centers for Disease Control and Prevention, National Center for Health Statistics (Washington, DC: U.S. Government Printing Office, 1993), p. 160; U.S. Bureau of the Census, *Statistical Abstract of the United States 1998* (Washington, DC: U.S. Government Printing Office, 1998); and Centers for Disease Control, National Center for Health Statistics, www/cdc.gov/nchs/fastats/hexpense.htm; U.S. Bureau of the Census, *Statistical Abstract of the United States: 2008* (Washington, DC: U.S. Government Printing Office, 2007), table 123.

What are the components of the health-care system? By way of introduction, let us briefly discuss four major components of the health-care system: physicians, nurses, hospitals, and patients (L. D. Weiss 1997). Each of these components is covered later in greater depth.

Although physicians constitute only about 10 percent of health-care workers in the United States, they establish the working framework for everyone else. Only physicians are authorized by most states to diagnose illness, prescribe medicine, and certify such events as birth and death. Hospitalized patients may be attended by many professionals and other employees, but decisions about their diagnosis and treatment are made by physicians. American physicians' responsibilities are matched by high levels of social prestige and monetary rewards.

Nursing became a recognized profession only in the late nineteenth century, and its identity continues to undergo change. When Florence Nightingale was sent (at her insistence) with nurses and supplies to help British army doctors in Turkey in 1854, she encountered substantial resistance. One colonel wrote: "The ladies seem to be on a new scheme, bless their hearts. . . . I do not wish to see, neither do I approve of, ladies doing the drudgery of nursing" (quoted in Mumford 1983:289). After finally winning acceptance, Nightingale influenced the establishment of nursing schools in England and elsewhere. Although this represented a major advance in medical care, Nightingale's vision of the nurse's role was narrower than most nurses would now accept:

1. The _____ embraces the professional services, organizations, training academies, and technological resources that are committed to the treatment, management, and prevention of disease.
2. Health-care expenditures now account for about _____ percent of the American GDP.
   a. 5
   b. 10
   c. 16
   d. 21

Answers: 1. health-care system; 2. c

She insisted that nurses should be "clean, chaste, quiet, and religious." She saw the nurse as providing wifely support, keeping the hospital going so the physician could do his important work. (Mumford 1983:290)

Possibly as a result of this initial orientation, nursing has experienced frequent controversy regarding education, professional roles, and compensation.

Training programs for nurses shifted from three-year diploma programs in hospitals to associate, baccalaureate, and masters programs in colleges and universities. Despite a need for increasingly sophisticated knowledge, college and university nursing programs were slow to develop because hospital administrators enjoyed the plentiful supply of cheap labor that was generated within their own hospital training programs. The shift of nurses' training to colleges and universities is now well established, but the fact that it took an interminable half a century symbolizes the professional and economic problems of nurses (Lynaugh and Brush 1996).

Hospitals provide specialized medical services to a variety of inpatients and outpatients. Hospitals range from small facilities specializing in short-term, uncomplicated care to large medical centers providing long-term care with complex technology. Typically, medical centers, as well as some large hospitals, combine research and training with patient care.

People usually enter the health-care system only because they have been defined by others as ill or injured. The definition of illness or injury is predicated on a complex social process involving many people. Symptoms that may be considered proof of illness in some circumstances are ignored in others. The elderly, for example, are labeled "ill" far more easily than young adults because of the common assumption that illness is a normal aspect of old age. Whether or not a patient's designation as sick is biologically justified, this social definition places the patient in a complex set of rights and responsibilities that Talcott Parsons calls the *sick role*. This leads directly to the theoretical perspectives, because the analysis of patients in relation to the sick role is part of the functionalist approach to the health-care system.

Theoretical Perspectives and the Health-Care System

Functionalism

Talcott Parsons (1951, 1964a, 1964b, 1975) proposed a view of sickness that was distinctively sociological, rather than merely medical. For the first time, sickness became relevant to sociological theory and research. Parsons assumed, first, that health problems are a threat to society. If people are sick and cannot fulfill their roles, society will not function smoothly. Either functions will not be performed at all, or they will be performed in an inferior manner by others who substitute for the sick. According to Parsons, society responds in two ways. As a deterrent to fake illness, society defines sham sickness as a form of deviant behavior. But society also institutionalizes legitimate patterns of behavior for a sick role (Weiss and Lonnquist 1996).

What is the sick role? The **sick role** is a confluence of appropriate behavior patterns for people who are ill. It serves to remove the sick from active involvement in daily routines, give these individuals special protection and privileges, and set the stage for their return to normal social roles. Parsons identifies four major aspects of the sick role:

1. *The sick are permitted to withdraw temporarily from other roles or at least reduce their involvement in them.* As long as others agree with the sick prescription, the ailing person can miss work or school or not perform usual household duties.
2. *It is assumed that the sick cannot simply will the sickness away.* Consequently, negative sanctions are avoided because others agree that the malady is not the patient's fault. Illness is defined as a physical rather than a moral matter.
3. *The sick are expected to define their condition as undesirable.* People are expected to prefer good health again. They are expected to neither resign themselves to illness nor take unfair advantage

of an ailment's benefits (relief from normal activities, displays of sympathy, service from others). Thus, individuals who feign an illness have no valid claim to the sick role.

4. *The sick are expected to seek and to follow the advice of competent health-care providers.* Although not blamed for the onset of their illness, the sick are responsible for their own improvement. If the sick fail to cooperate in efforts to cure themselves, they will be tagged as deviants.

If any of these elements are missing, the privileges of the sick role may be withdrawn. For example, the quality of care given to those who are "uncooperative" may be decreased. Some illnesses, including those linked to obesity and anorexia, may even be denied the usual privileges of the sick role if people judge the disorders to be self-imposed. The existence of the sick role and the relationships surrounding it demonstrate the social nature of illness.

What are the criticisms of the concept of the sick role? Like any important theoretical formulation, Parsons's concept of the sick role has been closely scrutinized. Several criticisms have been leveled. One criticism depicts the sick role as too narrow in scope: it speaks only to Western society; it does not capture all aspects of illness; and it does not seem to apply to such conditions as pregnancy. According to another objection, the sick role applies primarily to acute illnesses, such as viral infections, which are curable. In the case of an incurable disease, such as inoperable cancer, the patient is not expected to leave the sick role. Also, charge some critics, the sick role is not applied to stigmatized illnesses such as AIDS and mental illness. Some have also contended that the sick role is related to the professional health-care system and ignores other means of coping with sickness, such as self-treatment or the use of religious faith. Although analyzing sickness sociologically requires more than the concept of the sick role, Parsons's effort was pioneering and is still quite useful (Twaddle 1981; Twaddle and Hessler 1987; Aronowitz 1998).

Why has the American medical profession risen to its present heights? A second relevant issue to the functionalist perspective is the medical profession's rise to preeminence in American society. In his Pulitzer Prize–winning book, *The Social Transformation of American Medicine*, Paul Starr (1983) offers a functionalist explanation for the advancement of American physicians. Using a pluralist orientation, Starr contends that physicians gained the recognition of medicine as a profession through the efforts of the American Medical Association (AMA), especially after its reorganization around the turn of the twentieth century. As an interest group, the AMA was able to wield power vis-à-vis other competing interest groups such as patent medicine producers. The patent medicine companies were direct competitors with the doctors because early physicians, in addition to treating patients, often prepared their own drugs, and the patent medicine producers offered therapy and advice in addition to selling medicine.

According to Starr, the AMA persuaded Americans that they "needed" physicians to handle their health problems. The combination of a convincing lobby and a responsive American public elevated physicians in the eyes of the American public.

Conflict Theory

The conflict perspective challenges many health-care practices. Their concerns include an alternative explanation of the medical profession's prestige, an explanation of the high income and status of physicians, the professional and economic problems of nurses, and the inequality of health care (Freund and McGuire 1998; Annandale 1998; Pringle 1998; Waitzkin 2000; Scambler 2002).

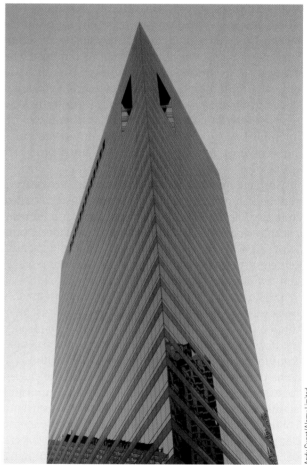

Andy Guest/Alamy Limited

While functionalists attribute the handsome income of physicians to their talent and training, advocates of conflict theory credit the medical establishment with limiting the number of available physicians. The American Medical Association building stands as a symbol of power in either case.

How do conflict theorists explain the prestige of the medical profession? According to Vicente Navarro (1976, 1986), the evolution of American medicine reflects the values and beliefs of American society. Navarro credits Starr with recognizing the role of the AMA in persuading American people of the need for physicians and the resulting preeminence of medicine. Nevertheless, Navarro offers a very different interpretation.

He begins with American power differentials based on class, race, and gender. The success of the medical profession, Navarro claims, is due to the power it possesses because of its alliances with the dominant capitalist class, the white race, and the male gender.

The same political and economic forces that determine the nature of capitalism determine the nature of the medical institution. Thus it was not merely persuasion that promoted the interests of the physicians but, more important, the coercive and repressive power physicians possess because of the capitalist power structure. Navarro would point, for example, to physician attempts at monopolization through repression of alternative (competing) approaches to health care—spiritual healing, acupuncture, osteopathy (based on the idea that disease results from loss of structural integrity), and homeopathy (treatment of disease by administering minute doses of a remedy that would in healthy persons produce symptoms of the disease being treated). In fact, a U.S. Supreme Court decision ruled that the American Medical Association was guilty of unfair restraint of trade in its efforts to keep osteopaths from practicing. It is these continuous conflicts and struggles, claims Navarro, that determine alterations in American society and in medicine.

Why do physicians receive a hefty reward? Conflict theorists reject the functionalist explanation for the high financial and prestige rewards of physicians. According to functionalist theory, the high rewards accorded any occupation, including physicians, are necessary for attracting talented people into demanding professions that require extensive training. But, according to the conflict perspective, occupational groups use various mechanisms (licensing, credential requirements) to limit competition. Thus, physicians' rewards are high in part because the medical establishment keeps medical school enrollments artificially low, thereby limiting price competition among physicians (Weeden 2002).

Whatever the reason, not all physicians have enjoyed the privileges accorded those in the United States today. In eighteenth-century England and early-twentieth-century France, physicians were poorly paid and were socially marginal. In the former Soviet Union, physicians received only about 75 percent as much income as the average industrial worker. Contemporary Great Britain has a few specialists who command very high incomes, but most general practitioners are only mod-erately well paid. Although physicians in most contemporary societies have higher status than was the case in ancient Rome (where physicians were primarily slaves, freedmen, and foreigners and medicine was considered a very low occupation), the situation in the United States is unusual (P. E. Starr 1983).

Can physicians maintain their rewards? Whether physicians in the United States will maintain their privileged and powerful position is not known. For one thing, the number of physicians for every 10,000 people in the United States has risen sharply since 1950 (Brownlee 2007). Also, because of declines in population growth, increasing costs of malpractice insurance, and the movement from general practice to specialized medicine, physicians are becoming less likely to work as independent professionals. Increasing numbers of physicians are taking salaried positions with organizations, or they are going into group practice and managed care organizations where expenses are shared. As these changes occur, individual physicians are beginning to lose control over their work:

> Perhaps the most subtle loss of autonomy for the profession will take place because of the increasing corporate influence over the rules and standards of medical work. Corporate management is already thinking about the different techniques for modifying the behavior of physicians, getting them to accept management's outlook and integrate it into their everyday work. (Starr 1983:447–448)

Medicine retains enough mystery that physicians are unlikely to lose all of their social status. As long as physicians use professional jargon, write illegible prescriptions, maintain control over patients' records, and force patients to wait in a waiting room "because the doctor's time is so valuable" (Mumford 1983:334), they will be held in awe by many members of society. It appears, however, that their ability to dictate the nature of the medical system is eroding.

One reflection of a growing sense of powerlessness is the 2003 strike of more than two dozen surgeons in West Virginia. They were protesting the rise in malpractice insurance premiums. A similar walkout in Pennsylvania was averted only because the state's governor proposed a $220 million fund to partially offset insurance premiums. In early 2003, protests, work slowdowns, and work stoppages occurred in New Jersey, Florida, and Mississippi. Other doctors are protesting rising malpractice costs by refusing to perform high-risk procedures, taking early retirement, or moving from states with the highest malpractice insurance premiums (L. Parker 2003; R. Stein 2003).

What are some of the professional and economic problems nurses face? Role expectations and authority relationships are problematic for nurses. Although nurses tend to patients far more frequently than do physicians, the nurses lack the authority to make any but the most routine decisions. In many situations, when nurses convey recommendations to physicians, they must appear to be passive. Leonard Stein illustrates this problem of domination in a hypothetical telephone conversation between a nurse and a physician:

This is Dr. Jones.

(An open and direct communication.)

Dr. Jones, this is Miss Smith of 2W—Mrs. Brown, who learned today of her father's death, is unable to fall asleep.

(This message has two levels. Openly, it describes a set of circumstances, a woman who is unable to sleep and who that morning received word of her father's death. Less openly, but just as directly, it is a diagnostic and recommendation statement; i.e., Mrs. Brown is unable to sleep because of her grief, and she should be given a sedative. Dr. Jones, accepting the diagnostic statement and replying to the recommendation statement, answers.)

What sleeping medication has been helpful to Mrs. Brown in the past?

(Dr. Jones, not knowing the patient, is asking for a recommendation from the nurse, who does know the patient, about what sleeping medication should be prescribed. Note, however, his question does not appear to be asking her for a recommendation. Miss Smith replies.)

Phenobarbital mg 100 was quite effective night before last.

(A disguised recommendation statement. Dr. Jones replies with a note of authority in his voice.)

Phenobarbital mg 100 before bedtime as needed for sleep, got it?

(Miss Smith ends the conversation with the note of a grateful supplicant.)

Yes I have and thank you very much doctor. (L. Stein 1967:700)

Despite the difficult tasks and inconvenient hours associated with nurses' work, nurses' pay ranks below that of most professionals. As noted in Chapter 10, women receive substantially lower pay for their work than men do. This is certainly the case for nurses compared with physicians. The low pay received by nurses may relate to the fact that more than 90 percent of registered nurses are female. In contrast, less than 30 percent of physicians are female.

Past efforts by nurses to improve their salaries were resisted by hospitals, but the nature of contemporary medical work could lead to an improvement. Medical work is increasingly defined as more than prescribing medicine and performing surgery. As demonstrated by Anselm Strauss and his colleagues (1985), successful treatment requires attention to safety, comfort, emotion, and the coordination of a variety of activities. Nurses, they note, play critical roles in these tasks.

Nursing itself is becoming increasingly specialized, and the need for coordination with other professionals is accelerating (Kovner 1990).

How do conflict advocates view inequality in health care in the United States? The concern of conflict advocates for health-care inequality begins with **epidemiology** — the study of the distribution of diseases within a population. Social epidemiologists study the relationship of illness to the social and physical environment (Rockett 1994, 1999). Conflict theorists are interested in the reasons minorities and the poor have shorter life expectancies and higher incidence of certain diseases than the general American population. (They are aware that certain illnesses are associated with some minorities more than others. For example, African Americans are more prone to heart disease, and alcoholism is more prevalent among Irish Americans.) Conflict proponents trace the proclivity for ill health among minorities and the poor to the nature of capitalist society. Specifically, they point to the role of the health-care establishment in maintaining unequal access to medical care (Navarro 1976, 1986; McKinlay 1985; Waitzkin 1986; Weitz 2007). In addition, underprivileged people more often live in areas with heavy industrial pollution.

How does the medical establishment promote unequal access to health care? The market philosophy promotes unequal access to medical care, making medical care a market commodity. Those without the money to purchase health care are likely to be ill more often, have more serious diseases, and die younger. Unable to purchase good health care, they suffer the consequences. Health-care reform will be discussed later in this chapter.

Why is affordable health care not provided to the poor in the United States? According to conflict theorists, medical care is a capitalistic industry, an industry that brings to its participants—hospitals, nursing homes, nurses, physicians, pharmaceutical companies, medical schools, medical suppliers, and pharmacies—almost $2 trillion dollars annually. According to conflict theorists, these combined vested interests ensure the perpetuation of health-care inequality (Weitz 2007).

Symbolic Interactionism

As we have seen repeatedly, symbolic interactionism is grounded in the concepts of socialization and symbols. This section examines the socialization of physicians, the personal attributes and socialization of nurses, and the influence of labeling.

The socialization process of most physicians begins early. Those who decide to become doctors tend to do so early in life, many by the time they are sixteen years old. This is not accidental. Early decision occurs in part because many of those who enter medical school have been influenced by relatives who are themselves phy-

The market philosophy of the American health-care system promotes unequal access to medical care. The poor more often turn to emergency rooms as a source of medical services.

sicians. This, of course, also helps explain why physicians tend to come from the higher social classes. This presocialization, however important in the decision to become a physician, does not prepare medical students for the socialization they experience in medical school.

What are the most common socialization experiences of medical students? That medical students acquire technical knowledge and skills and learn to diagnose and treat illness goes without saying. In addition, medical school socializes students to accept the beliefs, norms, values, and attitudes associated with the medical profession. They learn, for example, to be dispassionate and unemotional about illness, suffering, and death; to never criticize another physician publicly; and to avoid revealing a lack of knowledge. They also learn gradually to view themselves as doctors and to display a confident, professional appearance (Colombotos and Kirchner 1986; Twaddle and Hessler 1987; Kronenfeld 1998). Medical school also transforms students from eager neophytes who wish to absorb all medical knowledge to more experienced individuals who realize that their human limitations require a selective learning strategy. They must determine what is considered safe not to learn by assessing what they think the faculty wants them to learn, a social process involving meanings, symbols, and subjective interpretations (see Doing Research).

How are nurses socialized? Although nurses are both male and female and come from all social classes, they tend to be white, middle-class females. Compared with the general female college student population, nursing students are more altruistic, benevolent, and generous and are less interested in power, control, and self-advancement. They prefer to make their own decisions and they want to be treated empathetically. These characteristics are supported during the four years of undergraduate nursing education:

During their four years of nursing education, students are socialized to a nurse role that values individualized, direct patient care, rational knowledge, and innovation, and they

are taught to think of themselves as autonomous, professional persons. (Twaddle and Hessler 1987:215)

Upon graduation, however, nurses are faced with a rigid, bureaucratized work environment. Their work lives are spent in a setting that contradicts the values they brought to their education and that their school socialization reinforced. This creates the potential for many conflicts and difficulties because nurses are constantly attempting to interpret their occupational world with symbols and meanings that do not fit their work situation (Buresh et al. 2006; S. Gordon 2006).

How is the labeling perspective applied to illness? As discussed in Chapter 7, the labels and stigmas applied to people affect the way others behave toward them, which in turn affects the behavior of the labeled people. "Sickness" and "illness" are social labels that can be used to stigmatize people.

The nature of health-related labels is based on culture. Because culture varies, definitions associated with health and illness are also diverse. Mark Zborowski (1952, 1969) found Jews and Italians more likely to talk openly about their pain, whereas "old Americans" and the Irish, with objectively the same levels of pain,

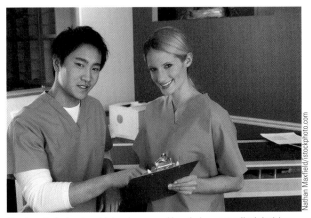

Nursing students place high value on making their own medical decisions and on their being treated with empathy. These expectations usually are not met in the inflexible organizational setting in which they work.

Howard S. Becker— Socialization of Students in White

Using participant observation, sociologist Howard Becker (1961) examined the socialization of medical students. Using a symbolic interactionist approach, his picture of student culture in medical school is not glamorous. Medical students grapple with a sense of eroding idealism, diminished concern for patient well-being, and bewilderment over the shortcomings of medicine.

Bright college graduates who enter medical school face a rude awakening. The gratifying thoughts of helping others through the practice of a high-status, honored profession are, in the first year in medical school, supplanted by the recognition of unforeseen barriers. Before becoming a physician, it is first necessary to learn to be a medical student. The medical student preoccupation blots out any images of how to perform as a doctor in the future.

Much of Becker's book concentrates on the ways students recognize and solve problems posed by their teachers. Medical students turn their eyes from the glorious distant mountain to the swamp that threatens to overwhelm them. Becker, in short, focuses on the social interaction between medical students and their instructors as they undergo a prolonged rite of passage from student to doctor.

Becker began his research without firm, testable hypotheses and without preset data-gathering research instruments. Because he wished to "discover" how the medical school as an organization influenced the basic perspectives of physicians in the making—something Becker did not know beforehand—he could not design formal research instruments in advance. This is not to say, however, that he lacked any research design. He states that his commitment to symbolic interactionism led him to use participant observation as the major research technique. Although he and his colleagues did not pose as medical students, they did "participate" in the life of medical students. The researchers attended classes and laboratories with students, listened to student conversations outside of class, accompanied students on rounds with attending physicians, ate with students, and took night calls with students.

In the end, Becker challenged the view (on the part of both the public and the medical school faculty) that medical students shed their idealism for cynicism as they progress through their training. Often cited evidence for this alleged shift from idealism to cynicism is the detached way medical students come to view their patients. Death and disease become depersonalized for medical students. Because medical students view death and disease unemotionally, as a medical responsibility rather than a tragedy, they are viewed by laypeople as cruel and heartless.

Becker does not deny that medical students appear to become cynical and detached in medical matters. He attributes this appearance not to the loss of an idealistic long-range perspective but to the need to survive the overwhelming demands of the training experience. A cadaver, for example, is not viewed as somebody's daughter or mother but as a necessary means for gathering information for a forthcoming examination. Successful medical students ask not what they can do for a cadaver but what a cadaver can do for them. It is, in short, the manifestations of the urgent and immediate need to survive medical training that give the appearance of cynicism to outsiders.

As graduation approaches, another change occurs. As medical students lose concern for the immediate situational demands of school, they begin to display the concern for service to others they had felt upon entering medical school. This new idealism, however, is more informed than the original.

The erroneous conclusions regarding cynicism are made more understandable within the framework of symbolic interactionism. The cynicism fails to acknowledge that individuals can hold conflicting values and rely on the relevant one in a particular situation. As the situation varies, so does an individual's definition of the situation. Thus, as the students' circumstances revert near graduation, the dormant values of idealism replace the expressions of cynicism. Corresponding alterations in behavior follow.

Evaluate the Research

1. Do you think the method of investigation is appropriate? Why or why not? What other methods could have been used?
2. Does the symbolic interactionist interpretation of medical school socialization ring true? Explain.
3. If you think they would be appropriate, explain how functionalism and conflict theory could be brought to bear on the study of the socialization of medical students. If you don't think one or both of these two theoretical perspectives are applicable, explain why.

tended not to complain. Similarly, Irving Zola (1966) reported that although the Irish did not see any relationship between their illnesses and their social relationships, the Italians presented relationship problems as a significant part of their illnesses.

Given this background, the existence of labels and stigma associated with certain illnesses is not surprising. The elderly are erroneously labeled "disabled" (e.g., forgetful) far more often than young adults because of the common assumption that illness is a normal aspect

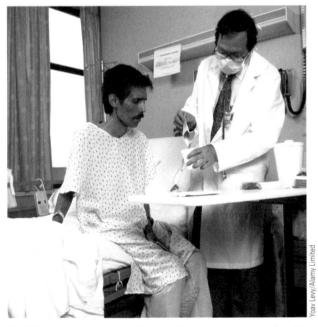

The high costs associated with AIDs make its victims particularly vulnerable in the American health-care system.

of aging. People with mental health problems are often stigmatized, which may lead to the adoption of a personal identity based on the properties of the stigma.

There is no more timely illustration of labeling and stigmatizing than the traditional view of AIDS in the United States. Because AIDS was initially discovered among homosexual men, who are already stigmatized in American society, people who acquire AIDS from such sources as blood transfusions or heterosexual partners are also labeled and stigmatized (Dennis Altman 1986). The taint of deviance and immorality is often attached to AIDS patients. This stigma is reflected in many responses. For example, a substantial number of Americans believe that AIDS is a punishment for the decline in moral standards.

The three theoretical perspectives provide very different insights into the health-care system; that is to be expected. In fact, in the absence of a single unifying theory, sociology is fortunate to have three powerful theoretical perspectives. Together, these theoretical approaches underscore the social nature of sickness and health care (see Table 15.1).

TABLE 15.1

FOCUS ON THEORETICAL PERSPECTIVES: Health Care in the United States

Each theoretical perspective opens unique avenues for research on health care. Can you think of other research questions for each theory?

Theoretical Perspective	Research Topic	Hypothesis
Functionalism	Sick role	• Society requires (needs) a timely exit from the state of illness.
Conflict theory	Power of physicians	• Physicians repress competing approaches to health care.
Symbolic interactionism	Stigmatization of illness	• AIDS victims are labeled as immoral and deviant.

FEEDBACK

1. The _____ serves to remove people from active involvement in everyday routines, give them special protection and privileges, and set the stage for their return to normal social roles.
2. If a sick person in American society does not attempt to recover, he or she will be labeled a _____.
3. According to the _____ perspective, physicians rose to dominance in the American health-care system by taking such actions as attempting to create a monopoly on medical services.
4. AIDS appears most frequently among homosexuals and intravenous drug users. This is an example of _____.
5. According to _____ theory, affordable health care is not provided to the poor in the United States because of the desire for profit.
6. Nurses are caught in a conflict between their values and their work environment. Which of the following perspectives best helps us understand this situation?
   a. functionalism
   b. conflict theory
   c. symbolic interactionism
   d. exchange theory
7. America's negative reaction to AIDS victims illustrates the effects of _____.

Answers: 1. sick role; 2. deviant; 3. conflict; 4. epidemiology; 5. conflict; 6. c; 7. labeling

Health-Care Reform in the United States

As mentioned in Using the Sociological Imagination at the beginning of the chapter, most modern societies provide free or inexpensive health care through their governments. The United Kingdom, although it has a system of private medical care, promotes equality of medical care through its National Health Service, which provides the vast majority of its services without charge. Canada also has a public health-care delivery system committed to the provision of equal access. As in the United Kingdom, health care in Canada is financed through the government.

Of the highly developed countries of the world, the United States is the only one without national health insurance for all its citizens. Moreover, except for the Veterans Administration hospital system, the military, and the Indian Health Service, the United States is without any direct mechanism for the delivery of health-care services (Waitzkin 2001).

George Bush's successful introduction of a new Medicare prescription drug plan in 2003 is one of the few efforts at health-care reform that has not been defeated in Congress. Consequently, health insurance in the United States is a privilege largely enjoyed, with the exception of the elderly, by those who have full-time jobs with well-established firms. Those with the power—including business interests and the medical establishment—have thus far successfully frustrated all efforts to reform the health-care system. By invoking threats of socialized medicine and profit reduction, contend conflict proponents, the elite in the United States have used capitalist ideology to prevent equality of access to medical care. In fact, say some conflict theorists, some form of socialized medicine will be required if equality of health care is to become a reality in the United States.

The Need for Health-Care Reform

Americans have conflicting ideas about the health-care system. Almost 60 percent of Americans say they are satisfied with their health care and health-care coverage. This is at odds with their attitudes regarding the need for health-care reform. Americans consistently identify health care as a top domestic issue for government. Three-fourths of the public believes that the present health-care system is either "in a state of crisis" or "has major problems." Support for replacing the current system jumped from 32 percent in 2004 to 41 percent in 2005 (Jones 2007b; Saad 2007b, 2007c).

Why is there a need for health-care reform? One motivation for health-care reform is economic. America annually devotes a larger share of its GDP to health care

than any other nation. As stated earlier, health care now accounts for 16 percent of America's GDP, or $2 trillion. This amounts to more than $5,000 annually for every man, woman, and child. And as you saw at the beginning of this chapter, the costs of health care are skyrocketing. The escalating costs of medical care also negatively affect the overall health of the economy. According to critics, economic resources of government—desperately needed in other areas of public life—are being siphoned off by a health-care system critics believe is out of control (Marmor and Barr 1992).

A second reason for reform is the substantial proportion of Americans who, except on an emergency basis, do not have access to medical care (see Figure 15.3). A large segment of the almost 45 million uninsured citizens are the most disadvantaged, those on the bottom of the stratification structure. Even with Medicaid, about half of America's poor are without medical coverage. Poor children are particularly vulnerable. This vulnerability extends beyond the poor. In 2005, one-fifth of those earning between $25,000 and $50,000 were uninsured as were one-fifth of those earning more than $50,000 (U.S. Bureau of the Census 2007c). Increasingly, either employers are requiring employees to pay both high premiums and deductibles for the coverage of their children and spouses, or they are providing health plans that simply do not cover dependents. Nearly one-half of uninsured children live with parents who are insured. Nearly 18 percent of children in the United States under 18 years of age were without health insurance in 2005.

Quality of life is a third reason for health reform. One obvious aspect of a lowered quality of life is the lack of health care or inferior health care. One-third

---

FIGURE 15.3

Percentage of Persons Not Covered by Health Insurance in the United States by Age: 2005

This figure displays the percentage of various age categories in the United States without health insurance. If you were running for the presidency of the United States, what would you say about this situation?

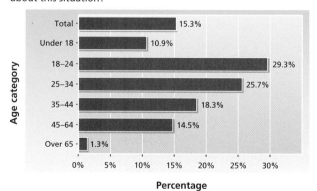

Source: U.S. Bureau of the Census, *Statistical Abstract of the United States: 2008* (Washington, DC: U.S. Government Printing Office, 2007) table 146.

Dynamic Graphics/Jupiter Images

Much of the skyrocketing cost of medical care in the United States is due to the changing nature of hospitals. This operating room displays the sophisticated technology that most hospital administrators and physicians appear to desire.

ever economic resources they have. Moreover, even the insured live with the prospect of losing coverage should they lose their jobs. As you saw in Chapter 13, more Americans, both blue- and white-collar, are experiencing job loss or, even if employed, lower compensation. In addition, companies are requiring employees to pay for more of their health-care coverage. In response, 17,000 General Electric workers went on strike in 2003 to protest the estimated $200 hike in their health-care costs (S. Greenhouse 2003).

A fourth reason for health care reform is the aging of the American population. As the baby boomers age, strains in the present health-care system are expected to increase dramatically. The remainder of this section dwells on options for health-care reform.

Health-Care Reform Options

Americans support a variety of health-care reforms, ranging from a national health-care system to government deregulation promoting competition among private health-care providers. Strength of support varies with different proposals. Seventy-seven percent of Americans support government deregulation to create competition. A narrow majority (54 percent) favor establishing a national health-care system (Saad 2007b).

What are the health-care reform options? There are a few basic options to delivering health-care services (Samuelson and Nordhaus 2004). The *modified competitiveness* approach is based on market principles such as consumer cost sharing. This approach, however, depends on universal health coverage as a precondition to health-care reform.

A second option, *managed competition*, is related to the first option, but is a combination of free-market forces and government regulation. Health care could be structured around several large plans modeled after

of Americans indicate they or a family member have postponed medical treatment because of cost (Rheault 2007). Perhaps less obvious is the fear and desperation that the uninsured feel. They know that an illness or accident experienced by themselves or a member of their family could either go untreated or absorb what-

---

FEEDBACK

1. Approximately _____ million Americans are without health insurance.
2. Which of the following highly industrialized countries does not have national health-care coverage?
   a. Germany
   b. France
   c. United States
   d. Canada
3. The major reason the United States does not have a national health-care program is because its citizens want different results from health-care programs than citizens of other developed societies. T or F?
4. Of those Americans without health-care protection, the poor and children comprise the majority. T or F?
5. The approach to health-care delivery under which medical services are financed governmentally is called
   a. managed competition.
   b. single payer.
   c. all payer.
   d. play or pay.

Answers: 1. 44; 2. c; 3. F; 4. T; 5. b

health maintenance organizations (HMOs). To increase their competitive edge, employers could form large purchasing networks, with health-care providers bidding on their business. President Bush's Medicare drug prescription plan enacted by Congress in 2003 reflects this approach through its reliance on competition among numerous health-care insurance providers. The idea is to create competition and to reward those health-care providers that have superior performance in cost, quality, and patient satisfaction.

A third option is the *single payer approach*, like the Canadian model. The government finances medical services. Canadians choose their doctors and hospitals and bill the government according to a standardized fee structure. A public opinion comparison of the U.S. and Canadian health-care systems favors the Canadian system. The Canadian approach engenders higher levels of consumer satisfaction with patient services, and it is regarded much more favorably than the U.S. system (H. Taylor 1994; Budrys 2001).

A fourth option—based primarily on the German model—uses a *play or pay mechanism*. Here, universal coverage is provided by employers that either offer employees health coverage (play) or pay into a public fund for covering the uninsured. Access to medical care in Germany, considered among the best in the world, is guaranteed for life (R. Atkinson 1994; Budrys 2001).

The Graying of America

America is not alone in its growing proportion of older persons. The situation in the United States, therefore, should be seen in perspective.

Aging of the World's Population

Graying is occurring around the world ("Why Population Aging Matters: A Global Perspective," 2007; see Worldview 15.1). Economic development carries with it a train of factors, not the least of which is improved medical care and knowledge of birth-control methods (see Chapter 17). These factors increase life expectancy while reducing the birth rate. For example, the proportion of the world's population sixty-five and older is increasing at a rate of about 6 percent annually, whereas the world's population as a whole has a growth rate under 2 percent. Consequently, the world in the twenty-first century will be an older world. The percent increase in the population aged sixty-five and over in China, for example, is expected to increase by more than 238 percent by 2025. If this occurs, the proportion of China's elderly population will increase to nearly 20 percent by the end of the first quarter of this century. By that same time, more than one-half of the world's elderly will inhabit East and South Asia. In Africa, where the most dramatic increases in the growth of the aged population are expected, the population aged sixty-five and over will have gone from 23 million persons in 1980 to more than 100 million in 2025 (Curran and Renzetti 2000).

Aging in the United States

An **age cohort** consists of persons born during the same time period in a particular population. The baby boomers, for example, constitute an age cohort born between roughly 1946 and 1964. The important thing about an age cohort is that its members pass through life

Population graying is a worldwide phenomenon. In China, the population sixty-five and over is projected to increase by over 238 percent by 2025.

Becky Reed/Alamy Limited

WORLDVIEW 15.1

Population Aging is Occurring Worldwide

Since 1950, the share of persons ages 65 and older has risen from 5 percent to 7 percent worldwide. As the map shows, Europe and Japan have led the way, with North America, Australia, and New Zealand close behind. However, older persons are now more than 5 percent of the inhabitants in many developing countries and by 2050 are expected to be 19 percent of Latin America's population and 18 percent of Asia's.

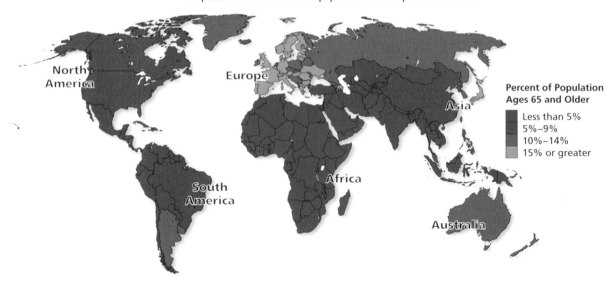

Percent of Population Ages 65 and Older

- Less than 5%
- 5%–9%
- 10%–14%
- 15% or greater

Source: *2007 World Population Data Sheet.* (Washington, D.C.: Population Reference Bureau, 2007).

Interpret the Map

1. Identify the major factors fueling worldwide population aging.

2. Do you think the elderly in the United States have too much political power? Why or why not?

3. Discuss some of the major implications of the data on this map for your generation.

together. Consequently, an age cohort not only exerts a certain influence on the rest of society, but experiences in common some effects of the particular nature of its society (Frey 2007).

The age cohort of persons sixty-five years and over—37 million strong—comprises 12 percent of the total U.S. population (U.S. Bureau of Census 2007c). By 2040 there will be about 82 million elderly Americans representing 20 percent of the nation's population (see Figure 15.4). The median age of the population will be forty-three years by 2080; it was about sixteen in 1800 and thirty-six in 2000. Due to the baby-boom generation, there will be an increasing proportion of elderly in the advanced age cohort. The number of Americans sixty-five to eighty-four will increase from about 30 million in 1990 to more than 50 million in the middle of the twenty-first century (Haupt and Kane 1998). Moreover, the number of Americans over age eighty-five will increase from about 4 million in 2000 to about 8 million in 2010 to more than 17 million by 2050.

Population pyramids are graphic representations illustrating the age and sex distribution of a population. Their use enhances understanding of an aging society. The pyramid on the left in Figure 15.5 shows the number of American males and females by age group in 1945. This is the classic pyramid shape, representing a population that has many younger people at the bottom, fewer adults in the middle, and a still smaller number of older people at the top. The pyramid on the right in Figure 15.5 portrays the projected age structure of the United States in 2050. Obviously, this is more of a rectangle or square than a geometric pyramid, with a similar number of people in all age groups from infancy to old age. This **age structure**—the distribution of people of different ages within a society—will be created by the combination of the low fertility rate (after the baby-boom years of 1946 to 1964), and increased longevity due to the low mortality rate.

What accounts for the rapid growth in the number of older Americans? The maximum size of the elderly population at any given time is due primarily to the number of births sixty-five or more years ago (Himes 2001). Between 2010 and 2030, those born in the post–World War II baby boom will dramatically increase the pool of the elderly. After this "senior boom," the rate of growth will fall significantly as those born in the so-called baby bust of the mid-1960s and 1970s reach retirement age. Around 2045,

FIGURE 15.4

Percentage of the U.S. Population 65 and Over

This graph charts the increase in elderly Americans from 1900 to 2040. Why has this trend occurred?

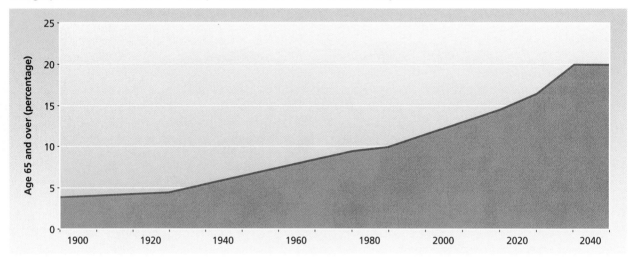

Source: U.S. Bureau of the Census.

another dramatic increase in the elderly will occur as the children of the baby boomers reach age sixty-five. This uneven growth rate will see the number of the elderly increase from 26 million in 1980 to more than 53.7 million in 2020 and then to almost 82 million in 2050.

The second most significant factor in the growth of the elderly population is the decline in the death rate. Thanks to medical advancements such as improved sanitation, antibiotics, and vaccines, the death rate among infants and children dropped precipitously between 1900 and 1954. Although only 40 percent of those Americans

born at the turn of the century had a life expectancy of sixty-five, today more than 80 percent are expected to reach this age. The life expectancy figures are, of course, even better than this. The average male life span, now seventy-five years, is expected to increase to seventy-six by 2015. Among females, the life span is projected to increase from the current eighty years to eighty-two years in 2015 (U.S. Bureau of the Census 2007c).

Immigration has had its effect through the massive pre–World War I immigration into the United States. Most of those immigrants were young adults with

FIGURE 15.5

Age and Sex Structure of the United States: 1945 and 2050

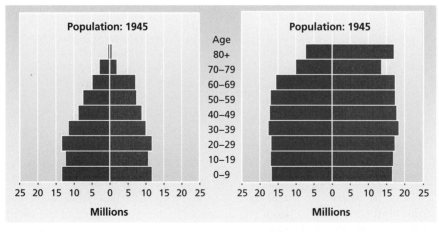

Source: U.S. Bureau of the Census.

a very high fertility rate. Their children, born in the United States, are now among the elderly.

The **dependency ratio** is the proportion of persons in the dependent ages (under fifteen and over sixty-four) in relation to those in the economically active ages (fifteen to sixty-four). A higher dependency ratio indicates that each worker has to support more dependent people; a lower ratio indicates that each worker has fewer people to support. Suppose a population of 200 persons was composed of 45 juveniles, 15 seniors, and 100 people of economically active ages. This would yield a dependency ratio of 60/100 or 0.60, denoting that each worker supports just over half a dependent person; to put it another way, there would be just over one dependent for every two working-age people. There are two subtypes of dependency: *youth dependency* and *old-age dependency*. The fictitious population just described would have a youth dependency ratio of 0.45 (45/100) and an old-age dependency ratio of 0.15 (15/100). That is, for every youthful dependent, there would be two-and-a-fraction workers; for every elderly person, there would be over six and a half workers. Less developed nations have much higher youth dependency than more developed nations. More developed nations have significantly higher old-age dependency.

Why is the dependency ratio important? For less developed countries such as Mexico, a high youth dependency means that national income must be diverted from savings (and the capital needed for economic development) to take care of its large population of children (food, housing, education). In more developed countries such as the United States, rising old-age dependency creates other challenges. There are fewer working-age people in the labor force to support the growing number of seniors. For example, in the United States in 1985, there were about 19 persons age sixty-five and over for every 100 Americans fifteen to sixty-four, an old-age dependency ratio of 0.19, or about five times as many in the economically active ages as there were elderly. By 1995, there were 25 dependent elderly Americans for every 100 economically active ones (ratio of 0.25), or just about four times as many Americans fifteen to sixty-four years of age as there were Americans sixty-five and older. In 2030, there will be 50 dependent elderly individuals for every 100 Americans of working age (ratio of 0.50). In other words, every two working-age Americans will have to support one elderly person. This shift will dramatically increase the burden on the economically active to pay for Social Security and Medicare (Kent and Mather 2002). This economic burden will be even heavier because of escalating health-care costs and the need for increasing health-care services, prescription drugs, and institutional arrangements for the long-term care of an aging population.

Perhaps, and only perhaps, the negative effects of the coming tide of some 80 million retiring baby boomers may be cushioned by their possible new view of retirement (Jacobe 2004; Carroll 2005; Saad 2006b, 2007a), which may be different than previous generations. Some surveys indicate that a majority of baby boomers intend

SNAPSHOT OF AMERICA 15.1

Projected Growth of Senior (Age 65+) Population by State, 2000-2030

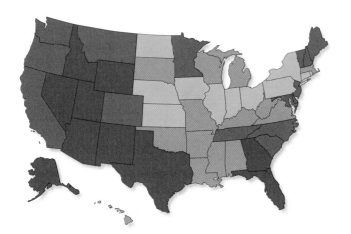

Percentage of Projected Growth

- 140% and up
- 100%–139.9%
- 70%–99.9%
- under 70%

Source: William H. Frey, "Mapping the Growth of Older America: Seniors and Boomers in the Early 21st Century," Washington, DC: The Brookings Institution, 2007, p.2.

The "graying of America" refers to the growing senior population. Improved medical care, better nutrition, and healthier lifestyles have all contributed to longer life expectancies in the United States. This map shows the projected growth of the population that is aged sixty-five years old and over in each state.

Interpret the Map

1. Describe the distribution pattern of senior people across the United States. Create a chart representing the distribution pattern.
2. Where is your state in this distribution?
3. Research the voting rate for seniors in your state. Is it higher or lower than the national average?

1. The aging of the population is confined to only the most highly industrialized societies. T or F?
2. By 2050, the size of the elderly population in the United States will constitute _____ percent of the population.
   a. 13
   b. 17
   c. 20
   d. 33
3. By 2050, the population pyramid of the United States is expected to resemble a
   a. triangle.
   b. a classic pyramid.
   c. a rectangle.
   d. an inverted pyramid.
4. By 2030, the old-age dependency ratio in the United States is expected to
   a. more than double its 1985 level.
   b. just about triple its 1985 level.
   c. remain relatively stable.
   d. be about half its 1985 level.

Answers: 1. F; 2. c; 3. c; 4. a

to retire at around age 64, but plan to reject either full-time leisure or full-time work. In a recent Gallup Poll, for example, about 35 percent of baby boomers express the hope to never work for pay again, while only 12 percent wish to work full-time (Saad 2006b). Fifty-one percent want to alternate periods of work and leisure by working part-time.

The new view of retirement is based on several factors. First, baby boomers are healthier, more highly educated, and living into their eighties. Second, the majority of baby boomers express the enjoyment provided by employment. Third, a significant proportion of baby boomers are worried about not having enough money in retirement because of soaring health-care costs and diminishing pensions (Chaddock 2004).

Health Care and Aging in the United States

We have just seen that the dependency ratio associated with the graying of America promises health-related economic burdens for America's working population. We explore this issue through topics of age inequality and ageism and health care.

Age Stratification

Thus far in our exploration of social inequality, we have examined race, ethnicity, and gender as major factors in the differential distribution of the scarce desirables (power, wealth, privileges) in a society. Because chronological age is another basis for social ranking, sociologists are interested in **age stratifi-** **cation**—the age-based unequal distribution of scarce desirables in a society. Although age can operate to the advantage or disadvantage of any age group in a society, sociologists are especially focused on the inequality existing among older Americans (Falk and Falk 1997; Pampel 1998; Schaie and Schooler 1998; Henretta 2001) because as the median age of the U.S. population grows older, this form of stratification affects more and more people.

What is the economic condition of the elderly? Because the economic situation among America's elderly has improved since 1960, the public thinks of the elderly as not only well-off, but as taking economic resources from the young (Duff 1995). It is true that the poverty rate for Americans over sixty-five years of age has declined since 1960 from 35 percent to 10 percent (see Figure 15.6). The conclusion that the elderly are therefore well-off economically, however, fails to capture the complexity of their economic condition. Among the factors masking the true economic situation of the elderly are distorted measures of poverty and factors of race, ethnicity, and gender.

What is being left out of this rosy economic picture? The measurement of poverty among older people distorts the reality: the poverty line is drawn at a higher dollar amount for the elderly than for younger Americans. Although the elderly spend proportionately more on health care and housing than younger people do, the federal government assumes a lower cost of living for the elderly. If the standard used for younger age categories were applied to elderly people, their poverty rate would increase from 10 to 15 percent. Gross poverty rates also fail to take into account the just over 7 percent of the elderly who are officially considered "near poor." Counting these at-risk elderly people, about one in six of those over age sixty-five is poor. Nor do official statistics include the "hidden poor" among the elderly population, who live either in institutions or with relatives because they cannot afford to live independently. Inclusion of these people would substantially raise the poverty rate for America's elderly.

Diversity is another factor complicating an evaluation of the economic situation of the elderly (Lee and Haaga 2002). There is a wide financial gap among the elderly. Some older people have moderate to high incomes based on dividends from assets, cash savings, and private retirement programs (Davis and Rowland 1991). Most elderly Americans, however, do not have sources of income beyond Social Security benefits. The existence of a small percentage of high-income older people gives the false impression that most elderly people are economically well-off (Himes 2001).

How do race and ethnicity compound age inequality? The elderly who are members of racial or ethnic minority groups are generally in worse condition than older white Americans. Whereas the median income of whites over age sixty-five was approximately $16,200 in 2005, the median income for elderly African Americans was $11,500; for elderly Latinos, it was about $10,400. The poverty rate among the African American elderly is nearly triple that of whites. For older Latinos, the poverty rate is two and one-half times that of white Americans. Unquestionably, this disparity is intimately linked to the

lifetime of prejudice and discrimination experienced by these racial and ethnic groups. The problem just becomes magnified in old age (Calisanti and Slevin 2001; L. Olson 2001).

How does gender heighten age inequality? Because women the world over live longer than males, there is a global trend toward the *feminization of poverty* among the elderly. Because the highest poverty rates exist among single women, the longer life expectancy of women means that, if poor, they will live longer than men will in that condition. Elderly women constitute one of the poorest segments of American society. Women over age sixty-five are almost twice as likely to live in poverty than are their male counterparts. Elderly women most likely to be poor are single women who have never married or who are divorced, separated, or widowed. Although the median income of elderly women is $12,100, it is $21,200 for males over sixty-five. Thus, the average income of elderly women is only about 60 percent that of elderly males, lower than the amount that American women under sixty-five earn relative to working males. This is not surprising, because the roots of poverty among older American women lie in the work-related experiences women have had throughout their lives (Sidel 1998). The consequences of gender inequality are compounded by the inequalities of age.

Are America's elderly a minority group? Jack Levin and William Levin (1980) contend that the field of **social gerontology**—the scientific study of the social

---

FIGURE 15.6

Poverty Rates Among Americans 65 and Over, 1960–2000

This figure documents the changing poverty rate among Americans 65 and over since 1960. Explain why it would be misleading to cite the current poverty rate as evidence that older people are economically well-off.

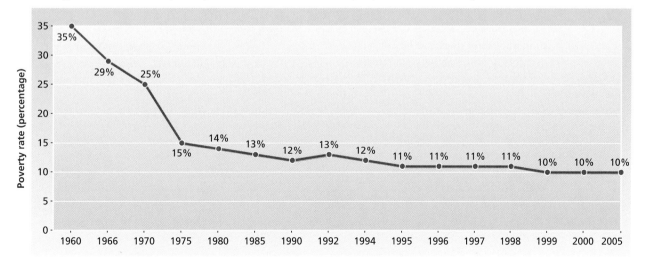

Source: U.S. Bureau of the Census, *Poverty in the United States: 2000* (Washington, DC: U.S. Government Printing Office, 2001)l; U.S. Bureau of the Census, *Statistical Abstract of the United States: 2008* (Washington, DC: U.S. Government Printing Office, 2007), table 691.

dimensions of aging—has in the past blamed the elderly for their situation in much the same way that Americans in general have blamed the poor for their plight. Because early researchers tended to study older people in institutions, their studies focused on people with diminished mental and physical capacities. This perspective coincides with the American public's negative view of the elderly. Many scholars, however, join Levin and Levin in their quest to avoid blaming the elderly for their situation and view them instead as a minority (M. L. Barron 1953; Comfort 1976; Palmore 1978; Hillier and Barrow 2007).

Richard Posner (1996) suggests that it is not necessary to consider the elderly (or women) as a minority to recognize them as victims of prejudice and discrimination. The existence of ageism alone substantiates the persistence of age-based inequality in America (Rubin 2007).

Age stratification, like most aspects of a society, must exist within a social and cultural context: it must make sense to those involved, and it must be socially justified. The justification for negative age stratification among the elderly in a society comes in the form of **ageism**—a set of beliefs, attitudes, norms, and values used to justify age-based prejudice and discrimination. Ageism should be understood in the context of aging and health care. We do so next by examining age-related stereotypes and their role in ageism and health care in the United States.

Ageism and Health Care

Stereotypes are ideas based on distortion, exaggeration, and oversimplification that are applied to all members of a social category. As with racism and sexism, ageism is based on stereotyping.

What are the stereotypes of the elderly? Robert Butler has succinctly summarized the many negative stereotypes of the elderly:

> An older person thinks and moves slowly. He does not think as he used to or as creatively. He is bound to himself and can no longer change or grow. He can learn neither well nor swiftly and, even if he could, he would not wish to. Tied to his personal traditions and growing conservatism, he dislikes innovations and is not disposed to new ideas. Not only can he not move forward, he often moves backward. He enters a second childhood, caught up in increasing egocentricity and demanding more from his environment than he is willing to give to it. Sometimes he becomes an intensification of himself, a caricature of a life-long personality. He becomes irritable and cantankerous, yet shallow and enfeebled. He lives in his past; he is behind times. He is aimless and wandering of mind, reminiscing and garrulous. Indeed, he is a study in decline, the picture of mental and physical failure. He has lost and cannot replace friends, spouse, job, status, power, influence, income. He is often stricken by diseases which, in turn, restrict his movement, his enjoyment of food, the pleasures of well-being. He has lost his desire and capacity for sex. His body shrinks, and so too does the flow of blood to his brain. His mind does not utilize oxygen and sugar at the same rate as formerly. Feeble, uninteresting, he awaits his death, a burden to society, to his family, and to himself. (Butler 1975:6–7)

Are these stereotypes accurate? By definition stereotypes are usually inaccurate because they do not apply to all members of a group. Stereotypes of older people are no exception, as much research has shown. Most elderly people are not senile. Old age is not a sexless period for the majority of those over sixty-five. There are few age differences in response to job-related challenges; most elderly people are able to learn new things and can enthusiastically adapt to change (Palmore 1980, 1981; Hendricks and Hendricks 1987; Atchley 2003).

In summary, there is enough evidence to challenge the popular stereotypes about elderly people. It is, of course, true that some older people fit one or more of these stereotypes (as do some younger people), and that more individuals are likely to fit one or more of them as they reach age seventy. This, however, does not justify applying these stereotypes to all senior people at any age or for mindlessly applying them to individuals in their fifties and sixties.

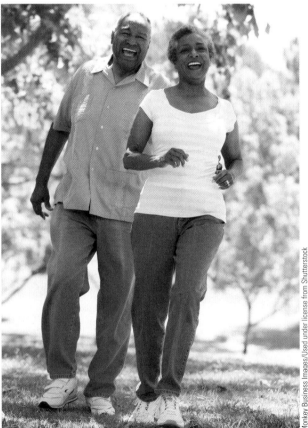

This healthy couple represents the legions of Americans who contradict the negative stereotypes of the aging.

Virtual Health Care?

No industry is structured like health care. The patient, in most cases, is not the person who directly pays most of the bill. The primary payer is almost always a third party, such as an insurance company or government agency. Knowledge that the patient is not paying most of the bill has been a disincentive to cost cutting. Therefore, doctors have provided as much care as possible at the highest cost. Given this setup, medical providers have had little reason to invest in information technology such as Internet services. It was not until the advent of managed care in the mid-1990s that concern over rising health insurance premiums became an issue in the health care industry (Francis 1999).

At the start of the twenty-first century, the Internet promises to make a significant impact on health-care costs. First, the Internet is selling medical "stuff" such as prescription drugs, medical supplies, and durable medical equipment. Insurance claims are also being filed, investigated, and paid online (Francis 1999; McTighe 1999). These mundane business tasks, however, do not directly affect the delivery of medical care.

Advances in the actual care of patients through the Internet have been longer in coming. Because the health-care industry is fragmented (thousands of individual physicians treat patients and gather information without a central means of communication), patients and doctors have not been able to access information easily. The Internet, with its standardized language and communication formats, is overcoming this obstacle and providing vast stores of easily accessed information. This central store of data makes analysis of medical research much easier and allows patients to obtain much more information about illnesses, diseases, and doctors. These improvements in health care are expected to produce better quality care and more informed patients (Nobel and Cherry 1999). Another potential benefit is the ability of patients to communicate with each other in chat rooms and to form support groups (McTighe 1999).

One example of improvement in quality and efficiency is the treatment of chronic illnesses such as diabetes, asthma, and heart disease, which are becoming more prevalent due to increased longevity of the American population. This field of medicine, know as disease management, seeks to keep patients away from expensive health care as much as possible while delivering high-quality service. The Internet permits this because patients can take medical readings from equipment provided by physicians and transmit them to health-care providers. The data, transmitted over the Internet,

are then analyzed by computer, and patients requiring special attention are made known to the physician (Nobel and Cherry 1999).

Although e-mail communication between doctors and patients is the exception, the American Medical Association states that more than one-quarter of physicians now use e-mail with some of their patients. Obvious advantages are less phone tag for doctors and fewer office visits and more rapid feedback for patients.

There are concerns about the use of the Internet in health care. The most notable of these worries is the possible breach of patients' privacy and information security. As you can imagine, the information about patients' health is very sensitive and must be well protected (McTighe 1999). Another concern is the relative lake of software programs to gather, compile, sort, analyze, and access medical data. This latter problem is expected to be solved in the near future, as software developers turn their attention to the health-care field (Francis 1999). Physicians also worry about reimbursement and increased liability. With respect to increased liability, many doctors fear harming patients through a misdiagnosis by e-mail and subsequent patient malpractice lawsuits (Kritz 2003).

Think About It
If the predicted changes in the health-care industry are realized, what effects will they have on the social status of your physician? Use one or more of the three sociological perspectives in your analysis.

Why are age-related stereotypes important in the context of health care? As with racism and sexism, ageism involves stereotyping. Prejudice and discrimination are justified in part through the creation and use of stereotypes. And when stereotypes are used to justify prejudice and discrimination, they harm the elderly.

In fact, stereotypes are used to justify prejudice and discrimination against the elderly in the context of health care. It is not a stereotype to say that health-care costs of the elderly in America will place a strain on the economy. But the elderly are unfairly harmed when age-related stereotypes are used to deny them proper health care.

Robert Butler (1989) has explained this issue, pointing out that stereotypes lead the younger generation to view the elderly as different from them. As we have seen in the study of racism and sexism, judgments of difference within a category of people usually become powerful reasons to treat them as less deserving of humane treatment. The younger generation is concerned that the impending avalanche of elderly will create a disproportionate economic burden on them. This concern may promote support for reductions in government spending on Medicare and other social-support programs for the elderly.

1. _____ is age-based in equal distribution of scarce desirables in a society.
2. Of the following, which is an *accurate* statement?
   a. Since 1960, the economic situation for the elderly in the United States has deteriorated.
   b. The poverty rate for Americans over sixty-five is lower than the official count indicates.
   c. The elderly who are members of racial or ethnic minorities are in much worse condition than other older people are.
   d. Only about one-fourth of American families in poverty have female heads of household.
3. _____ is a set of beliefs, attitudes, norms, and values used to justify age-based prejudice and discrimination.
4. Age-related _____ are used to justify prejudice and discrimination against the elderly in the context of health care.

Answers: 1. Age stratification; 2. c; 3. Ageism; 4. stereotypes

SUMMARY

1. All societies must develop ways of coping with the issues of health care and dying. Modern societies have evolved large and complex health-care systems. In the United States, for example, 14 percent of the gross domestic product is accounted for by health-care expenditures alone.
2. The major components of the health-care system are physicians, nurses, patients, and hospitals. Physicians control the health-care system in the United States and are highly rewarded in terms of power, prestige, and money. Partly because nursing began as a low-status, female-dominated occupation, it suffers from low power and pay.
3. Functionalists conceptualize a sick role, which carries with it prescriptions for the treatment and behavior of patients. Patients are expected to withdraw from or reduce their involvement in other roles, seek the advice of physicians, and attempt to recover. Failure to cooperate in recovery efforts may result in loss of accommodating privileges.
4. Functionalists and conflict theorists offer opposing views of the U.S. medical profession. The medical profession, functionalists contend, has become preeminent because Americans perceive it as essential. Conflict theorists attribute the dominance of the medical profession, the prestige of physicians, the professional and economic problems of nurses, and the inequality of health care to the power of the medical establishment.
5. Symbolic interactionism highlights the socialization of physicians and nurses. This perspective also deals with the labeling process in illness. It explains, for example, the attachment of a stigma to certain illnesses.
6. The public and politicians agree that the United States, the only major industrial power in the world without national health insurance for everyone, must undergo major health-care reform. There are several distinct options for such reform, but the surrounding political, social, and economic conflicts cloud the path.
7. The aging of the population is occurring in all highly industrialized societies. It is, therefore, not surprising that the graying of America is one of the most significant changes in the American population. Whereas persons over sixty-five years of age comprise 13 percent of the population today, they will constitute 20 percent in 2040.
8. One of the important consequences of the graying of America is the rising dependency ratio. Thus, as the population ages, there will be proportionately fewer employed people to support the growing retired population.
9. Age stratification—the distribution of scarce desirables based on chronological age—can affect any age group. The social inequality reflected in age stratification is always justified socially and culturally. The justification for the relatively low social standing among the elderly in a society is based on ageism—a set of beliefs, attitudes, norms, and values that permit age-based prejudice and discrimination.
10. Although the economic situation of America's elderly has improved over the last several decades, their poverty rate still stands at about 10 percent. Moreover, the gross poverty rate masks the true economic situation of most elderly Americans. The elderly who are members of racial or ethnic minorities or who are female are in the worst economic condition.
11. Age-related stereotypes have important implications for the health care of the elderly. They can be used to justify lowering the health-care benefits of seniors.

LEARNING OBJECTIVES REVIEW

- Define the concept of a health-care system, and identify its major components.
- Apply functionalism, conflict theory, and symbolic interactionism to the health-care system in the United States.
- Discuss health-care reform in the United States.
- Describe the aging of the world's population.
- Discuss the graying of America.
- Explain the importance of the dependency ratio.
- Distinguish between age stratification and ageism.
- Summarize the relationship between health care and aging in the United States.

CONCEPT REVIEW

Match the following concepts with the definitions listed below them.

_____ a. sick role
_____ b. ageism
_____ c. age cohort
_____ d. stereotypes

_____ e. population pyramids
_____ f. health-care system
_____ g. age stratification
_____ h. social gerontology

_____ i. dependency ratio
_____ j. epidemiology
_____ k. age structure

1. the professional services, organizations, training academics, and technological resources committed to the treatment, management, and prevention of disease
2. a set of beliefs, attitudes, norms, and values used to justify age-based prejudice and discrimination
3. the unequal distribution of scarce desirables in a society based on age
4. persons born during the same time period in a particular population
5. a social definition serving to remove people from active involvement in everyday routines, give them special protection and privileges, and set the stage for their return to their normal social roles

6. the distribution of people of different ages within a society
7. the proportion of persons in the dependent ages relative to those in the economically active ages
8. graphic representations illustrating the age and sex distribution of a population
9. the scientific study of the social dimensions of aging
10. ideas based on distortion, exaggeration, and oversimplification that are applied to all members of a social category.
11. the study of disease distribution patterns within a population

CRITICAL-THINKING QUESTIONS

1. Evaluate the idea that under certain circumstances sickness is seen as a form of deviant behavior. Draw on the concept of the sick role in forming your answer.

_____

_____

2. The medical profession in the United States enjoys a lofty position in the stratification structure. Compare the functionalist and conflict explanations of this fact.

_____

_____

3. Do you think the health-care system in the United States will be reformed in the next ten to fifteen years? Why or why not? If it is reformed, describe the basic direction you think it will take.

_____

_____

4. Discuss factors that led to the graying of America.

_____

_____

5. Based on information in this chapter, would you expect that the dependency ratio has increased, decreased, or remained stable over the past fifty years? What other factors influence the dependency ratio? How will America's dependency ratio affect you as you enter the labor force?

_____

_____

6. Examine the stereotypes commonly associated with the elderly. Based on your personal experience with older people, do you think those stereotypes are accurate? Why or why not?

_____

_____

MULTIPLE-CHOICE QUESTIONS

1. Although _____ constitute only about 10 percent of health-care workers in the United States, they establish the framework within which everyone else works.
   a. hospice personnel
   b. physicians
   c. registered nurses
   d. practical nurses
   e. medical consultants

2. Talcott Parsons uses the term _____ to describe the complex set of rights and responsibilities associated with biological illness.
   a. illness syndrome
   b. sick role
   c. doctor–patient nexus
   d. instrumental role nexus
   e. mutual contact

3. According to Marxist scholar Vicente Navarro, the success of the medical profession in the United States is due to
   a. clever lobbying by the American Medical Association.
   b. the high quality of American medical schools.
   c. the dedication of nurses and volunteers.
   d. its connection with dominant societal power blocs.
   e. its roots in religion.

4. Which of the following statements would conflict theorists be *most likely* to assert?
   a. Medical schools socialize students to accept the beliefs, norms, values, and attitudes associated with the medical profession.
   b. Those political and economic forces that determine the nature of capitalism also determine the nature of the health-care system.
   c. "Sickness" and "illness" are now seldom used social labels to stigmatize people in the United States.
   d. The experience of pain is not affected by cultural differences.
   e. Because it is now known that AIDS cannot be acquired through casual social interaction, this disease is on the way to losing its negative symbolic baggage.

5. Epidemiologists have demonstrated that health and illness are strongly influenced by _____ factors.
   a. social psychological
   b. demographic and social class
   c. anthropological
   d. genetic
   e. spiritual

6. With the exception of _____, racial and ethnic minorities in the United States have poorer health than the general white population.
   a. Native Americans
   b. Dominicans
   c. Asian Americans
   d. Aleuts
   e. Puerto Ricans

7. Which of the following statements regarding health-care reform in the United States is *accurate*?
   a. Few modern societies provide health care as a free or inexpensive service through their governments.
   b. The United States is one of the few highly developed countries with national health insurance.
   c. Only the poorest Americans are without access to medical care on a nonemergency basis.
   d. Only a little more than 1 million children in the United States are not covered by health insurance.
   e. A major reason for health-care reform in the United States is the substantial proportion of Americans who do not have access to medical care, except on an emergency basis.

8. Which of the following was *not* discussed as a basic health-care reform option for the United States?
   a. modified competitiveness
   b. managed competition
   c. all payer
   d. single payer
   e. play or pay

9. Age stratification is justified by
   a. scientific research.
   b. legal precedent.
   c. society.
   d. economic reality.

10. Graying (the aging of a population) is most common in
    a. developing countries.
    b. highly industrialized countries.
    c. preindustrial societies.
    d. African and Asian countries.

11. Which of the following factors does *not* affect the age structure of a population?
    a. fertility rate
    b. age stratification
    c. mortality rate
    d. life expectancy

12. The proportion of persons under age fifteen and over sixty-four to those between fifteen and sixty-four is known as the
    a. dependency ratio.
    b. age cohort.
    c. population pyramid.
    d. ageism.

13. Including the "near poor" in the number of at-risk elderly means that about one in _____ of those age sixty-five and older is poor.
    a. ten
    b. twenty
    c. two
    d. six

FEEDBACK REVIEW

True-False

1. The major reason the United States does not have a national health-care system is because its citizens want different results from health-care programs than citizens of other developed societies. T or F?

2. The aging of the population is largely confined to the most highly industrialized societies. T or F?

Fill in the Blank

3. The _____ embraces the professional services, organizations, training academies, and technological resources that are committed to the treatment, management, and prevention of disease.

4. If a sick person in American society does not attempt to recover, he or she will be labeled a _____.

5. According to _____ theory, affordable health care is not provided to the poor in the United States because of the desire for profit.

6. _____ refers to a set of beliefs, attitudes, norms, and values used to justify age-based prejudice and discrimination.

Multiple Choice

7. Health-care expenditures now account for about _____ _____ percent of the American GDP.
   a. 5
   b. 10
   c. 16
   d. 21

8. **Nurses are caught in a conflict between their values and their work environment. Which of the following perspectives *best* helps us understand this situation?**
   a. functionalism
   b. conflict theory
   c. symbolic interactionism
   d. exchange theory

9. **Of the following, which is an *accurate* statement?**
   a. Since 1960, the economic situation for the elderly in the United States has deteriorated.
   b. The poverty rate for Americans over sixty-five is lower than the official count indicates.
   c. The elderly who are members of racial or ethnic minorities are in much worse condition than other older people are.
   d. Only about one-fourth of American families in poverty have female heads of household.

10. **By 2050, the population pyramid of the United States is expected to resemble**
    a. a triangle.
    b. a classic pyramid.
    c. a rectangle.
    d. an inverted pyramid.

GRAPHIC REVIEW

Figure 15.5 depicts the age and sex structure of the United States as of 1945 and its projected age and sex structure in 2050. Answer the following questions to check your understanding of this information as it relates to the material in the chapter.

1. What was the largest age cohort in 1945?

2. What factors contribute to the change in the shapes of the population pyramids in 1945 and 2050?

3. Why is the seventy to seventy-nine age cohort markedly smaller than nearly all of the other age cohorts in the 2050 population pyramid? Check your answer against what you know about the baby boomers, Generation X, and Generation Y.

ANSWER KEY

Concept Review

a. 5
b. 2
c. 4
d. 10
e. 8
f. 1
g. 3
h. 9
i. 7
j. 11
k. 6

Multiple Choice

1. b
2. b
3. d
4. a
5. b
6. c
7. e
8. c
9. c
10. b
11. b
12. a
13. d

Feedback Review

1. F
2. F
3. Health care system
4. deviant
5. conflict
6. ageism
7. c
8. c
9. c
10. c

16 Sport

Diademimages/Dreamstime LLC

Chariot Race (gouache on paper), Jackson, Peter (1922–2003)/Private Collection. © Look and Learn/ The Bridgeman Art Library International

Is sport the way to upward social mobility for minorities? It may appear so on the surface.

Collegiate football and basketball have made it possible for many African Americans to go to college who would not have done so otherwise. There are many wealthy Latino baseball players. But there is evidence that sport actually impedes upward mobility for minorities. The chances of becoming a collegiate or professional player are very small. Minority youth, attracted by the lure of fame and wealth, often fail to obtain the education they need to advance in life.

The issue of social mobility and sport will be covered later in this chapter, along with sport and racism and sport and sexism. The topic of sport as a social institution, however, is our starting point.

The Nature of Sport

As a social institution, sport fulfills some important social needs. One of these is helping individuals identify with other members of society. Sport subcultures have developed around both team and individual sports. For this reason, sport is a reflection of society.

A Definition of Sport

For most people, sport consists of certain leisure activities, exercise, and spectator events. It is actually more complex than that. Sociologists define **sport** as a set of competitive activities in which winners and losers are determined by physical performance within a set of established rules. While sport is an important aspect of recreation, many forms of recreation do not involve sport. Sport sociologist Jay J. Coakley (2006) sees a spontaneous race between two skiers as more of a contest than a sport. Although a contest between skiers involves physical activity and competition, it does not involve definite rules or standardized conditions.

Sport as a Social Institution

Institutions fulfill certain basic needs and reflect the most important aspects of a society. The five most commonly recognized social institutions have been examined in preceding chapters: family, education, government, economy, and religion. Although these social institutions take different forms in different societies, they appear in every society because they fulfill needs common to all societies.

Because societies may have additional needs, there may be additional social institutions. In American society, sport is one of these additional institutions. Sport teaches some of the basic values of society. It also promotes attachment to society. For example, a society requires that its members identify with it. Members must feel that belonging to the society is an important part of who they are. Sport aids in this identification of self with society.

Chariot racing in ancient Rome is a clear illustration of this social identification. Athletes would risk their lives in this dangerous sport in part to reflect their self-identification as Romans.

The individual, even when free, did not belong to himself; he was strictly subordinated to the city. His life, his death, were only episodes in the history of the group. To confront death was not an act of exceptional heroism; it was the normal way of proving oneself a Roman. (Auguet 1972:198)

Sport, Culture, and Society

Sport plays a central role in American society in part because it reflects the culture's emphasis on achievement (Mandelbaum 2004a).

People who visit the United States from other countries are often amazed at the extent to which competition [in sport] is

The chariot races of the Roman Empire involved considerable skill and courage. Charioteers delivered the violence required to please the crowd.

used to distribute rewards and evaluate the work of human beings. (Coakley 2006:82)

The prevailing American view of sport is the one reportedly expressed by the late Vince Lombardi, coach of the Green Bay Packers of the National Football League during the 1960s: "Winning isn't everything. It's the only thing." For the most part, sport continues to be dominated by achievement-oriented values.

Sport reflects society, so it is not surprising that males dominate the sports world just as they do many other aspects of American society. Females are second to men overall in power, income, and job opportunities in sports just as they are in business, education, medicine, and law (Eitzen 2006).

Some progress toward equality is being made, however. The women's basketball teams at both the University of Tennessee and the University of Connecticut have, at different times, held the best single-season record in men's or women's basketball history. Women are making inroads in professional golf and tennis, and the women's professional basketball league (WNBA) has many fans. Women's place in athletics was given a healthy boost in 1999 when Mia Hamm and her teammates won soccer's World Cup. Equality of opportunity for women in sports, though, remains a distant goal.

Sport Subcultures

The relationship among sport, society, and culture can also be seen in sport subcultures. A **sport subculture** is a group within the larger context of sport that has some of its own distinct roles, values, and norms. These subcultures are organized around a sport activity, and beliefs vary widely. Sociologist Michael Smith (1979) wanted to know if violence among hockey players is due to involvement in a "subculture of violence." In this kind of subculture, violence is the expected response to a perceived challenge or insult—a jostle, a glance, or an offensive remark. Following this norm is essential in acquiring and maintaining honor, especially when challenges are associated with masculinity.

Smith found that hockey players favor violence more than nonplayers. Because of the expectations of coaches and teammates, many hockey players act violently during games. In fact, players criticize teammates who aren't violent. As one National Hockey League player put it:

I don't think that there's anything wrong with guys getting excited in a game and squaring off and throwing a few punches. That's just part of the game. It always has been. And you know if you tried to eliminate it, you wouldn't have hockey any more. You look at hockey from the time it was begun, guys gets excited and just fight, and it's always been like that. (Eitzen 2005a:165)

Kent Pearson (1981) researched subcultures involving water-related sports in Australia and New Zealand. He found major cultural differences between surfboard riders and surf lifesavers. Surfboard riders avoid formal organizations, work with loose and flexible definitions of the territory in which their sport will occur, place a heavy emphasis on physical prowess and individualism, and generally oppose the larger society. In contrast, surf lifesaving clubs are highly organized entities that stage competitions involving swimming, boating, and lifesaving. The territory for such competitions is precisely defined, and formal rules are employed.

Even in nonteam sports, subcultures emerge Thoroughbred jockeys have developed a subculture with a strong emphasis on displaying dignity, poise, and integrity. The ideal within the subculture of jockeys is a fiery animal with a cool rider.

The cool jockey can wait patiently with a horse in a pocket and get through on the inside, risking the possibility that there will be no opening. Coolness is waiting far back in the pack, risking the possibility that his horse will not "get up" in time. Coolness is sparing the whip on a front-running horse when another animal has pressed into the lead, risking the possibility that once his horse is passed he will not get started again. All these activities are taken by observers as instances of a jockey's character. In short, moral character is coolness in risky situations. (Scott 1981:146–147)

Jockeys take such chances partly because their subculture requires it. Jockeys who fail to display gallantry,

What values are at the center of the unique jockey subculture?

FEEDBACK

1. _____ is a set of competitive activities in which winners and losers are determined by physical performance within a set of established rules.

2. A group within the larger context of sport that has some of its own distinct roles, values, and norms is a _____.

3. Which of the following is *not* an example of sport?
   a. a baseball game between two major league teams
   b. a baseball game between two minor league teams
   c. a spontaneous race between two cyclists
   d. a swim meet involving amateur athletes

Answers: 1. Sport; 2. Subculture; 3. c.

integrity, and coolness—qualities expected of them by horse owners, trainers, and other jockeys—do not receive their choice of horses and therefore win few races. Failing to take risks leads to lost opportunities.

Theoretical Perspectives and Sport

Functionalists see sport positively, as a means for socializing young people, promoting social integration, providing a release for tensions, and developing sound character. Conflict theorists believe that organized sports can be harmful to character development. Symbolic interactionists focus on the self-concepts and relationships developed through sport activities.

Culture and Sport

Sport is a major social activity through which culture is created and reinforced. As noted earlier, sociologists recognize this important aspect of sport.

> *American sport embodies American values—striving for excellence, winning, individual and team competition, and materialism. Parents want their children to participate in sport because participation teaches them the basic values of American society and builds character.* (Eitzen 2005a:3)

Although sociologists agree that sport mirrors society, and that the relationship is complex, they disagree over the social implications of sport. Sport sociologist Stanley Eitzen has written a book on the paradoxes, or contradictions, of sport in America (see Table 16.1). Functionalists, who tend to concentrate on the benefits of sport, are represented in Eitzen's book. So are conflict theorists, who see a social downside to sport. Symbolic interactionists focus on personal meanings derived from sport.

Functionalism

How do functionalists view the role of sport in society? Functionalists think sport is important primarily because it helps society work more smoothly. It does this by performing the following functions (Eitzen and Sage 2002):

- *Sport teaches basic beliefs, norms, and values.* Sport readies us for adult roles. Games, for instance, prepare participating athletes for work in organizations. Young people who are exposed to competitive sport become more motivated to achieve than those who are not. And the earlier the exposure occurs, the higher the orientation toward achievement. This is important because achievement–motivation is essential to productivity in the modern economy.

- *Sport promotes a sense of social identification.* A team binds people to their community and nation. Clevelanders are united in their love of the Browns, Indians, and Cavaliers. The citizens of Green Bay, Wisconsin, seem to live or die on the fortunes of their beloved Packers.

- *Sport offers a safe release of aggressive feelings generated by the frustrations, anxieties, and strains of modern life.* It is socially acceptable to yell and scream for an athletic team. Similar behavior directed at a teacher, principal, parent, or employer can have negative consequences.

- *Sport encourages the development of character.* Coaches, school officials, and parents often draw a parallel between sport and "life." "When the going gets tough, the tough get going" is a sentiment expressed in most locker rooms. The hard work, discipline, and self-sacrifice demanded by team sports become part of an athlete's value system.

What are the social dysfunctions of sport? Functionalists have identified some drawbacks to sport. Because sport

TABLE 16.1

Sport Paradoxes

Stanley Eitzen, a highly respected sport sociologist, argues that sport is inherently contradictory. Here are a few of the paradoxes Eitzen identifies. Do you agree with Eitzen that these paradoxes exist?

Social Integration
Sport can unite different social classes and racial/ethnic groups
but
sport can heighten barriers that separate groups

Fair Play
Sport promostes fair play by teaching the importance of following the rules
but
sport's emphasis on winning tempts people to cheat

Physical Fitness
Sport promotes muscle strength, weight control, endurance, and coordination
but
sport can lead to the use of steroids adn other drugs, excessive weight loss or gain, and injuries.

Academics
Sport contributes to higher education through scholarships and fund raising
but
sport takes money away from academics and emphasizes athletic performance over learning and graduation.

Social Mobility
Sport allows athletes who might otherwise not attend college to obtain an education
but
only a few can achieve the promise of fame and wealth in the professional ranks.

Source: D. Stanley Eitzen, Fair and Foul (Lanham, MD: Rowman & Littlefield Publishers, Inc. 2006), pp. 4–7.

reflects society, it draws on achievement-oriented values that can be intensified to an extreme degree (Kohn 1992). When achievement and winning come to be seen as the primary goals of sport, any method of winning—including violence and cheating—may be encouraged.

We need not look far to see examples of violence in sport. Coaches and fans expect athletes to place their physical well-being on the line. Players in many sports are expected to resort to violence. In high school football, aggressive behavior is defended as preparation for "real-life" competition. Pressures are intensified at the professional level, where many sports have developed the informal role of *enforcer*—a team member whose major responsibility is to intimidate, provoke, and even injure opponents (Coakley 2006). Boston Bruins hockey player Marty McSorley used his hockey stick to deliver a vicious blindside slash to the head of opposing player Donald Brashear in February 2000. The attack was the culmination of a game marked by injuries and was the result of the long-standing rivalry between two "top enforcers."

Cheating may not be as easy as violent behavior to identify, but it is often present nonetheless. Cheating was no doubt involved when American Olympic skater Nancy Kerrigan was struck on the right leg with a metal rod by an assailant later linked to her competitor, Tonya Harding. In 2008, Marion Jones was sentenced to six months in prison for lying to federal agents investigat-

ing illegal steroid use (O'Connor 2008). Of course, the use of steroids by professional athletes in track, baseball, or any sport is cheating. And Bill Belichick, the coach of the New England Patriots, received a $500,000 fine by the National Football League for videotaping opponents' defensive signals. Cheating can even extend beyond players, as when a Utah state committee used illegal inducements to attract the 2002 Winter Olympics.

Sport promotes a sense of social identification. These Oakland Raiders fans are carrying social identification to the extreme.

WORLDVIEW 16.1

Olympic Success

Sport also plays an important role in today's global society. For some time, the winning of Olympic medals has been a source of regional and global prestige. This map shows the number of medals earned by each country in the 2004 Summer Olympic Games

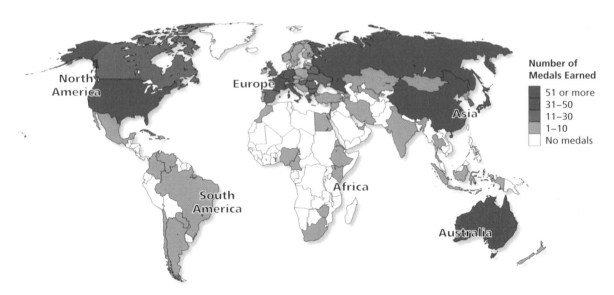

Number of Medals Earned

- 51 or more
- 31–50
- 11–30
- 1–10
- No medals

Source: *The World Almanac and Book of Facts, 2005*

Interpret the Map

1. Why do you think there is such wide variation in the number of Olympic medals earned?

2. Do the Olympics illustrate connection between sport and politics? Explain.

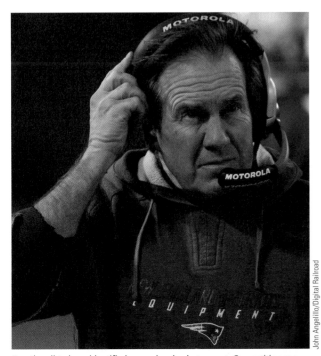

John Angelillo/Digital Railroad

Functionalists have identified some drawbacks to sport. Competition can lead to cheating, as when the New England Patriots under head coach Bill Belichick illegally taped opposing team signals during games.

Conflict Theory

Some sociologists have raised disturbing questions about the effects of sport on society. These questions are best understood through the conflict perspective. Conflict theorists are interested in who has the power and how elites use power to satisfy their own interests. To conflict theorists, sport is a social institution in which the most powerful oppress, manipulate, coerce, and exploit others. Conflict theorists highlight the ways in which sport mirrors the unequal distribution of power and money in society. They also emphasize the role of sport in maintaining inequality (Cozzillio 2005).

While functionalists see sport as contributing to the unification of society, conflict theorists do not. While people from all major segments of a community or society may join in cheering for the same team, their union is only temporary.

When the game is over, the enthusiasm dies, the solidarity runs short, and disharmony in other relations reasserts itself. Much as one hour a week cannot answer to the religious impulse, one game a week cannot answer to the solidarity needs of a racist, sexist, or elitist society. (Young 1986)

Basic social class divisions, in other words, will continue to exist and to affect social relationships in a community even if the local team has just won the World Series or the Super Bowl.

The contribution sport makes in forming good character is also questioned by conflict theorists. Among college athletes, studies have shown that the degree of sportsmanship apparently declines as athletes become more involved in the sports system. As sociologist Stanley Eitzen (2005b) notes, nonscholarship athletes display greater sportsmanship than those with athletic scholarships, and those who have not earned letters exhibit more sportsmanship than letter winners.

Conflict theorists can point to any number of past and present scandals in both the college and professional ranks. Americans are constantly reading in the sports section of the daily newspaper about athletes, in high school and college, who are taking drugs, cheating in school, or accepting illegitimate cash "gifts." One university after another is being investigated and penalized by the National Collegiate Athletic Association (NCAA). Coaches as well as players are involved in misconduct.

Athletes may use performance-enhancing drugs such as steroids and amphetamines to achieve a "competitive edge." . . . Big-time college coaches in their zeal to win have been found guilty of exploiting athletes, falsifying transcripts, providing illegal payments, hiring surrogate test takers, paying athletes for nonexistent summer jobs, and illegally using government Pell grants and work study monies for athletes. So much, I would argue, for the myth that "sport builds character." (Eitzen 2005a:189)

Symbolic Interactionism

Symbolic interactionism also contributes to our understanding of sport as a social institution. This theoretical perspective concentrates on personal meanings, social relationships, and self-identity processes. Symbolic interactionists are concerned with the symbols of sports. The meanings and interpretations of these symbols are important because they affect the self-concepts, as well as the relationships, of those involved.

The social context of Little League baseball illustrates this perspective. For three years, Gary Alan Fine (1987) studied American adolescent suburban males who played Little League baseball. He discovered and documented a variety of ways in which the boys assigned meanings to their team activities. In addition, he described how these meanings and interpretations influenced the boys' social interactions and affected their self-definitions.

What were these meanings? Much of the activity of coaches and parents centered on teaching the rules of the game and teaching values, such as team play, hard work, fair play, competition, and winning. But these ten- to twelve-year-old boys formed their own interpretation of these messages. The boys misinterpreted the adult values of hard work, competition, and so forth as the "masculine" values of dominance, "toughness," and risky behavior.

How were social interaction and self-concepts affected? In the first place, the boys' behavior convinced coaches and parents that the youngsters understood and accepted their values. For example, the aggressive behavior that the boys considered as evidence of their masculinity was seen by the coaches and parents as evidence of "hustle,"

Do you think participation in a team sport at an early age promotes or harms self-esteem?

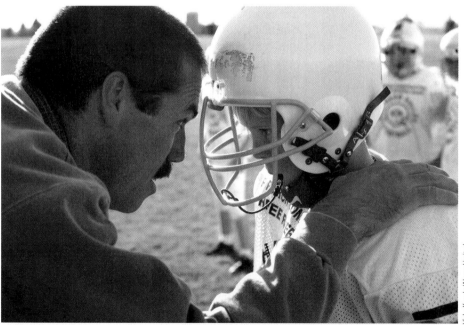

Vicky Kasala/Alamy Limited

a dedication to competition, and the desire to win. The boys were praised for this behavior, which encouraged them to continue it. "Weaker" peers, younger children, and girls in general frequently experienced the disdain of these Little Leaguers. This disrespect often led to a loss of self-esteem for children who suffered the brunt of the Little Leaguers' scorn.

What are some limitations of each perspective? The functionalist perspective makes important points regarding the positive and negative role of sport in society. Its critics, however, contend that many sports have become so closely tied to elite interests that they contribute more to private profit than to the general well-being of society. To investigate this point, the conflict perspective concentrates on some major concerns of sport, such as racism and sexism (discussed in the next section). On the other hand, conflict theorists tend to overlook the positive contributions of sport to society. They are accused of placing too much emphasis on the extent to which sport is manipulated and controlled by the elite. Their critics also claim that conflict theorists underestimate the character-building benefit of team sports. Symbolic interactionism contributes greatly to understanding the socialization process in sport. But, because it concentrates on social interaction, it fails to include the broader social and cultural context. For example, symbolic interactionism does not address the functions of sport in society or explore sport within the context of power and social inequity.

TABLE 16.2

Social Effects of Sport

This table illustrates how each theoretical perspective might study an issue involving sport. For each assumption, provide a specific example from your own experience or from a tem you folllow.

Theoretical Perspective	Concept	Power Elite Model
Functionalism	Social integration	Athletic teams promote togetherness and belonging in a community.
Conflict Theory	Social conflict	Deep social conflict exists within a community and persists despite widespread attachment to athletic teams.
Symbolic Interactionism	Social concept	Participation in a team sport may promote or harm self-esteem depending upon factors such as emphasis on winning and fair play.

FEEDBACK

1. Which of the following is *not* one of the functions of sport discussed in the text?
   a. Sport solidifies social class identification.
   b. Sport teaches basic beliefs, norms, and values.
   c. Sport provides a sense of identification.
   d. Sport offers a safe release of aggressive feelings.
   e. Sport encourages the development of character.
2. Several social aspects of sport are listed below. Indicate which would most interest functionalists (F), conflict theorists (C), and symbolic interactionists (SI).
   ____ a. Intensification of achievement-oriented values
   ____ b. Maintaining social inequality
   ____ c. Character building
   ____ d. Self-identification
   ____ e. Sport scandals

Answers: 1. a; 2. a. (F), b. (C), c. (F), d. (SI), e. (C)

Tough Guys, Wimps, and Weenies

Donna Eder researched the nature of middle-school sports using the case study method. Following the framework of symbolic interactionism, Eder assumes that the social world of teenagers is constructed through interaction with others. Thus, everyday exchanges—insults, greetings, gossip—give teenagers a sense of their self and the social world.

Middle-school coaches accented the value of toughness. In the world of athletics, having a "mean" attitude is masculine and being nice is effeminate. Wrestlers, for example, were told to make opponents "suffer." Football coaches did not tolerate fighting off the field, but as a means to handle conflict among athletes, these same coaches encouraged physical force on the field.

I said that I had heard that Coach Paulson wasn't pleased with the way the team played. Walter and Carl both agreed. Walter [the team manager] said that the team didn't hit like they should have and that made the coach mad. Carl said, "Yeah,

but I really socked that guy. Man, I threw him down on the concrete. Did you hear Coach James yelling, "Way to go, Orville"? (Eder 1995:62)

Evidence of weakness was greeted by derogatory names like "wuss," "wimp," and "girl." Ritual insults promoted stereotypically masculine behavior, particularly among higher-status boys. Stories of physical force in sports were repeated with pride. Even soccer players bragged about kicking opponents in the shins or throwing a ball into an opponent's face.

The most forcefully combative boys were the most respected. Although the coaches tried to curb physical violence outside of games and matches, many players considered fighting an appropriate way to handle all peer conflicts.

*[The] importance of being tough extended to behavior off the playing field as well as on it. Boys were continually challenged to develop more aspects of toughness, including the ability to deny pain and suppress feelings as well as respond combatively to verbal and physical attacks. Boys who rejected these messages were sometimes subject to ridicule by girls as well as boys, showing the difficulty boys faced when trying to escape the pressures of being masculine within this school setting. (*Eder 1995:72)

Insult exchanges could be won by getting another boy to become angry. By losing his cool, the other boy lost his image of toughness. Some boys

would insult another boy just to look good to others. An example is provided by one of the researcher's notes on Hank, the highest-status boy in the seventh grade, who had a reputation for verbal assault:

*Hank does seem to enjoy conflict or competition on a one-on-one basis. A couple of times today he left the table just to go down and abuse some kid at the end of the table, calling him a pud, a squirrel, or a wimp. Then he would come back and tell the group how the guy had done nothing when he had said this. Hank would get a big smile on his face and was really pleased. (*Eder 1995:73–74)

Insults and counter-insults delivered several messages. First, boys learned not to care about the feelings of others. Second, insulting, or even humiliating, their peers was a socially approved means of achieving or displaying higher status. Third, boys who humiliated low-status peers were rewarded with social recognition. This was true even if the target of ridicule had a disability or was overweight.

Evaluate the Research
1. Do you think this study describes sports at your school? Explain.
2. Do female athletes treat each other differently from the way boys treat each other? Explain.

Social Issues in Sport

Sport contributes to upward mobility among collegiate athletes, but the opportunities are too few. Minorities still face discrimination in sport. Women in sport suffer from gender-based stereotypes. Intercollegiate female athletes do not receive treatment equal to the treatment received by males, although this situation is slowly improving.

Sport and Social Mobility

The autobiographies of star athletes often point to sport as their way out of poverty. One educator once predicted that "football would enable a whole generation

of young men in the coal fields of Pennsylvania to turn their backs on the mines that employed their fathers" (Rudolph,1962:378). Many athletes do use sport as a means out of their equivalent "coal fields," and many minority members work their way out of poverty through sport. It is also true that the average salaries of professional athletes are very high. Even so, let's examine this alleged relationship between sport and social mobility.

Does sport really promote social mobility? Participating in sport increase the likelihood of improving a person's place in the stratification structure. Whatever sport they play, college athletes tend to be better educated, earn more money, and have higher occupational prestige than their fathers. This is the very definition of upward social mobility. And, in these terms, college athletes as a whole are more successful than college students who do not participate in sports (Leonard 1998). Although this finding is meaningful, it has not settled the debate regarding how much sport promotes upward mobility for minorities.

Does sport promote upward mobility for minorities? Some people argue that sport is a social class escalator for minorities. They point to Tiger Woods, Kobe Bryant, and Venus and Serena Williams, among others. A different viewpoint argues that the emphasis on sport is harmful because it diverts attention away from learning the academic and business-related skills necessary for success in mainstream American society. Because of the lure of high salaries and prestige, many aspiring minority athletes fail to develop alternative career plans. Minority members who spend their youth sharpening their athletic skills at the expense of their general education will very likely be casualties of an unrealizable dream of wealth and glory (Lapchick and Matthews 2004; Lapchick and General 2007).

Some convincing evidence supports those who see sport as a barrier to upward mobility for minorities. Table 16.3 shows that there are more than 1 million high school players. Fewer than 60,000 of these players become college football players. And only 1,600 of

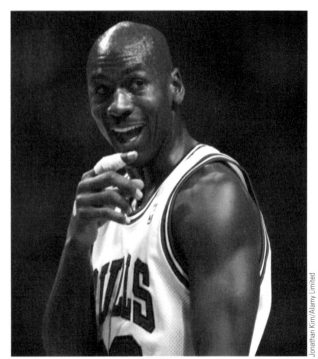

The phenomenal success of basketball star Michael Jordon is frequently used as proof that sport is a path to upward social mobility for minorities. How typical is Jordon's experience?

TABLE 16.3

Athletes' Chances of Advancing to the Pros

This table shows the slim chance that high school athletes have to play a professinal sport. Does this surprise you?

	Number of Players in High School	Percentage Advancing from High School to College	Number of Players in College	Percentage Advancing from College to Proffessional Level	Number of Players at Professional Level	Percentage Advancing from High School to Proffessional Level
MALES						
Football	1,002,734	6%	57,593	3%	1,643	0.16%
Basketball	541,130	3%	15,874	2%	348	0.06%
Baseball	451,701	6%	25,938	3%	750	0.17%
Ice hockey	27,245	13%	3,647	18%	648	2.38%
Total	2,022,811	5%	103,052	3%	3,389	0.17%
FEMALES						
Basketball	451,600	3%	14,445	1%	132	0.03%
Golf	49,690	6%	3,108	2%	52	0.10%
Tennis	159,740	5%	8,314	2%	150	0.09%
Total	661,030	4%	25,867	1%	334	0.05%
Grand Total	**2,683,841**	**5%**	**128,919**	**3%**	**3,723**	**0.14%**

Source: National Federation of State High School Associations, 1999–2000.

these college players become professional players. Thus, the probability that a high school football player will make it to the pros is less than two-tenths of 1 percent. Similarly a high school baseball player has a 0.2 percent chance of becoming a major leaguer. The odds are even worse for a high school basketball player, who has a 0.1 percent probability of making it to the National Basketball Association. Moreover, those who become professional athletes have short careers on the average: one to seven years for baseball players, four to six years for basketball players, and four-and-one-half years for football players.

Of course, this does not mean minority athletes should not enjoy the benefits of a collegiate sport. To be sure, some athletes, who may otherwise not have had the chance, have received good college educations. It does argue, however, that no high school athlete—minority or white, for that matter—should rely solely on sport as a ticket up the stratification structure.

Racial and Ethnic Inequality in Sport

One sign of systematic discrimination shows up in what is called stacking. In **stacking**, players are assigned to less central positions on the basis of race or ethnicity. "Central" positions are those that involve leadership and decision-making responsibilities and thus offer a greater likelihood of influencing the outcome of the game. Historically, minorities have more often been assigned to positions requiring relatively little interaction and coordination with other players. In football, for example, African American quarterbacks traditionally have been rare, while the proportions of African Americans in many defensive and other less central positions have been high (see Figure 16.1).

Such discrimination has important economic consequences, because the positions occupied by most African Americans have high injury rates that cut careers short. Both salaries and pension benefits are reduced as a result.

FIGURE 16.1

National Football League Positions, by Race

Do you think that these data support the presence of stacking in the NFL?

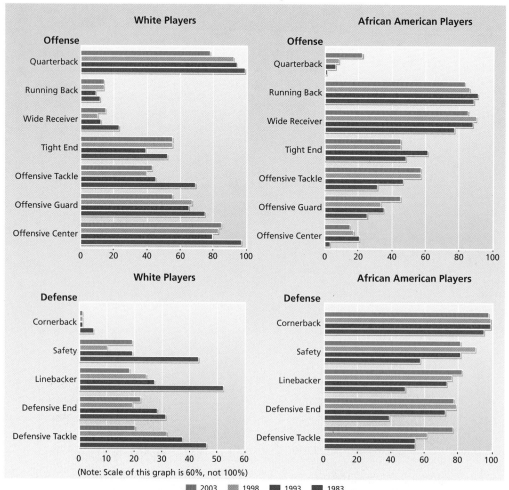

(Note: Scale of this graph is 60%, not 100%)

2003 1998 1993 1983

Source: The Institure for Diversity and Ethics in Sport, 2004.

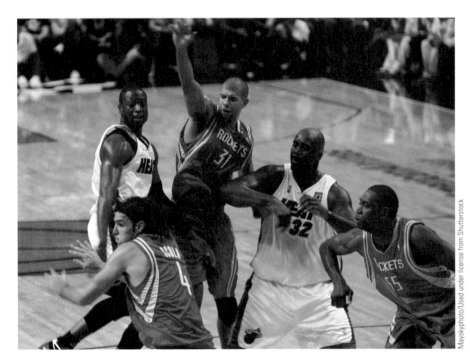

Minorities are well represented as players in major U. S. sports. However, after their playing days are over, minorities are underrepresented in positions within their sport.

Is there salary fairness in professional sports? Discrimination in salary at the professional level exists. African Americans in the major professional sports are, on the average, paid as much as or more than their white counterparts. It is only when level of performance is controlled that discrimination appears—African Americans have lower average salaries than whites for the same level of performance. In other words, African Americans must perform better than whites to avoid pay discrimination (Eitzen and Sage 2002).

What other areas of discrimination have been found? Minority former athletes profit much less than their white colleagues from personal appearances and commercial endorsements. They also lose out in sports-related careers when their playing days are over. While

SNAPSHOT OF AMERICA 16.1

Who Are the Biggest Baseball Fans?

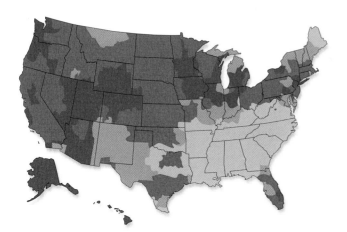

Regional Baseball Viewership on Television

- High
- Above average
- Below average
- Low

Adapted from *Latitudes and Attitudes: An Atlas of American Tastes, Trends, Politics, and Passions.* Boston: Little, Brown.

Baseball fans used to be young working-class white males. Today's fans are older and more affluent but still predominantly white and male. The relative lack of African American fans might be traced to baseball's traditional racist policies on the field and in the front office.

Interpret the Map

1. Do you see any regional patterns in the rates of baseball viewership? Describe.
2. How do you explain these patterns?
3. As a baseball fan, are you similar to or different from the general pattern in your state? Why?

approximately 76 percent of players in the National Basketball Association (NBA) are black, only about 17 percent of radio and television NBA sports announcers are African American, and only about 8 percent of the announcers are Latino.

At the professional level, there are few minorities represented in the power structure—head coaches, general managers, owners, executives, commissioners. In 1989, Bill White became the first African American to head a major professional sports league. As of 2004, however, only two major sport franchises in the United States were owned by minorities. And only four minorities in the NFL and Major League Baseball were general managers, presidents, or CEOs. There are only three African American head coaches in professional football, and only three African American baseball managers and four Latino baseball managers. However, almost 40 percent of NBA head coaches are members of minority groups.

What about racial and ethnic inequality in college sports? Much of what has been said about professional sports applies to collegiate sports. According to Richard Lapchick and Bente General (2007), positions of power at the collegiate level are overwhelming held by white men. Looking at Division 1A of the NCAA, they found that 92 percent of campus leadership positions are in the hands of whites, mostly males—94 percent of the presidents, 87 percent of the athletic directors, and 100 percent of the conference commissioners. (White women are in only about 15 percent of these 367 positions.) Only two Asian males, one Native American male, and one Latino woman are leaders.

Lapchick and General note some improvement in 2007, slight as it is. For example, head football coaches of color are up 2.4 percent (now six), and the percentage of female presidents increased by 3.2 percent (now seventeen). After recognition of this small improvement, the researchers stand by their overall conclusion that college sport is controlled by white males.

Gender Inequality in Sport

Racial and ethnic minorities have not been the only victims of prejudice and discrimination in sport. Women have experienced sexism in athletics. The cultural roots of sexism date back at least as far as the ancient Greeks. Greek gods were depicted as athletic, strong, powerful, competitive, rational, physical, and intellectual. Most Greek goddesses were passive, beautiful, physically weak, supportive, unathletic, and sexually attractive. (The few active strong goddesses were usually not attractive to nor attracted by men. To Greek males, women who were physically or intellectually superior to them were unfeminine.) These gender definitions have survived in large part for the past 2,500 years. Their influence is felt in sport just as it is in other aspects of social life.

What are some of the consequences of sexism? Stereotypes have traditionally discouraged females from playing sports. For centuries, the idea that playing sports makes females more masculine has been widespread. To be an athlete, females were told, is to be unfeminine. This stigma discouraged many females from participating in athletics and tyrannized many of those who did. Another barrier was the old, discredited argument that

Although sexism in sports has been decreasing, women athletes continue to suffer from inequalities.

Amy Sancetta/AP Photos

FIGURE 16.2

Percentage of College Women's Athletic Teams Coached by Women

What is most interesting to you about these data?

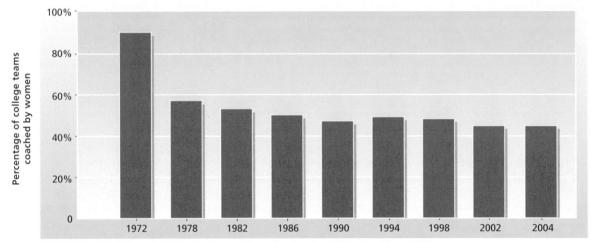

Source: R. Vivian Acosta and Linda Jean Carpenter, "Women in Intercollecgiate Sport." Brooklyn College, 2004.

sports harm a woman's health, particularly her ability to have children.

Sexism has denied females equal access to organized sports. At the local level, resistance to female participation in sports continues to exist. It was not until the mid-1970s that, under legal threat, the national Little League organization ended its males-only policy. Only when the 1972 Educational Amendment Act (Title IX) was passed were public high schools and colleges required to offer females equal access to sports. Originally, Title IX was interpreted as providing equal opportunity in "all" sport programs of institutions receiving federal funds. Ambiguities in Title IX have led to many legal suits. Important issues remain unresolved. Currently, the courts favor matching the ratio of males and females in a school's athletic programs to their proportionate numbers in the student body of that school (Hannon 2003; Orecklin 2003).

Why has the percentage of women coaching women's programs declined? Women are still denied equal access to the power structure of sport (Lapchick and Matthews 2004). What's more, although Title IX increased equality for female athletes, it led to a decrease in the number of coaching and administrative positions held by women. In the early 1970s, women's intercollegiate teams were headed almost entirely by women. Still, more than half of the NCAA women's teams are coached by men (see Figure 16.2). Only about 20 percent of all women's programs are headed by a female administrator, and females hold only 41 percent of all administrative jobs in women's programs (Acosta and Carpenter 2004).

Ironically, Title IX may be one reason for this decline. As the money and prestige associated with women's programs have increased, men have found these coaching jobs much more attractive. And conflict theorists believe that men, who are overwhelmingly in charge of athletic programs and who have the power to make hiring decisions, are more likely to choose men as coaches (Fazioli 2004).

Are women represented at the national level? Currently, professional sports for women include a Women's National Basketball Association (WNBA), a volleyball league, a golf tour, and a tennis circuit. As we have already seen, few women athletes make it to the professional ranks. Even those women who become professionals earn significantly less than their male counterparts (Levin 1996; Fazioli 2004). Golf, for example, is one of the few professional sports offering significant opportunities for women. Still, the leading money winner on the men's tour earns far more than twice as much as the leading money winner on the women's tour. This disparity is reflected in the 2007 prize money earned by the top men's Professional Golfers' Association (PGA) player and the top Ladies Professional Golf Association (LPGA) player—$11 million versus $4.3 million.

There are some positive, if small, signs of change. Tennis sensation Venus Williams has product endorsement contracts with American Express, McDonald's, and Wilson Racquet Sports. In 2003, Serena Williams signed a five-year contract with Nike worth close to $40 million. And shortly after sixteen-year old golfer Michelle Wie turned professional in 2005, she signed endorsement deals worth $10 million a year.

**F
E
E
D
B
A
C
K**

1. One positive thing about sports is that they eliminate discrimination. T or F?
2. Sport can impede the social mobility of minorities. T or F?
3. _____ involves the assignment of players to less central positions on the basis of race or ethnicity.
4. Minority former athletes benefit less than their white colleagues from personal appearances and commercial endorsements. T or F?
5. Females are denied equal access to organized sports because of _____.

Answers: 1. F; 2. T; 3. Stacking; 4. T; 5. sexism

How to Avoid Bigotry in Sports

Sports sociologist Jay Coakley supports the concerns of many Native Americans on the issue of team names. He wrote the following article about this issue.

Most of us are not very concerned about the use of Native American names by many athletic teams. But to Native Americans, war whoops and tomahawk chopping portray negative stereotypes.

Using stereotypes to characterize Native Americans in the United States is so common that most people don't even realize they are doing it .... When these stereotypes are used as a basis for team names, mascots, and logos, sports become a way of perpetuating an ideology that exploits, trivializes, and demeans the history and cultural heritage of Native Americans.

If teachers, administrators, and students in U.S. schools had a deep knowledge of the rich and diverse cultures of Native Americans and realized the discrimination native peoples currently face, they would not use names such as Indians, Redskins, Chiefs, Braves, Savages, Tribe, and Redmen for their teams; they would not allow Anglo students to entertain fans by dressing up as caricatures of Native Americans; and they would not allow fans to mimic Native American chants or act out demeaning stereotypes of war-whooping, tomahawk-chopping Native Americans.

Schools should not use any Native American name or symbol in connection with sport teams unless they do the following:

1. Sponsor a special curriculum to inform students of the history, cultural heritage, and current living conditions of the native group after which their sport teams are named. Unless 70 percent of the students can pass annual tests on this information, schools should drop the names they say are used to "honor" native people.
2. Publish two press releases per year in which information about the heritage and current circumstances of the native peoples honored by their team names is described and analyzed; publish similar materials annually in school newspapers and yearbooks.
3. Once per year, during homecoming or a major sport event, sponsor a special ceremony designed by and for native peoples in the local area, with the purpose of informing students and parents about the people they say they honor with their team names.

Source: Jay J. Coakley, *Sport in Society*, 6th ed. (Boston: McGraw-Hill, 2006, pp. 272–273).

Think About It
Is there a sport symbol in your community or state that might be offensive to Native Americans? Has the existence of this offensive symbol hurt your community or state economically? Explain.

SUMMARY

1. As a social institution, sport fulfills some important societal needs. One of these is helping individuals identify with others members of society.
2. Sport subcultures have developed around both team and individual sports. For this reason, sport is a reflection of society.
3. Functionalists see sport positively, as a means for socializing young people, promoting social integration, providing a release for tensions, and developing sound character. Conflict theorists believe that organized sports can be harmful to character development. Symbolic interactionists focus on the self-concepts and relationships developed through sport activities.

4. Sport contributes to upward mobility among collegiate athletes, but the opportunities are too few. Minorities still face discrimination in sport.
5. Women in sport suffer from gender-based stereotypes. Intercollegiate female athletes do not receive treatment equal to the treatment received by males, although this situation is slowly improving.

LEARNING OBJECTIVES REVIEW

• Justify sport as an American institution.
• Compare and contrast sport in America from a functionalist, conflict, and symbolic interactionist perspective.
• Understand the relationship between American sport and social mobility.
• Cite evidence of sexism and racism in American sport.

CONCEPT REVIEW

Match the following concepts with the definitions listed below them.

_____ a. stacking
_____ b. sport subculture
_____ c. sport

1. A set of competitive activities in which winners and losers are determined by physical performance within a set of established rules.
2. The assignment of players to less central positions on the basis of race or ethnicity.
3. A group within the larger context of sport that has some of its own distinct rules, values, and norms.

CRITICAL-THINKING QUESTIONS

1. Athletes with superior skills are often given extraordinary help in meeting school requirements. Many students feel this is unfair. Others point out that athletics bring in lots of money for colleges. They also say that athletes have skills as rare as high intelligence and so deserve their sports scholarships. What do you think?

_____

_____

2. Some schools have made efforts to change their school nicknames and mascots so as not to offend various groups that might have been negatively portrayed by these mascots and nicknames. Do you think that schools and teams have an obligation to take such actions? Or

should teams be allowed to retain their traditional nicknames and mascots?

_____

_____

3. In the National Basketball Association draft, the best players go to the teams that competed the previous season with the worst records. Why do you think the NBA uses this approach instead of allowing the best players to go to the teams with the most prestige, status, and monetary resources? Should the NBA use a different method?

_____

_____

MULTIPLE-CHOICE QUESTIONS

1. Which of the following is *not* a social institution?
   a. family
   b. government
   c. economy
   d. court system
   e. sport

2. Which of the following would *not* be included in the definition of sport?
   a. pick-up basketball game
   b. two vacationing skiers racing
   c. a round of golf at a country club
   d. boxing
   e. motorcycle racing

3. Sport plays a central role in American society in part because it reflects the culture's emphasis on
   _____
   a. equality.
   b. sexism.
   c. coolness.
   d. achievement.
   e. fame.

4. Professional football is an example of a(n) _____ ___ in sport.
   a. institution
   b. value
   c. culture
   d. counterculture
   e. subculture

5. Stanley Eitzen's book on the paradoxes of modern sports in the United States is an example of a(n) _____ _____ at work.
   a. functionalist
   b. subculturalist
   c. anthropologist
   d. conflict theorist
   e. symbolic interactionist

6. Extensive participation in sports is _____ __ related to achievement orientation.
   a. positively
   b. negatively
   c. curvilinearly
   d. hyperbolically
   e. parabolically

7. According to the text, violent behavior in football is often defended by adults in America because
   _____
   a. otherwise an attitude of losing is instilled.
   b. the prestige of schools is on the line.
   c. it prepares people to do well in a market society.
   d. coaches must win to keep their jobs.
   e. it demonstrates manliness.

8. Which of the following aspects of sport reflects the conflict perspective?
   a. equality
   b. integration
   c. inequality
   d. character development
   e. self-esteem

9. Cheating in sports is best accounted for by
   _____
   a. symbolic interactionism.
   b. conflict theory.
   c. structuration theory.
   d. deconstruction theory.
   e. structuralism.

10. The meaning of playing on a college basketball team would most likely be studied by _____
    a. functionalists.
    b. conflict theorists.
    c. subculturalists.
    d. symbolic interactionists.
    e. evolutionary theory.

11. Which of the following statement is *false?*
    a. College athletes are generally more successful than other students.
    b. Sport is an excellent avenue for upward social mobility for minorities.
    c. A high school player has a 0.2 percent chance of becoming a major league player.
    d. Professional athletes have short careers.
    e. Women still suffer from gender-based stereotypes.

12. _____ occurs when certain categories of players are assigned to less central positions.
    a. Success
    b. Failure
    c. Preemption
    d. Stacking
    e. Integration

13. The cultural roots of sexism in sport date to
    a. the Roman Olympics.
    b. Nazi Germany.
    c. the Ancient Greeks.
    d. the New Deal.
    e. the Gilded Age.

14. The overwhelming presence of males in athletic departments is best explained by
    a. conflict theory.
    b. symbolic interactionism.
    c. ground theory.
    d. evolutionary theory.
    e. functionalism.

FEEDBACK REVIEW

Multiple-Choice

1. Which of the following is *not* one of the functions of sport discussed in the text?
   a. Sport solidifies social class identification.
   b. Sport teaches basic beliefs, norms, and values.
   c. Sport provides a sense of identification.
   d. Sport offers a safe release of aggressive feelings.
   e. Sport encourages the development of character.

True-False

2. One positive thing about sports is that they eliminate discrimination. T or F?
3. Sport can impede the social mobility of minorities. T or F?
4. Minority former athletes benefit less than their white colleagues from personal appearances and commercial endorsements. T or F?

Fill in the Blank

5. _____ involves the assignment of players to less central positions on the basis of race or ethnicity.
6. _____ is a set of competitive activities in which winners and losers are determined by physical performance within a set of established rules.

7. A group within the larger context of sport that has some of its distinct roles, values, and norms is a _____.
8. Females are denied equal access to organized sports because of _____.

Matching

9. Several social aspects of sport are listed below. Indicate which would most interest functionalists (F), conflict theorists (C), and symbolic interactionists (SI).
   _____ a. intensification of achievement-oriented values
   _____ b. maintaining social inequity
   _____ c. character building
   _____ d. self-identification
   _____ e. sport scandals

GRAPHIC REVIEW

Table 16.3 shows an athlete's chances of making it to the next level in their respective sports.

1. Describe any patterns you see in the data in this table.

   _____
   _____

2. Discuss the implications of these data for the chances of upward social mobility for male and female athletes.

   _____
   _____

ANSWER KEY

Concept Review	Multiple-Choice Questions	Feedback Review
a. 2	1. d	1. a
b. 3	2. b	2. F
c. 1	3. d	3. T
	4. e	4. T
	5. a	5. stacking
	6. a	6. sport
	7. c	7. subculture
	8. c	8. sexism
	9. b	9. a. F
	10. d	b. C
	11. b	c. F
	12. d	d. SI
	13. c	e. C
	14. a	

17

Population and Urbanization

Rob Stuehrk/AP Photos

LEARNING OBJECTIVES

- Identify and distinguish among the three population processes.
- Discuss major dimensions of the world population growth problem.
- Relate the ideas of Thomas Malthus to the demographic transition.
- Predict world population trends.
- Discuss demographic trends of older Americans.

- Differentiate basic concepts of urbanization.
- Trace the historical development of preindustrial and modern cities.
- Describe some of the consequences of suburbanization.
- Discuss world urbanization.
- Compare and contrast four theories of city growth.

- Compare and contrast traditional and contemporary views on the quality of urban life.

Do you perceive urbanites as being part of the "lonely crowd," living amid a sea of strangers? If so, you can find support in the pioneering studies of urbanism. Subsequent research, however, presents another depiction: urban residents engage in personal relationships and these social ties contribute to a positive emotional state of mind. There are those without such social relationships, but they tend to be recently relocated residents who have not yet had time to form close human connections.

The quality of urban life is an important consideration. There are, of course, many other issues associated with population growth and urbanization. As a prelude to these issues, the chapter begins with the concept of demography.

The Dynamics of Demography

The Nature of Demography

Why is demography essential in the discussion of population change? **Demography,** the scientific study of population, encompasses all measures of population: size, distribution, composition, age structure, and change. Although basically a social science, demography draws from many disciplines, including biology, geography, mathematics, economics, sociology, and political science.

Demographic research can be divided into two subareas. **Formal demography** deals with gathering, collating, analyzing, and presenting population data. For example, formal demography looks at the data demonstrating changes in the entire American population. Another example of formal demography is the documentation of the dramatic increase in the Latino population of California and Texas. **Social demography,** on the other hand, is the study of population patterns within a social context. For example, social demography examines the relationship between population growth and congressional districting. Historically, social demographers have found that the projected growth of minorities in the United States benefited Democrats more than Republicans (Tilove 1999). Regardless of political affiliation, the growth of minorities affects the way congressional districts are drawn and is one reason census-taking can be a controversial topic. For a second

example, to help plan for hospitals and long-term nursing facilities, a social demographer might study trends in the population shifts of aging baby boomers.

Why study population trends? There is a demographer in all of us, observing and recording important events related to population change—births, deaths, the relocation of friends and family. The professional demographer considers these events in two ways: first, by gathering, organizing, and analyzing the patterns of population size, structure, composition, and distribution; and second, by attempting to identify and understand relationships between demographic and social processes. In other words, the demographer studies the link between the quantity of people and the quality of their lives.

Knowledge of population trends has never been more important. The study of population provides information about population growth, the characteristics of population, the location of population, and the probable long- and short-run effects of demographic trends. Demography plays a major role in policy formation, planning, and decision making in both public and private sectors of modern economies. Population information is used, for example, to plan for the health, education, transportation, and recreation needs of virtually every community in the United States. Demographers assist government policy makers and decision makers in meeting the needs of various socioeconomic groups, including the young, the poor, the unemployed, and the elderly. Demography generates information for identifying variable demands for products and services.

Three population processes—fertility, mortality, and migration—are responsible for population growth and decline. Any alterations in population characteristics originate from these three processes (Namboodiri 1996; McFalls 2007). Sociologists use these processes to chart population shifts, past and future. Figure 17.1 looks at the key demographic statistics by world regions.

Fertility

How is fertility related to population growth? **Fertility** measures the number of children born to a woman or to a population of women. Whereas fertility refers to the "actual" number of children women produce, **fecundity** is the maximum rate at which women can potentially produce children. The estimate for the upper limit of a society's average fecundity is fifteen births per woman. The record fertility rate for a group probably is held by the Hutterites, who migrated from Switzerland to North and South Dakota and Canada in the late nineteenth century. Hutterite women in the 1930s produced an average of more than twelve children each (Westoff and Westoff 1971). The Hutterites give us a good estimate

FIGURE 17.1

World Birth Rates, Death Rates, and Infant Mortality Rates

Would you always expect to see a correlation between crude birth rates and total fertility rates for a country? Between crude death rates and infant mortality rates?

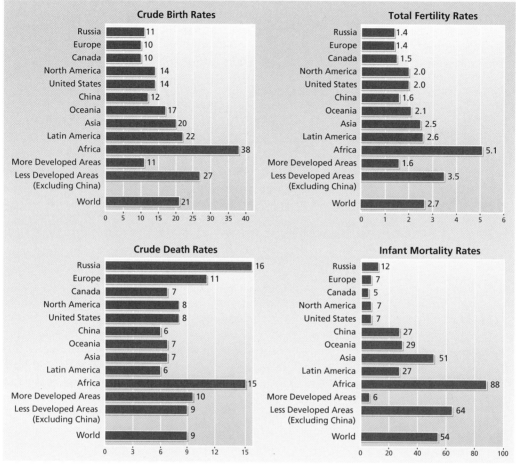

Source: Washington, DC: Population Reference Bureau, 2005.

of fecundity, because they are the best example of *natural fertility*—the number of children born to women in the absence of conscious birth control. No other known society has exhibited the fertility level of the Hutterites (Weeks 2008).

How is fertility measured? The **crude birth rate** is the annual number of live births per 1,000 members of a population. Birth rate varies considerably from one country to another. The birth rate for the United States is 14 per 1,000. Uganda, in East Africa, experiences a very high birth rate of 48 per 1,000; and Germany, a very low rate of 8 per 1,000 (*World Population Data Sheet* 2007).

To calculate birth rate, one simply needs to know the number of births in a year and the size of the population. One then divides the annual number of live births by the total population and multiplies that number by 1,000. As a formula, it would appear as

$$\text{Crude birth rate} = \frac{(\text{Live births} \times 1,000)}{(\text{Total population})}$$

This formula is "crude," however, because, being based on the entire population, it fails to identify those women in the population most likely to give birth; and it ignores the age structure of the population—two factors that affect the number of live births in any given year. Consequently, in addition to the crude birth rate, demographers use the **fertility rate**—the annual number of live births per 1,000 women ages fifteen to forty-four. **Age-specific fertility** is the number of live births per 1,000 women in a specific age group, such as twenty to twenty-four or thirty-five to thirty-nine. The rate easiest to comprehend is the **total fertility rate**—the average number of children born to a woman during her lifetime. Currently, total fertility rates in the world range from 5.2 in Africa to 1.4 in Europe. In the United States, the total fertility rate stands at 2.1.

SNAPSHOT OF AMERICA 17.1

Percentage of Population Under Age 18

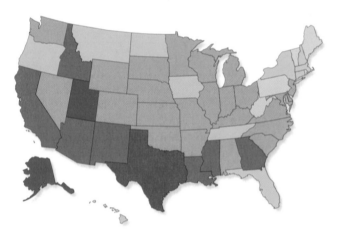

Percentage of Population Under Age 18

- 29% or more
- 28%–28.9%
- 26%–27.9%
- 24%–25.9%
- 23%–23.9%
- 22.9% or less

Source: U.S. Bureau of the Census, *Statistical Abstract of the United States, 2004–2005.*

Many students feel that as members of society they are not given enough respect by society. One reason could be that there are too few people in this age bracket to influence policy makers. This map shows the percentage of each state's population under age eighteen.

Interpret the Map

1. Which states have the smallest concentrations of young people? Can you explain why?
2. From this map, can you make any generalization about the American population? What additional information would help you to further describe the age structure of the U.S. population? Get that information for your state.

The birth rate of a population is influenced by both biological and social factors. A major biological influence is health. For example, widespread disease (especially rubella, or German measles) causes the birth rate to decline because many pregnancies end in miscarriage. Another biological factor, fecundity, influences fertility data: it still takes an average of nine months to give birth to a baby. Primary social factors affecting the birth rate include the average age at marriage, the level of economic development, the availability and use of contraceptives and abortion, the number of women in the labor force, the educational status of women, and social attitudes toward reproduction.

The U.S. birth rate has steadily declined over the past thirty years. It is slightly below the average of two children per woman of child-bearing age, which is required

How might fertility drugs affect the crude birthrate?

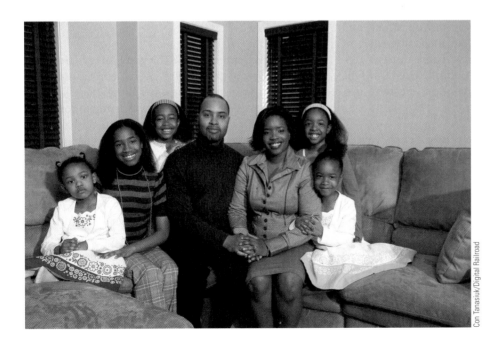

Con Tanasiuk/Digital Railroad

to keep the population size stable. An obvious precipitating factor here is the dramatic change in family planning. More Americans are using contraceptives and having abortions. A greater number of couples believe that two children—or even one child—is a desirable size for a family. And work patterns affect the birth rate as more American women are postponing having children until their late twenties and early thirties. As a result, fewer women are having children, and more women are having fewer children. Even though these factors remain in place, 1998 saw the first increase (1 to 2 percent) in the total fertility rate since 1990 (J. Martin et al. 2002). The significance—if any—of the rising birth rate for women in their twenties and thirties is not known yet.

Mortality

What are the dimensions of mortality? Mortality refers to deaths within a population. The important dimensions of mortality are life span and life expectancy. **Life span**—the most advanced age to which humans can survive—can be inferred only from the person with the greatest authenticated age, a Japanese man who lived nearly 121 years (Russell 1987). Although this reveals that people are biologically capable of living to be 120 years old, few even approach this age. **Life expectancy** is the average number of years that persons in a given population born at a particular time can expect to live. World life expectancy is sixty-eight years (*World Population Data Sheet* 2007).

Morbidity refers to rates of disease and illness in a population. For example, in early twentieth-century America, influenza, pneumonia, and tuberculosis accounted for nearly 25 percent of all deaths; but their share of all deaths has declined to less than 3 percent. As more people survive to older ages, chronic degenerative diseases such as heart disease, stroke, and cancer have come to represent an increasing share of deaths (nearly seven out of ten). These diseases represent the consequences of an industrialized lifestyle, including stress, alcohol consumption, cigarette use, and pollution.

How is mortality measured? The **crude death rate** is the annual number of deaths per 1,000 members of a population. Similar to the crude birth rate, the death rate is figured by dividing the annual number of deaths by the total population and multiplying by 1,000. Like the crude birth rate, the crude death rate varies widely throughout the world. The worldwide average death rate is 9 per 1,000 persons. Looking at specific regions of the world, the death rate varies from a low of 5 per 1,000 in Central America to a high of 18 per 1,000 in Middle Africa. The death rate in the United States is about 8 per 1,000 (*World Population Data Sheet* 2007).

Demographers are also interested in the variations in birth rates and death rates for specific groups. They have devised **age-specific death rates** to measure the number of deaths per 1,000 persons in a specific age group, such as fifteen to nineteen or sixty to sixty-four. This measure allows one to compare the risk of death to members of different groups. Although death eventually comes to everyone, the rate at which it occurs depends on many factors, including age, sex, race, occupation, social class, standard of living, and health care.

The **infant mortality rate**—the number of deaths among infants under one year of age per 1,000 live births—is considered a good indicator of the health status of any group, because infants are extremely susceptible to variations in food consumption, availability of medical care, and public sanitation. Infants in less developed countries are greater than seven times more likely to die before their first birthday than infants in the more developed nations are; this fact illustrates the wide variation in living standards among these nations.

Migration

In what two ways can we describe migration? **Migration** refers to the movement of people from one geographical area to another for the purpose of establishing a new residence. We can describe migration in relation to a move either within a country or from one country to another. A current example of international migration is the resettlement of Asian refugees from Vietnam and Cambodia to the United States and other countries around the world. Many of the refugees who settle in the United States in one particular city or region later move to another region, thus becoming internal migrants, as when people move from New York State to Arizona (Weeks 2008).

How is migration measured? The **gross migration rate** is the number of persons per 1,000 members of a population who, in a given year, enter (immigrants) or leave (emigrants) a geographical area. Net migration is the combined effect of immigration and emigration on the size of a population. Thus, the **net migration rate** is the annual increase or decrease per 1,000 members of a population resulting from movement into and out of the population. The United States, for example, has a net migration rate of about 4 per 1,000 population. That is, 4 more persons per 1,000 population enter the country than leave the country (*World Population Data Sheet* 2007).

When the U.S. Census Bureau reports migration rates, it refers only to the number of legal immigrants. Many people violate immigration quotas in the United States. In the 1970s, illegal entry into the United

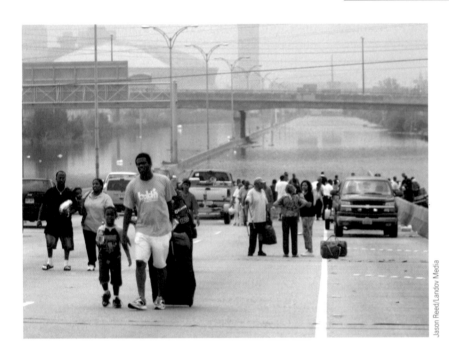

Not all migration is voluntary. Those leaving New Orleans because of Hurricane Katrina were being "pushed."

States—primarily from Latin American and Caribbean countries—became a major concern. It continues to be controversial. There are no precise statistics on either the illegal immigration rate or the total number of illegal immigrants living in the United States, but it is estimated that one-half of the foreigners in the United States entered illegally (Martin and Midgley 2006).

As of 2006, about 35 million legal and illegal immigrants were living in the United States, more than double the number (13.5 million) reached in 1910 during the last massive influx of immigrants. Thus, immigration has become the determinate factor in U.S. population growth, accounting for about 30 percent of the U.S. population increase between 1980 and 2000. In

FEEDBACK

Match the following terms and definitions.

_____ a. fertility   _____ e. fecundity   _____ i. migration
_____ b. crude death rate   _____ f. age-specific fertility _____ j. gross migration rate
_____ c. crude birth rate   _____ g. age-specific death rate _____ k. net migration rate
_____ d. demography   _____ h. infant mortality rate   _____ l. morbidity

(1) the number of deaths per 1,000 members of a population per year
(2) the movement of people from one geographical area to another for the purpose of establishing a new residence
(3) the increase or decrease per 1,000 members of a population per year as a result of people leaving and entering the population
(4) the number of live births per 1,000 women aged fifteen to forty-four years
(5) the number of deaths per 1,000 persons in a specific age group
(6) the study of the growth, distribution, composition, and change of a population
(7) the number of live births per 1,000 women in a specific age group
(8) the number of persons per 1,000 members of a population who enter or leave the population each year
(9) the number of deaths to infants under one year of age per 1,000 live births
(10) rates of disease and illness in a population
(11) the maximum rate at which women can physically produce children
(12) the number of live births per 1,000 members of a population per year

Answers: a. 4; b. 1; c. 12; d. 6; e. 11; f. 7; g. 5; h. 9; i. 2; j. 8; k. 3; l. 10

2005, the foreign-born comprised about 12 percent of the total population—the largest proportion in seventy years. Given current trends, the end of this decade will see the immigrant portion of the total population exceed the previous high of 15 percent, documented in 1890 (Camarota 2002; Martin and Midgley 2006).

Why do people migrate? The most general explanation for migration is the so-called push–pull theory. That is, people move either because they are attracted elsewhere or because they feel impelled to leave their present location. Often "push and pull" factors exist simultaneously. The United States was settled by people who were fleeing economic, religious, and social circumstances and who were attracted by the promise of a better future. People, of course, may not migrate despite significant pushing and pulling factors. Certain barriers, such as the cost of moving or bad health, may outweigh the push–pull factors.

In cases of voluntary movement, economic reasons are the most compelling. Voluntary migration is also associated with stages in the life cycle. The greatest movement is found among young people leaving home to attend college, take a job, marry, or join the military. Retired people may move from a metropolitan area to a rural community or from a colder climate to a warmer one.

Unfortunately, not all migration decisions are freely made. People may be forced to leave an area or prevented from leaving a locale, both of which occurred among Jews in Hitler's Germany and in Bosnia and Yugoslavia. An environmental crisis or an inhospitable climate producing famine or drought may give people no choice but to abandon their place of residence.

Demography for Businesses

Businesses have discovered that they can grow bigger by targeting smaller groups of consumers. These groups, called *generations*, or *cohorts*, are defined by important life experiences. Events occurring when people first become economic adults (usually between ages seventeen and twenty-one) affect their lifelong attitudes and values. These attitudes and values are unlikely to change as a person ages. So the kind of music that is popular during these formative years often remains the preferred type of music for life. Similarly, early lifetime experiences influence preferences in many other product and service categories.

Studies of the U.S. population have identified seven distinct groups described in the accompanying table. Which cohort are you? Your parents?

Cohort	Description	Born	Popular Music Styles
The Depression cohort	The G.I. generation	1912–1921	Big Band
The World War II cohort	The Depression generation	1922–1927	Swing
The postwar cohort	The silent generation	1928–1945	Frank Sinatra/Rat Pack
The boomers I cohort	The Woodstock generation	1946–1954	Rock and roll
The boomers II cohort	The zoomer generation	1955–1976	Grunge, rap, Country western
The boomlet cohort	The echo-boom generation	1977–	Retro-swing, Latin

Think About It

Conduct short interviews with members of at least two of the demographic business cohorts profiled here. Identify a number of differences in preferences for products between the members of different cohorts.

Source: Erin N. Berkowitz, Robert A. Kerin, Steven W. Hartley, and William Rudelius, *Marketing*, 5th ed. (Chicago: Irwin, 1997).

Population Growth

World Population Growth

On October 12, 1999, the United Nations officially declared that the world's population had reached 6 billion. How big is 6 billion? If you counted 100 numbers every minute for eight hours a day, five days a week, it would take you 500 years to reach 6 billion!

According to Population Connection, the world's population is growing at a rate of 80 million people per year. If asked about the cause of rapid world population growth, what would you say? Like most people, you would probably refer to the high birth rate in poorer countries. You could point out that 135 million new infants are born each year, enough people to more than duplicate Japan's population. You could note that every time you watch a thirty-minute television program, 4,860 infants are born. This explanation, however, is only half the story. It leaves out the death-rate side of the equation. Developing countries have passed through the stage when both birth rates and death rates were high. Now these countries are growing rapidly because their birth rates remain high while their death rates have dropped sharply, thanks to modern medicine, improved sanitation, and better hygiene. The growth and the distribution of the world's population—almost 7 billion—vary greatly among the nations of the world. Not only that, but population has grown at markedly different rates throughout history (Livi-Bacci 1997; Haupt and Kane 2004; McFalls, 2007; see Figure 17.2).

How fast is the world's population growing? When describing the growth of human population, it is impossible to avoid a certain amount of conjecture, because there has never been a complete worldwide numeration of the population. Although most countries now take a census, there are still national populations that are not calculated. Furthermore, the quality of census data varies a great deal from country to country. Nevertheless, it is possible to indicate world population growth patterns by using historical information and recently collected data (Gelbard, Haub, and Kent 1999).

Rapid world population growth—now standing at 1.2 percent, or about 80 million people per year—is a relatively recent phenomenon: "Those of us born before 1950 have seen more population growth during our lifetimes than occurred during the preceding four million years, since our early ancestors first stood upright" (L. R. Brown 1996:3). It is estimated that only about 250 million people were on the earth in A.D. 1. It was not until 1650 that the world's population doubled, to half a billion (see Figure 17.2). Subsequent doublings have taken less and less time. The second doubling occurred in 1850,

---

FIGURE 17.2

World Population Growth Through History

This figure shows estimated world population growth from the Old Stone Age to 2100. Population growth since 1800 has been phenomenal. The first doubling of the world's population occured between 8000 B.C. and 1650. The fourth doubling had occurred by 1976, and world population is on schedule to double for the fifth time in 2025.

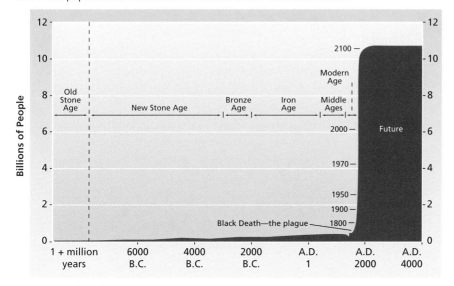

Sources: Population Reference Bureau and United Nations, *World Population Projections to 2100* (1998). Adapted from Joseph A. Falls Jr., *Population*, 5th ed. (Washington, DC: Population Reference Bureau, 2007), p. 25.

bringing the world population to just over 1 billion. By 1930, only eighty years later, another doubling had taken place. Only forty-five years later, in 1976, a fourth doubling raised the world's population to 4 billion. At the current growth rate, the world's population is expected to double again in approximately fifty-one years and is expected to reach almost 11 billion persons by the year 2100 (McFalls 2007; *World Population Data Sheet* 2007). Obviously, the time between each doubling of the population is getting shorter and shorter. Figure 17.3 breaks down world population projections by region.

Why is the world growing so rapidly? The population is increasing dramatically, in part because of the mathematical principle by which population increases. We are accustomed to thinking in terms of linear growth—arithmetical increases of a constant amount within a given time period (as in the progression 1, 2, 3, 4, 5 . . .). If you save $100 a year for ten years, you will end up with $1,000, accumulated in ten equal amounts. Population, however, does not grow linearly. It increases geometrically (as in 2, 4, 8, 16, 32 . . .), following the principle of **exponential growth**—the absolute growth that occurs within a given time period becomes part of the base for the growth rate in the next time period. This means that if the growth rate of a population remains the same for two successive years, the absolute growth will be larger in the second year.

Suppose that a city of 100,000 has an annual growth rate of 5 percent for two years. At the end of the first year, its population will increase by 5,000 (100,000 × 0.05). During the second year, the population will increase by an additional 5,250 (105,000 × 0.05). Consider a concrete case. The world population grew at about 2 percent per year between 1960 and 1970, which produced an increase of 650 million people. At the same rate, by 1980, the world population would increase by 800 million. This was nearly the case. In 1970, the world

---

FIGURE 17.3

Population Projections by Regions of the World

This graph displays population projections, by regions of the world, from 2005 to 2050. Note the dramatic difference in population doubling time between less developed areas and more developed areas.

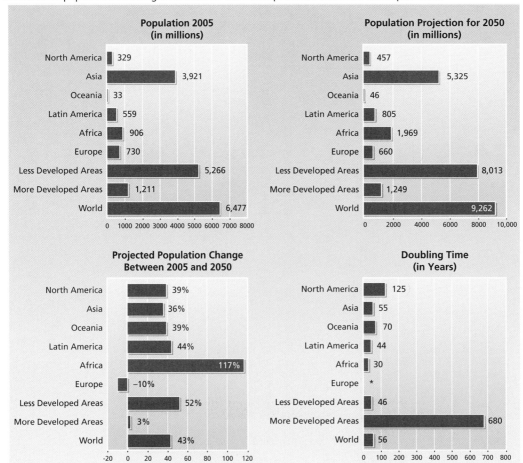

Source: Adapted from *World Population Data Sheet,* 2005.
* Doubling not projected to occur.

population was 3,261 million; in 1980, it stood at 4,414 million, an actual increase of 1,153 million.

A classic story offers an example of exponential growth. The story tells of a clever courtier who presented a beautiful chess set to his king and in return asked only that the king give one grain of rice for the first square on the chess board, two grains, or double the amount, for the second square, four (doubling again) for the third, and so forth. The king, being mathematically naive, agreed and ordered the rice brought forth. The eighth square required 128 grains, and the twelfth took more than a pound of rice. Long before reaching the sixty-fourth square, the king's coffers were depleted. Even today, the world's richest king could not produce enough rice to fill the final square. It would require more than 200 billion tons, or the equivalent of the world's current total production of rice for the next 653 years.

A term often used in analyzing population growth is **doubling time**—the number of years needed to double the original population size (given its current rate of growth). If a population is growing at 1 percent per year, it takes only seventy years to double. The number of people added each year becomes part of the total population, which then increases by another 1 percent in the following year.

Whereas the current world population growth rate of 1.3 percent may not seem high, it represents an increase of nearly 220,000 people every day. At this rate, the world's population grows every seven days as much as the size of the entire population of Utah (*World Population Data Sheet* 2007). To put it another way, the human population is growing at the rate of 152 per minute, nearly 220,000 per day, almost 80 million per year. The pace of growth is so rapid that additions to the world's population could equal the population of the United States in just over three years. At this rate, more people will be born in the year 2050 than were born in the 1,500 years after the birth of Christ.

On the encouraging side, the rate of world population growth has been declining since the 1970s. Data from fifty-six countries with an estimated population of 10 million or more indicate that the average annual rate of natural increase dropped from 2.1 percent in 1970 to 1.7 percent in the 1980s, to 1.2 percent in 2007 (*World Population Data Sheet* 2007).

Figure 17.2 shows the population in various regions of the world, along with the projected populations for 2050. Compare the projected population increase in the less developed regions (52 percent) with more developed nations (3 percent). Doubling time for less developed regions is 46 years, while the doubling time for more developed nations is 680 years. To put it another way, nearly 99 percent of the world's population growth will take place in less developed countries, while deaths exceed births annually in Europe (*World Population Data Sheet* 2007).

What difference does one child make? The importance of limiting family size, even by one child, can be illustrated by population projections for the United States (see Figure 17.4). Even though the United States is not going to a three-child average in the future, the hypothetical American case can help us understand the importance of population control. Figure 17.4 contrasts the projected

---

FIGURE 17.4

Projected Population of the United States, Comparing a Two-Child Average per Family (Replacement Level) to a Three-Child Average

This figure illustrates the importance of reaching the population replacement level (two children per family). Are you surprised with the difference in U.S. population growth caused by an average three children per family versus two children? Relate this difference to the principle of exponential growth.

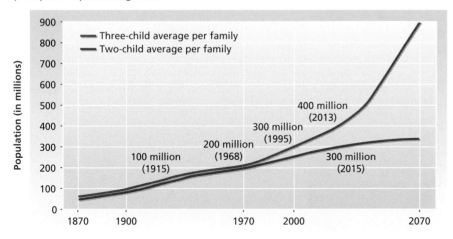

Source: *World Population Data Sheet*, Washington, DC: Population Reference Bureau, 2000.

population of the United States in the year 2070 for an average family size of two children and an average family size of three children. When small decreases in the crude death rate and a stable net migration are assumed, an average two-child family size would result in a population of 300 million in 2015. Taking the hypothetical average family size of three children, the U.S. population would grow to 400 million by 2013. As time passed, the difference of only one extra child per family would assume added significance. By 2070, the two-child family would produce a population of 350 million, but the three-child family would push the population to close to 1 billion! To say it another way, with an average family of two children, the U.S. population would not quite double itself between 1970 and 2070. But should the three-child family be the average, the population would double itself twice during this same period.

The importance of limiting population in developing regions becomes clear when the effect of even one child added to the average number of children in a family is recognized. Moreover, the addition of one child per family has a greater effect as the population base gets larger; not only is one extra person added, but theoretically that one person will be involved with the reproduction of yet another three; and on it goes. The largest populations are found in developing countries, which also have the largest average number of children per family.

The Malthusian Perspective

Although concern about population growth is strong, it is not new. In 1798, Thomas Robert Malthus, an English minister and economist, published *An Essay on the Principle of Population*, one of the most important works ever written on population growth and its repercussions. Thanks to the pessimism of Malthus's thesis, economics earned the label "dismal science."

In his essay, Malthus proposed a set of relationships between population growth and economic development:

1. Population, if left unchecked, will tend to exceed the available food supply. This is because population increases exponentially; the food supply does not.

2. Checks on population can be positive—those factors that increase mortality, such as famines, disease, wars—or preventive—those factors that decrease fertility, such as later marriages or abstaining from sexual relations in marriages. (At the time, there was no reliable birth control. For this conservative minister, sexual abstinence was the only morally acceptable—and practical—way to reduce the number of births.)

3. For the poor, any improvement in income is lost to additional births; this leads to reduced food consumption and lower standards of living and eventually to the operation of positive checks.

4. The wealthy and better educated already exercise preventive checks.

This last point leads to some of Malthus's suggestions for solving the "population problem." Because the wealthy and better educated had achieved controlled fertility, the method of extending this to the poor would be through universal education, which would offer all members of the population the chance of improving their situation: "The desire of bettering our condition, and the fear of making it worse, has been constantly in action and has been constantly directing people into the right road" (Malthus 1798:477). The beneficial effects of universal education on population control could be enhanced, Malthus suggested, by raising people's aspirations for a higher standard of living. This would be accomplished by raising wages above the minimum required for subsistence, thus providing the poor an opportunity to choose between more children at a minimal standard of living or smaller families with a higher quality of life.

English minister and economist Thomas Malthus wrote about the inability of the food supply to keep up with population growth.

Has the Malthus perspective had any effect? Malthus's writings have profoundly influenced the study of population change. Events during the late nineteenth and early twentieth centuries brought his predictions into question. The last half of the twentieth century, however, has again raised the specter of a Malthusian crisis: population growth outstripping both food resources and the ability of different societies to provide even minimal quantities of basic necessities (housing, health care, education, employment). Some social scientists—the neo-Malthusians—modified Malthus's propositions in an attempt to explain the current world situation and to predict possible futures. First, neo-Malthusians note that the development of reliable contraceptives has not distorted marital relations as Malthus feared that it might. Second, historical developments since Malthus's time indicate that values promoting, and norms supporting, smaller families are positively related to certain kinds of social and economic changes. In other words, values on family size and norms for fertility regulation are adapted to socioeconomic conditions, enabling society to avoid Malthusian checks through self-regulation. Third, neo-Malthusians argue that many nations have a rate of population growth that overloads this self-regulating process because population growth is excessive to the extent that resources are diverted from socioeconomic change to population maintenance (Humphrey and Buttel 1986).

A debate is taking place between those who believe that the world's population is exploding beyond control and those who are convinced that the population explosion will be defused. Those predicting a slackening of the current high population growth rate often base their optimism on the increasing worldwide acceptance of birth control. This debate is more easily understood when viewed within the context of the demographic transition.

The Demographic Transition

How does the demographic transition develop? To the extent that developed nations have defied Malthus's theory, they have done so through the **demographic transition**—the process by which a population, as a result of economic development, gradually moves from high birth rates and death rates to low birth rates and death rates (see Figure 17.5). The number of stages of population growth in the demographic transition ranges from three to five, depending on how finely the process is broken down. The description that follows is a four-stage model with examples:

Stage 1.   Both the birth rate and the death rate are high, and population growth is mod-

est. This is characteristic of preindustrial societies.

Stage 2.   The birth rate remains high, but the death rate begins to drop sharply because of modernization—sanitation, increased food production, medical advances. The rate of population growth is very high. Most sub-Saharan African countries are presently at this stage.

Stage 3.   The birth rate declines sharply; but because the death rate continues to drop, population growth is still rapid.

Stage 4.   Both the birth rate and the death rate are low, and the population grows slowly if at all. North America, Europe, and Japan are at this stage today. Europe, in fact, is in what is being called the "second demographic transition."

What is the second demographic transition? According to Figure 17.3, the expected population decrease for Europe between 2005 and 2050 is 10 percent. This reflects the "second demographic transition," which Europe alone is experiencing. Europe's first demographic transition began with a gradual death-rate decline in the early nineteenth century, followed by a birth-rate decline beginning around 1880. The transition to both low birth rate and low death rate was completed by the 1930s. Europe's second demographic transition began in the mid-1960s.

The primary demographic characteristic of the second demographic transition is the decline in fertility from slightly above **replacement level**—the birth rate at which a couple replaces itself in the population—to a fertility level of 1.7. Should Europe maintain this lower fertility rate, and if immigration does not interfere, Europe's population will decline, something that has already happened in Italy, Sweden, Germany, Hungary, and Russia. The original formulation of the demographic transition, as noted earlier, ended with stable or nongrowing population.

Does the demographic transition occur in developing societies? Traditionally, demographers assumed that developing societies would proceed through the same stages as the developed world. This assumption is being criticized for its ethnocentric bias. Why? Because developing societies did not begin the transition at the same point as Western societies. For example, less developed countries started the transition with birth rates and death rates higher than existed in the now-developed societies when they began the transition. And mortality declined in the West because of domestic industrialization rather than from the importation of sophisticated techniques of disease control.

FIGURE 17.5

Stages of the Demographic Transition

This figure illustrates the demographic transition. Stage 1 begins with smaller population growth due to a balance between birth rates and death rates (both at high levels). In Stage 2, population grows dramatically because the death rate decreases so much faster than the birth rate. Population growth begins to slow in Stage 3, when the birth rate belatedly drops sharply. Stage 4 is again a condition of smaller population growth because birth rates and death rates come into balance (both at low levels). Given this description, what would you call the gap between the two graphs representing birth and death rates?

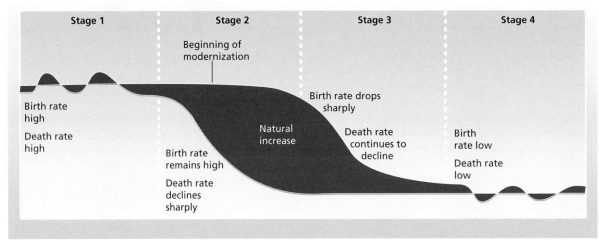

Source: Adapted from Joseph A. Falls Jr., *Population*, 5th ed. (Washington, DC: Population Reference Bureau, 2007), p. 27.
Note: Natural increase or decrease is produced from the difference between the number of births and deaths.

Population Control

Because death rates in both developing and developed nations have already dropped dramatically, efforts to curb world population growth must concentrate on lowering birth rates. Thus, **population control** is the conscious attempt to regulate population size through national birth control programs.

What is the history of birth rates? Historically, many societies were more concerned with increasing their population size than with overpopulation. High birth rates were needed to offset the high death rates from disease and poor hygiene. For some societies, larger populations enhanced security because of the ability to maintain larger armies. Agricultural societies needed large numbers of people to work the land. Aging parents wanted to be more secure in old age. High birth rates were also encouraged in countries with religious prohibitions against birth control.

Since the mid-twentieth century, however, more governments have come to view high birth rates as a threat to their national well-being. By 1990, most countries had in place formal programs to reduce birth rates—either voluntary or compulsory.

What is voluntary population control, and how successful has it been? Voluntary population control is generally known as **family planning**—making it technically possible for women to choose the number of children they will have. Beyond making contraceptive devices available, family planning programs provide birth prevention information and other services. Although women in the poorest countries still have the highest fertility, their overall fertility has been reduced by 50 percent since 1969 (Kluger and Dorfman 2002).

Even when effective, family planning programs merely enable families to achieve their *desired* family size. Unfortunately for effective population control, in many nations the desired family size is quite high. The World Fertility Survey reported that the average preferred family size of women in African nations is 7.1; in Middle Eastern nations, 5.1; in Latin American nations, 4.3; and in Asian Pacific nations, 4.0. In European countries, the average preferred family size ranges from 2.1 to 2.8.

Family planning has worked in Taiwan, where by the turn of the twenty-first century, the birth rate declined below replacement level. Taiwan's family planning efforts were launched under very favorable conditions. When the Japanese withdrew from Taiwan after World War II, they left behind a labor force trained for industrial work. Consequently, the Taiwanese were able to use this advantage to build an expanding economy. With economic development came a decline in both death rates and birth rates. In short, they went through the demographic transition fairly rapidly.

India was a different story. Family planning there got off to a very slow start, and the country has been unable to reduce the rate of population growth through voluntary means. Family planning efforts failed because

Family planning in Africa has failed to control population growth. Consequently, this scene of mothers and malnourished children waiting for medical treatment is all too familiar.

government officials and family planners underestimated the barriers to birth control—they did not take the broader social context into account. For one thing, India did not have Taiwan's advantage of relatively rapid economic development. In addition, the Indian officials and planners made insufficient efforts to overcome cultural and religious opposition to birth control. Nor did they address the lack of channels to effectively distribute birth control information and technology. Finally, the implementation of the national birth control policy was left in the hands of individual state governments.

Efforts at population control began to succeed in India only after the government turned to a sterilization program in 1976. Although the government did not use the force of law, a system of disincentives had the effect of compulsion. Those who could not produce official proof of a sterilization were denied such things as business permits, gun licenses, and ration cards for the purchase of basic goods (Weeks 2008). Still, India currently has a total fertility rate of 2.9 children per woman (the United States has a total fertility rate of about 2.1 children per woman). India is expected to have the largest population in the world by about 2050.

Have any countries successfully enforced compulsory population control methods? Both China and Singapore have required population control policies that seem to achieve their goals. In an attempt to reach zero population growth, the Chinese government first tried to discourage a third child in a family. The second step was to push for the one-child family. China was successful in reducing its total fertility rate from 7.5 in 1963 to 1.6 today through coercion and a system of rewards and punishments. In cities, couples with only one child can obtain a one-child certificate. This certificate ensures them a monthly allowance for the costs of children

up to age fourteen. One-child families receive a larger retirement pension and enjoy preference in housing, school admission for their children, and employment. Families with more than one child are subject to an escalating tax on each child, and they get no financial aid from the government for the medical and education costs of their extra children.

Singapore, an island city-state on the Malaysian peninsula, began formally discouraging large families in 1969, installing some economic penalties to encourage replacement-level fertility. The government included the following measures: denial of a paid eight-week maternity leave, loss of an income tax allowance, diminished access to public housing, increased maternity costs for each additional child, and lower likelihood of children entering the better schools.

These policies worked so well that the total fertility rate in Singapore dropped from 4.5 children per woman to 1.4 between 1966 and 1985. Worried over the reduction in population size, the government in 1987 switched to the guideline of three or more children for people able to afford them (Yap 1995). Despite this effort, Singapore's total birth rate is now 1.3, still below replacement level.

Future World Population Growth

World population growth has reached a watershed. After more than 200 years of acceleration, the annual population growth rate is declining. From 1950 to 1975, world population rose 64 percent; since 1975 it has increased 48 percent (Kluger and Dorfman 2002). The current growth rate of 1.3 percent compares favorably with the peak of 2.04 percent in the late 1960s. Moreover, the rate is projected to drop to zero by the end of the twenty-first century (United Nations Population Division 2000). The birth rate is already low in developed countries and, as noted earlier, is now declining in many developing countries. Similarly, the population growth rate is declining somewhat in both developed and developing areas of the world (Gelbard, Haub, and Kent 1999; Singer 1999).

Demographers are unsure of future world population growth in part because they do not know for sure how many children today's youth will have. Nor do they know what will happen to change life expectancy, particularly in developing countries.

Some analysts are now foreseeing a potential shift from concern about a "population explosion" to worry about a "population implosion" or decline due to lower fertility rates (Eberstadt 2001). Other demographers, however, point out that there are two very different worlds of population growth. In developed countries, fertility rates are at or below two children per couple (Europe; the United States; Canada; Japan; and some rapidly developing countries such as China, South

Korea, and Thailand). In developing societies, women average four children each (most of Africa, Asia, and Latin America). If the accelerating population growth that the world has known since the 1950s is to stop, there must be a dramatic decrease in fertility in developing countries (Gelbard and Haub 1998; Ashford 2001).

What is the future of world population growth? The United Nations offers three possible world population growth scenarios (see Figure 17.6). These three futures vary, depending on their assumptions regarding the average number of children women will bear. For the medium scenario to come true, women will have to average two children. If so, world population will rise to more than 9 billion by 2100. The medium scenario depicts **zero population growth**—when deaths are balanced by births so that the population does not grow. To make the high scenario a reality, women worldwide will have

FIGURE 17.6

The World's Population Has Exploded in Size

As this figure illustrates, the future growth of world population is uncertain. Which of these scenarios do you think is most likely to occur? Defend your answer with information from this chapter or from additional research.

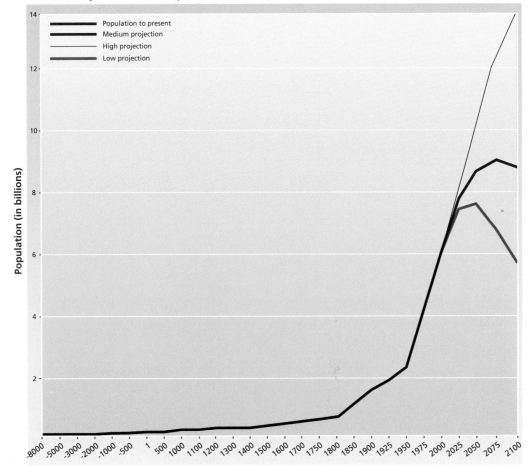

Sources: Data for years −8000 to 2000; projections are from United Nations Population Division, 2004, World Population to 2300 (New York; United Nations),Table 1. Adapted from John R. Weeks, *Population*, 10th ed. (Belmont, CA: Wadsworth, 2008), p. 35.

to average about 2.5 children. If this happens, world population will rise to 14 billion in 2100 and continue to grow indefinitely, exceeding 27 billion 100 years later. In the low scenario, fertility will have fallen to 1.6 children per woman or less. World population, under this condition, will peak at 8 billion in 2050 and drop to 6 billion by 2100.

But as we have seen, despite the reduction in the annual growth rate and birth rate, the world's population will continue to increase. Nearly 7 billion people are expected to inhabit the globe by 2010. Throughout the first half of the twenty-first century, the annual growth rate is expected to decline until world population stabilizes at about 12 billion people. At this point, the world will have reached zero population growth, when deaths are balanced by births so that the population does not increase.

Contrary to popular opinion, limiting the average family size to two children does not immediately produce zero population growth. This is because there is a time lag of sixty to seventy years, due to the high proportion of young women of childbearing age. Even if each of these women had only two children, the world population would grow because there are so many young women. Zero population growth will be reached only when this disproportionate number of young mothers disappears.

The time lag is what demographers call **population momentum**—a population continues to grow, regardless of a recent drop in the birth rate, because of the existing population base created by past growth. The increase in the world's population, like a huge boulder careening down a mountain, cannot be suddenly stopped. But the sooner the momentum of current population growth is halted, the better. The sooner the world fertility rate reaches the replacement level, the sooner zero population growth will be reached. The ultimate size of the world's population, when it does stop growing, depends greatly on the timing of the replacement level. To state it another way, each decade it takes to reach replacement level, fertility will increase the world's population by 15 percent.

Population Growth in the United States

The population of the United States has increased dramatically since the first census in 1790, growing steadily from less than 4 million in 1790 to 288 million in 2002. At the same time, the population growth rate has, with some deviations, tended to decline since 1790. Between 1990 and 2000, however, the American population surged by nearly 33 million, after growing more slowly between 1970 and 1990. The population of the United States is expected to continue growing despite the average American family reproducing at the replacement level of 2.1 children per family. In fact, contrary to other

industrialized nations, the United States is undergoing a *baby boomlet*. The 4.3 million births in 2006 was the largest number of children born in the United States in almost fifty years. In part, this trend is due to the baby-boom bulge in the age structure that produced the large number of women whose children and grandchildren are now part of the population (the "baby boomlet"). In addition to fertility, American population growth will be affected by its rates of mortality and migration (Kent and Mather 2002). If the current predictions of fertility, mortality, and immigration rates are accurate, the total U.S. population will reach 420 million by 2050. This would represent an almost 50 percent increase, considerably higher population growth than is expected in other more developed countries. In fact, as shown in Figure 17.7, U.S. population growth is projected to increase between now and 2050, while population growth in other more developed countries is expected to decline. Thus, the United States will remain the largest of the more developed countries.

Why will the U.S. population grow faster than other more developed countries? Generally, fertility is the primary factor in population change. The United States is no exception. Its population experienced **natural increase** (excess of births over deaths) each decade of the twentieth century. This trend continued between 2000 and 2001, when 4 million births were offset by only 2.4 million deaths. This natural increase of 1.6

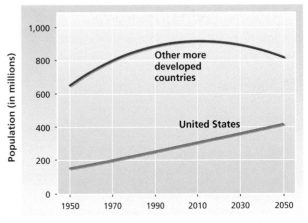

FIGURE 17.7

Population Growth in the United States and Other More Developed Countries: 1950–2050

This figure compares the curves of population growth in the United States with the rest of the more developed world. Describe any differences you see between 1950 and 1990. Describe the current differences in population growth between the two areas. Explain.

Note: "Other more developed countries" includes all European countries, Australia, Canada, Japan, and New Zealand.
Source: U.S. Census Bureau, International Data Base, October, 10, 2002, release; www.census.gov/ipc/www/idbagg.html (accessed October, 12, 2002).

million accounted for just over 60 percent of the total population growth that year (2.7 million).

Demographers often use the total fertility rate (the average number of children born to a woman during her lifetime) to predict population change. The total fertility rate ebbed and flowed during the twentieth century. In the first few decades of that century, native-born white women had an average of 3.5 children over their lifetimes. In part because of a lower total fertility rate, natural increase hit a low during the Great Depression of the 1930s. This declining natural increase was reversed after World War II when the baby boom occurred. Another total fertility rate low occurred in the early 1970s (1.7 children per woman), followed by an upswing beginning in the late 1970s. This small natural increase in the late 1970s and 1980s did not approach that of the baby boom of the 1950s, during which women were once again averaging 3.5 children.

A positive net migration rate has always, except during the Great Depression in the 1930s, played a larger role in the population growth in the United States than in most other nations. Among more developed countries, the United States has the largest foreign population, by a considerable margin (Camarota 2001; "The Longest Journey" 2002). The traditional importance of immigration to U.S. population growth has intensified over the last few decades because fertility among U.S.-born women has remained at or below replacement

level. During the 1980s, immigration was responsible for about one-third of U.S. population growth. The contribution of immigration to population growth increased in the 1990s, accounting for 40 percent of growth between 2000 and 2001. Because of high immigrant fertility, the contribution of immigration to population growth is even greater than suggested by the number of foreign-born people entering the United States. Approximately one-fifth of births in 2000 were among immigrant women. Should current immigration and fertility trends continue for 100 years, four in ten Americans will be post-1980 immigrants or their descendants.

During the 1950s, almost half of all immigrants came from northern and western Europe, whereas only two in ten came from less developed nations. Since 1960, the sources of immigration have shifted from Europe to Latin America, the Caribbean, and Asia. Less than one-fifth of the legal immigrants now come from Europe and Canada, whereas about three-fourths arrive from Latin American, the Caribbean, and Asia. Latin America and the Caribbean accounted for about 42 percent of the immigrants between 2000 and 2005 (Martin and Midgley 2006).

Although mortality rates can influence population growth rates, they have little effect among more developed countries, where mortality is already low (Kent and Mather 2002). At 9 deaths per 1,000 persons, the

FEEDBACK

1. According to the principle of _____, the absolute growth that occurs within a given time period becomes part of the base on which the growth rate is applied for the next time period.
2. Thomas Malthus saw famines as a _____ check on population growth.
3. The process by which a population gradually moves from high birth and death rates to low birth and death rates as a result of economic development is called the _____.
4. Europe's second demographic transition involves what kind of population growth?
   a. stable
   b. increasing
   c. declining
5. Which of the following figures is the world's population most likely to reach before it stops growing?
   a. 4 billion
   b. 8 billion
   c. 12 billion
   d. 25 billion
6. _____ exists when deaths are balanced by births so that the population does not increase.
7. Between now and 2050, the U.S. population is expected to
   a. remain stable.
   b. increase by almost 50 percent.
   c. decrease by 33 percent.
   d. dip in 2036 and start growing again.
8. A population with an excess of births over deaths will experience _____.

Answers: 1. exponential growth; 2. positive; 3. demographic transition; 4. c; 5. c; 6. Zero population growth; 7. b; 8. natural increase death rate in the United States is relatively low, but because of its aging of the population, the number of deaths has been increasing slightly since the early 1980s. Life expectancy, of course, has increased as the death rate has declined. On the average, an American born now can expect to live seventy-eight years, up from seventy-one years at the beginning of the 1970s. Life expectancy remains lower for males (seventy-five years) than for females (eighty years).

Actually, mortality rates could affect future U.S. population growth because they are higher at most ages than in other more developed countries, outside eastern Europe. U.S. mortality rates could decline further if preventable deaths due to such factors as poor medical care and poverty among younger Americans were lowered, particularly among African Americans. Death rates could also be reduced at later ages due to improved medical care and the adoption of healthier diets and lifestyles (Kent and Mather 2002).

Urbanization

For centuries, people have lived in large population settlements. The nature of those settlements has undergone important changes, and as the settlements have mutated, the language to discuss them has also been altered. As an aid to understanding the volatile conditions in urban areas, we first examine the terminology used to describe such areas.

Basic Concepts

What defines a community? Long before the rise of cities, people depended on communities for their common needs. In fact, a **community** is a concentration of people whose major social and economic needs are satisfied primarily within the area where they live. It is easy to identify the network of social relationships and organizations that fulfill basic social and economic needs in a small community of several hundred people, but the situation is far more complex when several million people live within a given geographical area. Some sociologists even doubt that such large concentrations of people can be described as communities. Consequently, social scientists are now concentrating on cities and metropolitan areas rather than on communities.

When does a village become a city? The census bureaus of most nations define cities as population aggregates exceeding a certain size. In Denmark and Sweden, an area with 200 inhabitants officially qualifies as a city. Japan uses a much higher cutoff point—30,000 inhabitants. The minimal cutoff point used by the U.S. Census Bureau to define a city is a population of 2,500. These relatively low definition points were set at a time when urbanization had just begun and population concentrations were small. Because 2,500 people is clearly low for modern times, the Census Bureau has come up with additional urban designations. Of course, using mere numbers of inhabitants to define a city overlooks other important aspects of cities. In addition to containing a reasonably large number of people, cities are characterized by permanence, density, heterogeneity, and occupational specialization of inhabitants. In brief, a city is a dense and permanent concentration of people who live in a limited geographical area and who earn their living primarily through nonagricultural activities.

The Census Bureau has several terms to distinguish urban areas of varying size (Palen 2007). An **urban area** is a census block with a population density of no less than 1,000 persons per square mile. **Urban clusters** are areas over 10,000 but under 50,000. An **urbanized area** is composed of a central city, or cities, of 50,000 or more and the surrounding densely inhabited territory. The 2000 Census identified 1,371 "urbanized areas" and "urban clusters." A **micropolitan statistical area (MicroSA)** has at least one urban concentration of no less than 10,000 but less than 50,000 population, plus adjoining territory that is highly integrated with the core. In 2005 10 percent of the population resided in 577 MicroSAs. A **metropolitan statistical area (MetroSA or MSA)** is a grouping that contains at least one city of 50,000 inhabitants or more or an urbanized area of at least 50,000 inhabitants and the surrounding integrated population.

The establishment of criteria for designating MicroSAs and MetroSAs allows the Census Bureau to identify large areas whose populations have interdependent relationships with a central city. This is important because the influence of any city extends far beyond its official boundaries. As of 2005, 83 percent of America's people lived in 361 MetroSAs. Even a MetroSA is too limited to identify some urban areas because many do not exist in isolation. For example, although Newark, New Jersey, qualifies as a MetroSA, it must be considered in relation to surrounding MSAs, including New York City. Because of such situations, the Census Bureau identifies a **combined statistical area (CSA)**—two or more adjacent MSAs or a MetroSA and one or more MicroSAs. As of 2005, there were 124 CSAs.

The Urban Transition

Urbanization is the process by which an increasingly larger portion of the world's population lives in urban areas (see Worldview 17.1). Urbanization has been such a prevalent trend that it is now taken for granted in many parts of the world. As you have just seen, this process has resulted in cities so large and interdependent that it is now necessary to think of strips of interlocking cities.

Urban Population as a Proportion of Total Population

As discussed in the text, the Industrial Revolution encouraged the rapid growth of cities. The map below shows that many countries now have urban populations that comprise 60 percent or more of their total populations.

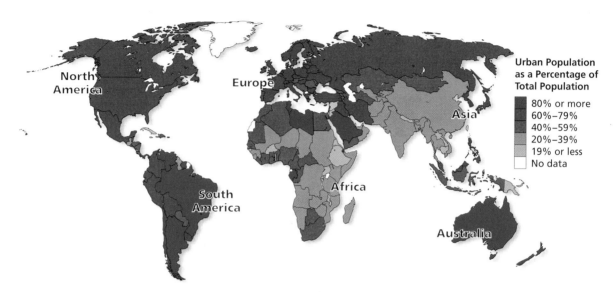

Urban Population as a Percentage of Total Population

- 80% or more
- 60%–79%
- 40%–59%
- 20%–39%
- 19% or less
- No data

Source: Population Reference Bureau, *2005 World Population Data Sheet.*

Interpret the Map

1. The map shows that countries such as the United Kingdom, Germany, and Sweden have urban populations that make up over 80 percent of their total populations. This can be explained by the effects of the Industrial Revolution, since these countries' economies are highly developed. However, other countries, such as Venezuela, Argentina, and Libya, which are not highly developed, also have urban populations that comprise over 80 percent of their totals. Can you think of reasons why this is so? Explain.

2. What effects will increased urbanization have on countries and the world?

At the beginning of the twenty-first century, almost as many people live in urban areas as in rural areas (Pacione 2005; Scripps-Howard 2002). This is a fairly recent development in human history.

What was the population of the early cities? The size of the first cities—founded only 5,000 years ago—was quite small by contemporary standards. Ur, located at the point where the Tigris and Euphrates Rivers meet (in modern-day Iraq), was one of the world's first major cities; but at its peak, it held only about 24,000 people. Later, during the time of the Roman Empire, it is unlikely that many cities had populations larger than 33,000. The population of Rome itself was probably under 350,000 (A. H. Hawley 1971).

In addition to their small size, the cities of ancient and medieval periods contained only a small portion of the world's population. As recently as 1800, less than 3 percent of the world's population lived in cities of 20,000 or more (K. Davis 1955). By contrast, 49 percent of the world's population now lives in urban areas (see Figure 17.8). In North America, 80 percent of the population lives in cities (*World Population Data Sheet* 2007). Only about 5 percent of the American population lived in urban areas in 1790 (U.S. Bureau of the Census 1993).

What accounts for this increase in urbanization? To find the answer to this question, we must take a closer look at urbanization before and after the Industrial Revolution. Although rapid urbanization is a fairly recent occurrence, its groundwork has been developing for several centuries.

Preindustrial Cities

Because people in cities hold primarily nonagricultural jobs, they require an agricultural surplus. Before 3500 B.C., few areas of the world were capable of producing a dependable supply of extra food for nonagriculturally productive people. For this reason, cities were initially slow to develop (Robert Adams 1965).

FIGURE 17.8

Urbanization Has Grown Dramatically Since the 1970s

As recently as 1975, one in three persons worldwide lived in urban areas, with most of those 1.5 billion urbanites living in areas of fewer than 1 million persons. (Many urban dwellers, in fact, live in communities of fewer than 5,000 residents.) Over the past 30 years, the urban population has increased so that more than half of the world population will be living in cities by 2008. Parallel to this overall urban growth (and the continued rise of smaller cities) is the emergence of "mega-cities—urban areas of at least 10 million persons. Between 1975 and 2005, the number of such large cities has mushroomed from three (two of which were in industrialized countries) to 20 (15 of which are in developing countries). Did you think that the world was more or les urban than shown below?

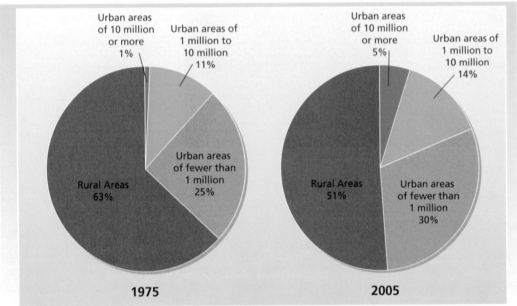

Source: United Nations Population Division, *World Urbanization Prospects: The 2005 Revision* (2006). Adapted from *World Population Data Sheet, 2007.*

What enabled the first cities to develop? According to anthropological research, the first urban settlements were established around 3500 B.C. in Mesopotamia (in southwestern Asia between the Tigris and Euphrates Rivers). The Mesopotamian region is among the world's most fertile areas, and at that time, its people were relatively advanced in the development of plant cultivation. So this area provided the necessary surplus of food for people in cities.

The existence of a surplus food supply clarifies the location of cities, but it does not explain people's attraction to them. Cities tended to attract four basic types of people: the elite, functionaries, craftspeople, and the poor and destitute. For the elite, who comprised but a small segment of the population, the city provided a setting for consolidating political, military, or religious power. The jewelry and other luxury items found in the tombs of the elite acknowledge the benefits these people gained from their consolidation of power and control. The functionaries were the political or religious officials who carried out the plans of the elite and whose lives were undoubtedly easier than those of the peasant-farmers in the countryside. Craftspeople, still lower on the stratification structure, came to the city to work and sell their products to the elite and functionaries. The poor and destitute, who were lured to the city for economic relief, were seldom able to improve their condition (Gist and Fava 1974).

Do preindustrial types of cities still exist? Many parts of Africa, Asia, and Latin America are not industrialized and therefore still have some preindustrial types of cities. This is particularly true of capital cities, which attract migrants from rural areas because of the symbolic importance of these cities and the promise of a better life. Unfortunately, those who migrate to many of these cities are disappointed, because the expected employment opportunities do not exist. Also, many cities in developing nations often lack the housing, sanitation, transportation, and communication facilities needed to accommodate their residents. As a result, many city dwellers in developing nations are forced to live in makeshift housing with few or none of the facilities that most people in modern societies consider necessities. Although cities in developing nations are unable to fulfill the social and economic needs of their residents, many people are driven to them because their life chances in the overpopulated rural areas and small towns are even worse.

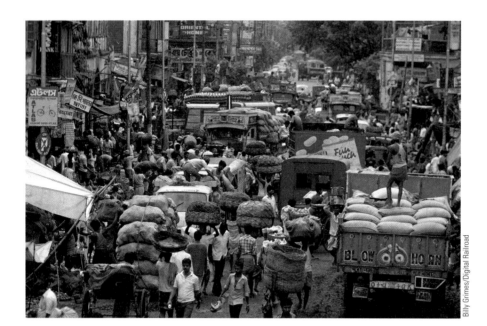

Calcutta, India remains essentially a preindustrial city.

Billy Grimes/Digital Railroad

Rise of the Modern City

As merchants gained power and wealth during the late medieval period, they often established small industries that would give them a stable supply of goods. As these industries gradually increased in size, the stage was set for rapid city growth.

How did the Industrial Revolution encourage the growth of cities? The Industrial Revolution, beginning in the eighteenth century, created major changes in transportation, agriculture, commerce, and industry. Technological developments during this period allowed for better food production and more efficient transportation systems with less human labor. As food production and transportation became more dependable, it became possible for more people with nonagricultural jobs to live in cities. The major reason for the growth of cities following the Industrial Revolution, however, was the increased use of factories.

Factories were not established to encourage the growth of cities, but they almost immediately had that effect; one factory tends to attract others to the same location. While taking advantage of natural features such as water power and river transport, factories located together can share raw materials and transportation costs. And at a time when transportation and communication were slow and undependable, it was desirable for producers of machinery and equipment to be near the factories they were supplying. The concentration of industry led to dense population settlements of factory workers, which in turn attracted retailers, innkeepers, entertainers, and a wide range of people offering services to city dwellers. Of course, the greater availability of urban services often enticed still more factories, keeping the cycle going. Although subse-

quent technological and social developments affected the growth of cities in different ways, the overall trend toward urbanization was set. The industrial world was becoming urbanized.

How does modern suburbanization differ from life on the outskirts of cities during preindustrial times? Life in preindustrial cities was directed toward the city's center. The walls, moats, and other protective devices found in these cities symbolized their role in defending the cities' inhabitants. It was safer to live in cities than outside them. The importance of the center of the city in preindustrial times is also reflected in residential patterns. Members of the upper class lived near the center of preindustrial cities, which gave them access to the most important government and religious buildings and, as a rule, the main market. The outskirts (including what we would now consider the suburban areas) were populated by the poor and those with low-status craft occupations, such as tanning and butchering. These people lived on the outskirts partly because more privileged residents wanted to have minimal contact with them. More important, though, the outskirts were the least desirable areas due to their lack of accessibility to the central city (Sjoberg 1966).

Modern cities, like preindustrial cities, located their major economic and government activities in their central core. Cities in industrialized nations, however, are connected by sophisticated communication and transportation networks, which have had the opposite effect of the walls and other protective devices of preindustrial cities. Thus, industrial cities tend to have an outward orientation. Most of their development has occurred away from the central city in the surrounding suburban areas. In industrialized nations, population movement from rural to urban areas has been accompanied by suburbanization.

In fact, many central cities are now losing population to the areas surrounding them (Brockerhoff 2000).

Suburbanization

Suburbanization—which occurs when central cities lose population to the area surrounding them—is a hallmark of contemporary American urban growth (Baldassare 1992; Gober 1993). By 1980, seventeen of the twenty-six central cities with a population of half a million or more in 1970 registered losses in population. Still, the population outside central cities is expanding several times as fast as the central cities themselves. According to the U.S. Census Bureau, the population of the central cities of the 255 MSAs increased by only 3.9 percent between 1990 and 1998, whereas the population of MSA areas outside the central cities (almost entirely suburbs) increased by 12.5 percent during the same period (U.S. Bureau of the Census 2000). During 1991–1992, the percentage of Americans moving from central cities to suburbs was twice the rate of movement from suburbs to central cities (U.S. Bureau of the Census 1993). This is a continuation of a trend that more than doubled the proportion of the population living in suburbs since 1950. Thus, the United States is now predominantly suburban.

What makes suburbanization possible? Suburbanization has become an important trend partly as a concomitant of technological developments. Improvements in communication—such as telephones, radios, television, and (later) computers, fax machines, and the Internet—allow people to live away from the central city without losing touch with its activities. Developments in transportation (especially highways, automobiles, trucks, and metro systems) make it possible both for people to commute to work and for many businesses to leave the central city for suburban locations (Baxandall and Ewen 2000).

Technology is not the only cause of suburbanization, however. Both cultural and economic pressures have encouraged the development of suburbs. Partly because of America's frontier heritage, American culture has always had a bias against urban living. Some Americans prefer urban life, but most report that they would rather live in a rural setting. Even those who, on balance, choose to live in the city believe they are giving up some advantages. Suburbs, with their low-density housing, allow many people to escape the problems of urban living without completely leaving the urban areas. Suburbs are attractive because of decreased crowding and traffic congestion, lower taxes, better schools, less crime, and reduced pollution.

Also, the scarcity and high cost of land in the central city encourage suburbanization. Developers of new housing, retail, and industrial projects often find suburban locations to be far less expensive than those near the central city.

Finally, government policy has often increased the impact of economic forces. Federal Housing Administration regulations, for example, favor the financing of new houses (which can be built most cheaply in suburban locations) rather than the refurbishing of older houses in central cities (Steinnes 1982).

As people move to suburban areas, increasing numbers of businesses and jobs follow. In fact, suburbanization is leading to the development of edge cities, and "boomburgs" that are changing the face of urban America.

What are edge cities and boomburgs? Originally, most suburbs were merely places where people lived. Suburbanites commuted daily from their "bedroom" communities to their jobs in the central city. Over the last decade, the nature of suburbanization has changed to include the movement of jobs and businesses to the suburbs as well. This change in suburbanization has created the **edge city**—a suburban unit specializing in a particular economic activity (Garreau 1991; Bingham and Bowen 1997). Employment in one edge city may focus on computer technology, another on financial services or health care. An edge city, of course, will have many other types of economic activities—industrial tracts, office parks, distribution and warehousing clusters, and home offices of national corporations. These are actually little cities in themselves with a full range of services, including schools, retail sales, restaurants, malls, medical facilities, hotels, motels, recreational complexes, and entertainment centers.

Edge cities do not have legal and physical boundaries separating them from the larger urban area in which they are located. This does not prevent names from being attached to several of them: Tyson's Corner is located in northern Virginia near Washington, D.C.; Las Colinas is close to the Dallas–Fort Worth airport; and King of Prussia is northwest of Philadelphia. Some edge cities bear the name of highways, such as Route 128 outside of Boston.

Boomburgs are places of more than 100,000 residents that are not the largest cities in their metropolitan areas and have experienced double-digit population growth in recent decades (Lang and Simmons 2001). According to the U.S. Census Bureau, the United States has 53 boomburgs, 41 of which exceed 100,000 people. Accounting for 51 percent of the 1990s growth are cities between 100,000 and 400,000 residents, boomburgs now contain one-fourth of all Americans living in cities of this size.

Despite developing into large cities, most boomburgs retain their suburban character. They are not without traditional urban problems, however, such as sprawl, overused government services, and traffic congestion.

Minorities are increasingly choosing to live in suburban communities. Despite this increase in residential integration, African Americans remain the most isolated minority in the United States.

Bill Bachmann/PhotoEdit

Consequences of Suburbanization

When suburbanization first became noticeable in the 1930s, only the upper and middle classes could bear the cost. It was not until the 1950s that the working class joined them. But until the 1970s, the suburbs have remained largely white, despite federal legislation prohibiting housing discrimination (Feagin and McKinney 2005).

Central-city minorities are beginning to move to the suburbs in greater numbers (Frey 1984, 1985; Massey and Denton 1987; R. Farley 1998; R. Farley and Haaga 2005). Although America's suburban counties remain predominantly white, minorities have poured into the suburbs over the past ten years. There are now 3 percent more Latinos in suburban counties than there were in 1990, 5 percent more Asians, and 5 percent more African Americans (Nasser 2001). A greater number of Latinos and Asian Americans, in fact, now reside in suburban counties than live in central cities. The percentage of African Americans living in suburbs increased almost 15 percent since 1980, standing now at 27 percent (Kleniewski 2006).

Despite this decline in the degree of residential segregation, African Americans remain the most isolated ethnic or racial minority in the United States (O'Hare et al. 1990; Farley and Frey 1994; South and Crowder 1997; Palen 2007). Although the percentage of African Americans living in suburbs increased from 16 percent to 26 percent between 1970 and 1990, the percentage of African Americans living in central cities declined only slightly. The situation is better for Latinos and Asians, as already noted. Not only are Latinos and Asians more suburbanized—and thus less concentrated in the central cities—than African Americans, they are also less segregated in suburban areas (R. Farley 1997; Palen 2007). Even this progress has a downside. As the more successful minority members move to the suburbs, the "underclass" accounts for a larger proportion of the remaining central-city population (Frey and Farley 1996; Massey and Denton 1996; Orfield 2001; see Chapter 9).

What is the central-city dilemma? Suburbanization created a major problem in the United States. The problem is not merely that minorities remain trapped in inner cities. Businesses have now followed the more affluent people to the suburbs, where they can find lower tax rates, less expensive land, less congestion, and their customers who have already left the city. Accompanying the exodus of the middle class, the manufacturers, and the retailers is the shrinking of the central-city tax base (Mouw 2000). As a result, the central city has become increasingly populated by the poor, the unskilled, and the uneducated. The result is a concentration of economic problems as well as a social dependency in cities that are financially unprepared to deal with them. This situation has been captured in what is known as the **central-city dilemma**—the concentration of a large population in need of public services (schools, transportation, health care) but without the money to provide for them (Wolfe 1999).

What are the consequences of the central-city dilemma? Poor inner-city African American children are handicapped by inferior educational facilities and teachers, and the dropout rate among them is considerably higher than among whites (Kozol 1992, 2006). During prosperous times, their unemployment rate is at least twice as high as it is for urban whites, and it skyrockets when the economy falters. Those who are employed generally must take dead-end, low-paying jobs because the more desirable jobs go to the better educated. A catalog of social ills flows from this socioeconomic condition in the inner city. African Americans in large-city slums exist in a world of poverty, congestion, prostitution, drug addiction, broken homes, and brutality. Crime rates, for example, appear to increase in central cities that have been affected by suburbanization, both because suburbanization creates stratification along city-suburban lines and because it contributes to an absolute decline in the socioeconomic status, the economic and tax base,

and the physical structure of the central city (Farley and Hansel 1981; Ganz 1985; Yinger 1995).

Can the central-city dilemma be solved? Some countertrends exist. There are city governments now requiring certain public employees to live in the city. Some parts of inner cities are being restored through **gentrification**—the development of low-income areas by middle-class home buyers, landlords, and professional developers ("Race, Place, and Segregation" 2002). Finally, there is a fairly significant movement of whites back to the central city. This movement is particularly evident among baby boomers who are remaining single or establishing childless or two-income families. Because these people are

John Hagedorn— Gang Violence

Gangs were a constant feature of the American urban landscape during most of the twentieth century. John Hagedorn (1998), however, contends that the extent and nature of gang violence has substantially changed with the emergence of postindustrial society in the 1960s. Hagedorn's conclusions are based on a combination of three methods: a review of the research of others, secondary analysis of precollected data (data collected by other researchers), and original data he gathered.

Gangs (mostly male) in the industrial period were tied to specific neighborhoods and new immigrant groups. These gangs emerged among unsupervised teens whose fathers worked in factories. Gang violence primarily centered on "turf" battles among neighborhood peer groups. Pride in violence came from defending the gang's space. Violence provided excitement and a sense of place in a group. Nevertheless, these working- and lower-class boys would eventually move on to hold decent

jobs, have families, and live in better neighborhoods.

Gangs today still tend to form around racial and ethnic groups and neighborhoods. Currently, gangs tend to be African American, Latino, or Asian, just as earlier gangs were mostly European immigrants, such as those from Ireland, Italy, or eastern Europe. Gangs form in the same poor urban areas, and members continue to come from dysfunctional families who do not supervise them. Juvenile gangs continue to engage in small-time crime, substance abuse, and periodic violence.

According to Hagedorn, however, the extent and nature of violence in postindustrial gangs underwent a major change. First, gang violence significantly increased. Second, gang-related homicides rose dramatically.

The question is, why? In answering this question, Hagedorn pinpoints important changes in postindustrial gangs, particularly gang functions. Gang violence, he notes, skyrocketed at the same time that American corporations were moving well-paying jobs from the central city. As legitimate work disappeared in inner cities, gangs turned from their earlier territorial emphasis to participation in the illegitimate drug market. The common outlook of gang members today is expressed by this gang member:

I got out of high school and I didn't have a diploma, wasn't no jobs, wasn't no source

of income, no nothing. That's basically the easy way for a . . . young man to be—selling some dope—you can get yourself some money real quick, you really don't have nothing to worry about, nothing but the feds. You know everybody in your neighborhood. Yeah, that's pretty safe just as long as you don't start smoking it yourself. (Hagedorn 1998:390)

Significantly, this gang member was not a teenager. More urban gang members today remain members into adulthood. Rather than getting a job, marrying, and settling down, more inner-city young adults hang out on the street, making a living on the drug trade.

Although drug selling was a small part of the gang scene before the 1980s, it was not a business. By contrast, Hagedorn found more young African American males in drug trafficking than any other area of employment.

There is still movement out of urban gangs, however. A minority of gang members remain committed to the drug economy, but most seek "legit" jobs as they approach their thirties.

Evaluate the Research

1. Relate Hagedorn's findings to Merton's strain theory in Chapter 7.
2. What do you think is the most important step in solving this problem? Justify your answer.

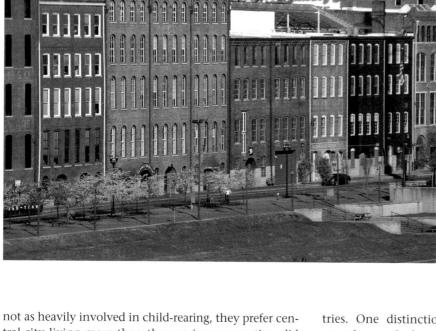

The movement of whites back to the central city has led to gentrification.

not as heavily involved in child-rearing, they prefer central-city living more than the previous generation did (Palen 2007). The importance of these countertrends for easing the central-city dilemma remains to be seen. They certainly have not been sufficiently important to stop the emergence of edge cities.

World Urbanization

Urbanization is a worldwide movement. From 1800 to the mid-1980s, the number of urban dwellers increased almost a hundredfold, whereas the population increased only about fivefold. Close to 50 percent of the world's population now lives in urban areas, and this is expected to reach 60 percent in 2025 (Weeks 2008). In more developed countries, 75 percent of the population lives in urban areas compared with 43 percent in less developed countries (Brockerhoff 2000; *World Population Data Sheet* 2007; Weeks 2008).

The pattern of urbanization is different in more developed countries than it is in less developed countries. One distinction is the rapidity of the urban growth rate. In less developed countries, most of the urban growth before the turn of the twentieth century occurred through colonial expansion. Western countries, which were involved in colonial expansion after the late fifteenth century, held half the world under colonial rule by the latter part of the nineteenth century. Only since World War II have many of these colonial countries become independent nations (Bardo and Hartman 1982). Since gaining independence, these former colonies have experienced rapid urbanization. In fact, urbanization in these areas is now proceeding faster than it did in the West, even during that area's biggest urban expansion period from 1850 to 1950 (Gelbard, Haub, and Kent 1999).

What are some additional differences in the pattern of urbanization in less developed and more developed countries? In the first place, industrial developments in less developed countries has not kept pace with urbanization. In the cities of less developed countries, the supply of labor from

Dennis MacDonald/PhotoEdit

**F
E
E
D
B
A
C
K**

1. A _____ is a concentration of people whose major social and economic needs are satisfied within the area in which they live.
2. _____ is the process by which an increasingly large proportion of the world's population lives in urban areas.
3. Thanks to worldwide urbanization, preindustrial cities no longer exist. T or F?
4. The _____ exists because diminishing public revenues are not sufficient to cope with the influx of the poor, unskilled, and uneducated.
5. _____ occurs when a city is unable to supply adequate jobs and housing for its inhabitants.

Answers: 1. community; 2. Urbanization; 3. F; 4. central-city dilemma; 5. Overurbanization

FIGURE 17.9

World's Largest Urban Areas: 1950, 2003

This figure compares the world's largest urban areas in 1950 and 2003. What is the most surprising aspect of these data to you?

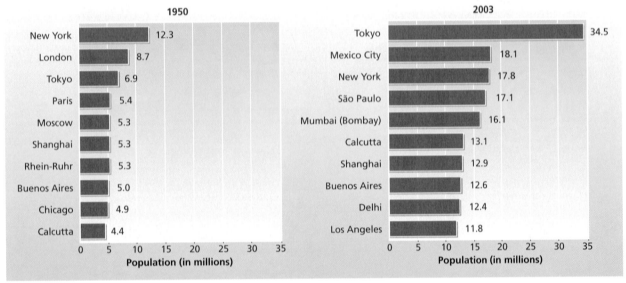

Source: *The World Almanac and Book of Facts, 2005.*

the countryside exceeds the demand for labor. A high rate of unemployment is the obvious result. The term **overurbanization** has been created to describe a situation in which a city is unable to supply adequate jobs and housing for its inhabitants. In the Western experience, on the other hand, industrial development kept pace with urbanization, and cities of North America and Europe had jobs for most migrants from rural areas.

Another difference between urbanization in more developed and less developed countries is the number and size of cities. When grouped by size, cities in developed countries form a pyramid: a few large cities at the top, many medium-sized cities in the middle, and a large base of small cities (Light 1983). In the less developed world, many countries have only one big city, a tremendously large city that dwarfs a large number of villages. Calcutta, India, and Mexico City are examples. In 1950, of the world's ten largest cities, only two—Shanghai and Calcutta—were in less developed areas. By 2000, seven of the top ten largest urban areas were in less developed countries. In 2005, there were twenty "megacities" with populations of 10 million or more, and that number is expected to increase to 22 million by 2015. Of these, fifteen were in developing countries, including the most impoverished societies in the world. Similarly, eight of the world's largest urban areas are in less developed countries (Weeks 2008; see Figure 17.9).

Why do people in developing countries move to large cities where they will find inadequate jobs and housing? Urban sociologists point to the operation of push–pull factors

in the rural-to-urban migration in less developed countries. Peasants are pushed out of their villages because expanding rural populations cannot be supported by the existing agricultural economy. Peasants are also attracted to cities by perceived opportunities for better education, employment, social welfare support, and availability of good medical care, even though they are likely to be disappointed (Firebaugh 1979; Sassen 2006).

Theories of City Growth

Every city is unique, but the patterns of all cities can be interpreted based on **urban ecology**—the study of the relationships between humans and their urban environments.

In the 1920s and 1930s, sociologists at the University of Chicago studied the effects of the city environment on urban residents. It turns out that the effects varied in different parts of the city. For example, Harvey Zorbaugh (1929) found significant differences between Chicago's wealthy Gold Coast and its slum areas. The realization that such differences existed led to additional questions: Why do differences exist between areas of a city? How do different areas affect one another? What are the processes through which areas are changed? To answer these and other questions, the University of Chicago sociologists developed theories of urban ecology, including theories of city growth (Micklin and Poston 1998; Flanagan 2001; Kleniewski 2006).

FIGURE 17.10

Theories of City Growth

This figure displays in graphic form the four major theories of city growth. Discuss one important advancement these theories make to understanding urban growth.

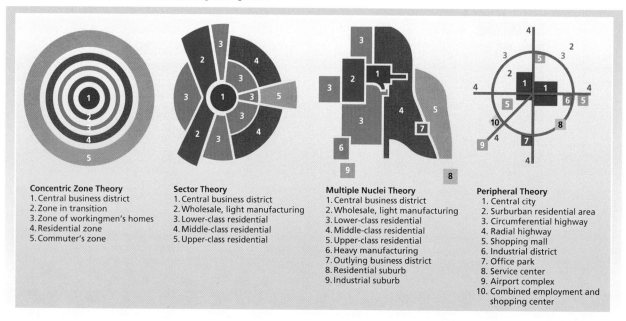

Concentric Zone Theory
1. Central business district
2. Zone in transition
3. Zone of workingmen's homes
4. Residential zone
5. Commuter's zone

Sector Theory
1. Central business district
2. Wholesale, light manufacturing
3. Lower-class residential
4. Middle-class residential
5. Upper-class residential

Multiple Nuclei Theory
1. Central business district
2. Wholesale, light manufacturing
3. Lower-class residential
4. Middle-class residential
5. Upper-class residential
6. Heavy manufacturing
7. Outlying business district
8. Residential suburb
9. Industrial suburb

Peripheral Theory
1. Central city
2. Surburban residential area
3. Circumferential highway
4. Radial highway
5. Shopping mall
6. Industrial district
7. Office park
8. Service center
9. Airport complex
10. Combined employment and shopping center

Source: Chauncy D. Harris and Edward L. Ullman, "The Nature of Cities," *The Annals* (1945):13. Copyright © 1945 by The American Academy of Political and Social Science; and Chauncy D. Harris, "'The Nature of Cities' and Urban Geography in the Last Half Century," *Urban Geography* 18(1997):17. Used by permission of V.H. Winston & Son, Inc.

There are four major theories of city growth. **Concentric zone theory** describes urban growth in relation to circular areas that grow from the central city outward. **Sector theory** emphasizes the importance of transportation routes in the process of urban growth. **Multiple-nuclei theory** focuses on specific geographical or historical influences. These three theories, developed more than fifty years ago, have been joined by a more contemporary model, the **peripheral theory,** which accents the growth of suburbs around the central city. These four theories lead to quite different images of urban space (see Figure 17.10). Indeed, all four tell us more when considered together than they tell us separately. To understand why this is so, we must first examine each theory.

Concentric Zone Theory

What is concentric zone theory? Ernest Burgess (1925), like other early sociologists at the University of Chicago, was interested in the causes and consequences of Chicago's growth. His work led to the concentric zone theory, which describes city growth in relation to distinctive zones that develop from the central city outward, in a circular pattern. As illustrated in Figure 17.10, the innermost circle is the *central business district*, the heart of the city, containing major government and private office buildings, banks, retail and wholesale stores, and entertainment and cultural facilities.

The central business district exerts an especially strong influence on other parts of a city. This influence is especially clear in the zone immediately surrounding it. Burgess called this the *zone in transition* because it is in the process of change. As new businesses and activities enter the central business district, the district expands by invading the next zone. This area may have been a residential area inhabited by middle- or upper-class families, who left because of the invasion of business activities. Most of the property in this zone is bought by those with little interest in the area. Rather than investing money in building maintenance, landowners simply extract rent from the property or sell it at a profit after the area is more commercialized. Until the zone in transition is completely absorbed into the central business district (which may never occur), it is used for slum housing, warehouses, and marginal businesses that are unable to compete economically for space in the central business district itself. In short, the invasion of business activities creates deterioration in the zone in transition.

Surrounding the zone in transition are three zones devoted primarily to housing. The *zone of workingmen's*

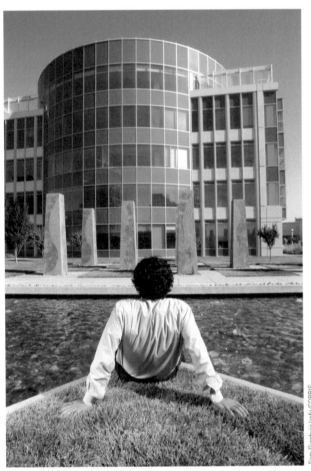

Explain which theory of urban growth best accounts for this suburban office building.

Sam Diephuis/zefa/CORBIS

homes contains modest but stable neighborhoods populated largely by blue-collar workers. In the northern United States, the zone of workingmen's homes is often inhabited by second-generation immigrants who have enough financial success to leave the deteriorating zone in transition. Next comes a *residential zone* containing mostly middle-class and upper-middle-class neighborhoods. Single-family dwellings dominate this zone, which is inhabited by managers, professionals, white-collar workers, and some well-paid factory workers. On the outskirts of the city, often outside the official city limits, is the *commuter's zone*, which contains upper-class and upper-middle-class suburbs.

Sector Theory

What is sector theory? Based on a study he performed for the Federal Housing Administration, Homer Hoyt (1939) developed sector theory. Hoyt's study indicated that patterns of land use do not necessarily change according to distance from the central business district. Instead, land is more strongly affected by major transportation routes. As Figure 17.10 shows, sectors tend to be pie-shaped wedges radiating from the central business district to the city's outskirts. Each sector is organized around a major transportation route. Once a given type of activity is organized around a transportation route, its nature tends to be set. Thus, some sectors will be predominantly industrial, others will contain stores and professional offices, others will be "neon strips" with motels and fast-food restaurants, and still others will be residential sectors, each with its own social class and ethnic composition.

As in concentric zone theory, sector theory depicts cities as generally circular in shape. But due to the importance of transportation routes extending from the central business district, the boundaries of many cities form a star-like pattern, rather than a uniformly circular shape. The exact shape of a city, however, is not a major issue in sector theory. Emphasis here is on land patterns around transportation routes.

Multiple-Nuclei Theory

What is multiple-nuclei theory? Many cities have areas that cannot be explained by either concentric zone or sector theory. For example, the Beacon Hill area of Boston defies concentric zone theory; it remains an expensive, prestigious neighborhood despite its location near the central business district. Because of such cases, Chauncy Harris and Edward Ullman (1945) have suggested that cities do not always follow a pattern contingent on a central district. (Figure 17.10 shows one example of a multiple-nuclei city pattern.) A city may have several separate centers, some devoted to manufacturing, some to retail trade, some to residential use, and so on. These specialized centers can develop because of the availability of automobiles and highways. They reflect such factors as geography, history, and tradition. Because the factors affecting land use are unique to each city, multiple-nuclei theorists do not predict any particular pattern of land use that would apply to all cities.

Peripheral Theory

What is peripheral theory? The preceding three theories of urban growth were developed more than fifty years ago. Despite their age, the insights of each theory help us to understand how cities have expanded from the center outward. This is especially the case for older cities such as Chicago and San Francisco. Many cities today, however, no longer have a central-city core to which other parts of the metropolitan area are oriented.

Dependence on shipping, railroads, and heavy manufacturing has been replaced by more flexible means of transportation, such as cars and trucks. And large urban areas are now encircled by highways. New technologies (fax machines, cell phones, computers, the Internet) are also loosening the ties of most parts of the city to the central-city core. As a result, many cities are now oriented *away* from the older urban core.

As mentioned earlier, many Americans have moved from the city to the suburbs. They have done so in part because many businesses—offices, factories, schools, retail stores, restaurants, health centers—are also in the suburbs. To encompass changes in urban areas today, urban geographer Chauncy Harris (1997) has formulated the peripheral theory. The dominant feature of this model, obviously, is the growth of suburbs (and edge cities) around and away from the central cities (see Figure 17.10). Peripheral theory brings urban growth research up to date.

Summing Up

Of what value are these diverse theories of city growth? Although each of these theories calls attention to the dynamics of city growth, none of them applies to all cities. Each theory, however, does emphasize the importance of certain factors that cannot be overlooked by anyone interested in city growth. Concentric zone theory underscores the fact that growth in any one area of a city is part of the processes of segregation, invasion, and succession. This theory also reveals the importance of economic power in city growth: the distribution of space in cities is heavily influenced by those with the money to buy the land they want for the purposes they have in mind. Concentric zone theory is in need of modification, however, because it is based on the assumption that cities will continue to grow and that change will occur as one area encroaches on another.

Many central cities, however, have ceased to grow. In those cases, the "push-from-the-center dynamic no longer applies" (Choldin and Hanson 1981:561).

Sector theorists also contributed to an understanding of urban dynamics. As they noted, transportation routes have a strong influence on cities. Decisions about the placement of railroad lines had important effects on the growth of cities in the nineteenth and early twentieth centuries. Highways and major routes have an even larger impact now.

Although multiple-nuclei theory is vague in its predictions, the types of geographical and historical factors it emphasizes are also important for understanding any specific city. Moreover, the apparent decline of many central business districts may increase the importance of specific local factors in the future growth of cities. Finally, peripheral theory brought urban growth research up to date by emphasizing the development of suburbs around the central city.

No single ecological theory describes the growth pattern of all cities; each is a contribution in itself. In recognition of this, some researchers explain patterns of land use by combining the insights of these theories. Several researchers, for example, have reported that people distribute themselves in cities in a way that can be predicted only by combining concentric zone theory and sector theory (Berry 1965; Timms 1971; Berry and Kasarda 1977).

For another example, different sectors attract people from different social classes. The geography and history associated with some transportation routes out of a city encourage people in higher social classes to settle there. Other sectors, which are less appealing, are left to lower-income people. Within any particular sector, however, family status and lifestyle seem to vary as concentric zone theory predicts. The newer single-family

---

FEEDBACK

1. Match the following zones of the concentric zone theory with the appropriate descriptions.
   - ____ a. central business district
   - ____ b. commuter's zone
   - ____ c. residential zone
   - ____ d. zone in transition
   - ____ e. zone of workingmen's homes

   (1) zone that is the next zone out from the zone in transition
   (2) site of major stores and offices
   (3) zone with physical deterioration stemming from few repairs
   (4) zone the wealthiest people live in
   (5) zone that is largely middle class in composition

2. The theory that emphasizes the importance of transportation routes is called _____ theory.
3. The _____ theory contends that the growth of cities does not necessarily follow any specific pattern.
4. _____ theory of city growth brings urban growth research up to date.
5. One of the ultimate contributions of the four major theories of city growth is their demonstration that cities do not grow in a random fashion. T or F?

Answers: 1. a. (2), b. (4), c. (5), d. (3), e. (1); 2. sector; 3. multiple-nuclei; 4. Peripheral; 5. T homes at the edge of the city attract different types of people than the apartment complexes and older homes that are likely to be found near the center of the city. Exact patterns of land use in any urban area, however, are likely to depend on the cultural values affecting that particular area (Abu-Lughod 1980).

Regardless of the ultimate fate of the concentric zone, sector, multiple-nuclei, and peripheral theories, these theories have performed an important service. By calling attention to the factors influencing land use in urban areas, these theories remind us that cities do not grow in a random manner.

The Quality of Urban Life

As just noted, sociologists from the University of Chicago contributed to our understanding of urban land use. Their interests were not, however, confined simply to the ecological characteristics of cities. These sociologists were also concerned with the effects of cities on people. Their basic conclusion: people in cities live a unique way of life. The researchers designated this distinctive way of life—**urbanism.**

Cities have always been thought to produce a wide variety of lifestyles. During medieval times, cities were viewed as places to escape the domination of powerful landowners, and it was said that "city air makes one free." Many of today's cities are also considered sources of intellectual and cultural diversity, sophistication, and freedom. Louis Wirth (1938), a University of Chicago sociologist and the first to develop the idea of urbanism as a way of life, agreed in part with the intellectual tradition that associated the city with freedom and tolerance. He believed that because urban residents are confronted with a variety of lifestyles, they are more tolerant of diversity and less ethnocentric than residents of small towns or rural areas (T. C. Wilson 1985, 1991). However, Wirth also drew on an intellectual tra-

dition that saw impersonality and social disorganization in urban areas, a tradition represented in the work of German sociologist Ferdinand Tönnies. Later sociologists, as noted in "Using the Sociological Imagination" at the beginning of the chapter, have a more positive view of the quality of urban social life.

The Traditional Perspective on Urbanism: Loss of Community

What was Tönnies's viewpoint? Although he did not use the term *urbanism*, Tönnies was the first sociologist to contrast the culture in rural and urban settings. As noted in Chapter 5, Tönnies distinguished between *gemeinschaft* (community) and *gesellschaft* (society). In the **gemeinschaft**—the type of society existing before the Industrial Revolution—social relationships were intimate because people knew each other. Daily life revolved around family and was based on tradition. There was a sense of a shared way of life and an interest in the common good of the community. Urbanization changes all that, according to Tönnies. In the **gesellschaft**, social relationships are impersonal because most people with whom one interacts are strangers. Although the immediate family is still important, the wider kinship network disappears. With the fading of tradition goes the sharing of common values. Interest shifts from the community's welfare to self-interest.

How did Wirth conceptualize urbanism? In Wirth's view, the urban conditions that promote sophistication, freedom, and diversity also involve a high cost. Urbanism, Wirth said, is marked by a depersonalization of human relationships. Rather than involving their total personalities in interaction, urban residents interact on the basis of rigidly defined roles. For example, a customer in a rural store may combine shopping with a lengthy

---

TABLE 17.1

FOCUS ON THEORETICAL PERSPECTIVES: Urban Society

This table illustrates how functionalism, conflict theory, and symbolic interactionisms might approach the study of urban society. State a research topic for the study of population for which any one of the theories would be appropriate.

Theoretical Perspective	Concept	Sample Research Topic
Functionalism	Urbanization	• Study of the relationship between population density and the suicide rate
Conflict theory	Overurbanization	• Investigation of the relationship between the distribution of scarce resources and social class
Symbolic interactionism	Urbanism	• Research on the effect of the degree of urbanization and the extent to which social interaction is based on shared meanings

conversation, but interactions in most urban stores are confined to the exchange of money and products. This emphasis on rigid role prescriptions, Wirth concluded, makes it difficult for urban residents to maintain a sense of their own individuality; it makes them indifferent and blasé. Heavy reliance on roles occurs partly because of sheer number: it is impossible to know everyone. Urbanites may actually know more people than do rural residents, but they know fewer people well enough to interact with them on a personal basis.

According to Wirth, urban residents behave as they do because of three demographic characteristics of cities: large size, high density (a large number of residents per square mile), and great diversity within the population. Wirth contended that as cities become larger, more densely settled, and more socially diverse, residents become more sophisticated and more reliant on impersonal relationships.

The Contemporary Perspective on Urbanism: Community Sustained

Wirth's theory of urbanism is criticized on several grounds, the most frequent being that it is an over-generalization. Some critics charge that Wirth's view of urbanism mistakenly assumes that all urban residents share a common lifestyle (I. M. Young 1990; Macionis and Parillo 2006). A leading spokesperson for this opposing line of thought is Herbert Gans (1962b, 1968).

Do all urban residents share the same way of life? Gans described five basic population groups found within inner-city areas, the first three of which live in the inner city by choice and are protected from the extreme negative consequences depicted by Wirth:

- *Cosmopolites.* People in this category are well educated and have high incomes. They live in inner-city areas because they are attracted to the area's cultural advantages, such as art museums, theaters, and symphony orchestras.
- *The unmarried and childless.* People in this category also live in the inner city by choice. They may not feel a strong commitment to the inner city, but living there allows them to meet other people, and the area is convenient to jobs and entertainment facilities.
- *Ethnic villagers.* People in this category live in inner-city neighborhoods with strong ethnic identities. Residents of such neighborhoods are suspicious of the city as a whole and participate little in its affairs. The neighborhood itself, however, is an important source of intimate, enduring relationships.
- *The trapped.* Some people—such as the elderly living on pensions or public assistance—cannot afford to

leave the inner city. They may identify with their own neighborhood, but this does not prevent them from becoming annoyed over changes in the central-city area that they consider undesirable.
- *The deprived.* Poor minorities, the psychologically and physically challenged, some divorced mothers, and other deprived people see the city as a source of increased opportunities for employment or welfare assistance. The poverty, however, prevents such individuals from choosing where they will live in the city. These people are therefore exposed to the most extreme effects of population size, density, and diversity.

According to Gans, then, no blanket statement (such as that formulated by Wirth) applies to every urban resident. Although the trapped and the deprived are unable to escape many of the adverse effects described by Wirth, members of the other three categories have chosen to live in the inner city. They would not have made this choice had they not seen many advantages.

Are city dwellers isolated, with no personal relationships? According to Wirth's theory, urban residents have relatively few opportunities for participation in informal relationships. Subsequent research shows, however, that urban residents do participate in informal relationships, and such relationships play an important role in maintaining a positive emotional state (Kadushin 1983; Creekmore 1985; C. S. Fischer 1982, 1984, 1995). In an intentional test of Wirth's theory, John Kasarda and Morris Janowitz (1974) found that population size and density do not necessarily produce a lack of involvement in local friendship, kinship, or other social relationships. Those who are not highly involved in such relationships in a large city tend to be those who have not lived there long enough to become a part of it. John Esposito and John Fiorillo (1974) found essentially the same thing in a study of working-class neighborhoods in New York City.

Other evidence indicates that families continue to provide urban residents with emotional support and mutual aid (see Chapter 12). Studies based on observations of actual behavior in areas as diverse as lower-class slums and middle-class suburbs are showing the importance of personal relationships among family members and friends. For example, Gans (1962a) observed the West End of Boston, where many aspects of Italian rural life continue to be honored. Personal relationships are encouraged there, some have argued, because the area is dominated by one ethnic group. However, neighborhoods in general appear to be important sources of interaction and support. Take, for example, research of a Chicago slum (Addams) by Gerald Suttles (1968): even in a low-income area populated by four different ethnic groups, a social order based on informal and personal relationships developed.

1. The distinctive way of life allegedly developed in urban areas is referred to as _____.
2. According to Ferdinand Tönnies, urbanism corresponds to _____ (gemeinschaft/gesellschaft).
3. Louis Wirth relied on three characteristics to explain the development of urbanism as a way of life. What were they?
    a. age of city, population density, political structure
    b. population size, economic base, political structure
    c. population size, population density, population diversity
    d. economic base, political structure, population structure
4. According to Herbert Gans, which of the following groups have relatively little choice within the inner city?
    a. cosmopolites, the unmarried and childless
    b. ethnic villagers, the deprived
    c. the unmarried and childless, ethnic villagers
    d. the trapped, the deprived
5. Research shows that urban residents are largely isolated from family and friends. T or F?

Answers: 1. urbanism; 2. gesellschaft; 3. c; 4. d; 5. F

SUMMARY

1. Demography is the scientific study of population. Population data are very important today, in part because of their use by government and industry.
2. Demographers use three population processes to account for population change: fertility, mortality, and migration.
3. Because population grows exponentially, it increases more each year in absolute numbers, even if the rate of growth remains stable. Most developed nations have undergone what is known as the demographic transition. This is the process accompanying economic development in which a population gradually moves from high birth and death rates to low birth and death rates. Europe seems to be experiencing a second demographic transition, which involves an eventual decline in population size.
4. In more developed societies, economic growth has led to a lower fertility rate for a number of reasons. But this type of economic development, along with continued inequalities, has not led to a reduction in fertility in less developed countries. These latter countries may or may not pass through the demographic transition.
5. Although the world's population growth rate is declining after 200 years of acceleration, the world's population will continue to grow. It is projected that world population growth will stop at just greater than 12 billion by the end of the twenty-first century.
6. The American population is expected to grow more rapidly than do those in other developed countries, due to a combination of the U.S. total fertility rate, low mortality rates, and high net migration rate.
7. As urban areas increased in size, sociologists gradually shifted their attention from communities to larger units,

such as metropolitan statistical areas (MetroSAs or MSAs) and combined statistical areas (CSAs). However, the basic unit of study is the city, which is characterized by a large number of inhabitants, permanence, density, and occupational specialization of inhabitants.

8. Although the process of urbanization is now commonplace throughout the world, the first cities developed only 5,000 to 6,000 years ago. The first preindustrial cities developed in fertile areas where plant cultivation was sufficiently advanced to support a nonagricultural population. A surplus food supply permitted relatively large numbers of people to live in cities and perform nonagricultural work. Preindustrial types of cities still exist in underdeveloped parts of the world.
9. The Industrial Revolution, with its technological innovations and alterations in social structure, caused a major increase in the rate of urbanization. The development of factories was an especially important influence on the location of cities. Although cities and metropolitan areas are important in industrialized societies, they are strongly affected by suburbanization and a recent tendency for medium-sized metropolitan and nonmetropolitan areas to grow more rapidly than large metropolitan areas.
10. Suburbanization has created severe problems for central cities and their residents. The problems are particularly serious for minorities, the poor, and the elderly in the inner city.
11. Less developed countries are having a different experience with urbanization than did the West. In less developed countries, the urban growth rate has been faster, industrialization has been slower, and a base of small and

middle-sized cities has not developed. Still, peasants in less developed countries migrate to cities both because they are forced off the land and because cities appear to provide unique opportunities.

12. Urban ecologists study the adjustment processes of human populations to their environments. Urban ecologists developed four major models of spatial development in cities: concentric zone theory, sector theory, multiple-nuclei theory, and peripheral theory. Each of these theo-

ries receives some empirical support, but it appears that combining insights from all of them is the most useful.

13. According to the traditional view of urbanism, cities provide a free and sophisticated environment but are characterized by impersonality and social disorganization. According to more recent empirical research, this description is not applicable to all urban residents; many people in urban areas are involved in personal and social relationships.

LEARNING OBJECTIVES REVIEW

- Identify and distinguish among the three population processes.
- Discuss major dimensions of the world population growth problem.
- Relate the ideas of Thomas Malthus to the demographic transition.
- Predict world population trends.
- Discuss demographic trends of older Americans.
- Differentiate basic concepts of urbanization.
- Trace the historical development of preindustrial and modern cities.
- Describe some of the consequences of suburbanization.
- Discuss world urbanization.
- Compare and contrast four theories of city growth.
- Compare and contrast traditional and contemporary views on the quality of urban life.

CONCEPT REVIEW

Match the following concepts with the definitions listed below them.

_____ a. gemeinschaft
_____ b. multiple-nuclei theory
_____ c. crude death rate
_____ d. urban ecology
_____ e. fecundity
_____ f. replacement level
_____ g. sector theory
_____ h. age-specific fertility

_____ i. social demography
_____ j. doubling time
_____ k. urbanization
_____ l. community
_____ m. gross migration rate
_____ n. combined statistical area (CSA)
_____ o. morbidity

_____ p. demography
_____ q. net migration rate
_____ r. metropolitan statistical area (MetroSA or MSA)
_____ s. fertility rate
_____ t. life span

1. the birth-rate level at which married couples replace themselves in the population
2. rates of disease and illness in a population
3. the maximum rate at which women can physically produce children
4. a description of the process of urban growth emphasizing the importance of transportation routes
5. the branch of demography that studies population in relation to its social setting
6. a grouping that contains at least one city of 50,000 inhabitants or more or an "urbanized area" of at least 50,000 inhabitants and the surrounding integrated population.
7. Tönnies's term for the type of society based on tradition, kinship, and intimate social relationships

8. a social group whose members live within a limited geographical area and whose social relationships fulfill major social and economic needs
9. the annual number of live births per 1,000 women ages fifteen to forty-four
10. the annual number of deaths per 1,000 members of a population
11. the annual increase or decrease per 1,000 members of a population resulting from people leaving and entering the population
12. the scientific study of population
13. a description of the process of urban growth that emphasizes specific and historical influences on areas within cities
14. the most advanced age to which humans can survive

15. the number of live births per 1,000 women in a specific age group
16. the study of the relationships between humans and their environments within cities
17. an urban area composed of two or more sets of neighboring primary metropolitan statistical areas (MetroSAs) or one or more MicroSAs.

18. the number of years needed to double the size of a population
19. the process by which an increasingly larger proportion of the world's population lives in urban areas
20. the annual number of persons per 1,000 members of a population who enter or leave a geographical area

CRITICAL-THINKING QUESTIONS

1. Suppose that the average number of children per family in the United States increased from 1.8 to 3. Discuss the short- and long-term effects this would have on population growth.

2. Should the slow-growing nations, such as the United States, have immigration policies that help relieve the population pressures on countries with high-growing populations? Defend your position using sociological concepts.

3. Explain and evaluate the usefulness of the following theories of city growth: concentric zone theory, sector theory, multiple-nuclei theory, and peripheral theory.

4. According to Louis Wirth, what are the major characteristics of urbanism as a way of life? Evaluate the accuracy of Wirth's explanation in view of more recent research findings.

MULTIPLE-CHOICE QUESTIONS

1. **The three population processes that are responsible for population growth and decline are**
   a. fecundity, mortality, and migration.
   b. fecundity, morbidity, and migration.
   c. fertility, morbidity, and migration.
   d. fertility, mortality, and migration.
   e. fecundity, mortality, and morbidity.

2. **Which of the following best describes contemporary Americans' behavior regarding reproduction?**
   a. Fewer American women are having abortions.
   b. Fewer Americans are using contraceptives.
   c. More American women are having children in their late teens and early twenties.
   d. More American women are now beginning to have larger families.
   e. Fewer American women are having children, and more American women are having fewer children.

3. **The demographic term for deaths within a population is**
   a. morbidity.
   b. mortality.
   c. terminality.
   d. finality.
   e. turnover.

4. **Rates of disease and illness in a population are known as**
   a. life expectancy.
   b. mortality.

   c. morbidity.
   d. life span.
   e. social pathology.

5. **The movement of people from one geographical area to another for the purpose of establishing a new residence is known as**
   a. fecundity.
   b. exponential transition.
   c. migration.
   d. morbidity.
   e. the demographic transition.

6. **_____ is the principle by which a population increases.**
   a. Linear growth
   b. Natural logarithms
   c. Multiplication
   d. Exponential growth
   e. Arithmetic explosion

7. **For Thomas Malthus, the solution to the population problem was**
   a. early marriage.
   b. improved contraception.
   c. universal education.
   d. federally financed abortions.
   e. infanticide.

8. _____ is the process by which a population gradually moves from high birth and death rates to low birth and death rates as a result of economic development.
   a. Exponential growth
   b. The demographic transition
   c. The dependency ratio
   d. The fertility–morbidity ratio
   e. The net migration rate

9. According to the text, three areas of the world are presently in the fourth stage of the demographic transition. These areas are
   a. China, Japan, and North America.
   b. Australia, Iran, and Europe.
   c. South America, Europe, and North America.
   d. Iran, North America, and Europe.
   e. Europe, North America, and Japan.

10. The primary characteristic of the second demographic transition is
    a. a decline in fertility from slightly above replacement level to slightly below.
    b. a reduction in the net migration rate.
    c. an increase in mortality.
    d. the progress achieved regarding adult literacy.
    e. the shift from gesellschaft to gemeinschaft patterns of social organization.

11. _____ has been reached when deaths are balanced by births so that a population does not increase.
    a. Zero population growth
    b. Exponential growth
    c. Equilibrium ratio
    d. The fecundity level
    e. Demographic transition

12. The primary resource needed by a preindustrial city was
    a. a strong trading center.
    b. a surplus food supply.
    c. adequate employment opportunities.
    d. a good system of waterways.
    e. a completed demographic transition.

13. Metropolitan dispersal refers to
    a. the spreading out of industry throughout a city.
    b. the dispersal of retail outlets outward due to inner-city deterioration.
    c. middle- and upper-class whites buying up downtown property after urban renewal.
    d. cities losing population to the areas surrounding them.
    e. cities gaining population over surrounding areas.

14. Which of the following theories of city growth explicitly deals with the zone in transition?
    a. concentric zone theory
    b. peripheral theory
    c. sector theory
    d. ecological invasion theory
    e. multiple-nuclei theory

15. Which of the following statements best describes the personal relationships of city dwellers?
    a. Urban residents conduct their social relationships on the basis of rigidly defined role prescriptions.
    b. Urban dwellers rely on impersonal relationships to protect themselves from the social claims of others.
    c. Kinship networks have disappeared from large cities.
    d. Urban residents know few people well enough to interact on a personal basis.
    e. Large population size does not necessarily produce a lack of involvement in local friendship, kinship, and other social relationships.

FEEDBACK REVIEW

True-False

1. Thanks to worldwide urbanization, preindustrial cities no longer exist. T or F?
2. One of the ultimate contributions of the four major theories of city growth is their demonstration that cities do not grow in a random fashion. T or F?

Fill in the Blank

3. Thomas Malthus sees famine as a _____ check on population growth.
4. _____ is the process by which an increasingly large proportion of the world's population lives in urban areas.
5. A population with an excess of births over deaths will experience _____.
6. _____ occurs when a city is unable to supply adequate jobs and housing for its inhabitants.
7. The distinctive way of life allegedly developed in urban areas is referred to as _____.

Multiple Choice

8. Which of the following figures is the world's population most likely to reach before it stops growing?
   a. 4 billion
   b. 8 billion
   c. 12 billion
   d. 25 billion

9. Europe's second demographic transition involves what kind of population growth?
   a. stable
   b. increasing
   c. declining

10. **Which of the following characteristics is not included in the definition of a city?**
    a. permanence
    b. dense settlement
    c. advanced culture
    d. occupational specialization in nonagricultural activities
11. **According to Herbert Gans, which of the following groups have relatively little choice within the inner city?**
    a. cosmopolites, the unmarried and childless
    b. ethnic villagers, the deprived
    c. ethnic villagers, the unmarried and childless
    d. the trapped, the deprived

Matching
12. Match the following zones of the concentric zone theory with the adjacent descriptions.
    ____ a.   central business district
    ____ b.   commuter's zone
    ____ c.   residential zone
    ____ d.   zone in transition
    ____ e.   zone of workingmen's homes
    (1) zone that is the next zone out from the zone in transition
    (2) site of major stores and offices
    (3) zone with physical deterioration stemming from few repairs
    (4) zone the wealthiest people live in
    (5) zone that is largely middle class in composition

GRAPHIC REVIEW

Population projections by world regions appear in Figure 17.3. Answer these questions to test your grasp of world population growth between 2005 and 2050.

1. Compare the projected population growth by world region between 2005 and 2050. Explain any surprises. If you were not surprised, explain why projections went just as you expected.

2. Compare the population doubling time in less and more economically developed areas. Explain the difference.

ANSWER KEY

Concept Review
a. 7
b. 13
c. 10
d. 16
e. 3
f. 1
g. 4
h. 15
i. 5
j. 18
k. 19
l. 8
m. 20
n. 17
o. 2
p. 12
q. 11
r. 6
s. 9
t. 14

Multiple Choice
1. d
2. e
3. b
4. c
5. c
6. d
7. c
8. b
9. e
10. a
11. a
12. b
13. d
14. a
15. e

Feedback Review
1. F
2. T
3. positive
4. Urbanization
5. natural increase
6. Overurbanization
7. urbanism
8. c
9. c
10. c

11. d
12. a. 2
    b. 4
    c. 5
    d. 3
    e. 1

18

Social Change and Collective Behavior

Dominic Favre/AP Photos

LEARNING OBJECTIVES

- Illustrate three social processes contributing to social change.
- Discuss, as sources of social change, the role of technology, population, the natural environment, conflict, and ideas.
- Compare and contrast the cyclical and evolutionary perspectives.
- Explain the why and how of social change within the context of the functionalist and conflict perspectives.
- Define modernization, and explain some of its major consequences.

- Define world-system theory, and summarize some of the consequences flowing from its description of the world economy.
- Describe social activities engaged in by dispersed collectivities.
- Describe the nature of a crowd, and identify the basic types of crowds.
- Contrast contagion theory, emergent norm theory, and convergence theory.

- Define the concept of social movement, and identify the primary types of social movements.
- Apply the concepts of relative deprivation and unfulfilled rising expectations to the emergence of social movements.
- Compare and contrast relative deprivation theory with value-added theory and resource mobilization theory.

Do successful revolutions bring radical social change? It seems so, almost by definition. There are dramatic examples of truly revolutionary changes in a society: France in the eighteenth century, China in the twentieth century, and Russia in the early twentieth century. Generally, however, victory celebrations are not followed by a new social order. Change after the American Revolution was not radical; it was gradual and long term. It is also a fact that the revolution was merely *one* of the precipitative factors creating change.

Revolution is only one factor contributing to the alteration of social structure. After defining social change, we turn immediately to several of its sources.

Social Change

Social Change a Constant

Imagine for a moment the history of the earth as a 365-day period, with midnight January 1 marking the starting point and December 31 being today. Within this time span, each "day" represents 12 million years. The first form of life, a simple bacterium, appeared in February. More complex life, such as fish, appeared about November 20. On December 10, the dinosaurs appeared; by Christmas they were extinct. The first recognizable human beings did not appear until the afternoon of December 31. *Homo sapiens* emerged shortly before midnight that last day of the year. All of recorded history occurred in the last 60 seconds of the year (Ornstein and Ehrlich 2000).

All societies change. In Egypt, where Islamic law permits a man to divorce his wife by declaring to her "You are divorced," one man did the deed by text messaging his wife, "I divorce you" (Knickmeyer 2008). The speed of social change may vary from the glacial to the mercurial. Nonetheless, social change is a constant. For sociologists, **social change** involves societal alterations with long-term and relatively important consequences (Abu-Lughod 2000).

Can social change be predicted? Predicting precise change is difficult, partly because a society absorbs change in ways consistent with its own culture. For instance, two societies that both adopt a democratic form of government may develop in very different ways. Both Britain and the United States are democracies. But their histories prior to becoming democracies (Britain had a royal tradition) led to divergent forms of government. In addition, people in a society can modify their behavior and consciously decide for themselves the ways in which change will occur. They can, for example, deliberately avoid a predicted state of affairs (Caplow 1991). These facts should not, however, discourage an attempt to understand alterations in society. But in doing so, one must base the accuracy of predictions on sound assumptions, just as Frenchman Alexis de Tocqueville did during the early 1800s in his remarkably penetrating study of American society (see Table 18.1).

Why do some societies change faster than others? This is a complex question with no easy answer. Technology,

---

TABLE 18.1

Key Assumptions in Predicting Social Change in America

The most accurate prognosticator of trends in American society has been the Frenchman Alexis de Tocqueville. Tocqueville's *Democracy in America*, which was published in the 1830s, displayed an amazing grasp of American society. Tocqueville's success has been attributed to several key assumptions he made.

- *Major social institutions would continue to exist.* Unlike many of his contemporaries—and many of ours—Tocqueville did not expect the family, religion, or the state to be greatly changed.
- *Human nature would remain the same.* Tocqueville did not expect men and women to become much better or worse or different from what history had shown them to be.
- *Equality and the trend toward centralized government would continue.*
- *The availability of material resources (such as land, minerals, and rich soils) limits social change.*
- *Change is affected by the past, but history does not strictly dictate the future.*
- *There are no social forces aside from human actions.* Historical events are not foreordained by factors beyond human control.

Sources: Adapted from Theodore Caplow, *American Social Trends* (New York: Harcourt Brace Jovanovich, 1991), p. 216.

discussed shortly, is a prime force in social change. Technologically complex societies change much faster than technologically simple societies (Nolan and Lenski 2005). This is because technology increases human control over the physical world. And as control over the environment increases, nontechnological changes occur in other parts of society—family, economy, education, medicine. For example, new technologies associated with the computer and information processing are radically changing health care. There are patients today who are diagnosed by physicians they will never meet in person.

Technology is, however, not the only reason some societies change faster than others. Sociologists have identified three important social processes and several more specific factors that influence the pace of social change.

Processes for Change

Three interrelated social processes lead to social change. Although related, the processes of discovery, invention, and diffusion are distinct.

How does each of these three social processes promote social change? In the **discovery** process, something is either learned or reinterpreted. Early ocean navigators, with skill and courage, demonstrated by experience the rounded shape of the earth. With this new interpretation, then, many worldwide changes followed, including new patterns of migration, intercontinental commerce, and colonization. Salt, another example, was first used to flavor food. Because it was so highly valued, it was subsequently used as money by Africans and as a religious offering among early Greeks and Romans.

More significantly, fire, which prehistoric peoples used for warmth and cooking, was later used (in about 7000 B.C.) by the first agriculturists to clear fields and to create ash for fertilizer. Fire, of course, was central to the ensuing transition to agricultural society.

Invention is the creation of a new element by combining two or more already existing elements, with a spawning of new rules for its use. Examples of inventions come easily to mind. It was not so much the materials Orville and Wilbur Wright used to make an airplane (most of the parts were available), but the way they manipulated and combined the elements, that led to their successful flight at Kitty Hawk.

The pace of social change through invention is closely tied to the complexity of the cultural and social base. As the base becomes more complex and varied, the number of elements, as well as the number of ways the elements can be combined into inventions, increases dramatically. Consequently, the more complex and varied a society, the more rapidly it will change. This helps explain why even though several million years passed between the evolution of the human species and the invention of the airplane, we reached the moon in less than seventy years after the Wright brothers' first flight. Social change, of course, occurs faster when an invention can be combined with many other aspects of a society. NASA (National Aeronautics and Space Administration) was able to reach the moon relatively quickly because the United States was already advanced in physics, aerodynamics, and the manufacturing of specialized materials.

When one group borrows something from another group—norms, values, roles, styles of architecture—change occurs through the process of **diffusion.** Whereas the extent and rate of invention depend on the

By which social process did this advertisement for a Mariah Carey concert reach Beirut, Lebanon?

Norbert Schiller/epa/CORBIS

complexity of a society, the extent and rate of diffusion depend on the degree of social contact. Borrowing from others may involve entire societies, as in the American importation of cotton growing that was first developed in India. Or diffusion may take place between groups within the same society, as in the social and cultural spin-offs of the jazz subculture, which was created by African American musicians.

As noted earlier, a borrowed element must harmonize with the group culture. Despite the upsurge of unisex fashions in America today, wearing a Scottish kilt on the job could get a construction worker laughed off the top of a skyscraper. Wearing kilts still clashes with the American definition of manhood. If skirts are ever to become as acceptable for American men as pants are for women, either their form will be altered or the cultural concept of masculinity will be modified.

Integrating a borrowed element may involve merely a selection of certain aspects. The Japanese, for example, accept capitalism but resist the American form of democratic government, the American style of conducting business, and the American family structure. Diffusion, then, almost always involves selectivity and modification.

In modern society, most aspects of culture are borrowed rather than created. The processes of discovery and invention are important, but usually far more elements enter a society through cultural diffusion.

Sources of Social Change

Discovery, invention, and diffusion are social processes through which social change occurs. A multitude of factors affect society through these processes. Most important among the factors encouraging social change are technology, population, the natural environment, revolution and war, and ideas.

Technology

Time magazine selected Albert Einstein as the man of the century, a choice reflecting the significance of science and technology in the twentieth century (Golden 1999). **Technology** is the combination of knowledge and hardware that is used to achieve practical goals. It has historically been viewed as a precursor to social change (Teich 1993; MacKenzie and Wajcman

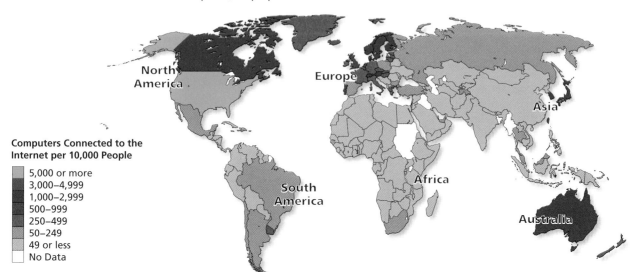

WORLDVIEW 18.1

Internet Connections

As this map shows, the number of people connected to the Internet varies widely from country to country. As of 2004, over 74 percent of Internet connections were on the North American continent. This map shows the number of computers connected to the Internet per 10,000 people.

Computers Connected to the Internet per 10,000 People

- 5,000 or more
- 3,000–4,999
- 1,000–2,999
- 500–999
- 250–499
- 50–249
- 49 or less
- No Data

Source: International Telecommunication Union, 2004.

Interpret the Map

1. Do you see a pattern in the number of connections to the Internet? Explain.

2. What implications might this distribution have for future social change?

1999; Volti 2003). For instance, Karl Marx placed the importance of technology in sharp perspective: "The windmill gives you society with the feudal lord; the steammill, society with the industrial capitalist" (Marx 1920:119). Although the relationship between technology and social change is not this simple and direct—and Marx knew it—their interconnectedness is firmly established.

To what extent can technological advances accelerate social change? Consider the social effects of even minor technological changes. The invention and diffusion of the stirrup altered medieval warfare. Soldiers on foot were easily defeated by knights on horseback because the stirrup permitted knights to continuously hold their lances, rather than throw them. This new military technology—a combination of the knight, the horse, and the lance—contributed to the rise of the feudal system, with the knightly class at the top (Lynn White 1972). Although originally thought to be a mere means of transportation, the automobile has had countless effects on modern society. It changed dating and courtship patterns, contributed to a new sexual morality, created new industries, promoted industrial expansion, produced suburbanization, and contributed immeasurably to environmental pollution. The creation of the silicon chip introduced technological change at an unprecedented rate. It took more than a century for telephone possession in the United States to reach 94 percent; but in less than a decade (from 1992 to 2000), the Internet reached about 30 percent of Americans; in just over twenty years, the personal computer market reached its present 40 percent household penetration; and in less than fifteen years, a cell phone was in almost 25 percent of American households (see Figure 18.1). In 2006, almost 80 percent of Americans were accessing the Internet (*World Almanac and Book of Facts* 2008).

The societal transformations wrought by the microprocessor are many and far-reaching. The workplace, by way of illustration, is undergoing a transformation, with effects as far-reaching as those of the Industrial Revolution (Fukuyama 1999). Telecommunications technology, for example, will allow many to work from their homes, but it will result in far less human interaction (McGinn and Raymond 1998). In the field of medicine, the microprocessor radically changed many surgical techniques, from microsurgeries to radio wave therapy (Cowley and Underwood 1998). On the autobahn in Germany, drivers using cell phones or electronic consoles in their cars can get real-time computer-generated information on traffic problems. Perhaps the most profound effect of the microprocessor was its contribution to the end of the Cold War. Soviet military personnel recognized the potential of computer guidance systems for long-range munitions:

The realization that the Soviet Union did not, and would not, possess the computer or sensor capabilities to produce such weapons—and so, in the long run, must inevitably face defeat at Western hands—was a significant factor behind Mikhail Gorbachev's decision to sue for peace. (Watson and Barry 1998)

The repercussions of sequencing the human genome promise even more dramatic social change. Now that scientists have cloned sheep, cows, pigs, domestic cats, a human embryo from adult cells, and artifical DNA, who can know where this capability will lead ("Hello, Dolly" 1997; R. Weiss 2000; J. Fischer 2001; Kluger 2002; Fukuyama 2003; "To Clone a Human" 2008)? In fact, Nobelist and Cal Tech president David Baltimore sees biology as a driving force in the transformation of society: "Instead of guessing about how we differ from one from another, we will understand and be able to tailor our life experiences to our inheritance" (Brave 2001:18). The 2001 national debate on stem-cell research policy is only the beginning of public debate; where will human genetic engineering take us?

Why should we be careful of technological determinism? Technological determinism assumes that technology is the primary determinate of the nature of a society. This is a seductive theory. By use of dramatic examples, it is easy to conclude that the nature of a society can be traced directly to the society's technological base.

FIGURE 18.1

Becoming Wired: Time It Took for 30 Percent of Americans to Acquire Selected Technologies

It took forty years for 30 percent of American homes to have telephones. How much time elapsed before 30 percent of Americans were online? How would you account for the difference in adoption time?

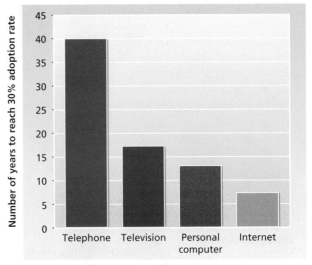

Source: David R. Francis, *The Internet in Healthcare* (Nashville, TN: J. C. Bradford, August 1999), p. 11.

Actually, the relationship between technology and social change is more complicated (Lauer 1991). First, social change can occur without technological developments. Ancient Greek commercial development was not based on improved production methods. Second, the introduction of technology does not necessarily lead to social change. A South Indian village did not change its community social structures following a switch from dry to wet cultivation of crops (Epstein 1962). Third, as just implied in the discussion of diffusion, the particular effects of technology will vary from society to society because the adoption and use of technology are always filtered through the nature of a given society. Although Russia, for example, has computer technology, it places much tighter controls on the use of the technology by individual citizens.

Although it seems inadvisable to see technology as a necessary and sufficient cause of social change, the contribution of technology to change should not be underestimated. It is best to view technology as an extremely important factor in a network of causes. Nor should we ignore the fact, as noted earlier, that technological innovation can occur very rapidly. David Freeman (1974) also makes this point by compressing the total lifetime of the earth (some 5 billion years) into an eighty-day period and calculating when some significant technological changes occurred: the Stone Age began six minutes ago, the agricultural revolution took place fifteen seconds ago, metals were first used ten seconds ago, and the Industrial Revolution started three-tenths of a second ago.

Population

Alterations in population size and composition have interesting effects. By way of illustration, population pressures in China led to a conservation of resources. This in turn produced a state policy replacing the more expensive, traditional burial rites with cremation ceremonies. Yet another consequence of population stress is found in the conflict and strain within Chinese families. Moreover, China instigated a national population control program to create the one-child family. This is already affecting Chinese social structure, as in the dishonesty engendered by some anxious local officials who underreport birth statistics to meet the government quotas. Less developed countries offer another illustration; they are undergoing many dramatic changes because their declining mortality is not being matched by lowered fertility.

America's baby boom following the return of American soldiers at the end of World War II is illustrative as well. The increased birth rate of Americans born between 1946 and 1964 required expanded medical personnel and facilities for child health care and then, in the 1950s and 1960s, created the need for more teachers and schools. On the other hand, the generation following the baby boomers, now in their thirties and in the labor market, is experiencing increased competition for jobs and fewer opportunities to move up the career ladder. As the baby boomers retire, problems of health care and Social Security loom large. Longer working hours, retraining programs, and reeducation for senior people will probably become prevailing patterns. As America's population continues to age, more attention is being paid to our senior citizens. Already, there are more extended-care homes for the aged, an increase in geriatric emphasis, and more television advertising and programming targeting the aging population.

The Natural Environment

Interaction with the natural environment, from the earliest times, has transformed American life. The vast territory west of the thirteen colonies permitted expansion, ultimately to the Pacific Ocean. This western movement helped shape our cultural identity and wrought untold changes, most tragically the destruction of many Native American cultures. Then there was the Great Depression of the 1930s, due in part to a long drought that hit the agricultural Midwestern plains states. Over-planting and -plowing had upset the fragile ecosystem, turning the prairies into a giant "dust bowl." The world economic slowdown in 2007 was due in part to the emergence of $100 a barrel for oil. Global warming, caused in part by human behavior, promises truly catastrophic consequences. Two important sources describe some implications of unchecked climate change, including rising conflict between rich and poor nations, tensions due to massive migrations, a proliferation of diseases, and realignment in global power driven by accessibility of natural resources (Campbell et al. 2007; United Nations Climate Change Conference 2007).

Revolution and War

From the Greek historian Polybius to Karl Marx, many early thinkers viewed conflict as the central form of human interaction. Marx and more contemporary conflict theorists, who see social conflict as the prime impetus for social change, are covered in the next major section. For now, let's focus on revolution and war as sources of social change that involve conflict.

What is a revolution? A **revolution** is a social movement that involves the violent toppling of a political regime. According to Charles Tilly (1978, 1997), a revolution results in the replacement of one set of power holders by another. Most revolutionaries, of course, expect more fundamental changes than this. Marx, for example, expected workers' revolutions to eliminate class-based inequality and therefore to profoundly affect the social and economic structures of the societies in which they occurred.

Are revolutions normally followed by radical changes? As suggested in "Using the Sociological Imagination" at the beginning of the chapter, revolutions are normally not followed by radical social change. In Crane Brinton's (1990) view, the changes introduced by revolutionary leaders are followed eventually by restoration of a social order similar to that existing in prerevolutionary days. In other words, even successful revolutions fail to tear down the old order and bring in an entirely new one. Radical changes are rarely permanent; after victory celebrations are over, a society operates with a great deal of continuity with the past—a compromise between the new and the old.

What sorts of changes do follow revolutions? There is a middle ground between Marx's position that revolutions produce major changes and Brinton's position that revolutions have little lasting effect. Tilly argues that revolutions *do* have lasting effects if there is a genuine transfer of power. Most revolutions, Tilly writes, *do not* have lasting effects because revolutionaries seize control of a government too weak to effect change. Eighteenth-century French revolutionaries, in contrast, took control of a powerful, centralized state, allowing them to seize and redistribute the property of aristocrats and churches. Despite a backlash, the French Revolution had lasting effects. Similarly, the Chinese Communists created a strong central state of their own, allowing them to reorganize village structure in a way that permanently altered Chinese society.

In the 1940s, a cornerstone of the communist revolution in China was sexual equality; liberation from sexism was a revolutionary plank. The situation for Chinese women has improved, but complete sexual equality is still a far-distant dream ("Closing the Gap" 1995). Revolutions, then, can be a powerful mechanism for social change, whether or not they create the total change promised by those who lead them and expected by those who participate in them.

What is the relationship between war and social change? Social change, as already seen, can be brought about by revolutionary conflict within societies. Change is also created by conflict in the form of **war**—armed conflict that occurs within a society or between nations. War and change are closely intertwined. The Golden Age of Athens (fifth century B.C.) started with the Persian War and ended with the Peloponnesian War. Rome's flowering in the first and second centuries cannot be separated from wars in Europe and Asia. War was a backdrop to the Age of Enlightenment. The founding of the United States is no exception; it required a war. The Civil War was instrumental in America's transition from an agri-cultural to an industrial society. Momentous technological, economic, and social changes also followed World War I, World War II, and the Vietnam War.

How does war promote social change? Robert Nisbet (1989) writes eloquently on the effects of war. Change is created through diffusion as war breaks down insulating barriers between societies. The crossbreeding and intermingling of people and cultures during and following a war leave participating societies different than they were before the conflict. Wars also promote invention and discovery. During the First World War (1914–1918), the U.S. government, due to the pressure of war, promoted and financed the development of such technologies as the airplane, automobile, nylon, and radio—each of which contributed to the social and cultural revolution that followed the war. And America's culture, both during and after World War I, was exported to societies all over the world.

Ideas

Indeed, ideas can lead to social change. Max Weber, for example, believed that the ideas inherent in Calvinism produced behavior conducive to the development of industrial capitalism.

Alfred North Whitehead wrote, "A general idea is always a danger to the existing order" (Whitehead 1933:22). We have seen the validity of this observation

Wars often bring social change because culturally dissimilar societies, such as the U. S. and Iraq, come into increased contact.

John Moore/AP Photos

F
E
E
D
B
A
C
K

1. _____ refers to alterations in social structures that have long-term and relatively important consequences.
2. The more complex a society, the more rapidly it will change through invention. T or F?
3. The process of _____ usually involves modification and selectivity.
4. The theory of social change that assumes technology to be the primary cause of the nature of social structure is called _____.
5. The baby boom in the United States is a good example of the effects of _____ on an entire society.
6. _____ is a type of social movement that involves the toppling of a political regime through violent means.
7. Contrary to conventional wisdom, war retards social change. T or F?
8. Was it Karl Marx or Max Weber who emphasized the role of ideas in creating social change? _____

Answers: 1. Social change; 2. T; 3. diffusion; 4. technological determinism; 5. population; 6. Revolution; 7. F; 8. Weber

in the earlier discussion on revolution. In the next section, we will see that two major theories of social movements also emphasize the crucial influence of ideas.

Do ideas necessarily lead to social change? Sometimes ideas retard social change. During the Middle Ages, Christian religious ideas forestalled change (Gay 1977). In fact, the Catholic Church based its strength on total resistance to change (Manchester 1993). For example, economic development was slowed in part because usury (charging interest on loans) was considered a sin. For many years, slavery in the United States was successfully defended on the basis of fallacious ideas that depicted African Americans as inferior. Similar unreasoned ideas have been used to deny women equal opportunity in many societies, including the United States.

How are these sources of change interrelated? Several, or all, of these factors can be operating in combination. Consider abortion. Abortions are possible because of medical inventions that have been widely diffused. Underlying the ability to perform abortions is medical technology. This technology affects the population by lowering the birth rate, which in turn affects family size and a number of other social conditions (child-free marriages, dual-employed couples, sexual freedom). A lowered birth rate also reduces demand on limited natural resources. Conflict between pro-life and pro-choice forces in the United States has existed ever since the Supreme Court legalized abortion in 1973 in the *Roe v. Wade* decision. This conflict, of course, has been based on disparate ideas regarding the moment at which a fetus constitutes a human being.

Theoretical Perspectives

It is through theory that we go beyond description of social change toward explanation. Before our theoretical discussion of functionalism, conflict theory, and symbolic interactionism, let's look, briefly, at some

early theories specific to social change. This will make the three contemporary theoretical approaches more meaningful.

The Cyclical Perspective

Scholars were baffled. Why would the most developed nations in the world spend $337 billion between 1914 and 1918 (World War I) to kill more than 8 million soldiers? This was very poor evidence indeed for a belief in the natural progression of civilization. Scholars began to ask whether change was necessarily progress. Was Western civilization on the road to improvement or on the way down?

How did advocates of the cyclical perspective describe the effects of social change? The dramatic upheaval during the early part of the twentieth century led scholars to consider the possibility that civilizations rise and fall rather than develop in a straight line. The rise and fall of Rome, after all, had been documented in six large volumes by British historian Edward Gibbon; his work was published between 1776 and 1788. Although many scholars disagreed with Gibbon's conclusions, they did not miss the point that social change may very well follow a pattern of growth and decay. Three of the most prominent advocates of the cyclical perspective were historians Oswald Spengler (1880–1936), Arnold Toynbee (1889–1975), and sociologist Pitirum Sorokin (1889–1968).

Spengler (1926–28) published *The Decline of the West*, in which he compared societies to living things, as did social Darwinist Herbert Spencer (1820–1903), but with a vital difference. Whereas Spencer saw advanced societies as superior survivors of the struggle of the fittest, Spengler believed that civilizations are born, ripen, decay, and perish. Specifically, he predicted that Western civilization—which had traded zest for materialism—was going to be replaced by a civilization from Asia. Spengler's work is now considered more literature than scientific explanation.

Almost twenty years later, British historian Arnold Toynbee (1946) published *A Study of History*, in which he presented his challenge-response theory of social change. According to Toynbee, civilizations rise only when a proper challenge in the environment is presented to a people who can respond successfully to it. If a challenge from the environment is not issued or if the society fails to meet a challenge, that society will either fail to rise in the first place or it will decline.

Like Spengler and Toynbee, Sorokin did not see in history any linear pattern of change. To him, sociocultural history involves a cyclical alternation among three reasonably homogeneous cultural types. *Ideational* cultures are based on the conviction that truth and value come from God. Emphasis is on the spiritual and the next world. In *sensate* cultures, truth is discovered through empirical observation by the senses. It is experience in the current world—gained through touching, smelling, hearing, seeing, and tasting—that reveals truth and reality. Sensate cultures are not absolutist but allow for reinterpretation of reality. People in ideational cultures focus on spiritual needs; members of sensate cultures place priority on physical needs and gratification of the senses. Whereas ideational cultures tend to be spiritual, sensate cultures are characterized by pleasure seeking and materialism. Sorokin's third cultural type, the *idealistic* culture, is a blend of the ideational and the sensate, with somewhat more accent on the ideational.

For Sorokin, then, history is best described as a continuous movement from the predominance of one type of culture to another. According to Sorokin, sensate culture dominated in Christ's time. Christian medieval Europe typified the shift from the sensate to the ideational. Idealistic culture emerged in thirteenth- and fourteenth-century Europe. Modern Western society is again grounded in sensate culture.

Regarding the cyclical approach to social change, what kinds of questions are left unanswered? Because history tells us that societies do, in fact, rise and fall, cyclical perspectives are especially appealing. The proof is there. Cyclical perspectives have a major problem, however. They are not so much explanations for change as descriptions of change. Spengler does not indicate why societies mature and decay; he just tries to show that they do. Toynbee does not state why a formerly successful society fails to meet a new challenge; he just writes that societies failing to respond to a challenge will decline. Sorokin provides no explanation for the cyclical pattern he describes; he just attempts to substantiate its existence with an amazing amount of historical information.

Is the cyclical perspective even worth considering? The cyclical perspective is worthy of consideration for several reasons. First, it represents an early, histori-

cally important attempt to understand social change. Second, it helps legitimate the scientific study of social change. Third, it is useful as a backdrop for distinguishing between description and explanation.

The Evolutionary Perspective

What was the nature of early evolutionary theory? "Every day in every way, things get better and better." In simple terms, this idea of progress, of societies constantly moving toward improvement, was at the heart of the evolutionary perspective that dominated nineteenth-century social thought. In its nineteenth-century form, the evolutionary perspective held that societies must pass through a set series of stages, each of which results in a more complex and advanced stage of development.

The evolutionary perspective became popular during the nineteenth century in part because of the times. During the middle and late nineteenth century, western Europeans prided themselves on their superiority over other peoples. Progress and territorial expansion were watchwords. During colonial expansion, contact with preliterate people only offered ethnocentric Europeans further proof that their societies were the most highly civilized ever to exist. "Primitives" simply represented early stages of social evolution.

Another important influence of the evolutionary perspective was the work of Charles Darwin (1809–1882), which provided an intellectual framework for interpreting the place of primitive and advanced societies in the world. Darwin's idea that all living things tend to improve and become more complex as they develop seemed to fit the fact that some societies were more advanced than others.

In the hands of the English sociologist Herbert Spencer, Darwin's biological ideas were transformed into social Darwinism. Spencer drew a parallel between living organisms and societies and coined the expression "survival of the fittest" to explain why some peoples had become "civilized" whereas others remained "savages" and "barbarians." Social Darwinism was so widely accepted in the Western world that governments in western Europe and America believed that they had the right to dominate, protect, and tutor less developed societies. Such was thought to be the destiny and burden of nature's hardiest people.

Social Darwinism and classical evolutionary perspectives survived into the first part of the twentieth century, but they are no longer considered valid. Social scientists no longer believe that societies necessarily improve as they change or that developing societies are destined to follow the Western model. Nevertheless, more recent perspectives on social change have overcome some of the faults of the classical evolutionary perspective (Sanderson 1990; Chirot 1994).

So how does modern evolutionary theory contribute to our view of social change? The contemporary evolutionary perspective rejects the unilinear, or single-direction, assumption of earlier perspectives. Given the social and cultural diversity around the world, modern evolutionary theorists ask how we can believe that all societies develop in a single direction. Because we know that societies can develop in many ways—some even toward greater deterioration and unhappiness—evolution must be multilinear, or multidirectional. Deliberately missing from the current evolutionary perspective, then, are the ideas of definite and orderly stages of development and of change inevitably producing progress and greater happiness. There are many paths and directions to evolutionary change (Steward 1979).

The most recent sociocultural evolutionary theory comes from Patrick Nolan and Gerhard Lenski (2005), whose initial assumption is that both change and stability must be considered if human societies are to be understood. Despite the dramatic changes that have taken place in the world over the past 10,000 years, they argue, most individual societies have successfully resisted change. This, of course, is a paradox; how can rapid social change occur when individual societies successfully resist change? To resolve this apparent paradox, Nolan and Lenski offer a proposition: a total system can change despite resistance to change in most of its parts, providing the parts that fail to change do not survive. This, they assert, is precisely what has happened in the "world system of societies." Of those societies that did not change or changed only a little, almost none are around today. Conversely, nearly all surviving societies are those that did alter their social structures. Thus, say Nolan and Lenski, in the world system of societies, a process of natural selection has operated that favors innovative societies over those that resist change.

For Nolan and Lenski, the key to societal survival is the accumulation of information, particularly information relevant to subsistence. In fact, improvements in subsistence technology are said to be a "necessary" precondition for a society to grow in complexity, size, wealth, or power. Subsistence technology is also important because it stimulates improvement in other types of technology, as in production, transportation, communication, and war. These further technological advances promote additional advances in subsistence technology. Nolan and Lenski are not technological determinists. For example, they think that a society's beliefs and values are crucial because, in part, they affect the extent to which a people are open to innovation and change.

Modern sociocultural evolutionary theory, then, helps us understand how societal changes are made. It shows that as subsistence technologies improve, societies have the means for gradually progressing from a hunting and gathering to a postindustrial form. As we

shall see, modern evolutionary theory is informed, in part, by both functionalist and conflict theory.

The Functionalist Perspective

Because functionalism emphasizes social stability and continuity, it may seem contradictory to refer to a functionalist theory of social change. However, two functionalist theories of social change—proposed by William Ogburn and Talcott Parsons—are especially interesting. Both of these theories are based on the concept of *equilibrium* discussed in Chapter 1.

What is the connection between functionalism and the concept of equilibrium? The word *equilibrium* implies balance and consistency. Concerned with equilibrium, the tightrope walker on a narrow rope inches his way across a deep chasm, continually shifting his body and using his long balancing pole to counterbalance the effects of the wind as well as the effects of his own motions. When applied to social life, equilibrium connotes attempts to

Social equilibrium was seriously shaken in New Orleans after Hurricane Katrina.

reestablish stability after some disturbance. According to functionalists, societies act as inherently stable wholes that react to changes by making adjustments and eventually assimilating change into a new state of equilibrium. A society in change, then, moves from stability to temporary instability and back to stability. Sociologists refer to this as a "dynamic," or "moving," equilibrium. For example, in 1972, a broken dam led to the destruction of the community of Buffalo Creek, West Virginia. The physical destruction of the community was accompanied by death and the rending of the old fabric of life. Despite the ensuing chaos, residents of the community pulled their lives together again. Although things were not the same as before, a new equilibrium was built out of the physical, social, and human wreckage (K. T. Erikson 1976). The 1960s were a time when the norms of sexual behavior changed radically. But, after skyrocketing during the sixties, teenage pregnancy is now declining. Although Americans do not follow the norms of the 1950s, a retreat from extremes is occurring as new norms of sexual behavior are being established.

Functionalism is based on the concept of equilibrium. Kingsley Davis writes:

> It is only in terms of equilibrium that most sociological concepts make sense. Either tacitly or explicitly, anyone who thinks about society tends to use the notion. The functional-structural approach to sociological analysis is basically an equilibrium theory. (Davis 1949:634)

Although not all sociologists would agree that most sociological concepts make sense only in relation to equilibrium, most would concur that functionalism involves the concept of equilibrium.

Before turning to two specific functionalist theories of social change, we should consider the similarity and difference between the evolutionary and functionalist approaches, because both of the functionalist theories of change we will examine draw on evolutionary theory. Both the evolutionary and functionalist perspectives view society as composed of many highly differentiated parts, all contributing to the maintenance of a smooth-running, stable society. However, whereas the traditional evolutionary perspective emphasizes a constant forward direction spurred by change, the functionalist perspective places emphasis on society's ability to recover its balance if equilibrium is upset by change. In functionalism, the various parts of society are seen as highly integrated so that if a change occurs in one part, other parts are affected. The chain reaction is supposed to eventually restore balance because change is absorbed and distributed among a society's elements until equilibrium is reestablished. The emphasis in the functionalist approach is always on a return to stability and order after some restructuring has taken place.

What is Ogburn's theory of social change? If a new equilibrium is achieved, the parts of society, asserted William Ogburn (1964; originally published in 1922), do not all reach the new balance at the same time. Some parts lag in time behind others. Ogburn applied the term **cultural lag** to any situation in which disequilibrium is caused by one segment of a society failing to change at the same rate as an interrelated segment. More specifically, Ogburn believed that changes in the nonmaterial dimensions of culture (norms, values, beliefs) lag behind alterations in the material culture (technology, inventions). Significant social change occurs when the nonmaterial culture is forced to change because of a prior change in the material culture. Ogburn's best-known example is the lag between the technological ability to cut down entire forests and the subsequent emergence of the conservationist movement. Cultural lag still exists. More recently, we can point to the sexual norms, values, and beliefs now in the process of attempting to catch up with the widespread distribution of birth control technology that occurred years ago. The Roman Catholic Church, officially, continues to oppose birth control, while many members of the church are waging personal struggles over contraception. The continuing conflict (sometimes violent) between pro-choice and pro-life advocates over the issue of abortion is a prime illustration of cultural lag.

How does Parsons approach social change? The concept of equilibrium is at the heart of Talcott Parsons's theory of social change. In his early work, Parsons (1937, 1951) did not emphasize social change. He depicted societies as systems attempting to resist change in order to maintain their current state of equilibrium. It was only later that Parsons (1966, 1971, 1977) began to depict change as a contribution to a new equilibrium with novel characteristics.

Consistent with his roots in evolutionary theory, Parsons was interested in the processes by which societies become more complex. In the first process, *differentiation*, aspects of a society are broken into separate parts. In simple societies, the family was responsible for nearly all functions—economic, educational, medical, emotional, and recreational. As societies became more complex, these functions began to differentiate from the family. Jobs were found in factories, education occurred in schools, medical needs were cared for by doctors and hospitals, emotional support could be found outside the home, and amusement was supplied by a variety of nonfamilial sources.

Differentiation brought on the need for the second process—*integration*. All the newly evolving social units formed workable links for a new equilibrium different from the previous one. Ways were developed for schools and families to mesh, for parents to accept their children's leaving for work in the factories, and for parents to cope with their offspring's independent lives outside their local communities. The process of integration not only leads to a new state of equilibrium but also, in conjunction with the process of differentiation, helps to produce a much more complex type of society.

What are the contributions and criticisms of the functionalist perspective on social change? A major contribution of the functionalist perspective is its attempt to explain both stability and change. Ogburn tries to show that a society's ways of thinking, feeling, and behaving are constantly attempting to catch up with prior technological change. Technology, for him, is an independent variable leading to further social change. Parsons identifies differentiation and integration as processes integral to the maintenance of a moving equilibrium. In addition, he emphasizes that change does not imply total change; continuity as well as change exists.

Functionalists have been criticized for having a conservative bias. The concept of equilibrium, argue some critics, assumes internal societal resistance to change. Also, the equilibrium model of functionalism is charged with depicting change as external to societies: societies are changing not because of their own internal dynamics, but as a result of outside forces. By failing to explore internal sources of change, it is argued, functionalists ignore many forces for change that influence a society. Finally, functionalists have been faulted for focusing on gradual change and failing to consider radical change (R. Collins 1975; Giddens 1979, 1987; Giddens 2007). The conflict perspective addresses these criticisms.

The Conflict Perspective

Whereas the functionalist perspective assumes society to be inherently stable, conflict theorists depict society as unstable. Whereas functionalism views society as an integrated whole whose various parts work harmoniously to achieve balance, the conflict perspective emphasizes the separate parts of society and the conflict that occurs between them. According to the conflict perspective, social change is the result of struggles between groups for scarce resources. Social change is created as these conflicts are resolved.

What are the origins of the conflict perspective? Many of the basic assumptions of the conflict perspective emerge from the writings of Karl Marx, who wrote that "without conflict, no progress: this is the law which civilization has followed to the present day" (Marx; in Feuer 1959:7). Ralf Dahrendorf, a modern advocate of the conflict perspective, summarizes Marx's view of society in this succinct passage:

For Marx, society is not primarily a smoothly functioning order of the form of a social organism, a social system, or a static social fabric. Its dominant characteristic is, rather, the continuous change of not only its elements, but its very structural form. This change in turn bears witness to the presence of conflicts as an essential feature of every society. Conflicts are not random; they are a systematic product of the structure of society itself. According to this image, there is no order except in the regularity of change. (Dahrendorf 1959:28)

More specifically, according to Marx, the struggle for scarce economic resources, particularly property, is the primary stimulus for change. By way of review, Marx claims that change in capitalist society is authored by class struggle, and as a capitalist society develops, it begins to divide into two classes: those who possess property and own the means for production (the bourgeoisie) and the exploited workers who sell their labor merely to survive (the proletariat). These two classes polarize as the rich get richer and the poor get poorer. At some point, a revolution permits the oppressed proletariat to seize power. With the proletarian revolution comes the development of a classless society. The absence of classes eliminates the source of conflict.

What are the contributions and criticisms of Marx's approach to social change? Defenders of Marx contend that contrary to the charge, Marx avoided economic determinism. Although Marx contended that the economic base of any society is the ultimate determinant of the society's nature and course, he also allowed for noneconomic influences on this economic base. Also, Marx is credited with effectively showing that how people think, feel, and behave reflects the basic underlying economic foundation of their society. In addition, Marx refined the view that social conflict is built into all stratification structures and that the distribution of power is rooted in the economic system (Lauer 1991).

According to critics, Marx placed too much emphasis on economic factors as the determinants of social change while downplaying relevant social and cultural forces. Others criticized Marx for failing to see that conflict can integrate societies as well as tear them apart and for not recognizing the prevalence of cooperation over conflict within societies. Marx's views on revolution were also challenged. Contrary to Marx's prediction, the critics assert, most revolutions of this century have come from the middle class rather than the working class. Also, the critics continue, communist revolutions did not occur in highly industrialized Western societies such as the United States, Great Britain, or Germany but in the agrarian societies of China, Russia, and Cuba. Finally, non-Marxists note the lack of a polarization of capitalist societies into ruling classes and working classes. They point to the emergence of a large middle class—composed of neither workers nor owners—in modern capitalist societies (Vago 2003).

Katherine S. Newman—The Withering of the American Dream?

After the economic recovery from the Great Depression, each subsequent generation of Americans expected a standard of living higher than that of their parents. A boost to this expectation was given by the unprecedented economic expansion following World War II. Upwardly spiraling affluence was assumed to be America's destiny.

Instead, according to Katherine Newman (1993), social and economic change are placing the American dream in jeopardy. The downscaling of jobs and pay that occurred during the 1980s and 1990s has replaced earlier optimism with anger, doubt, and fear. The traditional middle-class way of life is threatened by declining work compensation and a staggering national debt.

Newman analyzes this new reality among middle-class Americans as she tries to understand the effects on family life, political attitudes, and personal identities as well as the "rage, disappointment, and . . . sense of drift in communities across the land" (Newman 1993:x).

Bypassing the "bloodless" statistics, Newman talked with the Americans whose lives lie behind the graphs and dire popular media accounts. She spent two years conducting personal interviews with 150 Americans living in "Pleasanton," a representative suburban community. The population of Pleasanton corresponds to the hallmark of suburbia: a mix of skilled blue-collar workers and white-collar professionals from a variety of ethnic and religious origins. Her respondents were schoolteachers, guidance counselors, and sixty families whose children were then grown.

The promise of America had taken an unexpected wrong turn, and the residents of Pleasanton were trying to make sense of it. Newman attempted to understand the residents' view of this downward mobility. She looked for hidden meanings of culture as much as the more apparent behavior and norms. The stresses associated with changing economic conditions, she believed, would bring cultural expectations, disappointments, and conflicts close enough to the surface for a trained social scientist to see. In fact, she saw intergenerational conflict, marital discord, and racial and ethnic division. The following statement reveals a baby boomer's shattered confidence in the American dream:

I'll never have what my parents had. I can't even dream of that. I'm living a lifestyle that's way lower than it was when I was growing up and it's depressing. You know it's a rude awakening when you're out in the world on your own. . . . I took what was given to me and tried to use it the best way I could. Even if you are a hard worker and you never skipped a beat, you followed all the rules, did everything they told you you were supposed to do, it's still horrendous. They lied to me. You don't get where you were supposed to wind up. At the end of the road it isn't there. I worked all those years and then I didn't get to candy land. The prize wasn't there, damn it. (Newman 1993:3)

After a detailed and often personal exploration of what Newman calls the "withering American Dream," she turns to the larger social and political implications for society. She explores the transition from a society of upward mobility based on effort and merit to a society in which social classes of birth increasingly dictate future social and economic positions.

According to Newman, the soul of America is at stake. She raises these questions: Will Americans turn to exclusive self-interest, or will they care for others as well as themselves? Will suburbanites turn a blind eye to the rapidly deteriorating inner cities? Will the generational, racial, and ethnic groups turn inward, or will they attempt to bridge the divides that threaten to separate them further?

A partial answer to these questions is reflected in public opinion about federal, state, and local tax revenues. If the residents of Pleasanton are any guide, Americans do not wish to invest in the common good. Public schools, colleges, universities, and inner cities, for example, are receiving a rapidly declining share of public economic support. In conclusion, Newman states:

This does not augur well for the soul of the country in the twenty-first century. Every great nation draws its strength from a social contract, an unspoken agreement to provide for one another, to reach across the narrow self-interests of generations, ethnic groups, races, classes, and genders toward some vision of the common good. Taxes and budgets—the mundane preoccupations of city hall—express this commitment, or lack of it, in the bluntest fashion. Through these mechanistic devices, we are forced to confront some of the most searching philosophical questions that face any country: What do we owe one another as members of a society? Can we sustain a collective sense of purpose in the face of the declining fortunes that are tearing us apart, leaving those who are able to scramble for advantage and those who are not to suffer out of sight? (Newman 1993:221)

Evaluate the Research

1. Think about your experiences at home and in other social institutions (schools, churches). State the conception of the American dream these experiences provided. Critically analyze the ways in which society shaped your conception.

2. Newman's research was done in the early 1990s. Do you believe that she is right about the fate of the American dream? Explain.

3. If the American dream is withering, many social changes are in store. Describe the major changes you foresee.

4. Suppose Katherine Newman had decided to place her study in the context of sociological theory. Write a conclusion to her book from the theoretical perspective—functionalism, conflict theory, or symbolic interactionism—that you think is most appropriate.

Were there changes in the reemergence of the conflict perspective? Although the conflict perspective had little influence in sociology during the first half of the twentieth century, it has been revitalized. This reemergence was the work of sociologists, like Ralf Dahrendorf, who follow some of Marx's ideas while rejecting others (Dahrendorf 1958a, 1958b, 1959). Although Dahrendorf attributes change to the struggle over resources, the resources at stake are more than economic. Dahrendorf credits the quest for power—conflict over who controls whom—as the cause of social change. Whereas Marx envisioned two opposing social classes, Dahrendorf sees conflict between groups at all levels of society. Rather than change emerging from a single grand conflict, it comes from a multitude of competing interest groups. By rejecting Marx's contention that history is created by class conflict alone, Dahrendorf recognizes conflict between all types of interest groups—political, economic, religious, racial, ethnic, gender based. Thus, society changes as power relationships between interest groups change.

Historical events seem to support Dahrendorf's interpretation. Class conflict has not occurred in any capitalist society; social classes have not polarized into major warring factions. Rather, capitalist societies are composed of countless competing groups. In America, for example, racial groups struggle over the issue of equal economic opportunity; and environmentalists and industrialists debate the proper balance between environmental protection and economic development.

Some sociologists have concluded that no single perspective on social change is clearly superior to others in all respects. They assert that social and cultural change is too complex to be captured by any single theory, given our present state of knowledge. They believe that because each perspective has its strengths and weaknesses, the best way to explain social change is through the combination of various aspects of each theory. These efforts have been called functionalist conflict theory (R. Collins 1994).

Reconciling the Functionalist and Conflict Perspectives

What attempts have been made to accommodate the differences in these two perspectives? There has been no fully successful synthesis of the functionalist and the conflict perspectives addressing social change, but some interesting attempts have been made. Although he does not reconcile all the differences between these two perspectives, Pierre van den Berghe (1978) points to some commonalities. First, both perspectives observe society as systems with interrelated parts. Second, both perspectives are capable of viewing conflict as a contributor to social integration and integration as a producer of conflict. Third, both theoretical perspectives assume an evolutionary view of social change. Finally, van den Berghe sees both perspectives as equilibrium models.

Like Marx, Lewis Coser (1998) sees the pressure for social change coming from conflict: competing groups and classes rise and fall in the struggle for power and for protection of their interests. Unlike Marx, however, Coser does not believe that cataclysmic social change necessarily follows social conflict or that change produced by conflict must necessarily come about abruptly. In fact, Coser distinguishes between change *within* a system and change *of* a system. A society flexible enough to adjust to a new environment will experience change while maintaining its basic structure. Examples of change within a society would be the civil rights and women's movements in the United States. If a society is not flexible enough, a change of the system may

---

TABLE 18.2

FOCUS ON THEORETICAL PERSPECTIVES: Social Change

This table gives examples of how the functionalist, conflict, and symbolic interactionist perspectives view social change. Describe how a functionalist would look at an interest group and how a conflict theorist would view equilibrium.

Theoretical Perspective	Concept	Example
Functionalism	Equilibrium	• The nature of the presidency has continuity, despite scandals in the Nixon and Clinton administrations.
Conflict theory	Interest group	• Civil rights laws were enacted in the 1960s as a result of the struggle over racial equality.
Symbolic interactionism	Urbanism	• A smaller proportion of social interaction in a large city is based on shared meanings.

occur as a result of conflict. The Chinese Communist Revolution illustrates change of a society.

According to Coser, the behavior of a society's elite strongly influences the outcome of conflict, affecting whether conflict will produce a readjustment within the society or a breakdown and formation of a new society. If the dominant groups are flexible enough to allow free expression of complaints and make appropriate adjustments, change within the society is more likely to occur. Should those with the power choose to protect their interests by resisting change and stifling grievances, they run the risk of intensifying conflict and producing a change of the society.

Radical change of a society may occur suddenly, but it does not necessarily have to, according to Coser. Although basic institutions, values, and social relationships did change immediately following the Chinese Communist Revolution, change following the American War of Independence was gradual. The United States is a fundamentally different society than it was before 1776, but this restructuring cannot be linked to any specific event. In other words, for Coser, the fundamental restructuring of a society can take place as a result of gradual, cumulative changes within the society.

Nolan and Lenski (2005) also attempt to combine the functionalist and conflict perspectives. In the first place, they give equal weight to stability and change. Second, they explicitly incorporate conflict into their theory. For example, Nolan and Lenski contend that when societies have come into conflict with one another for territorial and other vital resources, those that have been technologically more advanced have usually been dominant. They make ample provision for conflict between societies, pointing out the need for societies to defend themselves and their territories from invasion by outsiders. Nor do they ignore conflict within societies, highlighting, for example, the struggle in industrializing agrarian societies between the elite in control of their society's resources and the masses who supply the labor for subsistence-level returns.

Regarding social change in particular, do these attempts at reconciliation mean that the functionalist and conflict perspectives can be abandoned? Although these attempts at reconciliation may be interesting, they do not eliminate the need for separate functionalist and conflict perspectives. Continued work by functionalists will tell us more about the processes involved in the maintenance of a dynamic equilibrium—how societies maintain stability and order while undergoing change. Conflict theorists continue to provide insights into how change occurs as a result of the struggles between those with differential power and opposing interests.

Modernization

The Nature of Modernization

It seems as if the entire world is either already modernized or attempting to move in that direction. **Modernization** refers to the updated social changes—there are a host of them—that accompany economic development. The trend toward modernization is one of the most significant trends in human history (Haferkamp and Smelser 1992; Chase-Dunn 1999; Fukuyama 1999, 2000; Giddens 2002).

What are some of the major consequences of modernization? Some significant demographic changes are associated with modernization. Population growth occurs as the death rate declines and life expectancy increases. Between the thirteenth and seventeenth centuries, for example, Europe had a life expectancy as low as twenty years; by 1930, life expectancy had increased to more than sixty years and now stands at seventy-eight years (*World Population Data Sheet* 2007). The population moves from rural to urban areas. Whereas most members of traditional societies work the land, most members of modern societies live and work in towns and cities, where industry and jobs are located.

FEEDBACK

1. According to the cyclical perspective on social change, societies rise and fall rather than move continuously toward improvement. T or F?
2. Cyclical theories explain the reason social change occurs. T or F?
3. Classical evolutionary theory assumed that change leads to improvement. T or F?
4. According to modern evolutionary theory, social change is _____. (unilinear/multilinear)
5. According to the _____ perspective, society is inherently stable, and any change that occurs is eventually assimilated so that a new state of equilibrium is achieved.
6. _____ refers to any situation in which disequilibrium is caused by one aspect of a society failing to change at the same rate as an interrelated aspect.
7. According to the _____ perspective, social and cultural change occur as a result of the struggles between groups representing different segments of a society.

Answers: 1. T; 2. F; 3. T; 4. multilinear; 5. functionalist; 6. Cultural lag; 7. conflict

Stratification structures are altered by modernization. Traditional society is characterized by a bipolar stratification structure—the wealthy at one end, the poor masses at the other. Modernization brings an expansion of the middle and upper classes. As more emphasis is placed on personal effort and achievement, social mobility increases and inequality generally declines.

The political institution is altered by modernization. The role of the state expands as it becomes more centralized and more involved in social and economic affairs. At the same time, modernization promotes political democracy, even though the extent of democratization varies considerably from society to society. Although political power is not equalized as a result of modernization—there is always a powerful elite—power is more widely dispersed.

Modernization transfers education from the family to a formal system of education. Education is no longer designed just for the privileged few but for the entire society; primary education, in fact, is intended for all members of a modernizing society. A literate population is absolutely necessary to create a workforce suitable for an industrialized economy.

Modernization also affects family life. The nuclear family replaces the extended family. Because the economy ceases to be based on a familial division of labor, people must move to cities to work. Extended family ties can be obliterated; they simply become much harder to maintain, and much of their intimacy can be lost. With modernization, functions formerly of the family's domain are adopted by other institutions. For example, government assumes more responsibility for the elderly, schools take care of children's educational needs, and the mass media provide entertainment.

How comparable are the social and cultural characteristics of modernizing societies? This question is debatable (Waters 2001). Advocates of the pattern of convergence foresee the development of social and cultural similarity among modernizing nations. They contend that as societies adopt the newest technological arrangements, they tend to develop social and cultural commonalities. This is said to occur because a particular technology carries in its wake a particular occupational structure, which in turn exercises similar wide-ranging effects on various social structures and social relationships. The necessity for an industrializing society to develop an educated populace is another major unifying force. Other pressures for uniformity include the enlarged role of government, the influence of multinational corporations within modernizing countries, and the accelerated growth of large-scale organizations (Inkeles 1998).

Movement toward globalization, for example, was reflected in the television coverage of millennium celebrations. Never before had the world shared in witnessing New Year's celebrations from Australia west to Hawaii. This international exposure—fueled by the almost universal adoption of computer technology and broadcast by commonly shared telecommunications technology—may foretell increased cross-cultural influence in the twenty-first century.

Supporters of the pattern of divergence, on the other hand, do not see social and cultural homogeneity as an inevitable result of modernization. Other

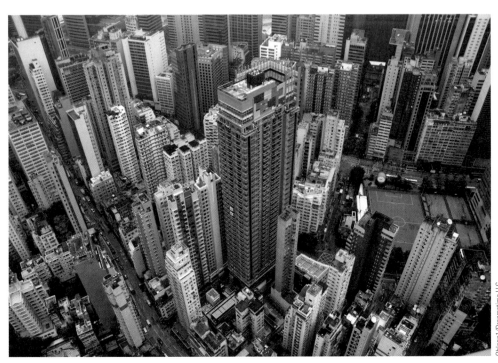

Hong Kong from the air looks like a modern Western city. Would social life on the ground look similar or different from that of a modern Western city?

Oblong1/Dreamstime LLC

sociologists emphasize the effects of idiosyncratic social and cultural forces to explain the emergence of new social patterns in modernizing societies. Advocates of divergence offer as evidence less developed countries that are now undergoing modernization along lines somewhat different from the path taken by Western societies. Many less developed countries that chose the Soviet economic model from the 1950s onward are, along with former Communist countries in eastern Europe, adopting the model of market economies (*World Development Report* 1991; Schnitzer 2000). Could this signal greater convergence in the future?

As in most complex debates, there is a middle ground that is probably closer to the truth. Social and cultural elements of traditional societies do not necessarily vanish in the process of modernization; traditional elements may blend with the new (Mazaar 1999; Guillen 2003). Changes in the family during modernization, for example, depend in part on the type of traditional family a society has. Some traditional family structures are more compatible with modernization than others are. To cite another example, democratization usually accompanies modernization, but its extent and form may vary due to the specific social and cultural traditions of a society. Although modern Japan is more democratic than it was in premodern times, it is less democratic than most Western societies. And although the Soviet Communist Party leaders voted overwhelmingly in July 1991 to endorse the principle of private property, freedom of religion, and a pluralistic political system, serious political opposition has since arisen. The lack of democratic progress is obvious when Russian President Vladimir Putin can choose the next president of Russia by simply endorsing his protégé, Dimitri Medvedev. Even before the 2008 election, Putin was labeled "the older czar" and Medvedev "the younger czar" (Schwirtz 2007).

World-System Theory

To this point, we have examined the process of modernization and its consequences *within* countries. When we consider modernization from an international perspective, quite different theories emerge. One of these is *world-system theory*, which also depicts social and cultural diversity among modernizing nations. Its approach is unique (Chase-Dunn and Hall 1997; Chase-Dunn 1999; Kardulias 1999; Guillen 2003).

What is world-system theory? According to **world-system theory,** the pattern of a nation's development hinges on its place in the world economy. The world economy is divided into segments: core and peripheral nations. Core nations, such as the United States, dominate the world economy by providing managerial expertise and technological innovation. This domination leads to the control and exploitation of the rest of the world; to benefit themselves, core nations use the cheap labor and natural resources of peripheral nations (less developed nations) (Wallerstein 1979, 2001, 2003a, 2003b, 2004).

What are the consequences of this arrangement? As a result of this world-system arrangement, the standard of living is much higher in core nations, whose skilled workers operate in a free labor market. Workers in peripheral nations provide unskilled, coerced labor and suffer a low standard of living. Peripheral nations do not develop as much as they could if they were not dominated and exploited by the core nations. In fact, world-system advocates contend, this deterrent to peripheral national development is the prime perpetuating impetus for the existing arrangement. Consequently, the economic, social, and cultural gap between core and peripheral nations is increasing (Rossides 1997; Wallerstein 2000; Robbins 2004; Stiglitz 2003, 2007).

According to many advocates of globalization, free markets promote democracy, prosperity, and peace (Mandelbaum 2004b; T. Friedman 2000, 2007). Free markets are said to promote enfranchisement of the masses, increase economic equality, and reduce conflict within and between societies. Looking at Russia and developing countries in Latin America, Africa, and Asia, Amy Chua (2004) questions this happy mix assumed by globalists. Conceding that free markets promote democracy, the results, Chua argues, do not include increased economic equality and reduced conflict. Instead, the introduction of the free market in a developing country creates a new class of extremely rich elites who tend to belong to the same minority group. This "market-dominant minority" is subsequently subjected to ethnic hatred and violence from the recently enfranchised majority. The majority government uses its newly acquired political freedom and power to wreak revenge on the politically weaker minority elites and confiscate or destroy their property. In Indonesia, to cite one of Chua's cases, the suppressed Muslim majority unleashed a vicious attack against the economically dominant ethnic Chinese, looting their businesses, breaking their store windows, and raping their women. Because Chua is going against the more conventional view, she is certain to be challenged by globalists. The same is true for Naomi Klein, who argues that Milton Friedman's free market movement (see M. Friedman 2002) has failed to deliver on its promise of a peaceful, prosperous global world. Rather, she argues, this movement's brand of capitalism has taken advantage of crises brought on by war to enrich those already rich and powerful (Klein 2007).

Collective behavior is popularly thought of as unstructured and irrational. What does this photo of volunteers unloading supplies after the World Trade Center disaster say about this conception?

Bill Kostroun/AP Photos

Collective Behavior

Another area of sociological study—collective behavior—often focuses on short-term behavior. Some forms of collective behavior are thought to be inconsequential; other forms have far-reaching consequences. We will examine both.

Collective behavior is the relatively spontaneous and unstructured social behavior of people who are responding to similar stimuli. It is *collective* because it usually occurs among a large number of people. The phrase "responding to similar stimuli" emphasizes collective behavior as a reaction by people to some person or event outside themselves, such as the attacks on the

World Trade Center and Pentagon on September 11, 2001. Collective behavior involves social interaction in which loosely connected participants influence one another's behavior.

The study of collective behavior poses some problems. In the first place, sociologists are accustomed to studying structured behavior. Second, how are researchers going to investigate a social phenomenon that occurs spontaneously? Despite these difficulties, sociologists have conducted fascinating research and formulated serviceable theories of collective behavior. It turns out that there is more structure and rationality to collective behavior than appears on the surface.

Initiating our discussion are the more disorganized, unplanned, and short-term forms of collective

FEEDBACK

1. _____ is a process involving all those social and cultural changes accompanying economic development.

2. Which of the following is *not* one of the major consequences of modernization discussed in the text?
   a. the social and cultural convergence of modernizing countries
   b. increased urbanization
   c. greater equality
   d. more political democracy
   e. widespread development of the nuclear family

3. According to _____ theory, the pattern of a nation's development largely depends on the nation's location in the world economy.

4. World-system theory would predict a (convergence/divergence) between core and peripheral nations.

Answers: 1. Modernization; 2. a; 3. world-system; 4. divergence behavior—rumors, mass hysteria, panics, fads, and fashions. Crowds are covered next. A discussion of social movements, the most highly structured, enduring, and rational form of collective behavior, closes the chapter.

Dispersed Collectivities

In the more structured forms of collective behavior, such as crowds and social movements, people are in physical contact. Other forms of collective behavior occur among dispersed members of a mass society. These **dispersed collectivities** engage in the less structured forms of collective behavior—rumors, mass hysteria, panics, fads, and fashions. Behavior among members of dispersed collectivities is not highly individualized. Members of dispersed collectivities are not physically connected but uniformly respond to some common object of attention; they are aware of being a member of a collectivity:

> When people are scattered about, they can communicate with one another in small clusters of people; all of the members of a public need not hear or see what every other member is saying or doing. And they can communicate in a variety of ways—by telephone, letter, Fax machine, computer linkup, as well as through second-, or third-, or fourth-hand talk in a gossip or rumor network. (Goode 1992:255)

Rumors

In the *Aeneid*, Virgil wrote these lines:

> Rumor! What evil can surpass her speed? In movement she grows mighty, and achieves strength and dominion as she swifter flies.

Rumor, as Virgil's words underscore, has a negative connotation. Rumors may be benign, as in the case of continual Elvis Presley sightings, or they can do considerable damage. They are often communicated as the truth when in fact they may be false. At best, rumors are usually inaccurate and misleading. In any event, the likelihood of an individual spreading a rumor depends in part on the degree of anxiety a person feels, the extent of uncertainty about events that the person is experiencing, the credibility of the person passing the rumor, and the relevance of the rumor to that particular person (Rosnow 1988, 1991).

Why is rumor considered an unstable form of collective behavior? A **rumor** is a widely circulating story of questionable truth. Rumors usually focus on people or events that are of great interest to others. Segments of the mass media exploit the public's fascination with rumors: entertainment magazines devoted exclusively to rock idols and movie stars; tabloid newspapers loaded with titillating guesswork, half-truths, and innuendos; even mainstream news publications catering to accounts of the rich, famous, and offbeat. As these examples suggest, rumors and gossip are closely related.

Following Hurricane Katrina, the national media were reporting a cholera outbreak, a dead baby in a trashcan, rampant rape, children with slit throats, and scores of bodies stacked inside a freezer in the New Orleans Superdome. Around the world in early 2006, rumors helped fuel Muslim riots, which were sparked by the publication of cartoon images caricaturing the prophet Muhammad. For example, leaflets distributed in Lebanon charged the West with declaring war on Islam. Another rumor about a cartoon, which pictured a hand grenade on the drawing of Muhammad's turban, claimed it also had hand grenades on each of his fingers. During the 2008 presidential campaign, Barack Obama fought the rumor that he was a part of a Muslim conspiracy against America, and that, if elected president, he would replace the Bible with the Koran at his inauguration. Obama, in fact, is not a Muslim (Bacon 2007). None of these rumors proved true; but they were spread and believed, in part because they touched on people's insecurities, uncertainties, and anxieties (Rosnow 1991; Fine and Turner 2004).

How is an urban legend dissimilar from a rumor? Akin to rumors are what Jan Harold Brunvand calls **urban legends** (Brunvand 2001, 2002). Although urban legends may incorporate current rumors, they tend to have a longer life and wider acceptance. They are moralistic tales passed along by people who swear the stories happened to someone they know, or to an acquaintance of a friend or family member (Heath, Bell, and Sternberg 2001). The tales often focus on current places, concerns, and fears such as AIDS and inner-city gangs. Consider the case of "The Baby in the Oven," in which a babysitter gets high on drugs while babysitting and does a tragic thing:

> This couple with a teenage son and a little baby left the baby with this hippie-type girl who was a friend of the son's. They went to a dinner party or something, and the mother called in the middle of the evening to see if everything was all right. "Sure," the girl says. "Everything's fine. I just stuffed the turkey and put it in the oven." Well, the lady couldn't remember having a turkey, so she figured something was wrong. She and her husband went home and they found the girl had stuffed the baby and put it in the oven. Now the son used a lot of drugs; she was his friend, so I guess they figure she took them, too. (Brunvand 1980:55)

Another typical story tells about a man who wakes up in a hotel room missing a kidney. Another describes alligators roaming the sewer systems of big cities. As cautionary tales, urban legends warn us against engaging in risky behaviors by pointing out what has supposedly happened to others who did what we might be tempted to try. Like rumors, urban legends permit us to play out some of our hidden fears and guilt feelings by being shocked and horrified at others' misfortune. "The Baby in the Oven" story, for example, may represent guilt feelings for parents who sometimes leave

their children with strangers or harbor a deep fear of outsiders entering their homes.

Mass Hysteria and Panics

What is mass hysteria? **Mass hysteria** is a collective anxiety created by the acceptance of one or more false beliefs. Orson Welles's famous "Men from Mars" radio

Terrorist Attacks and Disaster Myths

You fail a test, lose a boyfriend, have a minor auto accident, or suffer defeat by an archrival's basketball team. You might well describe each of these occasions as a "disaster." For sociologists, however, the term *disaster* is limited to events with the following characteristics.

- Extensive damage to property.
- Great loss of human life.
- Massive disruption to everyday living.
- Unpredictability and suddenness of a short-term event.

Researchers typically divide disasters into "natural disasters" such as floods, earthquakes, and hurricanes and "technological accidents" such as airline crashes, nuclear plant meltdowns, and ship sinkings. But how can we classify the September 11, 2001, attacks on the World Trade Center in New York City and the Pentagon in Washington, D.C.? They were neither natural nor an accident. But, they had all the characteristics of a disaster. In fact, terrorism is introducing a new type of disaster, one that involves technology and is intentionally caused by humans.

The World Trade Center and Pentagon attacks obviously met the criteria of a disaster. Less obviously, they also exposed as false many popular beliefs about human behavior in disasters. Let's con-

sider four such myths within the context of this national tragedy.

1. Victims of disasters panic. Contrary to this myth, disaster victims do not generally panic. While some individuals in disasters may panic and while mass panics may follow disasters, the prevailing response is one of general composure and problem-solving behavior. Some inside the World Trade Center did respond with incapacitating emotion. One secretary in shock, for example, had to be carried out by a fellow worker. Some people jumped from the towers. But the disaster failed to set off a widespread panic. Many who heeded the first building-wide instructions died after calmly remaining in their offices. And many of the survivors remained as interested observers, forcing police to broadcast an urgent plea for them to hurry away for their own safety.

2. Disaster victims respond as isolated individuals. Typically, we picture disaster victims as individuals trying to save only themselves. Actually, according to research, people immediately engage in group efforts to help others. People in the World Trade Center with cell phones offered them to other victims desperate to call family or friends. Scores of New York police and more than 300 firefighters died while working together to rescue trapped victims.

3. Disaster victims leave the scene as soon as possible. Contrary to this myth, the majority of victims remain near the disaster site. Rather than fleeing, most victims remained to help others, to witness the fire and rescue efforts, or to think about returning to their offices. Large numbers

of volunteers and off-site emergency personnel actually rushed to the scene. Bellevue Hospital at one point had five doctors for each emergency ward patient. Four firefighters who were playing golf on Staten Island saw the first plane hit the north tower. Three of those four lost their lives in rescue efforts, and they were just a few of the hundreds of firefighters who died after entering the disaster site. To help rescuers searching for survivors under the rubble, ironworkers, many of whom had built the World Trade Center, labored together in twelve-hour volunteer shifts clearing away twisted steel.

4. Crime is prevalent during disasters. Rather than increasing, crime actually decreases after a disaster. While some isolated instances of criminal behavior occur, the crime rate in a disaster falls. After the World Trade Center disaster, some looting in surrounding buildings was reported, and a Picasso drawing valued at $320,000 was stolen from a Madison Avenue art gallery. More importantly, the overall crime rate in New York City declined 34 percent in the week following the disaster. According to the NYPD, arrests were down 64 percent compared to the same seven days the previous years.

Think About It

1. Think of some event you formerly considered a disaster. Explain why it was not a disaster from a sociological viewpoint.

2. Do you think that the behavior following the terrorist attack on the World Trade Center is best explained by functionalism, conflict theory, or symbolic interactionism? Explain your choice.

broadcast in 1938, though based entirely on H. G. Wells's novel *The War of the Worlds*, caused nationwide hysteria. About 1 million listeners became frightened or disturbed and thousands of Americans hit the road to avoid the invading Martians. Telephone lines were jammed as people shared rumors, anxieties, fears, and escape plans (Houseman 1948; Cantril 1982; J. Barron 1988). Another example of mass hysteria is found in the responses to imagined witches in seventeenth-century Salem, Massachusetts. In Salem, twenty-two people who had been labeled witches died—twenty by execution—before the false testimony of several young girls began to be questioned.

A third example followed the death from AIDS of actor Rock Hudson in the mid-1980s. A 1987 Gallup poll showed that a substantial proportion of Americans held false beliefs regarding the spread of AIDS: 30 percent believed that insect bites could spread the disease; 26 percent related the spread to food handling or preparation; 26 percent thought AIDS could be transmitted via drinking glasses; 25 percent saw a risk in being coughed or sneezed on; and 18 percent believed that AIDS could be contracted from toilet seats (Gallup 1988). These mistaken ideas persisted on a widespread basis despite the medical community's conclusion that AIDS is spread through sexual contact, by sharing hypodermic needles, and by transfusion of infected blood. By the late 1990s, knowledge, toleration, compassion, and understanding of AIDS had increased substantially enough that the frequency of these rumors had dissipated.

What is the difference between mass hysteria and a panic?
A **panic** occurs when people react to a real threat in fearful, anxious, and often self-damaging ways. Panics usually occur in response to such unexpected events as fires, invasions, and ship sinkings. For example, 117 people died in two nightclubs in Chicago and Rhode Island in 2003, when panic reactions to a fire caused a jamming of escape routes (D. Johnson 2003a).

Interestingly enough, people often do not continue to panic after the initial chaos triggered by natural disasters such as earthquakes and floods. Although panic may occur at the outset, major natural catastrophes usually involve highly structured behavior (K. T. Erikson 1976, 1995; Dynes and Tierney 1994; Quarantelli 2001). An example of both initial panic and subsequent structured behavior occurred during the terrorist attacks of September 11, 2001. Some occupants of the buildings panicked, jumping from windows, whereas firefighters exhibited highly structured behaviors during their rescue efforts. Agreeing that rational behavior usually accompanies a catastrophe, Lee Clark writes:

> When the World Trade Center started to burn, the standards of civility that people carried with them everyday did not suddenly dissipate. The rules of behavior in extreme situations are not much different from rules of ordinary life. . . . When danger arises, the rule—as in normal situations—is for people to help those next to them before they help themselves. (Clark 2002:24)

Panics and episodes of mass hysteria are based on repulsion. Other forms of collective behavior, such as fads, crazes, and fashions, are rooted in attraction.

Fads, Crazes, and Fashions

Do fads have staying power? No. **Fads** are unusual behavior patterns that are spread rapidly, embraced zealously by a particular segment of society, and then disappear after a short time. The widespread popularity of a fad rests largely on its novelty. The "streaking" fad (running naked across college grounds or through occupied classrooms) delighted students in the early 1970s (Aguirre, Quarantelli, and Mendoza 1988). A reminder of that fad occurred in the summer of 1991, when a sixteen-year-old female, wearing only her eye shadow, ran down the first fairway at the British Open golf tournament in Southport, England. Ironically, on that same day, two male fans celebrated the Atlanta Braves' five-run sixth inning performance by baring it all on the field. One of these streakers impressed spectators and players alike with a head-first slide into home plate. Some fads in more recent years include tattoos, body piercing, and Scrabulous (played in the office and a virtual knockoff of the Scrabble board game) (Timmons 2008). Fads are not limited to hobbies, clothing, and entertainment; they also emerge regarding serious matters. Ron Insana (2002) documented past investment fads, including the bull market of the 1990s. By now you should be able to identify the latest fads.

What is the difference between a fad and a craze? A **craze** is a type of fad that can have serious consequences for its adopters. The fads previously mentioned are primarily recreational and harmless in nature. But consider the potential downside to fads such as losing weight through extreme diets, getting "high" from cold medication, and using the date rape drug "roofies."

Do fashions reflect a society's culture? Yes. **Fashions** are behavior patterns that evolve over time and are widely approved but expected to change periodically (Crane 2001). As societies modernize, fashion becomes more salient and may change more rapidly. Within contemporary American society, the "in" fashions for clothing are introduced seasonally. Women are "on top" of the defined dress or pant length, and men are questioning if pant cuffs are fashionable or passé. High school students wishing to be fashionable wear clothes with the hottest labels.

In modern societies, fashion becomes more important and changes more rapidly. Penelope Cruz arrives at the Oscars dressed in the latest creation of fashion designer Armani Prive.

groovy, tubular, tough, fine, rad, bad, phat, and *sick* were all slang terms of approval among young people at some point in time. Any American teenager who doesn't know what facebookable and teafragging mean, needs a computer (*Urban Dictionary* 2008). Anyone driving in Boston best not bang a Uey (make a U-turn).

Why is fashion particularly prevalent in modern society? In the first place, modern societies are based on growing economies fueled by mass consumption. Styles in clothing, automobiles, eyewear, and sporting equipment must change if profits are to be made and people employed. Second, without traditions to supply the brakes, people in modern societies are eager to respond to the tempting new fashions that are constantly being created by entrepreneurs and established corporations. Finally, the relative affluence in modern societies provides enough disposable income for people to indulge their desire for fashion novelty and change.

Crowds

Distinguishing Characteristics of Crowds

The most dramatic form of collective behavior is the crowd, because it involves intense emotions. Look at the content of local and national television news. If this is any indication, people are alternately fascinated and horrified by passions that are unleashed in crowds. We are engrossed by reports of crowds, mobs, and rioters, whether they are celebrating a joyous event or lamenting a sad one.

How different is a crowd from an aggregate of people? A **crowd** is a temporary collection of people who share an immediate common interest. The temporary residents of a large campground, each occupied with his or her own activities, constitute an *aggregate*. But if some event, such as the landing of a hot-air balloon or the appearance of a bear, serves as a common stimulus to draw the campers together, the aggregate becomes a crowd. What will happen next is highly unpredictable.

Although the most widely recognized examples of fashion are related to appearance (clothing, jewelry, hairstyles), fashions also come and go in such diverse arenas as automobile design, home decorating, architecture, and politics. Slang, another example, goes in and out of favor (J. Lofland 1993). What self-respecting teenager today would be caught calling something "neat"? *Cool,*

The specific subject of mutual interest—the common stimulus—is not important. What is significant is the collection of individuals into a situation that has the potential to stimulate an emotional reaction.

A crowd situation involves ambiguity and uncertainty; participants have no predefined ideas about the way they should behave toward one another or toward some target on which their attention is converging. Members of a crowd, however, are certain about one thing: they share the urgent feeling that something either is about to happen or should be made to happen (McPhail 1991; Marx and McAdam 1994).

How do crowds differ from one another? Although it is accurate to make generalizations about crowds, not all crowds are alike. Herbert Blumer (1969a) has distinguished four basic types of crowds. A *casual crowd* is the least organized, least emotional, and most temporary type of crowd. Although members of a casual crowd share some point of interest, it is minor and fades quickly. Members of a casual crowd may gather with others to observe the aftermath of an accident, to watch someone threatening to jump from a building, or to listen to a street-rap group.

A *conventional crowd* has a specific purpose and follows accepted guidelines for appropriate behavior. People watching a film, flying on a chartered flight to a university ballgame, or observing a tennis match are in conventional crowds. As in casual crowds, there is little interaction among members of conventional crowds. The fact that the activity of the conventional crowd follows some established procedures distinguishes this type of crowd from a casual crowd.

Expressive crowds have no significant or long-term purpose beyond unleashing emotion. Their members are collectively caught up in a dominating, all-encompassing mood of the moment. Free expression of emotion—yelling, crying, laughing, jumping—is the defining characteristic of this type of crowd. Hysterical fans at a rock concert and the people gathered at Times Square on New Year's Eve, are examples of expressive crowds. Following the assassination of presidential candidate Benazir Bhutto, hundreds of thousands of Pakistani mourners thronged the mausoleum of their countries' most famous political dynasty in an outpouring of emotion.

Finally, a crowd that takes some action toward a target is an *acting crowd*. This type of crowd concentrates intensely on some objective and engages in aggressive behavior to achieve it. The some 5 million demonstrators who congregated across the United States and the world in protest to the 2003 impending war on Iraq were acting crowds (Campo-Flores 2003; Tyler 2003). A conventional crowd may become an acting crowd, as when a crowd of European soccer fans attacked the officials. Similarly, an expressive crowd may become

an acting one, as in the case of celebrating sports fans who destroy property following their team's victory. In June 2000, a peaceful celebration of the Los Angeles Lakers' NBA championship deteriorated as hundreds of fans destroyed two police vehicles and two television vans by fire, looted businesses, and damaged more than seventy cars. One person was killed and at least twelve others were injured (Gildea 2000). Another example occurred in New York City in the summer of 2000 when a number of males at the annual National Puerto Rican Day Parade turned from participating in normal parade festivities to sexually attacking several females in Central Park (Campo-Flores and Rosenberg 2000). In 2008, Italian soccer fans raged against police in cities across their country in response to the police accidentally killing a fan while breaking up a brawl between fans of two rival teams. In Rome, a crowd set police cars on fire and stormed a police barracks, injuring scores of officers.

What is unique about mobs? A **mob** is an emotionally stimulated, disorderly crowd that is ready to use destructiveness and violence to achieve a specific purpose. Through group understanding, a mob knows what it wants to do and considers all other things as distractions. In fact, individuals who are tempted to deviate from the mob's purpose are pressured to conform. If a mob is storming a seat of government, for example, it is inappropriate for participants to waste their time raping or stealing, although that may come later. Concentration on the main event is maintained by strong leadership.

Mobs have a long and violent history. The French Revolution is a classic example of mob action. With the cry "To the Bastille! To the Bastille!" a Parisian mob stormed this symbol of oppression on July 14, 1789. The overthrow of royalty and the desire for political freedom were the larger objectives of the French Revolution, but this mob's more specific and limited goal was to destroy the symbol of tyranny, oppression, and fear. As in all revolutions, there were other mob actions (Charles Tilly 1992). In October 1789, a largely female mob forcibly entered the royal palace at Versailles, and peasant mobs throughout France overthrew their feudal masters. Ultimately, King Louis XVI and Queen Marie Antoinette were escorted to Paris for execution by yet another mob.

The formation of mobs is not limited to revolutions. During the mid-eighteenth century, American colonists mobbed tax collectors as well as other political officials appointed by the British. During the Civil War, more than a thousand people were killed or injured as armed mobs protested against the Union Army's draft. Beginning in the late nineteenth century, some mobs in the American South acted as judges, juries, and executioners in the lynching of African Americans (as well as some whites). A huge rally opposing Serbian President Slobodan Milosevic's continuation in office, after his 2000 election loss, turned into a popular uprising that forced his resignation (Hammer and Cirjakovic 2000). During a protest of the Kenyan presidential election in 2008, a mob burned a church containing hundreds of citizens seeking refuge. That same year, a group of protestors in Tibet went into a rampage, killing 19 people and injuring more than 600 (Drew 2008).

What is a smart mob? A "smart mob" is an acting collectivity organized through the mobile Internet. *Smart mob* is the label Howard Rheingold (2003) gives to the new technology created by the combination of mobile communications (including your cell phone), computing, and the Internet. Over the next decade, he predicts, mobile communications and pervasive computing technologies will "change the way people meet, mate, work, fight, buy, sell, govern, and create" (Rheingold 2003:xiii). These new technologies, he writes, permit people to join together in collective action not possible before. In 2003, 400,000 antiwar activists protested without hitting the streets. They jammed White House and congressional switchboards with phone calls, faxes, and e-mails. This is said to be the first national virtual demonstration (Taylor 2003).

Rheingold envisions both positive and negative social consequences of the new technologies. Negative repercussions include the potential loss of privacy in a surveillance state, the increased capacity for terrorists to operate undercover, and the creation of new opportunities for criminal behavior. On the positive side, smart mobs can overthrow dictators and organize mass protests.

How do riots differ from mobs? Some acting crowds, although engaging in deliberate destructiveness and violence, do not display a mob's sense of common purpose. These episodes of crowd destructiveness and violence are **riots.** Riots involve a much wider range of activities than mob action. Whereas a mob surges to burn a particular building, to lynch an individual, or to throw bombs at a government official's car, riots may include actions of several different crowds, with rioters often directing their violence and destructiveness at targets simply because they are convenient. People who participate in riots typically lack power and engage in destructive behavior as a way to express their frustrations. A riot, usually triggered by a single event, is best understood within the context of long-standing tensions. The 1967 summer ghetto riots tore through many large American cities, occurring against a background of massive unemployment, uncaring slum landlords, poverty, discrimination, and charges of police brutality. The rioters in the twenty-odd cities involved did not, however, lash out at the underlying cause of the riots—the enduring gap between black and white America (*Report of the National Advisory Commission on Civil Disorders* 1968). Although the rioters were protesting against discrimination and deprivation, and although white-owned businesses were damaged more frequently than those owned by African Americans, the fact remains that African Americans did more damage to themselves (the overwhelming majority of the dead were African American) and their neighborhoods than they did to the establishment. As further evidence of random behavior in riots, many participants saw the riots as a way to loot stores or to have some destructive, violent fun.

Riots, of course, did not end with the 1960s. The international publishing of Danish cartoon images associating Islam's prophet Muhammad with terrorism and ignorance sparked Muslim riots in 2006. Angry mobs burned embassies and staged protests inside and outside the Middle East, resulting in dozens of deaths. Following the disputed presidential election in 2008, gangs of Kenyan youth wielded machetes and clubs in battles with police. Some 800 ethnic rivals, including children, were burned alive or hacked to death during the riots.

Theories of Crowd Behavior

Theories have been developed to explain crowd behavior. The three most important are contagion theory, emergent norm theory, and convergence theory.

What is the focus for contagion theory? **Contagion theory** emphasizes the irrationality of crowds that is created by participants stimulating one another to higher and higher levels of emotional intensity. As emotional intensity in the crowd increases, people temporarily lose their individuality to the "will" of the crowd. This makes it possible for a charismatic or manipulative leader to direct behavior, at least initially.

This theory has its roots in the classic work of Gustave Le Bon (1960; originally published in 1895), a French aristocrat who disdained crowds composed of the masses. People in crowds, Le Bon thought, were reduced to a nearly subhuman level:

By the mere fact that he forms part of an organized crowd, a man descends several rungs in the ladder of civilization. Isolated, he may be a cultivated individual; in a crowd, he is a barbarian—that is, a creature acting by instinct. He possesses the spontaneity, the violence, the ferocity, and also the enthusiasm and heroism of primitive beings. (Le Bon 1960:32)

Herbert Blumer (1969a) offers another version of contagion theory, but he avoids Le Bon's elitist bias. Blumer's theory is more refined but still implies that crowds are irrational and out of control. For Blumer, the basic process in crowds is a "circular reaction"—people mutually stimulating one another. There are three stages to this process. In *milling*, the first stage, people move around in an aimless and random fashion, much like excited herds of cattle or sheep. Through milling, people become increasingly aware of and sensitive to one another.

The second stage, *collective excitement*, is a more intense form of milling. At this stage, crowd members become impulsive and highly responsive to the actions and suggestions of others. Individuals begin to lose their personal identities and take on the identity of the crowd.

The last stage, *social contagion*, is an extension of the other stages. Excitement begins to spread. Behavior in this stage lacks caution and judgment and is a nonrational transmission of mood, impulse, or behavior. For example, fans at soccer games in Europe have launched attacks on referees to such proportion that games have been interrupted and people killed or injured. Taking a less extreme case, people who observe an auction can find themselves buying white elephants because they have become immersed in the contagious excitement of bidding.

Blumer's theory is more refined than Le Bon's, but it still implies that people in crowds are irrational and out of control. Sociologists today, though, realize that much of crowd behavior, even within an acting crowd such as a mob, is actually very rational (McPhail 1991). Emergent norm theory illustrates the structure and rationality of crowd behavior.

Where is the emphasis for emergent norm theory? **Emergent norm theory** stresses the similarity between daily social behavior and crowd behavior. In both situations, norms guide behavior (R. H. Turner 1964; Turner and Killian 1987). So, even within crowds, rules develop. These rules, of course, are *emergent* norms, because the crowd participants are not aware of the rules until they find themselves in a particular situation. These norms develop on the spot as crowd participants pick up cues for expected behavior. Emergent norm theory contends, in short, that crowd behavior is no different from noncrowd behavior, except that crowds do not have ready-made norms.

Also, whereas contagion theory proposes a collective mind that motivates crowd members to action, emergent norm theory views people in a crowd as present for a variety of reasons; they do not all behave in the same way (McPhail and Wohlstein 1983; Zucher and Snow 1990). Conformity may be active (some people in a riot may take home as many watches and rings as they can carry) or passive (others may simply not interfere with the looters, although they take nothing for themselves). In Nazi Germany, for instance, some people destroyed the stores of Jewish merchants while others watched silently, either in support or afraid to disagree for fear that others would ridicule or hurt them.

What is convergence theory? Both the contagion and emergent norm theories of crowd behavior assume that individuals are merely responding to those around them. It may be a more emotional response (as in contagion theory) or a more rational response (as in emergent

Many Americans have visited the memorial site in Oklahoma City. It commemorates the 1995 terrorist bombing that destroyed a U. S. government complex, killing 168 people. How does this behavior relate to convergence theory for crowd behavior?

Jeff Mitchell/Landov Media

1. A _____ is a temporary group of people who are reacting to the same event or individual.
2. An _____ crowd has no purpose or direction beyond the unleashing of emotions.
3. *Mob* and *riot* are simply two terms for the same type of crowd. T or F?
4. Much crowd behavior is structured and rational. T or F?
5. Some individuals at a lynching do not participate or give verbal support, but do not attempt to stop it. Which of the following theories of crowd behavior best explains this response?
   a. contagion theory
   b. crowd decision theory
   c. emergent norm theory
   d. casual crowd theory

Answers: 1. crowd; 2. expressive; 3. F; 4. T; 5. c

norm theory). In other words, the independent variable in crowd behavior is the crowd itself. In contrast, in **convergence theory,** crowds are formed by people who deliberately congregate with others whom they know to be like-minded. According to convergence theory, the independent variable in crowd behavior is the desire of people with a common interest to come together.

There have been many instances of crowds gathering in front of clinics to discourage abortions. This behavior, say convergence theorists, does not simply occur because people happened to be at the same place and are influenced by others. Such a crowd is motivated to form because of shared values, beliefs, and attitudes (Berk 1974).

Contemporary sociologists view crowd behavior as a more structured and rational phenomenon than is apparent on the surface. Sociologists also agree that a social movement—discussed in the next section—is an even more highly structured, rational, and enduring form of collective behavior (Goode 1992; Marx and McAdam 1994).

Social Movements

The Nature of Social Movements

The next time you observe a **social movement,** you will be aware of its four defining elements: a large number of people, a common goal to promote or prevent social change, some degree of leadership and organization, and activity sustained over a relatively long period of time. It is the form of collective behavior that has the most structure, lasts the longest, and is the most likely to create social change (Lofland 1996). Most social movements are mounted to stimulate change. This was as much the case for Nazism and the American Revolution as it is for the U.S. militia movement (Cozic 1997). The

pro-life and pro-choice movements and the environmental movement are more contemporary examples.

Despite commonalities, various social movements have unique characteristics. It is difficult to compare the civil rights movement with the environmental movement. This has led sociologists to study differences among social movements (Crossley 2002; Goodwin and Jasper 2008).

What are the primary types of social movements? David Aberle (1991) has identified four basic types of social movements. A **revolutionary movement** attempts to change a society totally. An example is the revolutionary movement led by Mao Zedong in China; it entrenched a communist form of government. A **reformative movement** aims to effect only partial change in a society; it can either advocate change or resist change. Many reformative movements seek to promote change. The women's liberation movement illustrates this type of social movement (Baxandall and Gordon 2000, 2001). Other reformative social movements oppose change. Conservative political and fundamentalist religious organizations have mounted a concerted effort to oppose abortion (Chesler 2001). The movement against corporate-led globalization among liberal youth is another example (Ferguson 2002). A **redemptive movement** focuses on changing individuals. The religious cult of David Koresh (the Branch Davidians) was a redemptive movement. Finally, an **alternative movement** seeks only limited changes in individuals. Population Connection (formerly known as Zero Population Growth) illustrates such a movement. It attempts to persuade people to limit the size of their families, but it does not advocate sweeping lifestyle changes, nor does it advocate legal penalties for large families.

Social movements are the most highly structured form of collective behavior. Three illuminating theories of social movements are relative deprivation theory, value-added theory, and resource mobilization theory.

Relative Deprivation Theory

It is the frustrated and discontent who want change and who are the most willing to fight for it. Discontent with present conditions, in short, is necessary for collective action. People must see existing conditions as undesirable, unfair, and unjust (Rose 1982). Thus, American colonists protested taxation without representation; Fidel Castro pointed to the vast gap between rich and poor Cubans; and Iranian revolutionaries in 1979 felt strongly that the Shah had gone too far with the processes of modernization, Westernization, and secularization. Discontent is more likely to lead to a social movement if it is linked with relative deprivation and unfulfilled rising expectations.

When does one feel relative deprivation? Relative deprivation is felt when people compare themselves with others and believe that they should have as much as those others have. Women's liberationists compare the situation of women to men, and gays in the United States underscore the penalties they suffer when they reveal their sexual preference. Government statistics indicate that African Americans receive less income than whites of comparable educational background. African Americans who are aware of this fact are likely to experience relative deprivation. Because a comparison is made between one's own situation and the situation of others, deprivation of this type is purely relative. There is no absolute standard for comparison—only the conviction among certain people that they, wrongfully, have less than some specific others have.

What are unfulfilled rising expectations? Unfulfilled rising expectations occur when newly raised hopes for a better life either are not satisfied at all or are not satisfied as rapidly as people had expected. For example, newly industrializing countries are likely to experience some revolutionary discontent when people who have been poor all their lives are suddenly promised a better life. They revolt not just because of their poverty but because their expectations about their material well-being have changed more rapidly than their actual material condition. The phenomenon of unfulfilled rising expectations helps explain why many revolutionary situations arise only after people have experienced some economic and social improvement. Alexis de Tocqueville (1955; originally published in 1835) observed improvement in the French peasant's economic situation:

> It is a singular fact that this steadily increasing prosperity, far from tranquilizing the population, everywhere promoted a spirit of unrest. The general public became more and more hostile to every ancient institution, more and more discontented; indeed, it was increasingly obvious that the nation was heading for a revolution.

> Moreover, those parts of France in which the improvement in the standard of living was most pronounced were the chief centers of the revolutionary movement. Such records of the Ile-de-France region as have survived prove clearly that it was in the districts in the vicinity of Paris that the old order was soonest and most drastically superseded. In these parts the freedom and wealth of the peasant had long been better assured than in any other pays d'élection. (Tocqueville 1955:175)

James Davies (1979) has linked unfulfilled rising expectations to revolutionary social movements. According to Davies's J-curve theory, a revolutionary movement is most probable when a period of rising expectations accompanied by actual economic improvement is followed by a decline in the fortunes of the masses (see Figure 18.2). According to this model, once expectations begin to rise, they continue to do so. As long as expected need satisfaction and actual need satisfaction are reasonably close, people will tolerate the gap between what they want and what they are getting. It is when actual need satisfaction falls off sharply (note the upside-down J formed by the curve in Figure 18.2) that the gap between what people want and what they have becomes intolerable. At this point, a revolutionary social movement is most likely to occur.

What is the major shortcoming of relative deprivation theory? The theory cannot explain why some social discontent can exist without a subsequent social movement. Although African Americans had long been discontented and suffered from relative deprivation, a social movement was not mounted until the 1950s. We must conclude, then, that although discontent and deprivation are necessary conditions of social movements, they are not sufficient ones. Discontent and deprivation must precede a movement, but they cannot alone produce one. The other two theories of social movements point to other factors.

Value-Added Theory

One of the strengths of Neil Smelser's (1971) **value-added theory** is its applicability to many forms of collective behavior. Although the theory can be applied to such types of collective behavior as panics and riots, we describe it here only within the context of social movements.

What is value-added theory? Smelser's contribution is based on an economic theory. In the economic value-added process, each step in the creation of a product contributes (adds value) to the final entity. Smelser gives an example involving automobile production:

> An example of [the value-added process] is the conversion of iron ore into finished automobiles by a number of stages of

processing. Relevant stages would be mining, smelting, tempering, shaping, and combining the steel with other parts, painting, delivering to retailer, and selling. Each stage "adds its value" to the final cost of the finished product. The key element in this example is that the earlier stages must combine according to a certain pattern before the next stage can contribute its particular value to the finished product, an automobile. Painting, in order to be effective as a "determinant" in shaping the product, has to "wait" for the completion of the earlier processes. Every stage in the value-added process, therefore, is a necessary condition for the appropriate and effective condition of value in the next stage. The sufficient condition for final production, moreover, is the combination of every necessary condition, according to a definite pattern. (Smelser 1971:13–14)

According to Smelser's theory, six conditions are necessary and sufficient for the development of a social movement. That is, Smelser specifies conditions that must exist if a social movement is to occur and that will lead to a social movement if they are present.

What are Smelser's necessary conditions for the development of a social movement and how do these conditions apply to the anti-Vietnam war movement? *Structural conduciveness* refers to an environment that is social movement friendly. The antiwar movement in the 1960s and 1970s occurred because there was a war, yes, but also because most college campuses had convenient sites for

rallies and protest meetings. Without ready access to areas conducive for demonstrations, the antiwar movement could not have taken place.

A second condition promoting the emergence of a social movement is the presence of *structural strains*—conflicts, ambiguities, and discrepancies within a society. Without some form of strain, there is no impetus for change. Undoubtedly, conflict contributed to the 1960s antiwar movement: conflict between American credence in political self-determination, on the one hand, and the U.S. government's desire to save Southeast Asia from communism, on the other. Probably a key discrepancy in this case was the government's continued stance that there was no war (no legal war had been declared), despite the vast resources devoted to battle and the obvious combat casualties suffered.

The combination of structural conduciveness and structural strains increases the probability that a social movement will occur. When a third condition—*generalized beliefs*—is also present, a social movement is even more likely. Generalized beliefs include a general recognition not only that there is a problem, but that something should be done about it. Two shared beliefs were crucial to the antiwar movement. One was the belief that the Johnson and Nixon administrations were not telling the truth about the war. It was also believed that the Vietnam War was so morally wrong that it had to be stopped.

Even when structural conduciveness, structural strains, and generalized beliefs exist, a social movement

FIGURE 18.2

The J-Curve Theory of Revolution

According to the J-curve theory, a revolutionary social movement is most likely to occur when rising expectations cause expected need satisfaction to exceed actual need satisfaction by an unacceptable amount. Is this expected to be true even when total need satisfaction has actually increased? Explain.

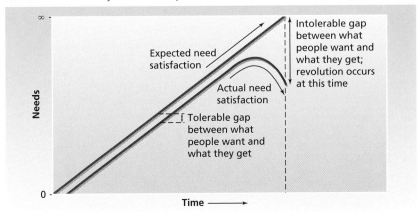

Source: James Chowning Davies, "The J-Curve of Rising and Declining Satisfactions as a Cause of Revolution and Rebellion," in Hugh Davis Graham and Ted Robert Gurr (eds.), *Violence in America*, rev. ed. (Beverly Hills, CA: Sage, 1979), p. 416. Reprinted by permission of J. C. Davies and the publisher, Sage Publications, Inc. (Beverly Hills/London).

SNAPSHOT OF AMERICA 18.1

Women in the Workplace

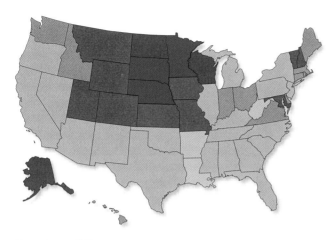

Percentage of Women in the Workforce, 2004

- 68% or more
- 66%–67.9%
- 62%–65.9%
- 60%–61.9%
- 55%–59.9%
- 54.9% or less

Source: U.S. Bureau of the Census, *Statistical Abstract of the United States, 2006.*

The number of women in the U.S. workplace is related to social change and social movements. The U.S. female workforce shot up during World War II. Once the soldiers returned home, however, a large percentage of those working women quit work to raise families. Owing in part to the women's rights movement, the U.S. has seen a resurgence of women entering the workforce. This map shows the percentage of women in each state who are active in the labor force.

Interpret the Map

1. Relate strain theory, the women's movement, and increased female labor force participation.
2. How does your state compare with other states in terms of female employment? Describe.

might not occur. One or more *precipitating factors* must occur to galvanize people into action. On April 30, 1970, President Nixon ordered the invasion of the neutral country of Cambodia. This event was a show of force to the North Vietnamese government, with whom the U.S. government was negotiating to end the war. President Nixon's announcement precipitated student antiwar protests and a chain of events that led to a confrontation between students and the Ohio National Guard, mobilized by the governor of Ohio. The encounter left many injured and four Kent State University students dead (G. H. Lewis 1972).

Once these first four conditions exist, the only remaining step to the emergence of a social movement is the *mobilization of participants for action*. It is at this point that leaders become very important. In the 1960s, radical student leaders, such as Mario Savio and Jack Weinberg, led efforts to mount an antiwar protest movement that began at the University of California, Berkeley. And the invasion of Cambodia fanned the flames of a student-led protest in California as well as across the country. Massive demonstrations were part of the political furor, with more than 100,000 opponents of the Vietnam War marching on Washington, D.C. Hundreds of colleges were forced to close as a result of strikes by 1.5 million students.

The sixth determinant of a social movement is ineffective *social control*—efforts on the part of society (media, police, courts, community leaders, political officials) to prevent, minimize, or interrupt the momentum for a social movement. If appropriate techniques of social control are applied, a potential social movement may be prevented, even though the first five determinants are present. Sometimes, social controls can be applied after a panic, riot, or mob action has already started. At that point, control efforts may block the social movement, minimize its effects, or make matters worse. Social control efforts in the antiwar movement were ineffective, however, in part because heavy-handedness on the part of politicians and law enforcement officials only stimulated further protest that hastened the ending of the war. In addition to the students killed on the Kent State University campus, two African American students were killed during an antiwar protest at Jackson State University in Mississippi. Repressive tactics (beatings, jailings) were also inflicted on antiwar protestors at the 1968 Democratic National Convention.

What are the strengths and weaknesses of value-added theory? According to value-added theory, more than discontent and deprivation are necessary for a social movement to emerge. Beyond structural strains there must be structural conduciveness, generalized beliefs, precipitating factors, and the mobilization of participants for action. Critics reject the traditional view of social movements as irrational, spontaneous, and initially unstructured (Oberschall 1973; Tilly 1978; Opp 1989; Gamson 1990; Opp and Roehl 1990). According to other critics, the value-added approach works best for crowd behavior, but even in that type of collective behavior, all six factors are not always present (Goode 1992). Value-added theory has also been criticized for failing to consider the importance of resources—funds, people, abilities—for the emergence and success of social movements (Olzak and West 1991).

Resource Mobilization Theory

What is involved in resource mobilization theory? According to resource mobilization theory, a key to galvanizing people for collective action is the mobilization of resources. **Resource mobilization** is the process through which members of a social movement secure and use the resources needed to advance their cause. Resources encompass human skills such as leadership, organizational ability, and labor power as well as material goods such as money, property, and equipment (Cress and Snow 1996; McCarthy and Wolfson 1996; Freeman and Johnson 1999). The civil rights movement of the 1960s succeeded because of the commitment of African Americans and because people of other races, and college students in particular, contributed money, energy, and skills necessary to stage repeated protests. In contrast, the gay movement in the United States experienced dif-

ficulty partially because of a relative shortage of money, foot soldiers, and affluent supporters.

John Lofland (1979) writes of the 1970s "white-hot" mobilization efforts of the Unification Church ("Moonies"). Upon taking up residence in the United States in 1971, the Reverend Sun Myung Moon was shocked by the scant resource mobilization work that had been done. He set out to change the situation, establishing a set of long- and short-term goals for the movement, including new-member and fund-raising quotas. Elaborate publicity campaigns and events were staged, including speaking tours and rallies. An organizational structure was established and a national training center created. The trainees were subsequently sent across the country in evangelistic teams. The United States was divided into ten regions, each headed by a regional director. Each regional director supervised several state directors who were in charge of local center directors. These efforts required about $15 million to be generated each year, some of it coming from Korean and Japanese branches of the movement. (Allegations were made that some of the money came from the South Koreans and American CIA.) Many local groups made money from their own businesses, such as housecleaning services, gas stations, and restaurants. Between 1971 and 1974, membership increased from 500 to 2,000. (For more detail on the current resources of the Unification Church, see Chapter 15.)

The case of the Unification Church makes a point central to resource mobilization theory: the collective action is organizationally based and led by rational people who calculate the likelihood of achieving their ends (A. D. Morris 1981, 1984). Resource mobilization advocates contend that although other theories of social

A human need to belong is a prime motivator for joining social movements such as represented by Reverend Sun Myung Moon's Unification Church. What other motivations might cause people to join a social movement?

Ahn Young-Joon/AP Photos

movements depict organization as something that emerges as the movement develops, resource mobilization theory sees preexisting organizational structure (and associated resources) as central to launching a social movement.

Several scholars have emphasized the role of "outsiders" in the creation of modern social movements (McCarthy and Zald 1977; Zald and McCarthy 1987; Burstein 1991). Outsiders may be volunteers who care about the plight of a category of people, or they may be professionals knowledgeable about the organization and management of social movements. For example, white students and adults from the North brought commitment and organizational skills to the civil rights movement in the 1960s.

How has resource mobilization theory been criticized?
Although critics praise resource mobilization theory for shedding light on the importance of resources in mounting and sustaining social movements, they fault it for deemphasizing the necessary social discontent and strain needed for a social movement to emerge. Resources are of little use if people are not sufficiently dissatisfied with current conditions (Jenkins 1983; Klandermans 1984). Although recognizing the importance of organization and planning, some social

movement theorists foresee a danger in dismissing the important role of spontaneity in the emergence and development of social movements. Social movement theorists, critics write, must not lose sight of the emotional factors that lead people to join a movement whose likelihood of success is small, and they must also consider the ways in which organization leaders are affected by the unpredictable, spontaneous actions of the people involved (Killian 1984; Opp 1988; Rule 1989; A. Scott 1995).

The Future Direction of Social Movement Theory

The three theories of social movements just discussed are more complementary than they are mutually exclusive. Relative deprivation theory emphasizes discontent from a social-psychological, or micro, viewpoint. Value-added theory implies discontent (within the concept of structural strain) but focuses on the operation of several factors at the macro level of analysis. Value-added theory, also at the macro level, implies the need for resources through its mobilization factor, but resource mobilization theory is needed to spell out this contributing factor to the rise of social movements.

TABLE 18.3

Major Forms of Collective Behavior

You can test your understanding of these forms of collective behavior by suggesting an additional example for each one.

Forms of Collective Behavior	Definition	Example
Rumor	• A widely circulating piece of information that is not verified as being true or false	• Continuously repeated prediction that airplanes would crash on a massive scale on January 1, 2000
Urban legend	• A moralistic tale that the teller swears happened to someone he or she knows or to an acquaintance of a friend or family member	• Fierce alligators in New York City's sewer system
Fad	• An unusual behavior pattern that spreads rapidly and disappears quickly	• Swing dancing
Fashion	• A widely accepted behavior pattern that changes periodically	• Wearing Nike shoes
Crowd	• A temporary collection of people who share an immediate interest	• New Year's celebrants at Times Square in New York City
Mob	• An emotional crowd ready to use violence for a specific purpose	• Lynch mob
Riot	• An episode of largely random destruction and violence carried out by a crowd	• Destructive behavior following the acquittal of police officers who were filmed using extreme force against Rodney King
Social movement	• Movement whose goal is to promote or prevent social change	• Civil rights movement

Social movement theory in the future will likely encompass both preexisting structure and spontaneity, both rationality and irrationality. It seems inevitable that the viewpoints of resource mobilization theory and other social movement theories will be considered by those on both sides of the debate.

A distinction is being made between "old" and "new" social movements (Eyerman 1992; Haferkamp and Smelser 1992; Buechler 2000). Old social movements, such as the American civil rights movements and the labor movement, are based more on the interests of the poor and the working class. They are centered on the struggle for power and control over economic conditions within the context of industrial capitalism. New social movements, such as the women's movement,

the ecology movement, the peace movement, the gay movement, and the animal rights movement, are not so embedded in economics (although women and gays seek economic equality) and opposition to capitalism (although environmentalists seek to moderate capitalism). Rather than expressing conflicts of industrial society and industrialization, new social movements rest more on conflicts unique to postindustrial society. The new social movements are fueled less by economic than by cultural conflicts, aimed at redefinitions of norms and values rather than at questions of economics and who gets what. New social movements are more global in focus, tend to center on quality-of-life issues, and are fueled by the interests of the middle and upper-middle class (Melucci 1980; Kriesi 1989; A. Scott 1995).

FEEDBACK

1. A _____ is the form of collective behavior that has the most structure, lasts the longest, and is the most likely to create social change.
2. Which of the following is an example of a reformative social movement?
   a. the French Revolution
   b. Zero Population Growth
   c. the Jesus People
   d. Women's Christian Temperance Union
3. _____ is felt when people compare themselves with others and believe that they should have as much as those others have.
4. _____ occur when newly raised hopes for a better life either are not satisfied at all or are not satisfied as rapidly as people expect them to be.
5. Once widespread discontent exists within a society, a social movement is bound to occur. T or F?
6. According to the _____ theory, several conditions are necessary and sufficient for the emergence of a social movement.
7. _____ is the process through which members of a social movement secure and use resources needed to press for social change.

Answers: 1. social movement; 2. d; 3. Relative deprivation; 4. Unfulfilled rising expectations; 5. F; 6. value-added; 7. Resource mobilization

SUMMARY

1. Social change refers to social structure alterations that have long-term and relatively important consequences. For most of the world, social change is accelerating at a dramatic rate.

2. Although predicting the precise nature of social change within a society is hazardous, several sources of social change are pretty well understood. Discovery, invention, and diffusion are the major social processes through which social change occurs. Important sources of social change are technology, population, the natural environment, conflict, and ideas. There are many interrelationships among these sources of social change.

3. Sociologists go beyond identifying sources of social change to developing theoretical perspectives of social change. The evolutionary, functionalist, and conflict perspectives view social change in very different ways. The cyclical perspective, short on explanatory power, envisions societies as changing through the process of growth and decay. Contemporary evolutionary theory emphasizes multilinear change (i.e., various societies may evolve in many diverse directions). The functionalist perspective depicts societies as relatively stable. Following a major change, these integrated systems seek a new equilibrium. According to the conflict perspective, societies are

REVIEW GUIDE

unstable systems that are constantly undergoing change. Attempts to combine the insights of the functionalist and conflict perspectives, although interesting, do not eliminate the need to retain separate perspectives.

4. With economic development comes modernization. This process involves numerous social and cultural changes. Consequences of modernization include rapid population growth, expanding urbanization, increased equality, greater political democracy, push for universal education, and a trend toward the nuclear family.

5. Even though modernization is a worldwide phenomenon, all modernizing societies are not on the road to social and cultural convergence. This is partly because modernization is filtered through existing social and cultural arrangements.

6. According to world-system theory, a nation's development hinges on its place in the world economy. So long as core nations control and exploit peripheral nations, the development gap between the former and the latter will increase.

7. Most areas of sociological study assume that social life is predictable, orderly, and recurrent. Collective behavior can be an important exception; much of this behavior is spontaneous, short term, and relatively unstructured. Some forms of collective behavior, though, are planned, structured, and enduring.

8. On the more unstructured end of the collective behavior continuum are dispersed collectivities—rumors, fads, crazes, panics, mass hysteria, and fashions. Even these forms of collective behavior are structured to some degree.

9. Rumors have a negative reputation, usually deserved, because they are composed of inaccurate, distorted, or false information. Rumors can be playful and harmless, or they can be damaging and hurtful.

10. Mass hysteria occurs when people accept false beliefs, then become anxious. Panics take place when there is a collective reaction to a real threat.

11. Unusual patterns of behavior that are adopted quickly, accepted enthusiastically, and disappear soon are called fads. Fashions are patterns of behavior that are widely approved but expected to change periodically. Fads and fashions are more central to modern than premodern societies.

12. Crowd behavior is fascinating to most people because it usually involves intense feelings and sometimes outrageous behavior. There are casual, conventional, expressive, and acting crowds.

13. Three quite different theories attempt to explain crowd behavior. Contagion theory stresses the irrationality of crowds and the buildup of intense emotion—a result of social interaction within a large collection of people. Emergent norm theory depicts crowd behavior as being more rational. According to emergent norm theory, crowd behavior is guided by norms that arise spontaneously. In convergence theory, crowds are formed by people with a common interest.

14. Social movements are closer to conventional social behavior than is crowd behavior because social movements are more permanent and more organized. Even so, they are not as permanent and structured as most aspects of social life.

15. There are three major explanations of social movements. According to relative deprivation theory, discontent and unfulfilled rising expectations create a breeding ground for social movements. Value-added theory outlines six necessary and sufficient conditions for the development of a social movement: structural conduciveness, structural strains, generalized beliefs, precipitating factors, mobilization of participants for action, and social control. According to the resource mobilization theory, resources are a crucial ingredient in the mounting of a social movement. Resources include human skills and economic assets, both of which often come from outside the movement.

LEARNING OBJECTIVES REVIEW

- Illustrate three social processes contributing to social change.
- Discuss, as sources of social change, the role of technology, population, the natural environment, conflict, and ideas.
- Compare and contrast the cyclical and evolutionary perspectives.
- Explain the why and how of social change within the context of the functionalist and conflict perspectives.
- Define modernization, and explain some of its major consequences.
- Define world-system theory, and summarize some of the consequences flowing from its description of the world economy.
- Describe social activities of dispersed collectivities.
- Describe the nature of a crowd, and identify the basic types of crowds.
- Contrast contagion theory, emergent norm theory, and convergence theory.
- Define the concept of social movement, and identify the primary types of social movements.
- Apply the concepts of relative deprivation and unfulfilled rising expectations to the emergence of social movements.
- Compare and contrast relative deprivation theory with value-added theory and resource mobilization theory.

CONCEPT REVIEW

Match the following concepts with the definitions listed below them.

____ a. diffusion
____ b. technology
____ c. invention
____ d. social change

____ e. collective behavior
____ f. mass hysteria
____ g. emergent norm theory
____ h. relative deprivation

____ i. panic
____ j. rumor
____ k. crowd

1. the social process that occurs when members of one group borrow social and cultural elements from another group
2. the creation of a new element by combining two or more already existing elements and creating new rules for their use as a unique combination
3. alterations in social structures that have long-term and relatively important consequences
4. the part of culture, including ideas and hardware, that is used to reach practical goals
5. a condition that exists when people compare themselves with others and believe that they should have as much as those others have
6. the relatively spontaneous and unstructured social behavior of people who are responding to similar stimuli

7. the form of collective behavior in which general social anxiety is created by acceptance of one or more false beliefs
8. a widely circulating story whose truth is questionable
9. the theory of crowd behavior that stresses the similarity between typical, everyday social behavior and crowd behavior
10. collective behavior that occurs when people react to a genuine threat in fearful, anxious, and often self-damaging ways
11. a temporary collection of people who share a common point of interest

CRITICAL-THINKING QUESTIONS

1. What do you think is the proper perspective on technological determinism? In the process of answering this question, be sure to incorporate the perspective of Karl Marx.

_____

2. Identify a major social change that has occurred in your lifetime. What do you think are the major sources of this change? Be careful to relate each source of change to the nature of the change itself.

_____

3. Is evolutionary theory compatible or incompatible with functionalism and conflict theory? Make clear the nature of all three perspectives on social change.

_____

4. Are you convinced that modernizing societies are becoming socially and culturally similar? Why or why not?

_____

5. Select a rumor, fad, craze, or fashion to demonstrate your understanding of a dispersed collectivity.

_____
_____

6. Sociologists take collective behavior to be a legitimate area of study on the grounds that it is more structured than it appears at first glance. Develop three examples of collective behavior of which you have some knowledge to support this claim by sociologists.

_____

7. You have participated in a crowd at some time. Think of an instance and identify it as one of the four types of crowds described in the text. Using your personal experience, provide examples of behavior within the crowd (yours or someone else's) that illustrate why you think it was a particular type of crowd.

_____

8. Which theory of social movements do you think best explains the women's movement? Link the theory to the nature of the women's movement.

_____
_____

MULTIPLE-CHOICE QUESTIONS

1. Alterations in social structures that have long-term and relatively important consequences are known as
   a. modernization.
   b. demographic change.
   c. cultural lag.
   d. social change.
   e. social movements.

2. Which of the following statements best describes the relationship between technology and social change?
   a. Social change cannot occur without technological developments.
   b. Technology does not necessarily lead to social change.
   c. The effects of technology are similar in all societies.
   d. Technology can be viewed as a single cause of social change.
   e. Technology has little effect on social change.

3. Max Weber, partially in response to Marx, emphasized the role of _____ in social change.
   a. technology
   b. revolution
   c. ideas
   d. inventions
   e. territorial discoveries

4. A major criticism of the cyclical perspective is that it
   a. assumes that society is inherently stable.
   b. assumes that society must pass through a set of progressive stages.
   c. fails to take into account the history of societies.
   d. describes rather than explains change.
   e. emphasizes inventions rather than discoveries.

5. According to the conflict theory, _____ is the major source of social instability in societies.
   a. technology
   b. revolution
   c. ideas
   d. politics
   e. scarcity of desired resources

6. _____ is the process involving all those social and cultural changes accompanying economic development.
   a. Modernization
   b. Postindustrialization
   c. Evolution
   d. Convergence
   e. Technological determinism

7. The development of social and cultural similarity among modernizing nations is known as
   a. convergence.
   b. world-system theory.
   c. congruence.
   d. consistency.
   e. modernization.

8. Which of the following statements *best* fits with world-system theory?
   a. The standard of living is much lower in core nations.
   b. Workers in peripheral nations experience a high standard of living.
   c. The economic gap between core and peripheral nations is increasing.
   d. Peripheral nations develop rapidly due to the aid given by core nations.
   e. The process of integration comes before the process of differentiation.

9. The study of collective behavior poses some particular problems for sociology because
   a. sociologists are accustomed to studying social interaction.
   b. collective behavior is difficult to define in sociological terms.
   c. sociologists are accustomed to studying structured behavior.
   d. sociological theories do not account for collective behavior.
   e. sociological research methods are inappropriate for the study of collective behavior.

10. When parents in a local school rebel against an AIDS victim attending school with their children, they are said to be exhibiting
    a. faddish behavior.
    b. mass hysteria.
    c. victimization.
    d. a public interest reaction.
    e. collective hypocrisy.

11. When a crowd of soccer fans abandons the guidelines for watching the match in order to attack the officials, a(an) _____ crowd has become a(an) _____ crowd.
    a. casual; acting
    b. conventional; acting
    c. expressive; acting
    d. conventional; expressive
    e. casual; expressive

12. An emotionally stimulated, disorderly crowd that is ready to use destructiveness and violence to achieve a specific purpose is known as a(an)
    a. mob.
    b. expressive crowd.
    c. unconventional crowd.
    d. revolution.
    e. interest group.

13. In his reformation of contagion theory, Herbert Blumer defined _____ as the stage in which people become increasingly aware and sensitive to one another and enter something akin to a hypnotic trance.

a. rising expectations

b. social contagion

c. milling

d. collective excitement

e. crowd decision

14. According to _____ theory, rules develop within crowds to tell people how they are expected to behave.

a. emergent norm

b. crowd decision

c. value-added

d. value-clarification

e. contagion

15. The role of "outsiders" in the creation of modern social movements is most prominent in _____ theory.

a. value-added

b. relative deprivation

c. emergent norm

d. contagion

e. resource mobilization

FEEDBACK REVIEW

True-False

1. According to the cyclical perspective on social change, societies rise and fall rather than move continuously toward improvement. T or F?

2. Classical evolutionary theory assumed that change leads to improvement. T or F?

3. *Mob* and *riot* are simply two terms for the same type of crowd. T or F?

4. Much crowd behavior is structured and rational. T or F?

Fill in the Blank

5. The theory of social change that assumes technology to be the primary cause of the nature of social structure is called _____.

6. Was it Karl Marx or Max Weber who most emphasized the role of ideas in creating social change? _____

7. According to the _____ perspective, social and cultural change occurs as a result of the struggles between groups representing different segments of a society.

8. Unusual patterns of behavior that spread rapidly, are embraced zealously, and disappear after a short time are called _____.

9. An _____ crowd has no purpose or direction beyond the unleashing of emotions.

10. A _____ is the form of collective behavior that has the most structure, lasts the longest, and is the most likely to create social change.

Multiple Choice

11. Which of the following is *not* one of the major consequences of modernization discussed in the text?

a. the social and cultural convergence of modernizing countries

b. increased urbanization

c. greater equality

d. more political democracy

e. widespread development of the nuclear family

12. Some individuals at a lynching do not participate or give verbal support, but do not attempt to stop it. Which of the following theories of crowd behavior best explains this response?

a. contagion theory

b. crowd decision theory

c. emergent norm theory

d. casual crowd theory

GRAPHIC REVIEW

Figure 18.1 shows the varying speeds with which Americans have adopted certain important technologies. Answer these questions to make sure you understand the information presented.

1. How would you explain the difference in adoption speed for the telephone and the Internet?

_____

_____

_____

2. Discuss some social implications of the increasing use of the Internet.

_____

_____

_____

ANSWER KEY

Concept Review

a. 1
b. 4
c. 2
d. 3
e. 6
f. 7
g. 9
h. 5
i. 10
j. 8
k. 11

Multiple Choice

1. d
2. b
3. c

4. d
5. e
6. a
7. a
8. c
9. c
10. b
11. b
12. a
13. c
14. a
15. e

Feedback Review

1. T
2. T
3. F

4. T
5. technological determinism
6. Max Weber
7. conflict
8. fads
9. expressive
10. social movement
11. a
12. c

Glossary

Absolute poverty The absence of enough money to secure life's necessities.

Achieved status A status within a social structure occupied because of an individual's efforts.

Age cohorts Persons born during the same time period in a particular population.

Ageism A set of beliefs, norms, and values used to justify age-based prejudice and discrimination.

Age-specific death rate The number of deaths per 1,000 persons in a specific age-group.

Age-specific fertility The number of live births per 1,000 women in a specific age-group.

Age stratification The unequal distribution of scarce desirables based on chronological age.

Age structure The distribution of people of different ages within a society.

Agricultural society A society whose subsistence relies primarily on the cultivation of crops with plows drawn by animals.

Alternative movement The type of social movement that seeks only limited changes in individuals.

Anomie A social condition in which norms are weak, conflicting, or absent.

Anticipatory socialization The process of preparing oneself for learning new norms, values, attitudes, and behaviors.

Ascribed status A status within a social structure that is not earned or chosen, but is assigned.

Assimilation The integration of a social or ethnic minority into a society.

Authoritarianism The type of political system controlled by nonelected rulers who generally permit some degree of individual freedom.

Authoritarian personality A personality characterized by excessive conformity; submissiveness to authority figures; inflexibility; repression of impulses, desires, and ideas; fearfulness; and arrogance toward persons or groups thought to be inferior.

Authority Power accepted as legitimate by those subjected to it.

Beliefs Ideas concerning the nature of reality.

Bilateral descent The familial arrangement in which inheritance and descent are passed equally through both parents.

Biological determinism The attribution of behavioral differences to inherited physical characteristics.

Blended family A family formed when at least one of the marriage partners has been married before and has one or more children from a previous marriage.

Boomburgs Places of more than 100,000 residents that are not the largest cities in their metropolitan areas and have experienced double-digit population growth in recent decades.

Bourgeoisie Members of a society who own the means for producing wealth.

Bureaucracies Formal organizations based on rationality and efficiency.

Capitalism An economic system founded on the sanctity of private property and the right of individuals to profit from their labor.

Case study A thorough, recorded investigation of a small group, incident, or community.

Caste system The type of stratification structure in which there is no social mobility because social status is inherited and cannot subsequently be changed.

Causation The idea that events occur in a predictable, non-random way and that one event leads to another.

Central-city dilemma The concentration of a large population in need of public services, but without the tax-generated money to provide them.

Charismatic authority Legitimate power based on an individual's personal characteristics.

Charter schools Publicly funded schools that are operated like private schools by public schoolteachers and administrators.

Church A life-encompassing religious organization to which all members of a society belong.

City A dense and permanent concentration of people living in a limited geographical area who gain their living primarily through nonagricultural activities.

Civil religion A public religion that expresses a strong tie between a deity and a culture.

Class conflict The conflict between those controlling the means for producing wealth and those laboring for them.

Class consciousness A sense of identification with the goals and interests of the members of one's own social class.

Closed-ended questions Questions a person must answer by choosing from a limited, predetermined set of responses.

Coercion Social interaction in which individuals or groups are forced to give in to the will of other individuals or groups.

Cognition The process of thinking, knowing, or mentally processing information.

Cognitive ability The capacity for thinking abstractly.

Cohabitation A marriage like living arrangement without the legal obligations and responsibilities of formal marriage.

Collective behavior The relatively spontaneous and unstructured social behavior of people who are responding to similar stimuli.

Combined statistical area (CSA) Two or more adjacent metropolitan statistical area (MSAs) or a MetroSA and one or more MicroSAs.

Communitarian capitalism The type of capitalism that emphasizes the interests of employees, customers, and society.

Community A concentration of people who live within a limited geographical area and whose social relationships fulfill major social and economic needs.

Competition A social process that provides rewards to people based on how their performance compares with that of others.

Concentric zone theory A description of the process of urban growth emphasizing circular areas that develop from the central city outward.

Conflict A form of social interaction in which individuals or groups work against one another to obtain a larger share of the valuables.

Conflict theory The theoretical perspective that emphasizes conflict, competition, change, and constraint within a society.

Conformity Behavior matching group expectations.

Conglomerate A network of unrelated businesses operating under a single corporate umbrella.

Conspicuous consumption The consumption of goods and services to display one's wealth to others.

Contagion theory The theory of crowd behavior that emphasizes the irrationality of crowds that is created by participants stimulating one another to higher and higher levels of emotional intensity.

Contingent employees Individuals hired on a part-time, short-term, or contract basis.

Continuity theory Presumes that most aging people maintain consistency with their past lives and use their life experiences to intentionally continue to develop in self-determined channels.

Control group The group in an experiment that is not exposed to the experimental variable.

Control theory The idea that conformity to social norms depends on the presence of a strong bond between individuals and society.

Convergence theory The theory that crowds are formed by people who deliberately congregate with others whom they know to be like-minded.

Cooperation A form of social interaction in which individuals or groups combine their efforts to reach some common goal.

Cooperative learning A nonbureaucratic classroom structure in which students study in groups, with teachers acting as guides rather than as the controlling agents determining all activities.

Core tier The tier of the occupational structure composed of large firms dominating their industries.

Corporate welfare Economic benefits government regularly gives to corporations.

Corporation An organization owned by shareholders who have limited liability and limited control over organizational affairs.

Correlation A statistical measure in which a change in one variable is associated with a change in another variable.

Counterculture A subculture that is deliberately and consciously opposed to certain central aspects of the dominant culture.

Craze A type of fad that can have serious consequences for its adopters.

Credentialism The idea that credentials (educational degrees) are unnecessarily required for many jobs.

Crime Acts in violation of the law.

Criminal justice system A system for controlling crime; comprised of police, courts, and corrections.

Crowd A temporary collection of people who share an immediate common interest.

Crude birth rate The annual number of live births per 1,000 members of a population.

Crude death rate The annual number of deaths per 1,000 members of a population.

Cult A religious organization whose characteristics are not drawn from existing religious traditions within a society.

Cultural bias In regard to intelligence tests, the unfair measurement of the cognitive abilities of people in some social categories.

Cultural lag Any situation in which disequilibrium is caused by one segment of a society's failing to change at the same rate as an interrelated segment.

Cultural particulars The widely varying, often distinctive ways societies demonstrate cultural universals.

Cultural relativism The idea that any given aspect of a particular culture should be evaluated in relation to its place within the larger cultural context of which it is a part rather than according to some alleged universal standard that is applied across all cultures.

Cultural transmission theory The theory that deviance is part of a subculture transmitted through socialization.

Cultural universals General cultural traits thought to exist in all known cultures.

Culture A people's way of life—consisting of material objects as well as the patterns of thinking, feeling, and behaving—that is passed from generation to generation among members of a society.

De facto subjugation Subjugation based on common, everyday social practices.

De jure subjugation Subjugation based on the law.

Democracy The type of political system in which elected officials are held responsible for fulfilling the goals of the majority of the electorate.

Democratic control The form of control in which authority is split evenly between husband and wife.

Demographic transition The process by which a population gradually moves from high birth rates and death rates to low birth rates and death rates as a result of economic development.

Demography The scientific study of population.

Denomination One of several religious organizations that members of a society accept as legitimate.

Dependency ratio The proportion of persons in the dependent ages relative to those who are economically active.

Dependent variable A variable in which a change (or effect) can be observed.

Desocialization The process of relinquishing old norms, values, attitudes, and behaviors.

Deterrence Intimidation of members of society into compliance with requirements of the legal system.

Deviance Any behavior that departs from societal or group norms.

Deviant A person who violates one or more of society's most highly valued norms.

Differential association theory The theory that deviant behavior is learned—that the more individuals are exposed to people who break the law, the more likely they are to become criminals.

Diffusion The process by which members of one group borrow social and cultural elements from another group, resulting in change.

Direct institutionalized discrimination Organizational or community actions intended to deprive a racial or ethnic minority of its rights.

Discovery The social process of learning something or reinterpreting something.

Discrimination Unequal treatment of individuals based on their minority membership.

Dispersed collectivities Participants in the less-structured forms of collective behavior such as rumors, panics, and fads.

Divorce rate The number of divorces in a particular year for every 1,000 members of the total population.

Divorce ratio The number of divorced persons per 1,000 persons in the population divided by the number of persons who are married and living with their spouses.

Doubling time The number of years needed to double the size of a population.

Downsizing The process by which companies reduce the size of their full-time workforce.

Dramaturgy The symbolic interactionist approach that depicts social life as theater.

Drives Impulses to do something to reduce discomfort.

Dual labor market The existence of a split between core and peripheral segments of the economy and the division of the labor force into preferred and marginalized workers.

Dynamic equilibrium The assumption by functionalists that a society both changes and maintains most of its original structure over time.

Dysfunction A negative consequence of some element of a society.

Economic determinism The idea that the nature of a society is based on the society's economy.

Economy The institution designed for the production and distribution of goods and services.

Edge city A suburban unit specializing in a particular economic activity.

Educational equality When schooling produces the same results (achievement and attitudes) for lower-class and minority children as it does for other children.

Elitism The theory of power distribution that sees society in the control of a relatively few individuals and organizations.

Emergent norm theory The theory of crowd behavior that stresses the similarity between typical, everyday social behavior and crowd behavior.

Endogamy Marriage within one's own group as required by social norms.

Epidemiology The study of the distribution of diseases within a population.

Ethnic minority A group of people who are socially identified by their unique characteristics related to culture or nationality.

Ethnocentrism The tendency to judge others in relation to one's own cultural standards.

Ethnomethodology The study of the processes people develop and use in understanding the routine behaviors expected of themselves and others in everyday life.

Euthanasia Deliberate termination of the life of a person suffering from a painful and incurable illness.

Exogamy Mate selection norms requiring individuals to marry someone outside their kind.

Experiment A laboratory procedure that attempts to eliminate all possible contaminating influences on the variables being studied.

Experimental group The group in an experiment exposed to the experimental variable.

Exponential growth When the absolute growth of a population that occurs within a given time period becomes part of the base for the growth rate in the next time period.

Extended family A family consisting of two or more adult generations that share a common household and economic resources.

Fads Unusual patterns of behavior that spread rapidly, are embraced zealously, and disappear after a short time.

False consciousness Acceptance of a system that works against one's own interests.

Family A group of people related by marriage, blood, or adoption.

Family of orientation The family into which an individual is born.

Family of procreation The family group established upon marriage.

Family planning The voluntary use of population control methods.

Family resiliency The capacity of the family to emerge from crises as stronger and more resourceful.

Fashions Behavior patterns that evolve over time and are widely approved, but are expected to change periodically.

Fecundity The maximum rate at which women can potentially produce children.

Feminism A social movement aimed at the achievement of sexual equality—socially, legally, politically, and economically.

Feminist theory A theoretical perspective that links the lives of women and men to the structure of gender relationships within society.

Feminization of poverty The trend toward more and more of the poor in the United States being women and children.

Fertility A measurement of the number of children born to a woman or to a population of women.

Fertility rate The annual number of live births per 1,000 women ages fifteen to forty-four.

Field research A research approach for studying aspects of social life that cannot be measured quantitatively and that are best understood within a natural setting.

Folk society A society that rests on tradition, cultural and social consensus, family, personal ties, little division of labor, and an emphasis on the sacred.

Folkways Norms without moral overtones.

Formal demography The branch of demography that deals with gathering, collating, analyzing, and presenting population data.

Formal organization A social structure deliberately created for the achievement of one or more goals.

Formal sanctions Rewards and punishments that may be given only by officially designated persons.

For-profit schools Schools run by private companies on government funds.

Functionalism The theoretical perspective that emphasizes the contribution (functions) made by each part of a society.

Fundamentalism The rejection of secularization and the adherence to traditional religious beliefs, rituals, and doctrines.

Game stage According to Mead, the stage of development during which children learn to consider the roles of several people at the same time.

Gemeinschaft Tönnies's term for the type of society based on tradition, kinship, and intimate social relationships.

Gender The expectations and behaviors associated with a sex category within a society.

Gender identity An awareness of being masculine or feminine, based on culture.

Gender roles Culturally based expectations associated with each sex.

Gender socialization The social process in which boys learn to act the way society assumes boys will act and girls learn to act in ways society expects of them.

Generalized other An integrated conception of the norms, values, and beliefs of one's community or society.

Genocide Politically motivated mass murder of most or all of a targeted population.

Gentrification The development of low-income areas by middle-class home buyers, landlords, and professional developers.

Geronticide The socially accepted killing of the elderly within a society.

Gerontocracy The type of society in which the elderly hold the greatest share of scarce desirables.

Gesellschaft Tönnies's term for the type of society characterized by weak family ties, competition, and impersonal social relationships.

Gesture Facial expression, body movement, or posture carrying culturally defined and shared symbolic meanings.

Goal displacement The situation that occurs when organizational rules and regulations become more important than organizational goals.

Government The political structure that rules a nation.

Gross migration rate The annual number of persons per 1,000 members of a population who enter or leave a geographical area.

Group A number of people who are in contact with one another; share some ways of thinking, feeling, and behaving; take one another's behavior into account; and have one or more interests or goals in common.

Group marriage The form of marriage in which two or more men are married to two or more women at the same time.

Groupthink A situation in which pressures toward uniformity discourage members of a group from expressing their reservations about group decisions.

Hate crime A criminal act motivated by prejudice.

Health-care system The combination of professional services, organizations, training academies, and technological resources that are committed to the treatment, management, and prevention of disease.

Health maintenance organizations (HMOs) Health plans in which members pay set fees on a regular basis and in return receive all necessary health care at little or no additional cost.

Heterogamy Marriage between people with differing social characteristics.

Hidden curriculum The educational curriculum that transmits to children a variety of nonacademic values, norms, beliefs, and attitudes.

Hidden unemployment Unemployment that is not measured because of discouraged workers who have stopped looking for work or part-time workers who would prefer full-time jobs.

Homogamy The tendency to marry someone similar to oneself based on personal preference.

Horizontal mobility A change from one occupation to another at the same general status level.

Horticultural society A society that solves the subsistence problem primarily through the domestication of plants.

Hospices Organizations designed to provide support for the dying and their families.

Hospitals Organizations providing specialized medical services to a variety of patients, including some who continue to live in their homes and others who will spend a prolonged period in the facility.

Humanist sociology The theoretical perspective that places human needs and goals at the center of sociology.

Hunting and gathering society A society that solves the subsistence problem through hunting animals and gathering edible fruits and vegetables.

Hypothesis A tentative, testable statement of a relationship between particular variables.

Hypothesis of linguistic relativity The idea that one's perception of reality is based on language.

"I" The spontaneous and unpredictable part of the self.

Ideal culture Cultural guidelines publicly embraced by members of a society.

Ideal-type method A method that involves isolating the most basic characteristics of some social entity.

Ideology A set of ideas used to justify and defend the interests and actions of those in power in a society.

Imitation stage The developmental stage during which, according to Mead, a child imitates the physical and verbal behavior of a significant other without comprehending the meaning of what is being imitated.

Incarceration The removal of criminals from society.

Income The amount of money received (within a given time period) by an individual or group.

Independent variable A variable that causes something to happen.

Indirect institutionalized discrimination Unintentional behavior that negatively affects a racial or ethnic minority.

Individualistic capitalism The type of capitalism founded on the principles of self-interest, the free market, profit maximization, and the highest return possible on stockholder investment.

Industrial society A society whose subsistence is based primarily on the application of science and technology to the production of goods and services.

Infant mortality rate The number of deaths among infants under one year of age per 1,000 live births.

Informal organization A group in which personal relationships are guided by rules, rituals, and sentiments not provided for by the formal organization.

Informal sanctions Rewards and punishments that may be applied by most members of a group.

In-group A group toward which one feels intense identification and loyalty.

Instincts Genetically inherited, complex patterns of behavior that always appear among members of a particular species under appropriate environmental conditions.

Institutionalized discrimination Discrimination that results from unfair practices that are part of the structure of society and have grown out of traditionally accepted behaviors.

Integrative curriculum An approach to education based on student–teacher collaboration.

Interest group A group organized to achieve some specific shared goal by influencing political decision making.

Intergenerational mobility Social mobility that occurs from one generation to the next.

Interlocking directorates Members of corporations sitting on one another's boards of directors.

Interorganizational relationship A pattern of interaction among authorized representatives of two or more formally independent organizations.

Intervening variable A variable that influences the relationship between an independent variable and a dependent variable.

Interview A set of questions asked by a trained interviewer.

Intragenerational mobility Social mobility that occurs from one generation to the next.

Invention The creation of a new element by combining two or more already existing elements and creating new rules for their use as a unique combination.

Invisible religion A private religion that is substituted for formal religious organizations, practices, and beliefs in a secular society.

Iron law of oligarchy The principle that power tends to become concentrated in the hands of a few members of any organized group.

Juvenile crime Violations of the law committed by those under eighteen years of age.

Labeling theory The theory that deviance exists when some members of a group or society label others as deviants.

Latent function An unintended and unrecognized consequence of some element of a society.

Laws Norms that are formally defined and enforced by officials.

Liberation sociology The theoretical approach to sociology that seeks to replace human oppression with greater democracy and social justice.

Life chances The likelihood of securing the "good things in life," such as housing, education, good health, and food.

Life expectancy The average number of years that persons in a given population born at a particular time can be expected to live.

Life span The most advanced age to which humans can survive.

Looking-glass self One's self-concept based on perceptions of others' judgments.

Macrosociology The level of analysis that focuses on relationships among social structures without reference to the interaction of the people involved.

Magnet schools Public schools that attempt to achieve excellence by specializing in a certain area.

Managed care Any health-care system based on cost containment, control of medical decisions, and regulation of treatments.

Manifest function An intended and recognized consequence of some element of a society.

Marriage A legal union based on mutual rights and obligations.

Marriage rate The number of marriages per year for every 1,000 members of a population.

Mass hysteria The form of collective behavior in which general social anxiety is created by acceptance of one or more false beliefs.

Mass media Those means of communication that reach large heterogeneous audiences without any personal interaction between the senders and the receivers of messages.

Master statuses Statuses that affect most other aspects of a person's life.

Matching Process in which participants in an experiment are matched in pairs according to all factors thought to affect the relationship being investigated.

Material culture The concrete, tangible objects within a culture.

Matriarchal control The form of control in which the oldest female living in a household has the authority.

Matrilineal descent The familial arrangement in which descent and inheritance are passed from the mother to her female descendants.

Matrilocal residence When a married couple is expected to live with the wife's parents.

"Me" The socialized part of the self.

Mechanical solidarity Social unity based on a consensus of values and norms, strong social pressure for conformity, and dependence on tradition and family.

Medical-industrial complex The network of physicians, medical schools, hospitals, medical equipment suppliers, pharmaceutical companies, and health insurance carriers.

Medicalization A process in which problems that were once considered matters of individual responsibility and choice are now considered a part of the domain of medicine and physicians.

Meritocracy A society in which social status is based on ability and achievement rather than background or parental status.

Metropolitan statistical area (MetroSA, or MSA) A grouping of counties that contains at least one city of 50,000 inhabitants or more or an "urbanized area" of at least 50,000 inhabitants and the surrounding integrated population.

Micropolitan statistical area (MicroSA) At least one urban concentration of no less than 10,000 but less than 50,000 population, plus adjoining territory that is highly integrated with the core.

Microsociology The level of analysis concerned with the study of people as they interact in daily life.

Migration Movement of people from one geographical area to another for the purpose of establishing a new residence.

Minority A people who possess some distinctive physical or cultural characteristics, are dominated by the majority, and are denied equal treatment.

Mob An emotionally stimulated, disorderly crowd that is ready to use destructiveness and violence to achieve a specific purpose.

Modernization The great number of social changes accompanying economic development.

Monogamy The form of marriage in which one man is married to only one woman at a time.

Monopoly A single company controlling a particular market.

Morbidity Rates of disease and illness in a population.

Mores Norms that have great moral significance and that should be followed by the members of a society.

Mortality Deaths within a population.

Multiculturalism An approach to education in which the curriculum accents the viewpoints, experiences, and contributions of women and diverse ethnic and racial minorities.

Multinational corporations Firms in highly industrialized societies with operating facilities throughout the world.

Multiple causation The idea that an event occurs as a result of several factors operating in combination.

Multiple-nuclei theory A description of the process of urban growth that emphasizes specific geographical and historical influences on areas within cities.

Nationalism In a nation-state, a people's commitment to a common destiny based on recognition of a common past and a vision of a shared future.

Nation-state The political entity that holds authority over a specified territory.

Natural increase Population growth due to an excess of births over deaths.

Negative correlation A statistical measure in which the independent and dependent variables change in opposite directions.

Negative deviance Behavior that underconforms to accepted norms.

Neolocal residence When a newly married couple establishes a residence separate from either of their parents.

Net migration rate The annual increase or decrease per 1,000 members of a population resulting from people leaving and entering the population.

Norms Rules defining appropriate and inappropriate behaviors.

Nuclear family The smallest group of individuals (mother, father, and children) that can be called a family.

Objectivity The principle stating that scientists are expected to prevent their personal biases from influencing the interpretation of their results.

Obligations Roles informing individuals of the behavior others expect from them.

Occupational sex segregation The concentration of men and women in different occupations.

Oligopoly A combination of companies controlling a market.

Open classroom A nonbureaucratic approach to education based on democracy, flexibility, and noncompetitiveness.

Open class system The type of stratification structure in which an individual's social status is based on merit and individual effort.

Open-ended questions Questions a person must answer in his or her own words.

Operational definition A definition of an abstract concept in terms of simpler, observable procedures.

Organic-adaptive systems Organizations based on rapid response to change rather than on the continuing implementation of established administrative principles.

Organic solidarity Social unity based on a complex of highly specialized roles that makes members of a society dependent on one another.

Out-group A group toward which one feels opposition, antagonism, or competition.

Overurbanization A situation in which a city cannot supply adequate jobs and housing for its inhabitants.

Panic Collective behavior that occurs when people react to a genuine threat in fearful, anxious, and often self-damaging ways.

Participant observation The type of field research technique in which a researcher becomes a member of the group being studied.

Pastoral societies Populations that depend primarily on raising and caring for domesticated animals for food.

Patriarchy The pattern in which the oldest man living in a household has authority over the rest of the family members. Also called *patriarchal control*.

Patrilineal descent The familial arrangement in which descent and inheritance are passed from the father to the male descendants.

Patrilocal residence When a married couple is expected to live with the husband's parents.

Peer group A group composed of individuals of roughly the same age and interests.

Peripheral theory Theory of city growth that emphasizes the growth of suburbs and edge cities around and away from central cities.

Peripheral tier The tier of the occupational structure composed of smaller, less profitable firms.

Personality The relatively organized complex of attitudes, beliefs, values, and behaviors associated with an individual.

Physician-hospital organizations (PHOs) Joint ventures between hospitals and organized groups of physicians to reach predetermined business objectives.

Play stage According to Mead, the stage of development during which children take on the roles of individuals, one at a time.

Pluralism The theory of power distribution that sees decision making as the result of competition, bargaining, and compromise among diverse special interest groups.

Political action committees (PACs) Organizations established by interest groups for the purpose of raising and distributing funds to selected political candidates.

Political institution The institution within which political power is supposed to be lodged and exercised.

Polyandry The form of marriage in which one woman is married to two or more men at the same time.

Polygamy Marriage of a male or female to multiple partners of the other sex.

Polygyny The form of marriage in which one man is married to two or more women at the same time.

Population All those people with the characteristics a researcher wants to study within the context of a particular research question.

Population control The conscious attempt to regulate population size through national birth control programs.

Population momentum When a population continues to grow, regardless of a recent drop in the birth rate to zero, because of the existing population base.

Population pyramids Graphic representations of the age and sex compositions of populations.

Positive correlation A statistical measure in which the independent and dependent variables change in the same direction.

Positive deviance Behavior that overconforms to social expectations.

Positivism The use of observation, experimentation, and other methods of the physical sciences in the study of social life.

Power The ability to control the behavior of others, even against their will.

Power elite A unified coalition of top military, corporate, and government leaders.

Preferred provider organizations (PPOs) Small groups of physicians providing a limited set of medical services to pre-enrolled patients at reduced prices.

Prejudice Widely held preconceptions of a group (minority or majority) and its individual members.

Presentation of self The ways that we, in a variety of social situations, attempt to create a favorable evaluation of ourselves in the minds of others.

Prestige Social recognition, respect, and admiration that a society attaches to a particular status.

Primary deviance Isolated acts of norm violation.

Primary group People who are emotionally close, know one another well, seek one another's company because they enjoy being together, and have a "we" feeling.

Primary sector That part of the economy producing goods from the natural environment.

Profane Nonsacred aspects of our lives.

Proletariat Members of a society who labor for the bourgeoisie at subsistence wages.

Protestant ethic A set of values, norms, beliefs, and attitudes stressing hard work, thrift, and self-discipline.

Public policy A broad course of governmental action expressed in specific laws, programs, and initiatives.

Qualitative variable A variable that consists of variation in kind rather than in number.

Quantitative variable A variable that can be measured and given a numerical value.

Questionnaire A written set of questions participants fill out by themselves.

Race A category of people who are alleged to share certain biologically inherited physical characteristics that are considered socially important within a society.

Racism Ideas that attempt to connect biological characteristics with innate racial superiority or inferiority.

Randomization Process by which subjects are assigned to the experimental or control group on a random or chance basis.

Random sample A sample selected on the basis of chance so that each member of a population has an equal opportunity of being selected.

Rationalism The solution of problems on the basis of logic, data, and planning rather than tradition and superstition.

Rationalization The tendency to use knowledge and impersonality in social relationships to gain increased control over society.

Rational-legal authority Authority based on rules and procedures associated with political offices.

Real culture Actual (subterranean) patterns of thinking, feeling, and behaving.

Recidivism A return to crime after prison.

Redemptive movement The type of social movement that focuses on changing individuals rather than society.

Reference group A group one uses to evaluate oneself, and from which one acquires attitudes, beliefs, values, and norms.

Reflexes Simple, biologically inherited, automatic reactions to physical stimuli.

Reformative movement The type of social movement that aims to bring partial change to a society.

Rehabilitation The resocialization of criminals into conformity with the legal code of society.

Relative deprivation A condition that exists when people compare themselves with others and believe that they should have as much as those others have.

Relative poverty The designation of poverty resulting from a comparison of the economic condition of those at the bottom of a society with that of other segments of the population.

Reliability The ability of a measurement technique to yield consistent results.

Religion A unified system of beliefs and practices relating to sacred things.

Religiosity The ways in which people express their religious interests and convictions.

Replacement level The birth rate at which a married couple replaces itself in the population.

Resocialization The process of learning to adopt new norms, values, attitudes, and behaviors.

Resource mobilization The process through which members of a social movement secure and utilize the resources needed to press for social change.

Retribution The public demand that criminals pay compensation equal to their offenses against society.

Revolution The type of social movement that involves the toppling of a political regime through violent means.

Revolutionary movement The type of social movement that attempts to change society totally.

Rights Roles informing individuals of the behavior that can be expected from others.

Riots Crowds that engage in episodes of destructiveness and violence without a mob's sense of common purpose.

Role conflict Conflict between the performance of a role in one status with the performance of a role in another status.

Role performance The actual conduct involved in putting a role into action.

Roles Culturally defined rights and obligations attached to social statuses indicating the behavior expected of individuals holding them.

Role strain Conflicting roles within a single status.

Role taking The process of mentally assuming the viewpoint of another individual and then responding to oneself from that imagined viewpoint.

Rumor A widely circulating story of questionable truth.

Sacred Entities that are set apart and given a special meaning that transcends immediate human existence.

Sample A limited number of cases drawn from a population.

Sanctions Rewards and punishments used to encourage conformity to norms.

Scapegoat A substitute object that serves as a convenient and less feared target on which to place the blame for one's own troubles, frustrations, failures, or sense of guilt.

Secondary analysis The use of information already collected by someone else for another purpose.

Secondary deviance Deviance as a lifestyle and personal identity.

Secondary group A group that is impersonal and task oriented and involves only a segment of the lives and personalities of its members.

Secondary relationships Relationships that are impersonal, that involve only limited aspects of others' personality, and that exist to accomplish a specific purpose beyond the relationship itself.

Secondary sector That part of the economy engaged in manufacturing goods from raw materials.

Sect A religious organization formed when members of an existing religious organization break away in an attempt to reform the "parent" group.

Sector theory A description of the process of urban growth emphasizing the importance of transportation routes.

Secularization The process whereby religion loses its influence over society as a whole.

Self-concept An image of oneself as an entity separate from other people.

Self-fulfilling prophecy The situation in which an expectation leads to behavior that causes the expectation to become a reality.

Sex The biological distinction between male and female.

Sexism A set of beliefs, norms, and values used to justify sexual inequality.

Sexual harassment The use of one's superior power in making unwelcome sexual advances.

Sick role A confluence of appropriate behavior patterns for people who are ill, serving to remove them from active involvement in everyday routines, give them special protection and privileges, and set the stage for their return to their normal social roles.

Significant others Those persons whose judgments are the most important to an individual's self-concept.

Social aggregate A number of people who happen to be physically located together.

Social category A group of persons who share a social characteristic.

Social change Alterations in social structures that have long-term and relatively important consequences.

Social class A segment of a population whose members have a relatively similar share of the desirable things and who share attitudes, values, norms, and an identifiable lifestyle.

Social control Means for promoting conformity to norms.

Social demography The branch of demography that studies population patterns within a social context.

Social dynamics The study of social change.

Social exchange A form of social interaction in which one person voluntarily does something for another, expecting a reward in return.

Social gerontology The scientific study of the social dimensions of aging.

Social interaction The process by which people influence one another's behavior as they relate.

Socialism An economic system founded on the belief that the means of production should be controlled by the people as a whole and that government should plan and control the economy.

Socialization The process of learning to participate in group life through the acquisition of culture.

Social mobility The movement of individuals or groups within a stratification structure.

Social movement A large number of people acting together with some degree of leadership and organization over a relatively long period with the goal of promoting or preventing social change.

Social network A web of social relationships that joins a person to other people and groups.

Social statics The study of stability and order in society.

Social stratification The creation of layers (strata) of people possessing unequal shares of scarce desirables.

Social structure Patterned, recurring social relationships.

Society People who live within defined territorial borders and participate in a common culture.

Sociobiology The study of the biological basis of human behavior.

Sociological imagination The set of mind that allows individuals to see the relationship between events in their personal lives and events in their society.

Sociology The scientific study of social structure.

Spirit of capitalism The preference to accumulate capital as a moral obligation (not as a necessity) to reinvest rather than to use for consumption.

Sport A set of competitive activities in which winners and losers are determined by physical performance within a set of established rules.

Sport subculture A group within the larger context of sport that has some of its own distinct roles, values, and norms.

Spurious correlation An apparent relationship between two variables that is actually produced by a third variable that affects both of the original two variables.

Stacking The assignment of players to less central positions on the basis of race or ethnicity.

Status The position that a person occupies within a social structure.

Status groups Groups composed of individuals who share a sense of status equality.

Status set All the statuses that an individual occupies at any particular time.

Stereotypes Ideas based on distortion, exaggeration, and oversimplification that are applied to all members of a social category.

Stigma An undesirable characteristic or label used by others to deny the deviant full social acceptance.

Strain theory The theory that deviance is most likely to occur when there is a discrepancy between culturally prescribed goals and legitimate (socially approved) means of obtaining them.

Stratified random sample A sample drawn from a population that has been divided into categories such as sex, race, or age. (The sample is selected at random from each category.)

Structural differentiation The division of one social structure into two or more social structures that operate more successfully separately than the one alone would under the new circumstances.

Structural mobility Mobility that occurs because of changes in the distribution of occupational opportunities.

Subculture A group that is part of the dominant culture but differs from it in some important respects.

Subjective approach A research method in which the aim is to understand some aspect of social reality through the study of the subjective interpretations of the participants themselves.

Suburbanization The process by which central cities lose population to the areas surrounding them.

Survey A research method in which people are asked to answer a series of questions.

Symbol Something that stands for, or represents, something else.

Symbolic interactionism The theoretical perspective that focuses on interaction among people based on mutually understood symbols.

Taboo A norm so important that its violation is considered repugnant.

Technology The combination of knowledge and hardware that is used to achieve practical goals.

Tertiary sector That part of the economy providing services.

Total fertility rate In a given population, the average number of children born to a woman during her lifetime.

Total institutions Places in which residents are separated from the rest of society and are controlled and manipulated by those in charge.

Totalitarianism The type of political system in which a ruler with absolute power attempts to control all phases of social life.

Tracking A process used in schools to place students in curricula consistent with the school's expectations of the student's eventual occupation.

Traditional authority Legitimate power rooted in custom.

Trained incapacity The situation that occurs when previous training prevents someone from adapting to new situations.

Underclass People typically unemployed who come from families with a history of unemployment for generations.

Unfulfilled rising expectations When newly raised hopes for a better life are either not satisfied at all or not satisfied as rapidly as people had expected.

Urban area A census block with a population density of no less than 1,000 persons per square mile.

Urban cluster An area with a population over 10,000 but under 50,000.

Urban ecology The study of the relationships between humans and their urban environments.

Urbanism The idea that urbanization involves a distinctive way of life.

Urbanization The process by which an increasingly large proportion of the world's population lives in urban areas.

Urbanized area A central city, or cities, of 50,000 or more and the surrounding densely inhabited territory.

Urban legends Moralistic tales passed along by people who swear the stories happened to someone they know, or to an acquaintance of a friend or family member.

Urban society A society in which social relationships are impersonal and contractual, kinship is de-emphasized, cultural and social consensus are not complete, the division of labor is complex, and secular concerns outweigh sacred ones.

Validity The ability of a measurement technique to actually measure what it is designed to measure.

Value-added theory A theory outlining six conditions that are necessary and sufficient for the emergence of a social movement.

Value-free research Research in which personal biases are not allowed to affect the research process and its outcomes.

Values Broad cultural principles that most people in a society consider desirable.

Variable Something that occurs in different degrees among individuals, groups, objects, and events.

Verifiability A principle of science by which any given piece of research can be duplicated (replicated) by other scientists.

Verstehen The method of understanding social behavior by putting oneself in the place of others.

Vertical mobility An upward or downward change in status level.

Victim discounting The act of seeing a crime as less serious if the victim is a member of a minority or a lower social class.

Vouchers Government money made available to families with school-age children for use toward public, private, or parochial schools.

War Armed conflict that occurs within a society or between nations.

Wealth All the economic resources possessed by an individual or group.

White-collar crime A crime committed by respectable and high-status people in the course of their occupations.

Working poor People employed in low-skill jobs with the lowest pay who do not earn enough to rise out of poverty.

World-system theory A theory of modernization that sees the pattern of a nation's development as largely dependent on that nation's location in the world economy.

Zero population growth When deaths are balanced by births so that a population does not increase.

References

Abbott, Andrew. *Department and Discipline: Chicago Sociology at One Hundred.* Chicago: University of Chicago Press, 1999.

Aberle, David. *The Peyote Religion among the Navaho.* Norman: University of Oklahoma Press, 1991.

Abramsky, Sasha. "Hard-Time Kids." *The American Prospect* (August 27, 2001):16–21.

Abu-Lughod, Janet L. *Rabat.* Princeton, NJ: Princeton University Press, 1980.

Abu-Lughod, Janet L. (ed.). *Sociology for the Twenty-First Century: Continuities and Cutting Edges* Chicago: University of Chicago Press, 2000.

Acosta, R. Vivian, and Linda Jean Carpenter. "Women in Intercollegiate Sport: A Longitudinal Study—Twenty Seven Year Update, 1977–2004." Department of Physical Education, Brooklyn College, 2004.

Adams, Bert N. *Kinship in an Urban Setting.* Chicago: Markham, 1968.

Adams, Bert N. "Isolation, Function, and Beyond: American Kinship in the 1960s." *Journal of Marriage and the Family* 32 (November 1970):575–597.

Adams, Rebecca G. "The Sociology of Jerry Garcia." *Garcia: Reflections* (1995):16–18.

Adams, Robert. *The Evolution of Urban Society.* Chicago: Aldine, 1965.

Addams, Jane. *Twenty Years at Hull House.* New York: New American Library, 1981.

Adelmann, P. K., T. C. Antonucci, S. E. Crohan, and L. M. Coleman. "Empty Nest, Cohort, and Employment in the Well-Being of Midlife Women." *Sex Roles* (1989):173–189.

Adelson, Joseph. "Why the Schools May Not Improve." *Commentary* 78 (October 1984):40–45.

Adler, Emily Stier, and Roger Clark. *How It's Done: An Invitation to Social Research,* 2nd ed. Belmont, CA: Wadsworth, 2003.

Adler, Jerry. "Spaced Out." *Newsweek* (January 13, 2003):48–51.

Adler, Patricia A., and Peter Adler. *Peer Power.* New Brunswick, NJ: Rutgers University Press, 1998.

Adorno, T. W., Else Frenkel-Brunswik, Daniel J. Levinson, and R. Nevitt Sanford. *The Authoritarian Personality.* New York: Harper, 1950.

Aftab, P., 2006. http://www.wiredsafety.net.

Aguirre, Adalberto, and Jonathan H. Turner. *American Ethnicity,* 2nd ed. New York: McGraw-Hill, 1997.

Aguirre, B. E., E. L. Quarantelli, and Jorge L. Mendoza. "The Collective Behavior of Fads: The Characteristics, Effects, and Career of Streaking." *American Sociological Review* 53 (1988):569–584.

Ahlburg, Dennis A., and Carol J. De Vita. "New Realities of the American Family." *Population Bulletin* 47 (August 1992):1–44.

Ahrens, Frank. "FCC Eases Media Ownership Rules." *Washington Post* (June 3, 2003):A1, A6, A7.

"Air Force Reports 54 Rapes, Assault." Associated Press, March 6, 2003.

Aizenman, N. C. "Influx of U.S. Inmates Slowing, Census Says." *The Washington Post* (September 27, 2007).

Alba, Richard D. "The Twilight of Americans of European Ancestry: The Case of Italians." *Ethnic and Racial Studies* 8 (January 1985):134–158.

Alba, Richard D. *The Transformation of White America.* New Haven, CT: Yale University Press, 1990.

Albert, Michel. *Capitalism vs. Capitalism.* New York: Four Walls Eight Windows, 1993.

Aldrich, Howard. *Organizations Evolving.* Thousand Oaks, CA: Sage, 1999.

Aldridge, Alan. *Religion in the Contemporary World.* Malden, MA: Blackwell, 2000.

Alexander, Jeffrey C., Bernard Giesen, Richard Munch, and Neil J. Smelser (eds.). *The Micro-Macro Link.* Berkeley: University of California Press, 1987.

Alfino, Mark, John S. Caputo, and Robin Wynyard. *McDonaldization Revisited.* Westport, CT: Praeger, 1998.

Alford, Robert R., and Roger Friedland. *Power of Theory.* Cambridge, UK: Cambridge University Press, 1985.

Allen, Jodi T. "A Nation of 'Haves' and 'Have-Nots.'" Washington, D.C.: Pew Research Center (September 13, 2007). http://pewresearch.org.pubs/593/haves-have-nots.

Allport, Gordon. *The Nature of Prejudice.* Reading, MA: Addison-Wesley, 1979.

Alter, Catherine, and Jerald Hage. *Organizations Working Together.* Newbury Park, CA: Sage, 1992.

Alter, Jonathan. "How America's Meanest Lobby Ran Out of Ammo." *Newsweek* (May 16, 1994):24–25.

Alterman, Eric. *What Liberal Media? The Truth about Bias and the News.* New York: Basic Books, 2003.

Altman, Daniel. "We Saved Your Job, But Gave You More Work." *New York Times* (October 29, 2002):G4.

Altman, Dennis. *AIDS in the Mind of America.* Garden City, NY: Anchor Books, 1986.

America 2000: An Educational Strategy. Washington, D.C.: U.S. Department of Education, 1991.

American Association of School Administrators. *Sex Equality in Educational Materials.* Executive Handbook Series, Vol. IV. Washington, D.C.: American Association of School Administrators, 1974.

American Jewish Year Book, 2006. New York: American Jewish Committee, 2006.

American Sociological Association. "Code of Ethics." Washington, D.C., 1997.

American Sociological Association. *The Health of Sociology,* Research and Development Department. Washington, D.C., 2007.

Amsden, Alice H. *The Rise of "the Rest": Challenges to the West from Late-Industrializing Economies.* Oxford, UK: Oxford University Press, 2003.

A Nation at Risk. The National Commission on Excellence in Education. Schaumburg, IL: Research USA Incorporated, 1989.

Andersen, Margaret L. *Thinking about Women,* 7th ed. Boston: Allyn & Bacon, 2005.

Andersen, Margaret L., and Patricia Hill Collins. *Race, Class, and Gender.* Belmont, CA: Wadsworth, 2007.

Anderson, Bernard E. "The Black Worker: Continuing Quest for Economic Parity." In Lee A. Daniels (ed.), *The State of Black America 2002.* New York: National Urban League, 2002, pp. 51–67.

Anderson, Craig A., and Brad J. Bushman. "The Effects of Media Violence on Society." *Science* 295 (29 March 2002):2377–2378.

Anderson, Curt. "Anti-Muslim Hate Crimes Soared Last Year, FBI Says." *The Philadelphia Inquirer* (November 26, 2002).

Anderson, Elijah, and Douglas S. Massey (eds.). *Problem of the Century: Racial Stratification in the United States.* New York: Russell Sage, 2001.

Anderson, Jack. *The Anderson Papers.* New York: Ballantine Books, 1974.

Anderson, T. B. "Widowhood as a Life Transition: Its Impact on Kinship Ties." *Journal of Marriage and the Family* 46 (February 1984):105–114.

Andrews, Edmund L. "Fed Acts to Rescue Financial Markets." *New York Times* (March 17, 2008).

Andrews, Lori B. *Black Power, White Blood.* New York: Pantheon Books, 1996.

Annandale, Ellen. *The Sociology of Health and Medicine: A Critical Introduction.* Malden, MA: Blackwell, 1998.

Antoun, Richard T. *Understanding Fundamentalism: Christian, Islamic, and Jewish Movements.* Thousand Oaks, CA: AltaMira Press, 2001.

Appel, Willa. *Cults in America.* New York: Holt, Rinehart and Winston, 1983.

Appiah, Kwame Anthony, and Amy Gutmann. *Color Conscious: The Political Morality of Race.* Princeton, NJ: Princeton University Press, 1998.

Appleby, Joyce. *Inheriting the Revolution: The First Generation of Americans.* Cambridge, MA: Harvard University Press, 2001.

Arenson, Karen W. "Applications to Colleges Are Breaking Records." *New York Times* (January 17, 2008).

Armour, Stephanie. "Faced with Less Time Off, Workers Take More." *USA Today* (October 29, 2002):1A.

Aron-Dine, Aviva. "New Data Show Income Concentration Jumped Again in 2005." Center on Budget and Policy Priorities, 2007, http://www.cbpp.org/3-29-07inc.htm.

Aronowitz, Robert A. *Making Sense of Illness.* New York: Cambridge University Press, 1998.

Aronowitz, Stanley, and Henry A. Giroux. *Education Still under Siege.* 2nd ed. Toronto: OISE Press, 1994.

Aronson, E., and A. Gonzalez. "Desegregation, Jigsaw, and the Mexican American Experience." In P. A. Katz and D. A. Taylor (eds.), *Eliminating Racism.* New York: Plenum Press, 1988, pp. 301–314.

Arrigo, Bruce A. *Social Justice/Criminal Justice: The Maturation of Critical Theory in Law, Crime, and Deviance.* Belmont, CA: Wadsworth, 1999.

Arrow, Kenneth Joseph, Steven N. Durlauf, and Samuel Bowles (eds.). *Meritocracy and Economic Inequality.* Princeton, NJ: Princeton University Press, 2000.

Asch, Solomon E. "Opinions and Social Pressure." *Scientific American* 193 (November 1955):31–35.

Ashford, Lori S. "New Population Policies: Advancing Women's Health & Rights." *Population Bulletin* 56 (March 2001):1–44.

Ashour, Ahmen Sakr. "The Contingency Model of Leadership Effectiveness: An Evaluation." *Organizational Behavior and Human Performance* 9 (June 1973):339–355.

Atchley, Robert C. *Social Forces and Aging.* 10th ed. Belmont, CA: Wadsworth, 2003.

Atchley, Robert C. *Continuity and Adaptation in Aging: Creating Positive Experiences.* Baltimore, MD: Johns Hopkins University Press, 2001.

Atchley, Robert C., and S. Miller. "Types of Elderly Couples." In T. H. Brubaker (ed.), *Family Relationships in Later Life.* Beverly Hills, CA: Sage, 1983, pp. 77–90.

Atkinson, Paul. "Ethnomethodology: A Critical Review." In W. Richard Scott and Judith Blake (eds.), *Annual Review of Sociology* 14 (1988):441–465.

Atkinson, Rick. "Germany's Health-Care System Such a Success It Needs Reform." *Washington Post* (June 22, 1994):A1.

Audi, Robert, and Nicholas Wolterstorff. *Religion in the Public Square: Debating Church and State.* Lanham, MD: Rowman & Littlefield, 1996.

Auguet, Roland. *Cruelty and Civilization.* London: George Allen and Unwin, Ltd., 1972.

Axtell, Roger E. *Gestures: The Do's and Taboos of Body Language around the World.* New York: Wiley, 1998.

Brubaker, T. H. "Family in Later Life: A Burgeoning Research Area." In A. Booth (ed.), *Contemporary Families.* Minneapolis: National Council on Family Relations, 1991, pp. 226–248.

Babbie, Earl R. *The Basics of Social Research.* 4th ed. Belmont, CA: Wadsworth, 2007.

Baca Zinn, Maxine, and D. Stanley Eitzen. "Economic Restructuring and Systems Inequality." In Margaret L. Andersen and Patricia Hill (eds.), *Race, Class, and Gender: An Anthology,* 6th ed. Belmont, CA: Wadsworth, 2006.

Bacon, Perry Jr. "Foes Use Obama's Muslim Ties to Fuel Rumors about Him." *Washington Post* (November 29, 2007).

Bagdikian, Ben. *The New Media Monopoly,* 10th ed. Boston: Beacon Press, 2004.

Bainbridge, William Sims. *The Sociology of Religious Movements.* New York: Routledge, 1997.

Baldassare, Mark. "Suburban Communities." *Annual Review of Sociology* 18 (1992):475–494.

Bales, Robert F. *Interaction Process Analysis.* Reading, MA: Addison-Wesley, 1950.

Ballantine, Jeanne H., and Joan Z. Spade (eds.). *Schools and Society: A Sociological Approach to Education,* 3rd ed. Thousand Oaks, CA: Sage, 2007.

Balrig, Flemming. *The Snow-White Image.* Oslo: Norwegian University Press, 1988.

Banner, Stuart. *The Death Penalty: An American History.* Cambridge, MA: Harvard University Press, 2003.

Banton, Michael. *Racial Theories.* New York: Cambridge University Press, 1998.

Bardes, Barbara A., Mack C. Shelley, and Steffen W. Schmidt. *American Government and Politics Start over Today: The Essentials,* 13th ed. Thousand Oaks, CA: Sage, 2007.

Bardo, John W., and John J. Hartman. *Urban Sociology.* Itasca, IL: Peacock, 1982.

Banerjee, Neela. "Poll Finds a Fluid Religious Life in U.S." *New York Times* (February 26, 2008).

Barker, Eileen. *The Making of a Moonie.* London: Basil Blackwell, 1984.

Barlett, Donald L., and James B. Steele. *America: Who Stole the Dream?* Kansas City, KA: Andrews and McMeel, 1996.

Barlett, Donald L., and James B. Steele. *The Great American Tax Dodge: How Spiraling Fraud and Avoidance Are Killing Fairness, Destroying the Income Tax, and Costing You.* Boston: Little, Brown and Company, 2002.

Barnes, Gill Gorell, et al. *Growing Up in Stepfamilies.* New York: Oxford University Press, 1998.

Barnes, Julian E. "The SAT Revolution." *U.S. News and World Report* (November 11, 2002).

Barnett, Ola W., Robin D. Perrin, and Cindy L. Miller-Perrin. *Family Violence Across the Lifespan,* 2nd ed. Thousand Oaks, CA: Sage, 2004.

Barr, C. "Pushing the Envelope: What Curriculum Integration Can Be." In E. Brazee and J. Capelluti (eds.), *Dissolving Boundaries: Toward an Integrative Curriculum.* Columbus, OH: National Middle School Association, 1995.

Barr, Stephen. "Women Make Gains in Parity." *Washington Post* (July 21, 1999):A19.

Barrett, Michael J. "The Case for More School Days." *The Atlantic Monthly* (November 1990):78–106.

Barrington, Linda. *Does a Rising Tide Lift All Boats? America's Full-Time Working Poor Reap Limited Gains in the New Economy.* New York: The Conference Board Inc., 2000.

Barron, James. "Fondly Recalling a Martian Invasion." *New York Times* (September 16, 1988):B1, B3.

Barron, Milton L. "Minority Group Characteristics of the Aged in American Society." *Journal of Gerontology* 8 (October 1953):477–482.

Barry, Herbert III, Margaret K. Bacon, and Irvin L. Child. "A Cross-Cultural Survey of Some Sex Differences in Socialization." *Journal of Abnormal and Social Psychology* 55 (November 1957):327–332.

Barry, Norman P. *Welfare,* 2nd ed. Minneapolis: University of Minnesota Press, 1999.

Baumohl, Bernard. "When Downsizing Becomes Dumbsizing." *Time* (March 15, 1993):55.

Baumol, William J., and Alan S. Blinder. *Economics Principles and Policy,* 10th ed. Belmont, CA: South-Western, 2007.

Baumol, William J., Alan S. Blinder, and Edward N. Wolff. *Downsizing in America.* New York: Russian Sage Foundation, 2005.

Baxandall, Rosalyn, and Elizabeth Ewen. *Picture Windows: How the Suburbs Happened.* New York: Basic Books, 2000.

Baxandall, Rosalyn, and Linda Gordon. "Second-Wave Soundings." *The Nation* (July 3, 2000):28–32.

Baxandall, Rosalyn, and Linda Gordon. *Dear Sisters.* New York: Basic Books, 2001.

Beals, Ralph L., Harry Hoijer, and Alan R. Beals. *An Introduction to Anthropology,* 5th ed. New York: Macmillan, 1977.

Bearman, Peter S., and Hannah Bruchner. "Opposite Sex Twins and Adolescent Same-Sex Attraction." *American Journal of Sociology* 107 (March 2002):1179–1205.

Becker, Howard S. *Art Worlds.* Berkeley: University of California Press, 1982.

Becker, Howard S. *Outsiders.* New York: Free Press, 1991.

Becker, Howard S., et al. *Boys in White.* Chicago: University of Chicago Press, 1961.

Beckford, James A. *Cult Controversies.* London: Tavistock, 1985.

Begley, Sharon. "Not Just a Pretty Face." *Newsweek* (November 1, 1993):63–67.

Begley, Sharon. "The 3–Million-Year-Old Man." *Newsweek* (April 11, 1994):84.

Begley, Sharon. "The Science Wars." *Newsweek* (April 21, 1997):54–57.

Behr, Peter. "Enron Skirted Taxes via Executive Pay Plan." *Washington Post* (February 14, 2003):E1, E3.

Bell, Daniel. "Equality and Equity." *The Seminar on Technology and Culture at MIT.* Boston: Massachusetts Institute of Technology, 1976.

Bell, Daniel. "Downfall of the Business Giants." *Dissent* (Summer 1993):316–323.

Bell, Daniel. *The Coming of Post-Industrial Society.* New York: Basic Books, 1999.

Bellah, Robert N. "Civil Religion in America." *Daedalus* 96 (Winter 1967):1–21.

Bellah, Robert N. *The Broken Covenant,* 2nd ed. Chicago: University of Chicago Press, 1992.

Bellah, Robert N., and Phillip E. Hammond. *Varieties of Civil Religion.* New York: Harper and Row, 1980.

Bellah, Robert N., et al. *The Good Society.* New York: Knopf, 1991.

Belton, Beth. "Despite Humming Economy, Workers Sweat Job Security." *USA Today* (March 2, 1999):A1.

Bendix, Reinhard. *Max Weber.* Garden City, NY: Doubleday, 1962.

Bennett, William J. "American Education: Making It Work." *Chronicle of Higher Education* 34 (May 4, 1988):29–41.

Bennis, Warren G. "Beyond Bureaucracy." *Transaction* 2 (July/August 1965):31–35.

Bennis, Warren G. "Response to Shariff: Beyond Bureaucracy Baiting." *Social Science Quarterly* 60 (June 1979):20–24.

Benokraitis, Nijole V. *Subtle Sexism.* Thousand Oaks, CA: Sage, 2004.

Benokraitis, Nijole V. *Marriages and Families: Changes, Choices and Constraints,* 6th ed. Upper Saddle River, NJ: Prentice Hall, 2007.

Benowitz, June Melby, and Yvonne M. Vowels. *Encyclopedia of American Women and Religion.* Santa Barbara, CA: ABC-CLIO Inc., 1998.

Bensman, Joseph, and Israel Gerver. "Crime and Punishment in the Factory: The Function of Deviancy in Maintaining the Social System." *American Sociological Review* 28 (August 1963):588–598.

Berger, Bennett M., Bruce Hacket, and R. Mervyn Millar. "The Communal Family." *The Family Coordinator* 21 (October 1972):419–427.

Berger, Joseph, M. Hamit Fisek, Cecilia L. Ridgeway, and Robert Z. Norman. "The Legitimization and Delegitimization of Power and Prestige Orders." *American Sociological Review* 63 (June 1998):379–405.

Berger, Peter L. *The Sacred Canopy: Elements of a Sociological Theory of Religion.* New York: Doubleday, 1990.

Berk, Richard A. *Collective Behavior.* Dubuque, IA: William C. Brown, 1974.

Berle, Adolph A. Jr., and Gardiner C. Means. *The Modern Corporation and Private Property.* New York: Harcourt Brace Jovanovich, 1968.

Berlin, Brent, and Paul Kay. *Basic Color Terms.* Berkeley: University of California Press, 1969.

Bernard, Thomas J. "Structure and Control: Reconsidering Hirschi's Concept of Commitment." *Justice Quarterly* 4 (1987):409–424.

Bernhardt, Annette D., et al. *Divergent Paths: Economic Mobility in the New American Labor Market.* New York: Russell Sage Foundation, 2001.

Bernstein, Aaron. "Inequality: How the Gap Between the Rich and Poor Hurts the Economy." *BusinessWeek* (August 15, 1994):78–83.

Berry, Brian J. L. "Internal Structure of the City." *Law and Contemporary Problems* 30 (Winter 1965):111–119.

Berry, Brian J. L., and John D. Kasadra. *Contemporary Urban Ecology.* New York: Macmillan, 1977.

Bertrand, Marianne, and Sendhil Mullainathan. "Are Emily and Brendan More Employable than Lakisha and Jamal? A Field Experiment on Labor Market Discrimination." Chicago: University of Chicago Graduate School of Business, Working Paper (November 18, 2002):3–28.

Bhalla, Surjit. *Imagine There's No Country: Poverty, Inequality, and Growth in the Era of Globalization.* Washington, D.C.: Institute for International Economics, 2002.

Bhargava, Deepak, and Joan Kuriansky. "Really Out of Line." *Washington Post* (September 15, 2002):B2.

Bianchi, Suzanne M., "America's Children: Mixed Prospects." *Population Bulletin* 45 (June 1990):3–41.

Bianchi, Suzanne M., and Lynne M. Casper. "American Families." *Population Bulletin* 55 (December 2000):1–44.

Bianchi, Suzanne M., and Daphne Spain. *Women, Work, and Family in America.* Washington, D.C.: Population Reference Bureau, 1996.

Bickman, Leonard, and Debra J. Rog (eds.). *Handbook of Applied Social Research Methods.* Thousand Oaks, CA: Sage, 1997.

Bielby, William T., and Denise D. Bielby. "Family Ties: Balancing Commitments to Work and Family in Dual Earner Households." *American Sociological Review* 54 (October 1989):776–789.

Bierstedt, Robert. "Sociology and General Education." In Charles H. Page (ed.), *Sociology and Contemporary Education.* New York: Random House, 1963, pp. 40–55.

Bigler, Rebecca, and Lynn Liber. "The Role of Attitudes and Interventions in Gender-Schematic Processing." *Child Development* 61 (1990):1440–1452.

Billson, Janet Mancini, and Bettina Huber. "Embarking Upon a Career with an Undergraduate Degree in Sociology." Washington, D.C.: American Sociological Association, 1993.

Bingham, Richard D., and William M. Bowen (eds.). *Beyond Edge Cities.* New York: Garland Publishing, 1997.

Birnbaum, Jeffrey. *The Money Men: The Real Story of Political Power in the USA.* New York: Crown, 2000.

Birnbaum, Jeffrey H., and Alan S. Murray. *Showdown at Gucci Gulch.* New York: Vintage, 1988.

Birren, James E., et al. *Handbook of the Psychology of Aging,* 6th ed. Amsterdam, Netherlands: Elsevier, 2005.

Black, Clifford. "Clinical Sociology and Criminal Justice Professions." *Free Inquiry in Creative Sociology* 12 (1984):117–120.

Black, Thomas R. *Doing Quantitative Research in the Social Sciences.* Thousand Oaks, CA: Sage, 2001.

Blackwell, John Reid. "Chance of Higher Cigarette Tax Dims." *Richmond Times Dispatch* (January 23, 2003).

"Black/White Relations in the United States." *The Gallup Poll Social Audit.* Princeton, NJ: The Gallup Organization, 1997.

Blank, Rebecca M., and Ron Haskins. *The New World of Welfare.* Washington, D.C.: Brookings Institution Press, 2001.

Blank, Rolf K., Roger E. LeVine, and Lauri Steel. "After 15 Years: Magnet Schools in Urban Education." In Bruce Fuller and Richard F. Elmore (eds.), *Who Wins, Who Loses?* New York: Teachers College Press, 1996, pp. 154–172.

Blau, Francine D., and Ronald G. Ehrenberg (eds.). *Gender and Family Issues in the Workplace.* New York: Russell Sage Foundation, 1997.

Blau, Peter M. *Exchange and Power in Social Life.* New Brunswick: Transaction Books, 1986.

Blau, Peter M., and Otis Dudley Duncan. *The American Occupational Structure.* New York: Wiley, 1967.

Blau, Peter M., and Marshall W. Meyer. *Bureaucracy in Modern Society,* 3rd ed. New York: Random House, 1987.

Blau, Peter M., and W. Richard Scott. *Formal Organizations.* London: Routledge and Kegan Paul, 1982.

Blee, Kathleen M., and Dwight B. Billings. *The Road to Poverty: The Making of Wealth and Hardship in Appalachia.* Cambridge, UK: Cambridge University Press, 1999.

Block, Jeanne H., Jack Block, and Per F. Gjerde. "The Personality of Children Prior to Divorce." *Child Development* 57 (1986):827–840.

Bloom, Allan. *The Closing of the American Mind.* New York: Simon & Schuster, 1987.

Blum, Deborah. *Love at Goon Park: Harry Harlow and the Science of Affection.* Cambridge, MA: Perseus Publishing, 2002.

Blum, Justin, and Jay Mathews. "Quality Uneven, Despite Popularity." *The Washington Post* (June 19, 2003):A1, A12, A13.

Blumer, Herbert. "Collective Behavior." In Alfred M. Lee (ed.), *Principles of Sociology,* 3rd ed. New York: Barnes and Noble, 1969a, pp. 65–121.

Blumer, Herbert. *Symbolic Interactionism.* Englewood Cliffs, NJ: Prentice Hall, 1969b.

Blumstein, Alfred, and Joel Wallman (eds.). *The Crime Drop in America,* 2nd ed. New York: Cambridge University Press, 2006.

Bonacich, Edna. "Advanced Capitalism and Black/White Race Relations in the United States: A Split Labor Market Interpretation." *American Sociological Review* 41 (February 1976):34–51.

Booth, Alan, and Ann C. Crouter (eds.). *Men in Families: When Do They Get Involved? What Difference Does It Make?* Mahwah, NJ: Erlbaum, 1998.

Booth, William. "1990's Social Reformer Now Toes the Line." *Washington Post* (July 12, 1994):A1, A4.

Boroditsky, Lera. "Does Language Shape Thought?: Mandarin and English Speakers' Conceptions of Time." *Cognitive Psychology* 43 (2001):1–22.

Boroughs, Don L. "Winter of Discontent." *U.S. News & World Report* (January 22, 1996):47–54.

Bottomore, Tom B. *Karl Marx.* New York: McGraw-Hill, 1963.

Bottomore, Tom B. *Political Sociology,* 2nd ed. Minneapolis: University of Minnesota Press, 1993.

Bouvier, Leon F., and Carol J. De Vita. "The Baby Boom— Entering Midlife." *Population Bulletin* 46 (November 1991):1–35.

Bowen, William G., and Derek Bok. *The Shape of the River: Long-Term Consequences of Considering Race in College and University Admissions.* Princeton, NJ: Princeton University Press, 2000.

Bowles, Samuel, and Herbert Gintis. *Schooling in Capitalist America.* New York: Basic Books, 1976.

Box, Steven. *Deviance, Reality and Society,* 2nd ed. New York: Holt, Rinehart and Winston, 1981.

Brajuha, Mario, and Lyle Hallowell. "Legal Intrusion and the Politics of Fieldwork: The Impact of the Brajuha Case." *Urban Life* 14 (January 1986):454–478.

Brauer, Jerald C. (ed.). *Religion and the American Revolution.* Philadelphia: Fortress Press, 1976.

Brave, Ralph. "Governing the Genome." *The Nation* (December 10, 2001):18–24.

Breed, Allen G. "Minority Vote Moves Center Stage." The Associated Press (January 12, 2008).

Bridges, George S., and Robert D. Crutchfield. "Law, Social Standing, and Imprisonment." *Social Forces* 66 (March 1988):699–724.

Brint, Steven, and Steve Brint. *Schools and Societies.* Thousand Oaks, CA: Pine Forge Press, 1998.

Brinton, Crane. *The Anatomy of Revolution.* New York: Knopf, 1990.

Brockerhoff, Martin P. "An Urbanizing World." *Population Bulletin* 55 (September 2000):1–44.

Brody, Jane E. "Personal Health; TV's Toll on Young Minds and Bodies." *New York Times* (August 3, 2004).

Brody, William. "Building a New Kind of Academy." *Planning for Higher Education* 25 (Summer 1997):72–74.

Bromley, David G. (ed.). *The Politics of Religious Apostasy: The Role of Apostates in the Transformation of Religious Movements.* Westport, CT: Praeger, 1998.

Bromley, David G., and Anson D. Shupe Jr. *New Christian Politics.* Macon, GA: Mercer University Press, 1984.

Bronfenbrenner, Urie. *Two Worlds of Childhood.* New York: Russell Sage Foundation, 1970.

Brouilette, John R. "The Liberalizing Effect of Taking Introductory Sociology." *Teaching Sociology* 12 (January 1985):131–148.

Brown, Elizabeth. "Lobbying FAQ: What is Permissible? Out of Bounds? Punishable?" Washington, D.C.: The Center for Public Integrity (January 18, 2006).

Brown, Lester R. *State of the World 1996.* New York: Norton, 1996.

Brown, Michael K. *Race, Money, and the American Welfare State.* Ithaca, NY: Cornell University Press, 1999.

Browne, Beverly A. "Gender Stereotypes in Advertising on Children's Television in the 1990s: A Cross-National Analysis." *Journal of Advertising* 27 (Spring 1998):83–96.

Brownlee, Shannon. "Overdose." *The Atlantic* (December 2007):36–38.

Bruce, Steve. *God Is Dead: Secularization in the West.* Malden, MA: Blackwell, 2002.

Bruner, Jerome. "Schooling Children in a Nasty Climate." *Psychology Today* (January 1982):57–63.

Brunner, H. G., M. Nelsen, X. O. Breakefield, H. H. Ropers, and B. A. van Oost. "Abnormal Behavior Associated with a Point Mutation in the Structural Gene for Monoamine Oxidase A." *Science* 262 (October 22, 1993):578–580.

Brunvand, Jan Harold. "Urban Legends: Folklore for Today." *Psychology Today* 14 (June 1980):50ff.

Brunvand, Jan Harold. *Too Good to Be True: The Colossal Book of Urban Legends.* New York: Norton, 2001.

Brunvand, Jan Harold. *Encyclopedia of Urban Legends.* New York: Norton, 2002.

Buckley, Stephen. "Shrugging Off the Burden of a Brainy Image." *Washington Post* (June 17, 1991):D1, D6.

Budrys, Grace. *Our Unsystematic Health Care System.* Lanham, MD: Rowman & Littlefield, 2001.

Buechler, Steven M. *Social Movements in Advanced Capitalism.* New York: Oxford University Press, 2000.

Bunzel, John H. (ed.). *Challenge to American Schools.* New York: Oxford University Press, 1985.

Burgess, Ernest W. "The Growth of the City." In Robert E. Park, Ernest W. Burgess, and Robert D. McKenzie (eds.), *The City.* Chicago: University of Chicago Press, 1925, pp. 47–62.

Burgess, J. K. "Widowers." In C. S. Chilman, E. W. Nunnally, and F. M. Cox (eds.), *Variant Family Forms.* Beverly Hills, CA: Sage, 1988, pp. 150–164.

Burke, Ronald, and Cary Cooper (eds.). *The Organization in Crisis: Downsizing, Restructuring and Privatization.* Cambridge, MA: Blackwell, 2000.

Burns, T., and G. M. Stalker. *The Management of Innovation.* 2nd ed. London: Tavistock, 1966.

Burros, Marian. "Even Women at the Top Still Have Floors to Do." *New York Times* (May 31, 1993):A1, A11.

Burstein, Paul. "Legal Mobilization as a Social Movement Tactic: The Struggle for Equal Employment Opportunity." *American Journal of Sociology* 96 (March 1991):1201–1225.

Buss, David M. "The Strategies of Human Mating: People Worldwide Are Attracted to the Same Qualities in the Opposite Sex." *American Scientist* 82 (May 1994a):238–249.

Buss, David M. *The Evolution of Desire: Strategies of Human Mating.* New York: Basic Books, 1994b.

Buss, David M. "Psychological Sex Differences: Origins Through Sexual Selection." *The American Psychologist* 50 (March 1995):164.

Buss, David M., Neil M. Malamuth, and Barbara A. Winstead. "Sex, Power, Conflict: Evolutionary and Feminist Perspectives." *Contemporary Psychology* 43 (April 1998).

Butler, Robert N. *Why Survive?* New York: Harper and Row, 1975.

Butler, Robert N. "Dispelling Ageism: The Cross-Cutting Intervention." *Annals of the American Academy of Political and Social Science* (1989):138–148.

Calasanti, Toni M. "Participation in a Dual Economy and Adjustment to Retirement." *International Journal of Aging and Human Development* 26 (1988):13–27.

Calasanti, Toni M., and Kathleen F. Slevin. *Gender, Social Inequalities, and Aging.* Walnut Creek, CA: AltaMira Press, 2001.

Calavita, Kitty, Henry N. Pontell, and Robert H. Tillman. *Big Money Crime.* Berkeley: University of California Press, 1999.

Califano, Joseph A. Jr. *America's Health Care Revolution.* New York: Random House, 1985.

Callan, Victor J. "The Voluntarily Childless and Their Perceptions of Parenthood and Childlessness." *Journal of Comparative Family Studies* 14 (Spring 1983):87–96.

Camarota, Steven A. "Immigrants in the United States— 2000: A Snapshot of America's Foreign-Born Population." *Backgrounder.* Washington, D.C.: Center for Immigration Studies, 2001.

Camarota, Steven A. "Immigrants in the United States— 2002." Washington, D.C.: Center for Immigration Studies, 2002.

Camic, Charles. *Reclaiming the Sociological Classics.* Malden, MA: Blackwell, 1998.

Campbell, Frances A., and Craig T. Ramey. "Effects of Early Intervention on Intellectual and Academic Achievement: A Follow-up Study of Children from Low-Income Families." *Child Development* 65 (1994):684–698.

Campbell, Kurt M., et al. "The Age of Consequences: The Foreign Policy and National Security Implications of Global Climate Change." New York: Center for Strategic and International Studies, 2007.

Campo-Flores, Adrian. "Giving Protest a Chance." *Newsweek* (February 3, 2003):28.

Campo-Flores, Adrian, and Yuval Rosenberg. "A Return to Wilding." *Newsweek* (June 26, 2000):28.

Canary, Daniel J., et al. "Diversity—Sex and Gender Differences in Personal Relationships." *Contemporary Psychology* 43 (1998):764.

Cancio, A. Silvia, T. David Evans, and David J. Maume Jr. "Reconsidering the Declining Significance of Race: Racial Differences in Early Career Wages." *American Sociological Review* 61 (August 1996):541–556.

Cantril, Hadley. *The Invasion from Mars.* Princeton, NJ: Princeton University Press, 1982.

Caplan, Gerald. "No Lessons Learned 12 Years after the Rwandan Genocide." *Pombazuka News* (April 14, 2006), http://pombazuka.org/en/category/features/33432.

Caplan, Nathan, Marcella H. Choy, and John K. Whitmore. "Indochinese Refugee Families and Academic Achievement." *Scientific American* 266 (February 1992):36–42.

Caplow, Theodore. *American Social Trends.* New York: Harcourt Brace Jovanovich, 1991.

Caplow, Theodore, Howard M. Bahr, and Bruce A. Chadwick. *All Faithful People*. Minneapolis: University of Minnesota Press, 1983.

Caplow, Theodore, Howard M. Bahr, Bruce A. Chadwick, Rueben Hill, and Margaret Holmes Williamson. *Middletown Families*. Minneapolis: University of Minnesota Press, 1982.

Cappelli, Peter, et al. *Change at Work*. New York: Oxford University Press, 1997.

Carnoy, Martin. *The Politics and Economics of Race in America*. Cambridge, UK: Cambridge University Press, 1994.

Carey, Benedict. "Study Quantifies Orphanage Link to I.Q." *New York Times* (December 2, 2007).

Carroll, Joseph. "Retirement Age Expectations and Realities." Gallup News Service (May 17, 2005).

Carroll, Joseph. "Hispanics' Views on Immigration Unchanged from Last Year." Gallup News Service (2007).

Carter, Bill, and Jacques Steinberg. "Off the Air: The Light Goes Out for Don Imus." *New York Times* (April 13, 2007).

Carter, Cynthia, and Kay C. Weaver. *Violence and the Media*. Philadelphia: Taylor and Francis, 2003.

Cashion, Barbara G. "Female-Headed Families: Effects on Children and Clinical Implications." *Journal of Marital and Family Therapy* (April 1982):77–85.

Casler, Lawrence. "The Effects of Extra Tactile Stimulation on a Group of Institutionalized Infants." *Genetic Psychology Monographs* 71 (1965):137–175.

Cassirer, Ernst. *The Philosophy of the Enlightenment*. Princeton, NJ: Princeton University Press, 1951.

Castro, Janice. "Disposable Workers." *Time* (March 15, 1993):47.

Cavanaugh, John C., and Fredda Blanchard-Fields. *Adult Development and Aging*, 4th ed. Pacific Grove, CA: Duxbury, 2001.

Cavanaugh, Stephanie. "The Trophy Kitchen." *Washington Post* (June 10, 1999):T14–T15.

Cavender, Gary, and Mark Fishman (eds.). *Entertaining Crime*. Hawthorne, NY: Aldine de Gruyter, 1998.

Center for Voting and Democracy. "International Voter Turnout, 1991–2000." 2002. Available online at www.fairvote.org/turnout/intturnout.htm.

Centers, Richard. *The Psychology of Social Classes*. New Haven, CT: Yale University Press, 1949.

Chaddock, Gail Russell. "Baby Boomers Face Retirement Squeeze." *Christian Science Monitor* (February 27, 2004).

Chafetz, Janet Saltzman. *Sex and Advantage*. Totowa, NJ: Rowman & Allanheld, 1984.

Chafetz, Janet Saltzman. *Gender Equity*. Newbury Park, CA: Sage, 1990.

Chagnon, Napoleon A. *Yanomamö*, 5th ed. Fort Worth, TX: Harcourt Brace College Publishers, 1997.

Chamberlain, Gary. "Metal Wars." *Design News* (October 4, 1999):115.

Chambliss, William J. "The Saints and the Roughnecks." *Society* 11 (November/December 1973):24–31.

Chambliss, William J., and Robert Seidman. *Law, Order, and Power*, 2nd ed. Reading, MA: Vantage, 1982.

Chandbasekaran, Rajiv. "Microsoft Trial Ends After 8 Months." *Washington Post* (June 25, 1999):A1, A20.

Chang, Iris. *The Rape of Nanking: The Forgotten Holocaust of World War II*. New York: Viking Penguin, 1998.

Chang, Kenneth. "Panel Says Bell Labs Scientist Faked Discoveries in Physics." *New York Times* (September 26, 2002):A1, A20.

Changing Public Attitudes toward the Criminal Justice System. Open Society Institute. Criminal Justice Initiative, 2002.

Chase-Dunn, Christopher. *Global Formation: Structures of the World Economy*. Lanham, MD: Rowman & Littlefield, 1999.

Chaves, Mark. "Abiding Faith." *Contexts* 1 (Summer 2002):19–26.

Cherlin, Andrew J., et al. "Longitudinal Studies of Effects of Divorce on Children in Great Britain and the United States." *Science* 7 (June 1991):1386–1389.

Cherlin, Andrew J. "A 'Quieting' of Change." *Contexts* (Spring 2002):67–68.

Chesler, Ellen. "New Options, New Politics." *The American Prospect* (Fall 2001):12–14.

Child Trends Data Bank. Washington, D.C.: Population Reference Bureau, 2002.

Children's Defense Fund. *The State of America's Children – 1991*. Washington, D.C.: Children's Defense Fund, 1991.

Chirot, Daniel. *How Societies Change*. Thousand Oaks, CA: Pine Forge Press, 1994.

Chirot, Daniel, and Martin E. P. Seligman (eds.). *Ethnopolitical Warfare: Causes, Consequences, and Political Solutions*. Washington, D.C.: American Psychological Association, 2001.

Choi, N. G. "Correlates of the Economic Status of Widowed and Divorced Elderly Men." *Journal of Family Issues* 12 (March 1992):38–54.

Choldin, Harvey M., and Claudine Hanson. "Subcommunity and Change in a Changing Metropolis." *Sociological Quarterly* 22 (Autumn 1981):549–564.

Christenson, James A., and Ronald C. Wimberley. "Who Is Civil Religious?" *Sociological Analysis* 39 (Spring 1978):77–83.

Christie, Nils. *Crime Control as Industry*, 3rd ed. London: Routledge, 2000.

Chu, Henry. "From Lowest Caste to High Office in India." *Los Angeles Times* (March 30, 2008).

Chua, Amy. *Day of Empire*. New York: Doubleday, 2007.

Chua, Amy. *World on Fire: How Exporting Free Market Democracy Breeds Ethnic Hatred and Global Instability*. New York: Knopf Publishing, 2004.

Cicourel, Aaron, and John Kitsuse. *The Educational Decision Makers*. New York: Bobbs-Merrill, 1963.

Citizens for Tax Justice. "Year-by-Year Analysis of the Bush Tax Cuts Shows Growing Tilt to the Very Rich." 2002. Available online at www.ctj.org/html/gwb0602.htm.

Citizens for Tax Justice. "The Bush Tax Cuts: The Latest CTJ Data March 2007." Available online www.ctj.org/pdf/gwb-data.pdf.

Clark, Charles S. "Cults in America." *CQ Researcher* 3 (May 7, 1993):385–408.

Clark, Lee. "Panic: Myth or Reality." *Contexts* 1 (Fall 2002):21–26.

Clausen, John A. *The Life Course*. Englewood Cliffs, NJ: Prentice Hall, 1986.

Clayton, Obie Jr. (ed.). *An American Dilemma Revisited*. New York: Russell Sage Foundation, 1996.

Clayton, Richard R., and James W. Gladden. "The Five Dimensions of Religiosity: Toward Demythologizing a Sacred Artifact." *Journal for the Scientific Study of Religion* 13 (June 1974):135–143.

Clemens, Elisabeth S. *The People's Lobby*. Chicago: University of Chicago Press, 1997.

"Closing the Gap." *The Economist* (September 2, 1995):74.

Cloward, Richard A., and Lloyd E. Ohlin. *Delinquency and Opportunity*. London: Routledge, 1998.

Coakley, Jay J (ed.). *Sports in Society,* 9th ed. Boston: Irwin/McGraw-Hill, 2006.

Cockburn, Alexander. "Beat the Devil." *The Nation* (January 17, 1994):42–43.

Cockburn, C. "The Relations of Technology." In R. Crompton and M. Mann (eds.), *Gender and Stratification.* Cambridge, UK: Polity, 1986.

Cohen, Adam, and Cathy Booth Thomas. "Inside a Layoff." *Time* (April 16, 2001):38–40.

Cohen, Albert K. *Delinquent Boys.* New York: Free Press, 1977.

Cohen, David, and Marvin Lazerson. "Education and the Corporate Order." *Socialist Revolution* 2 (March/April 1972):47–72.

Cohen, Jere. *Protestantism and Capitalism: The Mechanisms of Influence.* Hawthorne, NY: Aldine de Gruyter, 2002.

Coleman, James S. "The Concept of Equality of Educational Opportunity." *Harvard Educational Review* 38 (Winter 1968):7–22.

Coleman, James S. *Equal Educational Opportunity.* Cambridge, MA: Harvard University Press, 1969.

Coleman, James S. *Equality and Achievement in Education.* Boulder, CO: Westview Press, 1990.

Coleman, James S., Thomas Hoffer, and Sally B. Kilgore. "Questions and Answers: Our Response." *Harvard Educational Review* 51 (November 1981):526–545.

Coleman, James S., Sally B. Kilgore, and Thomas Hoffer. "Public and Private Schools." *Society* 19 (January/February 1982):4–9.

Coleman, James S., et al. *Equality of Educational Opportunity.* Washington, D.C.: U.S. Government Printing Office, 1966.

Coleman, James S., et al. "Cognitive Outcomes in Public and Private Schools." *Sociology of Education* 55 (April/July 1982a):65–76.

Coleman, James S., et al. "Achievement and Segregation in Secondary Schools: A Further Look at Public and Private School Differences." *Sociology of Education* 55 (April/July 1982b):162–182.

Coleman, James W. "Toward an Integrated Theory of White-Collar Crime." *American Journal of Sociology* 93 (September 1987):406–439.

Coleman, James W. *The Criminal Elite: The Sociology of White Collar Crime.* New York: St. Martin's Press, 1989.

Coleman, James W. *The Criminal Elite,* 4th ed. New York: St. Martin's Press, 1997.

College Board. *Trends in College Pricing 2001.* New York: College Entrance Examination Board, 2001.

College Board. *Trends in College Pricing 2007.* New York: College Entrance Examination Board, 2007.

Collier, Theresa J. "The Stigma of Mental Illness." *Newsweek* (April 26, 1993):16.

Collins, Chuck, and Felice Yeskel. *Economic Apartheid in America: A Primer on Economic Inequality and Insecurity,* 2nd ed. New York: New Press, 2005.

Collins, John J. *The Cult Experience.* New York: Charles C. Thomas, 1991.

Collins, Randall. "Functional and Conflict Theories of Educational Stratification." *American Sociological Review* 36 (December 1971):1002–1019.

Collins, Randall. *Conflict Sociology.* New York: Academic Press, 1975.

Collins, Randall. *The Credential Society: An Historical Sociology of Education.* New York: Academic Press, 1979.

Collins, Randall. *Four Sociological Traditions.* New York: Oxford University Press, 1994.

Collins, Randall. "Conflict Theory of Educational Stratification." In Jeanne H. Ballantine and Joan Z. Spade (eds.), *Schools and Society: A Sociological Approach to Education.* Belmont, CA: Wadsworth/Thomson, 2007.

Colombotos, John, and Corinne Kirchner. *Physicians and Social Change.* New York: Oxford University Press, 1986.

Comfort, Alex. "Age Prejudice in America." *Social Policy* 17 (November/December 1976):3–8.

Conason, Joe. *It Can Happen Here.* New York: St. Martin's Press, 2008.

Conrad, Peter, and Rochelle Kern (eds.). *Sociology of Health and Illness,* 3rd ed. New York: St. Martin's Press, 1990.

Cook, Karen S. "Exchange and Power in Networks of Interorganizational Relations." *Sociological Quarterly* 18 (Winter 1977):62–82.

Cooley, Charles Horton. *Human Nature and the Social Order.* New York: Scribner's, 1902.

Cooley, Charles Horton and Hans-Joachim Schubert. *On Self and Social Organization.* Chicago: University of Chicago Press, 1998.

Coon, Dennis, and John O. Mitterer. *Introduction to Psychology,* 3rd ed. Belmont, CA: Wadsworth, 2008.

Cose, Ellis. *The Rage of the Privileged Class.* New York: HarperCollins, 1995.

Cose, Ellis. "The Black Gender Gap." *Newsweek* (March 3, 2003): 48–51.

Coser, Lewis A. *The Functions of Social Conflict.* London: Routledge, 1998.

Cowley, Geoffrey, and Anne Underwood. "Surgeon, Drop That Scalpel." *Newsweek* (February 1998, Winter 1997–98 Special Edition):77–78.

Cox, Frank D. *Human Intimacy,* 10th ed. Belmont, CA: Wadsworth, 2005.

Cox, Harvey. *Turning East: The Promise and Peril of the New Orientalism.* New York: Simon & Schuster, 1977.

Cox, Harvey. *Religion in the Secular City.* New York: Simon & Schuster, 1992.

Cox, Harvey. *Fire from Heaven.* London: Cassell, 1996.

Cozic, Charles P. (ed.). *The Militia Movement.* San Diego, CA: Greenhaven Press, 1997.

Cozzillio, Michael J. *Sports and Inequality.* Durham, NC: Carolina Academic Press, 2005.

Craig, Steve. *Men, Masculinity and the Media.* Thousand Oaks: Sage, 1992.

Crain, Robert L. "School Integration and Occupational Achievement of Negroes." *American Journal of Sociology* 75 (January 1970):593–606.

Crain, Robert L., and Carol S. Weisman. *Discrimination, Personality and Achievements.* New York: Seminar Press, 1972.

Crane, Diana. *Fashion and Its Social Agendas: Class, Gender, and Identity in Clothing.* Chicago: University of Chicago Press, 2001.

Creekmore, C. R. "Cities Won't Drive You Crazy." *Psychology Today* (January 1985):46–53.

Cress, Daniel M., and David A. Snow. "Mobilization at the Margins: Resources, Benefactors, and the Viability of Homeless Social Movement Organizations." *American Sociological Review* 61 (December 1996):1089–1109.

Crewdson, John. "Fraud in Breast Cancer Study." *Chicago Tribune* (March 13, 1994):A1.

Crosby, F. E. *Juggling.* New York: Free Press, 1993.

Crosby, John F. *Reply to Myth: Perspectives on Intimacy.* New York: Wiley, 1985.

Crossley, Nick. *Making Sense of Social Movements.* Buckingham: Open University Press, 2002.

Crow, Barbara A. (ed.). *Radical Feminism: A Documentary Reader.* New York: New York University Press, 1998.

Cullen, John B., and Shelley M. Novick. "The Davis-Moore Theory of Stratification: A Further Examination and Extension." *American Journal of Sociology* 84 (May 1979):1424–1437.

"Cult Killings Exceed Jonestown Toll." *Los Angeles Times* (April 1, 2000):A14.

Cummings, Scott. "White Ethnics, Racial Prejudice, and Labor Market Segmentation." *American Journal of Sociology* 95 (January 1980):938–950.

Curran, Daniel J., and Claire M. Renzetti. *Social Problems: Society in Crisis,* 6th ed. Upper Saddle River, NJ: Prentice Hall, 2000.

Curtiss, Susan. *Genie.* New York: Academic Press, 1977.

Cushman, Thomas. *Notes from Underground: Rock Music Counterculture in Russia.* New York: St. Martin's Press, 1995.

Cuzzort, R. P., and E. W. King. *Humanity and Modern Social Thought,* 2nd ed. Hinsdale, IL: Dryden Press, 1976.

Cyert, Richard M., and James G. March. *A Behavioral Theory of the Firm.* Englewood Cliffs, NJ: Prentice Hall, 1963.

Dahl, Robert. *Who Governs?,* 2nd ed. New Haven, CT: Yale University Press, 2005.

Dahrendorf, Ralf. "Out of Utopia: Toward a Reorientation of Sociological Analysis." *American Journal of Sociology* 64 (September 1958a):115–127.

Dahrendorf, Ralf. "Toward a Theory of Social Conflict." *Journal of Conflict Resolution* 2 (June 1958b):170–183.

Dahrendorf, Ralf. *Class and Class Conflict in Industrial Society.* Stanford, CA: Stanford University Press, 1959.

Daly, Martin, and Margo Wilson. "Crime and Conflict: Homicide in Evolutionary Psychological Perspective." *Crime and Justice* 22 (1997).

Darder, Antonia, and Rodolfo D. Torres. *The Latino Reader.* Malden, MA: Blackwell, 1997.

Davidson, James D. "Glock's Model of Religious Commitment: Assessing Some Different Approaches and Results." *Review of Religious Research* 16 (Winter 1975):83–93.

Davies, Bronwyn. *Frogs and Snails and Feminist Tales.* New York: Pandora Press, 1990.

Davies, Christie. "Sexual Taboos and Social Boundaries." *American Journal of Sociology* 87 (March 1982):1032–1063.

Davies, James Chowning. "The J-Curve of Rising and Declining Satisfactions as a Cause of Revolution and Rebellion." In Hugh Davis Graham and Ted Robert Gurr (eds.), *Violence in America,* rev. ed. Beverly Hills, CA: Sage, 1979.

Davis, F. James. *Minority-Dominant Relations.* Arlington Heights, IL: AHM Publishing Corporation, 1978.

Davis, F. James. "Up and Down Opportunity's Ladder." *Public Opinion* 5 (June/July 1982):11ff.

Davis, Kingsley. "Final Note on a Case of Extreme Isolation." *American Journal of Sociology* 52 (March 1947):232–247.

Davis, Kingsley. *Human Society.* New York: Macmillan, 1949.

Davis, Kingsley. "The Origin and Growth of Urbanization in the World." *American Journal of Sociology* 60 (March 1955):429–437.

Davis, Kingsley, and D. Rowland. "Old and Poor: Policy Challenges in the 1990s." *Journal of Aging and Social Policy.* 2 (1991):37–59.

Dawkins, Richard. *The Selfish Gene,* 2nd ed. New York: Oxford University Press, 1990.

Deak, JoAnn. *How Girls Thrive.* Washington, D.C.: National Association of Independent Schools, 1998.

Death Penalty Information Center. "Race of Defendants Executed Since 1976." 2008. Available online at www.deathpenaltyinfo.org/dpicrace.html.

DeBarros, Anthony. "New Baby Boom Swamps Colleges." *USA Today* (January 2, 2003):1A, 2A.

Deegan, Mary J. *Women in Sociology: A Bio-Bibliographical Source Book.* Westport, CT: Greenwood, 1991.

Deegan, Mary J. *Jane Addams and the Men of the Chicago School, 1892–1918.* New Brunswick, NJ: Transaction Publishers, 2000.

Deegan, Mary Jo, and Michael Stein. "American Drama and Ritual: Nebraska Football." *International Review of Sport Sociology* 3 (1978):31–44.

DeFleur, Melvin L., and Sandra Ball-Rokeach. *Theories of Mass Communication,* 4th ed. New York: Longman, 1982.

de La Isla, Jose. *The Rise of Hispanic Political Power.* Santa Maria, CA: Archer Books, 2003.

del Pinal, Jorge, and Audrey Singer. "Generations of Diversity: Latinos in the United States." *Population Bulletin* 52 (October 1997):1–48.

Delgado, Richard, and Jean Stefancic (eds.). *The Latino Condition: A Critical Reader.* New York: New York University Press, 1998.

Deloria, Vine Jr., and Clifford Lytle. *The Nations Within.* New York: Pantheon Books, 1984.

De Luca, Rita Caccamo. *Back to Middletown: Three Generations of Sociological Reflections.* Stanford, CA: Stanford University Press, 2001.

Demaine, Jack (ed.). *Sociology of Education Today.* New York: Palgrave, 2001.

Demerath, N. J. III, and Rhys H. Williams. "Civil Religion in an Uncivil Society." *The Annals* 480 (July 1985):154–166.

Demo, D. H., and A. C. Acock. "The Impact of Divorce on Children." In A. Booth (ed.), *Contemporary Families.* Minneapolis: National Council on Family Relations, 1991, pp. 162–191.

DePaulo, Bella. *Singled Out.* New York: St. Martin's Press, 2007.

Devarics, Charles. "Is Welfare Reform Reforming Welfare?" *Population Today* 26 (October 1998):1–8.

DeSpelder, L. A., and D. L. Strickland. *The Last Dance.* Mountain View, CA: Mayfield Press, 1991.

Deutschman, Alan. "Pioneers of the New Balance." *Fortune* (May 20, 1991):60–68.

de Waal, Frans B. M. "Cultural Primatology Comes of Age." *Nature* 399 (June 1999):635–636.

Dewar, Helen. "Senate Approves Disclosure Measure." *Washington Post* (June 30, 2000):A1, A4.

Dewar, Helen, and Barbara Vobejda. "Clinton Signs Head Start Expansion." *Washington Post* (May 19, 1994):A1.

Dewey, John. *The School and Society: And the Child and the Curriculum.* Mineola, NY: Dover, 2001.

DeWitt, Karen. "Blacks Prone to Job Dismissal in Organizations." *New York Times* (April 20, 1995):A19.

DeYoung, Alan J. *Economics and American Education.* White Plains, NY: Longman, 1989.

Diamond, Jared. *The Third Chimpanzee: The Evolution and Future of the Human Animal.* New York: HarperPerennial, 1993.

Diamond, Jared. *Guns, Germs, and Steel: The Fates of Human Societies.* New York: Norton, 1999.

DiIulio, John J. Jr., and Anne Morrison Piehl. "Does Prison Pay?: The Stormy National Debate Over the Cost-

Effectiveness of Imprisonment." *The Brookings Review* (Fall 1991):28–35.

Dillon, Sam. "Ohio Goes after Charter Schools That Are Failing." *New York Times* (November 8, 2007).

DiMaggio, Paul (ed.). *The Twenty-First-Century Firm: Changing Economic Organization in International Perspective.* Princeton, NJ: Princeton University Press, 2001.

Dines, Gail, and Jean M. Humez (eds.). *Gender, Race, and Class in Media: A Text-Reader,* 2nd ed. Thousand Oaks, CA: Sage, 2002.

DiPrete, Thomas A. *Collectivist vs. Individualist Mobility Regimes?* Stockholm, Sweden: Stockholm University, 1996.

Directory of Corporate Affiliations. New Providence, NJ: LexisNexis Group, 2007.

Dobzhansky, Theodosius. *Mankind Evolving.* Toronto: Bantam Books, 1970.

Dollard, John, et al. *Frustration and Aggression.* New Haven, CT: Yale University Press, 1939.

Domhoff, G. William. *The Power Elite and the State.* Hawthorne, NY: Aldine de Gruyter, 1990.

Domhoff, G. William. *State Autonomy or Class Dominance.* New York: Aldine de Gruyter, 1996.

Domhoff, G. William. *Who Rules America?* New York: McGraw-Hill, 2005.

Dooley, Calvin M. *Income and Racial Disparities in the Undercount in the 2000 Presidential Election.* Washington, D.C.: U.S. House of Representatives Committee on Government Reform, 2001.

Doreian, Patrick, and Frans N. Stokman. *Evolution of Social Networks.* Amsterdam: Gordon and Breach Publishers, 1997.

Dornbush, Sanford M., et al. "Single Parents, Extended Households, and the Control of Adolescents." *Child Development* 56 (1985):326–341.

Douglas, Mary. "Accounting for Taste." *Psychology Today* 13 (July 1979):44ff.

Doyle, James A. *The Male Experience,* 3rd ed. Madison, WI: Brown and Benchmark, 1995.

Draut, Tammy. "New Opportunities?: Public Opinion on Poverty, Income Inequality and Public Policy: 1996–2002." New York: Demos, 2002.

Drew, Jill. "Eyewitnesses Recount Terrifying Day in Tibet." *Washington Post* (March 27, 2008):A1.

Dreyfuss, Robert, and Jason Vest. "The Lie Factory." *Mother Jones* 29 (January/February 2004):34–41.

Drucker, Peter F. "The Coming of the New Organization." *Harvard Business Review* 66 (January/February 1988):45–53.

Du Bois, William, and R. Wright. *Applying Sociology: Making a Better World.* Boston: Allyn & Bacon, 2000.

Dudley, Kathryn Marie. *The End of the Line: Lost Jobs, New Lives in Postindustrial America.* Chicago: University of Chicago Press, 1997.

Dudley, William. *Media Violence: Opposing Viewpoints.* San Diego, CA: Greenhaven Press, 1999.

Duff, Christina. "Profiling the Aged: Fat Cats or Hungry Victims." *The Wall Street Journal* (September 28, 1995):B1, B16.

Durkheim, Emile. *Suicide.* Translated by John A. Spaulding and George Simpson. Edited by George Simpson. New York: Free Press, 1964.

Durkheim, Emile. *The Rules of Sociological Method.* Translated by Sarah A. Solovay and John H. Mueller. Edited by George E. G. Catlin. New York: Free Press, 1966.

Durkheim, Emile. *The Elementary Forms of Religious Life.* New York: Free Press, 1995.

Durkheim, Emile. *The Division of Labor in Society.* New York: Free Press, 1997.

Dye, Robert R. *Politics in America,* 7th ed. Englewood Cliffs, NJ: Prentice-Hall, 2006.

Dye, Thomas R. *Who's Running America?,* 7th ed. Englewood Cliffs, NJ: Prentice Hall, 2001.

Dynes, Russell R., and Kathleen J. Tierney (eds.). *Disasters, Collective Behavior, and Social Organization.* Newark: University of Delaware Press, 1994.

Eagleton, Terry. *Ideology.* London: Longman, 1994.

Eagley, Alice H., and Wendy Wood. "The Origins of Sex Differences in Human Behavior: Evolved Dispositions versus Social Roles." *American Psychologist* 54 (1999):408–423.

Eaton, Susan E. *The Other Boston Busing Story: What's Won and Lost across the Boundary Line.* New Haven, CT: Yale University Press, 2001.

Ebenstein, Alan O., William Ebenstein, and Edwin Fogelman. *Today's Isms,* 11th ed. Upper Saddle River, NJ: Prentice Hall, 2000.

Eccles, Jacquelynne, et al. "Development during Adolescence: The Impact of Stage-Environment Fit on Young Adolescents' Experiences in Schools and in Families." *American Psychologist* 48 (February, 1993):90–101.

Eck, Diana L. *A New Religious America: How a "Christian Country" Has Become the World's Most Religiously Diverse Nation.* New York: HarperCollins Publishers, 2002.

Economist, The. "On the Right Track?" (March 2, 2002):59–60.

Eder, Donna. *School Talk.* New Brunswick, NJ: Rutgers University Press, 1995.

Edin, Kathryn, and Maria Kefalas. *Promises I Can Keep.* Berkeley, CA: University of California Press, 2006.

Edin, Kathryn, and Laura Lein. *Making Ends Meet: How Single Mothers Survive Welfare and Low-Wage Work.* New York: Russell Sage Foundation, 1997.

Edmonds, Patricia, and Mark Potok. "Followers 'See Themselves on Side of Jesus.'" *USA Today* (April 20, 1993): A1, A2.

Edsall, Thomas B. "New Ways to Harness Soft Money in Works." *Washington Post* (August 25, 2002a):A1, A6, A7.

Edsall, Thomas B. "Bush Has a Cabinet Full of Wealth." *Washington Post* (September 18, 2002b):A27.

Ehrenreich, Barbara. "Helping the Rich Stay That Way." *Time* (April 18, 1994):86.

Eisenstock, Barbara. "Sex-Role Differences in Children's Identification with Counterstereotypic Televised Portrayals." *Sex Roles* 10 (1984):417–430.

Eismeier, Theodore J., and Philip H. Pollock III. *Business, Money, and the Rise of Corporate PACs in American Elections.* Westport, CT: Greenwood Press, 1988.

Eitzen, D. Stanley. "Ethical Dilemas in American Sport: The Dark Side of Competition." In D. Stanley Eitzen (ed.), *Sport in Contemporary Society,* 7th ed. Boulder, CO: Paradigm Press, 2006a, pp. 165–174.

Eitzen, D. Stanley (ed.). *Sport in Contemporary Society: An Anthology,* 7th ed. Boulder, CO: Paradigm Press, 2006b.

Eitzen, D. Stanley, and George H. Sage. *Sociology of North American Sport,* 8th ed. New York: Paradigm Press, 2008.

Eitzen, D. Stanley, and Kelly Eitzen Smith. *Experiencing Poverty.* Belmont, CA: Wadsworth, 2003.

Elikann, Peter T. *The Tough-on-Crime Myth: Real Solutions to Cut Crime.* New York: Plenum, 1996.

Elkin, Frederick, and Gerald Handel. *The Child and Society,* 6th ed. New York: McGraw-Hill, 2001.

Elliott, John E. *Comparative Economic Systems,* 2nd ed. Belmont, CA: Wadsworth, 1985.

Elliott, Michael. "Crime and Punishment." *Newsweek* (April 18, 1994):18–22.

Ellis, Joseph J. *Founding Brothers: The Revolutionary Generation.* New York: Knopf, 2002.

Ellis, Joseph J. *American Creation.* New York: Knopf, 2007.

Elster, Jon, Clause Offe, and Ulrich Preuss. *Institutional Design in Post-Communist Societies.* Cambridge, UK: Cambridge University Press, 1998.

Ember, Carol R., and Melvin Ember. *Anthropology,* 9th ed. Upper Saddle River, NJ: Prentice Hall, 1999.

Empey, LaMar T. *American Delinquency,* 2nd ed. Homewood, IL: Dorsey, 1982.

Engels, Friedrich. *The Origin of the Family, Private Property, and the State.* New York: International Publishing, 2001.

Entwisle, Doris R., and Karl L. Alexander. "Early Schooling as a 'Critical Period' Phenomenon." In K. Namboodiri and R.G. Corwin (eds.), *Sociology of Education and Socialization.* Greenwich, CT: JAI Press, 1989, pp. 27–55.

Entwisle, Doris R., and Karl L. Alexander. "Entry into School: The Beginning School Transition and Educational Stratification in the United States." *Annual Review of Sociology* 19 (1993):401–423.

Epstein, T. Scarlett. *Economic Development and Social Change in South India.* New York: Humanities Press, 1962.

Erickson, Rosemary J., and Rita J. Simon. *The Use of Social Science Data in Supreme Court Decisions.* Urbana, IL: University of Illinois Press, 1998.

Erikson, Erik H. *Childhood and Society.* Harmondsworth, UK: Penguin Books, 1973.

Erikson, Erik H. *The Life Cycle Completed.* New York: Norton, 1982.

Erikson, Kai T. "A Comment on Disguised Observation in Sociology." *Social Problems* 14 (Spring 1967):366–373.

Erikson, Kai T. *Everything in Its Path.* New York: Simon & Schuster, 1976.

Erikson, Kai T. *A New Species of Trouble: Explorations in Disaster, Trauma, and Community.* New York: Norton, 1995.

Erikson, Kai T. *Sociological Visions.* Blue Ridge Summit, PA: Rowman & Littlefield, 1997.

Ermann, M. David, and Richard J. Lundman. *Corporate and Governmental Deviance,* 5th ed. New York: Oxford University Press, 1996.

Eshleman, J. Ross, and Richard A. Bulcroft. *The Family,* 11th ed. Boston: Allyn & Bacon, 2005.

Esping-Andersen, Gosta. *Welfare States in Transition.* Thousand Oaks, CA: Sage, 1996.

Espiritu, Yen Le. *Asian American Men and Women.* Thousand Oaks, CA: Sage, 1996.

Esposito, John, and John Fiorillo. "Who's Left on the Block? New York City's Working Class Neighborhoods." In Hans Spiegel (ed.), *Citizen Participation in Urban Development.* Fairfax, VA: Learning Resources, 1974, pp. 307–334.

Ethridge, Marcus E., and Howard Handelman. *Politics in a Changing World,* 4th ed. Belmont, CA: Wadsworth, 2008.

Etzioni, Amitai. *A Comparative Analysis of Complex Organizations,* rev. ed. New York: Free Press, 1975.

Etzioni, Amitai. "Education for Mutuality and Civility." *Futurist* 16 (October 1982):4–7.

Etzioni, Amitai. *The Monochrome Society.* Princeton, NJ: Princeton University Press, 2001.

Evans, Jean. *Three Men.* New York: Knopf, 1954.

Evans, Sara M. *Personal Politics.* New York: Vintage, 1980.

Evans, Sara M. *Born for Liberty: A History of Women in America.* New York: Free Press, 1989.

Eyerman, Ron. *Studying Collective Action.* Newbury Park, CA: Sage, 1992.

Eysenck, Hans. *Crime and Personality,* 3rd ed. London: Routledge and Kegan Paul, 1977.

Fagot, Joël, Edward A. Wasserman, and Michael E. Young. "Discriminating the Relation between Relations: The Role of Entropy in Abstract Conceptualization by Baboons *(Papio papio)* and Humans *(Homo sapiens).*" *Journal of Experimental Psychology: Animal Behavior Processes* 27 (October 2001): 316–328.

Faioloa, Anthony. "The New Economics of Hunger." *Washington Post* (April 27, 2008)

Falk, Erika. *Women for President.* Champaign, IL: University of Illinois Press, 2008.

Falk, Gerhard. *Stigma: How We Treat Outsiders.* Amherst, NY: Prometheus Books, 2001.

Falk, Ursula A., and Gerhard Falk. *Ageism, the Aged, and Aging in America.* Springfield, IL: C. C. Thomas, 1997.

Faltermayer, Charlotte. "Cyberveillance." *Time* (August 14, 2000):B22–B25.

Faludi, Susan. *Stiffed: The Betrayal of the American Man.* New York: Harper Collins Publishing, 2000.

Farhi, Paul. "TV Ratings Agreement Reached." *Washington Post* (July 10, 1999):A1, A16.

Farley, Christopher John. "The Butt Stops Here." *Time* (April 18, 1994):58–64.

Farley, John E. *Majority-Minority Relations,* 3rd ed. Englewood Cliffs, NJ: Prentice Hall, 1995.

Farley, John E., and Mark Hansel. "The Ecological Context of Urban Crime: A Further Exploration." *Urban Affairs Quarterly* 17 (September 1981):37–54.

Farley, Reynolds. *Blacks and Whites.* Cambridge, MA: Harvard University Press, 1984.

Farley, Reynolds, and William H. Frey. "Changes in the Segregation of Whites from Blacks During the 1980's: Small Steps Toward a More Integrated Society." *American Sociological Review* 59 (February 1994):23–45.

Farley, Reynolds, and John Haaga. "Modest Declines in U.S. Residential Segregation Observed." *Population Today* 25 (February 1997):1–2.

Farley, Reynolds, and John Haaga. *The New American Reality: Who We Are, How We Got There, Where We Are Going.* New York: Russell Sage Foundation, 1998.

Farley, Reynolds, and John Haaga (eds.). *American People.* New York: Russell Sage Foundation, 2005.

Farrell, Christopher. "Time to Prune the Ivy." *BusinessWeek* (May 24, 1993):112–118.

Fasteau, Marc. "The Male Machine: The High Price of Macho." *Psychology Today* 8 (September 1975):60.

Fattah, Hasson M. "Saudis Rethink Taboo on Women Behind the Wheel." *New York Times* (September 28, 2007).

Fausto-Sterling, Anne. *Myths of Gender.* New York: Basic Books, 1987.

Fazioli, Jennifer K. "The Advancement of Female Coaches in Intercollegiate Athletics." Department of Labor Studies and Industrial Relations, Pennsylvania State University, 2004.

Feagin, Joe R. *Subordinating the Poor.* Englewood Cliffs, NJ: Prentice Hall, 1975.

Feagin, Joe R. "The Continuing Significance of Race: Antiblack Discrimination in Public Places." *American Sociological Review* 56 (February 1991):101–115.

Feagin, Joe R. *Racist America: Roots, Current Realities and Future Reparations.* New York: Routledge, 2001.

Feagin, Joe R., and Clairece Booker Feagin. *Discrimination American Style.* Malabar, FL: Krieger, 1986.

Feagin, Joe R., and Karyn D. McKinney. *The Many Costs of Racism.* Lanham, MD: Rowman & Littlefield, 2005.

Feagin, Joe R., Hernán Vera, and Pinar Batur. *White Racism,* 2nd ed. New York: Routledge, 2001.

Fears, Darryl. "In Jena and Beyond, Nooses Return as a Symbol of Hate." *Washington Post* (October 20, 2007).

Featherman, David L. "The Socio-Economic Achievement of White Religio-Ethnic Sub-groups." *American Sociological Review* 36 (April 1971):207–222.

Feeley, Malcolm, and Jonathon Simon. "The New Penology: Notes on the Emerging Strategy of Corrections and Implications." *Criminology* 30 (1992):449–474.

Fein, Helen. *Genocide.* Newbury Park, CA: Sage, 1993.

Fenn, Richard. *Toward a Theory of Secularization.* Storrs, CT: Society for the Scientific Study of Religion, 1978.

Fenwick, Charles R. "Crime and Justice in Japan: Implications for the United States." *International Journal of Comparative and Applied Criminal Justice* 6 (1982):62–71.

Ferber, Abby L. *White Man Failing: Race, Gender, and White Supremacy.* Lanham, MD: Rowman & Littlefield, 1998.

Ferguson, Sarah. "Testing Protest in New York." *Mother Jones* (January 24, 2002). Available online at www.motherjones.com/web_exclusives/features/news/wef.html.

Fergusson, D. M., L. J. Horwood, and F. T. Shannon. "A Proportional Hazards Model of Family Breakdown." *Journal of Marriage and Family* 46 (August 1984):539–549.

Ferrar, Jane W. "Marriages in Urban Communal Households: Comparing the Spiritual and the Secular." In Peter J. Stein, Judith Richman, and Natalie Hannon (eds.), *The Family.* Reading, MA: Addison-Wesley, 1977, pp. 409–419.

Ferree, M., and B. Hess. *Controversy and Coalition: The New Feminist Movement Across Three Decades of Change.* Old Tappan, NJ: Twayne Publishers, 1994.

Feuer, Lewis S. (ed.). *Marx and Engels.* New York: Doubleday, 1959.

"Fewer Mothers Prefer Full-time Work." Washington, D.C.: Pew Research Center, Pew Social and Demographic Trends Project, July 12, 2007.

Fiedler, Fred E. *A Theory of Leadership Effectiveness.* New York: McGraw-Hill, 1967.

Fine, Gary Alan. *With the Boys: Little League Baseball and Preadolescent Culture.* Chicago: University of Chicago Press, 1987.

Fine, Gary Alan. "The Sad Demise, Mysterious Disappearance, and Glorious Triumph of Symbolic Interactionism." *Annual Review of Sociology* 19 (1993):61–87.

Fine, Gary Alan. *The Culture of Restaurant Work.* Berkeley and Los Angeles: University of California Press, 1996.

Fine, Gary Alan, and Patricia Turner. *Whispers on the Color Line: Rumor and Race in America.* Berkeley, CA: University of California Press, 2004.

Fineman, Howard. "Leadership? Don't Ask Us." *Newsweek* (May 11, 1992):42.

Fineman, Howard. "Shifting Racial Lines." *Newsweek* (July 10, 1995):38–39.

Finke, Roger, and Rodney Stark. *The Churching of America.* New Brunswick, NJ: Rutgers University Press, 1992.

Firebaugh, Glenn. "Structural Determinants of Urbanization in Asia and Latin America." *American Sociological Review* 44 (April 1979):199–215.

Fischer, Claude S. *To Dwell among Friends.* Chicago: University of Chicago Press, 1982.

Fischer, Claude S. *The Urban Experience,* 2nd ed. New York: Harcourt Brace Jovanovich, 1984.

Fischer, Claude S. "The Subcultural Theory of Urbanism: A Twentieth-Year Assessment." *American Journal of Sociology* 101 (November 1995):543–577.

Fischer, Claude S., et al. *Inequality by Design: Cracking the Bell Curve Myth.* Princeton, NJ: Princeton University Press, 1996.

Fischer, David Hackett. *Growing Old in America.* New York: Oxford University Press, 1977.

Fischer, Joannie. "Scientists Have Finally Cloned a Human Embryo." *U.S. News & World Report* (December 3, 2001): 51–63.

Flanagan, William G. *Contemporary Urban Sociology,* 4th ed. Boston: Allyn & Bacon, 2001.

Fletcher, Max E. "Harriet Martineau and Ayn Rand: Economics in the Guise of Fiction." *American Journal of Economics and Sociology* 33 (October 1974):367–379.

Fletcher, Michael A. "In Wisconsin, Vouching for Vouchers." *Washington Post* (March 20, 2001):A1, A14.

Florida, Richard, and Martin Kenney. "Transplanted Organizations: The Transfer of Japanese Industrial Organization to the U.S." *American Sociological Review* 56 (June 1991):381–398.

Flynn, C. P. "Relationship Violence by Women: Issues and Implications." *Family Relations* 39 (1990):194–198.

Fogarty, Thomas A., and Edward Iwata. "Regulators: Firms Traded Coverage for Business." *USA Today* (April 29, 2003):3B.

Fonda, Daren. "National Prosperous Radio." *Time* (March 24, 2003):49–51.

Francis, David K. *The Internet in Healthcare: The Final Frontier.* Nashville, TN: J. C. Bradford, 1999.

Frank, Robert H. *Luxury Fever: Money and Happiness in an Era of Excess.* Princeton, NJ: Princeton University Press, 2000.

Frank, Robert H. *Falling Behind.* Berkeley, CA: University of California Press, 2007.

Frankenberg, Erica, and Chungmei Lee. *Race in American Public Schools: Rapidly Resegregating School Districts.* Cambridge, MA: The Civil Rights Project, August 2002.

Frankenberg, Erica, and Gary Orfield. *Lessons in Integration.* Charlottesville, VA: University of Virginia Press, 2007.

Fraser, Jill Andresky. *White Collar Sweatshop: The Deterioration of Work and Its Rewards in Corporate America.* New York: Norton, 2002.

Freeman, David M. *Technology and Society.* Chicago: Rand McNally, 1974.

Freeman, Jo., and Victoria Johnson (eds.). *Waves of Protest: Social Movements Since the Sixties.* Lanham, MD: Rowman & Littlefield, 1999.

Freeman, Linton C. "The Sociological Concept of 'Group': An Empirical Test of Two Models." *American Journal of Sociology* 98 (July 1992):152–166.

Freeman, Richard B. *Black Elite.* New York: McGraw-Hill, 1977.

Freedman, Jane. *Feminism.* Buckingham, UK: Open University Press, 2001.

Freedman, Marc. *Encore.* New York: Perseus Publishing, 2007.

Freedman, Vicki A., Linda G. Martin, and Robert F. Schoeni. "Disability in America." *Population Bulletin* (September 2004): 1-32.

French, Howard W. "North Korea to Let Capitalism Loose in Investment Zone." *New York Times* (September 25, 2002): A1, A3.

Freund, Peter E., and Meredith B. McGuire. *Health, Illness, and the Social Body: A Critical Sociology,* 3rd ed. Englewood Cliffs, NJ: Prentice Hall, 1998.

Frey, William H. "Lifecourse Migration and Metropolitan Whites and Blacks and the Structure of Demographic Change in Large Central Cities." *American Sociological Review* 49 (December 1984):803–827.

Frey, William H. "Mapping the Growth of Older America: Seniors and Boomers in the Early 21st Century." Washington, D.C.: The Brookings Institution, 2007.

Frey, William H. "Mover Destination Selectivity and the Changing Suburbanization of Metropolitan Blacks and Whites." *Demography* 22 (May 1985):223–243.

Frey, William H., and Reynolds Farley. "Latino, Asian, and Black Segregation in U.S. Metropolitan Areas: Are Multiethnic Metros Different?" *Demography* 31 (February 1996):35–50.

Friedan, Betty. *The Feminine Mystique.* New York: Dell, 1963.

Friedman, Milton. *Capitalism and Freedom.* Chicago: University of Chicago Press, 2002.

Friedman, Thomas. *Lexus and the Olive Tree: Understanding Globalization.* New York: Knopf Publishing Group, 2000.

Friedman, Thomas. *The World Is Flat.* New York: Picador, 2007.

Friedrich, Carl J., and Zbigniew K. Brzezinski. *Totalitarian Dictatorship and Autocracy,* 2nd ed. Cambridge, MA: Harvard University Press, 1965.

Frieze, Irene H., Jacquelynne E. Parsons, Paula B. Johnson, Diane N. Rable, and Gail L. Zellman. *Women and Sex Roles.* New York: Norton, 1978.

From the Capital to the Classroom: State and Federal Efforts to Implement the No Child Left Behind Act. Washington, D.C.: Center on Education Policy, January 2003.

Fruch, Terry, and Paul McGhee. "Traditional Sex Role Development and Amount of Time Watching Television." *Developmental Psychology* 11 (1975):109ff.

Fukuyama, Francis. "The Great Disruption." *Atlantic Monthly* 283 (May 1999):55–80.

Fukuyama, Francis. *The Great Disruption: Human Nature and the Reconstitution of Social Order.* New York: The Free Press, 2000.

Fukuyama, Francis. *Our Posthuman Future: Consequences of the Biotechnology Revolution.* New York: Picador, 2003.

Fukuyama, Francis. *The End of History and the Last Man.* New York: Free Press, 2006.

Fuller, Bruce, and Richard F. Elmore (eds.). *Who Wins, Who Loses?* New York: Teachers College Press, 1996.

Fuller, John L., and William R. Thompson. *Behavior Genetics.* St. Louis: Mosby, 1978.

Furstenberg, Frank F. Jr., and Andrew J. Cherlin. *Divided Families.* Cambridge, MA: Harvard University Press, 1991.

Galloway, Joseph L. "In the Heart of Darkness." *U.S. News & World Report* (March 3, 1999):18.

Gallup, George H. Jr. *The Gallup Poll.* Wilmington, DE: Scholarly Resources, 1988.

Gallup, George H. Jr. *Religion in America 2002.* Princeton, NJ: The Princeton Religion Research Center, 2002.

Game, Ann, and Andrew W. Metcalfe. *Passionate Sociology.* Thousand Oaks, CA: Sage, 1996.

Gamoran, Adam. "Instructional and Institutional Effects of Ability Grouping." *Sociology and Education* 59 (1986):185–198.

Gamoran, Adam. "The Variable Effects of High School Tracking." *American Sociological Review* 57 (December 1992):812–828.

Gamson, William. *The Strategy of Social Protest,* 2nd ed. Belmont, CA: Wadsworth, 1990.

Ganong, Lawrence, and Marilyn Coleman. *Remarried Family Relationships.* Thousand Oaks, CA: Sage, 1994.

Gans, Herbert J. *The Urban Villagers.* New York: Free Press, 1962a.

Gans, Herbert J. "Urbanism and Suburbanism as Ways of Life." In Arnold M. Rose (ed.), *Human Behavior and Social Processes.* Boston: Houghton Mifflin, 1962b, pp. 625–648.

Gans, Herbert J. *People and Plans.* New York: Basic Books, 1968.

Gans, Herbert J. "Reconstructing the Underclass: The Term's Danger as a Planning Concept." *Journal of the American Planning Association* 56 (Summer 1990):271–277.

Ganz, Alexander. "Where Has the Urban Crisis Gone?" *Urban Affairs Quarterly* 20 (June 1985):449–468.

Garbarino, J. "The Incidence and Prevalence of Child Maltreatment." In L. Ohlin and M. Tonry (eds.), *Family Violence.* Chicago: University of Chicago Press, 1989, pp. 219–261.

Gardner, Carol Brooks. "A Family among Strangers: Kinship Claims among Gay Men in Public Places." In Don A. Chekki, Spencer E. Cahill, and Lyn H. Lofland (eds.), *Research in Community Sociology*, Supplement 1. Greenwich, CT: JAI Press, 1994, pp. 95–118.

Gardner, David Pierpont. *A Nation at Risk: The Imperative for Educational Reform.* Report of the National Commission on Excellence in Education. Washington, D.C.: U.S. Government Printing Office, 1983.

Gardner, Lytt. "Deprivation Dwarfism." *Scientific American* 227 (July 1972):76–82.

Garfinkel, Harold. *Studies in Ethnomethodology.* Cambridge, UK: Polity Press, 1984.

Garland, David (ed.). *Mass Imprisonment: Social Causes and Consequences.* Thousand Oaks, CA: Sage, 2001.

Garreau, J. *Edge City: Life on the New Frontier.* New York: Doubleday, 1991.

Garst, Jennifer, and Galen V. Bodenhausen. "Advertising's Effects on Men's Gender Role Attitudes." *Sex Roles* 36 (May 1997):551–572.

Gatewood, James V., and Min Zhou (eds.). *Contemporary Asian America: A Multidisciplinary Reader.* New York: New York University Press, 2000.

Gauntlett, David. *Media Gender and Identity: An Introduction.* New York: Routledge, 2002.

Gay, Peter. *The Enlightenment: The Rise of Modern Paganism.* New York: Norton, 1966.

Geary, David C. *Male, Female: The Evolution of Human Sex Differences.* Washington, D.C.: American Psychological Association, 1998.

Gecas, Viktor, and Michael L. Schwalbe. "Beyond the Looking-Glass Self: Social Structure and Efficacy-Based Self-Esteem." *Social Psychology Quarterly* 46 (1983):77–88.

Gehrig, Gail. "The American Civil Religion Debate: A Source for Theory Construction." *Journal for the Scientific Study of Religion* 20 (March 1981):51–63.

Geis, Gilbert, Robert Meier, and Lawrence Salinger (eds.). *White-Collar Crime: Offenses in Business, Politics and the Professions,* 3rd ed. New York: Free Press, 1995.

Gelbard, Alene, and Carl Haub. "Population 'Explosion' Not Over for Half the World." *Population Today* 26 (March 1998):1–2.

Gelbard, Alene, Carl Haub, and Mary M. Kent. "World Population Beyond Six Billion." Washington, D.C.: Population Reference Bureau (March 1999):1–44.

Gelles, Richard J. *Intimate Violence in Families,* 3rd ed. Thousand Oaks, CA: Sage, 1997.

Gelles, Richard J., et al. (eds.). *Current Controversies on Family Violence,* 2nd ed. Thousand Oaks, CA: Sage, 2004.

Gerber, Jurg. "Heidi and Imelda: The Changing Image of Crime in Switzerland." *Criminology* 8 (March 1991):121–128.

Gersoni-Stavn, Deane. *Sexism and Youth.* New York: Bowker, 1974.

Gerth, H. H., and C. Wright Mills (eds.). *From Max Weber.* New York: Oxford University Press, 1958.

Gest, Ted. "Are White-Collar Crooks Getting Off Too Easy?" *U.S. News & World Report* 99 (July 1985):43.

Gibbons, Tom. "Justice Not Equal for Poor Here." *Chicago Sun Times* (February 24, 1985):1,18.

Gibbs, Nancy. "Oh My God, They're Killing Themselves." *Time* (May 3, 1993a):27–43.

Gibbs, Nancy. "Angels among Us." *Time* (December 27, 1993b):56–65.

Gibbs, Nancy. "An American Family Goes to War." *Time* (March 24, 2003):26–33.

Gibson, Margaret A., and John V. Ogbu. *Minority Status and Schooling.* New York: Garland Publishing, 1991.

Giddens, Anthony. *Studies in Social and Political Theory.* London: Hutchinson, 1979.

Giddens, Anthony. *Social Theory and Modern Sociology.* Stanford, CA: Stanford University Press, 1987.

Giddens, Anthony. *Runaway World: How Globalisation Is Reshaping Our Lives.* New York: Routledge, 2002.

Giddens, Anthony, et al. *Introduction to Sociology,* 6th ed. New York: Norton, 2007.

Gilbert, Dennis A. *The American Class Structure in an Age of Growing Inequality,* 7th ed. Thousand Oaks, CA: Sage, 2008.

Gilbert, Lucia Albino. *Two Careers/One Family.* Newbury Park, CA: Sage, 1993.

Gildea, William. "Violence Mars Night of Victory." *Washington Post* (June 21, 2000):D1.

Gilens, Martin. *Why Americans Hate Welfare: Race, Media, and the Politics of Antipoverty Policy.* Chicago: University of Chicago Press, 1999.

Giles, Jeff. "Generation X." *Newsweek* (June 6, 1994):62–72.

Gill, Brian P., et al. *Rhetoric versus Reality: What We Know and What We Need to Know about Vouchers and Charter Schools.* Santa Monica, CA: The Rand Corporation, 2001.

Giroux, Henry A. "Theories of Reproduction and Resistance in the New Sociology of Education: A Critical Analysis." *Harvard Educational Review* 53 (August 1983):257–293.

Gist, Noel P., and Sylvia Fleis Fava. *Urban Society,* 6th ed. New York: Crowell, 1974.

Gittell, Marilyn J. (ed.). *Strategies for School Equity: Creating Productive Schools in a Just Society.* New Haven, CT: Yale University Press, 1998.

Glazer, Nathan. "The End of Meritocracy." *The New Republic* (September 27, 1999):26.

Gleckman, Howard. "Big Spending Wrapped in a Flag." *BusinessWeek* (February 18, 2002):44.

Gleckman, H., J. Carey, R. Mitchell, T. Smart, and C. Pousch. "The Technology Payoff." *BusinessWeek* (June 14, 1993):51ff.

Glick, P., and S. L. Lin. "More Young Adults Are Living with Their Parents: Who Are They?" *Journal of Marriage and the Family* (February 1986):107–112.

Glock, Charles Y., and Robert N. Bellah (eds.). *The New Religious Consciousness.* Berkeley: University of California Press, 1976.

Glock, Charles Y., and Rodney Stark. *Religion and Society in Tension.* Chicago: Rand McNally, 1965.

Glock, Charles Y., and Rodney Stark. *American Piety.* Berkeley: University of California Press, 1968.

Glueck, Sheldon, and Eleanor Glueck. *Unraveling Juvenile Delinquency.* Cambridge, MA: Harvard University Press, 1950.

Gober, Patricia. "Americans on the Move." *Population Bulletin* 48 (November 1993):1–40.

Goffman, Erving. *Encounters.* Indianapolis, IN: Bobbs-Merrill, 1961a.

Goffman, Erving. *Asylums.* Garden City, NY: Anchor Books, 1961b.

Goffman, Erving. *Stigma.* Englewood Cliffs, NJ: Prentice Hall, 1963.

Goffman, Erving. *The Presentation of Self in Everyday Life.* New York: Overlook Press, 1974.

Goffman, Erving. *Gender Advertisements.* New York: Harper and Row, 1979.

Goffman, Erving. "The Interaction Order." *American Sociological Review* 48 (February 1983):1–17.

Goldberger, Arthur S., and Glen G. Cain. "The Causal Analysis of Cognitive Outcomes in the Coleman Report." Madison, WI: Institute for Research on Poverty. University of Wisconsin, December 1981.

Goldberger, Arthur S., and Glen G. Cain. "The Causal Analysis of Cognitive Outcomes in the Coleman, Hoffer, and Kilgore Report." *Sociology of Education* 55 (April/July 1982):103–122.

Golden, Frederic. "Albert Einstein: Person of the Century." *Time* (December 31, 1999):62–65.

Goldfarb, William. "Psychological Privation in Infancy and Subsequent Adjustment." *American Journal of Orthopsychiatry* 15 (April 1945):247–255.

Goldscheider, Frances K., and Calvin Goldscheider. "Leaving and Returning Home in 20th Century America." *Population Bulletin* 48 (March 1994):1–35.

Goldscheider, Frances K., and Linda J. Waite. *New Families, No Families?* Chapel Hill: University of North Carolina Press, 1991.

Goldstein, Joshua R., and Catherine T. Kenney. "Marriage Delayed or Marriage Forgone? New Cohort Forecasts of First Marriage for U.S. Women." *American Sociological Review* 66 (August 2001):506–519.

Goleman, D. "An Emerging Theory on Blacks' IQ Scores." *New York Times Education Supplement* (April 10, 1988):22–24.

Goo, Sara Kehaulani. "American Airlines Chief Executive Ousted." *Washington Post* (April 25, 2003):E1.

Good, Mary-Jo DelVecchio, Paul E. Brodwin, Byron J. Good, and Arthur Kleinman (eds.). *Pain as Human Experience: An Anthropological Perspective.* Berkeley: University of California Press, 1994.

Goode, Erich. *Collective Behavior.* Fort Worth, TX: Sanders College Publishing, 1992.

Goode, Judith, and Jeff Maskovskyk (eds.). *The New Poverty Studies: The Ethnography of Power, Politics, and Impoverished People in the United States.* New York: New York University Press, 2001.

Goode, William J. *World Revolution and Family Patterns.* New York: Free Press, 1970.

Goodman, Ann B., Carole Siegel, Thomas J. Craig, and Shang P. Lin. "The Relationship between Socioeconomic Class and Prevalence of Schizophrenia, Alcoholism, and Affective Disorders Treated by Inpatient Care in a Suburban Area." *American Journal of Psychiatry* 140 (1983):166–170.

Goodstein, David. "Scientific Misconduct." *Academe* 88 (January/February 2002):28–31.

Goodwin, Jeff, and James Jasper (eds.). *Social Movements Reader.* New York: John W. Wiley, 2008.

Gordon, Milton M. *Assimilation in American Life.* New York: Oxford University Press, 1964.

Gordon, Milton M. *Human Nature, Class, and Ethnicity.* New York: Oxford University Press, 1978.

Gordon, Suzanne. *Nursing Against the Odds.* Ithaca, NY: Cornell University Press, 2006.

Gorer, Geofrey. *Grief and Mourning in Contemporary Britain.* London: Cresset Press, 1965.

Gottfredson, Michael R., and Travis Hirschi. *A General Theory of Crime.* Stanford, CA: Stanford University Press, 1990.

Gould, R. "Growth Toward Self Tolerance." *Psychology Today* (February 1975):47–48.

Gould, Stephen Jay. *The Mismeasurement of Man.* New York: Norton, 1981.

Gouldner, Alvin N. "Metaphysical Pathos and the Theory of Bureaucracy." *American Political Science Review* 49 (June 1955):496–507.

Grabosky, Peter N., and Russell G. Smith. *Electronic Theft: Unlawful Acquisition in Cyberspace.* New York: Cambridge University Press, 2001.

"Graduates Take Rites of Passage into Japanese Corporate Life." *Financial Times* (April 8, 1991):4.

Grandjean, B. "An Economic Analysis of the Davis-Moore Theory of Stratification." *Social Forces* 53 (June 1975): 543–552.

Granovetter, Mark. "The Strength of Weak Ties." *American Journal of Sociology* 78 (May 1973):1360–1380.

Grapes, Bryan J. *Prisons.* San Diego, CA : Greenhaven Press, 2000.

Graubard, Allen. "The Free School Movement." *Harvard Educational Review* 42 (August 1972):351–373.

Greeley, Andrew M. "Political Attitudes among American White Ethnics." In Charles H. Anderson (ed.), *Sociological Essays and Research,* rev. ed. Homewood, IL: Dorsey, 1974, pp. 202–209.

Greeley, Andrew M. *Ethnicity, Denomination, and Inequality.* Beverly Hills, CA: Sage, 1976.

Greeley, Andrew M. *Religious Change in America.* Cambridge, MA: Harvard University Press, 1996.

Green, Donald P., Laurence H. McFalls, and Jennifer K. Smith. "Hate Crime: An Emergent Research Agenda." *Annual Review of Sociology* 27 (2001):479–504.

Green, John. *U.S. Religious Landscape Survey.* Washington, D.C.: Pew Forum on Religion and Public Life, 2008.

Green, Mark. *Selling Out.* New York: HarperCollins Publishers, 2002.

Greenberg, Daniel S. *Science, Money, and Politics: Political Triumph and Ethical Erosion.* Chicago: University of Chicago Press, 2003.

Greenberg, David F., and Ronald C. Kessler. "The Effects of Arrests on Crime: A Multivariate Panel Analysis." *Social Forces* 60 (March 1982):771–790.

Greenhouse, Steven. "The Mood at Work: Anger and Anxiety." *New York Times* (October 29, 2002):G1, G9.

Greenhouse, Steven. "17,000 G.E. Workers Strike over Higher Health Costs." *New York Times* (January 15, 2003).

Greenwald, Elissa A., Hilary R. Persky, Jay R. Campbell, and John Mazzeo. *NAEP 1998 Writing: Report Card for the Nation and the States.* Washington, D.C.: National Center for Education Statistics, 1999.

Gregory, Paul R., and Robert C. Stuart. *Comparative Economic Systems,* 3rd ed. Boston: Houghton Mifflin, 1989.

Greider, William. *Who Will Tell the People?* New York: Simon & Schuster, 1993.

Greider, William. "Is This America's Top Corporate Crime Fighter?" *The Nation* (August 5, 2002):11–15.

Greider, William. *The Soul of Capitalism.* New York: Simon and Schuster, 2004.

Grescoe, P. *The Merchants of Venus: Inside Harlequin and the Empire of Romance.* Custer, WA: Orca Book Publishers, 1997.

Griffin, John Howard. *Black Like Me.* Boston: Houghton Mifflin, 1961.

Gross, Beatrice, and Ronald Gross. *The Great School Debate.* New York: Simon & Schuster, 1985.

Gross, L., and S. Jeffries-Fox. "What Do You Want to Be When You Grow Up, Little Girl?" In Gary Tuchman, Arlene Kaplan Daniels, and James Benet (eds.), *Hearth and Home.* New York: Oxford University Press, 1978, pp. 240–265.

Grubb, W. Norton, and Robert H. Wilson. "Trends in Wage and Salary Inequality, 1967–1988." *Monthly Labor Review* 115 (June 1992):23–39.

Guillen, Mauro F. *The Limits of Convergence: Globalization and Organizational Change in Argentina, South Korea, and Spain.* Princeton, NJ: Princeton University Press, 2003.

Guinzburg, Suzanne. "Education's Earning Power." *Psychology Today* 17 (October 1983):20–21.

Gur, R. C., et al. "Sex Differences in Regional Cerebral Glucose Metabolism During a Resting State." *Science* 267 (January 27, 1995):528–531.

Gutman, Henry G. *The Black Family in Slavery and Freedom, 1750–1925.* New York: Pantheon, 1983.

Hacker, Andrew. *Two Nations: Black and White, Separate, Hostile, and Unequal,* expanded and updated ed. New York: Ballantine Books, 1995.

Hadaway, C. Kirk, Penny Long Marler, and Mark Chaves. "What the Polls Don't Show: A Closer Look at U.S. Church Attendance." *American Sociological Review* 58 (December 1993):741–752.

Hadden, Jeffrey K., and Raymond R. Rymph. "The Marching Ministers." In Jeffrey K. Hadden (ed.), *Religion in Radical Transition,* 2nd ed. New Brunswick, NJ: Transaction Books, 1973, pp. 99–109.

Hadden, Jeffrey K., and Charles E. Swann. *Prime Time Preachers.* Reading, MA: Addison-Wesley, 1981.

Haferkamp, Hans, and Neil J. Smelser (eds.). *Social Change and Modernity.* Chapel Hill: University of North Carolina Press, 1992.

Hagan, Frank E. *Introduction to Criminology,* 6th ed. Thousand Oaks, CA: Sage, Nelson-Hall Publishers, 2007.

Hagan, John. "The Social Embeddedness in Crime and Unemployment." *Criminology* 31 (November 1993):464–492.

Hagan, John. *Crime and Disrepute.* Thousand Oaks, CA: Pine Forge Press, 1994.

Hagedorn, John M. "Gang Violence in the Postindustrial Era." In Michael Tonry and Mark H. Moore (eds.), *Youth Violence.* Chicago: University of Chicago Press, 1998, pp. 365–419.

Hall, Edward T. "How Cultures Collide." *Psychology Today* 10 (July 1976):66ff.

Hall, Edward T. "Learning the Arabs' Silent Language." *Psychology Today* 13 (August 1979):45ff.

Hall, Edward T. *The Silent Language.* New York: Anchor Books, 1990.

Hall, John. "President's Remarks Leave Settlement Up in the Air." *American Business Review* (September 18, 1997):1.

Hall, Richard H. "Intraorganizational Structural Variation: Application of the Bureaucratic Model." *Administrative Science Quarterly* 7 (December 1962):295–308.

Hall, Richard H., and Pamela S. Tolbert. *Organizations,* 9th ed. Upper Saddle River, NJ: Prentice Hall, 2004.

Hallinan, Maureen T. "Equality of Educational Opportunity." In W. Richard Scott and Judith Blake (eds.), *Annual Review of Sociology* 14 (1988):249–268.

Hallinan, Maureen T. (ed.). *Restructuring Schools: Promising Practices and Policies.* South Bend, IN: University of Notre Dame, 1995.

Hallinan, Maureen T. (ed.). *Handbook of the Sociology of Education.* New York: Springer-Verlag, 2006.

Hallock, Kevin F. "Reciprocally Interlocking Boards of Directors and Executive Compensation." *Journal of Financial and Quantitative Analysis* 32 (September 1997):331–344.

Hamilton, Peter, and Kenneth Thompson (eds.). *The Uses of Sociology.* Malden, MA: Blackwell Publishers, 2002.

Hamilton, Richard F. *The Social Misconstruction of Reality: Validity and Verification in the Scholarly Community.* New Haven, CT: Yale University Press, 1996.

Hammer, Joshua, and Zoran Cirjakovic. "Free at Last." *Newsweek* (October 16, 2000):24–31.

Hampden-Turner, Charles, and Fons Trompenaars. *The Seven Cultures of Capitalism.* London: Piatkus, 1995.

Hampton, Robert L., et al. (eds.). *Family Violence.* Newbury Park, CA: Sage, 1993.

Hancock, LynNell, and Pat Wingert. "A Mixed Report Card." *Newsweek* (November 13, 1995):69.

Hannan, Michael T., and John Freeman. *Organizational Ecology.* Cambridge, MA: Harvard University Press, 1989.

Hannon, Kelly. "Title IX: History of the Civil Rights Legislation." *Lynchburg News & Advance* (August 8, 2003).

Harlow, Harry F. "The Young Monkeys." *Psychology Today* 5 (1967):40–47.

Harlow, Harry F., and Margaret Harlow. "Social Deprivation in Monkeys." *Scientific American* 207 (November 1962): 137–146.

Harlow, Harry F., and Robert R. Zimmerman. "Affectional Responses in the Infant Monkey." *Science* 21 (August 1959):421–432.

Harmon, Amy. "In DNA Era, New Worries about Prejudice." *New York Times* (November 11, 2007).

Harris, Chauncy D. "The Nature of Cities and Urban Geography in the Last Half Century." *Urban Geography* 18 (1997):15–35.

Harris, Chauncy D., and Edward L. Ullman. "The Nature of Cities." *Annals of the American Academy of Political and Social Science* 242 (November 1945):7–17.

Harris, Gardiner. "Cigarette Company Paid for Lung Cancer Study." *New York Times* (March 26, 2008).

Harris, Judith Rich. *The Nurture Assumption: Why Children Turn Out the Way They Do.* New York: Free Press, 1998.

Harris, Marvin. *Cows, Pigs, Wars, and Witches.* New York: Random House, 1974.

Harris, Marvin. *Our Kind: Who We Are, Where We Came From, and Where We Are Going.* New York: HarperTrade, 1990.

Harrison, Bennett. *Lean and Mean.* New York: Basic Books, 1994.

Hartman, Laura P. "Technology and Ethics: Privacy in the Workplace." Waltham, MA: Center for Business Ethics (February 28, 2000):1–28.

Haupt, Arthur, and Thomas T. Kane. *Population Handbook.* 4th International ed. Washington, D.C.: Population Reference Bureau, 1998.

Haupt, Arthur, and Thomas T. Kane. *Population Handbook,* 5th ed. Washington, D.C.: Population Reference Bureau, 2004.

Hauser, Marc D. *Wild Minds: What Animals Really Think.* New York: Henry Holt, 2001.

Haviland, William A., et al. *Cultural Anthropology,* 12th ed. New York: Harcourt, 2007.

Hawkins, Asher. "Will the Silver State Go Green?" *Newsweek* (July 29, 2002):12.

Hawley, Amos H. "Human Ecology." In David L. Sills (ed.), *International Encyclopedia of the Social Sciences.* New York: Macmillan, 1968, pp. 329–337.

Hawley, Amos H. *Urban Society,* 2nd ed. New York: Wiley, 1971.

Hawley, Amos H. "The Logic of Macrosociology." *Annual Review of Sociology* 18 (1992):1–14.

Hawley, W. D. "Achieving Quality Integrated Education— With or Without Federal Help." In F. Schultz (ed.), *Annual Editions: Education 85/86.* Guilford, CT: Dushkin Publishing, 1985, pp. 142–145.

Hawley, W. D., and M. A. Smylie. "The Contribution of School Desegregation to Academic Achievement and Racial Integration." In P. A. Katz and D. A. Taylor (eds.), *Eliminating Racism.* New York: Plenum Press, 1988, pp. 281–297.

Hays, Kim. *Practicing Virtues: Moral Traditions at Quaker and Military Boarding Schools.* Berkeley and Los Angeles: University of California Press, 1994.

Heath, Chip, Chris Bell, and Emily Sternberg. "Emotional Selection in Memes: The Case of Urban Legends." *Journal of Personality and Social Psychology* 81 (December 2001): 1028–1041.

Heatherton, Todd F., Michelle R. Hebl, and Jay G. Hull (eds.). *The Social Psychology of Stigma.* New York: Guilford Publications, 2000.

Hechter, Michael, and Karl-Dieter Opp (eds.). *Social Norms.* New York: Russell Sage Foundation, 2005.

Heckert, Druann Maria. "Positive Deviance." In Patricia A. Adler and Peter Adler (eds.), *Constructions of Deviance: Social Power, Context, and Interaction.* Belmont, CA: Wadsworth, 2003, pp. 30–42.

Heckscher, Charles. *White-Collar Blues.* New York: Basic Books, 1995.

Heclo, Hugh, and Wilfred M. McClay (eds.). *Religion Returns to the Public Square: Faith and Policy in America.* Washington, D.C.: Woodrow Wilson Center Press, 2002.

Heelas, Paul. *The New Age Movement.* Cambridge, MA: Blackwell Publishers, 1996.

Hefner, Robert, and Rebecca Meda. "The Future of Sex Roles." In Marie Richmond-Abbott (ed.), *The American Woman.* New York: Holt, Rinehart and Winston, 1979, pp. 243–264.

Heilbrunn, Jacob. *They Knew They Were Right.* New York: Doubleday, 2008.

Heineman, Kenneth J. *Put Your Bodies upon the Wheels: Student Revolt in the 1960s.* Blue Ridge Summit, PA: Ivan R. Dee, 2002.

Helfer, Mary Edna, Ruth S. Kempe, and Richard D. Krugman (eds.). *The Battered Child,* 5th ed. Chicago: University of Chicago Press, 1999.

Helle, H. J., and S. N. Eisenstadt (eds.). *Macro-Sociological Theory,* vol. 1. Beverly Hills, CA: Sage, 1985a.

Helle, H. J., and S. N. Eisenstadt (eds.). *Micro-Sociological Theory,* vol. 2. Beverly Hills, CA: Sage, 1985b.

"Hello, Dolly." *The Economist* 18 (March 1, 1997):17–18.

Hendricks, Jon, and C. Davis Hendricks. *Aging in Mass Society,* 3rd ed. New York: HarperCollege, 1987.

Henig, Jeffrey R. *Rethinking School Choice.* Princeton, NJ: Princeton University Press, 1994.

Henretta, John C. *Age and Inequality.* New York: Perseus Publishing, 2001.

Henry, Stuart, and Mark Lanier (eds.). *What Is Crime? Controversies over the Nature of Crime and What to Do about It.* Lanham, MD: Rowman & Littlefield, 2001.

Henry, William A. III. "News as Entertainment: The Search for Dramatic Unity." In Elie Abel (ed.), *What's News.* San Francisco: Institute for Contemporary Studies, 1981, pp. 133–158.

Herbert, Wray. "When Strangers Become Family." *U.S. News and World Report* (November 29, 1999):58–67.

Herman, Arthur. *How the Scots Invented the Modern World.* New York: Crown Publishers, 2001.

Herring, Cedric. "Is Job Discrimination Dead?" *Contexts* (Summer 2002):13–18.

Herrnstein, Richard J., and Charles A. Murray. *The Bell Curve: Intelligence and Class Structure in American Life.* New York: Free Press Paperbacks, 1996.

Hertz, Rosanna, and Nancy L. Marshall (eds.). *Working Families: The Transformation of the American Home.* Berkeley, CA: University of California Press, 2001.

Hertz, Rosanna. *Single by Chance, Mothers by Choice.* New York: Oxford University Press, 2006.

Hesiod. *Theogony, Works and Days, Shield. Translation, Introduction, and Notes by Apostolos N. Athanassakis.* Baltimore: Johns Hopkins University Press, 1983.

Hettena, Seth. *Feasting on the Spoils.* New York: St. Martin's Press, 2007.

Hewitt, John P. *Self and Society: A Symbolic Interactionist Social Psychology,* 9th ed. Boston: Allyn & Bacon, 2002.

Hilbert, Richard A. "Ethnomethodology and the Micro-Macro Order." *American Sociological Review* 55 (December 1990):794–808.

Hill, Michael R., and Susan Hoecker-Drysdale (eds.). *Harriet Martineau: Theoretical and Methodological Perspectives.* New York: Routledge, 2001.

Hill, Patrice. "Clinton May Trim Some Budget Pork." *Washington Times* (August 5, 1997):A1, A14.

Hill, Steven. *Fixing Elections: The Failure of America's Winner Take All Politics.* Bristol, PA: Taylor and Francis, 2003.

Hillier, Susan, and Georgia M. Barrow. *Aging, the Individual, and Society,* 8th ed. Belmont, CA: Wadsworth, 2007.

Himes, Christine L. "Elderly Americans." *Population Bulletin* 56 (December 2001):1–40.

Hirsch, E. D. Jr. *Cultural Literacy.* Boston: Houghton Mifflin, 1987.

Hirsch, E. D. *The Schools We Need and Why We Don't Have Them.* New York: Doubleday, 1999.

Hirsch, E. D. *The New First Dictionary of Cultural Literacy,* 3rd ed. Boston: Houghton Mifflin Company, 2004.

Hirschfeld, Lawrence A. *Race in the Making.* Cambridge, MA: MIT Press, 1998.

Hirschi, Travis. *Causes of Delinquency.* Berkeley: University of California Press, 1972.

Hirschi, Travis, and Michael R. Gottfredson. "Substantive Positivism and the Idea of Crime." *Rationality and Society* 2 (October 1990):412–428.

Hirschi, Travis, and Hanan Selvin. *Principles of Survey Analysis.* New York: Free Press, 1973.

Hobbes, Thomas. *Leviathan.* Edited by Herbert W. Schneider. New York: Liberal Arts Press, 1958.

Hochschild, Arlie R. *The Second Shift.* With Anne Machung, contributor. New York: Avon, 1997.

Hochschild, Arlie R. *The Managed Heart: Commercialization of Feeling.* 20th anniversary ed. Berkeley: University of California Press, 2003.

Hochschild, Arlie R., and Anne Machung. *The Time Bind: When Work Becomes Home and Home Becomes Work.* New York: Penguin, 2003.

Hodge, Robert W., Paul M. Siegel, and Peter H. Rossi. "Occupational Prestige in the United States, 1925–1963." *American Journal of Sociology* 70 (November 1964):286–302.

Hodge, Robert W., and Donald J. Treiman. "Class Identification in the U.S." *American Journal of Sociology* 73 (March 1968):535–547.

Hodge, Robert W., Donald J. Treiman, and Peter H. Rossi. "A Comparative Study of Occupational Prestige." In Reinhard Bendix and Seymour Martin Lipset (eds.), *Class, Status, and Power,* 2nd ed. New York: Free Press, 1966, pp. 309–321.

Hodson, Randy, and Robert L. Kaufman. "Economic Dualism: A Critical Review." *American Sociological Review* 47 (December 1982):727–739.

Hodson, Randy, and Teresa A. Sullivan. *The Social Organization of Work,* 4th ed. Belmont, CA: Wadsworth, 2008.

Hoebel, E. Adamson. *The Law of Primitive Man.* New York: Atheneum, 1983.

Hoffer, Richard. "Bannister and Hillary." *Sports Illustrated* (December 27, 1999):96.

Hoffman, Lois W., and Francis I. Nye. *Working Mothers.* San Francisco: Jossey-Bass, 1975.

Hofstadter, Richard. *America at 1750.* New York: Vintage, 1973.

Hofstader, Richard. *Anti-Intellectualism in American Life.* New York: Knopf Publishing Group, 1966.

Hogan, Dennis P., and Nan Marie Astone. "Transition to Adulthood." In Ralph H. Turner and James F. Short Jr. (eds.), *Annual Review of Sociology* 12 (1986):109–130.

Hogg, Michael A., and Dominic Abrams (eds.). *Intergroup Relations: Essential Readings.* Philadelphia: Psychology Press, 2001.

Holland, Joshua. "The American Dream Is Alive and Well . . . in Finland." *Alternet* (December 11, 2007), http://www.alternet.org/story/70103/.

Hollingshead, August B. *Elmtown's Youth.* New York: Wiley, 1949.

Hollingshead, August B., and Frederick C. Redlich. *Social Class and Mental Illness.* New York: Wiley, 1958.

Holmberg, Allan R. *Nomads of the Long Bow.* Garden City, NY: Natural History Press, 1969.

Holt, John. *How Children Fail,* rev. ed. Reading, MA: Perseus Books, 1995.

Hook, Donald D. *Death in the Balance.* Lexington, MA: Heath, 1989.

Hoover, Stewart M. *The Social Sources of the Electronic Church.* Newbury Park, CA: Sage, 1988.

Hooyman, Nancy R., and H. Asuman Kiyak. *Social Gerontology,* 8th ed. Boston: Allyn & Bacon, 2007.

Hope, Keith. "Vertical and Nonvertical Class Mobility in Three Countries." *American Sociological Review* 47 (February 1982):99–113.

Horsley, Richard A. *Jesus and Empire: The Kingdom of God and the New World Disorder.* Minneapolis: Fortress Press, 2003.

Horwitz, Allan V. *The Logic of Social Control.* New York: Plenum, 1990.

Horwitz, Robert A. "Psychological Effects of the Open Classroom." *Review of Educational Research* 49 (Winter 1979):71–86.

Hotaling, G. T., et al. (eds.). *Family Abuse and Its Consequences.* Beverly Hills, CA: Sage, 1988.

Hougland, James G. Jr., and Willis A. Sutton Jr. "Factors Influencing Degree of Involvement in Interorganizational Relationships in a Rural County." *Rural Sociology* 43 (Winter 1978):649–670.

Houseman, John. "The Men from Mars." *Harper's Magazine* 197 (December 1948):74–82.

Hoyt, Homer. *The Structure and Growth of Residential Neighborhoods in American Cities.* Washington, D.C.: Federal Housing Authority, 1939.

Hrdy, Sarah Blaffer. *The Woman That Never Evolved,* rev. ed. Cambridge, MA: Harvard University Press, 1999. Available online at www.tvb.org/tvfacts/tvbacis/basics6.html.

Hrdy, Sarah Blaffer. *Mother Nature: Maternal Instincts and How They Shape the Human Species.* New York: Ballantine Books, 2000.

Hubbard, L. Ron. *What Is Scientology?* Los Angeles: Bridge, 1998.

Huber, Peter, and Jessica Korn. "The Plug-and-Play Economy." *Forbes* (July 7, 1997):268–272.

Hudak, Glenn M., and Paul Kihn (eds.). *Labeling: Pedagogy and Politics.* New York: Routledge, 2001.

Hudson, Ken. "Duality and Dual Labor Markets." Revision of a paper presented at the 93rd Annual Meetings of the American Sociological Association, 2000.

Huffington, Arianna. *Pigs at the Trough: How Corporate Greed and Political Corruption Are Undermining America.* New York: Crown Publishers, 2004.

Hughes, John A., Peter J. Martin, and W. W. Sharrock. *Understanding Classical Sociology.* Thousand Oaks, CA: Sage, 1995.

Hughes, Michael, and Melvin E. Thomas. "The Continuing Significance of Race Revisited: A Study of Race, Class, and Quality of Life in America, 1972 to 1996." *American Sociological Review* 63 (December 1998):785–795.

Huizinga, David, and Delbert S. Elliot. "Juvenile Offenders: Prevalence, Offender Incidence, and Arrest Rates by Race." *Crime and Delinquency* 33 (April, 1987):206–223.

Human Development Report 1999. New York: Oxford University Press, 1999.

Human Development Report 2002. New York: Oxford University Press, 2002.

Hummel, Ralph P. *The Bureaucratic Experience,* 3rd ed. New York: St. Martin's Press, 1987.

Humphrey, Craig R., and Frederick H. Buttel. *Environment, Energy and Society.* Malabar, FL: Krieger Publishing, 1986.

Humphreys, Laud. *Tearoom Trade.* New York: Aldine, 1979.

Hunt, Morton M. *The New Know-Nothings: The Political Foes of the Scientific Study of Human Nature.* New Brunswick, NJ: Transactions, 1999.

Hunt, Stephen, Malcolm Hamilton, and Tony Walter (eds.). *Charismatic Christianity: Sociological Perspectives.* New York: St. Martin's Press, 1998.

Hunter, Floyd. *Community Power Structure.* Chapel Hill: University of North Carolina Press, 1953.

Hunter, Floyd. *Community Power Succession.* Chapel Hill: University of North Carolina Press, 1980.

Hunter, James Davison. *Evangelicism.* Chicago: University of Chicago Press, 1987.

Hurley, Jennifer A. (ed.). *Racism.* San Diego, CA: Greenhaven Press, 1998.

Hurley, Jennifer A. (ed.). *American Values.* Guilford, CT: Greenhaven Press, 2000.

Hurn, Christopher J. *The Limits and Possibilities of Schooling,* 3rd ed. Boston: Allyn & Bacon, 1993.

Ikeda, Keiko. *A Room Full of Mirrors.* Stanford, CA: Stanford University Press, 1998.

"Impact of Racial and Ethnic Diversity on Educational Outcomes, The." Cambridge, MA: The Civil Rights Project, January 2002.

Inciardi, James A. *Criminal Justice,* 8th ed. New York: McGraw-Hill, 2007.

Infante, Esme M. "Panel: Nicotine Addictive." *USA Today* (August 13, 1994):A1.

Inglehart, Ronald, and Wayne E. Baker. "Modernization, Cultural Change, and the Persistence of Traditional Values." *American Sociological Review* 65 (2000):19–51.

Ingrassia, Michele, and Melinda Beck. "Patterns of Abuse." *Newsweek* (July 4, 1994):26–33.

Inkeles, Alex. *One World Emerging? Convergence and Divergence in Industrial Societies.* Boulder, CO: Westview Press, 1998.

Insana, Ron. *Trendwatching: Don't Be Fooled by the Next Investment Fad, Mania, or Bubble.* New York: HarperCollins, 2002.

Isaacs, Julia B. "Economic Mobility of Families across Generations." Economic Mobility Project. Washington, D.C.: Pew Charitable Trusts, 2007a.

Isaacs, Julia B. "Economic Mobility of Men and Women." Economic Mobility Project. Washington, D.C.: Pew Charitable Trusts, 2007b.

Isaacs, Julia B. "Economic Mobility of Black and White Families." Economic Mobility Project. Washington, D.C.: Pew Charitable Trusts, 2007c.

Isikoff, Michael. "Christian Coalition Steps Boldly into Politics." *Washington Post* (September 10, 1992):A1, A14.

Isikoff, Michael. "The Secret Money Chase." *Newsweek* (June 5, 2000):23.

Jackman, Mary R., and Robert W. Jackman. *Class Awareness in the United States.* Berkeley: University of California Press, 1983.

Jackson, Elton F., and Harry J. Crockett Jr. "Occupational Mobility in the United States: A Point Estimate and Trend Comparison." *American Sociological Review* 29 (February 1964):5–15.

Jackson, Stevi, and Sue Scott (eds.). *Gender: A Sociological Reader.* New York: Routledge, 2001.

Jacob, John E. "Black America, 1993: An Overview." In Billy J. Tidwell (ed.), *The State of Black America 1994.* New York: National Urban League, 1994, pp. 1–9.

Jacobe, Dennis. "Work is Part of Many Investors' Retirement Plans." Gallup News Service (November 16, 2004).

Jacobson, Neil S. and John Gottman. *When Men Batter Women.* New York: Simon & Schuster Trade, 1998.

Jacoby, Sanford M. *Modern Manors: Welfare Capitalism Since the New Deal.* Princeton, NJ: Princeton University Press, 1997.

Jacoby, Susan. "How Dumb Can We Get?" *The Week* (February 29, 2008a): 36-37.

Jacoby, Susan. *The Age of American Unreason.* New York: Knopf Publishing Group, 2008b.

Jaggar, Alison M., and Paula S. Rothenberg. *Feminist Frameworks,* 3rd ed. New York: McGraw-Hill, 1993.

James, Paul. *Nation Formation.* Thousand Oaks, CA: Sage, 1996.

Janis, Irving L. *Groupthink,* 2nd ed. Boston: Houghton Mifflin, 2006.

Janofsky, Michael. "Rustic Life of Amish Is Changing but Slowly." *New York Times* (July 6, 1997):7.

Janowitz, Morris. *Sociology and the Military Establishment,* 3rd ed. Beverly Hills, CA: Sage, 1974.

Janssen-Jurreit, Marie Louise. *Sexism.* New York: Farrar Strauss Giroux, 1982.

Jasinski, Jana L. (ed.). *Partner Violence: A Comprehensive Review of Twenty Years of Research.* Thousand Oaks, CA: Sage, 2000.

Jencks, Christopher. *Inequality.* New York: HarperCollins Publishers, 1981.

Jencks, Christopher, and Meredith Phillips (eds.). *The Black-White Test Score Gap.* Washington, D.C.: Brookings Institution, 1998.

Jenkins, J. Craig. "Resource Mobilization Theory and the Study of Social Movements." In Ralph H. Turner and James F. Short Jr. (eds.), *Annual Review of Sociology* 9 (1983):527–553.

Jenness, Valerie, and Ryken Grattet. *Making Hate a Crime: From Social Movement to Law Enforcement.* New York: Russell Sage Foundation, 2001.

Jensen, Arthur. "How Much Can We Boost IQ and Scholastic Achievement?" *Harvard Educational Review* 39 (Winter 1969):1–123.

Johnson, Charles Richard, and Patricia Smith. *Africans in America: America's Journey through Slavery.* San Diego, CA: Harcourt Brace and Company, 1998.

Johnson, Darragh. "SAT Scores Dip Slightly Overall: Widening Racial Gap Is Greater Concern." *Washington Post* (August 29, 2002):A03.

Johnson, David W., and Frank P. Johnson. *Joining Together,* 5th ed. Boston: Allyn & Bacon, 1994.

Johnson, Dirk. "A Leap of Fate." *Newsweek* (January 20, 2003a):34.

Johnson, Dirk. "The Night the Music Died." *Newsweek* (March 3, 2003b):40–42.

Johnson, Kevin. "DNA Tests Fuel Urgency to Free the Innocent." *USA Today* (February 2, 2008).

Johnson, Steven. *Everything Bad Is Good for You.* New York: Riverhead Books, 2006.

Johnson, Troy R. *Contemporary Native American Political Issues.* Thousand Oaks, CA: AltaMira Press, 2000.

Johnston, David Cay. "Enron Avoided Income Taxes in 4 of 5 Years." *New York Times* (January 17, 2002):A1.

Johnston, David Cay. *Perfectly Equal.* New York: Portfolio, 2005.

Johnston, David Cay. *Free Lunch.* New York: Penguin Books, 2007a.

Johnston, David Cay. "Report Says That the Rich Are Getting Richer, Faster, Much." *New York Times* (December 15, 2007b).

Johnstone, Ronald L. *Religion in Society,* 8th ed. Upper Saddle River, NJ: Prentice Hall, 2006.

Jones, A. H. M. *The Decline of the Ancient World.* New York: Holt, Rinehart and Winston, 1966.

Jones, Daniel (ed.). *Dylan Thomas: The Poems.* London: J. M. Dent and Sons, 1971.

Jones, James H. *Alfred C. Kinsey: A Public/Private Life.* New York: Norton, 1997.

Jones, James M. *Prejudice and Racism.* Reading, MA: Addison-Wesley, 1972.

Jones, James M. *Bad Blood,* expanded ed. New York: Free Press, 1993.

Jones, Jeffrey M. "Low Trust in Federal Government Rivals Watergate Era Levels." Gallup News Service (September 26, 2007a).

Jones, Jeffrey M. "Majority of Americans Satisfied with Their Healthcare Plans." Gallup News Service (November 29, 2007b).

Jones, Jeffrey M. "Trust in Government Declining, Near Lows for the Past Decade." Gallup News Service (September 26, 2006).

Jones, Robert Alun. *The Development of Durkheim's Social Realism.* New York: Cambridge University Press, 2005.

Juliano, Nick. "Income Inequality Worst Since 1920s, According to IRS Data." *The Raw Story* (October 12, 2007), http://rawstory.com/printstory.phpostory=7882.

Kadushin, Charles. "Mental Health and the Interpersonal Environment: A Reexamination of Some Effects of Social Structure on Mental Health." *American Sociological Review* 48 (April 1983):188–198.

Kalish, Carol B. *International Crime Rates.* U.S. Bureau of Justice Statistics. Washington, D.C.: U.S. Government Printing Office, 1988.

Kanter, Rosabeth Moss. *Commitment and Community.* Cambridge, MA: Harvard University Press, 1972.

Kanter, Rosabeth Moss. *Men and Women of the Corporation.* New York: Basic Books, 1993.

Kantrowitz, Barbara. "A Nation Still at Risk." *Newsweek* (April 19, 1993a):46–49.

Kantrowitz, Barbara. "The Group Classroom." *Newsweek* (May 10, 1993b):73.

Kaplan, David A. "Bobbitt Fever." *Newsweek* (January 24, 1994):52–55.

Kaplan, Fred. *Daydream Believers.* New York: John Wiley & Sons, 2008.

Kaplan, Jeffrey, and Helene Loow (eds.). *Cultic Milieu.* Thousand Oaks, CA: AltaMira Press, 2002.

Karatnycky, Adrian. "Democracies on the Rise, Democracies at Risk." *Freedom Review* 26 (January/February 1995):5–10.

Kardulias, P. Nick (ed.). *World-Systems Theory in Practice: Leadership, Production, and Exchange.* Lanham, MD: Rowman & Littlefield, 1999.

Karson, Jill (ed.). *Cults.* San Diego, CA: Greenhaven Press, 2000.

Kart, C. S. *The Realities of Aging.* Boston: Allyn & Bacon, 1990.

Kasarda, John D., and Morris Janowitz. "Community Attachment in Mass Society." *American Sociological Review* 39 (June 1974):328–339.

Kashef, Ziba. "Is DNA Research Giving New Life to the Idea That Race Exists?" *Alternet* (October 18, 2007), http://alternet.org/story/65484.

Katchadourian, Herant, and John Boli. *Cream of the Crop: The Impact of Elite Education in the Decade After College.* New York: Basic Books, 1994.

Kate, Nancy T. "Two Careers, One Marriage." *American Demographics* (April 1998):28.

Katel, Peter. "The Bust in Boot Camps." *Newsweek* (February 21, 1994):26.

Katsiaficas, George N., and Teodros Kiros (eds.). *The Promise of Multiculturalism: Education and Autonomy in the 21st Century.* New York: Routledge, 1998.

Katz, Irwin. "Desegregation or Integration in Public Schools?: The Policy Implications of Research." *Integrated Education* 5 (December 1967/January 1968):15–28.

Katz, Michael B. *Class, Bureaucracy and Schools.* New York: Praeger, 1975.

Katz, Michael B. (ed.). *The "Underclass" Debate.* Princeton, NJ: Princeton University Press, 1993.

Katznelson, Ira, Mark Kesselman, and Alan Draper. *The Politics of Power: A Critical Introduction to American Government,* 5th ed. Belmont, CA: Wadsworth, 2005.

Kaufman, Philip. *Dropout Rates in the United States: 1999.* Washington, D.C.: National Center for Education Statistics, November 2000.

Kaufman, Philip. *The Nation's Report Card: Fourth Grade Reading 2000.* Washington, D.C.: National Center for Education Statistics, 2001.

Kearl, Michael C., and Anoel Rinaldi. "The Political Uses of the Dead as Symbols in Contemporary Civil Religions." *Social Forces* 61 (March 1983):693–708.

Keen, Judy. "In Capital, Business and Politics Firmly Entwined." *USA Today* (July 31, 2002):1A, 2A.

Kellam, Sheppard G., Margaret E. Ensminger, and R. Jay Turner. "Family Structure and the Mental Health of Children." *Archives of General Psychiatry* 34 (September 1977):1012–1022.

Keller, Amy. "Roll Call's List of the 50 Richest Lawmakers." *Roll Call* (September 9, 2002).

Kelley, Harold H., and John W. Thibaut. "Group Problem Solving." In Gardner Lindzey and Elliot Aronson (eds.), *The Handbook of Social Psychology,* vol. 4. Reading, MA: Addison-Wesley, 1969, pp. 735–785.

Kelley, Maryellen R. "New Process Technology, Job Design, and Work Organization: A Contingency Model." *American Sociological Review* 55 (1990):191–208.

Kelly, Delos H. (ed.). *Deviant Behavior,* 5th ed. New York: St. Martin's Press, 1996.

Kendall, Diana. *Sociology in Our Times,* 7th ed. Belmont, CA: Wadsworth, 2008.

Kenneally, Christine. "When Language Can Hold the Answer." *New York Times* (April 22, 2008).

Kennedy, Randall. *Interracial Intimacies: Sex, Marriage, Identity, and Adoption.* New York: Knopf, 2003.

Kent, Mary M., and Mark Mather. "What Drives U.S. Population Growth?" *Population Bulletin* 57 (December 2002):3–40.

Kephart, William M., and Davor Jedlicka. *The Family, Society, and the Individual,* 7th ed. New York: HarperCollins, 1997.

Khouri, Norma. *Honor Lost: Love and Death in Modern-Day Jordan.* New York: Simon & Schuster, 2005.

Kibria, Nazli. *Becoming Asian American: Second-Generation Chinese and Korean American Identities.* Baltimore, MD: Johns Hopkins University Press, 2003.

Kiechel, W. "How Will We Work in the Year 2000?" *Fortune* (May 17, 1993):38.

Kilbourne, Jean. "Beauty and the Beast of Advertising." In James M. Henslin (ed.), *Down to Earth Sociology.* New York: Free Press, 2001, pp. 391–394.

Kilgore, Sally B. "The Organizational Context of Tracking in Schools." *American Sociological Review* 56 (April 1991):189–203.

Killian, Lewis M. "Organization, Rationality and Spontaneity in the Civil Rights Movement." *American Sociological Review* 49 (December 1984):770–783.

Kilson, Martin L. "The State of African-American Politics." In Lee A. Daniels (ed.), *The State of Black America 1998.* New York: National Urban League, 1998, pp. 247–270.

Kilson, Martin L. "African Americans and American Politics 2002: The Maturation Phase." In Lee A. Daniels (ed.), *The State of Black America 2002.* New York: National Urban League, 2002, pp. 147–180.

Kim, Alice. "N.J. Closes Death Row." *In These Times* (January 17, 2008).

Kimmel, Michael S. *Gendered Society.* New York: Oxford University Press, 2000.

Kinsey, Alfred C., and Paul Gebhard. *Sexual Behavior in the Human Female.* Philadelphia: Saunders, 1953.

Kinsey, Alfred C., Wardell Pomeroy, and Clyde Martin. *Sexual Behavior in the Human Male.* Philadelphia: Saunders, 1948.

Kinzie, Susan. "Internet Access Is Only Prerequisite for More and More College Classes." *Washington Post* (December 31, 2007).

Kirkpatrick, David. "Smart New Ways to Use Temps." *Fortune* (February 15, 1988):110–116.

Kissman, Kris, and Jo Ann Allen. *Single-Parent Families.* Newbury Park, CA: Sage, 1993.

Kitano, Harry H. *The Japanese Americans,* 2nd ed. New York: Chelsea House, 2005.

Kitano, Harry H., and Roger Daniels. *Asian Americans,* 4th ed. Englewood Cliffs, NJ: Prentice Hall, 2005.

Kitson, Gay C., and Helen J. Raschke. "Divorce Research: What We Know; What We Need to Know." *Journal of Divorce* 4 (Spring 1981):1–37.

Klandermans, Bert. "Mobilization and Participation: Social Psychological Explanations of Resource Mobilization Theory." *American Sociological Review* 49 (1984):583–600.

Klein, Naomi. *The Shock Doctrine: The Rise of Disaster Capitalism.* New York: Henry Holt & Company, 2007.

Klein, Stephen P., Susan Turner, and Joan Petersilia. *Racial Equity in Sentencing.* Santa Monica, CA: Rand, 1988.

Kleniewski, Nancy. *Cities, Change, and Conflict,* 3rd ed. Belmont, CA: Wadsworth, 2006.

Klineberg, Otto. "Race Differences: The Present Position of the Problem." *International Social Science Bulletin* 2 (Autumn 1950):460–466.

Klinker, Philip A., and Rogers M. Smith. *The Unsteady March: The Rise and Decline of Racial Equality in America.* Chicago: University of Chicago Press, 1999.

Kluegel, James R., and Eliot R. Smith. *Beliefs about Inequality: Americans' Views of What Ought to Be.* Hawthorne, NY: Aldine de Gruyter, 1990.

Kluger, Jeffrey. "Here, Kitty, Kitty!" *Time* (February 25, 2002):58–59.

Kluger, Jeffrey, and Andrea Dorfman. "The Challenges We Face." *Time* (August 26, 2002):A6–A12.

Knickmeyer, Ellen. "A New Text in Islamic Law." *Washington Post* (January 19, 2008).

Knox, David, and Caroline Schacht. *Choices in Relationships: Introduction to Marriage and the Family,* 9th ed. Belmont, CA: Wadsworth, 2007.

Kohl, Herbert. *Growing Minds.* New York: Harper Touchbook, 1988.

Kohn, Alfie. *No Contest: The Case against Competition,* rev. ed. Boston: Houghton Mifflin Company, 1992.

Kohn, Melvin L. "Social Class and Parent–Child Relationships: An Interpretation." *American Sociological Review* 68 (January 1963):471–480.

Kohn, Melvin L. *Class and Conformity,* 2nd ed. Chicago: University of Chicago Press, 1977.

Kohn, Melvin L. "Unresolved Issues in the Relationship between Work and Personality." In Kai Erikson and Steven Peter Vallas (eds.), *The Nature of Work.* New Haven, CT: Yale University Press, 1990, pp. 36–68.

Kohn, Melvin L., and Carmi Schooler. *Work and Personality: An Inquiry into the Impact of Social Stratification.* Norwood, NJ: Ablex, 1983.

Kohut, Andrew, et al. *The Diminishing Divide: Religion's Changing Role in American Politics.* Washington, D.C.: Brookings Institution, 2000.

Kolbe, Richard, and Joseph C. LaVoie. "Sex-Role Stereotyping in Preschool Children's Picture Books." *Social Psychology Quarterly* 44 (December 1981):369–374.

Konner, Melvin. "Darwin's Truth, Jefferson's Vision." *The American Prospect* (July/August 1999):30–38.

Koon, Jeff, Andy Powell, and Ward Schumaker. *You May Not Tie an Alligator to a Fire Hydrant: 101 Real Dumb Laws.* New York: Free Press, 2002.

Koppel, Ross. "American Public Policy: Formation and Implementation." In Roger A. Straus (ed.), *Using Sociology,* 3rd ed. Lanham, MD: Rowman & Littlefield, 2002, pp. 265–228.

Korgen, Kathleen Odell. *From Black to Biracial: Transforming Racial Identity Among Americans.* Westport, CT: Praeger, 1998.

Kornheiser, Tony. "You Get No Shot with a Bad Lie." *Washington Post* (December 15, 2001):D1,D6.

Kovner, Christine. "Nursing." In Anthony R. Kovner (ed.), *Health Care Delivery in the United States,* 4th ed. New York: Springer, 1990, pp. 87–105.

Kowalski, Robin M., Susan P. Limber, and Patricia W. Agatson. *Cyber Bullying.* Malden, MA: Blackwell Publishing, 2008.

Kozol, Jonathan. *Death at an Early Age.* New York: Penguin, 1985.

Kozol, Jonathan. *Savage Inequalities: Children in America's Schools.* New York: HarperCollins, 1992.

Kozol, Jonathan. *The Shame of the Nation.* New York: Crown Publishing Group, 2006.

Kraar, Louis. "The Real Threat to China's Hong Kong." *Fortune* 135 (May 26, 1997):84.

Kraditor, Aileen S. *The Ideas of the Woman's Suffrage Movement 1890–1920.* Garden City, NY: Anchor Books, 1971.

Kraybill, Donald B., and Marc A. Olshan (eds.). *The Amish Struggle with Modernity.* Hanover, NH: University Press of New England, 1994.

Kriesi, Hanspeter. "New Social Movements and the New Class in the Netherlands." *American Journal of Sociology* 94 (March 1989):1078–1116.

Kritz, Francesca Lunzer. "uncertainty@dr-mail.com." *The Washington Post* (April 1, 2003):F1, F5.

Kronenfeld, Jennie Jacobs (ed.). *Research in the Sociology of Health Care: Changing Organizational Forms of Delivering Health Care,* vol. 15. Greenwich, CT: JAI Press, 1998.

Kruschwitz, Robert B., and Robert C. Roberts. *The Virtues.* Belmont, CA: Wadsworth, 1987.

Kübler-Ross, Elisabeth. *Death's Final Stage.* New York: Simon & Schuster, 1986.

Kübler-Ross, Elisabeth. *Death.* New York: Simon & Schuster, 1997a.

Kübler-Ross, Elisabeth. *On Death and Dying.* New York: Touchstone, 1997b.

Kumar, Krishan. *From Post-Industrial to Post-Modern Society.* New York: Wiley, 2004.

Kurlaender, Michael, and John T. Yun. "Is Diversity a Compelling Educational Interest? Evidence from Metropolitan Louisville." Cambridge, MA: Civil Rights Project, August 2000.

Kuttner, Robert. "Competitiveness Craze." *Washington Post* (August 12, 1993a):A17.

Kuttner, Robert. "The Corporation in America: Is It Socially Redeemable?" *Dissent* (Winter 1993b):35–49.

Kuttner, Robert. *The Squandering of America.* New York: Knopf Publishing Group, 2007.

Labaree, David F. *How to Succeed in School without Really Learning: The Credentials Race in American Education.* New Haven, CT: Yale University Press, 1997.

Labich, Kenneth. "Can Your Career Hurt Your Kids?" *Fortune* (May 20, 1991):38–56.

Lacayo, Richard. "Wounding the Gun Lobby." *Time* (March 29, 1993):29–30.

Lachman, Margie E., et al. "Images of Midlife Development among Young, Middle-Aged, and Older Adults." *Journal of Adult Development* (1994):201–211.

Lachman, Margie E., and Jacquelyn Boone James. "Charting the Course of Midlife Development: An Overview." In Margie E. Lachman and Jacquelyn Boone James (eds.), *Multiple Paths of Midlife Development.* Chicago: University of Chicago Press, 1997, pp. 1–17.

Lamanna, Mary Ann, Agnes Riedmann, and Agnes Czerwinski Riedmann. *Marriages and Families: Making Choices and Facing Change,* 9th ed. Belmont, CA: Wadsworth, 2005.

Lamb, Michael B., M. Ann Easterbrooks, and George W. Holden. "Reinforcement and Punishment Among Preschoolers: Characteristics, Effects, and Correlates." *Child Development* 51 (1980):1230–1236.

Lampert, Leslie. "Fat Like Me." *Ladies Home Journal* (May 1993):154ff.

Landes, David S., and Charles Tilly (eds.). *History as Social Science.* Englewood Cliffs, NJ: Prentice Hall, 1991.

Landry, Bart. *The New Black Middle Class.* Berkeley: University of California Press, 1988.

Lane, Charles. "High Court Faces New Political Foray." *Washington Post* (May 3, 2003a):A1.

Lane, Charles. "Ruling Is Landmark Victory for Gay Rights." *Washington Post* (June 27, 2003b):A1,A16.

Lang, Robert E., and Patrick E. Simmons. "'Boomburbs': The Emergence of Large, Fast-Growing Suburban Cities in the United States." Fannie Mae Foundation *Census Note* 6 (June 2001).

Lanier, Mark M., and Stuart Henry. *Essential Criminology,* 2nd ed. Boulder, CO: Westview Press, 2004.

Lapchick, Richard, and Bente General. "The Buck Stops Here: Assessing Diversity among Campus and Conference Leaders for Division 1A Schools in 2007–08." University of Central Florida, DeVos Sport Business Management Program, 2007.

Lapchick, Richard E., and Kevin J. Matthews. *2004 Racial and Gender Report Card.* Center for the Study of Sport in Society: Northeastern University, 2004.

Lasswell, Harold. "The Structure and Function of Communication in Society." In Lyman Bryson (ed.), *The Communication of Ideas.* New York: Harper and Row, 1948.

Lauer, Robert H. *Perspectives on Social Change,* 4th ed. Boston: Allyn & Bacon, 1991.

Lawrence, Frederick M. *Punishing Hate; Bias Crimes under American Law.* Cambridge, MA: Harvard University Press, 2002.

Lawrence, Paul R., and Jay W. Lorsch. *Organization and Environment.* Boston: Division of Research, Graduate School of Business Administration, Harvard University, 1967.

Lazarsfeld, Paul. *Survey Design and Analysis.* New York: Free Press, 1955.

Le Bon, Gustave. *The Crowd.* New York: Viking, 1960.

Leach, Edmond. *Social Anthropology.* New York: Oxford University Press, 1982.

Leacock, Eleanor Burke. *Teaching and Learning in City Schools.* New York: Basic Books, 1969.

Least Developed Countries Report 2002, The: Escaping the Poverty Trap. United Nations Conference on Trade and Development. New York: United Nations, 2002.

Lee, Alfred McClung. *Toward Humanist Sociology.* New York: Oxford University Press, 1978.

Lee, Alfred McClung. *Sociology for People.* Syracuse, NY: Syracuse University Press, 1990.

Lee, Gary R. *Family Structure and Interaction.* Philadelphia: Lippincott, 1977.

Lee, Jennifer, and Frank D. Bean. "Beyond Black and White: Remaking Race in America." *Contexts* (Summer 2003):26–33.

Lee, Ronald, and John Haaga. "Government Spending in an Older America." Washington, D.C.: Population Reference Bureau, 2002.

Lee, Sharon M. "Asian Americans: Diverse and Growing." *Population Bulletin* 53 (June 1998):1–40.

"Left Out." *Newsweek* (March 21, 1983):26–35.

Lehrmann, Hartmut, and Guenther Roth. *Weber's Protestant Ethic: Origins, Evidence, Context.* New York: Cambridge University Press, 1993.

Leibenstein, Harvey. *Inside the Firm.* Cambridge, MA: Harvard University Press, 1987.

Lemann, Nicholas. *The Big Test: The Secret History of the American Meritocracy.* New York: Farrar, Straus & Giroux, 2000.

Lemert, Charles, and Ann Branaman (eds.). *The Goffman Reader.* Malden, MA: Blackwell, 1997.

Leming, Michael R., and George E. Dickinson. *Understanding Dying, Death, and Bereavement,* 5th ed. Fort Worth, TX: Harcourt Brace, 2002.

Lengermann, Patricia Madoo, and Jill Niebrugge-Brantley. *The Women Founders: Sociology and Social Theory, 1830–1930.* Boston: McGraw-Hill, 1998.

Lenski, Gerhard E. *Power and Privilege.* Chapel Hill: University of North Carolina Press, 1984.

Lenski, Gerhard E. "Rethinking Macrosociological Theory." *American Sociological Review* 53 (April 1988):163–171.

Leonard, Wilbert Marcellus II. *A Sociological Perspective of Sport,* 5th ed. Boston: Allyn & Bacon, 1998.

Leslie, Gerald R., and Sheila K. Korman. *The Family in Social Context,* 7th ed. New York: Oxford University Press, 1989.

Levin, Jack, and James Alan Fox. *Elementary Statistics in Social Research,* 9th ed. Boston: Allyn & Bacon, 2002.

Levin, Jack, and William C. Levin. *Ageism.* Belmont, CA: Wadsworth, 1980.

Levin, Jack, and Jack McDevitt. *Hate Crimes Revisited: America's War on Those Who Are Different.* Boulder, CO: Westview Press, 2002.

Levin, Susanna. "The Spoils of Victory: Who Gets Big Money from Sponsors, and Why?" In D. Stanley Eitzen (ed.), *Sport in Contemporary Society.* 5th ed. New York: St. Martin's Press, 1996, pp. 367–372.

Levine, Daniel U., and Rayna F. Levine. *Society and Education,* 9th ed. Boston: Allyn & Bacon, 1996.

Levine, Robert V. "The Kindness of Strangers." *American Scientist* (May–June 2003).

Levine, Sol, and Paul E. White. "Exchange as a Conceptual Framework for the Study of Interorganizational Relationships." *Administrative Science Quarterly* 5 (March 1961):583–610.

Levinson, Daniel J. *Seasons of a Man's Life.* New York: Random House, 1978.

Levinson, Daniel J. "A Conception of Adult Development." *American Psychologist* 41 (January 1986):3–13.

Lewis, Charles. *The Buying of Congress.* New York: Avon Books, 1998.

Lewis, Charles. *The Buying of the President 2004.* New York: Harper Collins, 2004.

Lewis, Charles, and Bill Allison. *Cheating of America: How Tax Avoidance and Evasion by the Super Rich Are Costing the Country Billions—and What You Can Do about It.* New York: Morrow, 2002.

Lewis, David Levering. *W. E. B. Du Bois: The Fight for Equality and the American Century, 1919–1963,* vol. 2. New York: Henry Holt, 2000.

Lewis, David Levering, and W. E. B. Du Bois. *W. E. B. Du Bois: Biography of a Race, 1868–1919.* New York: Henry Holt, 1993.

Lewis, Gordon H. "Role Differentiation." *American Sociological Review* 37 (August 1972):424–434.

Lewis, I. A., and William Schneider. "Hard Times: The Public on Poverty." *Public Opinion* 8 (June/July 1985):2ff.

Lewis, Lionel. "Working at Leisure." *Society* (July/August 1982):27–32.

Lewis, Neil A., and Richard A. Oppel Jr. "U.S. Court Issues Discordant Ruling on Campaign Law." *The New York Times* (May 3, 2003).

Lexis-Nexis Corporate Affiliations. New Providence, NJ: LexisNexis Group. 2008.

Liazos, Alexander. "The Poverty of the Sociology of Deviance: Nuts, Sluts, and Perverts." *Social Problems* 20 (Summer 1972):103–120.

Lichter, Daniel T., and Martha L. Crowley. "Poverty in America: Beyond Welfare Reform." *Population Bulletin* 57 (June 2002):1–36.

Lieberman, Myron. *Public Education.* Cambridge, MA: Harvard University Press, 1993.

Lieberman, Robert C. *Shifting the Color Line: Race and the American Welfare State.* Cambridge, MA: Harvard University Press, 1998.

Liebow, Elliot. *Talley's Corner.* Boston: Little, Brown, 2003.

Light, Ivan. *Cities in World Perspective.* New York: Macmillan, 1983.

Likert, Rensis. *New Patterns of Management.* New York: McGraw-Hill, 1961.

Likert, Rensis. *The Human Organization.* New York: McGraw-Hill, 1967.

Lilla, Mark. *The Stillborn God.* New York: Alfred A. Knopf, 2007.

Lin, Nan, Walter M. Ensel, and John C. Vaughn. "Social Resources and Strength of Ties: Structural Factors in Occupational Status Attainment." *American Sociological Review* 46 (August 1981):382–399.

Lincoln, Yvonna S., and Norman K. Denzin. *Turning Points in Qualitative Research: Tying Knots in the Handkerchief,* 3rd ed. Thousand Oaks, CA: AltaMira Press, 2005.

Linden, Eugene. "Can Animals Think?" *Time* (March 22, 1993a):55–61.

Linden, Eugene. *The Octopus and the Orangutan.* New York: Dutton, 2002a.

Linden, Eugene. "The Wife Beaters of Kibale." *Time* (August 19, 2002b):56–57.

Linder, Eileen W. (ed.). *Yearbook of American and Canadian Churches.* Nashville, TN: Abingdon Press, 2008.

Link, Bruce G. "Understanding Labeling Effects in the Area of Mental Disorders." *American Sociological Review* 52 (February 1987):96–112.

Link, Bruce G., and Francis Cullen. "Reconsidering the Social Rejection of Ex-Mental Patients: Levels of Attitudinal Response." *American Journal of Community Psychology* 11 (1983):261–273.

Link, Bruce G., Francis T. Cullen, James Frank, and John Wozniak. "The Social Rejection of Former Mental Patients: Understanding Why Labels Matter." *American Journal of Sociology* 92 (May 1987):1461–1500.

Link, Bruce G., Bruce P. Dohrenwand, and Andrew E. Skodal. "Socio-Economic Status and Schizophrenia: Noisome Occupational Characteristics as a Risk Factor." *American Sociological Review* 51 (April 1986):242–258.

Link, Bruce G., and Jo C. Phelan. "Conceptualizing Stigma." *Annual Review of Sociology* 27 (2001):363–385.

Linz, Juan J. "An Authoritarian Regime: Spain." In Erik Allardt and Yrjo Littunen (eds.), *Cleavages, Ideologies and Party Systems*. Helsinki, Finland: The Academic Bookstore, 1964.

Lipset, Seymour Martin. "Harriet Martineau's America." In Seymour Martin Lipset (ed.), *Harriet Martineau, Society in America*. Garden City, NY: Doubleday, 1962, pp. 5–41.

Lipset, Seymour Martin. *American Exceptionalism: A Double-Edged Sword*. New York: Norton, 1996.

Lipset, Seymour Martin, Martin Trow, and James Coleman. *Union Democracy*. New York: Free Press, 1956.

Liptak, Adam. "Challenges Remain for Lethal Injection." *The New York Times* (April 17, 2008a).

Liptak, Adam. "American Exception: Inmate Count in U.S. Dwarfs Other Nations." *New York Times* (April 23, 2008b).

Little, Suzanne. "Sex Roles in Faraway Places." *Ms.* (February 1975):77ff.

Litwak, Eugene. "Occupational Mobility and Extended Family Cohesion." *American Sociological Review* 25 (February 1960):9–21.

Litwak, Eugene. "Models of Bureaucracy which Permit Conflict." *American Journal of Sociology* 67 (September 1961):177–184.

Liu, Melinda. "Party Time in Beijing." *Newsweek* (November 25, 2002):36.

Livernash, Robert, and Eric Rodenburg. *Population Change, Resources, and the Environment*. Washington, D.C.: Population Reference Bureau, 1998.

Livi-Bacci, Massimo. *A Concise History of World Population*, 2nd ed. Malden, MA: Blackwell Publishers, 1997.

Livingston, E. *Making Sense of Ethnomethodology*. London: Routledge and Kegan Paul, 1987.

Lobo, Susan, and Kurt Peters (eds.). *American Indians and the Urban Experience*. Thousand Oaks, CA: AltaMira Press, 2001.

Locke, John. *Second Treatise of Government*. Edited by C. B. Macpherson. Indianapolis, IN: Hackett, 1980.

Lodge, George C. *The New American Ideology*. New York: Knopf, 1975.

Lodge, George C. *The American Disease*. Washington Square: New York University Press, 1986.

Lodge, George C., and Ezra F. Vogel (eds.). *Ideology and National Competitiveness*. Cambridge, MA: Harvard Business School, 1987.

Lofland, John. "White-Hat Mobilization: Strategies of a Millenarian Movement." In Mayer N. Zald and John D. McCarthy (eds.), *The Dynamics of Social Movements*. Cambridge, MA: Winthrop, 1979, pp. 157–166.

Lofland, John. *Polite Protesters*. Syracuse, NY: Syracuse University Press, 1993.

Lofland, John. *Social Movement Organizations*. Hawthorne, NY: Aldine de Gruyter, 1996.

Logan, Charles H. "Problems in Ratio Correlation: The Case of Deterrence Research." *Social Forces* 60 (March 1982): 791–810.

Lombroso, Cesare. *Crime*. Boston: Little, Brown and Company, 1918.

"Longest Journey, The." *The Economist* (November 2, 2002): 3–16.

Longshore, Douglas. "School Racial Composition and Blacks' Attitudes toward Desegregation: The Problem of Control in Desegregated Schools." *Social Science Quarterly* 63 (December 1982a):674–687.

Longshore, Douglas. "Race Composition and White Hostility: A Research Note on the Problem of Control in Desegregated Schools." *Social Forces* 61 (September 1982b):73–78.

Loo, Chalsa M. *Chinese American*, rev. ed. Thousand Oaks, CA: Sage, 1998.

Lopreato, Joseph. "From Social Evolutionism to Biocultural Evolutionism." *Sociological Forum* 5 (1990):187–212.

Lorber, Judith. *Gender Inequality: Feminist Theories and Politics*, 3rd ed. Los Angeles, CA: Roxbury Publishing, 2005.

Lorch, Donatella. "Ghosts of Tailhook." *Newsweek* (March 17, 2003):43.

Lord, George F. III, and William W. Falk. "Hidden Income and Segmentation: Structural Determinants of Fringe Benefits." *Social Science Quarterly* 63 (June 1982):208–224.

Lore, Diane. "Physicians Prescribe Unionizing." *Atlanta Constitution* (June 24, 1999):1A.

Lott, Bernice. *Women's Lives: Themes and Variations in Gender Learning*, 2nd ed. Pacific Grove, CA: Brooks/Cole, 1994.

Loury, Glenn C. *Anatomy of Racial Inequality*. Cambridge, MA: Harvard University Press, 2003.

Lowell, B. Lindsay. "Recession Pounds U.S. Hispanics." *Population Today* 30 (April 2002):1, 4.

Lucas, Samuel Roundfield. *Tracking Inequality: Stratification and Mobility in American High Schools*. New York: Teachers College Press, 1999.

Luckmann, Thomas. *The Invisible Religion*. New York: Macmillan, 1967.

Luhman, Reid. *Race and Ethnicity in the United States: Our Differences and Roots*. Fort Worth, TX: Harcourt College Publishers, 2002.

Lupyan, Gary. "The Conceptual Grouping Effect: Categories Matter (And Named Categories Matter More)." *Cognition* (Forthcoming 2008).

Lux, Kenneth. *Adam Smith's Mistake*. Boston: Shambhala, 1990.

Lyman, Stanford M. *Chinese Americans*. New York: Random House, 1974.

Lynaugh, Joan E., and Barbara L. Brush. *American Nursing*. Malden, MA: Blackwell, 1996.

Lyons, Linda. "Religiosity Measure Shows Stalled Recovery." Gallup News Service (January 11, 2005).

Maag, Christopher. "When Bullies Turn Faceless." *New York Times* (December 16, 2007).

Maccoby, Eleanor E. *The Two Sexes: Growing Up Apart, Coming Together*. Cambridge, MA: Belknap Press of Harvard University Press, 1997.

Machalek, Richard, and Michael Martin. "'Invisible' Religions: Some Preliminary Evidence." *Journal for the Scientific Study of Religion* 15 (December 1976):311–321.

MacIntyre, Alasdair. *A Short History of Ethics*, 1st Touchstone ed. New York: Simon & Schuster, 1998.

Macionis, John J., and Vincent N. Parrillo. *Cities and Urban Life*, 4th ed. Upper Saddle River, NJ: Prentice-Hall, 2006.

MacKay, Natalie J., and Katherine Corell. "The Impact of Women in Advertisements On Attitudes Toward Women." *Sex Roles* 36 (May 1997):573–583.

MacKellar, F. Landis, and Machiko Yanaghishita. "Homicide in the United States: Who's at Risk." Washington, D.C.: Population Reference Bureau, 1995.

MacKenzie, Donald A., and Judy Wajcman (eds.). *The Social Shaping of Technology*, 2nd ed. Bristol, PA: Taylor and Francis, 1999.

MacKinnon, Malcolm. "Calvinism and the Infallible Assurance of Grace: The Weber Thesis Reconsidered." *British Journal of Sociology* 39 (1988a):143–177.

MacKinnon, Malcolm. "Weber's Exploration of Calvinism: The Undiscovered Provenance of Capitalism." *British Journal of Sociology* 39 (1988b):178–210.

Maddox, Brenda. "Homosexual Parents." *Psychology Today* 16 (February 1982):62, 66–69.

Mahar, Maggie. "The Cholesterol Con—Where Were the Doctors?" *Alternet* (February 29, 2008).

Malinowski, Bronislaw. *The Sexual Life of Savages.* New York: Liveright, 1929.

Malthus, Thomas. *An Essay on the Principle of Population.* London: Reeves and Turner, 1798.

Manchester, William. *A World Lit Only by Fire.* Boston: Little, Brown and Company, 1993.

Mandelbaum, Michael. *The Meaning of Sports.* New York: Public Affairs, 2004a.

Mandelbaum, Michael. *The Ideas That Conquered the World: Peace, Democracy, and Free Markets in the Twenty-First Century.* New York: Public Affairs, 2004b.

Mann, Charles C. "Homeland Insecurity." *The Atlantic Monthly* (September 2002):81–102.

Mann, Patricia. *Micro-Politics: Agency in a Postfeminist Era.* Minneapolis: University of Minnesota Press, 1994.

Marchese, Aileen A. "What Companies Need to Know about Employment Practices Liability and Layoffs." 2002. Available online at www.plusweb.org/Downloads/Events/NeedtoKnowAboutEPLI&Layoffs-AileenMarchese.doc.

Marger, Martin N. *Race and Ethnic Relations: American and Global Perspectives*, 8th ed. Belmont, CA: Wadsworth, 2008.

Markle, Gerald E., and Ronald J. Troyer. "Cigarette Smoking as Deviant Behavior." In Ronald A. Farrell and Victoria Lynn Swigert (eds.), *Social Deviance*, 3rd ed. Belmont, CA: Wadsworth, 1988, pp. 82–91.

Marks, John. "A Hideous Déjà Vu from the Greatest Evil of All." *U.S. News & World Report* (April 26, 1999):30.

Marmor, Theodore R., and Michael S. Barr. "Health." In Mark Green (ed.), *Changing America.* New York: Newmarket Press, 1992, pp. 399–414.

Marriott, M. "Afrocentrism: Balancing or Skewing History?" *New York Times* (August 11, 1991):1, 18.

Marsden, Peter V. "Core Discussion Networks of Americans." *American Sociological Review* 52 (February 1987):122–131.

Marshall, Eliot. "DNA Studies Challenge the Meaning of Race." *Science* 282 (October 23, 1998):654–655.

Martin, Joyce A., et al. "Births: Final Data for 2000." *National Vital Statistics Reports* 50 (February 12, 2002).

Martin, Karin A. "Becoming a Gendered Body: Practices of Preschools." *American Sociological Review* 63 (August 1998):494–511.

Martin, Philip, and Elizabeth Midgley. "Immigration: Shaping and Reshaping America." 2nd ed. *Population Bulletin* 61 (2006):1–28.

Martineau, Harriet. *Society in America.* Paris: Bandry's European Library, 1837.

Martineau, Harriet. *How to Observe Manners and Morals.* London: C. Knight, 1838.

Martinelli, Alberto, and Neil J. Smelser. *Economy and Society: Overview in Economic Sociology.* Newbury Park, CA: Sage, 1990.

Martinez, Valerie J., R. Kenneth Godwin, Frank R. Kemerer, and Laura Perna. "The Consequences of School Choice: Who Leaves and Who Stays in the Inner City." *Social Science Quarterly* 76 (September 1995):485–501.

Martocchio, B. "Grief and Bereavement." *Nursing Clinics of North America* 20 (1985):327–346.

Marty, Martin E. "Fundamentalism Reborn: Faith and Fanaticism." *Saturday Review* (May 1980):37–42.

Marty, Martin E. *Pilgrims in Their Own Land.* New York: Penguin, 1985.

Marx, Gary T., and Douglas McAdam. *Collective Behavior and Social Movements.* Englewood Cliffs, NJ: Prentice Hall, 1994.

Marx, Karl. *The Poverty of Philosophy.* Translated by H. Quelch. Chicago: Charles H. Kerr, 1920.

Mason, Philip. *Patterns of Dominance.* New York: Oxford University Press, 1970.

Massey, Douglas S. "American Apartheid: Segregation and the Making of the Underclass." *American Journal of Sociology* 96 (September 1990):329–357.

Massey, Douglas S., and Nancy A. Denton. "Trends in the Residential Segregation of Blacks, Hispanics, and Asians: 1970–1980." *American Sociological Review* 52 (December 1987):802–825.

Massey, Douglas S., and Nancy A. Denton. *American Apartheid.* Cambridge, MA: Harvard University Press, 1996.

Massey, Douglas S., and Mitchell L. Eggers. "The Ecology of Inequality: Minorities and the Concentration of Poverty, 1970–1980." *American Journal of Sociology* 95 (March 1990):1153–1188.

Mauer, Marc. *Race to Incarcerate*, 2nd ed. New York: New Press, 2006.

Maxwell, Gerald. "Why Ascription: Parts of a More-or-Less Formal Theory of the Functions and Dysfunctions of Sex Roles." *American Sociological Review* 40 (1975):445–455.

May, Tim. *Social Research: Issues, Methods and Process.* Buckingham, United Kingdom: Open University Press, 2001.

Maynard, Douglas W. "On the Functions of Social Conflict among Children." *American Sociological Review* 50 (April 1985):207–223.

Mazaar, Michael J. *Global Trends 2005: An Owner's Manual for the Next Decade.* New York: St. Martin's Press, 1999.

Mazlish, Bruce. *A New Science.* University Park, PA: Penn State Press, 1993.

McArdle, Megan. "Body Counting." *The Atlantic* (April 2008):26, 28.

McCaghy, Charles H., Timothy A. Capron, J. D. Jamieson, and Sondra Harley Carey. *Deviant Behavior*, 8th ed. Boston: Allyn & Bacon, 2007.

McCall, Leslie. *Complex Inequality: Gender, Class and Race in the New Economy.* New York: Routledge, 2001.

McCarthy, John D., and Mark Wolfson. "Resource Mobilization by Local Social Movement Organizations: Agency, Strategy, and Organization in the Movement Against Drinking and Driving." *American Sociological Review* 61 (December 1996):1070–1088.

McCarthy, John D., and Mayer N. Zald. "Resource Mobilization and Social Movements: A Partial Theory." *American Journal of Sociology* 82 (May 1977):1212–1241.

McCrummen, Stephanie. "Tribal Rage Tears at Diverse Kenyan City." *Washington Post* (January 3, 2008).

McDonald, Scott. "Sex Bias in the Representation of Male and Female Characters in Children's Picture Books." *Journal of Genetic Psychology* 150 (December 1989):389–401.

McFalls, Joseph A. Jr. *Population: A Lively Introduction,* 5th ed. Washington, D.C.: Population Reference Bureau, 2007.

McGinn, Daniel, and Holly Peterson. "Playground of the Rich." *Newsweek* (November 25, 2002):34.

McGinn, Daniel, and Joan Raymond. "Workers of the World, Get Online." *Newsweek* (February 1998, Winter 1997–98 Special Edition):32–33.

McGrath, Ellie. "Confucian Work Ethic." *Time* (March 28, 1983):52.

McGuire, Meredith B. *Religion, The Social Context,* 5th ed. Belmont, CA: Wadsworth, 2001.

McIntyre, Robert S. "The Taxonomist." *The American Prospect* (November 20, 2000):12.

McIntyre, Robert S. "One for Oil." *The American Prospect* (March 11, 2002):17.

McKinlay, John B. (ed.). *Issues in the Political Economy of Health Care.* New York: Tavistock, 1985.

McLanahan, Sara, and Gary Fam Sandefur. *Growing Up with a Single Parent.* Cambridge, MA: Harvard University Press, 1996.

McLaughlin, Steven, and Associates. *The Changing Lives of Women.* Chapel Hill: University of North Carolina Press, 1988.

McManus, Ed. "The Death Penalty and the Race Factor." *Illinois Issues* 11 (March 1985):47.

McNamara, Justine M. "The Welfare Myth." *Journal of Sociology and Social Welfare* (December 1, 2004).

McNamee, Mike, and Joann Muller. "A Tale of Two Job Markets." *BusinessWeek* (December 21, 1998):38.

McNary, Sharon. "Neighbors Didn't Believe Norco Mother's Tale." *The Inland Southern California Press-Enterprise* (September 11, 1999):B1.

McPhail, Clark. *Myth of the Madding Crowd.* Hawthorne, NY: Aldine de Gruyter, 1991.

McPhail, Clark, and Ronald T. Wohlstein. "Individual and Collective Behaviors within Gatherings, Demonstrations, and Riots." In Ralph H. Turner and James F. Short Jr. (eds.), *Annual Review of Sociology* 9 (1983):579–600.

McQuail, Denis. *Mass Communication Theory,* 2nd ed. Beverly Hills, CA: Sage, 1987.

McTighe, Kathryn. "Internet-Healthcare: Ultimate Savior of the System?" *Health Management Technology* 20 (December 1999):24–27.

Mead, George Herbert. *Mind, Self and Society.* Chicago: University of Chicago Press, 1934.

Mead, Margaret. *Sex and Temperament in Three Primitive Societies.* New York: Mentor Books, 1950.

Mednick, Sarnoff A., William J. Gabrielli, and Barry Hutchings. "Genetic Influences in Criminal Convictions: Evidence from an Adoption Cohort." *Science* 224 (1984):891–894.

Melucci, Alberto. "The New Social Movements: A Theoretical Approach." *Social Science Information* 19 (May 1980):199–226.

Menaghan, E. G., and T. L. Parcel. "Parental Employment and Family Life: Research in the 1980s." In A. Booth (ed.), *Contemporary Families.* Minneapolis: National Council on Family Relations, 1991, pp. 361–380.

Mental Health: A Report of the Surgeon General. Washington, D.C.: Government Printing Office, 1999.

Mercader, Julio, et al. "4,300-Year-Old Chimpanzee Sites and the Origins of Percussive Stone Technology." *Proceedings of the National Academy of Sciences* 104 (February 27, 2007):3043–3048.

Merton, Robert K. *Social Theory and Social Structure,* enlarged ed. New York: Free Press, 1968.

Merton, Robert K. *On Social Structure and Science.* Chicago: University of Chicago Press, 1996.

Messner, Michael A. *Politics of Masculinities: Men in Movements.* Thousand Oaks, CA: Sage, 1997.

Messner, Michael A. *Taking the Field: Women, Men, and Sports.* Minneapolis, MN: University of Minnesota Press, 2002.

Meuller, Walt. *Understanding Today's Youth Culture.* Wheaton, IL: Tyndale House, 1999.

Meyer, John W., and W. Richard Scott. *Organizational Environments.* Beverly Hills, CA: Sage, 1985.

Michels, Robert. *Political Parties.* New York: Free Press, 1949.

Micklin, Michael, and Dudley L. Poston Jr. (eds.). *Continuities in Sociological Human Ecology.* New York: Plenum Press, 1998.

Miech, Richard, and Robert M. Hauser. *Social Class Indicators and Health at Midlife.* Madison: Center for Demography and Ecology, University of Wisconsin–Madison, 1998.

Milberg, William, and Robert L. Heilbroner. *Making of Economic Society,* 12th ed. Englewood Cliffs, NJ: Prentice Hall, 2007.

Milgram, Stanley. "Behavioral Study of Obedience." *Journal of Abnormal and Social Psychology* 67 (1963):371–378.

Milgram, Stanley. "Group Pressure and Action Against a Person." *Journal of Abnormal and Social Psychology* 68 (1964): 137–143.

Milgram, Stanley. "Some Conditions of Obedience and Disobedience to Authority." *Human Relations* 18 (1965): 57–76.

Milgram, Stanley. "The Small World Problem." *Psychology Today* 1 (1967):61–67.

Milgram, Stanley. *Obedience to Authority.* New York: Harper and Row, 1974.

Milkman, Ruth. *Farewell to the Factory.* Berkeley: University of California Press, 1997.

Mill, John Stuart. *On Liberty and Utilitarianism.* New York: Bantam Books, 1993.

Miller, Dale T. (ed.). *Cultural Divides: Understanding and Overcoming Group Conflict.* New York: Russell Sage Foundation, 1999.

Miller, Ellen, and Nick Penniman. "The Road to Nowhere." *The American Prospect* (August 12, 2002):14.

Miller, Joanne, and Howard H. Garrison. "Sex Roles: The Division of Labor at Home and in the Workplace." *Annual Review of Sociology* 8 (1982):237–262.

Miller-Perrin, Cindy L., and Robin D. Perrin (eds.). *Child Maltreatment.* Thousand Oaks, CA: Sage, 2006.

Millicent, Lawton. "Schools' 'Glass Ceiling' Imperils Girls, Study Says." *Education Week* (February 12, 1992):17.

Mills, C. Wright. *The Power Elite.* New York: Oxford University Press, 1956.

Mills, C. Wright, and Amitai Etzioni. *The Sociological Imagination,* 40th ed. New York: Oxford University Press, 1999.

Mills, Kay. *Something Better for My Children: The History and People of Head Start.* New York: NAL/Dutton, 1998.

Miner, Barbara. "Who's Vouching for Vouchers?" *The Nation* (June 5, 2000):23–24.

Mintz, Beth, and Michael Schwartz. "Interlocking Directorates and Interest Group Formation." *American Sociological Review* 46 (December 1981a):851–869.

Mintz, Beth, and Michael Schwartz. "The Structure of Intercorporate Unity in American Business." *Social Problems* 29 (December 1981b):87–103.

Mintz, Beth, and Michael Schwartz. *The Power Structure of American Business*. Chicago: University of Chicago Press, 1985.

Mintz, Morton, and Jerry Cohen. *America, Inc.* New York: Dial Press, 1972.

Mintzberg, Henry. *The Structuring of Organizations*. Englewood Cliffs, NJ: Prentice Hall, 1979.

Miron, Gary, and Brooks Applegate. "An Evaluation of Student Achievement in Edison Schools Opened in 1995 and 1996." Kalamazoo, MI: Western Michigan University Evaluation Center, 2000.

Mishel, Lawrence, Jared Bernstein, and Sylvia Allegretto (eds.). *The State of Working America, 2006–2007*. Ithaca, NY: Cornell University Press, 2006.

Mishel, Lawrence, John Schmitt, and Jared Bernstein. *State of Working America, 2000–2001*. Ithaca, NY: Cornell University Press, 2001.

Mitchell, B., and E. Gee. "Boomerang Kids and Midlife Parental Marital Satisfaction." *Family Relations* (October 1996):442–448.

Mitford, Jessica. *The American Way of Death Revisited*. New York: Alfred A. Knopf, 1998.

Mizruchi, Mark S., and Linda Brewster Stearns. "Getting Deals Done: The Use of Social Networks in Bank Decision-Making." *American Sociological Review* 66 (October 2001):647–671.

Moe, Terry M. *Schools, Vouchers, and the American Public*. Washington, D.C.: Brookings Institution Press, 2001.

Moen, Phyllis. "The Family Impact of Maternal Employment: Its Effects on Children." In Sameha Peterson, Judy Richardson, and Gretchen Kreuter (eds.), *The Two-Career Family*. Washington, D.C.: University Press, 1978, pp. 182–197.

Moen, Phyllis. *Women's Two Roles*. New York: Auburn House, 1992.

Moen, Phyllis. *It's About Time: Couples and Careers*. Ithaca, NY: Cornell University Press, 2003.

Moir, Anne. *Brain Sex: The Real Difference between Men and Women*. New York: Lyle Stuart, 1991.

Moller, David Wendell. *Confronting Death*. New York: Oxford University Press, 1995a.

Moller, David Wendell. *Death and Dying*. New York: Oxford University Press, 1995b.

Montagu, Ashley. "Don't Be Adultish." *Psychology Today* 11 (August 1977):46–50, 55.

Montagu, Ashley. *The Natural Superiority of Women*. 5th ed. Thousand Oaks, CA: AltaMira Press, 1997.

Montagu, Ashley. *Natural Superiority of Women*. Thousand Oaks, CA: AltaMira Press, 2000.

Montero, Darrel. "The Japanese Americans: Changing Patterns of Assimilation over Three Generations." *American Sociological Review* 46 (December 1981):829–839.

Moore, Joan, and Harry Pachon. *Hispanics in the United States*. Englewood Cliffs, NJ: Prentice Hall, 1985.

Moore, Michael. *Stupid White Men*. New York: HarperCollins, 2001.

Moore, Solomon. "Exoneration Using DNA Brings Change in Legal System." *New York Times* (October 1, 2007).

Moore, Thomas S. *The Disposable Work Force*. Hawthorne, NY: Aldine de Gruyter, 1996.

Moorhead, Gregory, Richard Ference, and Chris P. Neck. "Group Decision Fiascoes Continue: Space Shuttle Challenger and a Revised Groupthink Framework." *Human Relations* 44 (1991):539–550.

Mor, V., et al. (eds.). *The Hospice Experiment*. Baltimore: Johns Hopkins Press, 1988.

Morgan, C. "Adjusting to Widowhood: Do Social Networks Make It Easier?" *Gerontologist* 29 (1989):101–107.

Morgan, Dan. "Sugar Industry Expands Influence." *Washington Post* (November 3, 2007).

Morganthau, Tom. "IQ: Is It Destiny?" *Newsweek* (October 24, 1994):53–55.

Morganthau, Tom, and Seema Nayyar. "Those Scary College Costs." *Newsweek* (April 29, 1996):52–56.

Morin, Richard. "America's Middle-Class Meltdown." *Washington Post* (December 1, 1991):C1–C2.

Morin, Richard. "Crime Time: The Fear, The Facts." *Washington Post* (January 30, 1994):C1, C14.

Morris, Aldon D. "Black Southern Sit-in Movement: An Analysis of Internal Organization." *American Sociological Review* 46 (December 1981):744–767.

Morris, Aldon D. *The Origins of the Civil Rights Movement*. New York: Free Press, 1984.

Morris, Norval, and Michael Tonry. *Between Prison and Probation*. New York: Oxford University Press, 1990.

Morse, Jodie. "A Victory for Vouchers." *Time* (July 8, 2002):32–34.

Mouw, Ted. "Job Relocation and the Racial Gap in Unemployment in Detroit and Chicago, 1980–1990." *American Sociological Review* 65 (October 2000):730–753.

Mueller, D. P., and P. W. Cooper. "Children of Single Parent Families: How They Fare as Young Adults." *Family Relations* 35 (January 1986):169–176.

Muldavin, Joshua. "Commentary: Market Reforms Breed Discontent." *Los Angeles Times* (June 3, 1999):B9.

Muller, Eric. *American Inquisition*. Chapel Hill: University of North Carolina Press, 2007.

Muller, Jerry Z. *Adam Smith in His Time and Ours*. New York: Free Press, 1993.

Mumford, Emily. *Medical Sociology*. New York: McGraw-Hill, 1983.

Munger, Frank (ed.). *Laboring Below the Line: The New Ethnography of Poverty, Low-Wage Work, and Survival in the Global Economy*. New York: Russell Sage Foundation, 2002.

Murdock, George P. *Our Primitive Contemporaries*. New York: Macmillan, 1935.

Murdock, George P. "The Common Denominator of Cultures." In Ralph Linton (ed.), *The Science of Man in the World Crisis*. New York: Columbia University Press, 1945, pp. 123–142.

Murdock, George P. "World Ethnographic Sample." *American Anthropologist* 59 (August 1957):664–687.

"Muslim Americans: Middle Class and Mostly Mainstream." Washington, D.C.: Pew Forum, 2007.

Mussen, Paul H. "Early Sex-Role Development." In David Goslin (ed.), *Handbook of Socialization Theory and Research*. Chicago: Rand McNally, 1969, pp. 707–732.

Myers, David G. *Social Psychology*, 6th ed. Boston: McGraw-Hill College, 2001.

Myrdal, Gunnar. *An American Dilemma*. New York: Harper and Row, 1944.

Myrdal, Gunnar. *Objectivity in Social Research*. Middletown, CT: Wesleyan University Press, 1983.

Nader, Ralph, and Mark Green. "Crime in the Suites." *The New Republic* (April 29, 1972):17–19.

Nagata, Donna K. *Legacy of Injustice*. New York: Plenum, 1993.

Nagourney, Adam, and Megan Thee. "Young Americans Are Leaving Left, New Poll Finds." *New York Times* (June 27, 2007).

Naimark, Norman N. *Fires of Hatred: Ethnic Cleansing in Twentieth-Century Europe*. Cambridge, MA: Harvard University Press, 2002.

Nakao, Keiko, Robert W. Hodge, and Judith Treas. "On Revising Prestige Scores for All Occupations." General Social Survey Methodological Report Number 69, October 1990.

Nakao, Keiko, and Judith Treas. "Computing 1989 Occupational Prestige Scores." General Social Survey Methodological Report Number 70, November 1990.

Namboodiri, Krishman. *A Primer of Population Dynamics*. New York: Plenum, 1996.

Nanda, Serena, and Richard L. Warms. *Cultural Anthropology*, 9th ed. Belmont, CA: Wadsworth, 2006.

Nansel, R. et al. "Bullying Behavior Among U.S. Youth: Prevalence and Association with Psychosocial Adjustment." *Journal of American Medical Association* 285 (2001):2094–2100.

Nasser, Haya El. "Minorities Reshape Suburbs." *USA Today* (July 9, 2001):1A, 8A.

Nathanson, C. A. "Social Roles and Health Status among Women and the Significance of Employment." *Social Science and Medicine* 14 (1980):463–471.

National Survey of Latinos, 2002. Pew Hispanic Center/Kaiser Family Foundation. Los Angeles: University of Southern California, 2002.

"National Survey on Poverty in America." National Public Radio/Kaiser Family Foundation/Kennedy School of Government, 2001.

Navarro, Vicente. *Medicine Under Capitalism*. New York: Prodist, 1976.

Navarro, Vicente. *Crisis, Health, and Medicine*. New York: Tavistock, 1986.

Nee, Victor, and David Stark (eds.). *Remaking the Economic Institutions of Socialism*. Stanford, CA: Stanford University Press, 1989.

Nelan, Bruce W. "More Harm Than Good." *Time* (March 15, 1993a):40–45.

Nelan, Bruce W. "The Dark Side of Islam." *Time* (October 4, 1993b):62–64.

Nelson, Bryn. "Creation vs. Evolution." *Newsday* (March 11, 2002):B-6.

Nettler, Gwynn. *Explaining Crime*, 3rd ed. New York: McGraw-Hill, 1984.

Neubeck, Kenneth J., and Noel A. Cazenave. *Welfare Racism: Playing the Race Card against America's Poor*. London: Routledge, 2001.

Neugarten, B. L. "Adult Personality: Toward a Psychology of the Life Cycle." In B. L. Neugarten (ed.), *Middle Age and Aging*. Chicago: University of Chicago Press, 1968, pp. 137–147.

Neuman, William Lawrence. *Social Research Methods: Qualitative and Quantitative Approaches*, 5th ed. Boston: Allyn & Bacon, 2005.

Newman, Katherine S. *Declining Fortunes*. New York: Basic Books, 1993.

Newman, Katherine S. *No Shame in My Game: Working Poor in the Inner City*. New York: Alfred A. Knopf, 1999.

Newman, Katherine S., and Victor Tan Chen. *The Missing Class*. Boston: Beckon Press, 2007.

"New World of Work, The." *BusinessWeek* (October 17, 1994):76–86.

Newport, Frank. "Public Generally Satisfied with Rate of Organized Religion in America." Gallup News Service (January 30, 2007a).

Newport, Frank. "Same-Sex Marriage." Gallup News Service (May 10–13, 2007b).

Newport, Frank. "Questions and Answers about Americas' Religion." Gallup News Service (December 24, 2007c).

Newport, Frank, and Joseph Carroll. "Analysis: Impact of Personal Characteristics on Candidate Support." Gallup News Service (March 13, 2007).

Niebuhr, H. Richard. *The Social Sources of Denominationalism*. New York: World, 1968.

Nielsen, Niels C. Jr., et al. *Religions of the World*, 3rd ed. New York: St. Martin's Press, 1993.

Nierenberg, Danielle. *Correcting Gender Myopia: Gender Equality, Women's Welfare, and the Environment*. Washington, D.C.: Worldwatch Institute, 2002.

Nimkoff, Meyer F. (ed.). *Comparative Family Systems*. Boston: Houghton Mifflin, 1965.

Nisbet, Robert A. *The Sociological Tradition*. New York: Basic Books, 1966.

Nisbet, Robert A. *The Social Bond*. New York: Knopf, 1970.

Nisbet, Robert A. *The Present Age*. New York: Harper and Row, 1989.

Nobel, Jeramy J., and Julie C. Cherry. "It Helps Manage Patients with Chronic Illness." *Health Management Technology* 20 (December 1999):38–40.

"'No Child Left Behind' Gets Mixed Grades." Washington, D.C.: Pew Research Center for the People and the Press, 2008.

Nock, Steven L. "A Comparison of Marriages and Cohabiting Relationships." *Journal of Family Issues* 16 (January 1995):53–76.

Nolan, Patrick, and Gerhard Lenski. *Human Societies*, updated ed. Bolder, CO: Paradigm Publishers, 2005.

Nordland, Rod. "Learning the Killing Game." *Newsweek* (September 13, 1999):38–39.

Novak, Michael. *The Unmeltable Ethnics*, 2nd ed. New Brunswick: Transaction, 1996.

Nursing Homes: More Can Be Done to Protect Residents from Abuse. GAO-02-312. Washington, D.C.: U.S. General Accounting Office, 2002.

Nydeggar, Corinne N. "Family Ties of the Aged in Cross-Cultural Perspective." In Beth B. Hess and Elizabeth W. Markson (eds.), *Growing Old in America*. New Brunswick, NJ: Transaction, 1985.

Oakes, Jeannie. *Keeping Track*. New Haven, CT: Yale University Press, 1985.

Oakes, Jeannie, and Martin Lipton. In Laura I. Rendon and Richard O. Hope (eds.), *Educating a New Majority: Transforming America's Educational System for Diversity*. San Francisco: Jossey-Bass, 1996, pp. 168–200.

Oates, Stephen B. *Malice Toward None*. New York: Mentor Books, 1977.

Oberschall, Anthony. *Social Conflict and Social Movements.* Englewood Cliffs, NJ: Prentice Hall, 1973.

Obie, Clayton. *An American Dilemma Revisited: Race Relations in a Changing World.* New York: Russell Sage Foundation, 1996.

O'Brien, Joanne, and Martin Palmer. *The State of Religion Atlas.* New York: Touchstone, 1993.

O'Brien, Timothy L. "Officials Worried over a Sharp Rise in Identity Theft." *New York Times* (April 3, 2000):A1, A19.

O'Connell, Martin. "Where's Papa? Fathers' Role in Child Care." *Population Bulletin* 20 (September 1993):1–20.

O'Connor, Alice. *Poverty Knowledge: Social Science, Social Policy, and the Poor in Twentieth-Century U.S. History.* Princeton, NJ: Princeton University Press, 2002.

O'Connor, Timothy. "Disgraced Olympian Jones Given Six Month Sentence." *USA Today* (January 11, 2008).

O'Dwyer, Thomas. "The Taliban's Gender Apartheid." *Jerusalem Post* (October 27, 1999):6.

Ogburn, William F. *On Culture and Social Change.* Chicago: University of Chicago Press, 1964.

O'Hare, William P. "Can the Underclass Concept Be Applied to Rural Areas?" Population Reference Bureau. Staff Working Papers, January 1992a.

O'Hare, William P. "America's Minorities—The Demographics of Diversity." *Population Bulletin* 47 (December 1992b):1–47.

O'Hare, William P., and Brenda Curry-White. "The Rural Underclass: Examination of Multiple-Problem Populations in Urban and Rural Settings." Population Reference Bureau. Staff Working Papers, January 1992.

O'Hare, William P., et al. *Real Life Poverty in America: Where the American Public Would Set the Poverty Line.* Washington, D.C.: A Center on Budget and Policy Priorities and Families USA Foundation Report, 1990.

O'Harrow, Robert Jr. "Identity Thieves Thrive in Information Age." *Washington Post* (May 31, 2001):A1, A9.

Ollman, Bertel. *Market Socialism.* New York: Routledge, 1998.

Olson, Laura Katz (ed.). *Age through Ethnic Lenses.* Lanham, MD: Rowman & Littlefield, 2001.

Olson, Steve. *Mapping Human History: Discovering the Past through Our Genes.* Boston: Houghton Mifflin, 2002.

Olson, Steve. *Mapping Human History: Genes, Race, and Our Common Origins.* Boston: Houghton Mifflin, 2003.

Olzak, Susan, and Joane Nagel (eds.). *Competitive Ethnic Relations.* San Diego, CA: Academic Press, 1986.

Olzak, Susan, and Elizabeth West. "Ethnic Conflict and the Rise and Fall of Ethnic Newspapers." *American Sociological Review* 56 (August 1991):458–474.

Omi, Michael, and Howard Winant. *Racial Formation in the United States.* New York: Routledge and Kegan Paul, 1994.

O'Neill, Hugh M., and D. Jeffrey Lenn. "Voices of Survivors: Words that Downsizing CEOs Should Hear." *Academy of Management Executive* 9 (November 1995):23–34.

Opp, Karl-Dieter. "Grievances and Participation in Social Movements." *American Sociological Review* 53 (1988): 853–864.

Opp, Karl-Dieter. *The Rationality of Political Protest.* Boulder, CO: Westview, 1989.

Opp, Karl-Dieter, and Wolfgang Roehl. "Repression, Micromobilization, and Political Protest." *Social Forces* 69 (1990):521–547.

Orecklin, Michele. "Now She's Got Game." *Time* (March 3, 2003):57–59.

Orfield, Gary. "Schools More Separate: Consequences of a Decade of Resegregation." Cambridge, MA: The Civil Rights Project, July 2001.

Orfield, Gary, and Susan E. Eaton. *Dismantling Desegregation: The Quiet Reversal of* Brown v. Board of Education. New York: New Press, 1997.

Orfield, Gary, and Chungmei Lee. "Racial Transformation and the Changing Nature of Segregation." Cambridge, MA: Civil Rights Project, Harvard University, 2006.

Orfield, Gary A., et al. "Status of School Desegregation: The Next Generation." Report to the National School Board Association. Alexandria, VA: National School Board Association, 1992.

Ornstein, Robert, and Paul Ehrlich. *New World New Mind.* Los Altos, CA: Isak, 2000.

Orshansky, Mollie. "Who's Who among the Poor: A Demographic View of Poverty." *Social Security Bulletin* 28 (July 1965):3–32.

Orum, Anthony M. *Introduction to Political Sociology,* 2nd ed. Englewood Cliffs, NJ: Prentice Hall, 1983.

Osterman, Paul, et al. *Working in America: A Blueprint for the New Labor Market.* Cambridge, MA: MIT Press, 2001.

O'Sullivan, Christine Y., Clyde M. Reese, and John Mazzeo. *NAEP 1996 Science Report Card for the Nation and the States.* Washington, D.C.: National Center for Educational Statistics, 1997.

O'Toole, Patricia. *Money and Morals in America.* New York: Clarkson/Potter Publishers, 1998.

Paddock, Richard C. "Republic Stirs Debate by Allowing for Multiple Wives." *Los Angeles Times* (August 15, 1999):A1, A27–28.

Page, Susan. "Corporate Credentials Weigh Down Bush's Team." *USA Today* (August 6, 2002):1B.

Palast, Greg. *The Best Democracy Money Can Buy.* New York: Penguin, 2004.

Palen, John J. *The Urban World,* 8th ed. New York: McGraw-Hill, 2007.

Palmore, Erdman. "Are the Aged a Minority Group?" *Journal of the American Geriatrics Society* 26 (May 1978):214–217.

Palmore, Erdman. "The Facts on Aging Quiz: A Review of Findings." *Gerontologist* 20 (December 1980):669–672.

Palmore, Erdman. "More on Palmore's Facts on Aging Quiz." *Gerontologist* 21 (April 1981):115–116.

Pampel, Fred. *Aging, Inequality and Public Policy.* Thousand Oaks, CA: Pine Forge Press, 1998.

Parenti, Michael. *Democracy for the Few,* 8th ed. Belmont, CA: Wadsworth, 2007.

Parker, Laura. "Surgeons' Strike Forces Hospitals to Juggle Service." *USA Today* (January 20, 2003).

Parker, Laura, and Peter Eisler. "Ballots in Black Fla. Precincts Invalidated More." *USA Today* (April 6–8, 2001):1A, 3A.

Parker, Richard, and Delia Easton. "Sexuality, Culture, and Political Economy: Recent Developments in Anthropological and Cross-Cultural Sex Research." *Annual Review of Sex Research* 9 (1998):1–19.

Parkinson, C. Northcote. *Parkinson's Law.* Cutchogue, NY: Buccaneer Books, 1993.

Parrillo, Vincent N. *Diversity in America,* 2nd ed. Thousand Oaks, CA: Sage, 2005a.

Parrillo, Vincent N. *Strangers to These Shores,* 8th ed. Boston: Allyn & Bacon, 2005b.

Parsons, Talcott. *The Structure of Social Action.* New York: McGraw-Hill, 1937.

Parsons, Talcott. *The Social System*. New York: Free Press, 1951.

Parsons, Talcott. *Politics and Social Structure*. New York: Free Press, 1959.

Parsons, Talcott. "A Functional Theory of Change." In Amitai Etzioni and Eva Etzioni (eds.), *Social Change*. New York: Basic Books, 1964a, pp. 83–97.

Parsons, Talcott. "Definitions of Health and Illness in the Light of American Values and Social Structure." In Talcott Parsons, *Social Structure and Personality*. New York: Free Press, 1964b, pp. 257–291.

Parsons, Talcott. *Societies: Evolutionary and Comparative Perspectives*. Englewood Cliffs, NJ: Prentice Hall, 1966.

Parsons, Talcott. *The System of Modern Societies*. Englewood Cliffs, NJ: Prentice Hall, 1971.

Parsons, Talcott. "The Sick Role and the Role of the Physician Reconsidered." *Health and Society* (Summer 1975):257–278.

Parsons, Talcott. *The Evolution of Societies*. Englewood Cliffs, NJ: Prentice Hall, 1977.

Pascoe, C. J. *Dude: You're a Fag: Masculinity and Sexuality in High School*. Berkeley, CA: University of California Press, 2007.

Passas, Nilos and Robert Agnew (eds.). *The Future of Anomie Theory*. Boston: Northeastern University Press, 1997.

Passell, Peter. "'Bell Curve's Critics Say Early I.Q. Isn't Destiny." *New York Times* (November 9, 1994):B10.

Patchen, Martin. *Black-White Contact in Schools*. West Lafayette, IN: Purdue University Press, 1982.

Patterson, James T. *America's Struggle against Poverty in the Twentieth Century*. Cambridge, MA: Harvard University Press, 2000.

Patterson, Orlando. *Rituals of Blood: Consequences of Slavery in Two American Centuries*. Washington, D.C.: Civitas/Counterpoint, 1998.

Patterson, Thomas E. *Vanishing Voter: Public Involvement in an Age of Uncertainty*. New York: Knopf, 2003.

Patterson, Thomas E. *The American Democracy*, 8th ed. Boston: McGraw-Hill, 2006.

Pattillo-McCoy, Mary. *Black Picket Fences: Privilege and Peril among the Black Middle Class*. Chicago: University of Chicago Press, 2000.

Payer, Lynn. *Medicine and Culture*. New York: Henry Holt, 1996.

Payne, Richard J. *Getting beyond Race: The Changing American Culture*. Boulder, CO: Westview Press, 1998.

Pearlstein, Steven, and DeNeen L. Brown. "Black Teenagers Facing Worse Job Prospects." *Washington Post* (June 4, 1994): A1.

Pear, Robert. "Gap in Life Expectancy Widens for the Nation." *New York Times* (March 23, 2008).

Pearson, Kent. "Subcultures and Sport." In John W. Loy, Jr., Gerald S. Kenyon, and Barry D. McPherson (eds.), *Sport, Culture, and Society*. 2nd ed. Philadelphia: Lee and Febiger, 1981, pp. 131–145.

Pearson, Patricia, and Michael Finley. *When She Was Bad: Violent Women and the Myth of Innocence*. New York: Viking Penguin, 1997.

Peck, Dennis L., and J. Selwyn Hollingsworth (eds.). *Demographic and Structural Change: The Effects of the 1980s on American Society*. Westport, CT: Greenwood, 1996.

Pendleton, Brian F., Margaret M. Poloma, and T. Neal Garland. "Scales for Investigation of the Dual Career Family." *Journal of Marriage and the Family* 42 (May 1980):269–275.

Pennar, Karen. "The Ties That Lead to Prosperity." *BusinessWeek* (December 15, 1997):153–155.

Pennings, Johannes M. "The Relevance of the Structural-Contingency Model for Organizational Effectiveness." *Administrative Science Quarterly* 20 (September 1975):393–410.

Peoples, James, and Garrick Bailey. *Humanity: An Introduction to Cultural Anthropology*, 8th ed. Belmont, CA: Wadsworth, 2008.

Peplau, L. "Roles and Gender." In Harold H. Kelly et al. (eds.), *Close Relationships*. New York: Freeman, 1983, pp. 220–264.

Perlmann, Joel, and Robert A. Margo. *Women's Work? American Schoolteachers, 1650–1920*. Chicago: University of Chicago Press, 2001.

Perrow, Charles. *Complex Organizations*, 3rd ed. New York: McGraw-Hill, 1986.

Perrow, Charles. *Organizing America: Wealth, Power, and the Origins of Corporate Capitalism*. Princeton, NJ: Princeton University Press, 2005.

Perrucci, Robert, and Earl Wysong. *The New Class Society*, 3rd ed. Lanham, MD: Rowman & Littlefield, 2007.

Perry, Barbara. *In the Name of Hate: Understanding Hate Crimes*. Bristol, PA: Taylor and Francis, New York: Routledge, 2000.

Perry, Barbara. *Hate and Bias Crime: A Reader*. New York: Routledge, 2003.

Perry, Joellen, and Dan McGraw. "In Cleveland, It's a Back-to-School Daze." *U.S. News & World Report* (September 6, 1999):34.

Peter, Laurence J., and Raymond Hull. *The Peter Principle*. New York: William Morrow, 1996.

Peters, E. L. "Aspects of Family Life among the Bedouin of Cyrenaica." In Meyer F. Nimkoff (ed.), *Comparative Family Systems*. Boston: Houghton Mifflin, 1965, pp. 121–146.

Peters, Tom. "Restoring American Competitiveness: Looking for New Models of Organizations." *Academy of Management Executive* 2 (May 1988):103–109.

Petersen, Trond, Ishak Saporta, and Marc-David L. Seidel. "Offering a Job: Meritocracy and Social Networks." *American Journal of Sociology* 106 (November 2000):763–816.

Peterson, Karen S. "Changing the Shape of the American Family." *USA Today* (April 18, 2000):1D, 2D.

Peyser, Marc. "Tyranny of the Red Ribbon." *Newsweek* (June 28, 1993):61.

Pfeffer, Jeffrey. "The Micropolitics of Organizations." In Marshall W. Meyer and Associates (eds.), *Environments and Organizations*. San Francisco: Jossey-Bass, 1978, pp. 29–50.

Pfeffer, Jeffrey. *Power in Organizations*. Marshfield, MA: Pitman, 1981.

Pfeffer, Jeffrey. *New Directions for Organization Theory*. New York: Oxford University Press, 1997.

Phillips, Kevin. *Bad Money*. New York: Penguin Group, 2008.

Phillips, Derek. "Rejection: A Possible Consequence of Seeking Help for Mental Disorders." *American Sociological Review* 28 (December 1963):963–972.

Phillips, Derek. "Rejection of the Mentally Ill: The Influence of Behavior and Sex." *American Sociological Review* 29 (October 1964):679–687.

Phillips, John L. *The Origins of the Intellect*, 2nd ed. San Francisco: Freeman, 1975.

Phillips, Kevin. *Wealth and Democracy: How Great Fortunes and Government Created America's Aristocracy*. New York: Broadway Books, 2002.

Phu, Vu Duy. "Vietnam—Anticipating It." *Vietnam Economic News* (December 14, 1998).

Piaget, Jean. *The Psychology of Intelligence*. Totowa, NJ: Littlefield, Adams, 1981.

Piaget, Jean, and Bärbel Inhelder. *The Psychology of the Child.* London: Routledge, 1973.

Piaget, Jean, and Jane Valsiner. *The Child's Conception of Physical Causality.* New Brunswick, NJ: Transaction Publishers, 1999.

Pichanik, V. K. *Harriet Martineau.* Ann Arbor: University of Michigan Press, 1980.

Pickering, Michael. *Stereotyping: The Politics of Representation.* New York: Palgrave, 2001.

Pickering, W. S., and Geoffrey Walford (eds.). *Durkheim's Suicide.* New York: Routledge, 2000.

Pierre, Robert E., and Karl Lydersen. "Illinois Death Row Emptied." *Washington Post* (January 12, 2003):A1, A11.

Pillemer, K., and D. Finkelhor. "The Prevalence of Elder Abuse: A Random Sample Survey." *Gerontologist* 28 (February 1988):51–57.

Pillemer, V., and J. Suiter. "Will I Ever Escape My Children's Problems? Effects of Adult Children's Problems on Elderly Parents." *Journal of Marriage and the Family* (August 1991):585–594.

Pines, Maya. "The Civilizing of Genie." *Psychology Today* 15 (September 1981):28ff.

Pinker, Steven. *The Stuff of Thought.* New York: Penquin Press, 2007.

Pipher, Mary, and Susan Cohen. *The Middle of Everywhere: Helping Refugees Enter the American Community.* San Diego, CA: Harcourt Trade, 2002.

Plog, Fred, and Daniel G. Bates. *Cultural Anthropology,* 3rd ed. New York: McGraw-Hill, 1990.

Poddel, Lawrence. "Welfare History and Expectancy." *Families on Welfare in New York City.* Preliminary Report No. 5. New York: The Center for Social Research, City University of New York, 1968.

Pollard, Kelvin. "U.S. Diversity Is More than Black and White." Washington, D.C.: Population Reference Bureau, 1999.

Pollner, Melvin. "Left of Ethnomethodology: The Rise and Decline of Radical Reflexivity." *American Sociological Review* 56 (June 1991):370–380.

Polonko, Karen A., John Scanzoni, and Jay D. Teachman. "Childlessness and Marital Satisfaction: A Further Assessment." *Journal of Family Issues* 3 (December 1982):545–573.

Pomfret, John. "China Allows Its Capitalists to Join Party." *Washington Post* (July 2, 2001):A1, A14.

Pomfret, John. "Chinese Back New Breed of Capitalism." *Washington Post* (September 29, 2002):A1, A25.

Pope, Liston. *Millhands and Preachers.* New Haven, CT: Yale University Press, 1942.

Popenoe, David. *Life without Father.* Cambridge, MA: Harvard University Press, 1999.

Popenoe, David. "Essay: The Future of Marriage in America." National Marriage Project. Rutgers University, 2007, http://marriage.rutgers.edu/Publications/SOOU/TEXTSOOU2007.htm.

Posner, Richard A. *Aging and Old Age.* Chicago: University of Chicago Press, 1996.

Powell, Gary N., and Laura M. Graves. *Women and Men in Management,* 3rd ed. Thousand Oaks, CA: Sage, 2002.

Power, Samantha. "Bystanders to Genocide." *The Atlantic Monthly* (September 2001):84–108.

Power, Samantha. *A Problem from Hell: America and the Age of Genocide.* New York: Basic Books, 2003.

Pratt, Travis, "Race and Sentencing: A Meta-Analysis of Conflicting Empirical Research Results." *Journal of Criminal Justice* 26 (1998):513–525.

Pringle, Rosemary. *Sex and Medicine: Gender, Power and Authority in the Medical Profession.* New York: Cambridge University Press, 1998.

Prunier, Gerard. *The Rwanda Crisis: History of a Genocide.* New York: Columbia University Press, 1997.

Puente, Maria. "Election of Governor a Sign of Growing Political Clout." *USA Today* (November 19, 1996):1A.

Pulera, Dominic J. *Visible Differences: How Race Will Matter to Americans in the Twenty-First Century.* New York: Continuum International Publishing Group, 2002.

Purcell, Piper, and Lara Stewart. "Dick and Jane in 1989." *Sex Roles* 22 (1990):177–185.

Putnam, Robert. "Who Killed America?" *American Prospect* (March 1996):26–28.

Putnam, Robert. *Bowling Alone: The Collapse and Revival of American Community.* New York: Simon & Schuster, 2001.

Quarantelli, E. L. "Society of Panic." In *International Encyclopedia of the Social and Behavioral Sciences.* Oxford, UK: Pergamon Press, 2001, pp. 11020–11023.

Quarantelli, E. L., and Joseph Cooper. "Self-Conceptions and Others: A Further Test of the Meadian Hypothesis." *Sociological Quarterly* 7 (Summer 1966):281–297.

Quebedeaux, Richard. *The Worldly Evangelicals.* New York: Harper and Row, 1978.

Queen, Stuart A., Robert W. Habenstein, and Jill S. Quadagno. *The Family in Various Cultures,* 5th ed. New York: Harper and Row, 1985.

Quinn, Jane Bryant. "What's for Dinner, Mom?" *Newsweek* (April 5, 1993):68.

Quinney, Richard. *Class, State and Crime,* 2nd ed. New York: Longman, 1980.

"Race, Place, and Segregation." Cambridge, MA: The Civil Rights Project, May 2002.

Radin, Paul. *The World of Primitive Man.* New York: Henry Schuman, 1953.

Rae, John. *Life of Adam Smith.* Fairfield, NJ: Augustus M. Kelley, 1965.

Raine, Adrian (ed.). *The Psychopathology of Crime.* San Diego, CA: Academic Press, 1993.

Ramey, Craig T., and Frances Campbell. "Poverty, Early Childhood Education, and Academic Competence: The Abecedarian Experiment." In Aletha C. Huston (ed.), *Children in Poverty.* New York: Cambridge University Press, 1991, pp. 190–221.

Rank, Mark R. "As American as Apple Pie: Poverty and Welfare." *Contexts* (Summer 2003):41–49.

Rashid, Ahmed. *Taliban: Militant Islam, Oil and Fundamentalism in Central Asia.* New Haven, CT: Yale University Press, 2001.

Ravitch, Diane. "What Makes a Good School?" *Society* 19 (January/February 1982):10–11.

Ray, Julie. "Public Opinion Still 'Soft' on Voucher Issue." Gallup News Service (September 28, 2004).

Ray, Larry J. *Theorizing Classical Sociology.* Levittown, PA: Open University Press, 1999.

Redfield, Robert. *The Folk Culture of Yucatan.* Chicago: University of Chicago Press, 1941.

Redhead, Steve. *Subcultures to Clubcultures.* Malden, MA: Blackwell, 1997.

Reich, Robert B. *The Work of Nations.* New York: Knopf, 1997.

Reich, Robert B. *The Future of Success*. New York: Vintage, 2001.

Reich, Robert B. *Super Capitalism*. New York: Knopf Publishing Company, 2007.

Reich, Robert B. "America's Biggest Divide." *American Prospect* (January 4, 2008), http://www.prospect.org/cs/.

Reiman, Jeffrey H. *The Rich Get Richer and the Poor Get Prison*, 6th ed. Boston: Allyn & Bacon, 2000.

Reiss, Ira L. *Family Systems in America*, 2nd ed. Hinsdale, IL: Dryden Press, 1976.

Renner, Michael. *Vital Signs 2003: The Trends That Are Shaping Our Future*. New York: Worldwatch Institute, 2003.

Report of the National Advisory Commission on Civil Disorders. Washington, D.C.: U.S. Government Printing Office, 1968.

Reskin, Barbara. "Sex Segregation in the Workplace." *Annual Review of Sociology* 19 (1993):241–270.

Reskin, Barbara, and Irene Padavic. *Women and Men at Work*, 2nd ed. revised. Thousand Oaks, CA: Pine Forge Press, 2002.

Reverby, Susan M. *Tuskegee's Truth*. Chapel Hill: University of North Carolina Press, 2000.

Reynolds, Larry T., and Nancy J. Herman-Kinney (eds.). *Handbook of Symbolic Interactionism*. Thousand Oaks, CA: AltaMira Press, 2003.

Rheault, Magali. "Three in 10 Have Postponed Medical Treatment Due to Cost." Gallup News Service (December 14, 2007).

Rheingold, Howard. *Smart Mobs: The Next Social Revolution: Transforming Cultures and Communities in the Age of Instant Access*. Cambridge, MA: Perseus Publishing, 2003.

Richardson, James T. "Conversion, Brainwashing, and Deprogramming." *The Center Magazine* 15 (March/April 1982):18–24.

Richardson, Laurel, and Verta Taylor (eds.). *Feminist Frontiers*. Reading, MA: Addison-Wesley, 1983.

Riche, Martha Farnsworth. "America's Diversity and Growth: Signposts for the 21st Century." *Population Bulletin* 55 (June 2001):1–43.

Ridley, Matt. *The Origins of Virtue*. New York: Viking, 1996.

Ridley, Matt. *Nature Via Nurture: Genes, Experience, and What Makes Us Human*. New York: HarperCollins Publishers, 2003.

Riederer, Richard K. "Battered by the World Financial Crisis." *Metal Producing* 33 (March 1999):42–45.

Riesebrodt, Martin. *Pious Passion*. Berkeley: University of California Press, 1998.

Riesman, David. *On Higher Education*. San Francisco: Jossey-Bass, 1980.

Riesman, David, with Nathan Glazer and Reuel Denney. *The Lonely Crowd: A Study of the Changing American Character*. New Haven, CT: Yale University Press, 2001.

Rifkin, Jeremy. *The End of Work*. New York: Tarcher/Putnam, 1995.

Riley, Nancy E. *Gender, Power, and Population Change*. Washington, D.C.: Population Reference Bureau, 1997.

Risman, Barbara J. *Gender Vertigo: American Families in Transition*. New Haven, CT: Yale University Press, 1998.

Risman, Barbara, and Pepper Schwartz. "After the Sexual Revolution: Gender Politics in Teen Dating." *Contexts* (Spring 2002):16–24.

Rist, Ray C. "On Understanding the Process of Schooling: The Contributions of Labeling Theory." In Jerome Kagan and A. H. Halsey (eds.), *Power and Ideology in Education*. New York: Oxford University Press, 1977, pp. 292–305.

Ritzer, George. *The McDonaldization Thesis*. Thousand Oaks, CA: Sage, 1998.

Ritzer, George. *The McDonaldization of Society,* 4th ed. Thousand Oaks, CA: Sage, 2004.

Ritzer, George. *McDonaldization: The Reader,* 2nd ed. Thousand Oaks, CA: Sage, 2006.

Ritzer, George. *Sociological Theory,* 7th ed. Boston: McGraw-Hill, 2007.

Robbins, Richard Howard. *Global Problems and the Culture of Capitalism,* 2nd ed. Boston: Allyn & Bacon, 2004.

Robbins, Stephen P. *Organization Theory,* 3rd ed. Englewood Cliffs, NJ: Prentice Hall, 1990.

Robbins, Thomas, Dick Anthony, and James Richardson. "Theory and Research on Today's New Religions." *Sociological Analysis* 39 (Summer 1978):95–122.

Roberts, Sam. "In Name Count, Garcias Are Catching Up to the Joneses." *New York Times* (November 17, 2007).

Robinson, Greg. *By Order of the President: FDR and the Internment of Japanese Americans*. Cambridge, MA: Harvard University Press, 2001.

Rochon, Thomas R. *Culture Moves: Ideas, Activism, and Changing Values*. Princeton, NJ: Princeton University Press, 1998.

Rockett, Ian R. H. "Population and Health: An Introduction to Epidemiology." *Population Bulletin* 49 (November 1994): 1–47.

Rockett, Ian R. H. *Population and Health*. Washington, D.C.: Population Reference Bureau, 1999.

Rodriguez, Gregory. "From Minority to Mainstream, Latinos Find Their Voice." *Washington Post* (January 24, 1999): B1, B4.

Roediger, David R. *Colored White: Transcending the Racial Past*. Berkeley: University of California Press, 2003.

Roethlisberger, F. J., and William J. Dickson. *Management and the Worker*. New York: Wiley, 1964.

Romano, Renee Christine. *Race Mixing: Black–White Marriage in Postwar America*. Gainesville, FL: University Press of Florida, 2006.

Roof, Wade Clark. *A Generation of Seekers*. San Francisco: Harper San Francisco, 1994.

Roof, Wade Clark. *Spiritual Marketplace: Baby Boomers and the Remaking of American Religion*. Princeton, NJ: Princeton University Press, 2001.

Roozen, David, Jackson W. Carroll, and Wade Clark Roof. "Fifty Years of Religious Change in the United States" In Wade Clark Roof, Jackson W. Carroll, and David A. Roozen (eds.). *The Post-War Generation and Establishment Religion*. Boulder, CO: Westview Press, 1998, pp. 59–85.

Roper Organization. *Virginia Slims Poll*. New York: Richard Weiner, 1985.

Rose, Jerry D. *Outbreaks*. New York: Free Press, 1982.

Rosen, Jeffrey. *The Unwanted Gaze: The Destruction of Privacy in America*. New York: Random House, 2000.

Rosenbaum, Jill. "Social Control, Gender and Delinquency: An Analysis of Drug, Property and Violent Offenders." *Justice Quarterly* 4 (March 1987):117–132.

Rosenberg, Debra. "'Civil Unions' Cleared." *Newsweek* (April 25, 2000). Available online at www.newsweek.com/nw-srv/printed/us/na/a19076-2000apr25.htm.

Rosenfeld, Anne, and Elizabeth Stark. "The Prime of Our Lives." *Psychology Today* (May 1987):62–72.

Rosenfield, Sarah. "The Effects of Women's Employment: Personal Control and Sex Differences in Mental Health." *Journal of Health and Social Behavior* 30 (March 1989): 77–91.

Rosenthal, Robert, and Lenore Jacobson. *Pygmalion in the Classroom*. New York: Irvington Publishers, 1989.

Rosenwein, Rifka. "Why Media Mergers Matter." *Brill's Content* (December 1999/January 2000):92–111.

Rosnow, Ralph L. "Rumor as Communication: A Contextualist Approach." *Journal of Communication* 38 (Winter 1988): 12–28.

Rosnow, Ralph L. "Inside Rumor: A Personal Journey." *American Psychologist* 46 (May 1991):484–496.

Rossi, Alice S. (ed.). *The Feminist Papers.* New York: Bantam, 1974.

Rossides, Daniel W. *Social Stratification.* Upper Saddle River, NJ: Prentice Hall, 1997.

Roszak, Theodore. *The Making of a Counter Culture.* Berkeley and Los Angeles: University of California Press, 1995.

Rothschild, Joyce. "Alternatives to Bureaucracy: Democratic Participation in the Economy." *Annual Review of Sociology* 12 (1986):307–328.

Rothschild-Whitt, Joyce. "The Collectivistic Organization: An Alternative to Rational Bureaucratic Models." *American Sociological Review* 44 (August 1979):509–527.

Rothschild, Joyce, and J. Allen Whitt. *The Cooperative Workplace.* New York: Cambridge University Press, 1987.

Rotundo, E. Anthony. *American Manhood.* New York: Basic Books, 1993.

Rowe, David C. "Biometrical Models of Self-Reported Delinquent Behavior: A Twin Study." *Behavior Genetics* 13 (1983):473–489.

Rowe, David C. "Genetic and Environmental Components of Antisocial Behavior: A Study of 265 Twin Pairs." *Criminology* 24 (August 1986):513–532.

Roy, William G. *Socializing Capital: The Rise of the Large Industrial Corporation in America.* Princeton, NJ: Princeton University Press, 1999.

Rubenstein, Carin. "Guilty or Not Guilty." *Working Mother* (May 1991):53–56.

Rubin, Lillian B. *Worlds of Pain.* New York: Basic Books, 1992.

Rubin, Lillian B. *Families on the Faultline.* New York: HarperCollins, 1994.

Rubin, Lillian B. *60 and Up.* Boston: Beacon Press, 2007.

Ruesch, Hans. *Top of the World.* New York: Pocket Books, 1959.

Ruggiero, Vincent Ryan. *A Guide to Sociological Thinking.* Thousand Oaks, CA: Sage, 2001.

Rule, James B. "Rationality and Nonrationality in Militant Collective Action." *Sociological Theory* 7 (1989):145–160.

Russell, Alan. *Guinness Book of World Records.* New York: Sterling, 1987.

Ryan, William. *Blaming the Victim,* rev. and updated ed. New York: Vintage, 1976.

Saad, Lydia. "Future 'Retirees' Planning to Keep Busy on the Job." Gallup News Service (June 27, 2006b).

Saad, Lydia. "This is Not Your Father's Retirement." Gallup News Service (May 14, 2007a).

Saad, Lydia. "Any Healthcare Reform Plan Will Do for Americans." Gallup News Service (October 25, 2007b).

Saad, Lydia. "No Increase in Public Pressure for Healthcare Reform." (November 29, 2007c).

Saad, Lydia. "Anti-Muslim Sentiments Fairly Commonplace." Gallup News Service (2008).

Sadker, Myra, and David Sadker. "Sexism in the Schoolroom of the 80s." *Psychology Today* 19 (March 1985):54–57.

Sadker, Myra, and David Sadker. *Failing at Fairness.* New York: Simon & Schuster, 1995.

Sahlins, Marshall D. *Tribesmen.* Englewood Cliffs, NJ: Prentice Hall, 1968.

Sakamoto, Arthur, and Meichu D. Chen. "Inequality and Attainment in a Dual Labor Market." *American Sociological Review* 56 (June 1991):295–308.

"Salary Report, The 1997." *Working Woman* (January 1997):31–76.

Saletan, William. "Race, Genes, and Intelligence." *Slate* (November 19, 2007), http://www.slate.com/id/2178122.

Salholz, Eloise. "A Conflict of the Have-Nots." *Newsweek* (December 12, 1988):28–29.

Saltman, Kenneth J. *Collateral Damage: Corporatizing Public Schools: A Threat to Democracy.* Lanham, MD: Rowman & Littlefield, 2001.

Saltman, Kenneth J. *Capitalizing on Disaster.* Boulder, CO: Paradigm Publishers, 2007.

Sampson, Robert J., and John H. Laub. "Crime and Deviance over the Life Course: The Salience of Adult Social Bonds." *American Sociological Review* 55 (October 1990):609–627.

Samuda, Ronald. *The Psychological Testing of American Minorities.* New York: Dodd, Mead, 1975.

Samuelson, Paul A., and William D. Nordhaus. *Microeconomics,* 18th ed. New York: McGraw-Hill, 2004.

Samuelson, Robert J. "Welfare for Capitalists." *Newsweek* (May 5, 2003):54.

Sandberg, Jared. "Spinning a Web of Hate." *Newsweek* (July 19, 1999):28–29.

Sanders, Catherine M. *Surviving Grief.* New York: Wiley, 1992.

Sanderson, Stephen K. *Social Evolutionism.* New York: Basil Blackwell, 1990.

Santrock, John W. *Life-Span Development,* 8th ed. Boston: McGraw-Hill, 2002.

Sapir, Edward. "The Status of Linguistics as a Science." *Language* 5 (1929):207–214.

Sapolsky, Robert. "A Gene for Nothing." *Discover* (October 1997):40–46.

Sassen, Saskia. *Cities in a World Economy,* 3rd ed. Thousand Oaks, CA: Sage Publications, 2006.

Sawhill, Isabel, and John E. Morton. "Economic Mobility: Is the American Dream Alive and Well?" Washington, D.C.: The Pew Charitable Trusts, 2007.

Sawyer, R. Keith. "Emergence in Sociology: Contemporary Philosophy of Mind and Some Implications for Sociological Theory." *American Journal of Sociology* 107 (November 2001):551–585.

Sayles, Leonard R., and George Strauss. *The Local Union,* rev. ed. New York: Harcourt Brace Jovanovich, 1967.

Scambler, Graham. *Health and Social Change: A Critical Theory.* Buckingham, UK: Open University Press, 2002.

Schaefer, Richard T. *Racial and Ethnic Groups,* 11th ed. Upper Saddle River, NJ: Prentice Hall, 2007.

Schaie, K. Warner and Carmi Schooler (eds.). *Impact of Work on Older Adults.* New York: Springer, 1998.

Scheff, Thomas. "The Labeling Theory of Mental Illness." *American Sociological Review* 39 (June 1974):444–452.

Scheff, Thomas. *Microsociology.* Chicago: University of Chicago Press, 1990.

Scheff, Thomas. *Being Mentally Ill: A Sociological Theory,* 3rd ed. New York: Aldine de Gruyter, 1999.

Schell, Orville. "Deng's Revolution." *Newsweek* (August 22, 1997):21–27.

Schellenberg, James. A. *The Science of Conflict.* New York: Oxford University Press, 1982.

Schiff, Michel, and Richard Lewontin. *Education and Class.* New York: Oxford University Press, 1987.

Schlesinger, Arthur M. Jr. *The Disuniting of America: Reflections on a Multicultural Society.* New York: Norton, 1998.

Schnitzer, Martin C. *Comparative Economic Systems,* 8th ed. Mason, OH: South-Western College Publishing, 2000.

School Vouchers: Publicly Funded Programs in Cleveland and Milwaukee. U.S. General Accounting Office (GAO). Washington, D.C.: U.S. Government Printing Office, August 2001.

Schorr, Jonathan. "Giving Charter Schools a Chance." *The Nation* (June 5, 2000):19–22.

Schrag, Peter. "The Voucher Seduction." *The American Prospect* 11 (November 23, 1999):46.

Schriesheim, Chester A., and Steven Kerr. "Theories and Measures of Leadership: A Critical Appraisal of Current and Future Directions." In James G. Hunt and Lars L. Larson (eds.), *Leadership.* Carbondale: Southern Illinois University Press, 1977, pp. 9–45.

Schuman, Howard. "Sense and Nonsense about Surveys." *Contexts* (Summer 2002):40–47.

Schutt, Russell K. *Investigating the Social World,* 5th ed. Thousand Oaks, CA: Sage, 2006.

Schwartz, Barry. *The Battle for Human Nature.* New York: Norton, 1987.

Schwarz, John E. *Illusions of Opportunity: The American Dream in Question.* New York: Norton, 1997.

Schwartz, Martin D. *Researching Sexual Violence against Women.* Thousand Oaks, CA: Sage, 1997.

Schweinhart, Lawrence J. "How Much Do Good Early Childhood Programs Cost?" *Early Childhood Education and the Public Schools* 3 (April 1992):115–127.

Schwirtz, Michael. *The Lede in* The New York Times (December 11, 2007), http://thelede.blogs.nytimes.com/200712/11.

Scimecca, Joseph A. "Humanist Sociological Theory: The State of the Art." *Humanity and Society* 11 (1987):335–352.

Scott, Alan. *Ideology and the New Social Movements.* London: Routledge, 1995.

Scott, Martin B. "The Man on the Horse." In John W. Loy, Jr., Gerald S. Kenyon, and Barry D. McPherson (eds.), *Sport, Culture, and Society.* 2nd ed. Philadelphia: Lee and Febiger, 1981, pp. 146–156.

Scott, W. Richard. *Organizations,* 3rd ed. Englewood Cliffs, NJ: Prentice Hall, 1992.

Scripps-Howard. "Soon, Most of World Will Live in Cities." *Quad City Times* (May 16, 2002):A12.

Seale, Clive. *Constructing Death: The Sociology of Dying and Bereavement.* Cambridge, MA: Cambridge University Press, 1998.

Seligmann, Jean. "Husbands No, Babies, Yes." *Newsweek* (July 26, 1999):53.

Service, Elman R. *Origins of the State and Civilization.* New York: Norton, 1975.

Service, Elman R. *The Hunters,* 2nd ed. Englewood Cliffs, NJ: Prentice Hall, 1979.

Shaller, Elliot H., and Mary K. Qualiana. "The Family and Medical Leave Act—Key Provisions and Potential Problems." *Employee Relations Law Journal* 19 (Summer 1993):5–22.

Shanas, Ethel. *Old People in Three Industrial Societies.* New York: Arno Press, 1980.

Shapiro, Laura. "Guns and Dolls." *Newsweek* (May 28, 1990):54–65.

Sharot, Stephen. *A Comparative Sociology of World Religions: Virtuosos, Priests, and Popular Religion.* New York: New York University Press, 2001.

Shaw, Clifford, and Henry McKay. *Delinquency Areas.* Chicago: University of Chicago Press, 1929.

Sheehan, Molly O. "Poverty Persists." In *Vital Signs 2002: The Trends That Are Shaping Our Future.* New York: Norton, 2002, pp. 148–149.

Sheldon, William H. *Varieties of Delinquent Youth.* New York: Harper, 1949.

Shell, Adam. "Wall St. Firms Hit Over '90s Misdeeds." *USA Today* (April 29, 2003):1A.

Shepard, Jon M., and James G. Hougland Jr. "Contingency Theory: 'Complex Man' or 'Complex Organization'"? *Academy of Management Review* 3 (July 1978):413–427.

Shepard, Jon M., and Richard E. Wokutch. "The Problem of Business Ethics: Oxymoron or Inadequate Vocabulary." *Journal of Business and Psychology* 6 (1991):9–23.

Shepherd, William. "Legal Protection for Freedom of Religion." *The Center Magazine* (March/April 1982):30–37.

Sherman, Lawrence. "Defiance, Deference, and Irrelevance: A Theory of the Criminal Sanction." *Journal of Research in Crime and Delinquency* 30 (November 1993a):445–473.

Sherman, Lawrence. "A Brave New Darwinian Workplace." *Fortune* (February 25, 1993b):50.

Sherman, Arloc. "Income Inequity Hits Record Levels, New CBO Data Show." Center on Budget and Policy Priorities, 2007, http://www.cbpp.org/12-14-07inc.htm.

Shichor, David, and Dale K. Sechrest (eds.). *Three Strikes and You're Out.* Newbury Park, CA: Sage, 1996.

Shields, Vickie Rutledge, and Dawn Heinecken. *Measuring Up: How Advertising Images Shape Gender Identity.* Philadelphia: University of Pennsylvania Press, 2001.

Shils, Edward A., and Morris Janowitz. "Cohesion and Disintegration in the Wehrmacht in World War II." *Public Opinion Quarterly* 12 (Summer 1948):280–315.

Shils, Edward A., and Morris Janowitz. *Young Heroines.* Berkeley, CA: University of California Press, 2006.

Sidel, Ruth. *Keeping Women and Children Last.* New York: Penguin, 1998.

Siegel, Larry J., and Joseph J. Senna. *Introduction to Criminal Justice,* 11th ed. Belmont, CA: Wadsworth, 2007.

Siegel, Larry J. *Criminology,* 3rd ed. Belmont, CA: Wadsworth, 2008.

Sifford, Darrell. *The Only Child.* New York: Putnam, 1989.

Silberman, Charles E. *Crisis in the Classroom.* London: Wildwood House, 1973.

Simmel, Georg. *Conflict and the Web of Group Affiliation.* Translated by Kurt H. Wolff. New York: Free Press, 1964.

Simmons, J. L. *Deviants.* Berkeley, CA: Glendessary Press, 1969.

Simmons, Sally Lynn, and Amelia E. El-Hindi. "Six Transformations for Thinking about Integrative Curriculum." *Middle School Journal* 30 (November 1998): 32–36.

Simmons, Wendy W. "Majority of Americans Say More Women in Political Office Would Be Positive for the Country." Princeton, NJ: The Gallup Organization, 2001.

Simon, David R. *Elite Deviance,* 9th ed. Boston: Allyn & Bacon, 2007.

Simon, Herbert A. *Administrative Behavior,* 3rd ed. New York: Free Press, 1976.

Simpson, George Eaton. *Emile Durkheim.* New York: Crowell, 1963.

Simpson, Glenn R. "Lender Lobbying Blitz Abetted Mortgage Mess." *The Wall Street Journal* (December 31, 2007).

Sinclair, Karen. "Cross-Cultural Perspectives on American Sex Roles." In Marie Richmond-Abbott (ed.), *The American Woman*. New York: Holt, Rinehart and Winston, 1979, pp. 28–47.

Singh, B. Krishna, and J. Sherwood Williams. "Childlessness and Family Satisfaction." *Research on Aging* 3 (June 1981):218–227.

Singh, Sarban. "Shift in Focus." *Malaysian Business* (September 16, 1998):54.

Sizer, Theodore. *Horace's Hope: What Works for the American High School*. Boston: Houghton Mifflin, 1996.

Sjoberg, Gideon. *The Preindustrial City, Past and Present*. New York: Free Press, 1966.

Skinner, Denise A. "Dual-Career Families: Strains of Sharing." In Hamilton I. McCubbin and Charles R. Figley (eds.), *Stress and the Family*, vol. 1. New York: Brunner/Mazel, 1983, pp. 90–101.

Skocpol, Theda. "Bringing the State Back In: Strategies of Analysis in Current Research." In Peter B. Evans, Dietrich Rueschemeyer, and Theda Skocpol (eds.), *Bringing the State Back In*. Cambridge, UK: Cambridge University Press, 1985, pp. 3–37.

Skocpol, Theda, and Edwin Amenta. "States and Social Policies." In Ralph H. Turner and James F. Short Jr. (eds.), *Annual Review of Sociology* 12 (1986):131–157.

Skolnick, Arlene S., and Jerome H. Skolnick (eds.). *Family in Transition*, 14th ed. Boston: Allyn & Bacon, 2006.

Skolnick, Jerome H. "The Color of the Law." *American Prospect* 39 (July/August 1998):90–95.

Sloan, Allan. "The Hit Men." *Newsweek* (February 26, 1996):44–48.

Smaje, Chris. *Natural Hierarchies: The Historical Sociology of Race and Caste*. Malden, MA: Blackwell, 2000.

Small, Mario, and Katherine Newman. "Urban Poverty after *The Truly Disadvantaged*: The Rediscovery of the Family, the Neighborhood, and Culture." *Annual Review of Sociology* 27 (August 2001):23–45.

Smelser, Neil J. *Theory of Collective Behavior*. New York: Free Press, 1971.

Smelser, Neil J., William Julius Wilson, and Faith Mitchell (eds.). *America Becoming: Racial Trends and Their Consequences*, vols. 1 and 2. Washington, D.C.: National Academy Press, 2001.

Smith, Adam. *An Inquiry into the Nature and Causes of the Wealth of Nations*. Edwin Cannan (ed.). New York: Modern Library, 1965.

Smith, Adam. *The Theory of Moral Sentiments*. North Shadeland, IN: Liberty Press/Liberty Classics, 1976.

Smith, Adam, and Edwin Cannan. *The Wealth of Nations*. New York: Random House, 2000.

Smith, Christian. *Christian America*. Berkeley: University of California Press, 2000.

Smith, Christian, and Michael Emerson. *American Evangelicalism: Embattled and Thriving*. Chicago: University of Chicago Press, 1998.

Smith, Heather. "The Promotion of Women and Men." Ph.D. Dissertation, University of Delaware, 1999.

Smith, Lee. "Burned-Out Bosses." *Fortune* (July 25, 1994):44–52.

Smith, Michael D. "Hockey Violence: A Test of the Violent Subculture Hypothesis." *Social Problems* (27):235–247.

Smolowe, Jill. "When Violence Hits Home." *Time* (July 4, 1994):18–25.

Snarr, Brian B. "The Family and Medical Leave Act of 1993." *Compensation and Benefits Review* 25 (May/June 1993):6–9.

Snipp, C. Matthew. "A Demographic Comeback for American Indians?" *Population Today* 24 (November 1996):4–5.

Snizek, William E., and Joseph J. Kestel. "Understanding and Preventing the Premature Exodus of Mature Middle Managers from Today's Corporations." *Organization Development Journal* (Fall 1999):63–71.

Sobel, Michael E., Mark P. Becker, and Susan M. Minick. "Origins, Destinations, and Association in Occupational Mobility." *American Journal of Sociology* 104 (November 1998):687–721.

Sokoloff, Natalie J. *Between Money and Love*. New York: Praeger, 1980.

Solomon, Jolie. "Smoke From Washington." *Newsweek* (April 4, 1994):45.

Solomon, Jolie. "Texaco's Troubles." *Newsweek* (November 25, 1996):48–50.

Sorokin, Pitirim A. *Social Mobility*. New York: Harper, 1927.

Sourcebook of Criminal Justice Statistics Online. "Characteristics of Federal Prisioners." 2003. Available online at http://www.albany.edu/sourcebook/pdf/t654.pdf.

South, Scott J., and Kyle D. Crowder. "Escaping Distressed Neighborhoods: Individual, Community, and Metropolitan Influences." *American Journal of Sociology* 102 (January 1997):1040–1084.

Spain, Daphne, and Suzanne M. Bianchi. *Balancing Act: Motherhood, Marriage, and Employment among American Women*. New York: Russell Sage Foundation, 1996.

Spencer, Jon Michael. *The New Colored People*. New York: New York University Press, 1997.

Spengler, Oswald. *The Decline of the West*. New York: Knopf, 1926–1928.

Spickard, Paul R. "The Illogic of American Racial Categories." In Maria P. P. Root (ed.), *Racially Mixed People in America*. Thousand Oaks, CA: Sage, 1992.

Spindler, George D., and Louise Spindler. "Anthropologists View American Culture." *Annual Review of Anthropology* 12 (1983):49–78.

Spitz, René A. "Hospitalism." In Anna Freud et al. (eds.), *The Psychoanalytic Study of the Child*, vol. 2. New York: International Universities Press, 1946a, pp. 52–74.

Spitz, René A. "Hospitalism: A Follow-Up Report." In Anna Freud et al. (eds.), *The Psychoanalytic Study of the Child*, vol. 2. New York: International Universities Press, 1946b, pp. 113–117.

Spitze, G. "Women's Employment and Family Relations: A Review." In A. Booth (ed.), *Contemporary Families*. Minneapolis: National Council on Family Relations, 1991, pp. 381–404.

Spitzer, Steven. "Toward a Marxian Theory of Deviance." In Delos H. Kelly (ed.), *Criminal Behavior*. New York: St. Martin's Press, 1980, pp. 175–191.

Spotts, Peter N. "Science Labs, Too, 'Cooking the Books.'" *Christian Science Monitor* (July 19, 2002):1, 9.

Stacey, Judith. *Brave New Families*. New York: Basic Books, 1990.

Stolberg, Sheryl. "New Method Equalizes Stem Cell Debate." *The New York Times* (November 21, 2007).

Stangor, Charles. *Stereotypes and Prejudice*. Philadelphia, PA: Psychology Press, 2000.

Stark, David, and Laszlo Bruszt. *Postsocialist Pathways*. Cambridge, UK: Cambridge University Press, 1998.

Stark, Rodney, and William Sims Bainbridge. "Secularization and Cult Formation in the Jazz Age." *Journal for the Scientific Study of Religion* 20 (December 1981):360–373.

Stark, Rodney, and Roger Finke. *Acts of Faith: Explaining the Human Side of Religion.* Berkeley: University of California Press, 2000.

Stark, Rodney, Lori Kent, and Daniel P. Doyle. "Religion and the Ecology of a 'Lost' Relationship." *Journal of Research in Crime and Delinquency* 19–20 (January 1982):4–24.

Starr, Mark, and Allison Samuels. "Ear Today, But Gone Tomorrow." *Newsweek* (July 14, 1997):58–60.

Starr, Paul E. *The Social Transformation of American Medicine.* New York: Basic Books, 1983.

"Statement on Race and Intelligence." *Journal of Social Issues* 25 (Summer 1969):1–3.

State of Food Insecurity in the World 2002, The. Food and Agriculture Organization of the United Nations. Rome, Italy, 2002.

State of Our Nation's Youth, The. Alexandria, VA: Horatio Alger Association of Distinguished Americans, 2001.

Stavans, Ilan. *The Hispanic Condition,* 2nd ed. New York: Harper Collins, 2001.

Steedley, Gilbert. "The Forbes 500." *Forbes* 153 (April 25, 1994):195–230.

Stein, Joel. "The New View of Pot." *Time* (November 4, 2002):56–66.

Stein, Leonard I. "The Doctor-Nurse Game." *Archives of General Psychiatry* 16 (June 1967):699–700.

Stein, Rob. "Increase in Physicians' Insurance Hurts Care: Services Are Being Pared, and Clinics Are Closing." *Washington Post* (January 5, 2003):A1.

Steinberg, Jacques. "At 42 Newly Privatized Philadelphia Schools, Uncertainty Abounds." *New York Times* (April 19, 2002):A16.

Steinmetz, Suzanne K. *Duty Bound: Elder Abuse and Family Care.* Newbury Park, CA: Sage, 1988.

Steinnes, Donald N. "Suburbanization and the 'Malling of America': A Time-Series Approach." *Urban Affairs Quarterly* 17 (June 1982):401–418.

Stephens, Martha. *The Treatment: The Story of Those Who Died in the Cincinnati Radiation Tests.* Durham, NC: Duke University Press, 2002.

Stephens, W. Richard Jr., "What Now? The Relevance of Sociology to Your Life and Career." In Roger A. Straus (ed.), *Using Sociology,* 3rd ed. Lanham, MD: Rowman & Littlefield, 2002, pp. 347–363.

Stern, Kenneth S. *A Force Upon the Plane.* New York: Simon & Schuster, 1996.

Stern, Philip M. *The Best Congress Money Can Buy.* New York: Pantheon, 1988.

Steward, Julian H. *Theory of Culture Change.* Urbana: University of Illinois Press, 1979.

Stiglitz, Joseph E. *Globalization and Its Discontents.* New York: Norton, 2003.

Stiglitz, Joseph E. *Making Globalization Work.* New York: W.W. Norton & Company, 2007.

Stockard, Janice. *Marriage in Culture: Practice and Meaning across Diverse Societies.* Fort Worth, TX: Harcourt College Publishers, 2002.

Stockard, Jean, and Miriam M. Johnson. *Sex and Gender in Society,* 2nd ed. Englewood Cliffs, NJ: Prentice Hall, 1992.

Stoessinger, John. "The Anatomy of the Nation-State and the Nature of Power." In Michael Smith, Richard Little, and Michael Shackleton (eds.), *Perspectives on World Politics.* London: Croom Helm, 1981, pp. 25–26.

Stogdill, Ralph M. *Handbook of Leadership.* New York: Free Press, 1974.

Stone, Jon R. *On the Boundaries of American Evangelicalism.* New York: St. Martin's Press, 1997.

Stranger, Janice, Nicole C. Batchelder, William Brossman, and Gerald L. Uslarder. "What the Family and Medical Leave Act Means for Employers." *Journal of Compensation and Benefits* 9 (July/August 1993):12–19.

Strasburger, Victor C. *Adolescents and the Media: Medical and Psychological Impact.* Newbury Park, CA: Sage, 1995.

Straus, Murray A., and Suzanne K. Steinmetz. *Behind Closed Doors.* Piscataway, NJ: Tronaglion Publishers, 2006.

Straus, Murray A., David B. Sugarman, and Jean Giles-Sims. "Spanking by Parents and Subsequent Antisocial Behavior of the Child." *Archives of Pediatrics and Adolescent Medicine* (August 1997):761–767.

Straus, Roger A. *Using Sociology,* 3rd ed. Lanham, MD: Rowman & Littlefield, 2002.

Strauss, Anselm, and Juliet Corbin. *Basics of Qualitative Research,* 3rd ed. Thousand Oaks, CA: Sage, 2007.

Strauss, Anselm, Shizuko Fagerhaugh, Barbara Suczek, and Carolyn Wiener. *Social Organization of Medical Work.* Chicago: University of Chicago Press, 1985.

Strickland, S. S., and Prakash S. Shetty. *Human Biology and Social Inequality.* New York: Cambridge University Press, 1998.

Suarez-Orozco, Marcelo M., and Mariela Paez (eds.). *Latinos: Remaking America.* Berkeley: University of California Press, 2002.

Suchman, Edward A. "The 'Hang-Loose' Ethic and the Spirit of Drug Use." *Journal of Health and Social Behavior* 9 (June 1968):146–155.

Sullivan, Teresa A. "Work-Related Social Problems." In George Ritzer (ed.), *Handbook of Social Problems.* Beverly Hills, CA: Sage, 2004, pp. 193–208.

Sumner, William Graham. *Folkways.* Boston: Ginn, 1906.

Suro, Roberto. "Mixed Doubles." *American Demographics* 21 (November 1999):57–62.

Sussman, Marvin B. "The Help Pattern in the Middle Class Family." *American Sociological Review* 18 (February 1953):22–28.

Sussman, Marvin, Suzanne K. Steinmetz, and Gary W. Peterson. *Handbook of Marriage and the Family.* New York: Plenum, 1999.

Sutherland, Edwin H. "White Collar Criminality." *American Sociological Review* 5 (February 1940):1–12.

Sutherland, Edwin H. *White-Collar Crime.* New Haven, CT: Yale University Press, 1983.

Sutherland, Edwin H., and Donald R. Cressey. *Principles of Criminology,* 11th ed. Dix Hills, NY: General Hall, 1992.

Suttles, Gerald D. *The Social Order of the Slum.* Chicago: University of Chicago Press, 1968.

Swaaningen, René van. *Critical Criminology: Visions from Europe.* London: Sage, 1997.

Swain, Carol M. *New White Nationalism in America: Its Challenge to Integration.* New York: Cambridge University Press, 2002.

Swann, William B. Jr., Judith H. Langlois, and Lucia Albino Gilbert (eds.). *Sexism and Stereotypes.* Washington, D.C.: American Psychological Association, 1998.

Swatos, William H. Jr., and Daniel V. Olson (eds.). *The Secularization Debate.* Lanham, MD: Rowman & Littlefield, 2000.

Sweet, Jones, Larry Bumpass, and Vaughn Call. *National Survey of Families and Households*. Madison: University of Wisconsin, Center for Demography and Ecology, 1988.

Swidler, Ann. "Culture in Action: Symbols and Strategies." *American Sociological Review* 51 (April 1986):273–286.

Swinton, David H. "Economic Status of Blacks." In *The State of Black America 1989*. New York: National Urban League, 1989, pp. 129–152.

Symonds, William C. "For-Profit-Schools." *BusinessWeek* (February 7, 2000):64ff.

Symonds, William C. "Colleges in Crisis." *BusinessWeek* (April 28, 2003):72–79.

Szasz, Thomas. *The Myth of Mental Illness,* rev. ed. New York: Harper and Row, 1986.

Szymanski, Albert. "Racial Discrimination and White Gain." *American Sociological Review* 41 (June 1976):403–414.

Szymanski, Albert. *Class Structure*. New York: Praeger, 1983.

Tannenbaum, Frank. *Slave and Citizen*. New York: Knopf, 1947.

Taylor, Alexander L. "The Growing Gap in Retraining." *Time* (March 28, 1983):50–51.

Taylor, Chris. "Day of the Smart Mobs." *Time* (March 10, 2003):53.

Taylor, Humphrey. "The Canadian and U.S. Health Care Systems Compared." *The Harris Poll* (April 26, 1994):1–6.

Taylor, P. "Therapists Rethink Attitudes on Divorce: New Movement to Save Marriages Focuses on Impact on Children." *Washington Post* (January 29, 1991):A1, A6.

Taylor, Shelley E., Letitia Ann Peplau, and David O. Sears. *Social Psychology,* 9th ed. Englewood Cliffs, NJ: Prentice Hall, 1997.

Taylor, Verta. "Social Movement Continuity: The Women's Movement in Abeyance." *American Sociological Review* 54 (October 1989):761–775.

Teich, Albert H. (ed.). *Technology and the Future,* 6th ed. New York: St. Martin's Press, 1993.

Tellegen, Auke, D. T. Lykken, T. J. Bouchard Jr., and M. McGue. "Heritability of Interests: A Twin Study." *Journal of Applied Psychology* (August 1993):649–661.

Terry, James L. "Bringing Women In: A Modest Proposal." *Teaching Sociology* 10 (January 1983):251–261.

Therborn, Goran. *The Power of Ideology and the Ideology of Power*. London: New Left Books, 1982.

Therborn, Goran. "The Prospects of Labor and the Transformation of Advanced Capitalism." *New Left Review* 145 (May/June 1984):5–38.

Thernstrom, Stephan, and Abigail Thernstrom. *America in Black and White: One Nation, Indivisible*. New York: Simon & Schuster, 1999.

Thomas, Evan. "The Next Level." *Newsweek* (April 7, 1997):28–36.

Thomas, Evan. "The War over Gay Marriage." *Newsweek* (July 8, 2003):38–45.

Thomas, Evan, and Gregory L. Vistica. "A Question of Consent." *Newsweek* (April 28, 1997):41.

Thomas, Melvin E., and Michael Hughes. "The Continuing Significance of Race: A Study of Race, Class, and Quality of Life in America, 1972–1985." *American Sociological Review* 51 (December 1986):830–841.

Thomas, Susan L. *Gender and Poverty*. New York: Garland, 1994.

Thompson, Kenneth. "The Organizational Society." In Graeme Salaman and Kenneth Thompson (eds.), *Control and Ideology in Organizations*. Cambridge, MA: MIT Press, 1980, pp. 3–23.

Thompson, L., and A. J. Walker. "Gender in Families." In A. Booth (ed.), *Contemporary Families*. Minneapolis: National Council on Family Relations, 1991, pp. 76–102.

Thornton, Arland, and Linda Young-DeMarco. "Four Decades of Trends in Attitudes toward Family Issues in the United States: The 1960s through the 1990s." *Journal of Marriage and Family* 63 (2001):1009–1037.

Thurow, Lester C. *Head to Head*. New York: Harper Collins Publishers, 2003a.

Thurow, Lester C. *Fortune Favors the Bold: What We Must Do to Build a New and Lasting Global Prosperity*. New York: HarperCollins Publishers, 2003b.

Tilly, Charles. *From Mobilization to Revolution*. Reading, MA: Addison-Wesley, 1978.

Tilly, Charles. "Does Modernization Breed Revolution?" In Jack A. Gladstone (ed.), *Revolutions*. New York: Harcourt Brace Jovanovich, 1986, pp. 47–57.

Tilly, Charles. *Coercion, Capital, and European States, AD 990–1990*. Cambridge, MA: Blackwell, 1990.

Tilly, Charles. *Coercion, Capital, and European States, AD 990–1992*. Cambridge, MA: Blackwell, 1992.

Tilly, Charles. *Social Processes*. Lanham, MD: Rowman & Littlefield, 1997.

Tilly, Charles. *Durable Inequality*. Berkeley: University of California Press, 1999.

Tilly, Chris. "Beyond Patching the Safety Net: A Welfare and Work Survival Strategy." *Dollars and Sense* (January/February 1999):14, 36–38.

Tilly, Louise, and Joan Scott. *Women, Work, and Family*. New York: Holt, Rinehart and Winston, 1978.

Tilove, Jonathan. "The New Map of American Politics." *The American Prospect* (May/June 1999):34.

Timms, Duncan. *The Urban Mosaic*. Cambridge, UK: Cambridge University Press, 1971.

Timmons, Heather. "Online Scrabble Craze Leaves Sellers at Loss for Words." *The New York Times* (March 2, 2008).

Tobias, Sheila. *Sexual Politics*. Boulder, CO: Westview Press, 1997.

Toch, Thomas, and Warren Cohen. "Public Education: A Monopoly No Longer." *U.S. News & World Report* (November 23, 1998):25.

"To Clone a Human." *The Week* (February 15, 2008):13.

Tocqueville, Alexis de. *The Old Regime and the French Revolution*. Translated by Stuart Gilbert. Garden City, NY: Doubleday, 1955.

To Establish Justice, to Insure Domestic Tranquility: A Thirty Year Update of the National Convention on the Causes and Prevention of Violence. Washington, D.C.: The Milton S. Eisenhower Foundation, 1999.

Toland, John. *Adolph Hitler*. Garden City, NY: Doubleday, 1976.

Tomaskovic-Devey, Don, and Sheryl Skaggs. "Sex Segregation, Labor Process Organization, and Gender Earnings Inequality." *American Journal of Sociology* 108 (July 2002):102–128.

Tönnies, Ferdinand. *Community and Society*. Translated and edited by Charles P. Loomis. East Lansing: Michigan State University Press, 1957.

Tonry, Michael (ed.). *The Handbook of Crime and Punishment*. New York: Oxford University Press, 2000.

Tonry, Michael, and Mary Lynch. "Intermediate Sanctions." In Michael Tonry (ed.), *Crime and Justice: A Review of Research,* vol. 20. Chicago: University of Chicago Press, 1996, pp. 94–144.

Toynbee, Arnold. *A Study of History*. New York: Oxford University Press, 1946.

Treiman, Donald J. *Occupational Prestige in Comparative Perspective.* New York: Academic Press, 1977.

Trimble, J. E. "Stereotypical Images, American Indians, and Prejudice." In P. A. Katz and D. A. Taylor (eds.), *Eliminating Racism.* New York: Plenum Press, 1988, pp. 181–202.

Trister, Michael. "The Rise and Reform of Stealth PACs." *The American Prospect* (September 25–October 9, 2000): 32–35.

Troeltsch, Ernst. *The Social Teachings of the Christian Churches.* New York: Macmillan, 1931.

Troll, Lillian E. "The Family of Later Life: A Decade Review." *Journal of Marriage and the Family* 33 (May 1971):263–290.

Trueba, Enrique T. *Latinos Unidos: From Cultural Diversity to the Politics of Solidarity.* Lanham, MA: Rowman & Littlefield, 1999.

Tucker, Robert C. (ed.). *The Marx-Engels Reader.* New York: Norton, 1972.

Tumin, Melvin M. "Some Principles of Stratification: A Critical Analysis." *American Sociological Review* 18 (August 1953):387–394.

Turnbull, Colin. *The Mountain People.* London: Pimlico, 1994.

Turner, Jonathan H. *American Society,* 2nd ed. New York: Harper and Row, 1976.

Turner, Jonathan H. *The Structure of Sociological Theory,* 5th ed. Belmont, CA: Wadsworth, 2002a.

Turner, Jonathan H. (ed.). *Handbook of Sociological Theory.* New York: Kluwer Academic/Plenum Publishers, 2002b.

Turner, Jonathan H. *Face to Face: Toward a Sociological Theory of Interpersonal Behavior.* Stanford, CA: Stanford University Press, 2002c.

Turner, Ralph H. "Collective Behavior." In Robert E. L. Faris (ed.), *Handbook of Modern Sociology.* Chicago: Rand McNally, 1964, pp. 382–425.

Turner, Ralph H., and Lewis M. Killian. *Collective Behavior,* 3rd ed. Englewood Cliffs, NJ: Prentice Hall, 1987.

Twaddle, Andrew C. *Sickness Behavior and the Sick Role.* Cambridge, MA: Schenkman, 1981.

Twaddle, Andrew C., and Richard M. Hessler. *A Sociology of Health,* 2nd ed. New York: Macmillan, 1987.

Tyler, Patrick E. "Threats and Responses." *New York Times* (February 17, 2003):A1.

Ulbrich, Patricia M. "The Determinants of Depression in Two-Income Marriages." *Journal of Marriage and the Family* 50 (February 1988):121–131.

United Nations Climate Change Conference. United Nations, 2007. http://www.csis.org/.

United Nations. *Family.* New York: United Nations Publications, 1996.

United Nations Population Division. *World Population Prospects: The 1998 Revision.* New York: United Nations, 2000.

United Nations. *State of the World Population 2007.* New York: United Nations, 2007.

Urban Dictionary, 2008. http://www.urbandictionary.com/.

Urbina, Law. "Voter ID Laws Are Set to Face a Crucial Test." *New York Times* (January 7, 2008).

U.S. Bureau of the Census. *1990 Census of Population, Social and Economic Characteristics, United States.* CP-2-1. November 1993. Washington, D.C.: U.S. Government Printing Office, 1993.

U.S. Bureau of the Census. "Percentage Change in Metropolitan Population Inside and Outside Central Cities, by Region and Division," 2000. Available online at www.census.gov/population/estimates/metrocity/ma98-c2.gif.

U.S. Bureau of the Census. *Money Income in the United States: 2000.* Washington, D.C.: U.S. Government Printing Office, 2001.

U.S. Bureau of the Census. "Income, Poverty, and Health Insurance Coverage in the United States: 2006," 2007a. http://www.census.gov/prod/2007a.pubs/p60-233.pdf.

U.S. Bureau of the Census. *Statistical Abstract of the United States: 2008.* Washington, D.C.: U.S. Government Printing Office, 2007c.

U.S. Commission on Civil Rights. *Racial Isolation in the Public Schools,* vol. II. Washington, D.C.: U.S. Government Printing Office, 1967.

U.S. Commission on Civil Rights. *Success of Asian Americans: Fact or Fiction?* Clearinghouse Publication 64. Washington, D.C.: U.S. Government Printing Office, 1980.

U.S. Commission on Wartime Relocation and Internment of Civilians. *Personal Justice Denied.* Washington, D.C.: U.S. Government Printing Office, 1983.

U.S. Department of Health and Human Services. Administration on Child, Youth and Families. *Child Maltreatment 2005.* Washington, D.C.: U.S. Government Printing Office, 2007.

U.S. Department of Health and Human Services. "U.S. Birth Rate Reaches Record Low." Washington, D.C.: HHS News, July 25, 2003. http://www.hhs.gov.

U.S. Department of Justice. *White Collar Crime.* Washington, D.C.: U.S. Government Printing Office, 1987.

U.S. Department of Justice. "FBI Releases 2006 Hate Crime Statistics." Federal Bureau of Investigation. Press Release, November 19, 2007.

U.S. Department of Justice (2006a). *Criminal Victimization in the United States, 2006.* http://ojp.usdoj.gov/bJS/cvict.htm.

U.S. Department of Justice (2006b). Federal Bureau of Investigation. *Uniform Crime Reports, 2006.* http://www.fbi.ucr/ucr.htm.

U.S. Department of Justice. Federal Bureau of Investigation. *Uniform Crime Reports, 2006.* Washington, D.C.: U.S. Government Printing Office, 2007a.

U.S. Department of Justice. *Criminal Victimization in the United States, 2006.* Washington, D.C.: U.S. Government Printing Office, 2007b.

U.S. Department of Labor. Bureau of Labor Statistics. "Unemployed Persons by Marital Status, Race, Age, and Sex." 2001a. Available online at www.bls.gov/cps/cpsaat24.pdf.

U.S. Department of Labor. Bureau of Labor Statistics. "Issues in Labor Statistics." U.S. Department of Labor, 2006. http://www.bls.gov/opub/ils/pdf/opbils55.pdf.

U.S. Department of Labor. "Labor Force Statistics from the Current Population Survey." Bureau of Labor Statistics, 2008. http://www.bls.gov/newsrelease/empsit.t02.htm.

"U.S. Teen Sexual Activity." Menlo Park, CA: Kaiser Family Foundation, January 2005.

Useem, Michael. *The Inner Circle.* New York: Oxford University Press, 1984.

Vago, Steven. *Social Change,* 5th ed. Upper Saddle River, NJ: Prentice Hall, 2003.

Valdivieso, Rafael, and Cary Davis. "U.S. Hispanics: Challenging Issues for the 1990s." *Population Trends and Public Policy.* Washington, D.C.: Population Reference Bureau, 1988.

Valian, Virginia. *Why So Slow? The Advancement of Women.* Cambridge, MA: MIT Press, 1999.

Van Ausdale, Debra, and Joe R. Feagin. *The First R: How Children Learn Race and Racism.* Lanham, MD: Rowman & Littlefield, 2001.

VandeHei, Jim, and David S. Hilzenrath. "Hill Leaders Agree on Corporate Curbs." *Washington Post* (July 25, 2002):A-1.

VandeHei, Jim, and Tom Hamburger. "Businesses Prepare New Legislative Wish List for Bush." *The Wall Street Journal* (January 8, 2002):A20.

van den Berghe, Pierre L. "Dialectic and Functionalism: Toward a Theoretical Synthesis." *American Sociological Review* 28 (October 1963):695–705.

van den Berghe, Pierre L. *Race and Racism,* 2nd ed. New York: Wiley, 1978.

Vander Zanden, James W. *American Minority Relations.* New York: McGraw-Hill, 1990.

van Gennep, A. *The Rites of Passage.* Chicago: University of Chicago Press, 1960.

Vanhanen, Tatu. *Prospects of Democracy.* London: Routledge, 1997.

Van Zelst, Raymond H. "Sociometrically Selected Work Teams Increase Production." *Personnel Psychology* 5 (Autumn 1952):175–185.

Vazque, Francisco H., and Rodolfo D. Torres. *Latino/a Thought: Culture, Politics, and Society.* Lanham, MD: Rowman & Littlefield, 2002.

Veblen, Thorstein. *The Engineers and the Price System.* New York: Viking, 1933.

Veblen, Thorstein. *Theory of the Leisure Class.* New York: Penguin, 1995.

"Violence against Women: Research into Practice." Washington, D.C.: National Institute of Justice, 2002.

Vobejda, Barbara. "The Heartland Pulses with New Blood." *Washington Post* (August 11, 1991):A1,A18.

Volti, Rudi. *Society and Technological Change,* 4th ed. New York: Worth Publishers, 2003.

Voss, Kim, and Rachel Sherman. "Breaking the Iron Law of Oligarchy: Union Revitalization in the American Labor Movement." *American Journal of Sociology* 106 (September 2000):303–349.

Wagner, Tony. *Making the Grade: Reinventing America's Schools.* Bristol, PA: Taylor and Francis, 2003.

Wagster, Emily. "Many Workers Hurt abroad by Culture Shock." *USA Today* (December 17, 1993):1B.

Waitzkin, Howard. *The Second Sickness.* Chicago: University of Chicago Press, 1986.

Waitzkin, Howard. *The Second Sickness: Contradictions of Capitalist Health Care.* Lanham, MD: Rowman & Littlefield, 2000.

Waitzkin, Howard. *At the Front Lines of Medicine: How the Health Care System Alienates Doctors and Mistreats Patients . . . and What We Can Do about It.* Lanham, MD: Rowman & Littlefield, 2001.

Waldman, Steven. "The Stingy Politics of Head Start." *Newsweek.* Special Issue. *Education: A Consumer's Handbook* (Fall/Winter 1990–91):78–79.

Wallace, Walter J. *The Future of Ethnicity, Race and Nationality.* Westport, CT: Praeger, 1997.

Wallechinsky, David, and Irving Wallace. *The People's Almanac #3.* New York: Bantam, 1981.

Waller, Willard, and Reuben Hill. *The Family,* rev. ed. New York: Dryden Press, 1951.

Wallerstein, Immanuel. *The Capitalist World Economy.* New York: Cambridge University Press, 1979.

Wallerstein, Immanuel. *The Politics of the World Economy.* Cambridge, UK: Cambridge University Press, 1984.

Wallerstein, Immanuel. *The Essential Wallerstein.* New York: New Press, 2000.

Wallerstein, Immanuel. *The End of the World as We Know It: Social Science for the Twenty-First Century.* Minneapolis, MN: University of Minnesota Press, 2001.

Wallerstein, Immanuel. *The Modern World-System III.* Emerald Group Publishing, 2003a.

Wallerstein, Immanuel. *Decline of American Power: The U.S. in a Chaotic World.* New York: New Press, 2003b.

Wallerstein, Immanuel. *World-Systems Analysis.* Durham, NC: Duke University Press, 2004.

Wallich, Paul, and Elizabeth Corcoran. "The Discreet Disappearance of the Bourgeoisie." *Scientific American* 226 (February 1992):111.

Wallulis, Jerald. *The New Insecurity.* Albany: State University of New York Press, 1998.

Walsh, Edward. "In Midwest, Bush Calls Gore Obstacle to Reform." *Washington Post* (October 24, 2000):A12.

Ward, R., J. Logan, and G. Spitze. "The Influence of Parent and Child Needs on Coresidence in Middle and Later Years." *Journal of Marriage and the Family* (February 1992):209–221.

Ware, Leland, and Antoine Allen. "The Geography of Discrimination: Hypersegregation, Isolation and Fragmentation within the African-American Community." In Lee A. Daniels (ed.), *The State of Black America 2002.* New York: National Urban League, 2002, pp. 69–92.

Warner, R. Stephen. *Public Religions in the Modern World.* Chicago: University of Chicago Press, 1994.

Warner, W. Lloyd. *Social Class in America.* New York: Harper and Row, 1960.

Warner, W. Lloyd, and Paul S. Lunt. *The Social Life of a Modern Community.* New Haven, CT: Yale University Press, 1941.

Warr, Mark. *Companions in Crime: The Social Aspects of Criminal Conduct.* New York: Cambridge University Press, 2002.

Warren, Jenifer. "One in 100: Behind Bars in America." Washington, D.C.: Pew Center on the States and the Public Safety Performance Project, 2008.

Warren, Martin, and Sheldon Berkowitz. "The Employability of AFDC Mothers and Fathers." *Welfare in Review* 7 (July/August 1969):1–7.

Wasserman, Gary. *The Basics of American Politics,* 13th ed. New York: Longman, 2007.

Waters, Malcolm. *Globalization,* 2nd ed. New York: Routledge, 2001.

Waters, Mary C. *Ethnic Options.* Berkeley: University of California Press, 1990.

Watson, C. W. *Multiculturalism.* Bristol, PA: Taylor and Francis, 2000.

Watson, Russell, and John Barry. "Tomorrow's New Face of Battle." *Newsweek* (February 1998, Winter 1997–98 Special Edition):66–67.

Watterson, T. "Social Security: Many Widows Don't Get Benefits." *Baltimore Sun* (June 13, 1990):C24.

Watzman, Nancy, James Youngclaus, and Jennifer Shecter. *Cashing In.* Washington, D.C.: Center for Responsive Politics, 1997.

Wax, Emily. "In Times of Terror, Teens Talk the Talk." *Washington Post* (March 19, 2002):A1, A10.

Waxman, Chaim I. *The Stigma of Poverty.* New York: Pergamon, 1983.

Weakliem, David L. "Relative Wages and the Radical Theory of Economic Segmentation." *American Sociological Review* 55 (1990):574–590.

Webb, R. K. *Harriet Martineau, A Radical Victorian.* New York: Columbia University Press, 1960.

Weber, Max. *From Max Weber.* Edited by H. H. Gerth and C. Wright Mills. New York: Oxford University Press, 1946.

Weber, Max. *The Religion of China.* New York: Free Press, 1951.

Weber, Max. *Ancient Judaism.* New York: Free Press, 1952.

Weber, Max. *The Religion of India.* New York: Free Press, 1958.

Weber, Max. *The Sociology of Religion.* Boston: Beacon Press, 1964.

Webster, Murray Jr., and Stuart J. Hysom. "Creating Status Characteristics." *American Sociological Review* 63 (June 1998):351–378.

Weeden, Kim A. "Why Do Some Occupations Pay More than Others? Social Closure and Earnings Inequality in the United States." *American Journal of Sociology* 108 (July 2002):55–101.

Weeks, John R. *Population,* 10th ed. Belmont, CA: Wadsworth, 2008.

Weinberg, Meyer. *Desegregation Research,* 2nd ed. Bloomington, IN: Phi Delta Kappan, 1970.

Weingart, Peter, et al. (eds.). *Human by Nature: Between Biology and the Social Sciences.* Mahwah, NJ: Erlbaum, 1997.

Weinstein, Jay. "A (Further) Common on the Differences between Applied and Academic Sociology." *Contemporary Sociology* (March 2000):344–347.

Weisberg, Jacob. *The Bush Tragedy.* New York: Random House, 2008.

Weisbrot, Mark. "The Mirage of Progress: The Economic Failure of the Last Two Decades of the Twentieth Century." *American Prospect* (Winter 2002):A10–A12.

Weisburd, David, and Elin Waring. *White Collar Crime and Criminal Careers.* Cambridge, UK: Cambridge University Press, 2001.

Weiss, Gregory L., and Lynne E. Lonnquist. *The Sociology of Health, Healing, and Illness.* New York: Simon & Schuster, 1996.

Weiss, Lawrence D. *Private Medicine and Public Health.* Boulder, CO: Westview Press, 1997.

Wray, Herbert. "When Strangers Become Family." *U.S. News and World Report* (November 29, 1999):58–67.

Weiss, Rick. "Transplant Researchers Clone 5 Pigs." *Washington Post* (March 15, 2000):A1, A16.

Weiss, Rick. "In Laboratory, Ordinary Cells Are Turned Into Eggs." *Washington Post* (May 2, 2003):A1, A12.

Weiss, Rick. "Synthetic DNA on the Brink of Yielding New Life Forms." *Washington Post* (December 7, 2007a):A1.

Weiss, Rick. "Scientists Cure Mice of Sickle Cell Using Stem Cell Technique." *Washington Post* (December 7, 2007b):A2.

Weitz, Rose. *The Sociology of Health, Illness, and Health Care: A Critical Approach,* 4th ed. Belmont, CA: Wadsworth, 2007.

Weitzman, Leonore J., and Deborah Eifler. "Sex Role Socialization in Picture Books for Preschool Children." *American Journal of Sociology* 77 (May 1972):1125–1150.

Weitzman, Susan. *"Not to People Like Us": Hidden Abuse in Upscale Marriages.* New York: Basic Books, 2000.

Weller, Jack E. *Yesterday's People: Life in Contemporary Appalachia.* Lexington: University Press of Kentucky, 1980.

Wells, Spencer, and Mark Read. *The Journey of Man: A Genetic Odyssey.* Princeton, NJ: Princeton University Press, 2003.

Werhane, Patricia H. *Adam Smith and His Legacy for Modern Capitalism.* New York: Oxford University Press, 1991.

Wermuth, Laurie A. *Global Inequality and Human Needs: Health and Illness in an Increasingly Unequal World.* Boston: Allyn & Bacon, 2002.

West, Candace, and Don H. Zimmerman. "Doing Gender." *Gender and Society* 1 (June 1987):125–151.

West, Guida, and Rhoda Lois Blumberg (eds.). *Women and Social Protest.* New York: Oxford University Press, 1990.

Western, Bruce, and Becky Pettit. "Beyond Crime and Punishment: Prisons and Inequality." *Contexts* 1 (Fall 2002):37–43.

Westoff, Leslie Aldridge, and Charles F. Westoff. *From Now to Zero.* Boston: Little, Brown and Company, 1971.

Wetcher, Kenneth, Art Barker, and Rex McCaughtry. *Save the Males: Why Men Are Mistreated, Misdiagnosed, and Misunderstood.* Summit, NJ: PIA Press, 1991.

White, Leslie A. *The Science of Culture,* 2nd ed. New York: Farrar, Straus & Giroux, 1969.

White, Lynn. *Medieval Technology and Social Change.* New York: Oxford University Press, 1972.

White, Lynn, and David B. Brinkerhoff. "The Sexual Division of Labor: Evidence from Childhood." *Social Forces* 60 (September 1981):170–181.

White, Ralph K., and Ronald O. Lippitt. *Autocracy and Democracy.* New York: Harper and Row, 1960.

White, Robert, and Robert Althauser. "Internal Labor Markets, Promotions, and Worker Skill: An Indirect Test of Skill ILMS." *Social Science Research* 13 (December 1984):373–392.

Whitehead, Alfred North. *Adventures of Ideas.* New York: Mentor Books, 1933.

Whiten, A., et al. "Cultures in Chimpanzees." *Nature* 399 (June 1999):682–685.

Whorf, Benjamin Lee. *Language, Thought and Reality.* Edited by John B. Carroll. Cambridge, MA: MIT Press, 1956.

"Why Population Aging Matters: A Global Perspective." Washington, D.C.: National Institute on Aging, 2007.

Whyte, William Foote. *Street Corner Society,* 4th ed. Chicago: University of Chicago Press, 1993.

Whyte, William H. Jr. *The Organization Man.* New York: Simon & Schuster, 1972.

Whyte, William H., and Joseph Nocera. *The Organization Man: The Book That Defined a Generation.* Philadelphia: University of Pennsylvania Press, 2002.

Wiehe, Vernon R. *Sibling Abuse.* Lexington, MA: Heath, 1990.

Wiesel, Elie. "The Question of Genocide." *Newsweek* (April 12, 1999):37.

Wieviorka, Michel. *The Arena of Racism.* Newbury Park, CA: Sage, 1995.

Wildavsky, Ben. "Grading on a Curve: A New Controversy Erupts over Race, Class, and SAT Scores." *U.S. News & World Report* (September 13, 1999):53.

Wiley, Norbert (ed.). *The Marx-Weber Debate.* Beverly Hills, CA: Sage, 1987.

Wilkins, David E. *American Indian Politics and the American Political System.* Lanham, MD: Rowman & Littlefield, 2001.

Wilkinson, Doris Y. "Gender and Social Inequality: The Prevailing Significance of Race." *Daedalus* 124 (Winter 1995):167–178.

Wilkinson, Richard (ed.). *Class and Health.* New York: Tavistock, 1986.

Will, George. "The Nature of Human Nature." *Newsweek* (August 19, 2002):64.

Williams, J. Allen, Joetta A. Vernon, Martha C. Williams, and Karen Malecha. "Sex Role Socialization in Picture Books: An Update." *Social Science Quarterly* 68 (March 1987):148–156.

Williams, John E., and Deborah L. Best. *Measuring Sex Stereotypes,* rev. ed. Newbury Park, CA: Sage, 1990.

Williams, Mary E. *Minorities.* San Diego, CA: Greenhaven Press, 1997.

Williams, Mary E., Karin Swisher, and Brenda Stalcup (eds.). *Working Women: Opposing Viewpoints.* San Diego, CA: Greenhaven Press, 1997.

Williams, Robert L. "Scientific Racism and IQ: The Silent Mugging of the Black Community." *Psychology Today* 7 (May 1974):32–41, 101.

Williams, Robin M. Jr. *American Society,* 3rd ed. New York: Knopf, 1970.

Williamson, Elizabeth. "Getting around the Rules on Lobbying." *Washington Post* (October 14, 2007a).

Williamson, Elizabeth. "Industries Paid for Top Regulators' Travel." *Washington Post* (December 2, 2007b).

Williamson, Robert C. "A Partial Replication of the Kohn-Gecas-Nye Thesis in a German Sample." *Journal of Marriage and the Family* 46 (November 1984):971–979.

Wilson, Bryan. *Religion in Sociological Perspective.* New York: Oxford University Press, 1982.

Wilson, Craig. "Why People Will Do Almost Anything to Get on TV." *USA Today* (February 25–27, 2000):1A, 2A.

Wilson, Edward O. *On Human Nature.* Cambridge, MA: Harvard University Press, 1978.

Wilson, Edward O. *Biophilia.* Cambridge, MA: Harvard University Press, 1986.

Wilson, James Q. *The Moral Sense.* New York: Free Press, 1993.

Wilson, James Q. *Political Organizations.* Princeton, NJ: Princeton University Press, 1995.

Wilson, James Q., and Richard J. Herrnstein. *Crime and Human Nature.* New York: Simon & Schuster, 1985.

Wilson, Kenneth L., and W. Allen Martin. "Ethnic Enclaves: A Comparison of the Cuban and Black Economies in Miami." *American Journal of Sociology* 88 (July 1982):135–160.

Wilson, Stephen. *Informal Groups.* Englewood Cliffs, NJ: Prentice Hall, 1978.

Wilson, Thomas C. "Urbanism and Tolerance: A Test of Some Hypotheses Drawn from Wirth and Stouffer." *American Sociological Review* 50 (February 1985):117–123.

Wilson, Thomas C. "Urbanism, Migration, and Tolerance: A Reassessment." *American Sociological Review* 56 (February 1991):117–123.

Wilson, William Julius. *Power, Racism, and Privilege.* New York: Macmillan, 1973.

Wilson, William Julius. *The Declining Significance of Race,* 2nd ed. Chicago: University of Chicago Press, 1980.

Wilson, William Julius. "The Urban Underclass." In Leslie W. Dunbar (ed.), *Minority Report.* New York: Pantheon, 1984, pp. 75–117.

Wilson, William Julius. *The Truly Disadvantaged.* Chicago: University of Chicago Press, 1990.

Wilson, William Julius. "Studying Inner-City Social Dislocations: The Challenge of Public Agenda Research." *American Sociological Review* 56 (February 1991):1–14.

Wilson, William Julius (ed.). *The Ghetto Underclass.* Newbury Park, CA: Sage, 1993.

Wilson, William Julius. *When Work Disappears: The World of the New Urban Poor.* New York: Knopf, 1997.

Wilson, William Julius. *Visions of Social Inequality: Race, Class, and Poverty in Urban America.* Lanham, MD: Rowman & Littlefield, 2002.

Wilson, William Julius, et al. "The Ghetto Underclass and the Changing Structure of Urban Poverty." In Fred R. Harris and Roger W. Wilkins (eds.), *Quiet Riots.* New York: Pantheon, 1988, pp. 123–154.

Wimberley, Ronald C., and James A. Christenson. "Civil Religion and Church and State." *Sociological Quarterly* 21 (Spring 1980):35–40.

Wimberley, Ronald C., Donald A. Clelland, Thomas C. Hood, and C. M. Lipsey. "The Civil Religious Dimension: Is It There?" *Social Forces* 54 (June 1976):890–900.

Wind, James P., and James W. Lewis (eds.). *American Congregations,* vol. 2. Chicago: University of Chicago Press, 1994.

Wingert, Pat. "The Sum of Mediocrity." *Newsweek* (December 2, 1996):96.

Wingert, Pat. "The Report Card on Charter Schools." *Newsweek* (July 22, 2002):7.

Winship, Christopher, and Robert D. Mare. "Models for Sample Selection Bias." *Annual Review of Sociology* 18 (1992):327–350.

Winters, Rebecca. "The Philadelphia Experiment." *Time* (October 21, 2002):64–69.

Wirth, Louis. "Clinical Sociology." *American Journal of Sociology* 37 (July 1931):49–66.

Wirth, Louis. "Urbanism as a Way of Life." *American Journal of Sociology* 44 (July 1938):1–24.

Wirth, Louis. "The Problem of Minority Groups." In Ralph Linton (ed.), *The Science of Man in the World Crisis.* New York: Columbia University Press, 1945, pp. 347–372.

Wise, William. *Massacre at Mountain Meadows.* New York: Thomas Y. Crowell, 1976.

Wisensale, Steven K. *Family Leave Policy: The Political Economy of Work and Family in America.* Armonk, NY: M. E. Sharpe, 2001.

Witte, John F. *The Market Approach to Education: An Analysis of America's First Voucher Program.* Princeton, NJ: Princeton University Press, 2000.

Wodak, Ruth (ed.). *Gender and Discourse.* Thousand Oaks, CA: Sage, 1997.

Wolfe, Alan. "The Moral Meanings of Work." *The American Prospect* (September/October 1997):82–90.

Wolfe, Alan. *One Nation, After All.* New York: Penguin Books, 1999.

Wolfe, Alan. "And the Winners Is . . . ." *The Atlantic* (March 2008): 56–63.

Wolfinger, Raymond E., and Steven J. Rosenstone. *Who Votes?* New Haven, CT: Yale University Press, 1980.

Wolman, William, and Anne Colamosca. *The Judas Economy: The Triumph of Capital and the Betrayal of Work.* Reading, MA: Addison-Wesley, 1997.

Wood, James R. *Leadership in Voluntary Organizations.* New Brunswick, NJ: Rutgers University Press, 1981.

Wood, Gordon S. *The American Revolution: A History.* New York: The Modern Library, 2003.

Wooden, Wayne S. *Renegade Kids, Suburban Outlaws: From Youth Culture to Delinquency.* Belmont, CA: Wadsworth, 1995.

World Almanac and Book of Facts 2008, The. New York: World Almanac Education Group Inc., 2008.

World Bank Atlas, 1999, The. Washington, D.C.: World Bank, 1999.

World Development Report 1991. New York: Oxford University Press, 1991.

World Development Report 1994. New York: Oxford University Press, 1994.

World Development Report 2000–2001. New York: Oxford University Press, 2001.

World Development Report 2003. New York: Oxford University Press, 2003.

2007 World Population Data Sheet. Washington, D.C.: Population Reference Bureau, 2007.

World's Women 2005, The: Progress in Statistics. New York: United Nations Publications, 2005.

"World without Work." In Robert Staples (ed.), *The Black Family.* Belmont, CA: Wadsworth, 1999, pp. 291–311.

Wortzel, Larry M. *Class in China.* Westport, CT: Greenwood Press, 1987.

Wright, Charles R. "Functional Analysis and Mass Communication." *Public Opinion Quarterly* 24 (Winter 1960):606–620.

Wright, Erik Olin. *Classes.* London: Verse, 1985.

Wright, Erik Olin. *Class Counts.* Cambridge, UK: Cambridge University Press, 2000.

Wright, James D., and Sonia R. Wright, "Social Class and Parental Values for Children: A Partial Replication and Extension of the Kohn Thesis." *American Sociological Review* 4 (June 1976):527–537.

Wright, Kevin N. *The Great American Crime Myth.* New York: Praeger, 1987.

Wright, Lawrence. "Double Mystery." *The New Yorker* (August 7, 1995):45–62.

Wu, Jean Y., and Min Song (eds.). *Asian American Studies: A Reader.* Piscataway, NJ: Rutgers University Press, 2000.

Wuthnow, Robert. "Recent Patterns of Secularization: A Problem of Generations?" *American Sociological Review* 41 (October 1976):850–867.

Wuthnow, Robert. *Experimentation in American Religion.* Berkeley: University of California Press, 1978.

Wuthnow, Robert. "Review of All Faithful People." *Society* 22 (March/April 1985):87–88.

Wuthnow, Robert. *The Restructuring of American Religion.* Princeton, NJ: Princeton University Press, 1990.

Yamagata, Hisashi, et al. "Sex Segregation and Glass Ceilings: A Comparative Statistics Model of Women's Career Opportunities in the Federal Government over a Quarter Century." *American Journal of Sociology* 103 (November 1997):566–632.

Yang, Fenggang, and Helen Rose Ebaugh. "Transformations in New Immigrant Religions and Their Global Implications." *American Sociological Review* 66 (April 2001):269–288.

Yap, M. T. "Singapore's 'Three or More' Policy: The First Five Years." *Asia-Pacific Population Journal* 10 (1995):39–52.

Yinger, John. *Closed Doors, Opportunities Lost: The Continuing Costs of Housing Discrimination.* New York: Russell Sage Foundation, 1995.

Yinger, Milton J. "A Structural Examination of Religion." *Journal of the Scientific Study of Religion* 8 (Spring 1969): 88–89.

Yinger, Milton J. *The Scientific Study of Religion.* New York: Macmillan, 1970.

Yinger, Milton J., and Stephen J. Cutler. "The Moral Majority Viewed Sociologically." *Sociological Focus* 15 (October 1982):289–306.

Yochelson, Samuel, and Stanton E. Samenow. *The Criminal Personality,* vols. 1 and 2. London: James Aronson, 1994.

Young, Alford A. Jr., and Donald R. Deskins Jr. "Early Traditions of African-American Sociological Thought." *Annual Review of Sociology* 27 (2001):445–477.

Young, Iris Marion. *Justice and the Politics of Difference.* Princeton, NJ: Princeton University Press, 1990.

Young, Michael. *The Rise of the Meritocracy.* New York: Penguin, 1967.

Young, T. R. *The Drama of Social Life.* New Brunswick, NJ: Transaction Books, 1990.

Young, T. R. "The Sociology of Sport: Structural Marxist and Cultural Marxist Approaches." *Sociological Perspectives* 29 (January 1986):3–28.

Yunker, James A. "A New Statistical Analysis of Capital Punishment Incorporating U.S. Postmoratorium Data," *Social Science Quarterly* 82 (June 2001):297–311.

Zachary, G. Pascal. "Bell Labs Is Gone. Academia Steps In." *New York Times* (December 16, 2007).

Zald, Mayer N., and John D. McCarthy. "Introduction." In Mayer N. Zald and John D. McCarthy (eds.), *Social Movements in an Organizational Society.* New Brunswick, NJ: Transaction Books, 1987.

Zamble, Edward, and Vernon L. Quinsey. *The Criminal Recidivism Process.* New York: Cambridge University Press, 1997.

Zaret, David. "Calvin, Covenant Theology and the Weber Thesis." *British Journal of Sociology* 43 (1992):369–391.

Zborowski, Mark. "Cultural Components in Response to Pain." *Journal of Social Issues* 8 (1952):16–30.

Zborowski, Mark. *People in Pain.* San Francisco: Jossey-Bass, 1969.

Zelditch, Morris Jr. "Role Differentiation in the Nuclear Family: A Comparative Study." In Talcott Parsons and Robert F. Bales (eds.), *Family, Socialization and Interaction Process.* New York: Free Press, 1955, pp. 307–352.

Zellner, William M., and William M. Kephart. *Extraordinary Groups,* 6th ed. New York: St. Martin's Press, 2000.

Zellner, William W. *Countercultures: A Sociological Analysis,* 2nd ed. New York: St. Martin's Press, 1995.

Zellner, William W., and Marc Petrowsky. *Sects, Cults, and Spiritual Communities.* Westport, CT: Praeger, 1998.

Zepezauer, Mark. *Take the Rich off Welfare.* 2nd ed. Cambridge, MA: South End Press, 2004.

Zigler, Edward, and S. Muenchow. *Head Start.* New York: Basic Books, 1992.

Zigler, Edward, and Sally J. Styfco (eds.). *Head Start and Beyond.* New Haven, CT: Yale University Press, 1993.

Zimbardo, Philip G., S. M. Anderson, and L. G. Kabat. "Induced Hearing Deficit Generates Experimental Paranoia." *Science* (June 26, 1981):1529–1531.

Zimbardo, Philip. *The Lucifer Effect.* New York: Random House Publishing Group, 2008.

Zimring, Franklin E. *Contradictions of American Capital Punishment.* New York: Oxford University Press, 2004.

Zinn, Howard. *Declarations of Independence: Cross-Examining American Ideology.* New York: Harper Perennial, 2003.

Zinn, Howard. *A People's History of the United States: 1492–Present,* 20th anniversary ed. New York: HarperCollins Publishers, 2005.

Zogby, James J. *What Arabs Think: Values, Beliefs, and Concerns.* Utica, NY, and Beirut, Lebanon: Zogby International/The Arab Thought Foundation, 2002.

Zola, Irving K. "Culture and Symptoms—An Analysis of Patients Presenting Complaints." *American Sociological Review* 31 (October 1966):615–630.

Zorbaugh, Harvey. *The Gold Coast and the Slum.* Chicago: University of Chicago Press, 1929.

Zucher, Louis A., and David A. Snow. "Collective Behavior: Social Movements." In Morris Rosenberg and Ralph

H. Turner (eds.), *Social Psychology*. New Brunswick, NJ: Transaction Books, 1990, pp. 447–482.

Zuo, J., and S. Tang. "Breadwinner Status and Gender Ideologies of Men and Women Regarding Family Roles." *Sociological Perspectives* 43 (2000):29–43.

Zweig, Michael. *The Working Class Majority: America's Best Kept Secret*. Ithaca, NY: Cornell University Press, 2000.

Zweigenhaft, Richard L., and G. William Domhoff. *Diversity in the Power Elite*, 2nd ed. Lanham, MD: Rowan and Littlefield, 2006.

Zwerling, Craig, and Hilary Silver. "Race and Job Dismissals in a Federal Bureaucracy." *American Sociological Review* 57 (October 1992):651–660.

Credits

284 Chip East/Reuters /Landov

290 Patrick Clark/Getty Images

292 Chris Keane/Reuters /Landov

Chapter 11

301 Jim Craigmyle/Getty Images

303 *both* Monkey Business Images/Used under license from Shutterstock

305 Nruboc/Dreamstime LLC

308 Stephen Wilkes/Getty Images

317 arabianEye/Getty Images

323 FilmMagic/Getty Images

325 Michael Newman/PhotoEdit

327 Lise Gagne/istockphoto.com

329 Universal/Everett Collection

Chapter 12

336 Digital Vision/Getty Images

342 PhotoLink/Getty Images

344 Amy Conn-Gutierrez/AP Photo

345 Huy Lam/Getty Images

349 Mary Ann Chastain/AP Photo

352 Monkey Business Images/Index Open

358 Catherine Ledner/Getty Images

362 Corbis Premium RF/Alamy Limited

Chapter 13

367 Alex Wong/Getty Images

370 Nick Laham/Getty Images

374 AP Photo

377 Supplied by FilmStills.net/Digital Railroad

383 National Portrait Gallery, Smithsonian Institution/ Art Resource, NY

389 Lee Snider/Image Works

394 Ludwig Haskins/DIgital Railroad

395 Peter Sickles/Superstock

395 Dynamic Graphics/Jupiter Images

Chapter 14

405 *top* ARTEKI/Used under license from Shutterstock

405 *bottom left* Rolmat/Dreamstime LLC

405 *bottom right* Mikhail Nekrasov/Used under license from Shutterstock

407 Christineg/Dreamstime LLC

411 Hulton Archive/Getty Images

412 The Art Archive/Bibliothèque Universitaire Geneva/ Gianni Dagli Orti/Picture Desk

416 Chip Chipman/Digital Railroad

422 Jessica Kourkonis/AP Photo

424 Eduardo Sverdlin/Digital Railroad

426 Mark Mainz/Getty Images

429 Michael Blann/Getty Images

Chapter 15

436 Masterfile

437 iofoto/Used under license from Shutterstock

440 Andy Guest/Alamy Limited

443 *top* Alexandra Boulat/AP Photo

443 *bottom* Nathan Maxfield/istockphoto.com

445 Yoav Levy/Alamy Limited

447 Dynamic Graphics/Jupiter Images

448 Becky Reed/Alamy Limited

454 Monkey Business Images/Used Under license from Shutterstock

Chapter 16

460 Diademimages/Dreamstime LLC

461 *Chariot Race* (gouache on paper), Jackson, Peter (1922-2003)/Private Collection, © Look and Learn/ Bridgeman Art Library

462 Ed Betz/AP Photo

464 Paul Sakuma/AP Photo

465 John Angelillo/Digital Railroad

466 Vicky Kasala/Alamy Limited

469 Jonathan Kirn/Alamy Limited

471 Mayskyphoto/Used under license from Shutterstock

472 Amy Sancetta/AP Photos

Chapter 17

478 Rob Stuehrk/AP Photo

481 Con Tanasiuk/Digital Railroad

483 Jason Reed/Landov Media

488 Topham/Image Works

491 Peter Busomoke/Getty Images

498 Billy Grimes/Digital Railroad

500 Bill Bachmann/PhotoEdit

502 Dennis MacDonald/PhotoEdit

505 Sam Diephuis/zefa/CORBIS

Chapter 18

514 Dominic Favre/AP Photo

516 Norbert Schiller/epa/CORBIS

520 John Moore/AP Photo

523 Reuters/Landov Media

529 Oblong1/Dreamstime LLC

531 Bill Kostroun/AP Photo

535 Matt Sayles/AP Photo

538 Jeff Mitchell/Landov Media

543 Ahn Young-Joon/AP Photo

Name Index

Subject Index